ULLMANN'S
Agrochemicals

Volume 1
Plant Growth

1807–2007 Knowledge for Generations

Each generation has its unique needs and aspirations. When Charles Wiley first opened his small printing shop in lower Manhattan in 1807, it was a generation of boundless potential searching for an identity. And we were there, helping to define a new American literary tradition. Over half a century later, in the midst of the Second Industrial Revolution, it was a generation focused on building the future. Once again, we were there, supplying the critical scientific, technical, and engineering knowledge that helped frame the world. Throughout the 20th Century, and into the new millennium, nations began to reach out beyond their own borders and a new international community was born. Wiley was there, expanding its operations around the world to enable a global exchange of ideas, opinions, and know-how.

For 200 years, Wiley has been an integral part of each generation's journey, enabling the flow of information and understanding necessary to meet their needs and fulfill their aspirations. Today, bold new technologies are changing the way we live and learn. Wiley will be there, providing you the must-have knowledge you need to imagine new worlds, new possibilities, and new opportunities.

Generations come and go, but you can always count on Wiley to provide you the knowledge you need, when and where you need it!

William J. Pesce
President and Chief Executive Officer

Peter Booth Wiley
Chairman of the Board

ULLMANN'S
Agrochemicals

Volume 1
Plant Growth

WILEY-VCH Verlag GmbH & Co. KGaA

All books published by **Wiley-VCH** are carefully produced. Nevertheless, authors, editors and publisher do not warrant the information contained in these books, including this book, to be free of errors. Readers are advised to keep in mind that statements, data, illustrations, procedural details or other items may inadvertently be inaccurate.

Library of Congress Card No.:
applied for

British Library Cataloguing-in-Publication Data:
A catalogue record for this book is available from the British Library.

Bibliographic information published by The Deutsche Nationalbibliothek
The Deutsche Nationalibliothek lists this publication in the Deutsche Nationalbibliografie; detailed bibliographic data is available in the Internet at http://dnb.d-nb.de

© 2007 WILEY-VCH Verlag GmbH & Co KGaA, Weinheim

All rights reserved (including those of translation into other languages). No part of this book may be reproduced in any form – by photocopying, microfilm, or any other means – nor transmitted or translated into a machine language without written permission from the publishers. Registered names, trademarks, etc. used in this book, even when not specifically marked as such, are not to be considered unprotected by law.

Typesetting: Steingraeber Satztechnik GmbH, Ladenburg
Printing: betz-druck GmbH, Darmstadt
Binding: Litges & Dopf Buchbinderei GmbH, Heppenheim
Cover Design: Grafik-Design Schulz, Fußgönheim

Printed in the Federal Republic of Germany
Printed on acid-free paper

ISBN: 978-3-527-31604-5

Preface

Chemicals for plant growth and crop protection are a very important segment in the chemical industry world wide. The high standard of today's agricultural practices can be attributed to the great advances achieved in agricultural sciences since the second world war. This has led to a better understanding of plant physiology and the interaction between plants and their natural cohabitants, the latter of which can be direct competitors for the available nutrients and space, or pathogenic intruders of plants (fungi, nematodes, insects, rodents).

This ready reference presents an overview of all aspects of modern agricultural practices. Based on the latest electronic edition of Ullmann's Encyclopedia of Industrial Chemistry - and containing articles never before seen in print - the 2-volume set meets the need for a detailed survey of the fundamentals, industrial processes, effectiveness and the applications of agrochemicals as well as the environmental and toxicological aspects.

The detailed and meticulously edited articles have been written by renowned experts from industry and academia. Readers benefit from the clear and cross-indexed nature of the parent reference, and will find both broad introductory information and details of significance to industrial and academic environments. In each article practical and theoretical aspects are discussed, and each entry is accompanied by its chemical structure, physical properties, synthesis, agricultural use, trade names, toxicology, and its ecological significance. Deeper insight into any given area of interest is offered by highly referenced contributions.

We are convinced that this handbook will prove an invaluable source of information for all those involved in modern agricultural practices.

The Publisher

Contents

Volume 1

Symbols and Units	IX
Conversion Factors	XI
Abbreviations	XII
Country Codes	XVII
Periodic Table of Elements	XVIII
Fertilizers	3
Ammonia	143
Ammonium Compounds	299
Nitrates and Nitrites	325
Phosphate Fertilizers	353
Potassium Compounds	399
Urea	467
Plant Growth Regulators	503

Volume 2

Symbols and Units	IX
Conversion Factors	XI
Abbreviations	XII
Country Codes	XVII
Periodic Table of Elements	XVIII
Crop Protection	517
Acaricides	541
Fungicides, Agricultural	583
Insect Control	651
Molluscicides	715
Nematicides	723
Rodenticides	737
Weed Control	749
Biological Control	863
Author Index	885
Subject Index	887

Symbols and Units

Symbols and units agree with SI standards (for conversion factors see page XI). The following list gives the most important symbols used in the encyclopedia. Articles with many specific units and symbols have a similar list as front matter.

Symbol	Unit	Physical Quantity
a_B		activity of substance B
A_r		relative atomic mass (atomic weight)
A	m^2	area
c_B	mol/m^3, mol/L (M)	concentration of substance B
C	C/V	electric capacity
c_p, c_v	$J\,kg^{-1}K^{-1}$	specific heat capacity
d	cm, m	diameter
d		relative density (ϱ/ϱ_{water})
D	m^2/s	diffusion coefficient
D	Gy (= J/kg)	absorbed dose
e	C	elementary charge
E	J	energy
E	V/m	electric field strength
E	V	electromotive force
E_A	J	activation energy
f		activity coefficient
F	C/mol	Faraday constant
F	N	force
g	m/s^2	acceleration due to gravity
G	J	Gibbs free energy
h	m	height
$\hbar$	$W \cdot s^2$	Planck constant
H	J	enthalpy
I	A	electric current
I	cd	luminous intensity
k	(variable)	rate constant of a chemical reaction
k	J/K	Boltzmann constant
K	(variable)	equilibrium constant
l	m	length
m	g, kg, t	mass
M_r		relative molecular mass (molecular weight)
n_D^{20}		refractive index (sodium D-line, 20 °C)
n	mol	amount of substance
N_A	mol^{-1}	Avogadro constant (6.023×10^{23} mol^{-1})
p	Pa, bar *	pressure
Q	J	quantity of heat
r	m	radius
R	$J\,K^{-1}mol^{-1}$	gas constant
R	Ω	electric resistance
S	J/K	entropy
t	s, min, h, d, month, a	time

X   Symbols and Units

Symbols and Units (Continued from p. IX)

Symbol	Unit	Physical Quantity
t	°C	temperature
T	K	absolute temperature
u	m/s	velocity
U	V	electric potential
U	J	internal energy
V	m^3, L, mL, µL	volume
w		mass fraction
W	J	work
x_B		mole fraction of substance B
Z		proton number, atomic number
α		cubic expansion coefficient
α	$W\,m^{-2}K^{-1}$	heat-transfer coefficient (heat-transfer number)
α		degree of dissociation of electrolyte
$[\alpha]$	$10^{-2}\,deg\,cm^2\,g^{-1}$	specific rotation
η	Pa·s	dynamic viscosity
θ	°C	temperature
$\varkappa$		c_p/c_v
λ	$W\,m^{-1}K^{-1}$	thermal conductivity
λ	nm, m	wavelength
μ		chemical potential
ν	Hz, s^{-1}	frequency
ν	m^2/s	kinematic viscosity (η/ϱ)
π	Pa	osmotic pressure
ϱ	g/cm^3	density
σ	N/m	surface tension
τ	Pa (N/m^2)	shear stress
φ		volume fraction
χ	Pa^{-1} (m^2/N)	compressibility

* The official unit of pressure is the pascal (Pa).

Conversion Factors

SI unit	Non-SI unit	From SI to non-SI multiply by
Mass		
kg	pound (avoirdupois)	2.205
kg	ton (long)	9.842×10^{-4}
kg	ton (short)	1.102×10^{-3}
Volume		
m^3	cubic inch	6.102×10^4
m^3	cubic foot	35.315
m^3	gallon (U.S., liquid)	2.642×10^2
m^3	gallon (Imperial)	2.200×10^2
Temperature		
°C	°F	°C $\times 1.8 + 32$
Force		
N	dyne	1.0×10^5
Energy, Work		
J	Btu (int.)	9.480×10^{-4}
J	cal (int.)	2.389×10^{-1}
J	eV	6.242×10^{18}
J	erg	1.0×10^7
J	kW · h	2.778×10^{-7}
J	kp · m	1.020×10^{-1}
Pressure		
MPa	at	10.20
MPa	atm	9.869
MPa	bar	10
kPa	mbar	10
kPa	mm Hg	7.502
kPa	psi	0.145
kPa	torr	7.502

Powers of Ten

E (exa)	10^{18}		d (deci)	10^{-1}
P (peta)	10^{15}		c (centi)	10^{-2}
T (tera)	10^{12}		m (milli)	10^{-3}
G (giga)	10^9		μ (micro)	10^{-6}
M (mega)	10^6		n (nano)	10^{-9}
k (kilo)	10^3		p (pico)	10^{-12}
h (hecto)	10^2		f (femto)	10^{-15}
da (deca)	10		a (atto)	10^{-18}

Abbreviations

The following is a list of the abbreviations used in the text. Common terms, the names of publications and institutions, and legal agreements are included along with their full identities. Other abbreviations will be defined wherever they first occur in an article. For further abbreviations, see page IX, Symbols and Units; page XVI, Frequently Cited Companies (Abbreviations), and page XVII, Country Codes in patent references. The names of periodical publications are abbreviated exactly as done by Chemical Abstracts Service.

abs.	absolute	BAM	Bundesanstalt für Materialprüfung (Federal Republic of Germany)
a.c.	alternating current		
ACGIH	American Conference of Governmental Industrial Hygienists	BAT	Biologischer Arbeitsstoff-Toleranz-Wert (biological tolerance value for a working material, established by MAK Commission, see MAK)
ACS	American Chemical Society		
ADI	acceptable daily intake		
ADN	accord européen relatif au transport international des marchandises dangereuses par voie de navigation interieure (European agreement concerning the international transportation of dangerous goods by inland waterways)	Beilstein	Beilstein's Handbook of Organic Chemistry, Springer, Berlin – Heidelberg – New York
		BET	Brunauer – Emmett – Teller
		BGA	Bundesgesundheitsamt (Federal Republic of Germany)
		BGBl.	Bundesgesetzblatt (Federal Republic of Germany)
ADNR	ADN par le Rhin (regulation concerning the transportation of dangerous goods on the Rhine and all national waterways of the countries concerned)	BIOS	British Intelligence Objectives Subcommitee Report (see also FIAT)
		BOD	biological oxygen demand
ADP	adenosine 5′-diphosphate	bp	boiling point
ADR	accord européen relatif au transport international des marchandises dangereuses par route (European agreement concerning the international transportation of dangerous goods by road)	B.P.	British Pharmacopeia
		BS	British Standard
		ca.	circa
		calcd.	calculated
		CAS	Chemical Abstracts Service
		cat.	catalyst, catalyzed
AEC	Atomic Energy Commission (United States)	CEN	Comité Européen de Normalisation
		cf.	compare
a.i.	Active ingredient	CFR	Code of Federal Regulations (United States)
AIChE	American Institute of Chemical Engineers		
		cfu	colony forming units
AIME	American Institute of Mining, Metallurgical, and Petroleum Engineers	Chap.	chapter
		ChemG	Chemikaliengesetz (Federal Republic of Germany)
ANSI	American National Standards Institute	C.I.	Colour Index
		CIOS	Combined Intelligence Objectives Subcommitee Report (see also FIAT)
AMP	adenosine 5′-monophosphate		
APhA	American Pharmaceutical Association	CNS	central nervous system
		Co.	Company
API	American Petroleum Institute	COD	chemical oxygen demand
ASTM	American Society for Testing and Materials	conc.	concentrated
		const.	constant
ATP	adenosine 5′-triphosphate	Corp.	Corporation
		crit.	critical

Abbreviations

CTFA	The Cosmetic, Toiletry and Fragrance Association (United States)	FIAT	Field Information Agency, Technical (United States reports on the chemical industry in Germany, 1945)
DAB 9	Deutsches Arzneibuch, 9th ed., Deutscher Apotheker-Verlag, Stuttgart 1986	Fig.	figure
		fp	freezing point
d.c.	direct current	Friedländer	P. Friedländer, Fortschritte der Teerfarbenfabrikation und verwandter Industriezweige, Vol. 1–25, Springer, Berlin 1888–1942
decomp.	decompose, decomposition		
DFG	Deutsche Forschungsgemeinschaft (German Science Foundation)		
dil.	dilute, diluted	FT	Fourier transform
DIN	Deutsche Industrie Norm (Federal Republic of Germany)	(g)	gas, gaseous
		GC	gas chromatography
DMF	dimethylformamide	GefStoffV	Gefahrstoffverordnung (regulations in the Federal Republic of Germany concerning hazardous substances)
DNA	deoxyribonucleic acid		
DOE	Department of Energy (United States)		
		GGVE	Verordnung in der Bundesrepublik Deutschland über die Beförderung gefährlicher Güter mit der Eisenbahn (regulation in the Federal Republic of Germany concerning the transportation of dangerous goods by rail)
DOT	Department of Transportation – Materials Transportation Bureau (United States)		
DTA	differential thermal analysis		
EC	effective concentration		
EC	European Community	GGVS	Verordnung in der Bundesrepublik Deutschland über die Beförderung gefährlicher Güter auf der Straße (regulation in the Federal Republic of Germany concerning the transportation of dangerous goods by road)
ed.	editor, edition, edited		
e.g.	for example		
emf	electromotive force		
EmS	Emergency Schedule		
EN	European Standard (European Community)		
EPA	Environmental Protection Agency (United States)	GGVSee	Verordnung in der Bundesrepublik Deutschland über die Beförderung gefährlicher Güter mit Seeschiffen (regulation in the Federal Republic of Germany concerning the transportation of dangerous goods by sea-going vessels)
EPR	electron paramagnetic resonance		
Eq.	equation		
ESCA	electron spectroscopy for chemical analysis		
esp.	especially		
ESR	electron spin resonance		
Et	ethyl substituent ($-C_2H_5$)	GLC	gas-liquid chromatography
et al.	and others	Gmelin	Gmelin's Handbook of Inorganic Chemistry, 8th ed., Springer, Berlin–Heidelberg–New York
etc.	et cetera		
EVO	Eisenbahnverkehrsordnung (Federal Republic of Germany)		
		GRAS	generally recognized as safe
exp (...)	$e^{(...)}$, mathematical exponent	Hal	halogen substituent ($-F, -Cl, -Br, -I$)
FAO	Food and Agriculture Organization (United Nations)		
		Houben-Weyl	Methoden der organischen Chemie, 4th ed., Georg Thieme Verlag, Stuttgart
FDA	Food and Drug Administration (United States)		
FD & C	Food, Drug and Cosmetic Act (United States)	HPLC	high performance liquid chromatography
FHSA	Federal Hazardous Substances Act (United States)	IAEA	International Atomic Energy Agency
		IARC	International Agency for Research on Cancer, Lyon, France

XIV   Abbreviations

IATA-DGR	International Air Transport Association, Dangerous Goods Regulations
ICAO	International Civil Aviation Organization
i.e.	that is
i.m.	intramuscular
IMDG	International Maritime Dangerous Goods Code
IMO	Inter-Governmental Maritime Consultive Organization (in the past: IMCO)
Inst.	Institute
i.p.	intraperitoneal
IR	infrared
ISO	International Organization for Standardization
IUPAC	International Union of Pure and Applied Chemistry
i.v.	intravenous
Kirk-Othmer	Encyclopedia of Chemical Technology, 3rd ed., J. Wiley & Sons, New York – Chichester – Brisbane – Toronto 1978 – 1984; 4th ed., J. Wiley & Sons, New York – Chichester – Brisbane – Toronto 1991 – 1998
(l)	liquid
Landolt-Börnstein	Zahlenwerte u. Funktionen aus Physik, Chemie, Astronomie, Geophysik u. Technik, Springer, Heidelberg 1950 – 1980; Zahlenwerte und Funktionen aus Naturwissenschaften und Technik, Neue Serie, Springer, Heidelberg, since 1961
LC_{50}	lethal concentration for 50 % of the test animals
LCLo	lowest published lethal concentration
LD_{50}	lethal dose for 50 % of the test animals
LDLo	lowest published lethal dose
ln	logarithm (base e)
LNG	liquefied natural gas
log	logarithm (base 10)
LPG	liquefied petroleum gas
M	mol/L
M	metal (in chemical formulas)
MAK	Maximale Arbeitsplatz-Konzentration (maximum concentration at the workplace in the Federal Republic of Germany); cf. Deutsche Forschungsgemeinschaft (ed.): Maximale Arbeitsplatzkonzentrationen (MAK) und Biologische Arbeitsstoff-Toleranz-Werte (BAT), WILEY-VCH Verlag, Weinheim (published annually)
max.	maximum
MCA	Manufacturing Chemists Association (United States)
Me	methyl substituent ($-CH_3$)
Methodicum Chimicum	Methodicum Chimicum, Georg Thieme Verlag, Stuttgart
MFAG	Medical First Aid Guide for Use in Accidents Involving Dangerous Goods
MIK	maximale Immissionskonzentration (maximum immission concentration)
min.	minimum
mp	melting point
MS	mass spectrum, mass spectrometry
NAS	National Academy of Sciences (United States)
NASA	National Aeronautics and Space Administration (United States)
NBS	National Bureau of Standards (United States)
NCTC	National Collection of Type Cultures (United States)
NIH	National Institutes of Health (United States)
NIOSH	National Institute for Occupational Safety and Health (United States)
NMR	nuclear magnetic resonance
no.	number
NOEL	no observed effect level
NRC	Nuclear Regulatory Commission (United States)
NRDC	National Research Development Corporation (United States)
NSC	National Service Center (United States)
NSF	National Science Foundation (United States)
NTSB	National Transportation Safety Board (United States)
OECD	Organization for Economic Cooperation and Development
OSHA	Occupational Safety and Health Administration (United States)

p., pp.	page, pages		regulation in Federal Republic of Germany)
Patty	G. D. Clayton, F. E. Clayton (eds.): Patty's Industrial Hygiene and Toxicology, 3rd ed., Wiley Interscience, New York	TA Lärm	Technische Anleitung zum Schutz gegen Lärm (low noise regulation in Federal Republic of Germany)
PB report	Publication Board Report (U.S. Department of Commerce, Scientific and Industrial Reports)	TDLo	lowest published toxic dose
		THF	tetrahydrofuran
		TLC	thin layer chromatography
PEL	permitted exposure limit	TLV	Threshold Limit Value (TWA and STEL); published annually by the American Conference of Governmental Industrial Hygienists (ACGIH), Cincinnati, Ohio
Ph	phenyl substituent ($-C_6H_5$)		
Ph. Eur.	European Pharmacopoeia, 2nd. ed., Council of Europe, Strasbourg 1981		
phr	part per hundred rubber (resin)		
PNS	peripheral nervous system	TOD	total oxygen demand
ppm	parts per million	TRK	Technische Richtkonzentration (lowest technically feasible level)
q. v.	which see (quod vide)		
ref.	refer, reference	TSCA	Toxic Substances Control Act (United States)
resp.	respectively		
R_f	retention factor (TLC)	TÜV	Technischer Überwachungsverein (Technical Control Board of the Federal Republic of Germany)
R. H.	relative humidity		
RID	règlement international concernant le transport des marchandises dangereuses par chemin de fer (international convention concerning the transportation of dangerous goods by rail)		
		TWA	Time Weighted Average
		UBA	Umweltbundesamt (Federal Environmental Agency)
		Ullmann	Ullmann's Encyclopedia of Industrial Chemistry, 6th ed., Wiley-VCH, Weinheim, 2002, 5th ed., VCH Verlagsgesellschaft, Weinheim, 1985–1996; Ullmanns Encyklopädie der Technischen Chemie, 4th ed., Verlag Chemie, Weinheim 1972–1984
RNA	ribonucleic acid		
R phrase (R-Satz)	risk phrase according to ChemG and GefStoffV (Federal Republic of Germany)		
rpm	revolutions per minute		
RTECS	Registry of Toxic Effects of Chemical Substances, edited by the National Institute of Occupational Safety and Health (United States)		
		USAEC	United States Atomic Energy Commission
		USAN	United States Adopted Names
(s)	solid	USD	United States Dispensatory
SAE	Society of Automotive Engineers (United States)	USDA	United States Department of Agriculture
s.c.	subcutaneous	U.S.P.	United States Pharmacopeia
SI	International System of Units	UV	ultraviolet
SIMS	secondary ion mass spectrometry	UVV	Unfallverhütungsvorschriften der Berufsgenossenschaft (workplace safety regulations in the Federal Republic of Germany)
S phrase (S-Satz)	safety phrase according to ChemG and GefStoffV (Federal Republic of Germany)		
STEL	Short Term Exposure Limit (see TLV)	VbF	Verordnung in der Bundesrepublik Deutschland über die Errichtung und den Betrieb von Anlagen zur Lagerung, Abfüllung und Beförderung brennbarer Flüssigkeiten (regulation in the Federal Republic of Germany
STP	standard temperature and pressure (0° C, 101.325 kPa)		
T_g	glass transition temperature		
TA Luft	Technische Anleitung zur Reinhaltung der Luft (clean air		

	concerning the construction and operation of plants for storage, filling, and transportation of flammable liquids; classification according to the flash point of liquids, in accordance with the classification in the United States)
VDE	Verband Deutscher Elektroingenieure (Federal Republic of Germany)
VDI	Verein Deutscher Ingenieure (Federal Republic of Germany)
vol	volume
vol.	volume (of a series of books)
vs.	versus
WGK	Wassergefährdungsklasse (water hazard class)
WHO	World Health Organization (United Nations)
Winnacker-Küchler	Chemische Technologie, 4th ed., Carl Hanser Verlag, München, 1982-1986; Winnacker-Küchler, Chemische Technik: Prozesse und Produkte, Wiley-VCH, Weinheim, from 2003
wt	weight
$	U.S. dollar, unless otherwise stated

Frequently Cited Companies (Abbreviations)

Air Products	Air Products and Chemicals
Akzo	Algemene Koninklijke Zout Organon
Alcoa	Aluminum Company of America
Allied	Allied Corporation
Amer. Cyanamid	American Cyanamid Company
BASF	BASF Aktiengesellschaft
Bayer	Bayer AG
BP	British Petroleum Company
Celanese	Celanese Corporation
Daicel	Daicel Chemical Industries
Dainippon	Dainippon Ink and Chemicals Inc.
Dow Chemical	The Dow Chemical Company
DSM	Dutch Staats Mijnen
Du Pont	E.I. du Pont de Nemours & Company
Exxon	Exxon Corporation
FMC	Food Machinery & Chemical Corporation
GAF	General Aniline & Film Corporation
W.R. Grace	W.R. Grace & Company
Hoechst	Hoechst Aktiengesellschaft
IBM	International Business Machines Corporation
ICI	Imperial Chemical Industries
IFP	Institut Français du Pétrole
INCO	International Nickel Company
3M	Minnesota Mining and Manufacturing Company
Mitsubishi Chemical	Mitsubishi Chemical Industries
Monsanto	Monsanto Company
Nippon Shokubai	Nippon Shokubai Kagaku Kogyo
PCUK	Pechiney Ugine Kuhlmann
PPG	Pittsburg Plate Glass Industries
Searle	G.D. Searle & Company
SKF	Smith Kline & French Laboratories
SNAM	Societá Nazionale Metandotti
Sohio	Standard Oil of Ohio
Stauffer	Stauffer Chemical Company
Sumitomo	Sumitomo Chemical Company
Toray	Toray Industries Inc.
UCB	Union Chimique Belge
Union Carbide	Union Carbide Corporation
UOP	Universal Oil Products Company
VEBA	Vereinigte Elektrizitäts- und Bergwerks-AG
Wacker	Wacker Chemie GmbH

Country Codes

The following list contains a selection of standard country codes used in the patent references.

AT	Austria	ID	Indonesia
AU	Australia	IL	Israel
BE	Belgium	IT	Italy
BG	Bulgaria	JP	Japan*
BR	Brazil	LU	Luxembourg
CA	Canada	MA	Morocco
CH	Switzerland	NL	Netherlands*
DE	Federal Republic of Germany (and Germany before 1949)*	NO	Norway
		NZ	New Zealand
DK	Denmark	PL	Poland
ES	Spain	PT	Portugal
FI	Finland	SE	Sweden
FR	France	US	United States of America
GB	United Kingdom	ZA	South Africa
GR	Greece	EP	European Patent Office*
HU	Hungary	WO	World Intellectual Property Organization

* For Europe, Federal Republic of Germany, Japan, and the Netherlands, the type of patent is specified: EP (patent), EP-A (application), DE (patent), DE-OS (Offenlegungsschrift), DE-AS (Auslegeschrift), JP (patent), JP-Kokai (Kokai tokkyo koho), NL (patent), and NL-A (application).

Periodic Table of Elements

element symbol, atomic number, and relative atomic mass (atomic weight)

1A "European" group designation and old IUPAC recommendation
1 group designation to 1986 IUPAC proposal
IA "American" group designation, also used by the Chemical Abstracts Service until the end of 1986

1A 1 IA	2A 2 IIA											3B 13 IIIA	4B 14 IVA	5B 15 VA	6B 16 VIA	7B 17 VIIA	0 18 VIIIA
1 H 1.0079																	2 He 4.0026
3 Li 6.941	4 Be 9.0122											5 B 10.811	6 C 12.011	7 N 14.007	8 O 15.999	9 F 18.998	10 Ne 20.180
11 Na 22.990	12 Mg 24.305	3A 3 IIIB	4A 4 IVB	5A 5 VB	6A 6 VIB	7A 7 VIIB	8 8 VIII	8 9 VIII	8 10 VIII	1B 11 IB	2B 12 IIB	13 Al 26.982	14 Si 28.086	15 P 30.974	16 S 32.066	17 Cl 35.453	18 Ar 39.948
19 K 39.098	20 Ca 40.078	21 Sc 44.956	22 Ti 47.867	23 V 50.942	24 Cr 51.996	25 Mn 54.938	26 Fe 55.845	27 Co 58.933	28 Ni 58.693	29 Cu 63.546	30 Zn 65.409	31 Ga 69.723	32 Ge 72.61	33 As 74.922	34 Se 78.96	35 Br 79.904	36 Kr 83.80
37 Rb 85.468	38 Sr 87.62	39 Y 88.906	40 Zr 91.224	41 Nb 92.906	42 Mo 95.94	43 Tc* 98.906	44 Ru 101.07	45 Rh 102.91	46 Pd 106.42	47 Ag 107.87	48 Cd 112.41	49 In 114.82	50 Sn 118.71	51 Sb 121.76	52 Te 127.60	53 I 126.90	54 Xe 131.29
55 Cs 132.91	56 Ba 137.33		72 Hf 178.49	73 Ta 180.95	74 W 183.84	75 Re 186.21	76 Os 190.23	77 Ir 192.22	78 Pt 195.08	79 Au 196.97	80 Hg 200.59	81 Tl 204.38	82 Pb 207.2	83 Bi 208.98	84 Po* 208.98	85 At* 209.99	86 Rn* 222.02
87 Fr* 223.02	88 Ra* 226.03		104 Rf* 261.11	105 Db*^a 262.11	106 Sg 	107 Bh 	108 Hs 	109 Mt< br>	110 Uun* 	111 Uuu* 	112 Uub* 		114 Uuq* 		116 Uuh 		

^a provisional IUPAC symbol

57 La 138.91	58 Ce 140.12	59 Pr 140.91	60 Nd 144.24	61 Pm* 146.92	62 Sm 150.36	63 Eu 151.97	64 Gd 157.25	65 Tb 158.93	66 Dy 162.50	67 Ho 164.93	68 Er 167.26	69 Tm 168.93	70 Yb 173.04	71 Lu 174.97
89 Ac* 227.03	90 Th* 232.04	91 Pa* 231.04	92 U* 238.03	93 Np* 237.05	94 Pu* 244.06	95 Am* 243.06	96 Cm* 247.07	97 Bk* 247.07	98 Cf* 251.08	99 Es* 252.08	100 Fm* 257.10	101 Md* 258.10	102 No* 259.10	103 Lr* 260.11

* radioactive element; mass of most important isotope given.

Plant Growth

Fertilizers

HEINRICH W. SCHERER, Agrikulturchemisches Institut, Universität Bonn, Bonn, Federal Republic of Germany (Chap. 1 and 2)

KONRAD MENGEL, Institute for Plant Nutrition, Justus-Liebig-Universität Giessen, Giessen, Federal Republic of Germany (Chap. 1 and 2)

HEINRICH DITTMAR, BASF Aktiengesellschaft, Ludwigshafen, Federal Republic of Germany (Chap. 3 and 5)

MANFRED DRACH, BASF Aktiengesellschaft, Limburgerhof, Federal Republic of Germany (Chap. 4.1, 4.2 and 4.3)

RALF VOSSKAMP, BASF Aktiengesellschaft, Limburgerhof, Federal Republic of Germany (Chap. 4.1, 4.2 and 4.3)

MARTIN E. TRENKEL, Eusserthal, Federal Republic of Germany (Chap. 4.4 and 4.5)

REINHOLD GUTSER, Lehrstuhl für Pflanzenernährung, Technische Universität München-Weihenstephan, Freising, Federal Republic of Germany (Chap. 4.6)

GÜNTER STEFFENS, Landwirtschaftliche Untersuchungs- und Forschungsanstalt, Oldenburg, Federal Republic of Germany (Chap. 4.7)

VILMOS CZIKKELY, BASF Aktiengesellschaft, Ludwigshafen, Federal Republic of Germany (Chap. 6)

TITUS NIEDERMAIER, formerly BASF Aktiengesellschaft, Ludwigshafen, Federal Republic of Germany (Chap. 6)

REINHARDT HÄHNDEL, BASF Aktiengesellschaft, Limburgerhof, Federal Republic of Germany (Chap. 7)

HANS PRÜN, formerly BASF Aktiengesellschaft, Limburgerhof, Federal Republic of Germany (Chap. 7)

KARL-HEINZ ULLRICH, BASF Aktiengesellschaft, Limburgerhof, Federal Republic of Germany (Chap. 8)

HERMANN MÜHLFELD, formerly Chemische Fabrik Kalk GmbH, Köln, Federal Republic of Germany (Chap. 8)

WILFRIED WERNER, Agrikulturchemisches Institut der Universität Bonn, Bonn, Federal Republic of Germany (Chap. 9)

GÜNTER KLUGE, Bundesministerium für Ernährung, Landwirtschaft und Forsten, Bonn, Federal Republic of Germany (Chap. 10)

FRIEDRICH KUHLMANN, Institut für Betriebslehre der Agrar- und Ernährungswirtschaft der Justus-Liebig-Universität Giessen, Giessen, Federal Republic of Germany (Chap. 11.1)

HUGO STEINHAUSER, formerly Lehrstuhl für Wirtschaftslehre des Landbaues, Technische Universität München, Freising, Federal Republic of Germany (Chap. 11.1)

WALTER BRÄNDLEIN, BASF Aktiengesellschaft, Ludwigshafen, Federal Republic of Germany (Chap. 11.2)

KARL-FRIEDRICH KUMMER, BASF Aktiengesellschaft, Limburgerhof, Federal Republic of Germany (Chap. 11.3)

See also: Individual fertilizers are also described under the separate keywords → Ammonia, → Ammonium Compounds, → Nitrates and Nitrites, → Phosphate Fertilizers, → Potassium Compounds, and → Urea

1.	**Introduction**	5
2.	**Plant Nutrition and Soil Science**	7
2.1.	**Plant Nutrients**	7
2.1.1.	Definition and Classification	7
2.1.2.	Function of Plant Nutrients	8
2.2.	**Soil Science**	13
2.2.1.	Soil Classes, Soil Types, and Parent Material	13
2.2.2.	Nutrient Retention in Soils	15
2.2.3.	Soil pH, Buffer Power, and Liming	18
2.2.4.	Soil Water – Plant Relationships	19
2.2.5.	Organic Matter of Soils and Nitrogen Turnover	20
2.3.	**Nutrient Availability**	23
2.3.1.	Factors and Processes	23

2.3.2.	Determination of Available Plant Nutrients in Soils	24	4.4.5.3.	Polymer-Encapsulated Controlled-Release Fertilizers	54
2.4.	**Physiology of Plant Nutrition**	25	4.4.6.	Anti-Float Materials	55
2.4.1.	Nutrient Uptake and Long-Distance Transport in Plants	25	4.4.7.	Controlled-Release Fertilizers on Carriers	55
2.4.2.	Effect of Nutrition on Growth, Yield, and Quality	26	4.4.8.	Supergranules	55
			4.4.9.	Legislation	56
2.5.	**Nutrient Balance**	27	4.5.	**Nitrification and Urease Inhibitors**	56
2.5.1.	Gains and Losses of Plant Nutrients	27			
2.5.2.	Alternative Plant Nutrition	28	4.5.1.	Introduction	56
3.	**Standard Fertilizers**	28	4.5.2.	Types of Nitrification and Urease Inhibitors	57
3.1.	**Solid Fertilizers**	28			
3.1.1.	Straight Fertilizers	28	4.5.3.	Pyridines	58
3.1.2.	Multinutrient Fertilizers	28	4.5.3.1.	Nitrapyrin	58
3.1.3.	Lime Fertilizers	31	4.5.3.2.	Other pyridines	58
3.1.4.	Magnesium Fertilizers	32	4.5.4.	Dicyandiamide	59
3.2.	**Liquid Fertilizers**	32	4.5.5.	Pyrazoles	60
3.2.1.	Nitrogen Liquids	33	4.5.5.1.	1-Carbamoyl-3-methylpyrazole	60
3.2.2.	Multinutrient Liquids	36	4.5.5.2.	Outlook	61
3.2.2.1.	NP Liquids	36	4.5.6.	Neem/Neem-Coated Urea	61
3.2.2.2.	NPK liquids	39	4.5.7.	Urease Inhibitors	61
3.2.2.3.	UAS Liquids	39	4.5.8.	Environmental Aspects	62
3.2.3.	Suspensions	40	4.5.9.	Legal Requirements	62
4.	**Special Fertilizers**	42	4.6.	**Organic Fertilizers (Secondary Raw Material Fertilizers)**	63
4.1.	**Water-Soluble Nutrient Salts**	42			
4.2.	**Foliar Fertilizers**	42	4.6.1.	Fertilizers Based on Peat or Materials of Similar Stability	64
4.2.1.	Production	42			
4.2.2.	Application	43	4.6.2.	Fertilizers Based on Waste Materials of Animal Origin	64
4.2.3.	Combination with Agricultural Pesticides	44			
			4.6.3.	Fertilizers Based on Wastes of Plant Origin	65
4.3.	**Micronutrients**	44			
4.3.1.	Micronutrient Forms	44	4.6.4.	Fertilizers Based on Municipal Waste	66
4.3.2.	Production	45			
4.3.3.	Commercial Fertilizers	46	4.7.	**Manure**	67
4.3.4.	Use	46	4.7.1.	Composition	68
4.4.	**Slow- and Controlled-Release Fertilizers**	47	4.7.2.	Manure Nutrient Efficiency	68
			4.7.3.	Environmental Aspects	69
4.4.1.	Introduction	47	5.	**Fertilizer Granulation**	70
4.4.2.	Urea – Aldehyde Slow-Release Fertilizers	48	5.1.	**Introduction**	70
			5.2.	**Granulator Feedstocks**	72
4.4.2.1.	Urea – Formaldehyde Condensation Products	48	5.3.	**Granulation Equipment**	77
			5.3.1.	Pug Mill	77
4.4.2.2.	Other Urea – Aldehyde Condensation Products	49	5.3.2.	Drum Granulator	77
			5.3.3.	Pan Granulator	81
4.4.2.3.	Further Processing of Urea – Aldehyde Condensates	50	5.3.4.	The Granulator – Mixer	81
			5.3.5.	Roll Presses	82
4.4.3.	Other Organic Chemicals	51	5.4.	**Costs of Agglomeration**	82
4.4.4.	Inorganic Compounds	52	5.5.	**Bulk Blending**	83
4.4.5.	Coated and Encapsulated Controlled-Release Fertilizers	52	5.6.	**Quality Inspection**	84
			5.7.	**Fertilizer Conditioning**	85
4.4.5.1.	Sulfur-Coated Controlled-Release Fertilizers	52	5.8.	**Environmental Aspects**	86
			6.	**Analysis**	86
4.4.5.2.	Sulfur-Coated, Polymer-Encapsulated Controlled-Release Fertilizers	53	6.1.	**Sampling and Sample Preparation**	86
			6.2.	**Determination of Nitrogen**	86

6.3.	Determination of Phosphate	88	9.1.4.	Biosphere	107
6.4.	Determination of Potassium	89	9.1.5.	Pedosphere (Soil)	108
6.5.	Analysis of Calcium, Magnesium, and Trace Elements	89	9.1.6.	Countermeasures	108
			9.2.	**Phosphorus**	109
7.	**Synthetic Soil Conditioners**	89	9.2.1.	Eutrophication	109
7.1.	**Foams**	89	9.2.2.	Heavy Metals Buildup	111
7.1.1.	Closed-Cell Expandable Polystyrene Foam	90	10.	**Legal Aspects**	111
			11.	**Economic Aspects**	113
7.1.2.	Primarily Open-Cell Urea – Formaldehyde Resin Foams	90	11.1.	**Economics of Fertilization**	113
			11.1.1.	Input – Output Relationships: The Yield Function	113
7.2.	**Colloidal Silicates**	91			
7.3.	**Polymer Dispersions and Polymer Emulsions**	92	11.1.2.	Factors Controlling the Optimal Nitrogen Fertilization Level	114
7.4.	**Tensides**	93	11.1.3.	Factors Influencing the Optimal Nitrogen Fertilization Level	115
8.	**Storage, Transportation, and Application**	93	11.1.4.	Environmental Aspects of Fertilization	115
8.1.	**General Storage Requirements**	93			
8.2.	**Application**	96	11.2.	**World Consumption, Production, and Trade**	116
9.	**Environmental Aspects of Fertilizer Application**	98	11.3.	**Future Outlook**	119
			11.3.1.	Food Situation	120
9.1.	**Nitrogen**	99	11.3.2.	Development of Fertilizer Consumption	122
9.1.1.	Ground Water	100			
9.1.2.	Surface Waters	104	12.	**References**	123
9.1.3.	Atmosphere	104			

Fertilizers are products that improve the levels of available plant nutrients and/or the chemical and physical properties of soil. An overview is given over the chemical and physical aspects of plant nutrition uptake and soil properties. The different categories of fertilizers are discussed, and special interest is given on production processes and analyses, including storage and transportation as well as environmental, legal, and economic aspects.

1. Introduction

Fertilizers in the broadest sense are products that improve the levels of available plant nutrients and/or the chemical and physical properties of soil, thereby directly or indirectly enhancing plant growth, yield, and quality.

Fertilizers are classified as follows in terms of their chemical composition:

1) Mineral fertilizers consist of inorganic or synthetically produced organic compounds.
2) Organic fertilizers are waste products from animal husbandry (stable manure, slurry manure), plant decomposition products (compost, peat), or products from waste treatment (composted garbage, sewage sludge).
3) Synthetic soil conditioners are products whose main function is to improve the physical properties of soils, for example, friability and water and air transport.

The following categories are distinguished with respect to nutrient content:

1) Straight fertilizers generally contain only one primary nutrient.
2) Compound (complex or multinutrient) fertilizers contain several primary nutrients and sometimes micronutrients as well.
3) Micronutrient fertilizers contain nutrients required in small quantities by plants, as opposed to macronutrients; quantities range from 1 to 500 g $ha^{-1}a^{-1}$.

Finally, fertilizers can be classified as solid or liquid fertilizers and as soil or foliar fertilizers, the latter being applied exclusively by spraying on an existing plant population.

History. Fertilizing substances were applied even in antiquity. Their use can be attributed to the observation in nature that plants developed especially well in locations where human or animal excreta, ash residues, river mud, or dying plants were left. For example, the Egyptians knew about the fertility of the Nile mud, and the Babylonians recognized the value of stable manure; for example, HOMER mentions manure in the *Odyssey*. PLINY reports that the Ubians north of Mainz used "white earth," a calcareous marl, to fertilize their fields. The Romans acknowledged the advantages of green manuring, cultivating legumes and plowing them under. At the end of the first millenium, wood ash was much used as fertilizer in Central Europe. Not until the beginning of the 19th century did guano, at the suggestion of ALEXANDER VON HUMBOLDT (1800), and Chilean caliche, on the recommendation of HAENKES (1810), come into use as fertilizers. Up to that time, however, it was still believed that the organic matter of soil, humus, was the true source of plant nutrition.

Around 1800, the nutrition problem entered a critical phase in Europe. In 1798, MALTHUS set forth his pessimistic theses, saying that the quantity of food could increase only in arithmetic progression while the population grew geometrically. Combining results obtained by others (SPRENGEL, BOUSSINGAULT) with his own pathbreaking studies, J. VON LIEBIG set forth the theoretical principles of plant nutrition and plant production in *Chemistry in Its Application to Agriculture and Physiology* (1840). He took the view, now considered obvious, that plants require nitrogen, phosphate, and potassium salts as essential nutrients and extract them from the soil. LIEBIG's *mineral theory* was well supported by experimental data of J. B. BOUSSINGAULT (1802 – 1887) in France. He and also J. B. LAWES (1814 – 1900) and J. H. GILBERT (1827 – 1901) in England showed that plants benefit from inorganic N fertilizers. LIEBIG thus became the founder of the theory of mineral fertilizers, and his doctrines led to an increasing demand for them. A number of companies were subsequently founded in Europe to produce phosphate and potash fertilizers. Superphosphate was manufactured for the first time in 1846, in England.

In Germany, this industrial development started in 1855. The importation of saltpeter on a large scale began in the area of the German Federation (56 000 t in 1878). Peruvian guano soon came into heavy use (520 000 t in 1870).

Ammonium sulfate, a coke-oven byproduct, was later recognized as a valuable fertilizer, and the mining of water-soluble potassium minerals was undertaken in the 1860s [1].

The demand for nitrogen that developed at the end of the 19th century soon outstripped the availability of natural fertilizers. A crucial breakthrough came about with the discovery and large-scale implementation of ammonia synthesis by HABER (1909) and its industrial realization by BOSCH (1913).

Around the turn of the century, the technique of hydroponics led to the discovery of other essential plant nutrients. Research showed that plants in general require ten primary nutrients: carbon, hydrogen, oxygen. nitrogen, phosphorus, potassium, calcium, magnesium, sulfur, and iron. JAVILLIER and MAZE (1908) pointed out for zinc and AGULHON (1910) pointed out for boron the nutritional effects on plants. WARINGTON (1923) first described the symptoms of boron deficiency, and BRANDENBURG (1931) clearly recognized dry rot in the sugar beet as boron deficiency. Generally micronutrients were made available to the plant as liquid foliar fertilizer, a method first suggested for iron by GRIS in 1844. By 1950, this list of micronutrients had been expanded to include manganese, copper, and molybdenum.

Almost 70 years ago, serious research began on the best nutrient forms for individual plant species under various soil and climatic conditions. Besides the classical fertilizers, for example, controlled-release fertilizers, improved foliar fertilizers, nutrient chelates, and nitrification inhibitors have been developed in recent decades. This development of new nutrient forms is still in full swing in the special fertilizers sector.

In the developed market economies of Western Europe, the United States, and Japan, however, the level of mineral fertilizer use has not been increasing since the beginning of the 1980s. In some countries, genuine agricultural overproduction has occurred recently. Since better delivery of plant nutrients has led to increasing self-reliance even in the Third World economies (e.g., China, India, Brazil), these countries are not so important as purchasers of nutrients on the world market, so that surpluses cannot be exported without limit. The production of fertilizers is also on the increase in these countries.

Thus overproduction plus regional environmental problems (nitrates entering the ground water) are actually leading to a decrease in mineral fertilizer use in some areas. This decline will be limited by diminishing soil fertility in localities where fertility has been enhanced by decades of proper fertilization.

2. Plant Nutrition and Soil Science

The science of plant nutrition is situated between soil science and plant physiology. It comprises the definition of the elements nutritive for plants; the uptake of plant nutrients and their distribution in the plant; the function of the nutritive elements in plant metabolism; their effect on plant growth; yield formation and quality parameters in crops; soil nutrient exploitation by plant roots; factors and processes that control the plant nutrient availability in soils; toxic elements in soils and their impact on plant growth; the application of plant nutrient carriers (fertilizers) and their turnover in soils; nutrient balance; and the maintenance of soil fertility.

Plant nutrition is considered mainly from two aspects, an agronomic one and an ecological one. The former is focused on the question of fertilizing soil as an efficient means to increase crop yield and to maintain or even improve soil fertility. The latter, the ecological aspect of plant nutrition, is concerned with the nutritive condition of a soil and a location and with its effect on plant growth and plant communities. Since fertilizers are the topic of this article, the agronomic aspects of plant nutrition are treated with greater depth.

The science of plant nutrition is closely associated with the science of soils. The latter comprises a broad field of scientific activity and thus cannot be considered here in all its facets. In this article only those problems of soil science relevant to understanding plant nutrition are treated.

2.1. Plant Nutrients

2.1.1. Definition and Classification

From a scientific point of view, the term plant nutrient is not especially precise. More appropriate is to distinguish between nutritive elements of plants and nutritive carriers. Essential *nutritive elements* for plants are the chemical elements that are required for a normal life cycle and that satisfy the following criteria:

1) A deficiency of the element makes it impossible for the plant to complete its life cycle.
2) The deficiency is specific for the element in question.
3) The element is directly involved in the nutrition of plants because of either its chemical or its physical properties.

According to this definition, the following chemical elements are nutritive elements for plants: C, H, O, N, P, S, K, Ca, Mg, Fe, Mn, Cu, Zn, Mo, B. Further elements, such as Na, Cl, and Si, may affect plant growth positively, and there are particular plant species for which these elements are of great importance. Nevertheless, they are not essential nutritive elements for plants in the strict sense of the definition. Cobalt is required by some bacteria, e.g., by dinitrogen-fixing bacteria and thus may also benefit plant growth indirectly.

Generally it is not the element itself that is provided to and taken up by the plant, but an ion or a molecule in which the nutritive element is present, e.g., C is present in CO_2, P in $H_2PO_4^-$, N in NO_3^- or NH_4^+, and B in H_3BO_3. The particular molecule or ion in which the nutritive element is present is termed the *nutrient carrier*. In the case of metals, the corresponding ion or salts of ion species in question, e.g., K^+, Ca^{2+}, Zn^{2+}, can be considered the carrier. In this sense fertilizers are nutrient carriers.

Plant nutrients may be grouped into macronutrients and micronutrients. *Macronutrients* are required in high amounts and thus are present in plant tissues in much higher concentrations than the *micronutrients*. Carbon, H, O, N, P, S, K, Ca, and Mg belong to the macronutrients. Their concentration in the dry plant matter is in the range 1 – 50 mg/g, except for C, H, and O, which have much higher concentrations (see Table 2). The concentration of the micronutrients in the dry plant matter is in the range 1 – 1000 µg/g.

From the viewpoint of fertilization, those nutrients that are required by plants in high quantities and that must be regularly supplied by fertilization are of particular interest. These nutrients are N, K, P, and to a minor degree also Ca, Mg, and S. Calcium is a *soil nutrient*, which means that it is important for an optimum soil structure. Application of micronutrients is not a common practice, but they are applied at locations where soils are low in a particular micronutrient or where soils may bind this micronutrient very strongly. This is the case for heavy metals (Fe, Mn, Cu, Zn) and B in calcareous and alkaline soils (soils with a high pH value), while Mo is strongly fixed in acid soils. Acid organic soils are known for their low available Cu content.

According to the different quantitative requirements for macronutrients and micronutrients, the former are taken up in much higher quantities than the latter. Thus a wheat stand with a yield potential of 7 t of grain per hectare requires about 100 kg K but only 100 g Cu.

From a physiological point of view, plant nutrients are grouped into four groups, as shown in Table 1. The *first group*, comprising C, H, O, N, and S, includes all major elementary constituents of organic plant matter. Their carriers are present mainly in the oxidized form, and they must be reduced during the process of incorporation. The energy required for this reduction originates directly or indirectly from photosynthetically trapped energy. Assimilation of H is basically an oxidation process, namely, the oxidation of water with the help of light energy (photolysis):

$$H_2O \xrightarrow{h\nu} 2H^+ + 2e^- + 0.5\,O_2$$

The *second group* (P, B, Si) comprises elements that are taken up as oxo complexes in the partially deprotonated (P) and protonated (B, Si) form. The oxo complex is not reduced in the plant cell, but may form esters with hydroxyl groups of carbohydrates, thus producing phosphate, borate, and silicate esters.

The *third group* comprises metals that are taken up from the soil solution in ionic form. They are only partially incorporated into the organic structure of the plant tissue: Mg in the chlorophyll molecule, Mn in the electron donor complex of photosystem II, and Ca^{2+} as countercation of indiffusible anions in cell walls and particularly in biological membranes. Potassium is virtually not incorporated into the organic plant matter. It is only weakly adsorbed by Coulombic forces. There exist, however, some organic molecules that may bind K^+ very selectively (ionophores, see Section 2.4.1). These ionophores are likely to be involved in K^+ uptake.

The *fourth group* comprises heavy metals, of which Fe, Cu, and Zn are taken up as ions or in the form of soluble metal chelates, while Mo is taken up as molybdate. These molecules are easily incorporated into the organic structure, where they serve as essential elements of enzyme systems: Fe in the heme group and in ferredoxin, Mn in arginase [2], Cu in oxidases (polyphenol oxidase, cytochrome oxidase, ascorbate oxidase [3]), Zn in RNA polymerase [4], and Mo in nitrate reductase [5] and nitrogenase [6].

All nutritive elements of plants, therefore, are taken up in the form of inorganic complexes, mostly in oxidized form or as metal ions, i.e., in forms characterized by a low energy level. This is a unique feature of plants, and a feature in which they contrast sharply with animals and most kinds of microorganisms (bacteria and fungi). Animals and most microorganisms must take up food that is rich in chemical energy in order to meet their energy requirements. Plants, at least green plants, meet their energy requirement by converting radiation energy into chemical energy. This energy conversion process is manifest in the reduction of plant nutrient carriers (NO_3^-, SO_4^{2-}, CO_2) as already mentioned. Thus important processes of plant nutrition are closely linked with the unique function of plants in the great cycle of nature, i.e., the conversion of inorganic matter into organic form. Liebig [7] was correct in commenting on plant nutrition: "Die ersten Quellen der Nahrung liefert ausschließlich die anorganische Natur." (The primary source of nutrition is provided exclusively by the inorganic materials in nature.)

2.1.2. Function of Plant Nutrients

Most plant organs and particularly plant parts that are metabolically very active, such as young leaves and roots, are rich in water (ca. 80 – 90 wt % of the total fresh matter), while their organic material is ca. 12 – 18 wt % and their mineral content is 2 – 6 wt %. As shown in Table 2, in the dry matter of plant material O and C are by far the most abundant elements, followed by H, N, and K. The elements C, O, H, and, to some extent, N are mainly structural elements in plant matter. They can, however, form chemical

Table 1. Physiological classification of plant nutritive elements, nutrient carriers, and form in which the nutrient is taken up

Nutritive element	Nutrient carrier	Uptake
First group		
C	CO_2, HCO_3^-	CO_2 by leaves, HCO_3^- by roots
H	H_2O	H_2O by leaves, H_2O and HCO_3^- by roots
O	CO_2, HCO_3^-, O_2	O_2 and CO_2 by leaves, HCO_3^- and O_2 by roots
N	NH_4^+, NH_3, NO_3^-, NO_x	NH_4^+ and NO_3^- by roots, NH_3 and NO_x by leaves
S	SO_4^{2-}, SO_2, SO_3, H_2S	SO_4^{2-} by roots, SO_2, SO_3, and H_2S by leaves
Second group		
P	$H_2PO_4^-$, HPO_4^{2-}	$H_2PO_4^-$ and HPO_4^{2-} by roots
B	H_3BO_3, borates	H_3BO_3 and $B(OH)_4^-$ by roots
Si	silicates	$Si(OH)_4$ by roots
Third group		
K	K^+, K salts	K^+ by roots
Mg	Mg^{2+}, Mg salts	Mg^{2+} by roots
Ca	Ca^{2+}, Ca salts	Ca^{2+} by roots
Mn	Mn^{2+}, Mn salts	Mn^{2+} by roots
Fourth group		
Fe, Cu, Zn, Mo	ionic form or metal chelates, minerals containing these elements	by roots in ionic form or in the form of soluble metal chelates, Mo in the form of the molybdate

groups that are directly involved in metabolic processes, e.g., carboxyl groups, amino groups, hydroxyl groups.

Table 2. Mean content of chemical elements in the dry matter of green plant material

Element	Content, g/kg
O	440
C	420
H	60
N	30
K	20
P	4
All other elements	26

Since in many soils the available N is low, *nitrogen* [7727-37-9] is the most important fertilizer element, and for this reason its function in plant metabolism deserves particular interest. Nitrogen is an essential element for amino acids, proteins, nucleic acids, many coenzymes, and some phytohormones. Basic biochemical processes of meristematic growth, such as the synthesis of proteins and nucleic acids, require N. If this nutrient is not supplied in sufficient amounts, the growth rate is depressed and the synthesis of proteins affected. Nitrogen-deficient plants are characterized by low protein and high carbohydrate contents. This relationship is shown in Table 3 [8].

Nitrogen is also essential for the formation of chloroplasts, especially for the synthesis of chloroplast proteins. Hence N deficiency is characterized by low chlorophyll content; the leaves, especially the older ones, are pale and yellow; the stems thin and the plants small. Nitrogen-deficient plants senescence earlier, probably because of a deficiency of the phytohormone cytokinin. Abundant N supply increases the protein content, especially the content of free amino acids, and often also the content of NO_3^- in plants. An example of this is shown in Table 4 [9]. Excess nitrogen nutrition results in luxurious plants that frequently are susceptible to fungi attack.

Table 3. Effects of N supply on yield of dry matter and the content of organic N and carbohydrates in the dry matter of young timothy plants (*Phleum pratense*) [8]

Yield and content	N supply	
	Low	Sufficient
Yield, g/pot	15.7	20.2
Content, mg/g		
Organic N,	20.5	31.5
Sucrose	46.9	22.6
Fructans *	22.2	9.2
Starch	32.8	11.7
Cellulose	169	184

* Polysaccharides of fructose.

The ratio of N to S in plant matter is ca. 10:1. Hence *sulfur* [7704-34-9] is required in much lower quantities than N. Their functions are, however, similar. Sulfur is an elementary constituent of most proteins; the SH group in involved in various enzymatic processes and it is the reactive group of coenzyme A. Disulfide (S – S) bridges are essential structural elements in the tertiary structure of polypeptides and in many

Table 4. Relationship between N fertilizer rate and nitrogenous fractions in the dry matter of rye grass [9]

Nitrogen fertilizer rate, kg/ha	Nitrogenous fraction, g/kg			
	Total N	Protein N	Free amino acid N	NO_3^- and NO_2^- N
0	13.2	9.8	1.6	0.4
110	18.9	12.6	2.1	0.6
440	37.3	20.6	5.6	3.5

volatile S compounds, such as diallyl disulfide, which is the main component in garlic oil. Mustard oils occurring in many species of the Cruciferae contain a S-glycosidic bond and a sulfuryl group:

$$R-C\begin{array}{c}S-\text{glucose}\\ \\NO-S-OH\end{array}$$

Insufficient S supply results in a decrease of growth rate with extremely low levels of SO_4^{2-} and high concentrations of free amino compounds and NO_3^- in the leaves, which is due to hampered protein synthesis. Sulfur plays an important role in the baking quality of wheat, since the concentration of S compounds in the gluten fraction is responsible for the linkages between the protein molecules [10]. Sulfur deficiency may also affect N_2 fixation of legumes by causing unfavorable conditions in the host plant or because of the relatively high S content of nitrogenase and ferredoxin [11]. Deficiency symptoms of S appear at first in the youngest leaves, which turn light green to yellow. Abundant supply with S results in an accumulation of sulfate in plant tissues.

Sulfur oxide can be taken up by the leaves and metabolized and thus can contribute to the S nutrition of plants. Too high SO_2 concentrations in the atmosphere may be toxic. The toxicity symptoms are necrotic spots in the leaves. According to SAALBACH [12], the critical SO_2 level in the atmosphere for annual plants is 120 µg/m³. For trees and other perennials it is about half this level. The currently much discussed damage to trees in the forest of the Federal Republic of Germany (mainly spruce and silver fir) is not caused by toxic SO_2 levels.

Phosphorus [7723-14-0] is an essential element in nucleic acids and various phospholipids (phosphoglyceride and phosphosphingolipids). In both cases, phosphate is esterified with sugars (nucleic acid) or with alcohol groups of glycerol or sphingosine. Phosphate is also present in various coenzymes; the most prominent is adenosine triphosphate (ATP), which carries a kind of universal energy that is used in a number of biochemical processes. Metabolites and enzymes can be activated by phosphorylation, a transfer of the phosphoryl group from ATP to the metabolite according to the following reactions:

Activation of glucose
Glucose + ATP $\longrightarrow$ Glucose-6-phosphate + ADP

Phosphorylation of an enzyme
Enzyme – OH + ATP $\longrightarrow$ Enzyme–O^{P} + ADP

Such reactions demonstrate the essential role of P not only in plant metabolism but also in all living organisms. Undersupply with P results in a reduced growth rate, and seed and fruit formation is affected. The leaves of P-deficient plants often show a gray dark green color; the stems may turn red. The P reserve in seeds is the Mg (Ca) salt of the inositol hexaphosphate (phytic acid):

Myo–inositol hexaphosphate

$$\text{P} = -\overset{O}{\underset{OH}{\overset{\|}{P}}}-OH$$

Phosphoryl group

The physiological role of boron has remained obscure until now, and therefore various hypotheses with numerous modifications exist concerning the physiological and biochemical role of boron in higher plants. Depending on the pH of the soil, boron seems to be taken up mainly as undissociated boric acid or as the borate anion. Plant species differ in their boron uptake capacity, reflecting differences in boron requirements for growth. However, there is still some controversy about boron translocation in plants. At least in higher plants, a substantial proportion of the total boron content is complexed in the cell walls in in a *cis*-diol configuration [14]. According to BIRNBAUM et al. [13], B is involved in the synthesis of uracil and thus affects UTP formation. (UTP is an essential coenzyme for the synthesis of sucrose and cell-wall components.) Also the synthesis of ribonucleic acid is hampered in the case of B deficiency. Since uracil is an integral part of ribonucleic acid (RNA), the formation of RNA may also be related to the synthesis of uracil. POLLARD et al. [15] suggest that B has a specific influence on plant membranes by the reaction of borate with polyhydroxy compounds.

Boron deficiency appears as abnormal or retarded growth of the apical growing points. The youngest leaves are misshapen and wrinkled and show a darkish blue-green color. The fact that B deficiency primarily affects the apex is in accord with the impaired synthesis of ribonucleic acids required for meristematic growth. High levels of available B in the soil may cause B toxicity in plants. This is mainly the case in arid areas; however, B toxicity can also be the consequence of industrial pollution [16]. The toxicity is characterized by yellow leaf tips followed by progressive necrosis. The leaves take on a scorched appearance and drop prematurely.

Silicon [*7440-21-3*] is not an essential element for plants; however, it has a beneficial effect on various plant species, mainly grasses [17]. In plants well supplied with Si, cuticular water losses are diminished and resistance against fungal attack is improved [18]. The favorable effect of Si on rice growth is well known. Silicon-containing fertilizer is frequently applied in rice production.

Among the metal cation species, *the potassium* [*7440-09-7*] *ion*, K^+, is the nutrient plants take up from the nutrient medium at the highest rates. The K^+ concentration in the cytoplasm is about 100 mM and thus much higher than the concentration of other ion species [19]. Probably this high K^+ concentration has a favorable influence on the conformation of various enzyme proteins [20]. Potassium ions can easily penetrate plant membranes (see Section 2.4.1), which often leads to a depolarization of the membranes. Membrane depolarization, it is supposed, has a favorable effect on meristematic growth, photophosphorylation, aerobic phosphorylation, and phloem loading [21]. These basic processes are important for the long-distance transport of photosynthates, the synthesis of various organic compounds, and CO_2 assimilation.

The data in Table 5 show that with an increase of K^+ in alfalfa leaves (*Medicago sativa*), the CO_2 assimilation rate increased, while the mitochondrial respiration rate decreased [22]. In the case of low K, the respiration was about 2/3 of the CO_2 assimilation, while with high K the C gained by assimilation was about 11 times higher than the C lost by respiration. This typical behavior indicates that under the conditions of K^+ deficiency much of the stored carbohydrates must be respired in order to meet the ATP demand of the plant. Plants undersupplied with K^+ have therefore a low energy status. Such plants are highly susceptible to fungal attack, water stress, and frost damage.

Table 5. Relationship between K^+ concentration in the dry matter of alfalfa leaves, CO_2 assimilation, and mitochondrial respiration [22]

Concentration of K^+, mg/g	Carbon gain and loss, mg dm^{-2} h^{-1}	
	CO_2 assimilation	Mitochondrial respiration
13	11.9	7.56
20	21.7	3.34
38	34.0	3.06

Potassium is important in determining the osmotic pressure of plant fluids, and K^+-deficient plants are characterized by inefficient water use. Sodium ions may replace some K^+ functions, e.g., the less specific osmotic functions. Important counterions of K^+ in plant tissues are Cl^-, NO_3^-, and organic anions. The frequently observed favorable effect of Na^+ and Cl^- on plant growth is related to their osmotic functions.

Plants suffering from K^+ deficiency show a decrease in turgor, and under water stress they easily become flaccid. Plant growth is affected, and the older leaves show deficiency symptoms as necrosis beginning at the margins of tips and leaves. In K^+-deficient plant tissue, toxic amines such as putrescine and agmatine accumulate.

The most spectacular function of *magnesium* [*7439-95-4*] is its integral part in the chlorophyll molecule. Besides this function, Mg^{2+} is required in various other processes and, the Mg fixed in the chlorophyll molecule amounts only to about 20 % of the Mg present in green plant tissues. Magnesium is an essential ion in ribosomes and in the matrix of the cell nucleus. Here Mg^{2+} is bound by phosphate groups, since the Mg^{2+} is strongly electrophilic and thus attracts oxo complexes such as phosphate [23]. The magnesium ion activates numerous enzymatic reactions in which phosphate groups are involved. The activation is assumed to be brought about by bridging the phosphate group with the enzyme or with the substrate. This is an universal function of Mg^{2+} not only relevant for plant metabolism but also for practically all kinds of organisms.

Deficiency of Mg^{2+} affects chlorophyll synthesis: leaves turn yellow or red between the veins. The symptoms begin in the older leaves. Protein synthesis and CO_2 assimilation are depressed under Mg^{2+} deficiency conditions. Recent results [24] have shown that the yellowing of spruce needles in the Black Forest is due to a Mg^{2+} deficiency and can be cured by Mg^{2+} fertilizer application.

Calcium [7440-70-2] is the element of the apoplast (cell wall and "free space") and of biological membranes. Here it is adsorbed at the phosphate head groups of membrane lipids, thus stabilizing the membranes [25]. Most of the Ca^{2+} present in plant tissues is located in the apoplast and in the vacuole, some in the mitochondria and in the chloroplasts, while the cytoplasm is extremely low in Ca^{2+} (10^{-7} to 10^{-6} M). The maintenance of this low cytoplasmic Ca^{2+} concentration is of vital importance for the plant cell [26]. Higher cytoplasmic Ca^{2+} concentrations interfere with numerous enzymatic reactions and may even lead to a precipitation of inorganic phosphates. This low Ca^{2+} concentration suffices to form a complex with calmodulin, a polypeptide of 148 amino acids. The Ca – calmodulin complex is a universal enzyme activator. The activation is brought about by allosteric induction.

Direct Ca^{2+} deficiency in plants is rare, since most soils are relatively rich in Ca^{2+}. Physiological disorders as a consequence of an insufficient Ca^{2+} supply of particular plant parts, however, occur frequently. Calcium is mainly translocated by the transpiration stream. Hence plant parts such as fruits, which mainly feed from the phloem and less from the xylem sap, may suffer from an insufficient Ca^{2+} supply. Shear [27] cites a list of 35 such Ca^{2+}-related disorders in fruits and vegetables. Two of the most important ones involve storage tissues and result in poor crop qualities [28]: *bitter pit* in apples, characterized by small brown spots on the surface, and *blossom-end rot* in tomatoes, a cellular breakdown at the distal end of the fruit, which is then susceptible to fungal attack.

Manganese [7439-96-5] is an integral part of the superoxide dismutase and of the electron donor complex of photosystem II. Manganese may activate enzymes in the same way as Mg^{2+} by bridging the phosphate group with the enzyme or the substrate. Deficiency of Mn^{2+} leads to the breakdown of chloroplasts. Characteristic deficiency symptoms are smaller yellow spots on the leaves and interveinal chlorosis. Manganese toxicity may occur, especially on flooded soils, because of the reduction and thus solubilization of manganese oxides. Toxicity symptoms are generally characterized by brown spots of MnO_2 in the older leaves surrounded by chlorotic areas [29].

Iron [7439-89-6] is an essential element for haem and ferredoxin groups. Iron deficiency leads to chloroplast disorders; the synthesis of thylakoid membranes is disturbed and the photochemical activity affected [30]. Iron deficiency is characterized by yellow leaves. The symptoms are at first visible in the younger leaves. There is evidence that the deficiency, mainly occurring in plants growing on calcareous soils, is not induced by an insufficient Fe uptake from the soil but by a physiological disorder in leaves, affecting the Fe distribution in the leaf tissue [31].

Iron toxicity can be a problem under reducing soil conditions, which prevail in flooded soils. Under such conditions iron(III) oxides are reduced and the iron is rendered soluble. This may increase the Fe concentration in the soil solution by a factor of 10^2 to 10^3 [32] so that plants may suffer from Fe toxicity, characterized by tiny brown spots on the leaves, which later may turn uniformly brown. Iron toxicity is known as "bronzing."

Copper [7440-50-8] is an essential element of various enzymes, such as superoxide dismutase, polyphenol oxidases, plastocyanin of the photosynthetic transport chain, and cytochrome c oxidase, the terminal oxidase in the mitochondrial electron transport chain. Deficiency in Cu leads to pollen sterility and thus affects the fruiting of plants. Copper-deficient plants often are characterized by white twisted leaf tips and a tendency to become bushy.

Zinc [7440-66-6] is an integral part of carbonic anhydrase, superoxide dismutase, RNA polymerase, and various dehydrogenases. It is closely involved in the N metabolism of plants. In Zn-deficient plants, protein synthesis is hampered and free amino acids accumulate. There is evidence that Zn is involved in the synthesis of tryptophan, which is a precursor of indole acetic acid, an important phytohormone. Zinc deficiency is characterized by short internodes, small leaves, and chlorotic areas in the older leaves. Frequently the shoots die off and the leaves fall prematurely.

Molybdenum [7439-98-7] is present in the nitrate reductase and in the nitrogenase system that catalyzes the bacterial fixation (reduction) of dinitrogen. Deficiency of Mo frequently appears first in the middle and older leaves as a yellowish green coloration accompanied by a rolling of leaf margins. Cruciferae species are particularly susceptible to Mo deficiency. The most well-known Mo deficiency is the "whiptail" of cauliflower. For further information on the physiology of plant nutrition, see [3, 23, 33].

2.2. Soil Science

2.2.1. Soil Classes, Soil Types, and Parent Material

According to SCHROEDER [34], "soil is the transformation product of mineral and organic substances on the earth's surface under the influence of environmental factors operating over a very long time and having defined organisation and morphology. It is the growing medium for higher plants and basis of life for animals and mankind. As a space-time system, soil is four dimensional."

Soils are complex, quite heterogeneous, and may differ from each other considerably. Nevertheless, all soils have some common features. They possess a mineral, an organic, a liquid, and a gaseous component. In an ideal soil, the percentage proportions of these components are 45 %, 7 %, 23 %, and 25 %, respectively. The volumes of the liquid and gaseous components may change quickly. For example, in a water-saturated soil all pores are filled with water, and in a dry soil the soil pore volume is almost completely filled with air. The mineral and organic components contain plant nutrients and adsorb plant nutrients at their surfaces, and they are therefore of importance for the storage and retention of plant nutrients. The liquid phase of the soil is the soil solution. It contains dissolved plant nutrients and is the medium for the translocation of plant nutrients from various soil sites towards the plant roots. The gaseous soil component is essential for gas exchange, especially for the supply of plant roots with oxygen and for the release of CO_2 from the soil medium into the atmosphere.

For the description, comparison, and assessment of soils, a grouping according to general criteria is indispensable. There are two main grouping systems for soils: (1) soil classes or soil texture and (2) soil types. *Textural classes* are defined according to the particle size of soils. *Soil types* relate to the parent material of soils, to the pedological genesis, and to typical properties evident in the soil profile i.e., the horizontal layers of soils, called soil horizons.

Soil Classes. *Soil particle sizes* as a main characteristic of soil classes are grouped into four major groups as shown in Table 6. The major groups (sand, silt, and clay) are subdivided into coarse, medium, fine. Designation of the soil texture (soil class) depends on the percentage proportions of the sand, silt, and clay fraction in the total fine earth, which is sand + silt + clay. Soils in which the sand fraction dominates are termed *sandy soils*, soils consisting mainly of silt and clay are *silty clays*, and soils which contain all three fractions in more or less equal amounts are called *loams*. In the German terminology, abbreviations for the fractions are used (S = sand, U = silt, T = clay, L = loam). For example, if the major fraction is silt (U) and the next sand (S), the abbreviation is sU = sandy silt. Figure 1 shows the designations of the various soil classes according to the percentage proportion of the three main particle fractions.

In the farmer's practice, sandy soils are called light soils, soils rich in clay heavy soils. This distinction relates to the force required to work (plough, cultivate) a soil. Soils rich in clay, but also silty soils, tend to compaction when dried and hence are heavy to work.

Table 6. Particle size of soil fractions relating to soil texture

Diameter, mm	Designation	Abbrevation
> 2	pebbles, gravels	
0.06 – 2	sand	S
0.002 – 0.06	silt	U
< 0.002	clay	T

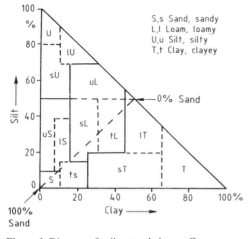

Figure 1. Diagram of soil textural classes, German system of SCHROEDER [34]
The vertical axis shows the percentage of silt, the horizontal the percentage of clay, and the dashed line the percentage of sand.

Although the grouping according to particle size is based on a physical factor, particle size is also associated with the chemical properties. This can be seen from Figure 2: the sand fraction consists mainly of quartz, which is a sterile material. Primary silicates (micas, feldspars) contain K, Ca, Mg, and other plant nutrients, which are released during the process of weathering. Clay minerals are less rich in plant nutrients than the primary silicates, but they possess large negatively charged surfaces that are of the utmost importance for the adsorption of plant nutrients and water.

The various soil particles form aggregates in which organic matter is also involved. This aggregation forming fine pores and holes in the soil is of relevance for soil structure. A good soil structure is characterized by a relatively high pore volume, ca. 50 % of the total soil volume. Soil structure depends much on the Ca saturation (see page 16). The richer the soil is in clay, the more important a good soil structure is.

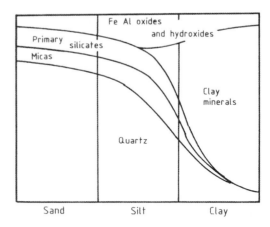

Figure 2. Mineral composition of the sand, silt, and clay fractions [34]

Soil Type. Soil type is related to the parent material from which a soil is developed and from the history of development, which is much influenced by climate and vegetation. Main groups of parent material are igneous rocks, sedimentary rocks, and metamorphic rocks. Also organic matter may be the main parent material. Content of plant nutrients, capacity to store plant nutrients, soil pH, and the rooting depth depend much on the parent material, but are also influenced by soil development.

In the following, a limited number of important soil types are considered according to the FAO World Soil Classification. Besides this system there are other systems, e.g., the U.S. Soil Taxonomy. The FAO classification comprises 26 classes.

A distinction can be made between young soils and old soils. The latter are generally highly weathered, their inorganic material consisting mainly of quartz and iron aluminum oxide hydroxides. Such soils are characterized by poor cation retention capacity (cation exchange capacity), low pH values, and a high phosphate fixing power. This soil type, called ferralsol, is frequent in the tropics, whereas in moderate climates highly weathered soils belong mainly to the podsols.

Young soils may be derived from the sedimentation of rivers and oceans (fluvisols) or from volcanic ash (andosols). These soils are generally rich in plant nutrients and thus form fertile soils. The most fertile soils belong to the black earths (chernozem). They are frequent in Russia, Central and East Europe as well as North America and are derived from loess. They are characterized by a neutral pH, by a well balanced content of clay and organic matter and by a deep rooting profile. They are naturally rich in plant nutrients and possess a high nutrient storage capacity. Soils in which the parent material loess is more weathered as compared with the chernozems belong to the luvisols. This soil type is common in Germany, Austria, and France where it represents the most fertile arable land. Gleysols are soils with a high water table, rendzinas are shallow soils derived from limestone, histosols are rich in only partially decomposed organic matter.

Under arid conditions salt may accumulate in the top soil layer. Solonchaks (white alkali soils) are saline soils with a pH of ca. 8 and with neutral anions as the most important anion component. Solonetz soils (black alkali soils) possess bicarbonate and carbonate as major anion component. Their pH is in the range 8 – 10. Crop growth on saline soils is extremely poor, and in many cases only a salt flora can grow under such conditions. This is particularly true for the solonetz soils.

For further information on soil texture and soil types, see [34 – 37], and the Soil Taxonomy of the Soil Conservation Service of U.S. Department of Agriculture [38].

2.2.2. Nutrient Retention in Soils

Nutrient retention is an important characteristic of fertile soil.

Cation Exchange. Cations are retained on soil colloids having a negative charge: the cations are bound at the surface of these particles by Coulomb forces. The most important cation species are Ca^{2+}, Mg^{2+}, K^+, Na^+, Al^{3+}, $Al(OH)^{2+}$, $Al(OH)_2^+$, and H^+. This is represented in Figure 3. A distinction can be made between inorganic and organic soil colloids capable of cation adsorption. Inorganic particles belonging to the clay fraction are known as *secondary clay minerals* because they are mainly derived by weathering of primary minerals such as orthoclase, plagioclase, and particularly mica. Organic soil colloids capable of cation adsorption belong to the *humic acids*. The negative charge of the inorganic soil colloids originates from the so-called isomorphic substitution and from deprotonation. Isomorphic substitution is the replacement of Si^{4+} in the crystal lattice by Al^{3+}, Fe^{2+}, or Mg^{2+}, thus leading to a surplus of negative charge, because the anionic groups of the lattice are not completely balanced by Al^{3+}, Fe^{2+}, or Mg^{2+}. Such a negative charge is a permanent charge, in contrast to labile charges that result from deprotonation. Labile charges are typical for organic colloids (humic acids): carboxylic groups and acid hydroxylic groups of phenols may be protonated or not depending on the pH of the environment.

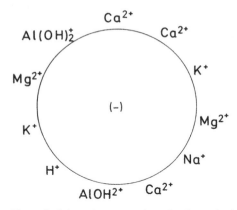

Figure 3. Schematic presentation of cations adsorbed to the negatively charged surface of a soil colloid

The secondary clay minerals are grouped into 1 : 1 clay minerals, in which a Si layer alternates with an Al layer, and 2 : 1 clay minerals, in which an Al layer is sandwiched by two Si layers. The most important representatives of the 1 : 1 clay minerals are the kaolinites. The 2 : 1 secondary clay minerals comprise the illite, transitional minerals, vermiculite, chlorite, and smectites ($\rightarrow$ Clays). Most of these 2 : 1 clay minerals possess inner surfaces. They are therefore characterized by a high cation retention (= cation exchange) capacity.

These negatively charged soil colloids, often also called sorption complexes, function like a *cation exchanger*. Adsorbed cations can be replaced by other cation species. The cation exchange is stoichiometric. Adsorption and desorption depend on the concentrations of the cation species in the surrounding solution. If a soil colloid completely saturated by K^+ is exposed to increasing Ca^{2+} concentrations, for example, adsorbed K^+ is more and more replaced by Ca^{2+} until eventually the sorption complex is completely saturated by Ca^{2+} (Fig. 4). In soils, such exchange and equilibrium reactions are complex as numerous cation species and sorption complexes with differing preferences for particular cation species are involved. The principle, however, is that cations adsorbed by Coulomb forces at soil colloids equilibrate with free cations in the soil solution. Thus adding cations to a soil by fertilization, e.g., the application of a potassium salt, results in replacing adsorbed cations with the newly added cations until a new equilibrium is reached. The adsorbed cations are protected against leaching. but they are available to plant roots. The strength of cation adsorption increases with the the charge of the cation species and with the thinness of the hydration shell. Provided that there are no specific adsorption sites, the strength of cation adsorption follows Hofmeister's cation sequence:

$Ca^{2+} > Mg^{2+} > K^+ > Na^+$

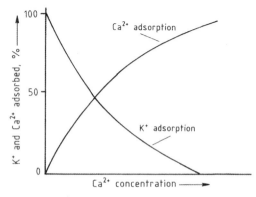

Figure 4. $Ca^{2+} - K^+$ exchange, K^+ desorption brought about by increasing Ca^{2+} concentration

At equilibrium, cation-exchange reactions are a helpful tool for predicting the distribution of ions between the adsorbed and solution phases of the soil as the amounts of cations present are changed. When a soil saturated with potassium is placed in a NaCl solution, the following equilibration occurs:

$K_{soil} + NaCl \rightleftharpoons Na_{soil} + KCl$

The exchange equation for this reaction is

$$\frac{[Na](K)}{[K](Na)} = k_1$$

Brackets refer to ions on the exchange site and parenthesis to the activity of ions in the solution. Since the proportionate strength of adsorption of the ions varies with the exchange site, values for k_1 differ for different exchange materials.

The divalent/monovalent system, which almost represents the situation in the soil, with K^+, Ca^{2+}, and Mg^{2+} as the dominant exchangeable cations, is more complex. The following equation, developed by GAPON [39], is widely used to describe monovalent/divalent exchange:

$$\frac{[K](Ca)^{1/2}}{[Ca](K)} = k_1$$

Cation exchange capacity (CEC) is defined as the quantity of cation equivalents adsorbed per unit soil or clay mineral. In Table 7 the exchange capacities of some soil classes are shown. The exchange capacity of the organic soil appears high if it is based on unit weight of soil. A more realistic picture is obtained, however, when the exchange capacity is based on soil volume, since under field conditions it is soil volume, not soil weight, that is related to a crop stand. Table 8 shows the cation exchange capacities of some important clay minerals and of humic acids in relation to the surface of these particles.

The cation exchange capacity of kaolinites and particularly of humic acids depends much on the pH of the medium. At low pH, most groups are protonated and hence the exchange capacity is low. Increasing soil pH, e.g., by liming, increases the cation exchange capacity if kaolinites and humic acids are the dominating exchange complexes.

Cation saturation of negatively charged soil colloids has some impact on soil structure, which is defined as the arrangement of soil particles

Table 7. Cation exchange capacity based on soil weight and soil volume as well as the specific weight of some soil classes

Soil class	Specific weight, kg/L	Cation exchange capacity *	
		cmol/kg	cmol/L
Sandy soil	1.5	3	4.5
Loam	1.5	15	22.5
Clay soil	1.5	30	45.0
Organic soil	0.3	75	22.5

* cmol = centimole.

Table 8. Cation exchange capacity and inner and outer surfaces of some soil colloids

	Total surface, m^2/g	Inner surface, %	Cation exchange capacity, mol/kg
Kaolinite	20	0	10
Illite	100	0	30
Smectite	800	90	100
Humic acids	800	0	200

into aggregates. High percentage of adsorbed Ca^{2+} favors the formation of aggregates. In well structured soils, such as in chernozems, 70 to 80 % of the total cation exchange capacity is occupied by Ca^{2+}. In acid solids, H^+ and Al cations (Al^{3+}, $Al(OH)^{2+}$, $Al(OH)_2^+$) and in saline soils Na^+ and Mg^{2+} are the dominating cation species adsorbed to soil colloids.

Anion Exchange. Soil particles may also adsorb anions. The adsorption occurs at the OH groups of aluminum and iron oxides as well as of some clay minerals. One may distinguish between a nonspecific adsorption and a specific anion adsorption. The nonspecific anion (A^-) adsorption originates from protonated hydroxylic groups.

$$M-O\begin{pmatrix}H\\H\end{pmatrix}^+ A^-$$

Protonation depends on soil pH and is particularly high under acid conditions. Hence nonspecific anion adsorption only plays a role in acid soils.

The specific anion adsorption is a ligand exchange. This is, for example, the case for phosphate. In step 1 $H_2PO_4^-$ replaces OH^-, resulting in a *mononuclear bond* between the phosphate and the iron oxide. In step 2, a further deprotonation of the phosphate occurs, followed by a second ligand exchange (step 3) to form a *binuclear bond* between the surface of the iron oxide and the phosphate.

The final structure is supposed to be very stable, and the phosphate so bound is hardly available to plant roots. This reaction sequence explains why anion (phosphate) adsorption is promoted under low pH conditions. In mineral soils with pH < 7, the adsorbed phosphate represents a major phosphate fraction. Increasing the soil pH, e.g., by liming, increases phosphate availability [40]. The relationship between free and adsorbed anions can be approximately described by the Langmuir equation:

$$A = A_{max}\frac{kc}{1+kc}$$

A	= surface concentration of adsorbed anions
A_{max}	= maximum surface concentration
c	= concentration of free anion
k	= constant related to adsorption energy, the adsorption strength increasing with k

Adsorption strength depends also on anion species decreasing in the order [41]:

phosphate > arsenate > selenite = molybdate > sulfate = fluoride > chloride > nitrate

Borate and silicate may also be adsorbed, but only at high pH. Under these conditions, boric acid and silicic acid may form anions according to the following equations:

$$H_3BO_3 + H_2O \rightarrow B(OH)_4^- + H^+$$

$$H_2SiO_3 + H_2O \rightarrow H_3SiO_4^- + H^+$$

This is why in neutral to alkaline soils boron can be strongly adsorbed (fixed) by soil particles, which may lead to boron deficiency in plants. The formation of a silicate anion can improve phosphate availability since $H_3SiO_4^-$ and phosphates compete for the same ligands at anion-adsorbing surfaces.

2.2.3. Soil pH, Buffer Power, and Liming

Proton concentration (pH) is of vital importance for all living organisms and also has an impact on soils and soil constituents. High H^+ concentrations (pH < 4) attack soil minerals, dissolving metal cations out of the crystal lattice, and eventually lead to mineral degradation. Under low soil pH conditions (pH < 5), bacterial life is suppressed while fungal life is relatively favored, which affects the decomposition of organic matter. Low soil pH (pH < 4) also affects root growth.

In many cases, however, it is not so much the H^+ but the toxic level of soluble Al species and Mn^{2+} associated with low pH conditions that considerably hamper root development and plant growth. A decrease in soil pH increases the solubility of aluminum oxide hydroxides and maganese oxides considerably. Under acid soil conditions Al^{3+}, $Al(OH)^{2+}$, $Al(OH)_2^+$ are dissolved in the soil solution or adsorbed to soil colloids. Of the three Al species, $Al(OH)^{2+}$ is considered to be the most toxic. *Aluminum toxicity* of plants is characterized by poor root growth: root tips and lateral roots become thickened and turn brown [42]. In the cell, Al interferes with the phosphate turnover and may even be adsorbed on the DNA double helix.

A distinction is made between the *actual acidity*, which is determined by the H^+ concentration in the soil solution, and the *potential acidity*, which is determined by both the H^+ of the soil solution and the adsorbed H^+. The actual acidity is measured by hydrogen-ion electrodes; the potential acidity, by titration with a base. Titration includes the Al species, which also consume OH^-

$$Al^{3+} + OH^- \rightarrow Al(OH)^{2+}$$

$$Al(OH)^{2+} + OH^- \rightarrow Al(OH)_2^+$$

$$Al(OH)_2^+ + OH^- \rightarrow Al(OH)_3 \text{ (gibbsite)}$$

Potential acidity may be high in Al-rich soils as well as in soils with a high cation exchange capacity. Such soils are characterized by a high *hydrogen-ion buffer power*: a pH change in the soil solution requires a relatively large amount of H^+ or OH^-.

In soils, different buffer power systems can be distinguished. Soils containing carbonates are buffered according to the following equation:

$$CaCO_3 + 2 H^+ \rightarrow Ca^{2+} + H_2O + CO_2$$

Reaction may be promoted by the presence of CO_2 originating from root and microbial respiration.

$$CaCO_3 + CO_2 + H_2O \rightarrow Ca(HCO_3)_2$$

$$Ca(HCO_3)_2 + 2 H^+ \rightarrow Ca^{2+} + 2 H_2O + 2 CO_2$$

At lower pH levels soils are buffered by adsorbed cations (Fig. 5) and by Al complexes, e.g.,

$$Al(OH)_3 + H^+ \rightarrow Al(OH_2)^+ + H_2O$$

Soil acidification results from different processes. Plant roots extrude H^+; net release of H^+ is especially high when plants are fed with NH_4^+, while NO_3^- nutrition results in a net release of OH^-. Leguminous species living in symbiosis with *Rhizobium* extrude H^+ at particularly high rates from their roots. Microbial oxidation of organic N, organic S, and elementary S leads to the formation of strong acids, such as HNO_3 and H_2SO_4, which have a marked acidifying effect on soils. Also SO_2 and NO_x as gases as well as the acid rain formed by these oxides have an acidifying effect. In Central Europe these pollutants carry ca. 3 – 5 kmol H^+ $ha^{-1}a^{-1}$ into soils.

Under anaerobic soil conditions, e.g., after flooding, protons are consumed by the reduction of Fe^{III} and Mn^{III} or Mn^{IV} as well as by the microbial reduction of NO_3^- (denitrification):

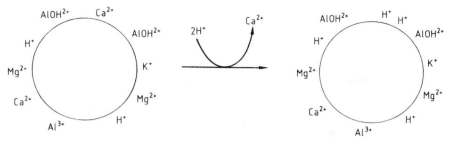

Figure 5. Principle of hydrogen-ion buffering by adsorbed cations

$$2\,NO_3^- + 10\,e^- + 12\,H^+ \rightarrow N_2 + 6\,H_2O$$

Some species of *Azalea, Calluna, Vaccinium*, and also tea (*Camellia sinensis*) are able to grow on acid habitats. These species can mask the Al with phenols and organic acids and thus avoid Al toxicity. Rye, potatoes, oats, and lupines tolerate weakly acid soils, whereas beets (*Beta vulgaris*), barley, rape, and most leguminous crops prefer more neutral soils. Wheat takes an intermediary position with regard to soil pH. In the case of leguminous crops, it is not so much the crop itself, rather it is the *Rhizobium* bacteria living in symbiosis with the crop that are affected by low soil pH. The mulitplication of the *Rhizobium* in the soil is depressed by soil acidity.

Low pH levels can be easily overcome by *liming*, the application of alkaline materials, mainly Ca/Mg oxides, carbonates, and silicates. They react with soil acidity as follows:

$$CaO + 2\,H^+ \rightarrow Ca^{2+} + H_2O$$

$$CaCO_3 + 2\,H^+ \rightarrow H_2O + CO_2 + Ca^{2+}$$

$$CaSiO_3 + 2\,H^+ \rightarrow H_2O + SiO_2 + Ca^{2+}$$

The quantity of lime required depends on the soil pH level and the buffer power. The lower the pH and the higher the buffer power, the more lime required.

Soil acidity is mainly a problem in humid zones, where the H^+ formed in the upper soil layer replaces the adsorbed metal cations (Ca^{2+}, Mg^{2+}, K^+), which are then leached. Under arid conditions, salts may accumulate in the top soil layer. If a major part of the anions accumulated are HCO_3^- and CO_3^{2-} (solonetz soils), high soil pH levels prevail, which affect plant growth and soil structure considerably. Such soils can be meliorated by heavy applications of elementary sulfur. Under aerobic conditions the S is oxidized to H_2SO_4 by soil microorganisms. The strong acid neutralizes the HCO_3^- and CO_3^{2-}

$$H_2SO_4 + 2\,HCO_3^- \rightarrow 2\,H_2CO_3 + SO_4^{2-} \rightarrow 2\,H_2O + 2\,CO_2 + SO_4^{2-}$$

$$H_2SO_4 + CO_3^{2-} \rightarrow SO_4^{2-} + H_2CO_3 \rightarrow H_2O + CO_2 + SO_4^{2-}$$

2.2.4. Soil Water – Plant Relationships

Plants continuously require water that is taken up from the soil by roots and transported from them to the upper plant parts, particularly to the leaves. From here water is released into the atmosphere. This last process is called *transpiration*. Water in the plant tissues is required to maintain optimum cell turgor, which is crucial for most metabolic processes.

Plants frequently have to overcome long dry periods during which they must feed from the soil water. Very crudely, the soil can be considered as a sponge that can store water in its pores and holes. The storage water in the soil must be retained against gravitation. The forces responsible for this retention are adsorption and capillary forces by which the water is sucked to the surface of the soil particles. This suction force can be considered as a negative pressure, and hence the strength of water binding in soils is measured in Pascals (Pa), the unit for pressure. The strength of water binding in soils is termed *water potential* (in older terminology, water tension). The higher the strength, the lower (more negative) the soil water potential. Water potentials in soils range from 0 to -1×10^6 kPa. Generally, however, soil water potentials of -10 to -1500 kPa prevail.

The total amount of water that can be adsorbed by a soil (all pores and holes filled with water) is called maximum water capacity. This is of minor importance; more relevant is the water quantity that can be retained against the gravitation force, the *field capacity*. Not all water of the field capacity fraction is available to plant roots. A proportion of the field capacity water is so strongly adsorbed that it can not be taken up by the roots. For most plant species this is water with a water potential < -1500 kPa: at such a low water potential, plants wilt. Therefore, this critical water potential is also called *wilting point*. The soil water fraction not available to plants is called *dead water*. The *available soil water* thus equals the difference between the field capacity and dead water.

Soils differ much in their capacity to store water. The higher the clay content of a soil, the larger the total surface of soil particles, and the more water that can be adsorbed. The water storage capacity of soils increases with the clay content. On the other hand, the water molecules are strongly adsorbed to the surface of clays and therefore the fraction of dead water increases with the clay content. For this reason, generally medium textured soils (loamy soils), and not the clay soils, possess the highest storage capacity for available soil water. Besides soil texture, also soil structure and the rooting depth of the soil profile determine the storage capacity for available soil water.

An important criterion of available soil water is the relationship between the percent water saturation of the soil and the water potential. This is shown in Figure 6 for a sandy, a loamy, and a clay soil. The section between field capacity (-10 kPa) and the wilting point (-1500 kPa) is the highest for the loamy soil.

The capability to use soil water economically differs considerably among plant species. A measure of this capability is the transpiration coefficient, the water quantity in kg (or L) required for the production of 1 kg plant dry matter:

Sorghum	277
Maize	349
Sugar beet	443
Spring wheat	491
Barley	527
Potatoes	575
Oats	583
Spring rye	634
Red clover	698
Flax	783
Alfalfa	844

Water loss under a vegetation cover results from evaporation (water release from the soil to the atmosphere) and transpiration (water release of the plant to the atmosphere). Evaporation is unproductive, transpiration productive. The relation between the two depends on plant nutrition, as can be seen in Table 9, which shows the favorable effect of N fertilizer on the productive use of soil water [43].

2.2.5. Organic Matter of Soils and Nitrogen Turnover

Organic matter of soils differs considerably. Soils can be classified according to the content of organic carbon in the soil (g/kg):

Low	<5
Medium low	5–10
Medium	10–20
High	20–40
Rich	40–80
Muck	80–150
Peat	>150

Soils with an organic carbon content of 50 g/kg are termed organic soils, in contrast to mineral soils. Enrichment of organic matter in soils depends on location and climatic conditions. Low temperature and a lack of oxygen favor the accumulation of organic matter in soils because these conditions hamper breakdown by soil microbes. Therefore, under cold continental climate conditions (frost, long winters) and under hydromorphic soil conditions (swamps, moors, bogs), organic matter accumulates.

The fertility status of organic soils differs considerably. Moors located on the tops of mountains are generally poor in nutrients, especially in N and K, since they are fed mainly from rain. Moors in lowland fed from rivers and streams may be rich in plant nutrients, especially

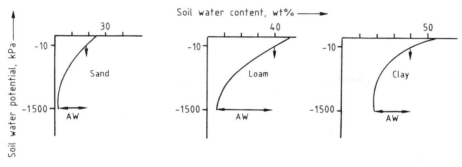

Figure 6. Relationship between the soil water content and the soil water potential and the resulting available water (AW) for three soil classes
Field capacity is -10 kPa; wilting point is -1500 kPa.

Table 9. Relationship between N fertilizer rate, grain yield of barley, evaporation, and transpiration [43]

Nitrogen fertilizer rate, kg/ha	Grain yield, t/ha	Transpiration,* L/m^2	Evaporation,* L/m^2
30	1.02	85	235
125	1.65	121	278
225	2.69	217	212

* Transpiration and evaporation in liters per square meter of soil surface.

in N. The C:N ratio of the organic matter and soil pH are suitable indicators of the fertility status. Fertile soils possess C:N ratios of ca. 20 in their organic matter and pH values in the weak acid to weak alkaline range. Acid organic soils (highland moors) have much higher C:N ratios, ca. 50 or more in their organic matter. The C:N ratio has direct impact on the decomposition of organic N by soil microbes; the higher the ratio the lower the net release of mineral N by microbial activity.

Nitrogen turnover in soils is related not only to biological processes but also to physicochemical processes. In addition, there is a rapid exchange of N between the biosphere, the soil, and the atmosphere. The main processes of this nitrogen cycle are shown in Figure 7. Inorganic nitrogen, mainly NO_3^- and NH_4^+, including fertilizer N, can be easily assimilated by higher plants as well as by microbes (fungi, bacteria). Also dinitrogen (the N_2 of the atmosphere) can be reduced to NH_3 by some soil bacteria. The N_2 fixation capacity of the so-called free living bacteria, mainly species of *Azotobacter, Beijerinckia, Azospirillum,* and some species of the Cyanobacteria (*Anabaena, Nostoc, Rivularia*) is moderate, amounting to ca. $5-50$ kg ha^{-1}a^{-1}. Symbiontic N_2-fixing bacteria, mainly species of *Rhizobium* and *Actinomyces,* have a fixation capacity about 10 times higher: for pulses (grain legumes) $50-100$ kg/ha per growth period and for forage legumes even $200-500$ kg ha^{-1}a^{-1}. They are of utmost importance in the N turnover and N availability in soils.

Inorganic nitrogen (N_2, NH_4^+, NO_3^-) assimilated by living organisms is mainly used for the synthesis of proteins, amino sugars, and nucleic acids. As soon as these organisms die, the organic N can be attacked by other microorganisms, which are able to convert the organic N into an inorganic form, a process called nitrogen mineralization. This starts with ammonification, and under aerobic conditions and favorable soil pH ammonification is followed by nitrification in the sequence:

Organic N $\rightarrow NH_4^+ \rightarrow NO_2^- \rightarrow NO_3^-$

Ammonification is carried out by a broad spectrum of heterotrophic organisms; nitrification only by a small number of autotrophic bacteria [44]. The microbial oxidation of NH_4^+ to NO_2^- and NO_2^- to NO_3^- requires oxygen and hence proceeds only under aerobic conditions. The oxidation of NH_4^+ to NO_2^- is brought about by species of *Nitrosomonas, Nitrosolobus,* and *Nitrospira,* oxidation of NO_2^- to NO_3^- by species of *Nitrobacter.* They all require weak acid to neutral soil conditions; in acid soils nitrification is more or less blocked.

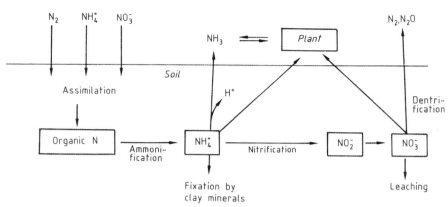

Figure 7. N cycle in nature. Transfer of N between soil, plant, and atmosphere

Ammonium ions produced by microbial breakdown of organic N, including urea, as well as NH_4^+ fertilizer can also be fixed by 2:1 clay minerals. In this form, NH_4^+ is protected against nitrification and leaching, but, depending on the type of clay minerals, may still be available to plant roots [45]. This fixed NH_4^+ fraction is of major significance for plant nutrition in soils derived from loess.

The concentration of ammonium ion in the soil solution is governed by the equilibrium

$$NH_4^+ \rightleftharpoons NH_3 + H^+ \quad pK = 9.25$$

At pH < 6, there is virtually no NH_3 present; with an increase in pH the deprotonation of NH_4^+ increases, and so does the risk of NH_3 loss by volatilization. In alkaline and calcareous soils considerable amounts of N can thus be lost by the soil system [46]. High losses of NH_3 may also occur from the application of slurries, which generally have an alkaline pH [47].

Ammonium as well as NO_3^- are taken up by plant roots at high rates, and vigorous crop stands can deplete the NO_3^- concentration in the soil to a great extent. Nitrate is very mobile in soils since it is virtually unadsorbed on soil colloids. It can be leached by rainfall to deeper soil layers or even into the ground water (see Section 2.5.1). Nitrate losses may also occur under anaerobic soil conditions, for some bacterial species are able to use the oxygen of the NO_3^- as e^- acceptor for respiration. Nitrate is thus reduced to volatile NO, N_2O, and N_2 [49]:

$$NO_3^- \rightarrow NO_2^- \rightarrow NO_{(g)} \rightarrow N_2O_{(g)} \rightarrow N_{2(g)}$$

This process, brought about mainly by species of *Pseudomonas*, *Alcalignes*, *Azospirillum*, *Rhizobium*, and *Tropionibacterium*, is called denitrification. It may cause considerable soil N losses particularly in flooded rice soils, in which anaerobic conditions prevail [50].

Loss of NO_3^- by leaching or denitrification can be reduced by blocking the NO_2^- formation by application of nitrification inhibitors such as Nitrapyrin (2-chloro-6-trichloromethylpyridine), AM (2-amino-4-chloro-6-trimethylpyrimidine), or terrazole (5-ethoxy-3-trichloromethyl-1,2,4-thiadizole). There are also some natural compounds that are nitrification inhibitors [51]. The most important is neem, which occurs in the seeds of *Azadirachta indica*, a tree common in the tropics.

Some of the organic N in the soil may be incorporated into a very stable organic form. This nitrogen, which mainly occurs in humic acids, is hardly mineralized. In most soils the humus-N fraction is by far the largest, comprising 80–90 % of the total soil N. It is of great importance for soil structure, but has hardly any relevance as a nutrient reserve.

The fraction of hydrolyzable soil N can be mineralized and thus may serve as a source for N absorbed by plants. The most important fraction in this respect is the N of the biomass, which comprises ca. 40–200 kg/ha, thus only a small fraction of total soil N, which may be 2000–8000 kg/ha in arable soils.

2.3. Nutrient Availability

2.3.1. Factors and Processes

From the total amount of N present in the soil, only a small proportion can be made available for plants (see above). This is also true for other plant nutrients. For example, in a clay-rich soil ca. 200 000 kg K^+ may be present in 1 ha within the rooting depth of 80 – 100 cm, but only 1 % may be available to plant roots.

Plant nutrient availability depends on physicochemical and biological factors. A young root pushing into a soil directly contacts only a small amount of macronutrients, which would contribute only a few percent of the total nutrient demand. By far the greatest proportion of nutrients (NO_3^-, NH_4^+, K^+, Ca^{2+}, Mg^{2+}) required by the plant must be transported towards the plant roots. This transport can be brought about by mass flow and/or diffusion. In *mass flow*, the nutrients are moved with the water flow from the soil towards the roots. Therefore mass flow depends much on water uptake and transpiration conditions. At zero transpiration (100 % relative humidity, rainy, or foggy weather) mass flow is also zero. Mass flow plays a major role for the transport of Ca^{2+}, Mg^{2+}, and also for NO_3^- in cases where the NO_3^- concentration of the soil solution is high, e.g., after fertilizer application.

Nutrients that are taken up at high rates by plant roots (K^+, NH_4^+; NO_3^-, phosphate) but that have relatively low total concentration in the soil solution are mainly transported by *diffusion*. Uptake of nutrients by roots decreases the nutrient concentration near the root surface and establishes a concentration gradient, which drives the diffusive flux of nutrients from the soil towards the plant roots. Absorbing roots thus act as a sink for plant nutrients. Typical depletion profiles are shown in Figure 8 for phosphate [52]. Phosphate and K^+ concentrations in the soil solution at the root surface may be as low as 1 µM, whereas in the bulk soil solution concentrations of 50 to 300 µM phosphate and 500 to 1000 µM K^+ may prevail. Fertilizer application increases the nutrient concentration of the soil solution and hence the concentration gradient that drives the nutrients towards the roots. Therefore, the level of the nutrient concentration in the bulk soil solution is an important factor of nutrient availability.

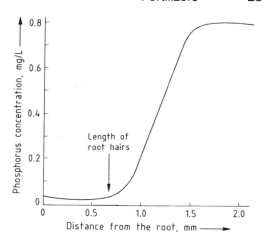

Figure 8. Phosphate depletion around a corn root: P concentration in the soil solution as a function of the distance from the root surface [52]

Diffusive flux and mass flow in the soil depend much on soil moisture. The dryer a soil, the smaller the water cross section (the size of soil pores that are still filled with water) and the more the nutrient flux is hampered. Therefore, soil moisture is another important factor of nutrient availability [53]. A third important factor of nutrient availability is the *nutrient buffer power* of a soil, a factor of particular relevance for phosphate, K^+, and NH_4^+. Here buffer power means the capability of a soil to maintain the nutrient concentration level in the soil solution (in analogy to the hydrogen-ion buffer power, see Section 2.2.3). In well-buffered soils nutrients absorbed from the soil solution by roots are replenished by nutrient desorption, e.g., by cation exchange.

The most important biological factor for nutrient availability is root growth [54]. For most plant nutrients, only the soil volume around the root can be exploited by the plant. For phosphate the depletion zone extends only a few millimeter from the root surface and depends much on the length of root hairs, as can be seen from Figure 8. For K^+ and NH_4^+, the depletion zone is more extended, ca. 1 – 4 cm from the root surface. Root mass, root length, and root hairs therefore are of great importance for the portion of soil volume that can be exploited by a crop stand.

Roots excrete organic materials such as organic acids, sugars, and slimes from which bacteria feed. As a result, the bacterial colonization in the rhizosphere (the volume around the root) is much denser than in the soil apart from the roots. These bacteria are involved in the N turnover, e.g., for the N_2 fixation of free-living bacteria and for denitrifying bacteria [55].

Proton excretion of roots affects bacterial activity in the rhizosphere, the dissolution of calcium phosphates, and the cation exchange. Net proton release of plant roots is strongly affected by the type of N supply. Ammonium nutrition results in a high net release of H^+, nitrate nutrition in release of OH^- and HCO_3^-. Leguminous species living in symbiosis with *Rhizobium* are known for a high H^+ release by roots and therefore have a strong acidifying effect on soils.

Nutrient deficiency, e.g., phosphate or Fe deficiency, also increases net release of H^+, which may contribute to the dissolution of iron oxides and calcium phosphates. The release of avenic acid and mugineic acid by plant roots is of particular importance for the mobilization of Fe in the rhizosphere [56].

Plant growth may be enhanced after infection with mycorrhizal fungi, which leads to increased nutrient uptake due to increases in the effective absorptive surface of the root, mobilization of sparingly available nutrient sources, or excretion of ectoenzymes or chelating compounds. Furthermore, mycorrhizal fungi may protect roots from soil pathogens [57] and and in this way enhance root growth and nutrient acquisition of the host plant. This is particularly important when considering the nutrition of plants with immobile nutrients such as phosphorus, as fungal hyphae are known to absorb P and translocate it into the host plant [58].

2.3.2. Determination of Available Plant Nutrients in Soils

The level of available plant nutrients in soils can be assessed by means of plant analysis and soil analysis. In the case of *plant analysis*, the nutrient content of a particular plant organ at a certain physiological stage may reflect the nutritional status of the plant, hence also the nutrient status of the soils [59]. Such diagnostic plant analysis is particularly common for perennial crops, including fruit trees and forest trees.

In *soil analysis*, soil samples are extracted with special extractants. The quantity of nutrients extracted reflects the level of available nutrients in the soil. Numerous soil extractants have been developed. In Central Europe the DL method (double lactate method) and the CAL method (calcium acetate lactate method) are widely used for the determination of available soil K and P. In the Netherlands, available P is extracted by water (P-water method). In the United States, the Olsen method (extraction with NH_4F + HCl) is used for the determination of available soil phosphate [60]. Ion-exchange resins are also useful tools for the determination of available plant nutrients [61].

Of particular interest is the determination of available soil N early in spring before the first application of N fertilizer. Mineral N (NO_3^-, exchangeable NH_4^+) is extracted with a $CaCl_2$ or K_2SO_4 solution [62]. This technique, called the N_{min} method, provides reliable information on the level of directly available soil N. The easily mineralizable N in the soil, which frequently is the important fraction for the release of available N during the growth period, is not obtained by the N_{min} method.

Electro-ultrafiltration (EUF method) has been used for the determination of available soil nutrients [63]. This method uses an electric field to separate nutrient fractions from a soil suspension. Most plant nutrients can be extracted. The advantage of this method is that, besides inorganic nitrogen, the readily mineralizable organic N fraction is extracted [64].

In the last decade S has attracted interest as a plant nutrient. Numerous procedures have been proposed for the determination of plant-available S in soils. The procedures include extraction with water, various salts and acids, and S mineralization by incubation [65].

The relationship between soil analysis data and the response of crops to fertilizer application is not always satisfying since other factors may interfere, such as rooting depth, root morphology, soil moisture, and particularly the clay content. These factors should be taken into consideration in interpreting soil test data.

2.4. Physiology of Plant Nutrition

2.4.1. Nutrient Uptake and Long-Distance Transport in Plants

Oxygen and CO_2 are mainly taken up by aboveground plant parts. The process of uptake is a diffusion of CO_2 and O_2 into the plant tissue. For the entry of these gases into the plant, the stomata are of major importance. Water and other plants nutrients are mainly absorbed from the soil solution. The rate of nutrient uptake increases with the concentration of the particular nutrient in the soil solution, the rate of uptake leveling off at higher soil-solution concentrations.

Plant nutrients in the soil solution, mainly present in ionic form, diffuse into the root tissue. The outer plasma membrane of the cells (plasmalemma) is a great diffusion barrier. The transport of nutrients across this barrier is the proper process of ion uptake. This transport is not mere diffusion, but is related to specific membrane components and to metabolic processes that allow selective uptake of the plant nutrients, which is often associated with an accumulation of the nutrient in the cell. For example, the K^+ concentration in the cell (cytoplasm) may be higher by a factor of $10^2 - 10^3$ than the K^+ concentration in the soil solution.

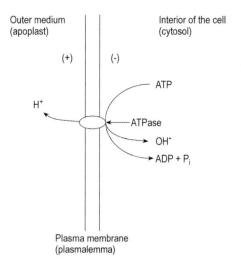

Figure 9. Scheme of plasmalemma-located ATPase, hydrolyzing ATP and pumping H^+ into the apoplast (proton pump)

Nutrient uptake is initiated by an enzyme located in the plasma membrane called ATPase (ATP hydrolase). Its substrate is ATP. Hydrolyzation of ATP results in the splitting of H_2O into H^+ and OH^-, from which the latter remains in the cytosol of the cell while H^+ is extruded into the outer medium (Fig. 9). Thus an electrochemical potential is created between the two sides of the membrane. The proton motive force (p.m.f.) obtained in this way is described by the following equation [66]:

$$\text{p.m.f.} = -50\,\Delta\text{pH} + \Delta\varphi \quad \varphi = \text{electrical charge}$$

The p.m.f. is the driving force for ion uptake. Cations are directly attracted by the negatively charged cell. Since the plasma membrane, however, represents a strong barrier, the entry of cation species must be mediated by particular carriers and ion channels. Little is known about these carriers and channels in plant membranes. These are assumed to be ionophores like valinomycin, nonactin, or gramicidin, which bind selectively to cation species and hence mediate a selective cation transport across the membrane. Such a type of carrier transport is shown in Figure 10. The carrier is hydrophobic and therefore quite mobile in the membrane, which consists mainly of lipids. At the outer side of the membrane it combines selectively with a cation species, e.g., K^+. The cation carrier complex then diffuses to the inner side of the membrane, where the K^+ is released. Release and combining with K^+ depend on the electrochemical difference between the two sides of the membrane. High K^+ concentration and a positive charge favor the combining process; low K^+ concentration and a negative charge favor the release of K^+. Net transport rate becomes zero as soon as the electrochemical equilibrium is attained, which is governed by the Nernst equation:

$$E = \frac{RT}{zF} \ln \frac{a_o}{a_i}$$

E = electrical potential difference between the two sides of the membrane
F = Faraday constant
z = oxidation state of the cation
a_o = activity of the cation species in the outer solution
a_i = activity of the cation species in the inner solution

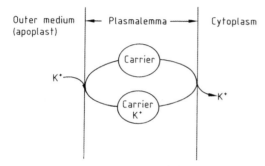

Figure 10. Scheme of K^+ carrier transport across the plasmalemma

The uptake of anions (NO_3^-, $H_2PO_4^-$) is also assumed to be driven by the plasmalemma-located ATPase. The anions presumably form protonated carriers at the outer side of the membrane and then are selectively tranported across the membrane. The protonated carrier – anion complexes are positively charged. Hence the electropotential difference between either side of the plasmamembrane represents the driving force for anion uptake.

Ion absorbed by cells of the root cortex are translocated via the symplasm in centripetal direction towards the central cylinder, where they are secreted into the xylem vessels. The actual process of this secretion is not yet understood. In the xylem the ions are translocated to the upper plant parts with the transpiration stream. They thus follow the vascular system of the plant and are distributed along the major and minor vein system of leaves from where they diffuse into the pores and intercellular spaces of cell wall (apoplast). The transport from the apoplast across the plasmamembrane into the cytoplasm of leaf cells is a process analogous to the nutrient uptake of root cells.

Some plant nutrients, such as N, P, K, and Mg, but not Fe and Ca, may also be translocated against the transpiration stream via the phloem tissue. These nutrients therefore may be transported from the tops to the roots or from older leaves to younger leaves.

2.4.2. Effect of Nutrition on Growth, Yield, and Quality

Meristematic growth requires plant nutrients: N and P for the synthesis of proteins and nucleic acids, K and Mg for the activation of enzymes and for membrane potentials, and all the other nutrients for various processes. The quantities required differ greatly but for practical purposes mainly N, K, and P, in some cases also Mg and Ca, limit plant growth. The growth rate is controlled by the nutrient with the lowest availability (LIEBIG's law of minimum). Application of this particular nutrient results in a growth response. This response is not linear, but rather follows a saturation type of curve (Fig. 11). Also crop yields as a function of increasing rates of fertilizer applicaton reflect this curve, which is also called the *Mitscherlich curve* since MITSCHERLICH [67] investigated these relationships extensively. The curve is described by the following equation:

$$\log(A-y) = \log(A-cx)$$

A = maximum yield
y = obtained yield
x = growth factor, e.g., fertilizer rate
c = constant

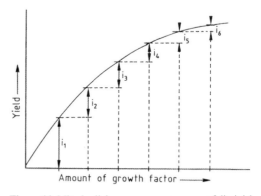

Figure 11. Mitscherlich curve, response curve of diminishing increments (i) with each further unit of growth factor

The term $(A - y)$ is the increment that is required to attain the maximum yield. This increment becomes smaller as the variable x (growth factor) becomes greater. Such growth factors include not only plant nutrients, including CO_2, but also light intensity and temperature. The growth response obtained by these factors also follows the saturation type of curve shown in Figure 11.

Growth and metabolic processes in plant tissues depend not only on the rates of plant nutrients supplied but also on the ratio in which the nutrients are provided. If the N supply is relatively high as compared with the supply of other nutrients, the synthesis of N-containing compounds, such as amino acids and proteins, is promoted. This may have a favorable effect on the protein synthesis in grains of cereals and hence improve their baking quality and their nutritional value. In grains for malting purposes and in particular barley, however, low protein content and high starch content are required. In this case, relatively high N rates have a detrimental effect on grain quality. An analogous case is the sugar beet, which should be high in sugar and low in N-containing compounds, especially amino acids. Relatively high N supply favors the synthesis of vitamins of the vitamin B group and the synthesis of carotenes in green plant tissue but has a negative effect on the content in vitamin C. High N rates may also increase the sensitivity of leaves and culms to fungi attack.

Phosphate and especially K^+ have a favorable impact on the energy status of plants. A relatively high supply of both nutrients promotes the synthesis of carbohydrates and the development of cell wall material, which increases the resistance against fungi attack.

2.5. Nutrient Balance

2.5.1. Gains and Losses of Plant Nutrients

In order to maintain the level of available plant nutrients in soils, the quantity of nutrients lost from the soil must be replenished. Nutrients may be lost by harvesting of plant material, leaching, transition into a non-available form in the soil, and volatilization. Nutrient gains result from fertilizer application, soil weathering, rainfall, and microbial dinitrogen fixation (see Section 2.2.5). In most soils, weathering provides enough of those nutrients that are only required in minor amounts, such as Fe, Mn, Cu, Zn, Mo, and B. Young soils being rich in primary minerals also may weather appreciable amounts of Mg, K, and phosphate. Generally, the contribution of plant nutrients released by weathering is more important if the yield level is lower. Up to the mid-1980s, sulfur supply to crops was to a large extent brought about by SO_2 in the atmosphere and by SO_2 dissolved in precipitation. However, in the last two decades increased global environmental concern has prompted reductions in SO_2 emissions, which in past years were beneficial for crop growth in many countries [68]. With the reduction in the sulfur supply from atmospheric emissions, the greater use of high-analysis S-free fertilizers, new high-yielding crop varieties, and declining use of sulfur as a fungicide, sulfur deficiency became widespread in various parts of the world.

The quantity of nutrients taken up by crops depends mainly on the yield level and on the kind of plant organs. For a rough calculation of *nutrient uptake by crops* the following figures may be used:

1 t of grain (cereals) contains
   20 – 25   kg N
   5   kg K
   4   kg P
1 t of dry matter of green plant material (leaves, stems) contains
   15 – 30   kg N
   15 – 30   kg K
   2 – 4   kg P
1 t of roots (beets) or tubers contains
   1.5 – 3   kg N
   4 – 5   kg K
   0.3 – 0.6   kg P

Nutrient losses only occur from plant parts removed from the field. For example, if only grain is harvested and the straw remains on the field, only the plant nutrients present in the grain are lost from the system.

High nutrient losses may occur by *leaching*. The quantities lost depend much on climate, weather conditions, and soil properties. In Central Europe, with a rainfall of about 700 mm/a, ca. 25 – 50 % of the precipitation passes through the soil profile to a depth >1 m. Plant nutrients dissolved in this drainage water are lost from the soil. Table 10 shows maximum and minimum leaching rates obtained during a period of eight years by Vömel [69]. Generally, leaching rates are low if the soil clay content is high. High leaching rates prevail in fallow soils, low rates in soils with a permanent plant cover, e.g., grassland. Phosphate is effectively not leached. Leaching occurs mainly with the winter or monsoon rainfall. Under such humid conditions plant nutrients present in plant residues (stubble, straw, roots, leaves) also may be leached.

Table 10. Leaching rates of plant nutrients from soils [69]

Soil	Plant nutrient, kg ha^{-1} a^{-1}			
	N	K	Ca	Mg
Sand	12 – 52	7 – 17	110 – 300	17 – 34
Sandy loam	0 – 27	0 – 14	0 – 242	0 – 37
Loam	9 – 44	3 – 8	21 – 176	9 – 61
Clay	5 – 44	3 – 8	72 – 341	10 – 54

Volatile losses only play a role for nitrogen, which may be lost in form of NH_3, N_2, and N_2O (see Section 2.2.5). Nutrient loss by conversion into a non-available form is only relevant for phosphate and in some cases for K^+.

2.5.2. Alternative Plant Nutrition

The terms "ecological agriculture" and "biological agriculture" are used as synonyms for "organic agriculture". One of the main approaches of organic agriculture is a mixed farm system within a more or less closed nutrient cycle. With regard to plant nutrition the production system can be characterized by the following principles:

- Nearly closed cycles of nutrients and organic matter within the farm
- Predominant use of farmyard manure and compost
- Slowly soluble P minerals, if necessary
- K fertilizers not in the form of chloride

Besides nitrogen, which may be imported with manure, symbiotically bound nitrogen is the main N source. For this reason N_2 fixation by legumes is of great importance. Synthetic fertilizers are renounced. In the last few years major efforts were undertaken to quantify the cycles of nutrients as well as the organic matter cycles.

Organic agriculture has to deal with limited amounts of nutrients. Nutrient management, defined as systematic target-oriented organization of nutrient flow, is therefore considered as the optional combination of resources that are restricted or have to be released. Strategies must be adopted that make nutrients in the system internally available by achieving optimized utilization or which keep nutrients potentially available in the long term. The main nutrient flows in organic farm are fixed for the long term by organizing and optimizing the site-adapted crop rotation [70].

3. Standard Fertilizers

Standard fertilizers include the products used in large quantities worldwide. They are applied to agricultural and large-scale garden crops. They can be classified as solid and liquid fertilizers.

3.1. Solid Fertilizers

Solid fertilizers are the most important group of fertilizers. Worldwide, nearly 90 % of all the nitrogen applied in fertilizers is in solid form (season or fertilizer year 1996/97). In Germany about 83.5 % of total nitrogen is applied in the solid form, and the corresponding figure in the United States is 44 %. Phosphate and potash are applied mainly in solid form. Solid fertilizers include granular, prilled, and compacted products (see Chap. 5).

A summary of the most important fertilizers can be found in [71].

3.1.1. Straight Fertilizers

Straight fertilizers contain only one nutrient, for instance, urea (N) or triple superphosphate (P_2O_5). Straight solid fertilizers are listed in Table 11.

3.1.2. Multinutrient Fertilizers

Multinutrient or compound fertilizers contain two or more nutrients. The term complex fertilizer refers to a compound fertilizer formed by mixing ingredients that react chemically. In bulk-blend or blended fertilizers (see Chap. 5), two or more granular fertilizers of similar size are mixed to form a compound fertilizer.

There are several routes for manufacturing fertilizers. An important route is the Odda process, in which phosphate rock is digested with

Table 11. Straight fertilizers

Type	Production; main constituents	Analysis, wt %	Remarks	Ullmann keyword
Nitrogen fertilizers				
Calcium nitrate [13477-34-4]	synthetic	15.5 % N (Ca(NO$_3$)$_2$), NH$_4$ N, max. 1.5 %	fast acting; chiefly for special crops and refertilizing	→ Nitrates and Nitrites
Sodium nitrate [7631-99-4]	synthetic	16 % N (NaNO$_3$)	fast acting; chiefly for special crops and refertilizing	→ Nitrates and Nitrites
Chile saltpeter	from caliche; sodium nitrate	≥ 15 % N	fast acting; chiefly for special crops and refertilizing	
Ammonium nitrate [6484-52-2]	synthetic; may contain additives such as ground limestone, calcium sulfate, dolomite, magnesium sulfate, kieserite	34 % N(NH$_4$NO$_3$)	special storage provisions (see Chap. 8)	→ Ammonium Compounds, Chap. 1
Calcium ammonium nitrate (CAN)	synthetic; may contain additives such as ground limestone, calcium sulfate dolomite, magnesium sulfate, kieserite	26 – 28 % N, ≥ 2 % CaCO$_3$	versatile fertilizer, both fast and slow acting	→ Ammonium Compounds
Ammonium sulfate nitrate (ASN)	synthetic	added as limestone or dolomite 26 % N (NH$_4$NO$_3^+$ and (NH$_4$)$_2$SO$_4$), ≥ 5 % NO$_3$ N	versatile fertilizer with both fast and slow action	→ Ammonium Compounds
Magnesium ammonium nitrate	synthetic; ammonium nitrate and magnesium compounds (dolomite, magnesium carbonate, or magnesium sulfate)	20 % N (NH$_4$NO$_3$)	for soils and crops requiring magnesium	
Ammonium sulfate [7783-20-2]	synthetic, often as byproduct	≥ 6 % NO$_3$ N, 5 % MgO 21 % N ((NH$_4$)$_2$SO$_4$)	slow acting, for soils containing lime and/or high-precipitation areas	→ Ammonium Compounds, Chap. 2
Ammonium chloride [12125-02-9]	synthetic	≥ 25 % N (NH$_4$Cl)	rice fertilization, chiefly in Japan and India	→ Ammonium Compounds, Chap. 3
Urea [57-13-6]	synthetic	46 % N (CO(NH$_2$)$_2$) biuret ≤ 1.2 %	optimal only when soil lime, heat, and moisture are appropriate; also suitable for foliar application; volatile losses are possible	→ Urea
Calcium cyanamide [156-62-7]	synthetic	20 – 22 % N; ≥ 75 % of total N from cyanamide; ca.40 % CaO 66 % N	slow acting, special requirements, also for weed control nitrification	→ Cyanamides, Chap. 1
Dicyandiamide [461-58-5]	synthetic		nitrification inhibitor, added pure to liquid manure; also as additive to straight and compound fertilizers containing ammonium	→ Cyanamides, Chap. 3

Table 11. (continued)

Type	Production; main constituents	Analysis, wt %	Remarks	Ullmann keyword
Phosphate fertilizers				
Basic slag (Thomas phosphate)	byproduct of steel production from ores containing phosphate; calcium silicophosphate	$\geq 12\%$ P_2O_5 soluble in mineral acid, $\geq 75\%$ soluble in 2% citric acid	phosphate containing up to 45% CaO, versatile phosphate fertilizer	→ Phosphate Fertilizers
Superphosphate	digestion of ground phosphate rock with sulfuric acid; monocalcium phosphate with calcium sulfate	16 – 22% P_2O_5 soluble in neutral ammonium citrate solution, 93% water soluble	for all soils	→ Phosphate Fertilizers
Triple superphosphate	digestion of ground phosphate rock with phosphoric acid; monocalcium phosphate	45% P_2O_5; soluble in neutral ammonium citrate solution, 93% water soluble	for all soils	→ Phosphate Fertilizers
Partly digested phosphate rock	partial digestion of ground phosphate rock with sulfuric or phosphoric acid; monocalcium phosphate, tricalcium phosphate, calcium sulfate	$\geq 20\%$ P_2O_5; soluble in mineral acid, 40% water soluble	for all acidic soils (pH < 7)	→ Phosphate Fertilizers
Soft phosphate rock	grinding of soft phosphate rock; tricalcium phosphate and calcium carbonate	$\geq 25\%$ P_2O_5 soluble in mineral acid, 55% soluble in 2% formic acid	for acidic soils and low-precipitation areas	→ Phosphate Fertilizers
Dicalcium phosphate [7789-77-7]	neutralization of phosphoric acid with calcium hydroxide dicalcium phosphate dihydrate	$\geq 38\%$ P_2O_5 soluble in alkalized ammonium citrate; not water-soluble	for all soils	→ Phosphoric Acid and Phosphates
Thermal (fused) phosphate	thermal digestion of ground phosphate rock with alkali-metal compounds and silicic acid; alkali-metal calcium phosphate and calcium silicate	$\geq 25\%$ P_2O_5 soluble in alkalized ammonium citrate solution and in citric acid	for all soils and locations	
Aluminum calcium phosphate	thermal digestion and grinding; amorphous aluminum and calcium phosphates	29 – 33% P_2O_5 soluble in mineral acid, 75% soluble in alkalized ammonium citrate solution	for all soils and locations	
Potassium fertilizers				
Potash ore	run-of-mine potash	$\geq 10\%$ K_2O, $\geq 5\%$ MgO, contains sodium	for soils and crops where Na and Mg are important, especially cattle fodder	→ Potassium Compounds
Beneficiated potash ore	beneficiation of run-of-mine potash and mixing with potassium chloride	$\geq 18\%$ K_2O	for all soils	→ Potassium Compounds
Potassium chloride [7447-40-7]	beneficiation of run-of-mine potash	$\geq 37\%$ K_2O (up to 60%)	for all crops not sensitive to chloride	→ Potassium Compounds
Potassium chloride with magnesium	beneficiation of run-of-mine potash and addition of Mg salts	$\geq 37\%$ K_2O, $\geq 5\%$ MgO	on Mg-deficient soils for all crops not sensitive to chloride	→ Potassium Compounds
Potassium sulfate [7778-80-5]	synthetic	$\geq 47\%$ K_2O, $\leq 3\%$ Cl	for all crops sensitive to chloride	→ Potassium Compounds
Potassium sulfate with magnesium	synthetic, from potassium sulfate, with addition of Mg salts; potassium sulfate and magnesium sulfate	$\geq 22\%$ K_2O, $\geq 8\%$ MgO, $\leq 3\%$ Cl	deficient soils for all crops sensitive to chloride	→ Potassium Compounds

nitric acid. Calcium nitrate is crystallized by cooling and removed, and the mother liquor is neutralized by addition of gaseous ammonia (see → Phosphate Fertilizers, Chap. 11). Other important processes are based on digestion of phosphate rock with sulfuric acid and the so-called mixed-acid process [355]. For the production of NPK fertilizers, potassium compounds are added in the desired amount to the corresponding slurries. Magnesium (as kieserite or dolomite) and micronutrients may also be added.

NPK Fertilizers. According to the EEC Guidelines, NPK fertilizers must contain at least 3 % N plus 5 % P_2O_5 plus 5 % K_2O and at least 20 % total nutrients. The most commonly used grades (N-P_2O_5-K_2O, each in wt %) are

Nutrient ratio 1 : 1 : 1
15–15–15, 16–16–16, 17–17–17
Nutrient ratios 1 : 2 : 3 and 1 : 1.5 : 2
5–10–15, 6–12–18, 10–15–20
Nutrient ratio 1 : 1 : 1.5 – 1.7
13–13–21, 14–14–20, 12–12–17
Nutrient ratios 3 : 1 : 1 and 2 : 1 : 1
24–8-8, 20–10–10
Low-phosphate grades
15–5-20, 15–9-15

If additional numbers are given in a fertilizer grade, the fourth is the wt % MgO and the fifth is wt % S. Micronutrient contents may also be stated. In some countries the grade is expressed in terms of the elements rather than oxides.

The NPK fertilizers have the important advantage of simplified application, since all the important nutrients can be distributed in one operation. Each grain of fertilizer has the same content of nutrients. Serious errors are prevented by the harmonic nutrient ratio, provided the metering rate is correct. The nitrogen is usually present as nitrate and ammonium N in roughly equal parts. The phosphate is 30 – 90 % water-soluble, the rest being soluble in ammonium citrate solution. Most of the potassium is present as the chloride; the sulfate is used for chloride-sensitive crops.

In bulk-blended products (see Section 5.5), the individual fertilizers are combined in the desired nutrient ratio. The operational advantage is the same as that of synthetic NPK fertilizers. Because, however, the individual granular products may differ in grain-size spectrum, granule surface characteristics, and density, segregation may occur during handling, storage, packaging and even during application. These products are therefore suitable for practical use only when the individual components have similar physical qualities.

NP Fertilizers. The minimum analysis for NP fertilizers under the EEC Guidelines is 3 % N and 5 % P_2O_5 and at least 18 % total nutrients. Common grades are 20–20, 22–22, 26–14, 11–52, 16–48, and 18–46. These products are appropriate for potassium-rich soils or where potash is supplied as a separate fertilizer.

NK Fertilizers. The minimum analysis for NK fertilizers under the EEC Guidelines is 3 % N and 5 % K_2O and at least 18 % total nutrients. These products are suitable for phosphate-rich soils or where phosphate is distributed separately.

PK Fertilizers. In the group of PK fertilizers, all combinations of the straight phosphate and potassium components listed in Table 11 are possible. In general, the materials are first milled and then mixed and granulated, so that a fairly homogeneous mixture is obtained. Some products are also made by bulk blending. The EEC Guidelines set forth a minimum analysis of 5 % P_2O_5, 5 % K_2O, and at least 18 % nutrients. Magnesium and micronutrient boron can be added.

3.1.3. Lime Fertilizers

Solid fertilizers also include lime fertilizers. The main purpose of using lime is to optimize soil pH; a secondary purpose is to supply calcium as a plant nutrient. The use of large amounts of lime to increase the pH is referred to as soil-improvement liming. Use to maintain the present pH is called maintenance liming.

The starting materials for lime fertilizers are limestone [*1317-65-3*] and dolomite [*17069-72-6*]. These are marketed in various forms. National fertilizer regulations (e.g., the type list in the Federal Republic of Germany's fertilizer law, see Chap. 10) govern the fineness of grinding. Calcination at 900 – 1400 °C yields quicklime CaO; hydration with water gives slaked lime

Ca(OH)$_2$. The most immediately effective form is finely divided quicklime with a particle size of ca. 0.15 mm or less. If a lime grade contains at least 15 % MgO or MgCO$_3$, it can be called magnesium lime. The content of MgO depends on the starting material (proportion of dolomite) and may be up to 40 %.

Calcium carbonates [471-34-1], CaCO$_3$, act slowly. They are recommended above all for lighter soils. Unslaked and slaked limes are faster acting. They are suitable mainly for loamy and clayey soils.

Besides these lime products from natural rock, there are also limes from industrial processes. They include smelting lime and converter lime, which are steel-industry byproducts. Converter lime with phosphate must contain 40 % CaO + MgO and also at least 5 % P$_2$O$_5$. These products must also satisfy minimum requirements on fineness to ensure sufficiently fast action.

Finally, the refining of beet sugar yields a product containing lime. According to the fertilizer law in the Federal Republic of Germany, this lime form must contain at least 30 % CaO and 5 % MgO.

3.1.4. Magnesium Fertilizers

The following magnesium fertilizers are offered in commerce: kieserite [14567-64-7] (27 % MgO) for soil application, and Epsom salts [7487-88-9] (17 % MgO) for foliar application in liquid form.

3.2. Liquid Fertilizers

Only solid fertilizers were produced and used up to ca. 1950: lower production costs, higher nutrient concentrations, and the ease of making compound fertilizers were for a long time the reasons that development was limited to the solid fertilizers. Since then, liquid fertilizers have also been developed and supplied to agriculture. Particularly in the United States, liquid fertilizers have come into heavy use. Modern production methods have lowered production costs, even for liquid fertilizers with high nutrient concentrations.

Homogeneous liquid fertilizers, in contrast to solid fertilizers, present no special problems during application by the farmer. Furthermore the storage of liquid fertilizers is less difficult than that of solid ones. Small amounts of herbicides and insecticides can be mixed with liquid fertilizers far more easily. Production plants for solid fertilizers are much more costly than those for liquid fertilizers. Liquid fertilizers have the drawbacks that they usually have lower nutrient content and are sensitive to impurities, as well as to precipitation and crystallization, especially caused by magnesium and fluorine.

Liquid fertilizers are classified as (1) anhydrous ammonia [7664-41-7] with 82 wt % N at high pressure, (2) aqueous ammonia with up to 24 wt % N, (3) ammoniates (ammoniated ammonium nitrate and/or urea solutions) with up to 50 wt % N at moderate to atmospheric pressure, and (4) nonpressure urea – ammonium nitrate solutions with up to 32 wt % N [72]. Compound liquid fertilizers include both clear liquids and suspensions.

The growing interest in liquid fertilizers results in large part from the fact that field spraying can be used for specific, exactly meterable, inexpensive, clean application with an injector. Even low nitrogen rates of 15 – 30 kg/ha can be exactly and uniformly metered out in liquid form.

Liquid-fertilizer plants are classified according to type of operation as hot mix or cold mix. Hot-mix plants use phosphoric acid and ammonia, whereby hot mix refers to the heat of reaction. Cold-mix plants mix ammonium phosphate solution with other ingredients; no heat of reaction is evolved. Cold-mix plants are popular in the United States, because they are simple and inexpensive. They are essentially a blending and mixing operation, the liquid counterpart of the bulk-blend plants (see Section 5.5).

Table 12 lists the consumption of nitrogen in liquid fertilizers and their share of total nitrogen consumption by countries. In Germany consumption of nitrogen in liquids has risen since the season 1985/86, but total nitrogen consumption is decreasing [72]. The use of liquid ammonia plays no role in Germany and has strongly declined in Denmark. In the United States more nitrogen is consumed in liquid than in solid form.

Liquid fertilizers can be easily mixed with a variety of micronutrients and agricultural pesticides. The dissolving of micronutrients in clear liquid fertilizers is promoted by gelatinizing agents or by complexing with polyphosphates.

Table 12. Consumption of liquid fertilizers and their share of total nitrogen consumption

Country and fertilizer year *	Liquid ammonia direct application		Nitrogen solutions		Total liquids		Total N
	10^3 t N	%	10^3 t N	%	10^3 t N	%	10^3 t N
World							
1985/86	5022.2	7.13	3811.1	5.41	8833.3	12.54	70 461.6
1993/94	5052.4	6.97	3801.9	5.24	8854.3	12.21	72 497.7
1994/95	4122.8	5.69	3794.3	5.24	7917.1	10.93	72 454.8
1995/96	4649.2	5.92	4013.6	5.11	8662.8	11.03	78 592.5
1996/97 **	4688.1	5.67	4097.0	4.96	8785.1	10.63	82 645.6
United States							
1985/86	3400.5	35.96	1887.6	19.96	5288.1	55.92	9457.0
1993/94	4116.3	35.89	2543.0	22.17	6659.3	58.06	11 469.4
1994/95	3308.7	31.12	2514.0	23.65	5822.7	54.77	10 632.1
1995/96	3693.6	33.09	2642.8	23.68	6336.4	56.77	11 161.5
1996/97 **	3651.6	32.65	2651.3	23.71	6302.9	56.36	11 184.5
Germany							
1985/86	15.0	0.66	65.0	2.84	80.0	3.50	2285.7
1993/94	0.0	0.00	205.0	12.72	205.0	12.72	1612.0
1994/95	0.0	0.00	226.0	12.65	226.0	12.65	1787.0
1995/96	0.0	0.00	282.0	15.93	282.0	15.93	1770.0
1996/97 **	0.0	0.00	290.0	16.53	290.0	16.53	1754.0
France							
1985/86	31.0	1.29	482.0	20.02	513.0	21.31	2408.0
1993/94	37.0	1.67	560.0	25.20	597.0	26.87	2222.0
1994/95	35.7	1.55	566.4	24.54	602.1	26.09	2308.3
1995/96	42.7	1.79	577.0	24.13	619.7	25.92	2391.7
1996/97 **	44.8	1.78	643.2	25.48	688.0	27.26	2523.9
Denmark							
1985/86	102.0	26.69	0.0	0.00	102.0	26.69	382.1
1993/94	0.0	0.00	4.0	1.23	4.0	1.23	325.0
1994/95	0.0	0.00	6.0	1.90	6.0	1.90	316.0
1995/96	18.0	6.19	7.0	2.41	25.0	8.60	291.0
1996/97 **	17.0	5.94	7.0	2.45	24.0	8.39	286.0

* The fertilizer year (season) runs from July 1 to 30 next year.
** Preliminary.

The pulverized additions such as copper sulfate, sodium molybdate, sodium borate, zinc oxide, or manganese oxide, which are mixed with suspensions [73], can be suspended in water by stirring.

Nonpressure liquid fertilizers are generally stored in concrete or mild steel tanks [308, p. 129]. A corrosion inhibitor is required in the case of nonpressure urea – ammonium nitrate solution. Concrete tanks have the shape of vertical cylinders. The tanks should be lined with a film of a durable plastic such as poly(vinyl chloride). Before lining, the concrete tank must be protectively coated [74]. For pressurized liquid fertilizers, double-walled vessels with leak warning systems should be employed. When single-walled tanks are used, a retaining basin must be erected.

3.2.1. Nitrogen Liquids

Liquid Ammonia. Since 1950, liquid ammonia has found increasing use as a direct-application fertilizer, especially in the United States, since it can be produced in large amounts at low cost. Because of its high vapor pressure (6 bar at 10 °C, 9 bar at 20 °C, and 12 bar at 30 °C), anhydrous ammonia must be stored and transported in pressure vessels and applied with special equipment.

Ammonia is injected into the soil to a depth of roughly 15 cm with injection prongs. In general, the equipment suitable for this purpose deposits the fertilizer in a pipe whose diameter depends on the soil type and soil moisture content. Figures 12 and 13 show the ammonia loss as a function of the soil type, the soil water content, and the depth of application [75]. If the soil contains 15 % water, the NH_3 loss is virtually independent of depth and, at about the 1 % level, can be neglected. Despite some advantages, the application of liquid ammonia is too expensive for most farmers in the EC. A publication for farmers on the safe use of liquid ammonia is available [76]. For storage and transportation, see → Ammonia and [308, p. 73]

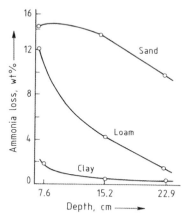

Figure 12. Ammonia losses as a function of application depth for three types of air-dry soils

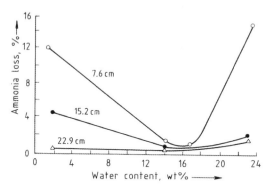

Figure 13. Ammonia loss as a function of soil water content for three depths of application

Pressurized Nitrogen Liquids. Together with ammonia, ammonium nitrate and/or urea can be converted into pressurized aqueous nitrogen fertilizers. The following categories are distinguished: low-pressure solutions (up to 1.3 bar) for direct application in agriculture, medium-pressure solutions (up to 7 bar), mostly used for ammoniating superphosphates or phosphoric acid, and high-pressure solutions (over 7 bar), used only for ammoniation. Table 13 lists some nitrogen solutions with their saturation temperatures and vapor pressures at 40 °C (313 K) [77].

Nonpressure Nitrogen Liquids. Nitrogen solutions that contain little or no ammonia have no significant vapor pressure at ambient temperatures. The most common nonpressure liquid nitrogen fertilizers comprise mixtures of ammonium nitrate, urea, and water (UAN), which are the most popular nitrogen fertilizers in the United States. A commercial solution might have the following composition:

Ammonium nitrate	39.5 % = 14 % N
Urea	30.5 % = 14 % N
Water	30.0 %
UAN	100.0 % = 28 % N

The density at 15 °C is ca. 1.28 g/cm^3, corresponding to 36 kg of nitrogen in 100 L of the product.

Figure 14 gives solubility isotherms and phase boundaries for the ammonium nitrate – urea – water system. In order to optimize the amount of UAN as a function of temperature during transportation and in the field, the UAN composition is adjusted to suit the conditions [77, 79]. This is especially important for winter application. Among almost-nonpressure nitrogen solutions, mixtures containing added NH_3 (see Table 14) should also be mentioned [80].

A urea – ammonium nitrate solution can be produced by dissolving solid urea in an ammonium nitrate solution. If the raw materials are solid urea and ammonium nitrate, UAN can be produced in a slightly modified dissolver. Figure 15 gives a simplified diagram of such a production unit.

The desired ammonium nitrate : urea ratio is obtained by mixing a 75 – 80 % urea solution at 120 °C with an 80 – 85 % ammonium nitrate solution at 40 °C, the quantities being controlled. After the addition of water, the liquid fertilizer is transferred to a storage tank after passing via a cooler. The mixer and cooler are made of stainless steel, and the equipment downstream of the cooler are made of carbon steel [94]

Table 13. Nomenclature and physical properties of some nitrogen solutions

% N × 10	Composition *	Vapor pressure at 40 °C, bar **	Saturation temperature, °C
410	22–65–0	0.7	− 6
410	26–56–0	1.2	−32
444	25–55–10	1.5	−29
453	31–0–43	3.4	+ 8
454	37–0–33	4.0	− 9
490	33–45–13	3.6	−27
490	34–60–0	3.4	−47
530	49–36–0	7.3	−73
580	50–50–0	10.1	−34

* $HN_3 - NH_4NO_3 - (NH_2)_2CO$, wt %.
** Multiply by 0.1 to convert to MPa.

Table 14. Composition, nitrogen concentration (N), and crystallization temperature (T_c) of nitrogen solutions [80]

Composition, wt %				N, wt %	T_c, °C	Pressure at 10 °C, bar *
NH_3	NH_4NO_3	Urea	H_2O			
24	56	10	10	44	−26	0.3
20	45	15	20	39.2	−32	0.2
19	58	11	12	41	−14	0.2

* Multiply by 0.1 to convert to MPa.

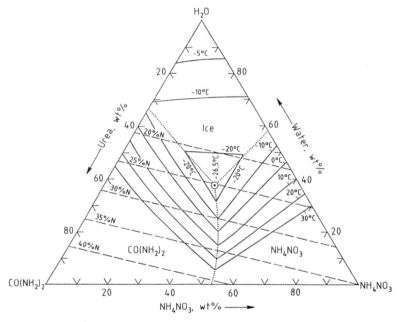

Figure 14. The ammonium nitrate – urea – water system [77, 78]
Solid lines are solubility isotherms; dotted lines show phase boundaries; dashed lines show constant nitrogen content.

Since urea – ammonium nitrate solutions attack ordinary steels, steel storage tanks and tank cars should be coated with Derakane-470, a poly(vinyl ester) [82], or lined with polypropylene or polyethylene [83]. This practice also prevents stress-corrosion cracking of the steels, especially in the presence of $(NH_4)_2CO_3$ [84]. Poly(vinylidene fluoride) provides excellent corrosion protection up to 120 – 130 °C [83]. The most common inhibitor for these solutions is anhydrous ammonia, which is used to adjust the pH to 7. Ammonium thiocyanate and ammonium phosphate are also effective [308, p. 129]

In the production and use of UAN, such solutions can explode if they are evaporated to <5 % water and simultaneously heated, especially when the two components have become segregated [85].

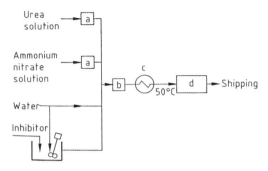

Figure 15. Production of urea – ammonium nitrate solution [112]
a) Ratio controller; b) Mixer; c) Cooler; d) Storage tank

3.2.2. Multinutrient Liquids

Liquid mixed fertilizers consist of aqueous solutions and dispersions of the nutrients nitrogen, phosphate, and potash. These liquid mixed fertilizers are produced mainly from phosphoric acid, anhydrous ammonia or aqueous ammonia, ammonium nitrate, urea, and potassium salts, chiefly KCl but also K_2SO_4 (hot mix). In most cases, fine-grained potassium chloride, with a somewhat higher potash content (62 % K_2O) than the normal fertilizer salt, is employed because it dissolves better. If only solid raw materials are used, the N-P-K grade of the liquid mixed fertilizer can be varied only over a narrow range. The following solid nitrogenous components are mainly employed: ammonium nitrate, urea, monoammonium phosphate, and diammonium phosphate (cold mix).

For intensive field and greenhouse cultivation of vegetables, etc., the nutrients can be introduced into the watering system in the form of a low-percentage nutrient solution. Because this solution is often delivered directly to the roots through hoses, it should have the lowest possible level of impurities.

3.2.2.1. NP Liquids

In the early 1950s the commercial production of mixed liquid fertilizers was started in the United States. Pure electric-furnace orthophosphoric acid was used as the phosphate source. The acid was ammoniated to give an 8–24–0 grade liquid which does not salt out at − 8 °C. At mixing plants, this base solution was mixed with urea – ammonium nitrate solution to give a liquid containing ca. 28 – 32 % nitrogen. An additional potassium content was obtained by dissolving potassium chloride in the liquid. Maximum-mix grades attainable in this way included 13–13–0 and 7-7-7. A problem was the low concentrations of the 8–24–0 base solution and the final liquid mixes. Moreover, use of the cheaper wet-process orthophosphoric acid was not possible, because on ammoniation of such acids, iron, aluminum, magnesium and other impurities form voluminous, gelatinous, and crystalline precipitates that are difficult to handle.

In 1957 TVA introduced superphosphoric acid and base solutions made by ammoniation thereof [86]. The superphosphoric acid, initially produced by the electric-furnace process, contained ca. 76 % P_2O_5 as compared to 54 % P_2O_5 in commercial wet-process acid. At this higher P_2O_5 concentration, ca. 50 % of the P_2O_5 is in the form of pyrophosphoric, tripolyphosphoric, and tetrapolyphosphoric acids [88]. Because of higher solubility of the ammonium salts of these polyphosphoric acids, ammoniation of superphosphoric acid allowed production of higher grade base solution (10–34–0 instead of 8–24–0) and higher grade mixed fertilizer solutions. A further advantage was that the polyphosphates are effective sequestering agents that prevent precipitation of many metal ions. The kinetics of formation of ammonium polyphosphates are reported in [87].

Although the introduction of superphosphoric acid was an effective solution to the problem of low-grade liquids, there were still the high costs of the electric-furnace acid process. Studies were undertaken to produce usable superphosphoric acid by concentration of the cheaper (54 % P_2O_5) wet-process phosphoric acid to about 72 % P_2O_5 and a polyphosphate content of 45 – 50 %. Problems of the wet-process superphosphoric acid of this concentration were its high viscosity and the poor storage properties of the 10–34–0 base solution derived from it. Although the polyphosphate content of the liquid initially keeps impurities in solution in complexed form, precipitates may still form during storage if the acid used in ammoniation had too high an impurity level and the polyphosphate concentration is too low. A frequent problem with NP liquids is the precipitation of insoluble magnesium compounds. If 0.4 wt % magnesium (calculated as oxide) is present, the storage time before appearance of cloudy precipitates is only about one month. For magnesium oxide concentrations of 0.2 % and 0.1 %, the storage life is about 1 – 2 years [89].

The usual raw material for phosphoric acid productions, apatite, with its impurities and its variable composition, is not a simple raw material. In the production of phosphoric acid, however, most of the impurities (e.g., aluminum, iron, and magnesium compounds) remain behind in the acid slurry.

The removal of the impurities from the wet-process phosphoric acid (Fe, Al, Mg, Ca, SO_3, F) is necessary for storage and application of the liquid fertilizer. Two typical processes reduce the concentrations of the most troublesome impurities significantly (Table 15) [95]. Process B, which is used in Europe, reduces the impurity concentrations to lower levels than the U.S. Process A, while not reducing P_2O_5 concentration as much; however, Process B is more expensive.

The cleaning of the wet-process phosphoric acid by means of continuous extraction with ammonia and acetone has been described for the "green acid" from North Carolina phosphate rock and the "black acid" from Florida phosphate rock [90]. An extraction process is described in [91]. For the use of chelating agents, see [92, p. 58].

Addition of ammonium fluoride to aqueous aluminum phosphate-containing slurries precipitates aluminum as $(NH_4)_3 AlF_6$ and forms ammonium and diammonium phosphate:

$$6\,NH_4F + AlPO_4 \rightarrow (NH_4)_3AlF_6 + (NH_4)_2HPO_4 + NH_3$$

resulting in a filtrate that can be subsequently worked up to give a liquid fertilizer.

Figure 16 shows how the polyphosphate content and the $N : P_2O_5$ ratio affect the solubility of ammonium phosphates at 0 °C. The bottom curve represents the solubility of the orthophosphates. The middle and top curves demonstrate the increase in solubility as the polyphosphate content increases [93].

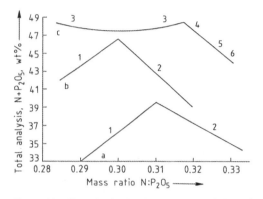

Figure 16. Effect of polyphosphate content and $N : P_2 O_5$ ratio on solubility of ammoniated phosphoric acids at 0 °C
Polyphosphate contents as percent of total P_2O_5: a) 0; b) 45; c) 70
Crystallizing phases: 1) $(NH_4)H_2PO_4$; 2) $(NH_4)_2HPO_4$; 3) $(NH_4)_3HP_2O_7 \cdot H_2O$; 4) $(NH_4)_5P_3O_{10} \cdot 2\,H_2O$; 5) $(NH_4)_2HPO_4 \cdot 2\,H_2O$; 6) $(NH_4)_4P_2O_7 \cdot H_2O$

Table 16 characterizes NP base solutions made from superphosphoric acid. The two solutions are 10 – 34 from wet-process superacid and 11 – 37 from furnace-grade superacid. The 10 – 34 NP solution is used predominantly [77].

A large number of experimental data points were subjected to regression analysis. With the resulting equation, the total nutrient concentration in NP solutions made from ammonium

Table 15. Purification of wet-process phosphoric acid

	Raw acid	Purified acid	
		Process A	Process B
P_2O_2	52.5	40.4	48.9
Fe	0.6	0.2	0.07
Al	0.9	0.3	0.01
Mg	0.3	0.03	0.001
Fe	0.8	0.3	0.06

Table 16. Composition of NP base solutions [77]

Grade	Phosphoric acid, wt % P_2O_5	Percentage composition *				
		Ortho-phosphate	Pyro-phosphate	Tripoly-phosphate	Tetrapoly-phosphate	Higher phosphates
10–34–0	76	49	42	8	1	0
11–37–0	79 – 80	29	42	21	5	3
	80 – 81	20	37	23	10	10

* Distribution of P_2O_5 sources: of the total P_2O_5 present (34 or 37 wt %), the percentages given in the table are present in the form of orthophosphate, polyphosphate, etc.

phosphates and urea at 0 – 50 °C can be calculated with a relative error of 1 % or else determined with a nomograph [80].

Figure 17 shows schematically a plant for producing NP base solutions (tank reactor process).

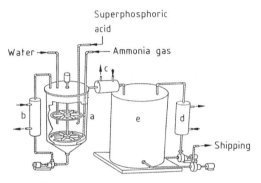

Figure 17. Plant for producing NP solution from superphosphoric acid [77]
a) Reactor; b, c, and d) Coolers; e) Storage tank

Superphosphoric acid, usually containing 40 to 50 % of the P_2O_5 as polyphosphates, ammonia, and water are continuously fed to a stirred-tank reactor. A heat exchanger operating in closed cycle holds the temperature in the reactor at roughly 70 °C. The solution concentration is controlled through density measurement, and the N : P_2O_5 ratio is controlled by means of the pH. The NP solution outlet from the reactor is cooled to 35 °C in another heat exchanger (c) before being transferred to a storage tank. To minimize hydrolysis of polyphosphate to orthophosphate, the product is further cooled to ca. 20 °C and the storage tank is protected from sunlight. The polyphosphate content of the product is always slightly lower than that of the feed acid. For example, a base solution containing about 50 % polyphosphate, which is about the minimum polyphosphate content of the product with acceptable storage properties, requires a feed acid containing 55 % polyphosphate.

Because of the expensive heat-exchange problems — heat must be supplied in the production of superphosphoric acid and removed in the subsequent ammoniation step — other techniques have been developed for producing ammonium polyphosphate solution from orthophosphoric acid in a *direct process*, thereby largely eliminating the expensive heat exchange. Two examples are the Swift process and the Tennessee Valley Authority (TVA) process.

In the Swift process (Swift Agricultural Chemical Corporation), ammonia gas reacts in a jet reactor (Fig. 18) with preheated wet-process phosphoric acid to produce ammonium polyphosphate. The phosphate melt inlet to the mixing tank is converted to liquid fertilizer by the addition of water; the grade produced is usually 10–34–0. The extent of polymerization depends on the temperature that prevails in the process reactor. At 300 °C, it is about 60 %. In addition to clear 10–34–0 liquid fertilizer, a 12–40–0 suspension can also be obtained with this process [94].

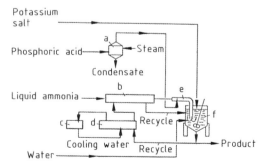

Figure 18. Swift process for producing NP (or NPK) solution from orthophosphoric acid [94]
a) Acid heater; b) Ammonia vaporizer; c) Cooling tower; d) Cooler; e) Pipe reactor; f) Mixing tank

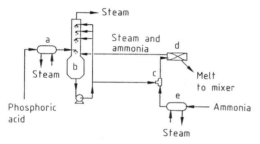

Figure 19. TVA process for producing NP solution from orthophosphoric acid [94]
a) Acid heater; b) Spray tower; c) T reactor; d) Rotating disengager; e) Ammonia preheater

The TVA process uses a T reactor (Fig. 19) to produce ammonium polyphosphate. Wet-process phosphoric acid (52 – 54 wt % P_2O_5) is preheated and then partly neutralized by ammonia in a simple spray washer (spray tower). The solution is then fed into the T reactor, which is also supplied with heated anhydrous ammonia from the other side. At the adjusted temperature of 232 – 243 °C, the reaction with ammonia goes to completion while the residual water is simultaneously vaporized and a foamy polyphosphate melt is formed [95]. After the product passes through a rotary separator (disengager), excess ammonia and steam are removed; the ammonia is scrubbed out, and the steam is discharged. From the melt, with a polyphosphate content of roughly 50 %, either clear liquid fertilizers such as 10–34–0 or suspensions such as 10–37–0 can be produced [95, 96].

The T reactor process has enabled reasonable equipment costs, process simplicity, and efficiency in the production of high-quality liquid fertilizers. The unique feature of the process is the conversion of a low-polyphosphate (20 – 35 %) superphosphoric acid to an high-polyphosphate (ca. 80 %) 10–34–0 or 11–37–0 base solution. The high polyphosphate content substantially improves the stability, storage life, and handling properties of the base solutions and NPK liquids. In contrast to the tank-reactor process the reaction is accomplished in a limited volume of the pipe reactor with minimal heat losses [97].

In the Ugine Kuhlmann process [98], clear NP liquid fertilizers with good storage properties are produced from *unpurified wet-process phosphoric acid* [90].

The crude acid is first treated with urea to yield urea phosphate, which is crystallized and separated. The mineral impurities remaining in the mother liquor can be used in the production of solid fertilizers. Thus there is no waste disposal problem. In the second reaction step, the urea phosphate is reacted with ammonia, yielding a urea – ammonium polyphosphate melt, which is dissolved in water to obtain a 16–30–0 liquid fertilizer.

3.2.2.2. NPK liquids

Liquid NPK fertilizers are produced by adding a potassium component, usually potassium chloride, to UAN and NP solutions. The amount of water needed to dissolve 100 g of a mixture of UAN, NP solution, and potassium chloride at a specific temperature, as well as the solid phase, can be shown conveniently in the form of three-component phase diagrams [80]. Other triangular diagrams show the solubility relationships in mixed liquid fertilizer systems [97]. While the clear NP solutions with polyphosphate have compositions corresponding to those of solid fertilizers, the contents must be kept lower in clear ternary NPK solutions because potassium nitrate, which is formed from ammonium nitrate and potassium chloride, has a low solubility.

Table 17 presents several three-component liquid fertilizers obtained by mixing a 10–34–0 base solution with potassium chloride powder. The added component lowers the nutrient content from almost 39 % to ca. 29 – 30 % [99].

Production takes place in either cold-mix or hot-mix plants [100], [312, p. 456]. Because cold-mix plants offer low investment costs and are easy to operate, they are in more widespread use. In a stirred vessel, the appropriate quantities of potassium chloride are added to the base solutions (e.g., 28–0-0 and 10–34–0) and mixed. Hot-mix plants, especially small-capacity ones, require relatively large investments and, because of their troublesome operation, must have well-trained personnel.

3.2.2.3. UAS Liquids

The use of high-analysis fertilizers containing little or no sulfur has led to soil sulfur deficiencies in areas far from industries. The critical level

Table 17. Three-component liquid fertilizers produced by mixing 10–34–0 base solution with KCl [99]

Fertilizer type	Salt composition, wt %						Total nutrient analysis, wt % *	Freezing point, °C
	Ammonium orthophosphate	Ammonium polyphosphate	Urea	NH$_4$NO$_3$	KCl	H$_2$O		
1:1:1	7.93	10.11	7.8	16.3	15.6	42.2	28.85	−21
1:2:2	9.20	12.4	5.0	10.4	18.7	44.4	29.54	−18.5
1:1.5:1	10.70	13.65	7.3	15.3	14.1	38.4	30.57	−23.5
2:1:1	5.75	7.33	11.9	24.8	10.7	39.5	27.89	−24.5
1:1:0	16.10	20.5	15.9	33.2	0	14.3	38.60	−33

* N + P$_2$O$_5$ + K$_2$O.

for sulfur in soil is 0.1 – 0.3 wt %. Sulfur deficiencies in the soil of Kuwait led to the development of a sulfur-containing liquid fertilizer: 21.2 wt % (NH$_4$)$_2$SO$_4$ + 37.6 wt % urea + 41.2 wt % water [101]. The total nitrogen content is 22 wt %, and the saturation temperature is 4 °C.

The reaction of urea with sulfuric acid

$$(NH_2)_2CO + H_2SO_4 \rightarrow (NH_2)_2CO \cdot H_2SO_4$$

yields liquid fertilizers with pH = 1. Because of corrosion, the plant must have polyethylene linings or else use expensive chrome – nickel – molybdenum steel. The following are typical UAS solutions [102]:

Sulfuric acid, wt %	N, wt %	S, wt %
28	9	27
17	16	49
15	17	52
10	18	55

Liquid fertilizers can be applied even in winter if ammonium thiosulfate is added in order to lower the salting-out temperature to − 18 °C. From the standpoint of nutrient content, this compound is better than water. Ammonium thiosulfate itself is a clear liquid fertilizer (12–0-0 + 26S) with a salting-out temperature of − 5 °C [102].

3.2.3. Suspensions

Suspension fertilizers contain solid nutrient salts, especially potassium chloride, suspended in a concentrated ammonium phosphate solution. Their development has made it possible to attain nutrient contents corresponding to those of solid NPK fertilizers:

Solutions	Suspensions
8–8–8	15–15–15
3–9–9	7–21–21
2–16–12	5–15–30
10–5–5	20–10–10

What is more, suspension fertilizers have the advantage that low-priced phosphates that are difficult to dissolve in water can also be employed.

The storage and application of suspension fertilizers, however, is more difficult than that of clear liquid fertilizers, since the former tend to undergo sedimentation. If an easily handled suspension is desired, the product must be stirred vigorously before use. Swellable clays such as palygorskite (attapulgite), a magnesium silicate-containing clay, have proved suitable as suspending agents. Suspensions thus make special demands on technique. The methods of storage, transportation, and spraying cannot be transferred directly from clear liquid fertilizer solutions. As a rule, substantial changes in equipment are needed, and these require additional investments. The quality of spray distribution does not match that for clear fertilizer solutions either. For all these reasons, suspensions have scarcely become established in Europe, despite efforts to introduce them. In the United States, on the other hand, a larger scale of farming, together with a well-developed distribution service sector, has allowed the successful introduction of suspensions.

A suspension fertilizer can be produced from *solid monoammonium phosphate* by intensive stirring in water and simultaneous adding of ammonia. The addition of swellable clay (1 – 2 %) ensures a fairly stable suspension. By dilution, a product of grade 9–27–0 is obtained. Applied in a comparable way, it behaves much like the commercial 10–30–0 liquid fertilizer [103].

Anhydrous ammonia is used to improve the solubility and mixing of monoammonium phosphate. Ammonium orthophosphate displays the highest solubility at a molar ratio $NH_3 : H_3PO_4$ of 1.5. A turbine mixer is suitable for homogenizing the suspension obtained by the addition of KCl and 1 or 2 wt % palygorskite [104].

A stable suspension is obtained by ammoniating *wet-process orthophosphoric acid* (54 wt % P_2O_5) at 71 °C and a pH between 7.5 and 8.5. The product is then adjusted to the desired pH with a small amount of acid. Because of the fineness of the solid particles, the suspension (8–24–0) remains stable. After cooling, it can be blended with urea – ammonium nitrate solutions (32 % N) and potassium chloride to produce various compositions of liquid fertilizers [105].

A two-step process developed by TVA yields low-priced, high-analysis suspension fertilizers in which the phosphate component is obtained by ammoniating orthophosphoric acid (Fig. 20). In the first reaction step, the acid is partly ammoniated with ammonia, and the solution is held for about 30 min at 106 °C. In the second step, ammoniation is completed. Then the product is cooled to 60 °C, and the suspension of fine diammonium phosphate crystals is stabilized with 1.5 wt % clay. A fluid NPK fertilizer of 15–15–15 grade can be produced from the 12–40–0 suspension [106]. Raising the acid concentration from 54 wt % to 70 – 73 wt % P_2O_5 (wet-process superphosphoric acid) gives useful suspension fertilizers. In general, however, the Mg content is so high that prolonged storage is excluded because of the precipitation of struvite, $MgNH_4PO_4 \cdot 6\,H_2O$.

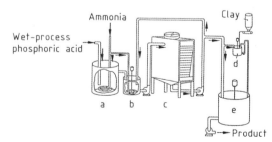

Figure 20. Production of orthophosphate suspensions
a) First reaction stage; b) Second reaction stage; c) Cooler; d) Slurry vessel for suspending agent; e) Product tank

This difficulty is avoided if *triple superphosphate* is employed as the sole phosphate source. Palygorskite is slurried with water in a mixing vessel, aqueous ammonia and urea – ammonium nitrate solution are added, and then finely-ground triple superphosphate and potassium chloride are incorporated. The suspensions made by this process have a typical grade of 12–12–12 [107]. The addition of palygorskite can be omitted if part of the triple superphosphate is first treated with aqueous ammonia to yield a voluminous precipitate, which then acts as a suspending agent for the other solid particles [108].

Suspensions from the Nitrophosphate Process. A low-cost process for producing a suspension fertilizer involves digesting crude phosphate with nitric acid. The slurry of phosphoric acid and calcium nitrate is inlet to a hot-mix plant and ammoniated to form an NP suspension fertilizer. Such processes have been developed in the United States, the USSR, and the Federal Republic of Germany [81, 109, 110]. Aside from the cheap raw material, the fixed costs of the fertilizer are relatively high. Since nitrophosphate processes yield suspension fertilizers with large amounts of calcium, the dilution effect of the calcium holds the nutrient content relatively low (e.g., 9–9–9 or 6–12–12).

High-Nitrogen Suspension Fertilizer. TVA has devised a promising method for producing a high-nitrogen suspension fertilizer (36 wt % N) [102]. This product is a mixture of microspray-crystallized urea and ammonium nitrate solution (76 wt % AN) with palygorskite. The composition is 57.5 wt % urea + 31.5 wt % ammonium nitrate + 10.0 wt % water + 1 wt % palygorskite [105, 108, 111].

Sulfur-Containing Suspensions. In regions where the effects of sulfur deficiency in the soil are pronounced in certain regions, sulfur in fertilizers not only increases yield but also can improve product quality. TVA has manufactured a 29–0–0–5S urea ammonium sulfate (UAS) suspension fertilizer in a pilot plant. The feedstocks are urea (70 % aqueous solution), sulfuric acid (93 %), anhydrous ammonia, and water. Clay (ca. 2 %) is added to the UAS solution to give the final product [102].

4. Special Fertilizers

The special fertilizers include water-soluble nutrient salts, foliar fertilizers, micronutrient fertilizers, slow- and controlled-release fertilizers, nitrification inhibitors, and organic fertilizers, including sewage sludge, compost and manure.

4.1. Water-Soluble Nutrient Salts

Especially for fertilization in commercial horticulture, but also for specialty field crops, water-soluble nutrient salts (e.g., ammonium nitrate, potassium nitrate, ammonium phosphate, magnesium sulfate) are preferably used with water-saving irrigation systems (drip irrigation/fertigation) and with soilless cultivation systems. Various crop-specific fertilizer recipes can be formulated with these salts. A convenient form for the consumer are multinutrient fertilizers made by mixing various water-soluble salts, which generally also include micronutrients. Such salt mixtures are available worldwide from most major fertilizer manufacturers.

From these salts and salt mixtures, the grower prepares highly concentrated master solutions, which for application are further diluted to render them plant-compatible and then fed into the irrigation system. By using appropriate automatic fertilizer systems it can be assured that irrigation intervals and concentration of the nutrient solution are adjusted to the crop's needs.

A distiction is made between urea-containing and urea-free nutrient salt mixtures. Since the application rate of nutrient salts is frequently monitored by means of conductivity, products containing urea cause problems as urea has no ionic conductivity and is therefore not detected. Therefore, products containing urea are mainly used for foliar fertilization (see Section 4.2) where these calibration problems do not exist and because urea nitrogen is the N form with the best uptake rate via the leaves.

A prerequisite for all nutrient salts is rapid dissolution in water without any residues. This is achieved by producing the salt mixtures from appropriate ingredients and by properly grinding them. Moreover, for special fields of application in horticulture (flood irrigation and soilless cultivation systems) they are rated by their nitrogen forms. For these fields of application, nitrate-based nutrient salts are preferred.

In principle, if appropriately diluted, all nutrient salts are suitable both for soil fertilization and for foliar fertilization.

4.2. Foliar Fertilizers

For over 100 years it has been known that nutrients can be taken up through the leaves. However, with some exceptions, plants cannot be provided with nutrients through the leaves alone, since even leafy plants such as potatoes cannot absorb sufficient nutrients through the leaves. Foliar fertilization, therefore, is usually a supplement to soil fertilization. However, it is often the case that proper foliar fertilization puts the plants in the position to utilize soil-derived nutrients better. Recent decades have seen foliar fertilizers come into use not only for special crops or in certain localities, but throughout agriculture.

Foliar fertilizers are substances that contain primary nutrients and/or micronutrients, are applied to the leaves, and are absorbed into the leaves. The most important foliar fertilizer is urea (46 % N) which is highly soluble in water and rapidly absorbed by plants via the leaves. Therefore, urea is frequently used as a component of fertilizer suspensions and solutions. In addition to nitrogen, other macro- and micronutrients can be added, frequently in a nutrient ratio tailored to the demand of specific target crops. Special raw materials and formulations ensure a good plant compatibility and optimum foliar nutrient uptake. Also used for foliar fertilization are solutions with organic ingredients (e.g., amino acids), readily soluble salts (e.g., potassium nitrate or micronutrients, mostly based on sulfate) and salt mixtures containing macronutrients and special micronutrient mixtures.

4.2.1. Production

Solid foliar fertilizers are produced by mixing salts that are readily soluble in water. Before application, the salts are dissolved for spraying. Liquid foliar fertilizers, which are made from readily soluble salts or solutions and adjusted to a given analysis, are also on the market. The production of suspensions for foliar application involves the use of dispersing agents, and in some cases ionic surfactants, to stabilize the spray.

4.2.2. Application

In order to achieve optimal utilization of nutrient uptake through the leaves, the nutrient salts must remain as long as possible in the dissolved liquid state on the leaf surface. For this reason, spraying in overcast weather or toward evening ("evening dew") gives better results than spraying on dry days in bright sunlight. Depending on the sensitivity of the leaves and the concentration of the spray liquid, application in hot, sunny weather may cause burning. Table 18 gives the time required for 50 % absorption of nutrients applied to the leaves of bean plants [113].

Table 18. Nutrient absorption rates [113]

Nutrient	Time for 50 % uptake
Nitrogen	1 – 6 h
Phosphate	2.5 – 6 days
Potash	1 – 4 days
Calcium	1 – 4 days
Sulfur	5 – 10 days
Magnesium	1 h (20 % uptake)
Iron	24 h (8 % uptake)
Manganese	1 – 2 days
Zinc	1 day
Molybdenum	10 – 20 days

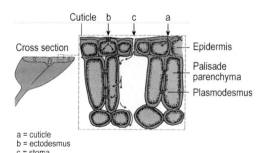

a = cuticle
b = ectodesmus
c = stoma

Figure 21. Access and transport paths for nutrients in leaves

Nutrients placed on the leaf surface follow three paths to the leaf cells [114 – 116]. After the cuticle has been wet by the nutrient spray, the cutin swells, which increases the distances between the wax plates in the cuticle. As a result, the nutrients are able to diffuse through the cuticle and into the cell wall (Fig. 21). The nutrients can then diffuse into the space around the epidermal cell and to the cell membranes, where they can be absorbed, or they can diffuse further along the cell wall of the epidermal cell and adjoining cells, deeper into the leaf, where they can be absorbed by a parenchymal cell. Nutrients that have been absorbed by an epidermal cell can be transferred to other cells by cytoplasmic strands connecting the cells, the plasmodesmata, a second path. If surfactants are included in the nutrient spray, a third path becomes available: the nutrients can pass through the stomata into the air spaces of the leaf.

Reasons for Using Foliar Fertilizers. Foliar fertilizers are applied to remedy obvious nutrient deficiencies. If a deficiency is recognized, the missing nutrient is supplied by spraying. If the deficiency is not well defined, complete foliar fertilizers or a mixed micronutrient fertilizer can be employed.

Another reason for foliar application is to remedy hidden (latent) nutrient deficiencies, which can seriously impair crop yield and quality.

Foliar fertilizers are also used to protect yield and quality, to meet peak nutrient demands (i.e., remedy latent deficiencies and stimulate the plants to absorb more nutrients from the soil), and to improve quality. The last point is especially important for special crops (wine, fruits, vegetables).

Advantages and Drawbacks of Foliar Fertilizers. Foliar application has a number of advantages over soil application: Nutrients act rapidly, nutrient utilization is high, nutrients are placed where they are needed, the risk of leaching losses and groundwater contamination is reduced. micronutrients are supplied to meet much of the demand for them, micronutrients are not fixed in the soil, application does not entail extra costs if fertilizers are combined with agricultural pesticides, and some nutritional problems can be cleared up only through application to the leaves or fruit (bitter pit in apples, stem necrosis in grapes).

The main disadvantages of foliar fertilization are that the action time is often short, the quantity of primary nutrients delivered is relatively

4.2.3. Combination with Agricultural Pesticides

Foliar fertilizers are usually applied along with agricultural pesticides for the following reasons [117]: to lower the cost of application, since pesticide spraying is required anyway; to reduce the stress due to the application of pesticide (e.g., certain herbicides); to improve the quality of pesticide spray liquids (lower the pH or increase the electrolyte content) and thus stabilize them; to reduce evaporation at low air humidity, so that most of the fine droplets reach the plant in liquid form; and to lower the surface tension of the water.

4.3. Micronutrients

The most important micronutrients (trace elements) are iron, manganese, boron, zinc, copper, and molybdenum. The delivery of micronutrients to plants is just as indispensable today as that of macronutrients or primary nutrients. Diseases resulting from an undersupply of micronutrients are seen more and more often as agriculture and gardening become more intensive, with a corresponding increase in plant nutrition [118]. For this reason, better and more specific micronutrient fertilizers have now been developed.

The term micronutrient also signifies that these substances are needed only in small quantities, often just 1 – 100 g per hectare and year. These small amounts are, however, absolutely necessary. But if the optimal dose is much exceeded, serious crop damage and loss of yield may occur. For boron, in particular, the difference between an adequate and an excessive amount is slight.

To work in the proper way, the micronutrients must be absorbed by the plants. Uptake is mainly in the form of ions: cations or metal–chelate ions of iron, manganese, copper, and zinc and anions in the case of boron and molybdenum (borates, molybdates). Absorption is through the leaves (foliar fertilizer) or the roots (soil solution).

4.3.1. Micronutrient Forms

Micronutrients are used in the following inorganic forms:

1) Various metal salts (sulfates, chlorides, nitrates)
2) Milled oxides
3) Metal flours and metal slags

Table 19 lists only those micronutrient compounds that are in common use. The table also states the nutrient analyses, the solubilities in water, and the various types of fertilizers in which the compounds appear.

In addition to these, the nitrates and chlorides of iron, manganese, copper, and zinc are also used, especially in concentrated liquid fertilizers. The trace-element nitrates and chlorides have high solubilities (e.g., 268 g per 100 g H_2O for $Fe(NO_3)_2 \cdot 6 H_2O$ or 400 g per 100 g H_2O for $FeCl_3 \cdot 6 H_2O$; some of the nitrates and chlorides of manganese, copper, and zinc have still higher values). Because of the price and the handling qualities in the solid state, however, the sulfates are preferred. Many crops are sensitive to trace-element chlorides in large amounts. At the levels usual in foliar application, however, this is certainly not a major problem. Trace-element sulfates and nitrates provide two nutrient elements each.

Because soluble iron compounds are fixed as phosphates or oxides in calcareous soils, organic complexes were first developed for this element. One successful complexing agent is ethylenediamine di(o-hydroxyphenylacetic acid) (EDDHA):

$$\underset{\underset{\text{COOH}}{|}}{\text{C}_6\text{H}_4(\text{OH})-\text{CH}}-\text{NH}-\text{CH}_2-\text{CH}_2-\text{NH}-\underset{\underset{\text{HOOC}}{|}}{\text{CH}-\text{C}_6\text{H}_4(\text{OH})}$$

Other compounds of trace metals with N-carboxylalkylamino acids are less effective in the soil, especially with soil pH values > 7, but if applied on the leaves are superior to the EDDHA compounds and to the simple salt solutions.

The N-carboxylalkylamino acids have also come into wide use today as chelating agents for manganese, copper, and zinc:

Table 19. Micronutrient fertilizers

Substances and products	Formula	Analysis, wt %	Solubility, g/100 g H_2O	Uses
Salts (technical)				
Iron(II) sulfate heptahydrate	$FeSO_4 \cdot 7H_2O$	19.5	62.3	liquid fertilizers,
Manganese(II) sulfate monohydrate	$MnSO_4 \cdot H_2O$	32	63.0	nutrient-salt
Copper(II) sulfate pentahydrate	$CuSO_4 \cdot 5H_2O$	25	36.0	mixtures, solid
Zinc(II) sulfate monohydrate	$ZnSO_4 \cdot H_2O$	35.5	63.7	straight and
Zinc(II) sulfate heptahydrate	$ZnSO_4 \cdot 7H_2O$	22	165.5	multinutrient
Boric acid	H_3BO_3	17		fertilizers,
Colemanite (boron mineral)	$Ca_2B_6O_{11} \cdot 5H_2O$	11		micronutrient and
Disodium octaborate tetrahydrate *	$Na_2B_8O_{13} \cdot 4H_2O$	20.8	9.5	multimicronutrient
Hexammonium heptamolybdate tetrahydrate	$(NH_4)_6Mo_7O_{24} \cdot 4H_2O$	54		fertilizers
Oxides				
Manganese dioxide (mineral)	MnO_2	45		solid straight and
Zinc oxide (residual zinc)	ZnO	69		multinutirent
Molybdenum trioxide	MoO_3	60		fertilizers
Slags				
Copper slag flour		2.5 Cu, 0.05 Co		micronutrient fertilizers
Metal flours				
Manganese flour		20 Mn, 10 Fe, 0.5 Zn, 0.25 Cu		micronutrient
Copper flour		2.6 – 25 Cu		fertilizers

* Disodium octaborate is not a homogeneous substance. It has a $Na_2O : B_2O_3$ ratio of 1 : 4 and can be produced as a concentrated solution. It is sold by US Borax under the trade name Solubor.

EDTA ethylenediaminetetraacetic acid
HEDTA hydroxyethylethylenediaminetriacetic acid
DTPA diethylenetriaminepentaacetic acid
EDDHA ethylenediamine di(o-hydroxyphenylacetic acid)
EDDHMA ethylenediamine di(o-hydroxy-p-methylphenylacetic acid)

Other chelating agents such as lignosulfonates and citric acid have so far been used only to a lesser extent.

4.3.2. Production

The list of fertilizers containing micronutrients comprises straight micronutrient fertilizers, combination micronutrient fertilizers, and ordinary fertilizers with micronutrients. All these types can be manufactured in either liquid or solid form. Among the solid micronutrient fertilizers, the powdered and fine-crystalline ones predominate, since nearly all of them can be dissolved in water and applied through the leaves. They generally consist of the several substances in Table 1 as well as the N-carboxyalkylamino acid chelates cited.

Solid micronutrient fertilizers for soil application include metal flours, metal slags, and micronutrient frits. Metal flours come from fabrication waste in the manganese production and processing industry and from specially prepared alloys. The source material is ground and granulated to between 0.25 and 1.5 mm. Metal slags such as copper slag flour are also won from waste products. The metal slags are quenched in water and comminuted to fine, powdery products.

Frits are made by melting glasses and incorporating the desired micronutrients (single-element and multi-element frits). The glasses are then finely milled [119].

Solid straight micronutrient fertilizers and multi-micronutrient fertilizers with complexing can be produced in various ways. One technique is to grind a metal sulfate with chelating agents, or several metal sulfates with chelating agents (in many cases borates, molybdates, and magnesium sulfate are added) and blending [120]. Chelation takes place when the material is dissolved in water before use. A second approach is to stir and thereby dissolve chelating acids and metal oxides (including magnesium oxide) in water at elevated temperatures; additives that cannot be chelated, such as borates and molybdates, are added; the product is further stirred, filtered, and spray-dried [121] or crystallized.

Liquid micronutrient fertilizers and combination micronutrient fertilizers can be prepared by the same technique but without the spray-drying step.

In the production of fertilizers with micronutrients, the micronutrients are usually added before granulation; in the case of liquid fertilizers, micronutrient compounds or chelates are dissolved along with the primary nutrient forms.

4.3.3. Commercial Fertilizers

Several firms in Western Europe manufacture pure micronutrients or combination micronutrient fertilizers, most of them in chelate form. Table 20 lists selected firms and examples of their products.

Urania (Germany) uses metal slags as raw material to produce granular copper fertilizers containing 2.5 % and 5 % Cu, as well as a copper – kieserite granular fertilizer with 2.5 % Cu and 21 % MgO. Similar micronutrient products and the same chelating agents are made in the United States and marketed as commercial fertilizers [119, pp. B68 to B83].

There are also fertilizers with micronutrient supplements, namely: granulated straight and multinutrient fertilizers, water-soluble nutrient salts, suspensions and solutions with micronutrients, which are chosen and added according to crop demand.

4.3.4. Use

In order to combat micronutrient deficiencies, either fertilization far exceeding actual depletion or directed foliar fertilization must be carried out [117, p. 128]. The annual depletion of micronutrients can be partly offset by the use of macronutrient fertilizers with micronutrients. On the basis of average micronutrient contents and average crops (rotation 70 % cereals, 20 % root crops, 10 % fodder), the following are withdrawn from the topsoil per hectare and year in central Europe [117, p. 122]: 400 – 700 g manganese, 260 – 400 g zinc, 150 – 200 g boron, 80 – 120 g copper. The amounts for grassland (per hectare and year) are 800 g manganese, 300 g zinc, 80 g copper, 50 g boron, and 0.5 g cobalt. In addition, leaching removes an average of 250 g manganese, 250 g boron, 100 g zinc, and 30 g copper per hectare and year [117, p. 122].

Table 20. Commercial fertilizers

Company	Products
ABM Chemical (United Kingdom)	Nevanaid Fe, finely powdered solid product with 9 % Fe (HEDTA)
	Nevanaid Fe, finely powdered solid product with 7 % Fe (DTPA)
	Nevanaid Mn, finely powdered solid product with 10 % Mn (EDTA)
Allied Colloids (United Kingdom)	Librel Fe-Dp, finely powdered solid product with 7 % Fe (DTPA)
	Librel Mn, finely powdered solid product with 13 % Mn (EDTA)
BASF (Germany)	Fetrilon 13 %, spray-dried solid fertilizer with 13 % Fe (EDTA)
	Fetrilon-Combi, spray-dried solid multimicronutrient fertilizer with 9 % MgO, 4 % Fe, 4 % Mn, 1.5 % Cu, 1.5 % Zn, 0.5 % B, 0.1 % Mo (all heavy metals as EDTA, boron as borate, Mo as molybdate)
	Nutribor, crystalline powder with 8 % B, 1 % Mn, 0.1 % Zn, 0.04 % Mo, 6 % N, 5 % MgO, 12 % S (EDTA)
	Solubor DF, microgranular with 17.5 % B
Ciba Geigy (Switzerland)	Sequestrene 138, solid fertilizer with 6 % Fe (EDDHA), soil application
	Ferrogan 330, solid fertilizer with 10 % Fe (DTPA)
Grace Rexoline (Sweden)	Rexene 224 Fe, powdered or granular fertilizer with 6 – 7 % Fe (EDDHMA), soil application
	Rexenol Cu, powdered solid fertilizer with 9 % Cu (HEDTA)
	Rexene Zn, liquid fertilizer with 6.5 % Zn (EDTA)

Normally, fertilization must more than make up for this consumption. Per hectare, boron-deficient soils must receive 1 – 2 kg of boron, and manganese-deficient soils must receive as much as 12 – 24 kg of manganese.

In order to prevent the development of latent deficiencies, micronutrients are often used in small quantities, especially for foliar application. This applies in particular to intensive cereal farming, where — even on well-supplied soils — the freely available micronutrients are soon consumed or else the nutrient flux at peak demand times cannot be covered from soil reserves. If relatively small amounts of a multi-micronutrient fertilizer are used in specific treatments during tillering and in the shoot and ear phases, latent deficiencies are generally eliminated and the yield potential is fully utilized [122, 123].

4.4. Slow- and Controlled-Release Fertilizers

4.4.1. Introduction

In plant nutrition, soil and plants are two antagonistic systems that compete for the nutrients available in or applied to the soil. This competition is the main reason why only a portion of nutrients is taken up and used by the plants and crops grown, while another portion is (temporarily) immobilized in the soil or lost by denitrification/volatilization and leaching (particularly of nitrogen) [124, p. 124]. The fertilizer industry has developed special types of fertilizers and fertilizer modifications which avoid or at least reduce such losses, such as:

– Foliar fertilizers
– Slow and controlled-release fertilizers
– Nitrification and urease inhibitors; stabilized fertilizers

The utilization rate of nutrients is improved considerably by leaf application. However, in practice it is impossible to supply all the necessary nutrients via plant leaves [124, pp. 125 – 127], [125, pp. 14 – 16]. A more practical route is the use of nitrogen fertilizers which release the nutrients according to the plants' requirements, that is, slow- and controlled-release fertilizers.

Terminology. The Association of American Plant Food Control Officials (AAPFCO) gives the following definition: "Delay of initial availability or extended time of continued availability may occur by a variety of mechanisms. These include controlled water solubility of the material (by semipermeable coatings, occlusion, or by inherent water insolubility of polymers, natural nitrogenous organics, protein materials, or other chemical forms), by slow hydrolysis of water-soluble low molecular weight compounds, or by other unknown means" [126].

There is no official differentiation between slow-release and controlled-release fertilizers. However, the microbially degradable N products, such as urea – formaldehydes (UFs and other urea – aldehyde compositions), are commonly referred to in the trade as slow-release fertilizers, and coated or encapsulated products as controlled-release fertilizers [127, p. 12].

Advantages and Disadvantages. Slow- and controlled-release fertilizers reduce toxicity, especially in use with seedlings. The toxicity of conventional soluble fertilizers is caused by the high ionic concentrations resulting from quick dissolution. Consequently, these slow and controlled-release fertilizers permit the application of substantially larger fertilizer dressings (depot fertilization). This results in significant savings in labor, time, and energy. They also allow the full nutrient requirements of crops grown under plastic cover to be met (protected crop cultivation). These fertilizers significantly reduce possible losses of nutrients due to the gradual nutrient release (particularly losses of nitrate nitrogen). They also reduce evaporative losses of ammonia. They further contribute to a reduction in environmentally relevant gas emissions (N_2O) [127, pp. 15 – 16].

There are no standardized methods for reliable determination of the nutrient release pattern available as yet due to the lack of correlation between laboratory tests and field conditions. With urea – formaldehyde fertilizers, a proportion of the nitrogen content may be released extremely slowly or not at all. With sulfur-coated controlled-release fertilizers the initial nutrient release may be too rapid and cause damage to turf or the crop. Repeated use of sulfur-coated urea may also increase the acidity of the soil. Polymer-coated or encapsulated controlled-release fertilizers can cause an environmental problem since undesirable residues of the coating material may accumulate in the fields. However, the main disadvantage is that the cost of manufacturing slow- and controlled-release fertilizers is still considerably higher than that of conventional mineral fertilizers. At present their cost/benefit ratio prevents their wider use in general agriculture. Consequently, the vast majority is applied in nonagricultural sectors such as nurseries and greenhouses, golf courses, professional lawn care, as well as by consumers (home and garden) and landscape gardeners [127, pp. 17 – 18]. Total world consumption of slow- and controlled-release fertilizers is estimated at 562 000 t (1995/96), amounting to only 0.15 % of world total consumption of $N + P_2O_5 + K_2O$ in the form of fertilizer material (ca. 380×10^6 t) [128].

Types of Slow- and Controlled-Release Fertilizers. The two most important groups are:

1) Condensation products of urea such as urea – formaldehydes (slow-release)
2) Coated or encapsulated fertilizers (controlled-release)

Of lesser importance are other organic chemicals, ion-exchange materials, and supergranules.

4.4.2. Urea – Aldehyde Slow-Release Fertilizers

Three types of urea – aldehyde condensation products (see also → Urea, Chap. 8.3) have gained practical importance:

1) Urea – formaldehyde (UF)
2) Urea – isobutyraldehyde (IBDU/Isodur)
3) Urea – acetaldehyde/crotonaldehyde (CDU/Crotodur)

At pH values below 2, crotonaldehyde and acetaldehyde can form cyclic condensation products [131]. For production of urea – aldehyde condensates, see [129, pp. 3 – 87], [130, pp. 1 – 137], [132, pp. 153 – 156], [133], [134, pp. 247 – 279].

The urea – formaldehyde products have the largest share of the slow- and controlled-release fertilizer market (40 % of world consumption in 1995/96); IBDU- and CDU-based products are less widely used (15 % in 1995/96), since their manufacturing costs are even higher than that of urea – formaldehydes [127, pp. 61 – 63].

4.4.2.1. Urea – Formaldehyde Condensation Products.

Ureaform, as defined by the American Association of Plant Food Control Officials (AAPFCO) is the oldest type of urea – formaldehyde condensate. As early as 1924, Badische Anilin- & Soda-Fabrik (now BASF) registered the first patent (DRP 431 585) on urea – formaldehyde condensation fertilizers [135]. In the United States they were patented for use as fertilizers in 1947; commercial production began in 1955.

Ureaforms are a mixture of methylene – urea oligomers of various molecular masses, polymer chain lengths, and hence varying water solubilities, such as methylene diurea (MDU), and dimethylene triurea (DMTU). They also contain a certain amount of unchanged urea.

Manufacture. The manufacture of urea – formaldehyde products is a two step process [136, p.13], [137 – 143]:

$$NH_2CONH_2 + CH_2O \rightleftharpoons HOCH_2NHCONH_2$$

$$CH_2O + HOCH_2NHCONH_2 \rightleftharpoons HOCH_2NHCONHCH_2HO$$

First, urea and formaldehyde are combined to give the intermediates monomethylol- [*1000-82-4*] and dimethylolurea [*140-95-4*]. Under acidic conditions these methylolureas react with further urea to give various oligomers of methylene urea:

$$NH_2CONH_2 + HOCH_2NHCONH_2 \rightarrow$$
$$NH_2CONHCH_2NHCONH_2 + H_2O$$

In the production of granular urea – formaldehyde products, water must be removed by evaporation. The main problem in the manufacture of urea – formaldehyde slow-release fertilizers is the production of condensation oligomers in the desired proportions. A number of processes can be used to meet this target (e.g., dilute- and concentrated-solution processes). The procedure commonly applied is to use suspensions of methylene urea or solutions of urea and methylol urea as well as solid product. In this way the condensation reactions take place in the granulator itself (in situ process) [144].

Properties. The urea – formaldehyde products are separated into the following three fractions:

1) Fraction I: cold water (25 °C) soluble (CWS), containing residual urea, methylene diurea (MDU), dimethylene triureas (DMTU), and other soluble reaction products. Depending on soil temperature the availability of Fraction I nitrogen is slow.
2) Fraction II: hot water (100 °C) soluble (HWS), containing methylene ureas of intermediate chain length: slow-acting nitrogen.
3) Fraction III: hot water insoluble (HWI) containing methylene ureas of very long chain length, insoluble in both cold and hot water; extremely slow-acting or ineffective in plant nutrition.

How the proportion of the different methylene ureas affects the release of nitrogen and the nitrogen efficiency are expressed by the activity index (AI). The AI is calculated from the solubility fractions of the fertilizer under various conditions [127, p. 22]. In the past urea – formaldehydes had an AI of about 40 – 50; more recent formulations have AI values of 55 – 65.

In general, the nitrogen content of urea – formaldehyde condensation products ranges from 35 to 42 % N. The American Association of Plant Food Control Officials (AAPFCO) specifies a minimum AI of 40, with at least 60 % of the nitrogen as cold water insoluble nitrogen (CWI N), a total N content of at least 35 % N, and an unreacted urea nitrogen content of less than 15 % of total nitrogen. In the United States, Western Europe, the Former Soviet Union, and Israel research has been carried out to reduce the fraction of HWI nitrogen [148]. In the 1980s research resulted in the development of MDU/DMTU compositions which consist of shorter chain polymers with at least 60 % CWS polymer nitrogen. Although they have higher contents of CWS nitrogen, they still have safer agronomic and environmental properties than conventional nitrogen fertilizers. Commercial products are white, colorless powders or granules. In wet granulation, the pH value and temperature must be controlled to avoid hydrolysis and thus losses of formaldehyde. Under normal conditions the finished products are stable in handling and storage. Typical properties of urea – formaldehyde products are given in Table 21.

Application. The release of plant-available nitrogen from urea – formaldehyde products mainly involves decomposition through microbial activity and dissolution by hydrolysis. Consequently, factors affecting microbial activity, such as higher temperature, moisture, pH value and oxygen availability, also affect the release of nitrogen. These products are therefore widely used in warmer climates (in the Mediterranean region of Europe and in the southern and southwestern United States).

Table 21. Properties of typical ureaforms

Property	Nitroform*	Azolon**
Total N, wt %	37.4	38.0
Insoluble in cold water, wt %	26.9	26.0
Soluble in hot water, wt %	15.4	10.4
Activity index	43	60
Bulk density, t/m^3		0.75
Granule size, mm	0.5 – 2.0	1.0 – 4

* Trademark of Nor-AM; BASF product analysis 1980.
** Trademark of Aglukon; analysis from [149].

4.4.2.2. Other Urea – Aldehyde Condensation Products

Due to the higher costs involved in the combination of urea with higher aldehydes, only two products have gained commercial importance. These are IBDU or Isodur (urea + isobutyraldehyde) and CDU or Crotodur (urea + acetaldehyde or crotonaldehyde).

IBDU (Isodur). Products derived through the combination of urea and isobutyraldehyde consist of mainly isobutylidene diurea [6104-30-9] (with small quantities of slow-acting by-products).

$$\begin{array}{c} H_3C \\ \diagdown \\ CH-CH \\ \diagup \\ H_3C \end{array} \begin{array}{c} NH-CO-NH_2 \\ \\ \\ NH-CO-NH_2 \end{array}$$

Isobutylidene diurea

Manufacture. Isobutylidene diurea is manufactured by condensation of liquid isobutyraldehyde with urea (either in solution or solid form). In contrast to urea – formaldehyde, the reaction of urea with isobutyraldehyde results in a single oligomer. To obtain an optimal proportion of IBDU it is important to stop the reaction by neutralization when the IBDU yield is at a maximum.

Isobutylidene diurea is produced in Japan and Germany [130, 150 – 152] (Mitsubishi, BASF).

In the BASF process urea reacts with isobutyraldehyde in an aqueous solution to give a high proportion of slow-release N [153]. A Mitsubishi-developed process is operated by IB Chemicals in Alabama.

Properties and Application. IBDU is a white crystalline solid with a theoretical N content of 32.18 wt % N. The official definition (AAPFCO) requires 30 wt % N, of which 90 % must be cold water insoluble prior to grinding. Since it is almost completely soluble in hot water, the N content is therefore nearly all slow-release N. It has a calculated AI of 90 – 99 (ureaform: 55 – 65). Nitrogen is released from IBDU by hydrolysis, which is affected by soil moisture and temperature. Both urea molecules of IBDU can be liberated. The rate of N release is mainly a function of particle size: the finer the particles, the more rapid the rate of N release. IBDU is unstable in an acid media, whereby it decomposes into the starting materials. Therefore, it tends to release its nitrogen more rapidly in strongly acid soils. This can also be counteracted by using larger, well-compacted granules [158, 159].

The safety margin and agronomic response from IBDU is good with turf; occasional phytotoxicity has been observed in greenhouse use. Since it is independent of microbial activity, IBDU is particularly suited to low-temperature application. Properties of IBDU are given in Table 22.

CDU (Crotodur). Crotonylidene diurea [*1129-42-6*] is a mixture of 75 – 80 wt % crotonylidene diurea (2 mol urea + 2 mol acetaldehyde) and 15 – 20 wt % 5-oxyethylcrotonylidene diurea [*23048-84-2*] (2-oxo-4-methyl-5-oxyethyl-6-ureidohexahydropyridine), 2 mol urea + 3 mol acetaldehyde), [145] and 5 – 7 wt % K_2SO_4.

Crotonylidene diurea

Manufacture. Crotodur, which was patented as a slow-release fertilizer in 1959 [153], is produced by acid-catalyzed reaction of urea with acetaldehyde (Chisso Corp., Japan) or crotonaldehyde (BASF, Germany) [154, 155]. The mother liquor is neutralized, and the CDU isolated as a white powder by spray drying or filtration.

Properties (see Table 22). CDU or Crotodur contains 85 wt % to > 90 wt % pure crotonylidene diurea. It is almost completely insoluble in cold water, but soluble in hot water with a calculated AI of 90 – 99. In contrast to isobutylidene diurea, N release from crotonylidene diurea depends on hydrolysis and microbial activity. Only the urea molecule from the side chain (6-position) can be liberated by hydrolysis. The urea which forms part of the ring can only be released by microbial action [146]. The particle size also influences the N release; with large particles, release is strongly delayed.

Application. In Japan and Europe Crotodur is mainly used on turf and in speciality agriculture, either as a straight N fertilizer or in granulated NPK fertilizers.

4.4.2.3. Further Processing of Urea – Aldehyde Condensates

All commercial urea – aldehyde condensates such as ureaform, IBDU, Isodur, CDU, and Crotodur can be further processed by compaction or moist granulation with other conventional fertilizers with rapidly available plant nutrients, thus producing straight and compound fertilizers with both slow-release and rapidly available nitrogen. However, it is necessary to keep the pH between 5.4 and 6.2 [147], or preferably between 5.0 and 5.5 [156], the temperature below 90 °C, and the dwell time as well as the quantity of recycled material as low as possible [130, pp. 1 – 15], [156] to prevent hydrolysis and, particularly in the case of ureaform, further condensation to higher oligomers.

Commercial Products and Trade Names. There are a large number of various slow-release fertilizers based on ureaform 38–0-0, methylene ureas 40–0-0, IBDU 31–0-0, and CDU 31–0-0, which are formulated as straight fertilizers and in combination with P, K, and secondary and trace elements in solid form, as well as in solutions or suspensions: Scotts Granuform, Scotts MU-40, ProGrow, ProTurf (The Scotts

Table 22. Properties of CDU, IBDU, and Isodur

Property	CDU	IBDU	Isodur
mp, °C	245a	207 – 208	203 – 204b
Bulk density, kg/m^3	ca. 600	600 – 700	500 – 600
Solubility in water at 25 °C, g/L			
fertilizer grade		0.3 – 3.0	2.7
pure substance	1.3 (pH 2); 0.9 (pH 7)		1
Total N, wt %	31	31	31
pH (10 wt % suspension)	4 – 6	5 – 8	6 – 8
LD$_{50}$ (rat, oral), g/kgb	10.0		10.0

a 259 – 260 °C for crotonylidene diurea (BASF).
b 236 °C for isobutylidene diurea (BASF).
c Toxicological data from BASF.

Company); Nitroform, Nutralene (Omnicology, Inc./AgrEvo); Plantosan, Nitroform, Nutralene, Azolon (Aglukon Spezialdünger). Hydroform, Hydrolene (Hydroagri US). Folocron, CoRoN (CoRoN Corp.); Isodur (Floranid), Crotodur (Triabon) (BASF); Azorit (EniChem); Urea – formaldehyde (Mitsui Toatsu Fertilizers); IBDU (Mitsubishi Kasei); CDU (Chisso).

Analyses of some typical fertilizers containing Isodur or Crotodur are listed in Table 23.

4.4.3. Other Organic Chemicals

Some other organic compounds such as oxamide, triazones, and melamine have also been used as slow-release fertilizers but have not obtained the commercial importance of urea – aldehyde products.

Oxamide [*471-46-5*], the diamide of oxalic acid, N content 31.8 wt %, M_r 88.08, mp 419 °C, ϱ^{20} 1.667 g/cm^3, is a nonhygroscopic, colorless compound that forms needle-shaped crystals.

Its solubility in water is only 0.4 g/L [41, p. 320]. By hydrolysis in the soil, it is transformed first into oxamic acid with liberation of ammonia, and then into oxalic acid. Oxalic acid is toxic to plants if it is not further converted into carbon dioxide by microbial activity. Because the nitrogen is released by hydrolysis, the slow-release effect is primarily a function of particle size. Oxamide is manufactured by oxidation of hydrogen cyanide with hydrogen peroxide to give cyanogen, which is then partially hydrolyzed [161]. It can also be produced directly in one step from hydrogen cyanide, oxygen, and water, with copper nitrate as the catalyst [162 – 164].

Table 23. Analysis and physical properties of typical fertilizers containing Isodur or Crotodur *

Property	Floranid permanent 15 + 9 + 15 (+ 2)**	Floranid N 32**	Triabon 16 + 8 + 12 (+ 4)**
Nitrogen content, wt %			
Total	15	32	16
IBDU	5	29	
CDU			11.0
Ammonium	5.8		4.0
Nitrate	4.2		
Carbamide		3.0	1.0
Phosphate (P$_2$O$_5$)	9		8
Potassium (K$_2$O)	15		12
Magnesium (MgO)	2		4
Granule size***	0.7 – 2.8	0.6 – 2.6	1.0 – 3.5
Bulk density, kg/m^3	960	600	820

* Floranid and Triabon are trademarks of BASF.
** Floranid Permanent and Triabon also contain trace elements.
*** 90 % within given range [157].

In Japan, production of oxamide as a slow-release fertilizer amounts to about 1000 t/a (1995, Ube Industries) [165].

Symmetrical Triazones. Some symmetrical triazones are used as slow-release nitrogen fertilizers [160], particularly in urea – triazone solutions with 28 % N for application to soil. Triazones are synthesized by condensation reactions of formaldehyde or other aldehydes with urea, organic amines, or ammonia [131, p. 251].

Commercial Products and Trade Names. Hickson Kerley, United States: N-Sure 28–0-0, N-Sure-Lite 30–0-0, Trisert 13–3-4.

Melamine, triaminotriazine [*108-78-1*], is a high-nitrogen (66 wt % N) crystalline powder. It is produced through heating urea under pressure in the presence of a catalyst. Melamine slow-release fertilizer material is available as a powder (Nitrazine 66 % N). Currently only small amounts are used in the fabrication of slow-release spikes and stakes for shrubs and trees.

4.4.4. Inorganic Compounds

Some sparingly soluble inorganic compounds such as metal ammonium phosphates and metal potassium phosphates [130, pp. 256 – 284] are also used as slow-release fertilizers. Their solubility in water at 25 °C is [166, 167]:

$MgNH_4PO_4 \cdot H_2O$	0.014 g/100 mL
$MgNH_4PO_4 \cdot 6H_2O$	0.018 g/100 mL
$MgKPO_4 \cdot H_2O$	0.21 g/100 mL
$MgKPO_4 \cdot 6H_2O$	0.23 g/100 mL

The release of nitrogen, particularly from the two ammonium compounds, again depends on particle size (slow release only with larger particles or granules).

The manufacturing processes for magnesium ammonium and magnesium potassium phosphates are described in [130, pp. 256 – 284], [168, 169].

Other sparingly soluble inorganic compounds which have been used as slow-release fertilizers are based on crystalline ammonium potassium polyphosphates [170], and on glassy melts of ammonium dihydrogenphosphate, potassium hydrogenphosphate, and dicalcium phosphate [171].

Commercial Products and Trade Names. MagAmp (magnesium ammonium phosphate; Grace Sierra Horticultural Products); EnMag (magnesium ammonium phosphate + potassium sulfate; ICI).

4.4.5. Coated and Encapsulated Controlled-Release Fertilizers

Coated fertilizers are conventional soluble fertilizer materials whose plant nutrients are rapidly available and which after granulation, prilling, or crystallization are given a protective coating to control water penetration and hence the rate of dissolution and nutrient release in the soil. The AAPFCO definition is: "A product containing sources of water soluble nutrients, release of which in the soil is controlled by a coating applied to the fertilizer".

Coated fertilizers are the fastest growing group of slow- and controlled-release fertilizers in the United States and in Japan. They accounted for 24 % of total world consumption of slow- and controlled-release fertilizers in 1995/96 [127, pp. 61 – 63]. The growth is due to improved economics in production, the possibility of controlling the release of nutrients other than nitrogen, and the greater flexibility in determining the nutrient release pattern.

Many condensation polymers, drying oils, waxes, and bitumen were tested for their suitability as coating materials [129, pp. 102 – 263], [172, 173]. However, only three categories of coated/encapsulated controlled-release fertilizers have gained commercial importance:

1) Sulfur coatings
2) Polymer coatings (e.g., PVDC copolymers, polyolefins, polyurethanes, urea – formaldehyde resins, polyethylene, polyesters, alkyd resins)
3) Sulfur – polymer coatings (hybrid products with a multilayer coating of sulfur and polymer) [174 – 186]

4.4.5.1. Sulfur-Coated Controlled-Release Fertilizers

The only inorganic coating material which has achieved any importance is sulfur [129, pp. 102

– 263], [132, pp. 151 – 169], [193 – 196]. The sulfur coating can be regarded as an impermeable membrane which slowly degrades in the soil through microbial, chemical, and physical processes. Nutrient release depends on the thickness of the coating in relation to the size of the granule or prill and the quality of the urea substrate. The total N content of sulfur coated ureas varies with the amount of coating applied; products currently available contain 30 – 42 % N, 6 – 30 % S, and various sealants and conditioners. Commercial production started in 1972 when ICI commissioned a pilot plant in the United Kingdom.

Manufacture. Most of the modern industrial processes are still based more or less on the technology developed in the 1960s and 1970s by TVA [193, 194, 197] (now National Fertilizer and Environmental Research Center). Preheated (71 – 82 °C) urea granules (1.7 – 2.9 mm) are introduced into a horizontal rotating cylindrical drum. Molten sulfur (143 °C) is sprayed onto the urea granules and quickly solidifies on contact. The average target thickness is 40 µm, but there are various random proportions of granules having thin (< 30 µm), medium (30 – 50 µm) and thick (> 50 µm) sulfur coatings. Any pores and cracks are closed in a second step by addition of a polymeric hydrocarbon/petroleum-based wax or a high-viscosity polymeric paraffin oil with a polyethylene sealant (2 – 3 % of total weight). A flow conditioner such as diatomaceous earth, talc, clay, or silica (2 – 3 % of total weight) is added to give a dust-free, free-flowing product with good handling and storage properties.

Other straight and compound fertilizers can also be coated with sulfur. However, ammonium nitrates and fertilizers with high contents of nitrate nitrogen are excluded due to the risk of explosion.

Agronomic Properties. Generally, sulfur-coated urea products have good slow-release properties. However, resistance of the coating to impact and abrasion is low. The quality of sulfur-coated urea (SCU) is characterized by the rate of N release into the soil solution within seven days (TVA method). SCU-30 indicates a product that releases 30 % of its nitrogen within seven days, resulting in a rather rapid initial effect. If coated too thickly they may exhibit lock-off, i.e., no effective nutrient release. These disadvantages of sulfur-coated conventional fertilizers were the reason for the development of sulfur-coated, polymer encapsulated fertilizers.

Commercial Fertilizers and Trade Names. Sulfur-coated fertilizers are mainly distributed as straight nitrogen grades, e.g., Enspan 39 % N (Hydro Agri, North America), sulfur-coated urea, and sulfur-coated potassium (Nu-Gro Canada).

4.4.5.2. Sulfur-Coated, Polymer-Encapsulated Controlled-Release Fertilizers

Sulfur – polymer hybrid coatings combine the controlled-release performance of polymer-coated fertilizers with the lower cost of sulfur-coated fertilizers. Figure 22 [198] shows the flow diagram of the process developed by RLC Technologies. The process yields a sulfur-coated controlled-release granular urea fertilizer with a uniform, durable polymer coating over the sulfur coating. The hot-melt polymer liquid sealant of the conventional process is replaced by specific liquid monomers. When applied sequentially onto the surface of the hot sulfur-coated urea granules, they copolymerize to form a firm, tack-free, water-insoluble polymer coating sealant. The liquid monomers used in this process are diisocyanates, such as MDI (4,4-diphenylmethane diisocyanate), and a polyol mixture of DEG (diethylene glycol) and TEA (triethanolamine); the TEA acts both as a reactive polyol and a catalyst. The resulting product has improved impact resistance. The RLC sealant provides a substantial improvement over polymeric wax and solvent-dispersed sealants [198].

Commercial Products and Trade Names. The commercial products (in the United States) generally contain 38.5 to 42 % N, 11 to 15 % S and ca. 2 % polymer sealant: TriKote PCSCU 39–42N (Pursell Technologies), Poly-S PCSCU 38.5–40N (Scotts), and POLY PLUS PCSCU 39N (Lesco).

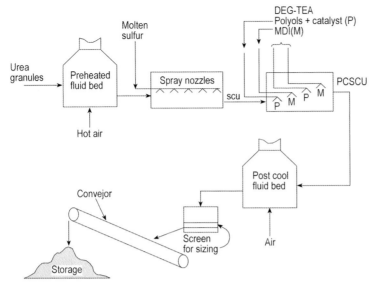

Figure 22. Flow diagram of the RLC Technologies process

4.4.5.3. Polymer-Encapsulated Controlled-Release Fertilizers

Application of controlled-release fertilizers to high-value crops requires precise control of nutrient release geared to plant requirements. Hence, a range of polymer-coated controlled release fertilizers has been developed, in which the rate of nutrient release can be altered by means of the composition and thickness of the coating, giving longevities from one to 24 months. The release pattern of these controlled-release fertilizers is significantly more linear than that of PCSCU.

Polymer coatings can be semipermeable membranes or impermeable membranes with tiny pores. Most polymers used in coating conventional fertilizers decompose extremely slowly or not at all in the soil and leave residues of up to 50 kg ha^{-1} a^{-1}). This may be considered as an environmental disadvantage even though the accumulation of 10 years (500 kg) only represents 200 ppm in dry soil. Nevertheless, extensive research is being carried out into the development of polymer coatings that are biodegraded after application. In the case of polyethylene, polypropylene, and ethylene copolymers, incorporation of ethylene – carbon monoxide copolymer promotes photochemical degradation of the coating, and coatings that contain a poly(3-hydroxy-3-alkylpropionic acid) as active ingredient are biodegradable [178, 179], [185, p. 14].

Manufacture. In the *Sierra Process* [187] the coating material is a copolymer of dicyclopentadiene with drying or semidrying oils in an organic solvent (glycerol ester of linseed oil). The granules are coated with at least two layers in a coating drum operating at 65 – 70 °C. Maleic acid, for example, is added to improve the drying of the oil. Coating weights vary from 10 to 20 wt %; the commercial products are mainly blends of different coating weights.

The *Chisso Asahi process* [188, 189] produces a particularly smooth coating. The incorporation of finely powdered inorganic materials such as talc and silica into the coating [185, pp. 15, 16] makes it possible to produce controlled-release fertilizers in which the rate of release of nutrients in the soil varies with temperature. A hot 5 wt % solution of the coating material in a hydrocarbon or chlorinated hydrocarbon solvent is sprayed onto the warm granules of fertilizer in a fluidized bed, and the solvents are immediately evaporated with hot air (60 – 70 °C).

Polyolefins, ethylene – vinyl acetate copolymers, poly(vinylidene chloride), and mixtures thereof are used as coating material. The moisture permeability is modified by means of the ratio of ethylene – vinyl acetate (high permeability) to polyethylene (low permeability) [185, pp. 12 – 16].

The release pattern is determined by a water-leach test at 25 °C; for example, T-180 indicates that the product releases 80 % of its nutrient over 180 d at 25 °C in water. Products are produced from T-40 to T-360 [185, pp. 19 – 21].

The *Pursell Technologies Reactive Layers Coating (RLC)* process [190, 191] produces attrition resistant controlled-release fertilizers by coating a plant nutrient with a coating material that chemically bonds to it. Solvent-free polyols and polyisocyanates are pumped to nozzles, located along the entire length of a horizontal, cylindrical rotating drum. The two reactive liquids are applied sequentially in ultrathin layers onto the surface of preheated urea granules, where they polymerize directly to form a continuous polyurethane encapsulating membrane. The first layer of polyisocyanate can also react with the urea granule, chemically bonding the polymer coating to the surface of the urea granule. The thickness of the layer depends on the number of reactive layers and allows controllable release durations of up to 6 months to be achieved.

The polyols used in the process are predominantly polyester polyols. The isocyanates used are also a major factor in influencing the properties of the polyurethane encapsulation. Preferred isocyanate raw materials include polymeric diphenylmethane diisocyanates. Though the process had been primarily defined for the use of urea, most other plant nutrients/fertilizers can also be utilized.

Commercial Products and Trade Names. The main polymer-coated substrates are urea (40–0–0 and 38–0–0), potassium chloride/potassium sulfate, and NP, NK, and ammonium nitrate containing NPK fertilizers. Various grades also contain secondary and minor elements.

Trade names: The Scotts Company, US: Osmocote, Osmocote Plus (NPKs + MgO + TEs), High-N, Sierra, and Sierrablen; Prokote, Scottkote. Chisso Corp., Japan: Meister (urea, KCl and K_2SO_4) Lp and Long, Nutricote (NP, NK, NPK). Pursell Technologies Inc., US: POLYON lines (PC-U, PC-SOP, PC-NPK, PC-MAP, PC-KNO_3). Agrium, Canada: Duration and ESN (clay-coated PCU). Aglukon Spezialdünger, Germany (subsidiary of AgrEvo): Plantacote Depot, Plantacote Control (NPK), and Plantacote Mix. BASF, Germany: Basacote (NPK + TE). Haifa Chemicals, Israel: Multicote 4 (PC-NPK, PC-U, PC-KNO_3). Asahi Chemical Ind., Japan: Nutricote. Mitsubishi Chemical, Japan: M cote (PCU).

4.4.6. Anti-Float Materials

Special controlled-release fertilizers which sink immediately on application have been developed for application to irrigated crops. Diatomaceous earth or wetting agents such as liquid surfactants are applied to the external surface of the coated fertilizer to achieve this antifloat effect by breaking the surface tension between the water and the coated fertilizer [192].

Commercial Products and Trade Names. Pursell Technologies Inc. POLYON PCU-AF/AntiFloat, marketed in Japan by Sumitomo; Haifa Chemicals resin-coated anti-floating urea MULTICOTE.

4.4.7. Controlled-Release Fertilizers on Carriers

Bayer manufactures controlled-release NPK fertilizers based on water-insoluble synthetic ion-exchange resins [199, 200]. The nutrients are released from the resin granules by reaction with salts in the ground water.

Commercial Products and Trade Names. Bayer AG Lewatit HD 5 NPK 18–7-15 and Lewaterr 80 NPK (31–12–33).

4.4.8. Supergranules

A slow release of nutrients can also be achieved by granulation or compaction of conventional fertilizers with a relatively small surface to volume ratio (supergranules, briquettes, tablets or sticks or stakes). Some of these formulations also contain urea – formaldehyde or IBDU.

4.4.9. Legislation

In the United States, 50 states regulate their own agricultural policies, including fertilizers. There are some guidelines and Federal EPA regulations which can be imposed on the individual states (mainly concerning registration of pesticides under RECRA — Resource Conservation And Recovery Act — in the EPA). However, fertilizers are excluded [201]. In Western Europe there are not as yet general regulations of the EU Commission on slow- and controlled-release fertilizers, and there are no coated controlled-release fertilizers in the EU type list. There are regulations concerning definitions and classification in the individual member states. These cover fertilizer types such as coated urea, coated NPK, and partly coated NPK.

In Germany, CDU, IBDU and urea – formaldehydes (UFs) are classified as individual fertilizers. Legislation also includes the group of N, NPK, NP, and NK fertilizers containing UF, CDU or IBDU; other legislation covers coated and encapsulated fertilizers [202, 203].

To achieve European standardization, a Task Force slow-release fertilizers (TFsrf) has been formed with the aim of defining the conditions under which type of fertilizer included in the EU fertilizer type list can be newly categorized as a slow- or controlled-release fertilizer [127, p. 11]: A fertilizer may be described as slow-release if the nutrients declared as slow- (controlled-) release meet, under defined conditions including that of a temperature of 25°C, each of the following three criteria:

1) No more than 15 % released in 24 h
2) No more than 75 % released in 28 d
3) At least about 75 % released in the stated release time

This European Task Force has close contacts to the Controlled Release Task Force formed in the United States by AAPFCO (Association of American Plant Food Control Officials) and TFI (The Fertilizer Institute). In Japan registration of slow- or controlled-release fertilizers requires a dissolution test in water under well defined conditions [204], [127, p. 105].

4.5. Nitrification and Urease Inhibitors

4.5.1. Introduction

Ammonium ions in the soil — whether from decomposition of organic material or from the application of ammonia-containing mineral fertilizers — are oxidized to nitrite and nitrate. Bacteria of the species *Nitrosomonas* spp. are responsible for the transformation into nitrite. The nitrite is relatively rapidly oxidized to nitrate by *Nitrobacter* and *Nitrosolobus* spp., so that there is normally no toxic accumulation of nitrite in the soil [205, pp. 156, 157], [206, pp. 287, 289]. The process is known as nitrification. The formation of the environmentally relevant gases N_2O and NO may be regarded as a side-reaction of the nitrification process [207].

Nitrate is readily soluble in water and in the aqueous soil medium, and so is completely mobile in the soil, in contrast to ammonia, which is strongly adsorbed in soil colloids and base-exchange complexes. Therefore, it can be leached readily from the soil [208 – 212]. Under unfavorable (anaerobic) conditions nitrate can be reduced by denitrification to N_2 [206, p. 289]. This can result in further considerable losses of nitrogen [213, pp. 90 – 95].

Addition of a nitrification inhibitor to ammonia containing fertilizers or urea retards nitrification and minimizes leaching of nitrogen as nitrate and losses of nitrogen due to denitrification. Furthermore, nitrification inhibitors also suppress methane emissions and lower nitrous oxide emissions [207, 212, 214 – 218].

On application to the soil, amide nitrogen, as in urea, UAN (urea – ammonium nitrate solution), and some NPK fertilizers, is transformed by the enzyme urease via the unstable ammonium carbamate ($H_2NCOONH_4$) to ammonia, and CO_2 [205, p. 156].

This transformation has two major drawbacks:

1) It results in sometimes very high volatilization losses of ammonia when urea is applied to the surface [219, 220] or under flooded conditions.
2) It can produce severe seedling damage by ammonia and nitrite [221].

Urease inhibitors inhibit or reduce the formation of urease, which is ubiquitous in surface soils.

This slows down the rate of urea hydrolysis in the soil and prevents or at least depresses the transformation of amide nitrogen into ammonia.

Definitions. *Nitrification inhibitors* are compounds that delay bacterial oxidation of the ammonium ion by depressing the activity of *Nitrosomonas* bacteria in the soil over a certain period of time. Thus, they control leaching of nitrate by keeping nitrogen in the ammonium form longer, and preventing denitrification of nitrate [222, p. 12]. *Urease inhibitors* prevent or depress transformation of the amide nitrogen of urea into ammonium hydroxide and ammonia over a certain period of time by inhibiting hydrolytic action on urea by urease; thus they avoid or reduce volatilization losses of ammonia.

There is considerable confusion concerning the terms nitrogen stabilizers, nitrification inhibitors, urease inhibitors, and stabilized fertilizers. The terms nitrogen stabilizers and nitrification inhibitors have been used interchangeably. Strictly speaking, stabilized fertilizers refers only to those which are modified during production with a nitrification inhibitor, such as ALZON and BASAMMON. In all other cases, fertilizers and nitrification and urease inhibitors are sold separately [222, p. 12].

Advantages and Disadvantages of Nitrification and Urease Inhibitors. Nitrification inhibitors significantly reduce leaching losses of nitrate by stabilization of ammonia [219, 220] and reduce emissions of the environmentally relevant gases N_2O and NO [207, 214 – 216, 223]. Nitrification inhibitors indirectly improve the mobilization and the uptake of phosphate in the rizosphere [224]. Urease inhibitors reduce ammonia volatilization losses, particularly from top-dressed agricultural fields and under reduced tillage conditions [225, p. 9, 10], [226 – 228]. Urease inhibitors furthermore reduce seedling damage where seed-placed levels of urea-containing fertilizers are too high [219, 221].

Possible disadvantages include the fact that fertilizers containing ammonia and a nitrification inhibitor may result in increased ammonia volatilization if they are not incorporated into the soil immediately after application. Depending on the type of nitrification inhibitor, the activity of soil bacteria may not only be interrupted for a certain time period, but the soil bacteria may actually be killed. This can be regarded an undesirable interference in a natural soil process [213, p. 219], [229, pp. 37 – 44].

4.5.2. Types of Nitrification and Urease Inhibitors

Extensive research on nitrification and urease inhibitors has been carried out mainly in Europe, Japan, Russia, and the United States. Various chemical fumigants and pesticides have also been tested to establish their possible effectiveness in inhibiting nitrification (nematicides; soil-insecticides and herbicides) [230, pp. 547 – 554].

Until the late 1960s research was carried out in the United States and Japan on N-Serve (2-chloro-6-trichloromethylpyridine; Dow Chemical Company), AM (2-amino-4-chloro-6-methylpyrimidine; Mitsui Toatsu Chemicals), Terrazole (5-ethylene oxide-3-trichloromethyl-1,2,4-thiodiazole; Olin Mathieson); ASU (1-amide-2-thiourea; Nitto Chemical Industry), and ATC (4-amino-1,2,4-triazole hydrochloride); substituted phenyl compounds (DCS), and compounds of the *s*-triazine line (MAST) [231, pp. 64 – 82]. In Eastern Europe and the former Soviet Union, CMP (1-carbamoyle-3-methylpyrazole) and its main metabolite MP (3-methylpyrazole) were tested extensively. However, only products based on pyridines, dicyandiamide, pyrazoles have gained practical agronomic importance as nitrification inhibitors. Terrazole, AM, and ASU (thiourea) had some regional importance, particularly in the United States and in Japan.

Reseach on urease inhibitors has concentrated on phosphoric triamides. Limited research has been carried out with PPD/PPDA (phenyl phosphorodiamidate) and ATS (ammonium thiosulfate) [222, p. 32].

There are no reliable statistics publicly available on the use of nitrification inhibitors due to the unique production structure. Estimates of the acreage treated with fertilizers containing nitrification inhibitors for the United States are 1860 $\times 10^6$ ha (1995/96), 1660 $\times 10^6$ ha thereof with nitrapyrin and 200 000 ha with dicyandiamide. For Western Europe a very rough estimate is 200

000 ha of arable cropland treated with fertilizers containing dicyandiamide [222, p. 63, 64].

4.5.3. Pyridines

4.5.3.1. Nitrapyrin

Nitrapyrin consists of 2-chloro-6-trichloromethylpyridine [*1929-82-4*] and related chlorinated pyridines, such as 4,6-dichloro-2-trichloromethylpyridine [*1129-19-47*].

Cl—N—CCl$_3$

Manufacture. The (trichloromethyl)pyridine compounds are manufactured by photochlorination of methyl-substituted pyridines. The desired amount of chlorine gas is passed through the appropriate methyl-substituted pyridine, generally in the presence of its hydrochloride. The product is recovered by conventional procedures such as filtration or distillation [232]. A flow diagram for the production of chlorinated picolines from readily available raw materials such as α-picoline is given in [233].

Nitrapyrin is produced exclusively by DowElanco in the United States and distributed under the trade name N-Serve (NS) [234 – 236].

Properties. Nitrapyrin is a white crystalline solid with a mild sweetish odor, *mp* 62 – 63 °C; *bp* 101 °C at 133 Pa. For use in agriculture nitrapyrin is formulated as a liquid product.

N-Serve 24 nitrogen stabilizer consists of 22.2 wt % 2-chloro-6-trichloromethylpyridine, 2.5 wt % related chlorinated pyridines including 4,6-dichloro-2-trichloromethylpyridine, and 75.32 wt % Xylene-range aromatic solvent [*64742-96-6*].

N-Serve 24E nitrogen stabilizer consists of 21.9 wt % 2-chloro-6-trichloromethylpyridine, 2.4 wt % related chlorinated pyridines, 4,6-dichloro-2-trichloromethylpyridine, and 75.7 wt % Xylene-range aromatic solvent.

Both formulations contain 2 lb of active ingredients per gallon ($\approx$ 240 g/L); NS24 is recommended for use with anhydrous ammonia (82 % N) and impregnation onto urea; NS24E for use with liquid fertilizers (aqueous ammonia, solutions) and with manure (slurry).

Nitrapyrin has a very selective effect on *Nitrosomonas* bacteria. However, this effect is not only bacteriostatic but also bactericidal, so that part of the population in treated soil is killed [213, p. 219], [236, 237].

Toxicity. The single dose toxicity of nitrapyrin is low; the technical material has a LD$_{50}$ of ca. 1000 mg/kg of body weight in laboratory animals. It is slightly irritating to the eyes and skin and has a low vapor toxicity. The equivalent LD$_{50}$ (oral, female rat) for the two formulations is 2140 mg/kg (N-Serve 24) and 3300 mg/kg (N-Serve 24E). In soil and in plants, nitrapyrin is chemically and biologically rapidly degraded into 6-chloropicolinic acid, the only significant chemical residue from its use, and further to N_2, Cl^-, CO_2, and H_2O.

Application. The recommended application rate is 1.4 – 5.6 L/ha. In warm soils the nitrification inhibiting period is normally 6 – 8 weeks; it can be 30 weeks or more in cool soils.

However, in agronomic use this long-term standard nitrification inhibitor has two drawbacks:

1) Loss by volatilization from treated fertilizers during storage
2) The bactericidal effect, which may be regarded as an undesirable interference in a natural soil process

The loss by volatilization restricts it to simultaneous application with anhydrous or aqueous ammonia or fertilizer solutions which are injected directly into the soil at a depth of at least 5 – 10 cm. This limits acceptance in regions where nitrogen fertilizer is not commonly injected. Therefore, N-Serve is available commercially only in the United States.

4.5.3.2. Other pyridines

3,6-dichloro-2-trichloromethylpyridine is a proven nitrification inhibitor [238]. However, it is not readily obtained by ring chlorination of a 2-substituted pyridine because such chlorinations are not sufficiently selective to produce a preponderance of the desired isomer.

Hence, 3,6-dichloro-2-methylpyridine is prepared by the addition reaction 1,1-dichloro-2-propanone and acrylonitrile to give 4,4-dichloro-5-oxohexanenitrile. Subsequent cyclization in the presence of hydrogen chloride gives 3,6-dichloro-2-methylpyridine, which is chlorinated to obtain 3,6-dichloro-2-trichloromethylpyridine [239].

Commercial Products and Trade Names. N-Serve 24 Nitrogen Stabilizer, N-Serve 24E Nitrogen Stabilizer, DowElanco.

4.5.4. Dicyandiamide

Dicyandiamide [*461-58-5*] exists in two tautomeric forms (→ Cyanamides, Chap. 3). Its nitrification-inhibiting property was first reported in 1959 [240].

Manufacture. All large-scale production processes are based on calcium cyanamide. In the first step cyanamide is liberated from calcium cyanamide by carbonation in aqueous solution at pH 7 – 8. The cyanamide is then dimerized to dicyandiamide at pH 9 – 10.

$CaNCN + H_2O + CO_2 \rightarrow H_2NCN + CaCO_3$

$2 H_2NCN \rightarrow H_2N-C=NCN$

The dicyandiamide is isolated by filtration or centrifugal crystallization.

The disadvantage of this production process is the high energy input for the production of calcium carbide, the raw material for the production of $CaCN_2$; this is reflected in the relatively high price for the product as a nitrification inhibitor. There are only three major producers worldwide: SKW Trostberg and SKW Stickstoffwerke Piesteritz in Germany; Odda Smelteverk in Norway; and Nippon Carbide in Japan. The former Canadian producer, Cyanamid Canada, has ceased production.

Properties. Dicyandiamide is generally supplied as white or colorless crystals in paper or jute bags. Under dry conditions it can be stored for an unlimited period. It has low water solubility (3.2 g/100 g water at 20 °C) and contains at least 65 % nitrogen. When applied to the soil it is decomposed (partly abiotically and partly biotically by specific enzymes) and converted via guanylurea and guanidine to urea, a conventional fertilizer [241 – 244].

Dicyandiamide has a bacteriostatic effect on the *Nitrosomonas* bacteria. Depending on the amount of nitrogen applied, soil moisture, and temperature, the nitrification-inhibiting effect of dicyandiamide lasts 6 – 8 weeks.

Toxicity. With an LD_{50} of 10 000 mg/kg oral (female rat), dicyandiamide is practically nontoxic. The Ames test did not reveal any mutagenic activity. Furthermore, long-term studies have shown that dicyandiamide has no cancerogenity. Therefore, any risks for human health can be excluded. This also refers to its residues [237, 245].

Processing. In Western Europe the majority of ammonium-containing fertilizers are applied in solid form. However, a satisfactory nitrification-inhibiting effect can not be obtained by simple mechanical mixing of dicyandiamide with the solid fertilizer. Dicyandiamide added to the fertilizer prior to granulation, it decomposes at the usual granulation temperatures of 100 – 140 °C. When dicyandiamide is spread onto the finished fertilizer granules it does not adhere firmly to the granule surface. The use of vegetable, animal, or mineral oils as adhesion promoters has drawbacks with regard to storage and spreading. BASF has developed a process for applying dicyandiamide to ammonium- and sulfate-containing fertilizers without decomposition and with good adherence to the surface of the granules [246]. The fertilizer granules at 60 – 130 °C are mixed in a rotary drum with 2 – 4 wt % of finely ground dicyandiamide with simultaneous addition of a 40 – 50 wt % aqueous solution of $Ca(NO_2)$. The surface of the fertilizer material is rapidly cooled, thus preventing a decomposition of the applied dicyandiamide, which adheres well to the fertilizer granules.

Freeport-McMoRan Resource Partners in the United States has developed two processes for incorporating dicyandiamide into urea in combination with other nitrification- or urease-inhibiting compounds. In the first, dicyandiamide is incorporated into urea in combination with ammonium thiosulfate and a phosphate. Controlled release of N is claimed [247]. In the second, dicyandiamide is incorporated into urea in combination with *N*-(*n*-butyl)thiophosphoric triamide (NBPT) [248].

Application. Dicyandiamide is applied directly with N or NPK fertilizers and fertilizer solution, in which it is incorporated in correct proportion for the ammonia content. These stabilized fertilizers are recommended for all agricultural crops, particularly when grown on light-textured soils where heavy precipitation is expected within 6 – 8 weeks after application or where crops have a relatively slow growth rate during the early growing stages (e.g., potatoes).

Commercial Products and Trade Names. Ensan (BASF technical product); Basammon stabil (27 % total N, of which 1.6 % is dicyandiamide N (Ensan) and 13 % S); Nitrophoska stabil 12–8-17 (12 % total N, of which 1.1 % is dicyandiamide N (Ensan), plus 2 % MgO and 7 % S). Didin (SKW technical product); Alzon 27 (27 % total of N, of which 1.6 % is dicyandiamide N (Didin) and 31 % S); Alzon 47 [47 % total N, of which 3 % is dicyandiamide N (Didin)]; Piadin (SKW Stickstoffwerke Piesteritz) liquid mixture of DCD and 3MP (3-methylpyrazole, the main metabolite of CMP) in a proportion of 15 : 1, contained in Piasin 28/Alzon-flüssig (a urea – ammonium nitrate solution).

4.5.5. Pyrazoles

The group of pyrazole compounds has the largest number of compounds showing a nitrification-inhibiting effect [249, 250]. They include 3-methylpyrazole; 3,4-dimethylpyrazole; 4-chloro-3-methylpyrazole; 3-methylpyrazole phosphate; 3,4-dimethylpyrazole phosphate; 4-chloro-3-methylpyrazole phosphate; compounds resulting from the addition of polyacrylic acid to 3,4-dimethylpyrazole, 4-chloro-3-methylpyrazole, 3,4-dimethylpyrazole phosphate, 4-chloro-3-methylpyrazole phosphate; and from the addition of polyphosphoric acid to 3,4-dimethylpyrazole (1:20) and (1:1). In the 1980s and 1990s various compositions were patented [251 – 256, 324].

Properties. The nitrification-inhibiting effect of pyrazole compounds is better than that of dicyandiamide. However, they have one or more of the following disadvantages: they are liable to hydrolysis, which lowers the stability in storage and the activity period in the soil. They are too toxic and they are highly volatile. 3-Methylpyrazole is so volatile that significant amounts are lost when it is applied onto the surface of fertilizer granules or on fertilizer storages. Metal salts and metal complexes of 3-methylpyrazole [257] are also relatively volatile.

The most extensively tested pyrazole is 1-carbamoyl-3-methylpyrazole (CMP; in Russia: KMP). This compound is not as volatile as 3-methylpyrazole, but its manufacture is difficult and high evaporation losses may still occur when it is applied to fertilizers and on storage.

4.5.5.1. 1-Carbamoyl-3-methylpyrazole

1-Carbamoyl-3-methylpyrazole (and its metabolite 3-methylpyrazole, MP) has the molecular formula $C_5H_7N_3O$.

CMP was developed by VEB Agrochemie Piesteritz (now SKW Stickstoffwerke Piesteritz, Germany). Further research and practical use was mainly carried out in the former German Democratic Republic, in Central Eastern Europe, and in the former Soviet Union.

Manufacture. Carbamoyl-3-methylpyrazole (CMP) is produced from 3(5)-methylpyrazole (MP) by carbamoylation with sodium cyanate in aqueous HNO_3. After 1 h, the resulting CMP is removed by filtration, washed with water, and dried in a vacuum dryer. The yield is 90 %.

Properties. Bulk density 630 kg/m^3, solubility in water 0.56 g/100 g, *mp* 123 – 125 °C, evaporation pressure 4.41×10^{-2} Pa (20 °C). Therefore, the technical solid product has to be transported and stored in closed containers. When mixed with solid or liquid fertilizers, these have to be applied and incorporated into the soil immediately after preparation of the mixture. CMP has a bacteriostatic but not bactericidal effect on *Nitrosomonas* bacteria.

Toxicity. CMP has an LD_{50} of 1580 mg/kg (rat, oral) and its metabolite 3MP and LD_{50} of 1312 mg/kg (rat, oral).

Application. For use in field testing and agriculture CMP was formulated as a 50 % CMP-preparation for mixing with solid ammonium-containing fertilizers or solutions. However, because of its liability to hydrolysis and to prevent evaporation losses the CMP formulation could only be added at the time of applying the fertilizer and, like nitrapyrin, had to be incorporated into the soil immediately.

Commercial Products and Trade Names. At present (1998) there is no registration for 1-carbamoyl-3-methylpyrazole or any other pyrazole compound as a nitrification inhibitor in Europe. In accordance with German fertilizer law, Stickstoffwerke Piesteritz has registered a 15 : 1 mixture of dicyandiamide with 3MP under the name Piadin, recommended for use with UAN fertilizer solutions.

4.5.5.2. Outlook

In recent years research has concentrated on overcoming the high evaporation losses on adding pyrazole compounds to fertilizer granules and solutions, on storing such modified fertilizers, and on applying them onto the soil. BASF [258] has reported a process in which fertilizer granules are covered with salts of 3-methylpyrazole and made resistant against abraision and evaporation by addition of an inorganic or organic polyacid [259]. This treatment significantly lowers the volatility of the nitrification inhibitor and allows smaller amounts to be used. The storage stability of fertilizers treated by the BASF process is also improved. In a test for 4 weeks at 30 °C, 40 – 50 % R. H., and 1.2 m/s air velocity on 3,4-dimethylpyrazole/polyphosphoric acid 1/20 and 1/1 (DMPP), losses of the applied nitrification inhibitor were 0 and 12 %, respectively. Thus, DMPP is of future interest as a nitrification inhibitor.

4.5.6. Neem/Neem-Coated Urea

The extract from the press cake of the seed of the neem tree (*Azadirachta indica*) exerts a nitrification-inhibiting effect on *Nitrosomonas* spp. This effect is reversible.

The active compound in this biological nitrification inhibitor consists of various terpenes/triterpenes (epinimbin, nimbin, deacetylnimbin, salanin, azadirachtin, deacetylsalanin).

Toxicity. The formulated product is nontoxic.

Application. For practical use, a product containing 5.0 % active ingredient is to be mixed with urea in a ration of 1 : 100 before application.

Commercial Products and Trade Names. Nimin (5 – 5.5 % active ingredient) produced by Godrej Agrovet, Pirojshanagar (India).

4.5.7. Urease Inhibitors

Although it has long been known that substantial evaporation losses in form of ammonia from urea may occur, it was only in the 1980s that the first chemical substances with effective urease-inhibiting properties in soils were disclosed, namely the phosphoric triamide compounds. Losses of between 3 and 40 % and more of applied nitrogen are possible when urea is not incorporated immediately after application and stays on dry soil (particularly on soil with a high pH and with high temperature). Also if it is applied with conservative tillage, particularly no-till, and on crops which are not tilled, such as bananas, sugar cane, oil palms, and rubber, as well as on flooded rice [206, p. 290], [213, p. 211], [218, 260 – 263], [264, pp. 52, 59]. Reference [265] lists 146 triamide compounds, of which *N*-(*n*-butyl)thiophosphoric triamide (NBPT) showed the best inhibition values.

N-(n-butyl)thiophosphoric triamide [*94317-64-3*], NBPT urease inhibitor is the best developed of the *N*-alkyl thiophosphoric triamide for commercial applications.

Manufacture. NBPT is prepared by a two-step synthesis in THF:

$PSCl_3 + n\text{-BuNH}_2 + NEt_3 \rightarrow n\text{-BuNHPSCl}_2 + HNEt_3Cl$

$n\text{-BuNHPSCl}_2 + 4\,NH_3 \rightarrow n\text{-BuNHPS(NH}_2)_2 + 2\,NH_4Cl$

The approximate product composition from this manufacturing process is as follows [337]:

– *N-n*-butylthiophosphoric triamide (NBPT), min. 85 wt %
– *N,N*-di-*n*-butylthiophosphoric triamide (DNBPT) 0 – 3 wt %
– *N,N,N*-tri-(*n*-butyl)thiophosphoric triamide (TNBPT) 0 – 3 wt %
– Thiophosphoric triamide (TPT) 0 – 3 wt %
– Others 0 – 10 wt %
– THF 0 – 2 wt %
– Triethylamine 0 – 2 wt %

Properties. Pure NBPT is a white crystalline solid, industrial grade NBPT is a waxy, sticky, heat- and water-sensitive material, which render this material difficult to handle. This material is susceptible to decomposition during storage and distribution. The vapor pressure is 1.1 kPa at 40 °C. The compound is an excellent urease inhibitor that inhibits the activity of urease for 12 – 14 d on dry soil. When incorporated into the soil in combination with urea or urea-containing fertilizers, it significantly reduces seedling damage when seed-placed levels of such fertilizers are too high [219, 221]. For satisfactory use on flooded crops, further research is needed. *N*-methyl-pyrrolidone (NMP) is a good carrier for NBPT [268]. However, cheaper formulations with better long-term stability are obtained with glycols and glycols with a liquid amine cosolvent [269].

Toxicity. The acute oral LD_{50} of NBPT is 1000 – 4000 mg/kg. The Ames tests were negative [266, p. 23]. The compound poses a very low acute toxicity hazard to workers and has received EPA approval. Because it inhibits free urease in the soil without affecting bacterial growth, it is not classified as a pesticide and hence not regulated under TSCA. In the soil, the product degrades into the fertilizer elements nitrogen, phosphorus and sulfur.

Application. For practical use in agriculture NBPT (Agrotain) is formulated as a green clear liquid containing [267] 25 % *N*-(*n*-butyl)thiophosphoric triamide, as active ingredient, 10 % N-methylpyrrolidone, and 60 – 65 % other nonhazardous ingredients. The recommended rate of application depends exclusively on the quantity of amide-nitrogen applied as urea, UAN, or in the form of NPK-fertilizers (1.4 kg per tonne of urea).

Agrotain is primarily recommended for preplanting surface application of urea and urea-containing fertilizers, but may be used in pre-emergence, side-dress, top-dress, or other postplanting applications. It is not recommended for use if rain is imminent [222, pp. 38, 39].

Commercial Products and Trade Names. The only commercial product for use in agriculture is Agrotain (IMC-Agrico).

4.5.8. Environmental Aspects

In assessing the value of nitrification and of urease inhibitors not only must the better utilization of the applied nitrogen be taken into account, but also the possibility of maintaining clean ground water, as well as reduced emissions of ammonia and other environmentally relevant gases [222]. In Germany, where, in water catchment areas with restrictions or for other reasons, a reduction in applied nitrogen is required, the recommendation is that nitrogen application can be reduced by approximately 20 kg/ha without loss of yield.

Of equal importance are the positive environmental properties of nitrification inhibitors in significantly reducing emissions of climatically relevant gases such as N_2O and methane [207].

The application of urea or UAN-solution modified with a urease inhibitor such as NBPT would permit a substantial reduction in nitrogen losses to the atmosphere, and consequently also in the application rates, without affecting growth and yield of fertilized crops.

4.5.9. Legal Requirements

In Western Europe, there is no uniform regulation for DCD, the leading nitrification inhibitor (urease inhibitors are not yet in use). As in the case of slow- and controlled release fertilizers, individual countries [270, 271] have established national classification and legislation [272]. The pyrazole compound CMP (in combination with DCD) also comes under fertilizer legislation.

In the United States only fertilizers, but not additives to fertilizers, are excluded from RECRA (Resource Conservation And Recovery Act) regulations of the EPA. Therefore nitrapyrin — the active ingredient in N-Serve, the leading nitrification inhibitor in the United States — is classified as a pesticide in the EPA registration. In 1996 it was decided that all nitrification inhibitors have to be EPA registered as pesticides in the United States [222]. However, in spite of the new regulation, DCD will not require registration as a pesticide. NBPT, which does not affect soil organisms, is not classified as a biocide, but is regulated under the TSCA.

4.6. Organic Fertilizers (Secondary Raw Material Fertilizers)

At present an EC catalogue of commercial organic fertilizers is not available. Since these fertilizers mainly contain exploitable plant and animal wastes and in future will contain more waste of organic origin (with or without pretreatment such as composting, anaerobic fermentation, etc.), comprehensive legal provisions are in preparation. In Germany these organic fertilizers are designated secondary raw material fertilizers (*Sekundärrohstoffdünger*). According to the German Fertilizer Law their main components are organic waste materials for agricultural and horticultural utilization (secondary raw materials). These fertilizers are also subject to the German Waste Law (analysis of pollutant content, description of subsequent utilization) and the German Fertilizer Law (control of nutrient content) [273 – 275]. Depending on the quality standard, monitoring systems varying in analytic sophistication are specified. The main argument for the application of these fertilizers is the preservation of natural (and therefore limited) nutrient resources (e.g., P) by means of systematic recycling. The applied amount of these fertilizers is mainly limited by their nutrient content. In the case of low quality, the pollutant load can also determine the applied quantity. In Germany the permitted pollutant load conforms to the threshold values fixed in the decrees for the utilization of sewage sludge (AbfKlärV), which are likely to be lowered substantially for reasons of soil protection [276]. According to the pollutant content or the ratio of nutrients to pollutants, organic wastes can be ranked as to their applicability as secondary raw material fertilizers (Fig. 23). For fertilizers containing several organic waste materials, each component must meet the legal requirements for secondary raw material fertilizers. Blending of unapproved material with compounds of low pollutant content is prohibited.

Another group of organic – mineral fertilizers are mixtures of fertilizer salts (N, P, K) with peat, composted bark, lignite dust, or, occasionally, dried slurry. The difference between these fertilizers and secondary raw material fertilizers is that the primary product of organic origin is not classified as waste. Nevertheless, threshold values for heavy metal concentrations have been set [e.g., 3 mg Cd/kg dry weight (dw); 750 mg Zn/kg dw]. In the case of peat, different combinations of fertilizer salts (N, NP, NPK, PK) are added. In the case of composted bark, lignite, or slurry, only mixtures with NPK fertilizers are supplied [274].

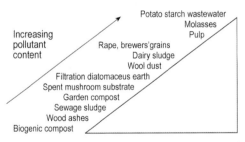

Figure 23. Applicability of wastes for agricultural use [277]

Organic fertilizers are also classified according to their effects:

– Fertilizers with an improving effect on soil condition have a stable organic substance as well as a slow effect on N supply (low concentration of readily available nitrogen)
– Fertilizers with short-term effects on nutrient supply, especially of nitrogen, have high contents of mineral nitrogen and/or readily available organic N compounds

Examples for the first category are composts (i.e. biogenic composts, garden composts) and fertilizers based on peat. The second category includes sewage sludge (high NH_4 concentration), blood meal, and potato starch wastewater (readily degradable organic N compounds).

Organic fertilizers with or without the addition of mineral fertilizer salts are classified as follows:

1) Fertilizers based on peat (or materials of similar stability): peat, composted bark, lignite
2) Fertilizers based on waste materials of animal origin: horns, bone meal, blood meal, hide meal, feather meal, guano
3) Fertilizers based on waste materials of plant origin (selected examples of waste materials from the food- and feedstuff industries): castor cake, cacao waste, brewer's grains, rape (marc), vinasse, spent mash, potato starch wastewater, filtration diatomaceus earth
4) Fertilizers based on municipal waste: sewage sludge, biogenic and garden composts

4.6.1. Fertilizers Based on Peat or Materials of Similar Stability

Peat is an organic material of stable structure and low nutrient content that has no effect on nutrient supply without the addition of mineral fertilizers, but improves soil condition (e.g., aeration, water-retention capacity). The raw material for peat fertilizers is mainly highly decomposed upland moor peat (black peat) or acidic low moor peat (pH < 6). These peats (pH 3 – 5, volume weight 50 – 200 g/L, salt content 50 – 1500 mg/L, ash 2 – 15 %, organic matter 85 – 98 %) have high sorption capacity (cation sorption capacity 300 – 600 mval/L).

Depending on the kind of application, the crop, and the nutrient status of the soil, several mixtures of peat with mineral fertilizers (N, NP, NPK, and PK salts) are available. The following minimum requirements for nutrient content are specified: 1 % N, 0.5 % P, 0.8 % K, 30 % organic substance [274].

Similar mixtures of organic and mineral fertilizers are based on composted bark or lignite (partial substitute for peat) and other organic materials such as dried slurry or spent mushroom substrate (SMS). The minimum fraction of organic substance is set at 15 %. The organic substance of composted bark is highly stable against biological degradation and resembles the organic substance of peat. After complete decomposition, composted bark shows no N immobilization and is therefore suitable for replacing peat in mixtures with other fertilizers (pH 5 – 7, volume weight 150 – 300 g/L, salt content 100 – 1500 mg/L, ash 12 – 45 %, organic substance 55 – 88 %). Threshold values for heavy metal concentrations are specified for these fertilizers [274, 278].

4.6.2. Fertilizers Based on Waste Materials of Animal Origin

Only waste materials that pose no health risk may be converted into fertilizers. The raw materials arise as byproducts at butchers' shops, slaughterhouses, and carcass-disposal plants. Horn, bone meal, and blood meal are the most widely used in agriculture. In some cases, the primary products, containing mainly N and P, are upgraded by the addition of K salts [279 – 282].

Horns. Horn materials (e.g., horns, hoofs, claws) consist mainly of the filament protein keratin. Since keratin decomposes slowly, horns represent a slow N source (22 % of the protein is in the form of cystein). The rate of N release increases with the extent of grinding (chips < grit < meal). Nutrient concentrations are listed in Table 24.

Bone Meal. Bones are ground, partly degreased, and cleaned. The main constituent of bone protein is the filament protein collagen (90 % of bone N is bound in collagen). By treatment with hot water and steam, collagen is converted into glutine and is removed. Therefore, steamed bone meal contains only 0.8 % N (untreated bones 4 – 6 % N). Due to the high P content (7 – 12 % P), bone meals are mainly applied as P fertilizers (Table 24). Occasionally, tricalcium phosphate is converted into monocalcium phosphate by acid treatment.

Blood Meal. Besides blood, blood meals often contain other slaughterhouse wastes such as intestine contents. These mixtures are dried and ground (N content 9 – 13 %; Table 24). Sometimes, mixtures of fresh blood (3 % N, 0.1 – 0.2 % P, 0.1 % K, 80 % water content) with solid wastes are used. Like meat meal, blood meal has a short-term effect on N supply.

Table 24. Nutrient concentrations in fertilizers based on animal wastes (% in dry matter)

Fertilizer	N	P	K	Ca	Mg	C/N
Horn	10 – 14	0.4 – 4.0	0.2 – 0.8	1.5 – 7.5	0.5 – 1.0	3 – 4
Blood meal	9 – 13	0.2 – 1.0	0.2 – 1.5	1.5 – 3.0	0.4	2.4
Feather meal	12	0.3	0.2			4
Bone meal	4 – 6 (0.8)	7 – 12	0.2	18 – 25	0.6	4.5
Carcass meal	6 – 11	2.4 – 7.0	0.3 – 0.5	4 – 10	0.2	3.5
Guano	8 – 16	2 – 7	1 – 3	18	3 – 5	3 – 4

Hide Meal. The main constituents of hide meal, a waste material from leather production, are skins and hair. The effect on N supply is quite slow (8 – 11 % N dw, C/N 5). In Germany, the threshold value for Cr^{3+} is set at 0.3 % in dry matter (for agricultural utilization, Cr^{6+} is prohibited in hide meal).

Feather Meals. These fertilizers show slow N release, comparable to that of horns. The N content of 13 – 14 % (Table 24) is mainly bound as keratin.

Meat and Carcass Meal. Occasionally, meat meal and carcass meal are converted into organic fertilizers. These slaughterhouse wastes have a high proportion of protein and hence a short-term effect on N supply (meat meal: 11 % N, 2.4 % P; meat-and-bone meal: 10 % N, 2.5 – 7 % P; carcass meal 6 – 10 % N; Table 24).

Guano consists of partially mineralized excrements of seabirds and can also contain feathers and carcasses. It is obtained from deposits in arid coastal regions of South America. Guano is converted to guano fertilizers by acid treatment. Guano fertilizers consist mainly of inorganic substances and contain 8 – 16 % N, 2 – 7 % P (ammonium and calcium phosphates) and 1 – 3 % K (Table 24). Since up to 50 % is in the form of ammonium, guano has a short-term effect on N supply.

4.6.3. Fertilizers Based on Wastes of Plant Origin

Provided the waste materials do not pose any health risk, conversion into fertilizers can be performed without special pretreatment. Both aerobic (composting) and anaerobic methods (fermentation, generation of biogas) are suitable for the treatment of plant wastes. The pretreatment and conditioning of some wastes are regulated. All substances described have low pollutant contents. Therefore, the applied quantities are only limited by nutrient contents [274, 281, 282]. The nutrient concentrations in these fertilizers are summarized in Table 25.

Castor cake is the residue of castor beans after oil has been pressed. Owing to its high protein concentration, castor cake contains 5 – 6 % N, which is ammonified in soil quite rapidly and thus becomes plant-available. The fertilizers are only traded in sealed packages due to their allergenic properties. Furthermore, the toxic substances ricin and ricinin must be destroyed by steam treatment at 120 °C for several hours, and dust has to be absorbed by treating the castor cake with linseed oil.

Cacao waste is a residue of cocoa production and has a lower N and P content and higher K content than castor cake.

Brewers' grains (water content 75 %) and *rape* (water content 40 – 60 %) result as waste from brewing and from production of fruit juices and wine. They are used as fertilizers in fresh and composted form and have only a minor short-term effect on N supply.

Vinasse. Sugar-containing molasses is a waste material of sugar production from sugar beet. It is used for alcohol, yeast, and glutamate-production. The remaining sugar-free molasses is concentrated to 60 – 70 % and is then used as sugar-beet vinasse, an organic N and especially K fertilizer. Vinasse has a medium-term effect on N supply.

Pulp remains after alcohol production (distillery) from corn or potatoes (5 – 7 % dry matter). Potato pulp contains more K than corn pulp.

Table 25. Nutrient concentrations (% dw) in fertilizers based on wastes of plant origin

Fertilizer	N	P	K	Ca	Mg	org. S	C/N
Castor cake	5.5	0.8	0.8 – 1.6	0.4	0.3	80	8
Cacao waste	2.3 – 3	0.4 – 0.5	2.5 – 3.0	0.6	0.5	90	17 – 24
Brewers grains	4	0.4 – 0.7	0.5 – 1.0	0.3	0.2	65 – 75	10 – 12
Fruit pulp	1.0	0.3	1.3	0.8	0.1	85 – 95	40 – 50
Rape	1.5 – 2.5	0.4 – 0.8	3 – 3.5	1.0 – 1.8	0.2	75	20 – 30
Vinasse	3 – 4	0.15	6 – 7	0.6 – 1.2	0.3	50	8
Pulp	5 – 6	0.7 – 1.0	0.8 – 6.0			75 – 85	8 – 10
Potato starch wastewater	4 – 8 *	0.7 – 0.9 *	10 * – 12				5 * – 7
Filtration diatomaceus earth	0.7	< 0.1	< 0.1	0.2	< 0.1	6	4 – 8

* In fresh potato starch wastewater.

Pulp has a medium-term effect on N supply, because potato and corn protein must first be decomposed by microorganisms. Ten cubic meters of pulp is equivalent to 30 – 40 kg N, 4 – 6 kg P, 5 – 40 kg K, and 200 – 300 kg C. Pulp is applied for fertilization as fresh material. Storage without addition of preservatives leads to unpleasant odors [283].

Potato Starch Wastewater. The production of potato starch generates large quantities of potato starch wastewater, which is used as a fertilizer. Depending on the production technique, 3 – 50 % of the total N is present in the ammonium form. Furthermore, the protein N is mineralized rapidly. Therefore, potato starch wastewater has a short-term effect on N supply. It also has a high K concentration. Potato starch wastewater is classified as follows:

– Fresh potato starch wastewater (ca. 6 % dry matter; 1 L is equivalent to 60 g dry matter, 5 g N, 025 g P, 6 g K)
– Treated potato starch wastewater (e.g., protein precipitation, evaporation, partly mixed with wastewater), 1 – 5 % dry matter (in the case of storable syrup, up to 55 % dry matter [284, 285])

Filtration diatomaceous earth is the filtration residue in breweries (75 % water content). Yeast proteins are the main N-containing substances. Owing to the narrow C/N ratio, this waste material has a short-term effect on N supply.

4.6.4. Fertilizers Based on Municipal Waste

This group is represented by biogenic and garden composts as well as by sewage sludge. In Germany the produced nutrient quantity of secondary raw material fertilizers is estimated at 270 000 t N per annum and 90 000 t P per annum, of which sewage sludge accounted for 60 % of N and 75 % of P. Composts accounted for 20 – 25 % of N and P. This demonstrates the importance of nutrient recycling by utilizing these waste materials as fertilizers. The precondition for recycling these wastes in agriculture and horticulture is a high quality standard, which strongly depends on the content of pollutants, especially heavy metals, dioxines, and furans. In Germany legal provisions for the utilization of sewage sludge and biocomposts have already been set or are in preparation (Waste Law, Fertilizer Law). These rules are likely to reduce environmental risks to a tolerable minimum and to promote the acceptance of the secondary raw material fertilizers [275, 281, 282]. Lately, sewage sludge and biocomposts have been classified as organic NPK fertilizers in the group of secondary raw material fertilizers [274].

Sewage sludge is mainly produced in municipal purification plants. Sewage sludge is classified into primary (residue after mechanical purification, deposits in settling tanks) and into secondary sludge (residue after biological purification by biomass and chemical purification by precipitation). Generally, the two kinds of sewage sludge are mixed and are subjected to various treatments. A precondition for the utilization of sewage sludge as fertilizer is an acceptable health standard. In Germany pollutant concentrations of sewage sludge have been considerably reduced over the last 20 years (e.g., Cd, Figure 24). At present, the average Cd concentration is 80 % lower than the threshold value specified in the decrees (AbfKlärV) for the utilization of sewage sludge (10 mg Cd/ kg dry matter) [276]. In 1993 dioxin and furan concentrations averaged 27 ng/kg, well below the threshold value of 100 ng/kg dry matter. Adjustment of the threshold values to the present quality standard is in preparation. In 1994 the average concentrations of heavy metals in sewage sludge applied in agriculture (in mg/kg dry matter) were Pb 84, Cd 1.8, Cr 56, Cu 251, Ni 32, Hg 1.6, Zn 977, all well below the permitted maximum loads set in the decrees for the utilization of sewage sludge (AbfKlärV) [276, 286].

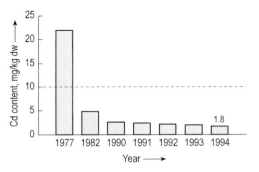

Figure 24. Cadmium concentration in sewage sludge from 1977 to 1994 [286] (the dashed line marks the threshold value for agricultural use [276])

Depending on the pretreatment (dewatering; precipitation of P with Ca, Fe, Al salts; lime addition) nutrient concentrations in sewage sludge vary considerably (Table 26). Therefore, the nutrient contents of these fertilizers must be routinely monitored and adjusted if necessary. The short-term effect on N supply depends on the NH_4 content (fraction of total N). The effect on P supply depends mainly on prior precipitation and pretreatment (lime addition, drying). The applied amount of these fertilizers is determined by nutrient concentrations. For example, the input of P should correspond to the P removed by harvest. Thus the applied quantity remains under the maximum amount of 5 t dry weight of sewage sludge per hectare in three years.

Biogenic and Garden Composts. Biogenic compost originates from biologically degradable waste, separated and collected in households. The portion of biologically usable waste in total domestic waste is 30 – 40 wt %. Garden compost consists of waste materials of plant origin such as litter or remains from pruning, collected in gardens, parks, and embankments and in National Parks. Composts represent the final product of a controlled rotting process in piles or bioreactors. According to the extent of rotting, composts are characterized as fresh or mature composts. Biogenic wastes are subjected to biogas production and subsequent composting.

As a result of the separate collection of the organic wastes, biogenic composts have lower pollutant concentrations than composts based on total domestic wastes (Table 27). Heavy metal concentrations of biogenic and especially garden composts remain under the threshold values set in the decrees for the utilization of biogenic composts (BioAbfV, in preparation). Average concentrations of organic pollutants (e.g., dioxins, furans) of 10 – 15 ng TE/kg dry weight lie within the range of the unavoidable natural background load from the atmosphere. Nutrient concentrations of composts vary less than those of sewage sludge but still have to be routinely monitored (Table 28). Garden composts have lower nutrient contents than biogenic composts. Compared to sewage sludge (high N and P levels) composts have high K contents and mainly a soil-meliorating effect. The short-term effect on N supply is quite low (only 10 % of total N is represented by mineral N). An effect on N supply is barely detectable after applications for several years. The maximum quantity of compost application is likely to be restricted to 20 t (30 t in case of high quality , i.e., minor contents of heavy metals) dry weight over three years per hectare [i.e., 100 (150) kg ha^{-1} a^{-1} of N) [275, 287, 289].

4.7. Manure

There are four types of animal manures: solid manure, liquid manure, slurry, and dry manure from poultry. Solid manure consists of feces, urine, and bedding. Usually, *solid manure* contains 100 % of the feces; the fraction of the urine retained depends on amount and kind of bedding material, animal type, and the way of housing. The most widely used bedding material is straw, but it can also be peat litter or sawdust. *Liquid manure* contains the urine of the animal to-

Table 26. Dry matter and nutrient contents of sewage sludge (in wt %) [275]

	Dry matter	N	P	K	Mg	CaO
Range	6 – 23	2 – 6	1.5 – 7.0	0.2 – 0.5	0.4 – 0.7	0 – 15
Maximum	75	25	15			40
Average	12	3.8	1.6	0.3	0.6	

Table 27. Heavy metal concentrations in composts originated from domestic, biogenic and garden wastes (mg/kg dry matter) [287, 288]

Element	Domestic waste	Biogenic waste	Garden waste	Maximum concentration (BioabfV *)
Pb	286	55	59	100 (150) **
Cd	3.9	0.4	0.4	1 (1.5)
Cr	60	27	28	70 (100)
Cu	261	50	36	70 (100)
Ni	40	15	14	35 (50)
Hg	2.6	0.3	0.2	0.7 (1)
Zn	1020	201	133	300 (400)

* BioabfV: Bioabfallverordnung (German Biowaste Regulations), August 1998.
** Maximum quantity of compost application: 30 (20) t dry weight over three years per hectare.

Table 28. Nutrient concentrations in composts (% dry matter) [275]

Compost from	N	P	K	Mg	CaO	Salts (g/L)
Biogenic wastes	1 – 1.8	0.2 – 0.5	0.5 – 1.3	0.3 – 1.6	1.7 – 9.5	3 – 9
Garden wastes	1.0	0.2	0.6	0.5	8.5	(1 – 6)

gether with some rain water and cleaning water and small amounts of feces and bedding material. *Slurry* is a mixture of feces and urine with some additional water and some bedding and feed material. *Layer dry manure* consists of the excrement layers from battery farming.

Solid manure production is the most frequent animal manure handling system for raising young layers, broilers, turkeys, and cattle, the latter especially on smaller farms. For fattening pigs and cattle on larger farms, slurry systems have been favored in the last 30 years because production costs, especially labor costs, are lower.

4.7.1. Composition [290 – 296]

A major fraction of the nutrients consumed by animals is excreted with the feces and urine. For nitrogen, this is ca. 70 – 80 % of the intake, for phosphorus ca. 80 %, and for potassium 90 – 95 %. Table 29 lists the nutrient amounts produced per year by various animals.

The nutrient contents of manures are affected by type of feed, kind and amount of bedding material, amount of water supply, and kind and length of storage. Even for the same type of manure, the nutrient content can therefore show a wide range; deviations of more than ± 50 % from the average contents are possible. Table 30 lists the average content of organic matter, nitrogen, phosphorus, and potassium in various types of manures.

4.7.2. Manure Nutrient Efficiency [291, 294, 297 – 303]

Availability of Nutrients. In animal manures, phosphorus, potassium, magnesium, calcium, and the micronutrients are predominantly present as inorganic compounds. Their plant availability is comparable to that of mineral fertilizers. The fraction of inorganic nitrogen depends on the type of manure (Table 31). The organic nitrogen becomes partly available to the crop after mineralization in the soil. In the long run and under optimal conditions, liquid manure reaches a nitrogen fertilizer equivalent of up to 100 %, slurry 70 – 90 %, poultry excrement 60 – 70 %, and solid manure 50 – 60 %.

As for all fertilizers, especially with regard to nitrogen, the nutrient efficiency of manures is highly dependent on the application conditions. Especially for slurry and liquid manure, spring application usually results in a much better nitrogen efficiency than autumn or winter application, especially in areas where nitrate leaching

Table 29. Nutrient amounts excreted per year from various animals

Animal	Nutrient amount, kg		
	N *	P_2O_5	K_2O
Milking cow (6000 kg/a milk yield)	110	38	140
Fattening cattle (125 – 600 kg live weight)	42	18	44
Sow with piglets	36	19	16
Fattening pig	13	6	6
Laying hens (100)	74	41	33
Fryers (100)	29	16	16
Turkeys (100)	164	81	71

* Excluding gaseous losses from storage or application.

Table 30. Contents of organic matter and nutrients in various manures

Manure type and amount	Organic matter, t	Nutrient amount, kg		
		N	P_2O_5	K_2O
Solid manure, cattle, 10 t	2	54	32	70
Solid manure, pig, 10 t	2	80	80	60
Solid manure, fryer, 1 t	0.2	28	21	23
Solid manure, turkey, 1 t	0.2	23	17	16
Dry poultry manure, 1 t (50 wt % solids)	0.4	28	21	15
Slurry, cattle, 10 m^3 (10 wt % solids)	0.7	47	19	62
Slurry, pig, 10 m^3 (6 wt % solids)	0.5	56	31	30
Slurry, poultry, 10 m^3 (14 wt % solids)	0.9	98	83	48

Table 31. Percentage of ammonium N and organic N in various manures

Type of animal	Type of manure	Percentage nitrogen as	
		Ammonium	Organic
Various animals	solid manure	15	85
Poultry	dried feces	30	70
Poultry	fresh feces	45	55
Cattle	slurry	55	45
Poultry	slurry	60	40
Pigs	slurry	70	30

occurs over winter. Ammonia losses can be kept low if slurry is incorporated into the soil as soon as possible after application.

Humus Effect. As shown in Table 30, animal manures contain an considerable amount of organic matter. This organic matter may increase the organic matter content of the soil, depending on the application rate and the cropping system. Application of solid organic manures usually leads to a larger increase in organic matter content than slurry due to the bedding material, which is low in N and hence less readily decomposable. The application of organic matter improves soil quality in terms of water-retention capacity, turnover nutrient availability, and nutrient pore volume, and resistance to soil erosion.

4.7.3. Environmental Aspects [294, 295, 298, 304 – 307]

Environmental problems may result from inappropriate application of animal manures, especially at times when the crop does not require nutrients, at rates which exceed the nutrient demand, or if animal manures are not incorporated into the soil immediately after application. The consequences can be:

– Nitrate losses due to leaching after nitrification of ammonium N or mineralization of organic N
– Ammonia losses from the soil or crop surface
– Potassium losses due to leaching on light soils
– Phosphorous accumulation in the top soil and in the subsoil; phosphorus leaching from peat soils

With the aim of decreasing nitrate leaching into ground water, some countries have established legal regulations that restrict the amount of manure and/or timing of application (see Chap. 9). A regulation to reduce ammonia losses from manure application and storage has been implemented in the Netherlands.

5. Fertilizer Granulation

5.1. Introduction

The granulation of fertilizers was one of the most significant advances in fertilizer technology, affording considerable advantages to both manufacturer and user. Today, a well-defined grain size distribution is specified just as nutrient contents and good application properties are. Although the first granular fertilizers came on the market between 1920 and 1930, a stronger trend toward granulation developed — especially in the United States — only after the end of World War II.

In 1976, both granular fertilizers and bulk blends enjoyed shares of somewhat more than 40 % in the U.S. mixed-fertilizer market. Granular fertilizers were losing ground against bulk-blend products and liquid fertilizers [313]. In 1990, the corresponding figures are about 63 % for bulk blends, 22 % for liquid fertilizers and 15 % for granular fertilizers. In Europe, Africa, and Asia, granular fertilizers are the most frequently used form, far ahead of bulk blends and fluids.

Advantages of Granular Fertilizers. Forming and subsequent conditioning are indispensable for the production of fertilizers suitable for use. It was recognized at an early stage that fertilizers in powdered or finely divided form readily cake during storage. This is less of a problem with low-surface-area granules. Only free-flowing materials allow mechanized handling and distribution. Granules often require less storage space because of their greater bulk density: they are stored and transported more economically. A further advantage of granular fertilizers over powdered and crystalline products is that they tend to produce less dust, so that product losses are reduced. A granular product with a definite grain-size spectrum is a prerequisite for uniform mechanical application with field equipment (see Section 8.2): granules with diameters between 1 and 5 mm are most suitable. At the same time, losses caused by the wind, and the accompanying environmental problems, are dramatically reduced. Moreover, granules produced from various feedstocks (solids, slurries, melts) by granulation do not segregate, in contrast to bulk-blended products (Section 5.5).

The use of granular instead of powdered fertilizers delays nutrient delivery to the plant until the granules have disintegrated completely (controlled delivery to the plant, diminished leaching losses). In the case of some controlled-release fertilizers, larger granules release nitrogen more slowly (see Section 4.4). Field studies in Swedish soils have shown that granular superphosphate with a grain diameter of $1-3.5$ mm was twice as effective as finely-divided fertilizers [314], since the granular form retards phosphate fixation in the soil [315]. This reported effect varies with the soil type, the pH, the proportion of water-soluble P_2O_5 and the type of plant [316]. In the case of mineral fertilizers not containing P_2O_5 (N, NK, and NMg fertilizers), however, the grain size has only a slight effect.

Definitions [312, Chap. 1].

– Straight fertilizer: a fertilizer containing only one nutrient.
– Compound fertilizer: a fertilizer containing two or more nutrients.
– Complex fertilizer: a compound fertilizer formed by mixing ingredients that react chemically.
– Granular fertilizer: a fertilizer in the form of particles between two screen sizes usually within the range of $1-4$ mm.
– Prilled fertilizer: a granular fertilizer of near-spherical form made by solidification of free-falling droplets in air or other fluid medium (e.g., oil).
– Coated fertilizer: a granular fertilizer that has been coated with a thin layer of some substance to prevent caking or to control dissolution rate.
– Conditioned fertilizer: a fertilizer treated with an additive to improve physical condition or prevent caking. The conditioning agent may be applied as a coating or incorporated in the granule.

– Bulk-blend fertilizer: two or more granular fertilizers of similar size mixed together to form a compound fertilizer.

Granulation Loop Granulation may be coupled with a production step, such as the manufacture of ammoniated triple superphosphate, or on the other hand it may be only a forming step in a production process, for example, granulation in the nitrophosphate process [317]. But other production operations also come under the heading of granulation: the preparation of feed materials and, after forming, the steps of drying (Fig. 26), cooling, screening, comminution of material with too large a grain diameter (oversize), recycling of this comminuted material and of material with too small a grain diameter (undersize) to the granulator, and finally conditioning of the particles with the desired grain size (product fraction). The processing steps, linked into a loop by the recycle, are called the granulation loop (Fig. 25).

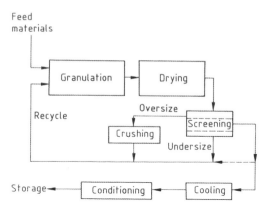

Figure 25. Granulation loop

Recycling is carried out for the following reasons:

1) The size distribution leaving the granulator differs from the required distribution
2) The ratio of liquids to solids in the available feed is in excess of the requirements for the desired size enlargement
3) Granulated material is recycled to provide nuclei for the granulation process

Processes which correspond to (1) may be described as granulation efficiency limited; an example is the agglomeration of low-solubility fertilizers. Condition (2) is generally encountered where readily soluble or high water content feeds are agglomerated and can be described as liquid phase balance limited. Liquid phase balance frequently leads to high recycle rates with consequently high processing costs (Chap. 5.4). For the control of granulation see [318, p. 280]

The *granulation efficiency* often is defined as the mass fraction of particulate material that leaves the granulator as finished product, that is, with grain sizes in the desired range (assuming 100 % sieve efficiency) [317]. It is also possible, however, to state the granulation efficiency as the mass fraction of finished product at the dryer outlet [319]. This definition allows for some regranulation in the dryer. The mass ratio of material not withdrawn (recycled material) to product is often referred to as the recycle ratio. For example, a 20 % granulation efficiency implies a recycle ratio of 4 : 1 if other losses are disregarded.

The *recycle ratio* is important to the process of granulation. Recycle is necessary because, after the product has passed once through the granulator, a certain quantity of particles lies outside the desired region of the grain-size spectrum (off-size material) and must be run through again. For a given mixture and a given temperature, optimal granulation takes place only within a narrow range of the solid-to-liquid ratio. The quantity of recycled fines depends not only on the chemical properties of the materials but also on the water content of the slurry and on the granulation device [320]. Recycle is also needed to generate nuclei for agglomeration and to stabilize the granulation conditions in the granulator.

The quality of the granules is influenced by the following factors:

– Type and fineness of the feedstock
– Moisture content of the granules
– Surface tension of the wetting liquid and wettability of the particles
– Mode of motion in the granulator
– Inclination and speed of the granulator
– Type and properties of the binder

Granulation Processes. Granulation processes can be classified by the nature of the feed materials to be granulated (i.e., granulation

Figure 26. Drying drum, showing granulated product
Courtesy of BASF Aktiengesellschaft

of solids, slurries, melts [308, pp. 250 – 260], [313]) and by the type of granulation equipment used. The most important types of equipment for granulating fertilizers are shown schematically in Figure 27. For the most important fertilizer materials, both straight and multinutrient, Table 32 offers an overview of the main commercial granulation processes, along with further possibilities.

5.2. Granulator Feedstocks

Granulation of Solids with Water or Aqueous Solutions. A solid phase (dry mixed nongranular or powdered material) and a liquid phase or steam (granulation aid) are required. Steam is discharged under the bed of material at the feed end, and water is sprayed on the bed through spray nozzles. For each mixture there is a percentage of liquid phase at which granulator efficiency is optimum. The higher the temperature, the less water and hence less drying is required [308, p. 251]. The system is granulation-efficiency controlled. Granulation takes place by agglomeration of the particles. The granulation efficiency is high and the recycle ratio is low (roughly 1 : 1). Examples are the granulation of superphosphate (with and without simultaneous ammoniation), the granulation of superphosphate in the presence of $(NH_4)_2SO_4$ and K_2SO_4, and the granulation of monoammonium phosphate together with other nutrients [313] to yield high-analysis formulations. While the granulation of solids has proved to be a flexible and economical process, it has the drawbacks of diminished quality as to physical properties and appearance. What is more, the P_2O_5 component (monoammonium phosphate or superphosphate) has to be prepared in a separate plant [313]. For an example of granulation with a solid P_2O_5 component, see [358, 359]; for granulation of NPK fertilizers containing urea, see [360].

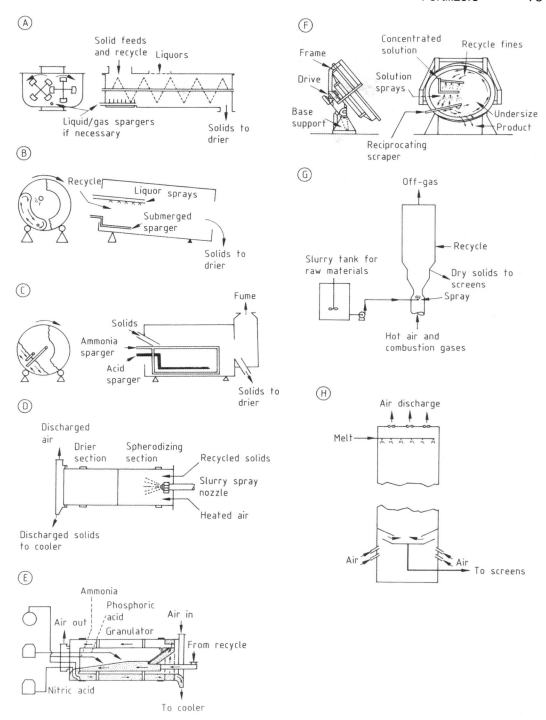

Figure 27. Granulation equipment [321]
A) Pugmill (blunger); B) Rotary drum; C) TVA ammoniator – granulator drum; D) Spherodizer process; E) SAI-R drum granulator; F) Inclined pan granulator; G) Fluidized-bed granulator/drier; H) Air-cooled prilling tower

Table 32. Equipment for granulation of fertilizer materials * [317]

Fertilizer material	Granulation equipment: main commercial techniques are in boldface, while possibilities are in normal typeface
Calcium nitrate	prilling, flaking, pan granulator [322], drum granulator, compaction, pugmill/blunger [320]
Ammonium nitrate	**prilling** [308, p. 103], [323], **cold spherodizer** [323, 324], **pan granulator** [322, 323, 325, 326], **drum granulator**, fluidized-bed granulation [327, 328], spouted-bed granulation, TVA falling-curtain drum granulation
Calcium ammonium nitrate	**prilling** [309, p. 195], **pugmill/blunger** [308, p. 104], [309, p. 196], **drum granulator, pan granulator** [322, 329], **hot spherodizer**, cold spherodizer, fluidized-bed granulation [327, 330], spouted-bed granulation, TVA falling-curtain drum granulation
Ammonium sulfate nitrate	**pugmill/blunger** [309, p. 205], **drum granulator** [331], prilling [309, p. 205], pan granulator [332]
Ammonium sulfate	**crystallization**, hot spherodizer, compaction, pipe reactor-drum [333]
Urea	**prilling** [323, 328, 334] cold spherodizer [323, 324], pan granulator [322, 323, 325, 326, 335], crystallization, drum granulator [336, 337], compaction [338], fluidized-bed granulation [328], spouted-bed granulation [339], TVA falling-curtain drum granulation
Urea with ammonium sulfate	**prilling** [341, pp. 71 – 73], cold spherodizer, pan granulator [342], fluidized-bed granulation [340], spouted-bed granulation, TVA falling-curtain drum granulation
Superphosphate	drum granulator [343], pan granulator [309, p. 234], pugmill/blunger
Triple superphosphate	**drum granulator** [308, p. 191], [344], **pan granulator** [309, p. 348], **pugmill/blunger** [308, p. 191], [344], compaction
Monoammonium phosphate	**drum/ammoniator-granulator** [311, pp. 6 – 8], [345, 346], **pugmill/blunger** [311, p. 30], [347], prilling [311, p. 8], [347], crystallization, compaction
Diammonium phosphate	**ammoniator-granulator** [311, pp. 6 – 8], [345, 346, 348], crystallization, **pugmill/blunger** [346 – 348], compaction
Ammonium polyphosphate	**ammoniator-granulator**[311, p. 217], **pugmill/blunger** [311, p. 217]
Nitrophosphate (NP)	**hot spherodizer, prilling** [349, 350], **pugmill/blunger** [320, 351], **pan granulator** [332, 351], **drum granulator** [351, 352], fluidized-bed granulation
Potash	compaction [353], crystallization
PK	drum granulator [343], pan granulator, pugmill/blunger, fluidized-bed granulation [354]
Nitrophosphate (NPK)	**hot spherodizer** [324, 355, 356], **prilling** [349, 350], **pugmill/blunger** [320, 346, 355, 356], **pan granulator** [357], **drum granulator** [345, 346, 351, 355, 356], fluidized-bed granulation, compaction [317]
Compounds on ammoniacal base	drum/ammoniator-granulator, pugmill/blunger, hot spherodizer, prilling
Compounds with ammonium nitrate	ammoniator-granulator, pugmill/blunger, hot spherodizer, drum granulator, prilling
Compounds with urea	drum/ammoniator-granulator, pugmill/blunger, hot spherodizer, compaction, prilling
Compounds with micronutrients	ammoniator-granulator, pugmill/blunger, hot spherodizer, drum granulator, compaction

* Compaction is illustrated in Figure 34. The equipment for the other processes is illustrated in Figure 27.

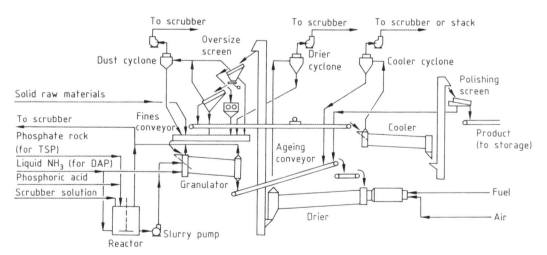

Figure 28. Slurry granulation process [317, p. 23]

Slurry Granulation. The materials to be granulated are in the form of a slurry, usually derived from reaction of sulfuric, nitric, or phosphoric acid with ammonia, phosphate rock, or a combination thereof. In some process modifications, solid materials may be added to the slurry before or during granulation (Fig. 28). Slurry granulation is liquid-phase controlled. Usually a thin film of a slurry having the fertilizer composition is sprayed onto small solid particles. The granules are built up in layers (layering process). The process is mainly controlled through the recycle and the slurry water content (the recycle ratio may be 5 : 1 or more [313]). Granulation is aided by various impurities (Al/Fe compounds, organic substances); see [320]. Drying can be combined with granulation into one processing step. A modification of the slurry process is the Spherodizer process developed by C & I Girdler (Section 5.3.2). Slurry granulation is widely practiced in Europe for the production of N, NP, and NPK fertilizers. In the United States, the process has been modified so that acids, phosphoric and/or sulfuric, or partly neutralized acids are completely ammoniated in the granulator (ammoniator – granulator, Section 5.3.2) [361]. For example, $(NH_4)_2SO_4$ can be granulated in a drum by this method [414].

Granulation with solutions or slurries includes fluid-bed spray granulation (mechanism of agglomeration [362, 363]) and spray drying. In the continuous fluid-bed spray granulation process, solutions, suspensions, or melts can be converted to a granular product in a single step [364]. In contrast to spray (flash) drying, this process can be made to yield granules with a particle size of up to 5 mm [365]. The liquid for granulation is sprayed through nozzles located in or above the fluid bed onto the particles, which comprise comminuted oversize or undersize from the cyclone separator. Warm air in the fluid bed promotes the drying of the particles, and the sprayed particles increase in size. If melts are sprayed into the bed, cold fluidizing air carries off the heat of solidification. Fluidization is accomplished by blowing air through a plenum with a grid. Agglomeration with urea, NH_4NO_3, and K_3PO_4 has been reported [362].

Spray or flash drying represents a direct path from the liquid product to granules. The end product ranges from a powder to a fine grit. The feed liquid is atomized hydraulically, through feed nozzles, or pneumatically, with two-fluid nozzles or atomizer disks. The solution is sprayed into a tower-like vessel with a hot air stream and thus solidified into the fine granules. The dry product is removed pneumatically and collected in a cyclone system, or it can be removed with a bucket wheel at the bottom of the tower [366]. A few special fertilizer products are made by spray drying.

Fluidized-bed methods include the NSM process (Fig. 29) [327, 328], [370, pp. 277 – 288]. The granulator is a rectangular vessel with a perforated plate at the bottom to provide a uniform distribution of air. The fluid bed, which is initially made up of fines, has a height of 0.5 – 2 m and an area of several square meters. It is subdivided into chambers to obtain a narrow gradation in the end product. The granulation liquid or melt is sprayed into the fluid bed in each section by air. Heavier particles, which remain in the bottom portion of the fluid bed, can pass into the next section or to the outlet. In this way, the granular product migrates through the fluid bed from the first to the last section. The NSM plant has a capacity of 800 t/d for urea, and the production costs are less than those for prilling. For the properties of slurry granulation processes, see [367, 368]; for studies on the layering process, [369]. For fluid-bed granulation, see [363].

A modification combining granulation and drying in a variant of the fluid-bed process is the "spouting-bed granulator" [313, 371, 372]. The conical vessel stands on end, with a Venturi tube at the bottom, the narrowest section; into this section, either hot air propels a hot saturated solution or cold air propels a melt. The fluid bed is set up in the cone. The fast-moving particles require no perforated distributor plate. Drying or cooling takes place rapidly, and the material builds up in onion fashion. The dust collected in the cyclone is recycled (2 : 1 recycle ratio) [313]. Granulation tests on ammonium sulfate in the spouted bed have been reported [373]. Despite successful pilot tests, no full-scale unit has yet been built [371].

Melt Granulation. Spherical agglomerates produced from the melt (e.g., urea and ammonium nitrate) are called prills. These are usually obtained by spraying a salt melt or a highly concentrated solution into the top of a tower. The melt should have a very low viscosity (< 5 cP)

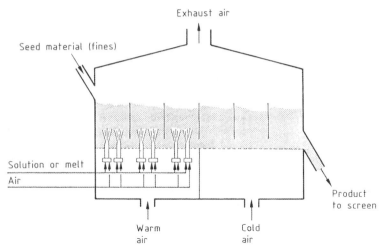

Figure 29. NSM fluidized-bed granulator [327, p. 7]

but a high surface tension at temperatures just a few degrees above the melting point [374]. The liquid jet breaks up into droplets in the free air space. As they fall in contact with counter-flowing cool air, the droplets solidify. The tower height (and thus the falling distance) and the velocity of the cool air are adjusted so that the prills are sufficiently hard when they strike the bottom of the tower [329]; tower heights are typically 45 – 55 m for ammonium nitrate [396]. The prills can be removed with scrapers or belt conveyers, or they can be cooled in a fluid-bed cooler located at the bottom end of the prilling tower [375]. Alternatively, the heat of crystallization can be carried away by spraying the droplets of melt into an oil bath. This is done, for example, with calcium nitrate [376], which is subsequently centrifuged and screened [377].

The recycle ratio in prilling is ca. 0.1 – 0.2 [315]. The prill, with a diameter of 1 – 3 mm, is usually smaller and rounder than the particle obtained by granulation. Because of the high air throughputs in the prilling space and the resulting off-gas problems, and also because of the smaller particle size, prilling has lost importance [315]. For small capacities, such as 250 t/d, granulation is economically superior to prilling; for high capacities, from 1000 t/d up, conditions determine which process is more economical [315, 378]. Schoemaker and Smit present a comparison between granulation and prilling in the manufacture of fertilizers [379].

For the prilling of NPK mixtures consisting of NH_4NO_3, $NH_4H_2PO_4$, and KCl, see [374].

For tests of oil prilling of a urea – ammonium sulfate mixture (34–0–0–9S), see [341, pp. 71 – 73], [342]; for tests of oil prilling of urea – ammonium polyphosphate mixtures, see [382]. Monoammonium phosphate can be obtained in melt form with a pipe reactor and sprayed into a prilling tower (Swift process) [347].

Multinutrient fertilizers mostly have high melting points and are very viscous [313]. One exception is a mixture of monoammonium phosphate and ammonium nitrate, which melts at a low temperature and has a low viscosity. The melt is granulated in a drum (recycle ratio 1 : 1). Depending on whether KCl is added in the granulator, formulations such as 24–24–0 and 17–17–17 are obtained. In a TVA process [380], phosphoric acid and NH_3 are reacted in a T reactor to yield an anhydrous melt; this can be granulated by itself to an ammonium phosphate/ammonium polyphosphate mixture (from 11–55–0 to 12–57–0) [313], or urea can be added to obtain a 35–17–0 or 28–28–0 granular product. If KCl is added, 19–19–19 can be produced [313, 381]. Granulation takes place in a drum or a pug mill (cf. Sections 5.3.1 and 5.3.2).

Melt-granulation processes have the advantage that a dryer, which is otherwise the largest and most expensive unit in granulation plant, can be dispensed with [308, p. 256]

5.3. Granulation Equipment

The condition for granulation is that a bed of solid particles moves, with simultaneous intensive mixing, in the presence of a liquid phase. This motion provides the particle collisions and bonding needed for granule growth. There are many types and models of granulating equipment, most of which use one of three basic intensive mixing mechanisms [321]:

1) Rotation of one or more shafts carrying staggered paddles in a fixed trough (pug mill, blunger).
2) Rotation of the whole device, such as drum or pan.
3) Movement of particles by a third phase, as by blown air in a fluid-bed granulator. In slurry granulation the third phase is usually hot air or hot combustion off-gases, which can serve as a drying medium at the same time. In this way, two processing steps in the granulation loop can be carried out in a single apparatus [365].

In order to improve pelleting conditions or pellet qualities, binding agents can be added along with the granulation liquid. The binding agents may be solid or liquid, may form films, crusts, or crystals, and may harden at standard temperature or at higher temperature [383].

Because granules can also be obtained by dry compaction, the compactor should be considered as a granulator here.

Various authors have reported data on granulators [309, 317, 321, 352, 384 – 388]. Ries has attempted to classify granulating equipment and processes [366, 389 – 391].

The granulating devices used most often in the fertilizer industry are drums, pans, and pug mills. While fluid-bed granulation has come into use in the fertilizer field, mixer – granulators and compactors are more frequently employed to form fertilizer granules. Spray drying and extrusion processes are used only for special fertilizer products.

5.3.1. Pug Mill

A pug mill (Fig. 27 A) consists of a U-shaped trough and, inside it, one or two shafts bearing strong paddles staggered in a screw-thread fashion. In frequency of use, two-shaft pug mills are dominant [317]. The shafts rotate at equal speeds in opposite directions in the horizontal or slightly inclined trough. The solid particles (fresh feed plus recycle) are fed in at one end of the trough and are thrown up in the middle of the trough, where they are wetted with the granulation liquid. In the trough, the paddles move, knead, and transport the moistened particles toward the discharge end. The particles can be given a better-rounded external shape either in a downstream tumbling drum or in the feed zone of the drying drum. Placement of a perforated NH_3 inlet pipe (sparger) at the bottom of the trough makes it possible to ammoniate and agglomerate the fertilizer at the same time. The pug mill is sturdy and can adapt to variable service conditions; it produces hard granules of uniform composition [321]. If the angle between the paddles and the shafts is optimized, the energy consumption can be reduced. The paddles are usually provided with a wear-resistant coating to prevent abrasion [320]. Processes have been described for granulating in a pug mill an ammonium polyphosphate melt (12–57–0), and the same melt with urea (28–28–0) [380], and the same with KCl (19–19–19) [341]. For tests on 35–17–0, see [392]. The combination of a pipe reactor with a pug mill for the granulation of NPK has been reported [393].

5.3.2. Drum Granulator

The drum granulator (Fig. 27 B and Fig. 30), which is the type of granulator in widest use for fertilizers, is an inclined rotating cylinder. The rotation speed is usually adjustable. For a given drum and a given granular product, there is an optimal peripheral speed that gives the highest yield of granules.

An inclination of up to 10° from the horizontal ensures adequate movement of product toward the discharge end. Because, however, this inclination is not enough to effect classification, the discharged product has a fairly broad grain-size distribution, in contrast to the pan-granulator product (Fig. 31A).

Drums in which the lengthwise axis is inclined upward from the feed end to the discharge [395] give a narrower particle-size spectrum. Such an upward inclination also increases the drum fillage.

Figure 30. Drum granulator
Courtesy of BASF Aktiengesellschaft

In drums currently used in the fertilizer industry, the length-to-diameter ratio is ≥ 1 and may reach 6 : 1. The feed end may have empirically designed distributing elements on the inside wall to spread the feed material. In the adjacent part of the drum, where the granulation liquid is fed in, a good tumbling motion should be ensured. This can be achieved with light lifting flights, but they must not lift the granulate too high. In the remainder of the drum, the pregranulated material should be tumbled to a round shape and further compacted. This is also achieved with light flights on the wall and an appropriate fillage. The fillage in the spray and tumbling areas can be controlled by means of internal ring dams. The cylinder may be either open ended or fitted with ring dams at the ends [396] to prevent overflowing at the feed end and to control the bed depth and thus residence time. Fixed or movable scrapers inside the drum or hammers or other rapping devices outside on the drum can be used to remove or reduce excessive product caking inside the drum. Some material buildup on the drum wall may promote granulation [397]. Cylindrical drums are used for continuous granulation with and without internals.

As in the case of the pug mill, recycled product (undersize) generates a moving bed of material in the drum; a slurry containing, say, 3 – 8 % water can be sprayed onto the bed [309].

Powdered feed materials (mixed and wetted in an upstream mixer if necessary to provide granule nuclei) can be granulated in the drum through spraying with water, solutions, suspensions, and highly concentrated slurries, or through blowing with steam. The bed volume should be 20 – 30 % of the cylinder capacity [398]. The recycle ratios for drum granulation are generally between 1 : 3 and 1 : 6. Optimization of these plant parameters for each product class is done by trial and error.

The drum granules are better rounded but less dense than the pug-mill granules [321]. Drums 4.5 m in diameter and 16 m long are in use in the fertilizer industry.

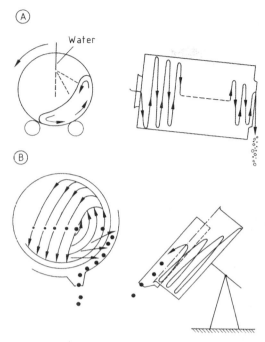

Figure 31. Schematic representation of granule development in the drum granulator (A) and pan granulator (B) [394]

An important modification of drum granulation is the TVA ammoniator – granulator (Fig. 27C and Fig. 32) [308, pp. 250 –260], [399 – 401]. This is a drum roughly equal in length and diameter, with ring dams at the ends but no lifting flights. Ammonia reacts with phosphoric and sulfuric acids below the surface of the tumbling bed of fresh feed and recycle. The reaction generates heat, which vaporizes the water at the same time that granulation takes place. The heat is removed by injected air. The ammonia and the acids are supplied to the bed through perforated distribution pipes mounted parallel to the drum axis. The requisite bed depth is ensured by the ring dam at the drum discharge. In a modern process, a mixture of phosphoric and sulfuric acids and ammonia is neutralized in a pipe-cross reactor situated upstream of the granulating drum (Fig. 33). The slurry is then fed to the drum along with recycle. While more phosphoric acid is sprayed onto the tumbling bed, ammonia is fed into the bed [313]. In this way, NPK fertilizers can also be produced [335, pp. 44 – 48]. In the SA CROS process for monoammonium phosphate production, phosphoric acid and ammonia are mixed and reacted in a pipe reactor. The slurry is distributed over the tumbling bed together with the steam generated; no subsequent ammoniation takes place in the bed [402]. For granule improvement with an interstage pan, see [403]. The use of the pipe reactor in combination with the granulating drum for the manufacture of granular ammonium phosphates was introduced by TVA in 1973 [311, p. 45] and was later incorporated in many plants [404]. A possible improvement in the drum granulator is represented by the double-pipe granulator, which is especially well-suited to fertilizer mixtures with a high proportion of recycle (Scottish Agricultural Industries system, Fig. 27 E) [309, 405]. For example, by virtue of the high recycle ratio with corresponding residence times, a hard ammonium nitrate – ammonium phosphate mixture can be granulated.

Another modification of the drum granulation process described is the spray-drum process (Spherodizer, Fig. 27D). In a rotating drum, preneutralized slurry is sprayed onto a dense curtain of granules cascading from baffles inside the drum. The water content of the slurry must be, say, 12 – 18 % to allow good spray dispersion [309]. During granulation, hot combustion gases flow through the drum in cocurrent [406], so that drying takes place at the same time. The dried particles are then sprayed again, redried, and so forth. The grains grow in shell fashion with an onion structure and are very hard. Spherodizer units are built in capacities of up to 650 t/d. Such an apparatus has a diameter of 4.5 m and a length of 12 m [309]. The Spherodizer, developed by C & I Girdler [407, 408], was first used on an industrial scale in 1959. The *cold* and *hot* used for the versions of the Spherodizer describe the condition of the air that flows through the drum. The cold version is used with melt feeds, especially ammonium nitrate and urea, while the hot version serves for granulation and spraying with slurries (NPK fertilizers, nitrophosphates, ammonium phosphate – nitrate, urea – ammonium phosphate) [317]. Granulation and drying thus take place in the same device. Under optimal service conditions, the recycle ratio is approximately 1 : 1.

A combination of drum granulation and fluid-bed technology is embodied in the Kaltenbach-Thuring Fluidized Drum Granulation (FDG) process [325, p. 39], [409, 410]. The technique

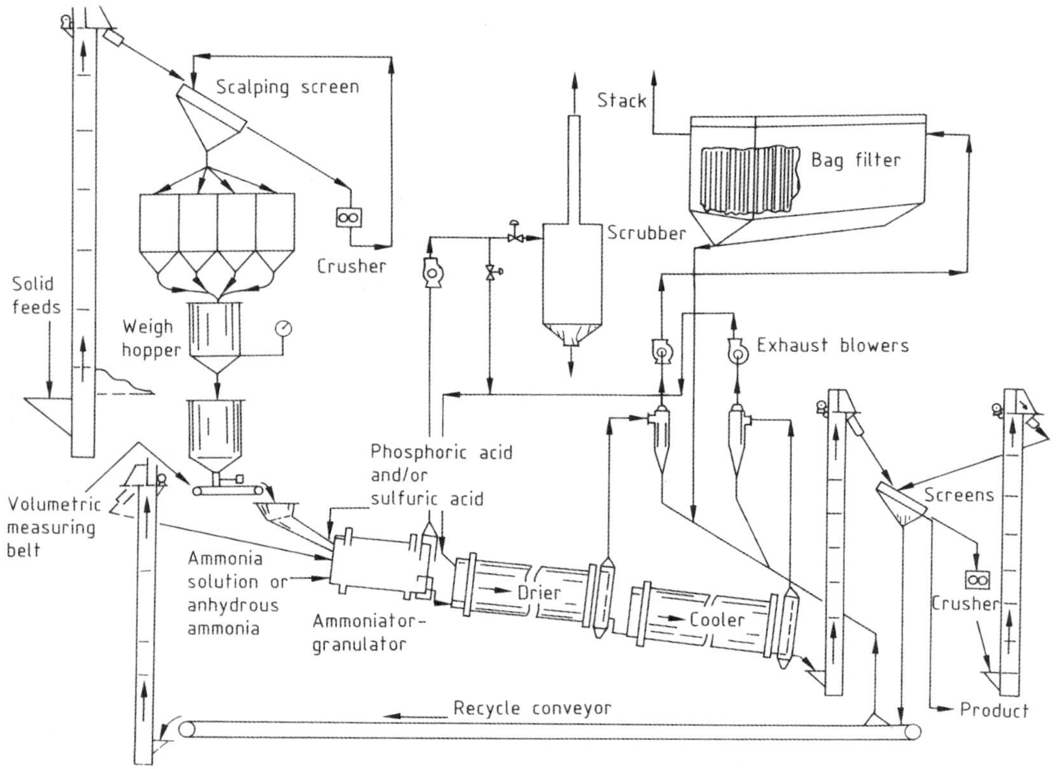

80   Fertilizers

Figure 32. Ammoniator – granulator plant for NPK mixtures [317, p. 23]

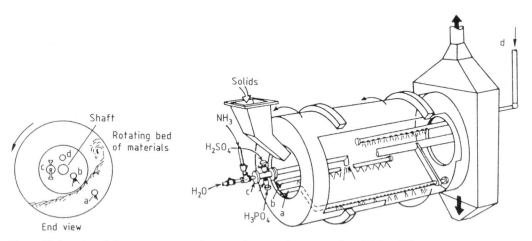

Figure 33. Location of pipe-cross reactor and spargers in ammoniator – granulator [313, p. 87]
a) Ammonia sparger; b) Phosphoric acid sparger; c) Pipe-cross reactor; d) Scrubber liquor
Reprinted by permission of John Wiley & Sons, Inc.

is suitable for both melt and slurry granulation (e.g., size enlargement for urea and ammonium nitrate prills).

In a drum, the best granulation takes place at 25 – 45 % of the critical rotation speed [361], which is the rotation speed at which the weight of the granules and the centrifugal forces are in balance [386], [318, p. 204]:

$$N_{crit} = 42.3\sqrt{\frac{\sin\beta}{D}}$$

D = drum diameter, m
β = drum inclination
N_{crit} = critical rotation speed, rpm

5.3.3. Pan Granulator

The tumbling motion of granules during agglomeration can also take place on a rotating inclined pan (Figs. 27 and 31 B).

For a given pan size, if the inclination of the pan axis to the horizontal is increased, the granules roll upward less steeply but have a longer residence time in the pan. The granulation nuclei and the small granules initially move in the vicinity of the pan bottom. During granulation, the rotation of the pan and the force of gravity cause them to take up a spiral path. The particles grow and eventually reach the bed surface. The spiral diameter decreases continuously until the granules, finally becoming large enough, run over the rim of the pan (Fig. 31 B). Melts or slurries are often sprayed onto the bed, but water or solutions can also be used as granulation aids, and steam can be injected into the bed. If water is employed, it should be applied in the region of the largest spiral diameter [320]. Experience has shown that the spray liquid must be dispersed more finely, the finer the solids for granulation [413]. Because the overflow product has a rather uniform grain size, downstream classification can often be dispensed with. By means of an advancing and retreating scraper blade, the pan bottom can be kept fairly clean and the formation of crusts can be avoided. Here, as in the drum, some material coating the pan prevents wear and promotes the correct tumbling action [396]. The pan can also be made in the shape of a truncated cone or can have at its periphery a tumbling ring, onto which the granules fall from the pan rim; surface coating agents can be applied. Pan granulators are manufactured with diameters of up to 7.5 m [317]. Typically, the height of the rim is one-fifth of the diameter.

Concentrated salt melts of urea, ammonium nitrate, or calcium nitrate can be processed in the pan granulator; the products are easily applied fertilizer pellets, and an alternative to prills [322, 323, 383, 415].

For the production of granular triple superphosphate, phosphoric acid is added to digest finely milled crude phosphate in a granulator – mixer; this step yields a moist, crumbly product, which is directly processed in a subsequent pan to the required pellet size with the injection of steam and the addition of hot phosphoric acid [383]. For the pan granulation of urea – ammonium sulfate mixtures, see [342].

The relationship between the critical rotation speed and the pan diameter and inclination is the same as for the drum granulator [396].

In general, pan rotation speeds are $n \approx 0.6 - 0.75 n_{crit}$, where n_{crit} is the critical rotation speed; the pan axis is usually inclined at 45 – 55 ° to the horizontal [411]. According to the TVA [412], the optimal angle is ca. 65 °. The throughput of a pan granulator can be calculated roughly as follows [386]:

$$\dot{m} = k \cdot 1.5\, D^3$$

$\dot{m}$ = throughput, t/h
k = factor ca. 0.95 – 1.1 for mixed fertilizers
D = pan diameter, m

5.3.4. The Granulator – Mixer

While the pan granulator must be fed with powdered or pre-pelleted material, the granulator – mixer can accept friable, plastic, pasty, or crumbly feeds. If the mixing elements move at the proper speed, the material is comminuted to the desired grain sizes [383]. The disintegration of hard agglomerates is made possible by cutter rotors mounted at the sides of the mixing space [416]. Granulator – mixers are often used in batch operation, while pan granulators are run continuously. Process engineers in the fertilizer industry have also combined the two kinds of apparatus with the mixer upstream to improve product quality. The mixing vessel itself either has a fixed position or may rotate, while the movable mixing elements (e.g., mixing stars, spirals, shafts with attached vanes, plowshare mixing elements) effect intimate mixing and thus granulation by virtue of their rotation. The shape and rotation speed of the mixing elements are usually variable and can be adapted to a range of mixing and granulation jobs. Wear of the mixing elements must be expected. The mixing vessel proper can have a variety of shapes: smooth pipe, zigzag rotary pipe, pan, cone, Y, tub, and so forth. The mixer is often provided with external auxiliary heating. The liquid used for granulation can be fed to the mixing space and distributed by means of a hollow shaft [385], but suitable openings and feed pipes on the vessel can also feed in the liquid. For granulation – mixing of fertilizers, the device has a specific energy consumption of 2 – 6 kW per 100 kg of product [417]. The granulation time can be taken as 5 – 10 min. Grain sizes between 0.1 and 5 mm are achievable. The capacity is up to 30 000 kg/h per mixer [366]. RIES compares the grain-size distribution curves of fertilizer granules from granulator – mixers with those of the starting fineness [418].

5.3.5. Roll Presses

The size enlargement of a finely dispersed charge material by external compression is implemented in press agglomeration (Fig. 34). The charge is gripped by two counterrotating rolls, nipped in the gap, and compressed. As the void volume decreases, the material generally undergoes a two- to threefold compaction. While a charge hopper is adequate for a free-flowing material, a material that is not sufficiently free-flowing can be transported to the nip by screw feeders, with some precompaction. If the rolls are smooth, the material exiting from the nip ("shell") has a smooth surface. If the rolls have mating depressions, briquetts are produced. The shells are next reduced to the desired grain size (in a crusher or mill) and screened. The fines and oversize are recycled (Fig. 35). Rolls are manufactured in diameters up to 1.4 m and widths up to 1.5 m [390]. They may be mounted side-by-side or over-and-under.

For the specific compressive forces for urea, KCl, and $(NH_4)_2SO_4$, see [353]; for data on the compaction of special fertilizers, [420, 421]; of K_2SO_4, [320]; of calcium cyanamide, [422]. For a general description of fertilizer compaction, see [423, 424]; for a monograph on roll pressing, [425]; for the principles of pressure agglomeration, [419].

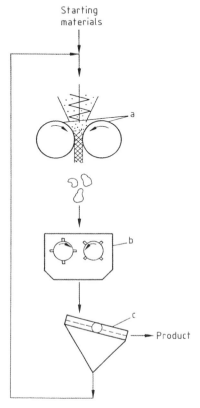

Figure 34. Press agglomeration with smooth rolls [386, p. 215]
a) Rolls; b) Crusher; c) Screen

5.4. Costs of Agglomeration

The costs incurred for granulation depend not just on the agglomeration properties, but — for equal or nearly equal agglomeration properties — on the size and type of equipment used in the process. For equipment and investment costs for pan granulators, drums, mixers, and roll-compaction equipment, and on hourly operating costs versus equipment size, see [426]. With regard to personnel, mixers and roll presses are

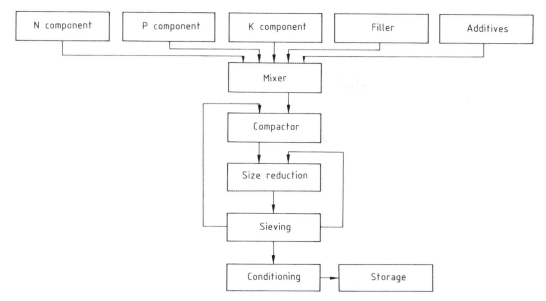

Figure 35. Compacting of a multinutrient fertilizer [419, p. 34]

considered to require one-half man (year-round, rotating shifts), while pans and drums are figured at one man each. In contrast to dry compaction (roll presses), drying costs have to be added in for mixer granulation. Up to a certain moisture content, mixer granulation with drying is quite competitive with dry compaction. For processes and costs of agglomeration, see also [427]. In comparing the granulation of solids and slurries, the investment costs are one-third higher for slurry granulation, and the operating and utility costs are likewise greater [313].

For production costs with various granulating equipment, see [428, 429]. For a cost and process comparison between prilling and granulating, see [378, 430]; for the costs of granulating monoammonium phosphate and diammonium phosphate, [311]; for the costs of NPK granulation in the Norsk Hydro nitrophosphate process, [349]; for the costs of fertilizer compaction, [431]. For economic aspects and comparative estimates of manufacturing costs, see [308, pp. 138, 266].

5.5. Bulk Blending

The mechanical mixing of single components in granular form, called bulk blending, is a special way of producing multinutrient fertilizers. Bulk blending was introduced in the USA at the beginning of the fertilizer industry [447]; it is not nearly so widespread in Europe. In the process, several of the usual starting components, such as superphosphate, triple superphosphate, monoammonium phosphate, diammonium phosphate, urea, and potassium chloride, are combined in an uncomplicated device such as a rotating drum. The nutrient ratio can be adjusted as desired. The components are briefly mixed (up to 15 t/h) and made available to the farmer in batches that are usually loaded directly into the distributor.

The precondition for this process is that the components in the mixture be physically and chemically compatible [317, 448]. For example, urea and ammonium nitrate must not be present together, since a mixture of these is very hygroscopic and tends to deliquesce. Further, stoichiometric mixtures of urea and ammonium nitrate are sensitive to impact, and even a solution of the two can form an explosive mixture [449]. Mixtures of urea or diammonium phosphate with normal or triple superphosphate have only limited compatibility. If an aqueous salt mixture has a somewhat elevated pH and simultaneously contains NH_4^+ ion, NH_3 may be liberated.

Bulk blending has the drawback that segregation can occur during silo filling, packaging, transportation, and application. A uniform grain size or grain-size distribution is essential for reducing segregation, even if the particles differ in density. Design measures at the silo inlets and inside the silos can prevent segregation [335, pp. 37–39]. Drum mixing generates dust, which may necessitate cleanup measures depending on the amount of dust and the size of the mixing equipment.

The process has the advantages that the N : P_2O_5 : K_2O ratio desired by the farmer can easily be obtained; micronutrients, insecticides, and herbicides can easily be metered in, and the dealer requires less storage space.

For the production of bulk-blended urea with an appropriate gradation, see [370, pp. 277–288]. For the bulk-blending system in the United States, see [447, 450–452]; for the use of mechanical fertilizer mixing in Germany, see [453].

5.6. Quality Inspection

For successful handling and application, certain ranges of physical properties must be specified for the fertilizer particles. Quality control, which includes chemical analysis as well, is performed by the fertilizer manufacturer. For the determination of the physical properties of mineral fertilizers, see [308, 432–434].

Grain Shape. Fertilizer particles should have the least possible surface area, since irregular shapes lead to increased abrasion and a tendency to cake.

Grain-Size Spectrum. The diameter of ordinary commercial fertilizer grains is in the range of 0.5–6 mm. U.S. products generally have a somewhat finer size spectrum (primarily 1–3.35 mm) [317] than European products (primarily 2–5 mm). In exceptional cases, the product grains may be coarser, as in the case of a woodland fertilizer applied from the air (6–12 mm), or finer, as in the case of ammonium sulfate and crystalline mixed-salt fertilizers (< 2 mm) and other special fertilizers. The grain-size distribution is important for the intended application, for example to ensure uniform distribution of fertilizer nutrients by field equipment. The grain-size spectrum is determined by screen analysis (ISO 8397 standard screening). For the grading curves of granular fertilizers in comparison to the starting fineness, see [418].

Settled Density. The settled density is important for the sizing of packaging equipment and storage areas. For a given fertilizer grade, it should fluctuate as little as possible. The settled densities of granular fertilizers can be determined in accordance with ISO 3944; those of finely divided fertilizers (with a high content of <0.5 mm), in accordance with ISO 7837.

Compacted Density. The compacted density is generally as much as 10 % higher than the settled density. It represents the maximum bulk density that can be achieved through vigorous shaking. The compacted density can be measured in accordance with ISO 5311.

Dumping Angle. The dumping angle (angle of slope) of a fertilizer is important for the design of storage areas and for transportation. The dumping angle should be as large as possible and can be measured in accordance with ISO 8398.

Grain Hardness. The grain hardness is a measure of the fracture strength of fertilizer grains and their mechanical stability in storage. As a rule it is measured by placing grains of a definite size between two parallel plates and compressing [435].

Abrasion Resistance. The abrasion resistance is a measure of the mechanical stability of fertilizer grains moving against one another and of their stability in free fall (wear due to tumbling and dropping). Abrasion causes dusting of the fertilizers during storage operations, transportation, and application. The abrasion resistance should be as great as possible and can be determined with, for example, the TVA method [308].

Caking Tendency. One of the most important properties of a fertilizer is its storability, which can be determined through measurement of its tendency to cake. By careful drying and effective surface treatment (see Section 5.7), the caking tendency can be significantly reduced and thus the storage qualities improved. This property is measured by, for example, a shear test on caked fertilizer.

Hygroscopicity. The hygroscopicity of a fertilizer characterizes its sensitivity to atmospheric humidity. Grains of a highly hygroscopic fertilizer exposed to sufficiently high air humidity take up moisture, which impairs their initial grain hardness and abrasion resistance. The hygroscopicity of a fertilizer is assessed from the water-vapor adsorption isotherm. The rate of water uptake and the critical relative humidity of the salt system can also be determined [434, 436]. Critical humidities are listed in [437]. Corresponding to the critical relative humidity is the partial pressure of water vapor over a saturated salt solution forming a very thin skin of liquid over the salt surface. If the humidity of the ambient air is less than critical, the liquid skin gives up water; if greater, the product gains moisture.

5.7. Fertilizer Conditioning

A conditioner is a material added to a fertilizer to promote the maintenance of good physical condition (flowability) during storage and handling. The use of conditioners is essential with many products, but is not required with all fertilizers. It is preferable to use other means, such as good drying, to avoid caking. Even if the fertilizer grains are dried adequately from an economic standpoint, caking may occur and impede storage, transportation, and field application.

Hardening and caking result from the crystallization of water-soluble salts and the formation of bridges between the grain surfaces during storage. The surfaces also suffer plastic deformation under pressure, and the reduction of the water vapor pressure in the joint between the new contact surfaces causes the particles to adhere to one another [438]. The reaction

$NH_4NO_3 + KCl \rightarrow NH_4Cl + KNO_3$

during storage may also be indirectly important in poor storage qualities [315, p. 370], [439].

The internal conditioning of fertilizers means incorporating additives in the granules before or during granulation to improve the physical properties and the anticaking qualities. Internal conditioners usually act as hardeners or crystal modifiers, for example to improve the storage properties of ammonium nitrate fertilizers. Internal conditioners inhibit or modify the effects of crystal phase inversions due to temperature changes during storage. The inversion at 32 °C can cause uninhibited ammonium nitrate granules and prills to shatter and cake. In the case of urea prills and granules, 0.2 – 0.5 % of formaldehyde or urea – formaldehyde is added to the urea melt as a hardener and anticaking additive [308, p. 301]. The addition of 1.8 % Mg$(NO_3)_2$ protects ammonium nitrate from caking. The destructure effect of the phase change at about 32 °C is avoided [440, p. 200].

External conditioning, also called coating and surface treatment, means applying to the granule surface a thin layer of powders or surfactants to reduce the caking tendency. The addition of wax and/or oil enhances the action by suppressing dust formation. This process step is carried out in a rotating drum or a fluid bed. Although coating with a fine, inert powder (kieselguhr, talcum, lime, kaolin) has long been practiced as an external form of inorganic conditioning, surface treatment with nonionic organic sealants (polyethylene waxes, paraffins, urea – aldehyde resins) and coating with surfactants to make the grain surface hydrophobic came into use later. The surfactants employed are, above all, fatty amines and sulfonates. These are also used [315] in combination with powders and waxes and/or oils [370, pp. 289 – 303]. For the use of special oils as anticaking agents, see [441]; for an example of a surfactant, [442]. For special coatings to prevent caking, see [443]. Because of the many anticaking agents and fertilizers in production, no overall recommendation can be made as to special additives for general use [438].

Intentional aging of fertilizer in a storage pile prior to bagging or bulk shipment is referred to as curing. In products that benefit from curing, chemical reactions that cause caking bonds apparently proceed to near completion during the curing period. The heat of reaction retained in the curing pile speeds the completion of the reactions. After curing there is reduced tendency for additional bonds to develop [308, p. 302]. In the manufacture of superphosphates, pile curing for about 30 d frequently is employed to improve physical properties.

5.8. Environmental Aspects

In fertilizer plants, the gaseous effluents from all equipment handling solid materials, including screens, have to be cleaned owing to their content of fine dust (and harmful gases). Although the most serious dusting occurs during the drying of granular fertilizers, dust is also formed in the granulators. These units are therefore operated under a partial vacuum. The dust in the off-gas is usually collected in cyclones and recycled. When dry dust collection is inadequate, wet separation in scrubbers, which also absorb gases like NH_3, is employed. Recycling of the scrub waters is implemented in the AZF process [444]. For NPK fertilizers, experience has shown that a soft granular product has a stronger dusting tendency than a hard one. For certain fertilizer formulations, less dust is produced from the drum than from the pug mill [445].

Gaseous effluents like ammonia, nitrogen oxides, and fluorine compounds are evolved in the production of NPK fertilizers and feedstocks. Normally these exhaust gases are scrubbed, and the resulting scrubber liquors are recycled to the process.

Liquid or aqueous effluents from fertilizer plants are usually of smaller volume compared to those vented to the atmosphere [308, p. 322]. They generally result from scrubbing equipment and can be concentrated and recycled. Spills and washings are usually collected in floor sumps and also concentrated and recycled.

For the removal of emissions from fluid-bed granulators, see [370, pp. 277 – 288]; for fluorine emission in triple superphosphate production, [446]; for a summary of environmental practices in the fertilizer industry, [308, pp. 319 – 328].

6. Analysis

The point and purpose of fertilizer analysis is to check the declared fertilizer grade and confirm that the contents of grade-determining components, nutrients, nutrient forms, nutrient solubilities, and incidental components are as printed on the bag. In the Western countries, all fertilizers are subject to official inspection, and the laws permit only small deviations within given limits.

In nearly every country, only products with established compositions and analyses are allowed in trade; therefore, inspections must be performed during production, and factory laboratories must use testing methods in compliance with the established official techniques of analysis (Table 33)

6.1. Sampling and Sample Preparation

Sampling must produce a sample that corresponds to an exact, representative average of a large product batch. The most appropriate time for fertilizer sampling is during bagging or when the bulk product is being loaded into conveyances. Only when sampling at this time is impossible should samples be taken from closed bags or from a bulk product stockpile. As a rule, the sample taken should make up at least 0.1 % by weight of a batch for delivery. Fertilizer regulations also contain provisions for sampling methods. A minimum requirement is given in EN 1482.

A suitable sample divider is next used to reduce the sample quantity to some 1 – 2 kg. The final sample must be prepared for analysis by comminution, screening, and homogenizing. The smallest portion called for by the analytical technique must still be representative of the whole final sample. At the same time, this preparation must not alter the gradation of the fertilizer in such a way as to have a marked effect on typical properties such as the solubility in extractants.

6.2. Determination of Nitrogen

The most important nitrogen compounds are nitrates, ammonium salts, urea, urea – aldehyde condensates, and cyanamide. Farm-produced fertilizers (e.g., liquid manure) have most of their nitrogen in organic form (protein).

Total Nitrogen. The analytical method for determining total nitrogen is dictated by the components present in the fertilizer. For straight N fertilizers that contain only ammonia nitrogen, e.g., ammonium sulfate, it is sufficient to alkalize the sample solution and distill the liberated

Table 33. National and international standards for fertilizer analysis

Method	European Economic Community [454]	Federal Republic of Germany [455]	International Organization for Standardization [456]	United States (AOAC) [457]
Sampling and sample preparation (see Section 6.1)	No. L 213 on pp. 2 – 4 EN 1482	No. 1.1, 1.2, 1.3 DIN/EN 1482	ISO 3963; ISO 5306; ISO 7410; ISO/DIS 7742	No. 2.1.01 – 2.1.05
Nitrogen analysis (see Section 6.2)				
Total nitrogen	No. 2.3; § 7.1 in No. 2.6.1; § 7.2 in No. 2.6.2	No. 3.5 (3.5.1 – 3.5.5)	ISO 5315	No. 2.4.02 – 2.4.06
Ammonia and nitrate N	2.2 (2.2.1 – 2.2.3)	No. 3.3 (3.3.2 – 3.3.4)	ISO 11 742	No. 2.4.09 and 2.4.10
Ammonia N	No. 2.1; § 7.2.5 in No. 2.6.1 and § 7.5 in No. 2.6.2	No. 3.2.1 – 3.2.6	ISO 5314, ISO 7408	No. 2.4.07 and 2.4.08
Nitrate N	§ 7.2.4 in No. 2.6.1 and § 7.4 in No. 2.6.2	No. 3.4.1 and 3.4.2	ISO 4176	No. 2.4.11 and 2.4.12
Urea N	§ 7.2.6 in No. 2.6.1, § 7.6 in No. 2.6.2	No. 3.8 (3.8.1 – 3.8.5)	ISO 8603	No. 2.4.20
Controlled-release N	EN 13 266	No. 3.10	in preparation	
Phosphate analysis (see Section 6.3)				
Extraction	No. 3.1.1 – 3.1.6	No. 4.1.1 – 4.1.7	ISO 5316, ISO/DIS 7497	No. 2.3.01, 2.3.06, 2.3.11, 2.3.14,
Dissolved phosphate	No. 3.2	No. 4.2, 4.3	ISO/DIS 6598	No. 2.3.02 – 2.3.05, 2.3.07 – 2.3.09, 2.3.12 – 2.3.14
Potassium analyses (see Section 6.4)	No. 4.1	No. 5.1, 5.2	ISO 5317, ISO 7407, ISO 5310	No. 2.5.01 – 2.5.08
Other analyses (see Section 6.5)				
Calcium		No. 6.1 and 6.2	ISO/CD 10 151	No. 2.6.01, 2.6.05 – 2.6.07
Magnesium	No. 8.1, 8.3, 8.6, 8.7	No. 7.1 and 7.2	ISO/CD 10 152	No. 2.6.01, 2.6.17 – 2.6.21
Sulfur	No. 8.1, 8.2, 8.4, 8.5, 8.9		ISO 10 084	No. 2.6.28
Micronutrients	No. 9.1 – 9.11	Chap. 8 (8.1.1 – 8.9.2)		No. 2.6.01 – 2.6.04, 2.6.10 – 2.6.16, 2.6.22 – 2.6.25, 2.6.29 – 2.6.32

ammonia into an acid of known concentration in the receiving flask. For samples containing nitrate, on the other hand, the nitrate content must be reduced first. Examples of ways to reduce the nitrate are with iron or chromium powder in an acid solution, with ARND alloy in a neutral solution, or with DEVARDA alloy in an alkaline solution. Nascent hydrogen reacts with the nitrate to give ammonia in quantitative yield. Distillation from the alkalized reaction solution gives the sum of ammonia N and nitrate N.

For fertilizers that contain other forms of N besides ammonia and nitrate, the reduction step must be followed by the conversion of the other N forms (urea, cyanamide, protein) to ammonia by Kjeldahl digestion.

The final determination of total N in every case is by distillation of ammonia from the alkalized pretreated solution into a receiving flask with an acid standard solution and backtitration of the residual acid excess.

Ammonia Nitrogen Apart from Other N Compounds. In theory, the ammonia component is easily isolated from commercial mixed-salt fertilizers by distillation from sample solutions alkalized with NaOH or KOH. Urea and other organic N forms readily give additional ammonia in alkaline solution. If such compounds are present, the distillation conditions must be kept as mild as possible. The solutions are alkalized with soda, calcined magnesium oxide, or freshly precipitated calcium carbonate, and the ammonia liberated is distilled under reduced pressure at room temperature or with the help of a strong air purge.

For pure ammonium salts of strong acids, the formaldehyde method offers a quick procedure for ammonia determination. The reaction of ammonium ion with formaldehyde

$$4\,NH_4^+ + 6\,CH_2O \rightarrow (CH_2)_6N_4 + 6\,H_2O + 4\,H^+$$

liberates protons equivalent to the ammonium ions present; these can be directly titrated with sodium hydroxide standard solution.

Nitrate Nitrogen Apart from Other N Compounds. In many cases, nitrate N is determined as total N minus ammonia and urea N. The most versatile direct method is precipitation of the nitrate with nitron and final determination by gravimetry. Another sensitive method, recommended for specially formulated mixed fertilizers with low nitrate contents, is a colorimetric technique based on the nitration reactions with 2,4- or 3,4-xylenol. The nitration products give intensely colored compounds in alkaline solution.

Urea Apart from Other N Compounds. Three different methods are available for the selective determination of urea:

1) Separation of urea by precipitation with xanthydrol(9-hydroxyxanthene), yielding dixanthylurea; gravimetric final determination
2) Enzymatic hydrolysis of urea, controlled by ureases; distillation of produced ammonia into a receiving flask with acid standard solution, where it is absorbed
3) Colorimetric method (for low urea contents) based on the reaction of urea with p-dimethylaminobenzaldehyde

Nitrogen Fertilizers with Controlled-Release Urea – Aldehyde Condensates. Ureaform, Isodur, and Crotodur are nitrogen compounds that are only slightly soluble in cold water but are nonetheless effective as fertilizers. The delayed water solubility provides a basis for the analytical characterization of these compounds: extraction with cold water isolates the readily available N components (ammonium, nitrate, free urea). Only when the extraction residue is treated with hot water do isobutylidene diurea, crotonylidene diurea, and the short-chain components of the urea – formaldehyde condensates go into solution.

In an acid medium, especially at high temperature, condensates of urea with formaldehyde and with isobutyraldehyde tend to hydrolyze to urea and the aldehyde. In mixed fertilizers with an inherent acid tendency, there is thus a danger of altering the solubility characteristics in the course of the analysis. For this reason, the separations are carried out in buffered solutions of pH 7.5 [455].

To calculate the percent N contents, either the total N is determined in the extraction solutions or else the AOAC procedures for urea – formaldehyde condensates are carried out and a nitrogen activity index is determined.

$$\frac{\%CWIN - \%HWIN}{\%CWIN} \times 100 = AI \text{ in}\%,$$

where CWIN is cold-water-insoluble nitrogen and HWIN is hot-water-insoluble nitrogen.

The nitrogen activity index for urea – formaldehyde condensates is around 50; crotonylidene diurea and isobutylidene diurea have values of more than 90.

6.3. Determination of Phosphate

Because phosphorus fertilizers are rated by their contents of phosphate available to plants, various extraction methods have been devised.

Extraction with Water. Most countries define *water-soluble phosphates* as those components that go into solution at room temperature when an aqueous suspension is made with 1 g of sample to 50 mL of suspension volume (in the USA, 1 g of sample to 250 mL).

Extraction with Citrate Solution. The technique of citrate extraction determines phosphates, such as $CaHPO_4$, that are insoluble in water but soluble in complex-forming salt solutions.

The most common methods in Europe use ammonium citrate solutions with a prescribed citrate concentration. The officially approved extraction processes differ in the choice of pH. Extraction is performed in a strong alkaline medium with an ammoniacal ammonium citrate solution, or in a neutral ammonium citrate solution. Most often, extraction is carried on for 30 min at room temperature and is followed by a treatment at 40 °C; for the EEC process with neutral ammonium citrate solution, 65 °C is prescribed.

Extraction with Citric Acid Solution. Originally, the citric acid solubility was stated only for Thomas phosphate, but the lack of agreement between the plant availability of P_2O_5 and the solubility in ammoniacal citrate solution led to other straight and mixed phosphate fertilizers being rated by citric acid solubility. Treatment with 2 wt% citric acid solution (5 g of sample is agitated in 500 mL of citric acid solution for 30 min at 20 °C) dissolves phosphorus – silicon heteropolyacids, which are hardly soluble in complex-forming salt solutions.

Extraction with Formic Acid Solution. Soft and hard phosphate rock can be distinguished by their behavior in 2 % formic acid. A 5-g ground rock sample is shaken at 20 °C for 30 min with 500 mL of the formic acid solution. In the case of soft phosphate rock, 50 – 70 % of the total phosphate dissolves; in the case of hard phosphate rock, only ca. 25 % dissolves.

Extraction for Total Phosphate Determination. A technique applicable to all fertilizers as well as crude phosphates is digestion with sulfuric acid, nitric acid, and copper sulfate. Mixtures of hydrochloric and nitric acids or nitric and perchloric acids have also been proposed for quantitative extraction.

Determination of Phosphate. After hydrolysis of polyphosphates in acid solution, if necessary, phosphate is precipitated as quinolinium molybdatophosphate and weighed. Possible substitutes for the gravimetric final determination are methods in which the precipitate, washed acid-free, is dissolved with an excess of sodium hydroxide standard solution and back-titrated with nitric acid standard solution. Reaction of the dissolved phosphate with a molybdate – vanadate solution yields a soluble, colored heteropolyacid complex, which can be determined by colorimetry.

6.4. Determination of Potassium

For almost all fertilizers that contain potassium, this component is dissolved in water. In complete fertilizers, it is advisable to perform the extraction in acid solution so that, if gypsum is present, none of the potassium will take part in double salt formation and remain undissolved as syngenite. For the determination, the potassium is precipitated with sodium tetraphenylborate from a weakly alkaline or acidic solution; the product, slightly soluble potassium tetraphenylborate, is weighed. Another proposal, mainly in the USA, is to precipitate the potassium as the chloroplatinate (AOAC). Flame photometry is also recognized as an official test method.

6.5. Analysis of Calcium, Magnesium, and Trace Elements

Complexometric methods with EDTA have come into widespread use for determining calcium and magnesium. Photometric techniques for the determination of other metal ions effective in growth (Cu, Fe, Mn, Zn, Co, Mo) have been described. To an increasing extent, however, atomic absorption spectrophotometry has gained in importance. Emission spectral analysis with plasma excitation is also used in fertilizer analysis. But this procedure is new, and as yet no official analytical specifications exist for it.

Boron must be isolated by distillation in the form of methyl borate before it can be determined. The boric acid is reacted with 1,1-dianthrimide or azomethine H, and the colored product is determined by photometry. For boron contents of 0.1 % and higher, boric acid and polyhydric alcohol (usually mannitol) are reacted to give didiolboric acid, which is determined volumetrically.

7. Synthetic Soil Conditioners

Soil conditioners are substances with which the soil and substrate properties, seldom ideal for the growth of plants, can be optimized and stabilized. Their purpose is to exert biotic, chemical, or physical influences on soils in such way as to improve the soil structure and water regime [462]. Synthetic soil conditioners have well-defined composition, stable quality, and properties suited to the requirements they must meet. They may, for example, supplement or replace natural substances that are unsuitable from an environmental and ecological standpoint, are in scant supply, or must be conserved.

7.1. Foams

Foamed polystyrene and foamed urea – formaldehyde resin are used as soil conditioners; foamed phenolic and polyurethane resins are used as florists' mounting media or plant growth media [458].

7.1.1. Closed-Cell Expandable Polystyrene Foam

Expandable polystyrene (PS) foams consist of polystyrene pre-expanded by blowing agents at temperatures of about 100 °C and, after holding, expanded with steam. The forms used for soil conditioners are flakes, beads, or raspings [458 – 460, 463, 464].

Physical and Chemical Properties. Expandable PS foams have closed cells. They contain 98 % air by volume and can hold 1 – 2 % water on their surface [465, 466]. They do not, however, absorb water. Air exchange does take place [467]. Styromull foam flakes are 4 – 16 mm in diameter with a loose density of 12 – 20 kg/m^3 [459]. The permeability to water between expandable PS flakes without compression is roughly the same as that of fine gravel; under 25 % compression, it matches that of coarse sand; at 75 % compression, that of fine sand.

Expandable PS foams are largely resistant to attack by acids and alkalies; they are odorless, chemically neutral, and unobjectionable to plants. These foams have been considered resistant to degradation by soil bacteria. Physical degradation takes place under the action of ultraviolet radiation, and biodegradation, especially by soil fungi, yields carbon dioxide and water [467]. This process is slow, however, so that the foams are effective soil conditioners with a relatively long lifetime.

Use. Expandable PS foams in comminuted form for soil application have been described [468, 469]. Uses range from incorporation into soils and garden substrates to the conditioning of weak-structured peat [470].

Incorporated, they have soil-loosening, aerating, and draining actions. Foams are also employed as soil conditioners in crop farming and landscaping, and they have proved useful as drainage aids in slit drainage and as filter materials for covering drain pipes [458, 471, 472].

For soil aeration and structural stabilization around the roots of urban trees, expandable PS foam beads (1 – 2 mm diameter) are recommended for use with soil aeration equipment (e.g., Terralift) [473, 474].

Trade Names. Hygropor, Styromull (BASF, Ludwigshafen, Germany); Styrofoam (Dow Chemical, Midland, Mich.).

7.1.2. Primarily Open-Cell Urea – Formaldehyde Resin Foams

Urea – formaldehyde foams consist of condensates of urea with formaldehyde in a well-defined molar ratio. They are solid foams with rigid walls and little flexibility. The materials used as soil conditioners are mainly urea – formaldehyde foams modified for plant compatibility.

Physical Properties. Urea – formaldehyde foams can be produced with densities of 2 – 50 kg/m^3 (as water-free uncomminuted material); foams with a density of 22 kg/m^3 are used for mixing into soils and substrates. Packed moist from processing, urea – formaldehyde flakes (trade name Hygromull) weigh approximately 35 – 40 kg/m^3.

The cell walls are partly open. They are thicker, and the cell volumes smaller, the higher the density. Cell diameters lie between 100 and 300 µm [460, 475, 476]. The capillarity is low (1 – 2.5 %); the capillary elevation is 300 – 400 mm [460]. The capillary elevation for a garden substrate treated with foam is described as advantageous [477].

The grades used as soil conditioners are made water-absorbing by modification of the condensate resin and comminution to flakes measuring 4 – 20 mm. Under vacuum, the water capacity is more than 90 vol % [478]; for Hygromull flakes having a density of 22 kg/m^3, this corresponds to 4100 wt %. The water capacity can be determined appropriately by the method of DIN 11542 (1967). At atmospheric pressure, the initial water uptake is slow but rewetting is very fast [478].

Water release is uniform and goes to completion without losses due to evaporation, so that the water stored in the foam flakes can be economically utilized by plants. The water is held in place by suction forces of pF 1.2 – 2.54 (cf. Section 2.2.4). Urea – formaldehyde foams in a loose bed have the permeability of moderately fine sand; under 25 – 35 % compression, that of fine sand [479].

Chemical Properties. As the condensation product of urea and formaldehyde, the foams are synthetic organic substances. Incorporated into moist soils and substrates in flake form, they are known to be biodegradable. At pH 6 – 7, the process goes at 3 – 5 % per year [480]; at pH 3.9 – 4.1, degradation is speeded up (15 – 20 %), especially at temperatures over 40 °C. The stability falls off rapidly above 80 °C. In horticultural substrates, steam sterilization is tolerated up to a maximum of 1 h.

The components of the urea – formaldehyde foams are readily biodegradable: The urea component mineralizes to ammonium and is then converted microbially to nitrate. The formaldehyde component is liberated upon mineralization; it undergoes microbial degradation in 48 – 72 h in soil or water, even more quickly in air. Bioaccumulation does not occur [481].

A theoretical effective lifetime of some 20 – 30 years is desirable for soil conditioning [462, 482].

Raw Materials and Manufacture. A process, part of which has been patented, is used to produce foams from modified urea – formaldehyde condensation products with compressed air, foaming agents, and hardeners in fixed or mobile equipment [483].

In Germany, the maximum allowable workplace concentration (MAK) in production is 1 ppm formaldehyde; this level may be exceeded only for short periods and within limits. In the United States, the absolute MAK is 3 ppm.

Use. The use of urea – formaldehyde foams as soil conditioners has been described by DOEHLER [484]. BAUMANN advanced their use (1953 – 1968) and developed the Plastopanik technique [459]. Foams were brought into reproducible use through extensive research [458, 460, 462, 470, 485, 486].

For reasons of cost, applications cluster in the areas of garden crops, such as pot and container plants, cut flowers (e.g., carnations, chrysanthemums), flower bulbs, fruits (including strawberries), vegetables, and eucalyptus, and in landscaping, where foams are used in newly seeded lawns and woods, in the transplanting of large trees, and in sports areas and golf courses.

In terms of soil physics, urea – formaldehyde foams optimize the air and water regimes in soils and substrates. They enhance the pore volumes, the maximum, minimum, and plant-available water capacity, and the aeration, and they lower the soil density. Plants respond with improved shoot and root growth, better early growth and development rates, and increased crop yields and qualities [462].

Trade Names. Hygromull, Hygropor (mixtures of Hygromull with Styromull) (BASF Aktiengesellschaft, Ludwigshafen); Agricon (Intrinco, Vaduz, Liechtenstein).

Official licenses have been issued for Hygromull and Hygropor in Austria, Belgium, France, Luxembourg, the Netherlands, Norway, Spain, and Switzerland and also for Agricon in Saudi Arabia and other Middle Eastern countries.

7.2. Colloidal Silicates

Colloidal silicates include compounds of polysilicic acid, produced and stabilized mainly by synthetic means, with a high content of reversibly soluble silicic acid and added flocculating electrolytes [460, 462, 487 – 489].

Agrosil colloidal silicate consists of (1) partly dehydrated sodium silicate, precipitated (neutralized) with acids, (2) electrolytes (phosphate, sulfate), and (3) an organic additive to retard aging. Official licenses have been issued for it in Austria, Belgium, Denmark, France, Luxembourg, the Netherlands, Norway, Spain, Sweden, Switzerland, and some Middle Eastern countries.

Physical and Chemical Properties. Colloidal silicates are solid, fine-grained, and thus easily distributed substances that can be dispersed with water to form silica gels and silica sols. Silica gels, examined under the electron microscope, are described as porous; they exhibit surface activity for the addition and inclusion of water and nutrients [490, 491]. Incorporated into soils, they fill the voids between soil particles [492] and bind these together with organic and inorganic complexing agents to form water-stable crumbs [462, 493 – 496]. Pore redistribution takes place in this process [490, 497]. Sorptive salt buffering has been described in soil mixtures [488, 491, 498].

The chemical activity is ascribed in particular to the low-molecular-mass silica sols. In the soil, they enable the holding of phosphate in solution or its activation by desorption [490]; depending on the pH, they can irreversibly fix heavy metals [461, 462, 499 – 504].

Raw Materials and Manufacture. Raw materials for the production of Agrosil colloidal silicate are spray-dried and liquid sodium silicate, phosphoric and sulfuric acids, and an organic aging retardant such as urea, humic substances, pectins, or proteins. The manufacturing process is patented [505, 506].

Use. Colloidal silicates are conceived as soil supplements or soil conditioners [462, 488, 493]. To a certain extent, they make it possible to return soils affected by salts and heavy metals to agricultural use [462].

Their action as soil conditioners depends on their incorporation in the soil and their ability to take up water (precipitation, irrigation). Quantities used per are ($= 10^2$ m^2) are 7 – 20 kg, preferably 10 – 15 kg. In gardening and landscaping, they have an effective lifetime of 3 – 5 years; aftereffects in unworked soils have been seen for 10 and 14 years.

Direct effects due to soil conditioning include improved water and sorption capacities, the activation of plant nutrients in the soil, physical improvement of soil structure through pore redistribution and crumb formation; and the fixing of heavy metals (such as Cd, Cu, Pb, and Zn) in the presence of alkaline earths. Indirect effects include improved soil life (respiratory activity and nodule bacteria), the release of excess water (a consequence of crumb formation), and enhanced phosphate mobility. The growth of new shoots and roots is encouraged, both organic and inorganic substances are accumulated in the vegetation in greater quantities, less water is needed, wilting of grass is reduced, and fungus and bacteria resistance is increased [462, 507].

In the Swiss classification, the product is not classed as toxic.

Trade Names. Agrosil (Guano-Werke AG, BASF).

7.3. Polymer Dispersions and Polymer Emulsions

Poly(vinyl acetate), poly(vinyl propionate), butadiene – styrene copolymer, *cis*-butadiene, and various acrylic acid polymers have been described and employed as soil conditioners with actions in and on the soil [460].

Physical and Chemical Properties. The products are mainly applied to the soil surface in aqueous solutions. They cross-link the particles of the uppermost soil layer and, depending on the concentration of the active agent, form a closed film or networklike coatings. These are permeable to precipitation but diminish evaporation. Incorporated into the soil, they promote crumb formation [494].

As organic substances with a relatively high dilution, polymer dispersions are degraded relatively quickly by UV radiation when on the surface of the soil and by microbial action when in the soil. Their action is therefore limited in time, and their stability depends on ambient conditions such as cold and heat.

Use. Polymer dispersions are employed chiefly for seed protection in landscaping, and also in vegetable and flower-bulb growing. They are applied at planting time by spraying, usually along with fertilizers and soil conditioners (e.g., Agrosil) or mulches (cellulose, straw). They can also be applied after planting by area spraying or, in the case of vegetables planted in rows, by stripe spraying [508].

The quantities used depend on the product and vary between 10 and 50 g/m^2; the dilution with water depends on the purpose and varies between 1 : 1 and 1 : 60 (product : water).

The duration of the structure-stabilizing or protective action depends on the quantity used [509] and on environmental conditions such as weathering and insolation.

Polyacrylates and polyacrylamides are recommended for use alone or with, e.g., starch to promote water storage in soils [460, 510 – 515]. The waterholding effect is, however, greatly diminished by pH values higher or lower than 7.0, water hardness, and dissolved substances such as urine or soil nutrients. No satisfactory practical solution has been found for these limitations [516], but more recent generations of these so called superabsorbers are claimed to be more tolerant towards dissolved salts. Various applications to gardening substrates and flower-growing soils have been described [517].

Polymer formulations are used for protection against erosion, especially in conjunction with the sprouting of seeds. They perform an environmental function by offering limited protection to the seeds or small plants against wind and rain erosion until the plants are large enough to protect themselves. They are of increasing importance in furrow irrigation systems in North America for preventing soil erosion and increasing water infiltration into the soil. A concentration of 10 ppm in the irrigation water is necessary to achieve the desired effects [518]. Biodegradability is a prerequisite for the soil and plant application of selected polymers. Biodegradability is not assured in the case of some polyacrylates.

Trade Names. Curasol for poly(vinyl acetate) (Hoechst Aktiengesellschaft, Frankfurt); Hüls 801 for *cis*-butadiene (Chemische Werke Hüls, Marl); Aqua-Gel for polyacrylate plus starch (Miller Chemical and Fertilizer Corp.); Hydrogel Viterra for poly(ethylene oxide) (Union Carbide Corp., New York); SGP Absorbent for sodium polyacrylate plus starch (General Mills Chemicals, Inc., Minneapolis, Minn.); Soiltex G1 for polyacrylamides (Allied Colloids).

7.4. Tensides

Many tensides have been tested as soil conditioners, including ammonium laureth sulfate, ethoxylated alkyl phenols, polyoxyethylene esters of alkylated phenols, polyoxyalkylene glycols and their polymers, polyoxyethylene, polypropoxypropanol glycol butylesters, alkyl polyglycosides, sulfosuccinates, and poly(propylene oxide)s. The hydrophobic ends of the molecules of the these so-called wetting agents are adsorbed by water-repellent organic matter, and the hydrophil ends link the soil particles with water.

Use. In turfgrass culture thatch accumulation may occur due to reduced decay of the organic matter of the grass sward. Under dry conditions, this material becomes hydrophobic, and irrigation results in water losses because of poor infiltration rates and run-off [519, 520]. Spraying of wetting agents onto the affected areas at a rate of 10 – 20 L/ha increases water absorption and efficiency. Solid formulations are also available.

Trade Names. Primer (Aquatrols, USA); wetx'tra (Rhône Poulenc, France); Turf Ex (Service Chemicals, UK), Saturaid (Debco, Australia).

8. Storage, Transportation, and Application

Fertilizers are produced continuously, year-round, in large capital-intensive plants, but they are sold only in a few months at the beginning of and during the vegetation period. Large quantities must therefore be stored for long times. The logistic of delivering the product to the user at the proper time leads to a subdivision of storage into plant storage, dealer storage, and user storage. In this way, loading and transport are spread evenly through the year, with additional control by seasonally varied sales rebates.

The construction and operation of storage facilities depend on the size, the location, the type of fertilizer, and the danger level, all of which are dictated by the product.

8.1. General Storage Requirements

Fertilizers are classified as follows by danger level:

– Group A. Explosive fertilizers, such as those with a high ammonium nitrate content or with a low ammonium nitrate content and > 0.4 wt % organic matter.
– Group B. Fertilizers in which self-sustaining, progressive thermal decomposition (low-temperature decomposition) is possible. In the past this group comprised mainly multinutrient fertilizers which contain ammonium nitrate. Today, the ammonium nitrate based multinutrient fertilizers which are marketed are almost exclusively rated Group C.
– Groups C and D. Fertilizers that do not explode and are not susceptible to self-sustaining, progressive thermal decomposition.

The precise classification is set forth by law [521]. With its technical regulations, the German regulation concerning hazardous substances (*Gefahrstoffverordnung*) prescribes in detail how to construct and to operate storage facilities for ammonium nitrate based fertilizers. How many regulations have to be observed depends on the classification of the fertilizer. The strictest standards apply for Group A, followed by Group B. For nonflammable fertilizers (Groups C and D), the normal storage conditions for bulk materials apply.

Most fertilizers come in granular form and readily absorb moisture. As a result, they experience caking or granule disintegration; either effect makes application more difficult. The storage areas must therefore be dry. Not only the roof and walls must be tight, but the floor must be safe against rising moisture. Care must also be taken that moisture does not get into the storage building because of too much air circulation. Doors and windows must be kept closed when the relative humidity is over 65 %. The surface : volume ratio in storage should be as small as possible.

Fertilizers containing ammonium nitrate, especially calcium ammonium nitrate, must be protected from direct sunlight and from repeated temperature changes at 32 °C, because otherwise repeated recrystallization leads to granule disintegration.

When ammonium nitrate fertilizers are heated above 130 °C, low-temperature decomposition sets in, generating gases that may be toxic, depending on concentration. Practices must therefore be employed that safely prevent the action of fire or external heating. Such practices include general fire-protection measures such as no smoking, no use of fire or open light, and no unauthorized entry. Ammonium nitrate fertilizers must not be mixed with flammable materials (sawdust, coal dust, sulfur, petroleum, etc.) or with acidic or basic substances (lime, basic slag phosphate, acid salts, etc.). Welding and burning work may be carried out only with the most stringent safety precautions. Electrical equipment must be located outside the fertilizer pile. Cables and wiring must be located at least 0.5 m above the highest possible pile level. Special care should be taken that the pile does not cover up lights, especially portable lights connected by cable to the power supply.

With regard to moving mechanical parts (belt, screw, and bucket conveyors, etc.), care must be taken that these devices cannot run hot or experience fertilizer buildup at critical points. Diesel and gasoline trucks that go directly into the storage area represent a danger due to hot exhaust gases and hot mufflers. Hardened fertilizers must be loosened mechanically, not with explosives.

Low-temperature decomposition can be detected by white or brown smoke or a piercing smell. If this process develops and is recognized promptly, the sintered reaction zone can be separated from the unreacted fertilizer. Good access is needed, and precautions are needed to keep the sizable amounts of gas evolved from being a health hazard. If safe access is not possible, the decomposition must be brought to a stop by cooling with water. If any low-temperature decomposition takes place, an alarm must immediately be sent to the fire-fighting service.

Fertilizers affect most of the conventional building materials, especially when wet. Steel, aluminum, and zinc suffer more rapid corrosion and must be protected by several coats of paint (total thickness > 180 µm). Fertilizer dust can cause brittle fracture at welds in steel. Subsequent welding must be carried out with special care.

Mineral fertilizers attack concrete more or less severely, causing scaling and corrosion of the steel reinforcement. For this reason, concrete floors contacted by fertilizer should be protected by filled coal-tar pitch or epoxy-based plastic.

Wood, especially in the new form of glued lamellae, is a suitable material. Care should be taken to use corrosion-resistant fasteners.

Plant Storage. Storage at the fertilizer plant is almost always in bulk form. Large warehouses, with capacities of up to and beyond 50 000 t for one or more grades, are employed. The total storage capacity is usually several months' production. The size, shape, and outfitting of the warehouses depend on the frequency of shipping, the number of grades, the required loading capacity, and the area available. In order to reduce the risk, some grades are limited as to amount stored in a single warehouse, which may be accomplished by dividing the building. Large quantities are stored in elevators (tower silos) only under some restrictions, because of the risk of caking and the resulting high reclaiming cost.

If enough space is available and shipments are not frequent, a flat-floor warehouse in which material can be piled more or less high against the walls is the most economical design. Trough-shaped or basin-shaped storage structures with appropriate reclaiming equipment are used when space is limited and intervals between shipments are short.

The material used for the foundation and walls is usually reinforced concrete; the roof is most often made of wood. Glued lamellar wood construction makes it possible to span wide buildings without posts. Fertilizer can be piled against the walls to a height of 10 m or more. However, if the pile is too high, the material can segregate during placement. Slides during reclaiming are also a problem if material is piled too deep. The plant warehouses generally need not be heated, since the fertilizer leaves the production facility at a temperature of 35 – 45 °C (not over 50 °C). On the other hand, overhead conveyors and loading apparatus spanning moderate distances should be heated to keep them dry.

Piling is done mainly with overhead belts at the center or side, with transverse distribution by belts, chutes, or discharge pipes. Newer facilities have a distribution control room with closed-circuit television units.

Reclaiming capacity must be several times piling capacity. Because requirements vary, several methods have been devised. For high capacity and short shipment intervals, track-borne scrapers capable of automation are used. Bridge cranes are also employed. Many wheeled tractor loaders are in service; these can be shifted from one location to another, easily replaced if damaged, and reinforced if necessary. All warehouse equipment moves material by way of transfer hoppers to conveyor belts, troughs, or bucket conveyors, which carry the material to the loading point. Crucial factors in the choice of equipment are the loading capacity required, the storage qualities of the fertilizer, the cost of structures and equipment, the safety of the operation, and the expected operating costs.

Warehouses are not usually operated on the first in, first out principle. For practical reasons, however, the warehouse should be emptied completely from time to time, since varying raw materials may result in slight color variations, while going to zero aids precise inventorying, and for other reasons.

Shipping from the Plant. Fertilizers at present are usually shipped in bulk, in self-discharging railroad cars, trucks, silo cars with pneumatic loading of elevators, and canal boats and barges.

Before shipping, many plants screen the fertilizer again to remove dust, lumps, and impurities.

Shipment of fertilizers in bags is declining. The product is usually put in 50-kg open bags (pillow bags), which are then sewed or welded shut, or in valve bags. Valve bags are better suited to stacking on pallets because of their boxlike shape. Most bags are made of plastic — extruded polyethylene tubing or woven polypropylene strips coated or lined with polyethylene. Paper and jute bags are now rarely used. The machinery industry has developed largely automated equipment for bagging and loading or palleting, and much work is still under way to improve these devices and adapt them to various requirements. Because of the associated high investment costs, bagging takes place mostly at the fertilizer plant. Palleting, which simplifies subsequent transfer of the bagged product, has gained in importance. In many countries, fertilizers are shipped in large 500 – 1000-kg bags, likewise made of woven polyethylene or polypropylene.

Dealer Storage. Fertilizer is stored chiefly in buildings, which are usually divided boxwise. Because of the caking tendency and the resulting problems with reclaiming, especially in the case of ammonium nitrate fertilizers, elevators find use only for free-flowing grades, for a limited span of time, and for quantities up to 30 t.

In building storage, special care should be taken to protect against moisture, since the fertilizer no longer has its heat from production, and it often sits for several months under cool, moist conditions. Before receiving material, the warehouse must be completely dry. Fertilizer should be placed continuously, without long breaks. The pile height should not exceed 5 m. Immediately after placement, the fertilizer surface must be carefully covered with film not less than 25 μm thick, because of the mechanical stress. Joints in the film must have an ample overlap, usually 1 m. A tight seal should be made at the walls (e.g., with wood lath). Air circulation should be minimized (closed windows and doors, airtight roof). If product is reclaimed in stages, the cover must be restored each time.

A belt conveyor is suitable for placement of fertilizer. Mechanical abrasion restricts the use of screw conveyors (not over 5 – 6 m, 100 rpm).

User Storage. The interim storage of mineral fertilizers by the consumer is declining to the same extent that the dealers become more willing to undertake this kind of storage. Bulk fertilizer is kept mainly in flat storage boxes, just as in the trade, and the boxes are similar to those used for dealer storage. Elevator storage is practiced on a very limited basis.

Bagged product can be stored by the user without major precautions. To simplify transfer operations, between 20 and 30 50-kg bags are assembled on a pallet, with or without shrink-wrap; in some countries, large bags (0.5 – 1.0 t) are also employed. Bagged fertilizer can be stored outdoors for a limited time under certain conditions (no temperatures over 32 °C, white UV-stabilized covers).

Transportation. Group B fertilizers can generally be transported without special precautions. Only in exceptional cases are they considered hazardous goods as defined in the transportation regulations. Specific provisions for Group B, as well as Group A fertilizers, are set forth in international regulations for various kinds of carriers. The most important rules are recognized by nearly every industrial nation and have been incorporated into the national regulations.

Sea transport: IMDG (International Maritime Dangerous Goods Code)
Rail transport: CIM (International Convention on the Transport of Merchandise by Rail)
Road: ADR (European agreement concerning the international transportation of dangerous goods by road)
Air: IATA-RAR (IATA Restricted Articles Regulations)

The CFR contains rules for all modes of transportation in the USA.

8.2. Application

Solid mineral fertilizers are generally applied to agricultural sites in the form of granules, coarse or fine crystals, and powders. Fine crystals and powders are becoming less and less important, with the exception of lime which is used almost exclusively in powdered form.

To achieve a good nutrient efficiency and to avoid environmental pollution, the equipment used for the application of mineral fertilizers has to meet the following requirements:

1) Exact calibration of the amount of fertilizer applied and uniform distribution, both in the direction of travel and in the transverse direction, and independent of traveling speed and application rate.
2) In order to attain a high area of application, the effective band must be wide and the distributor design must permit a high rate of operation. The rate must be variable between 50 and 1000 kg/ha. Special distributors for liming must offer rates of up to 5000 kg/ha.

The uniformity of fertilizer application can be determined exactly on a test stand. Boxes with areas of 0.25 m^2 catch the fertilizer over the whole application width, and the product is then weighed. The mean is calculated from the individual values. The smaller the mean deviation from the average value, the more uniform the fertilizer distribution. Internationally, the test method is largely unified, so that the results are also easily comparable from country to country.

Application Equipment. Mineral fertilizers in the form of granules or coarse crystals are generally applied by broadcasting-type fertilizer distributors. Spreader rigs with pneumatic transverse distribution are also utilized, but their use is of declining importance. Powdered fertilizers such as lime are preferably applied by fixed-width fertilizer spreaders with mechanical transverse distribution to avoid dust formation during operation.

Broadcasting fertilizer spreaders, also called spinner spreaders, operate with working widths of 6 – 36 m. They are characterized by simple construction and easy handling, together with a high area performance. They constitute the absolute majority of all fertilizer distributors.

Fertilizers 97

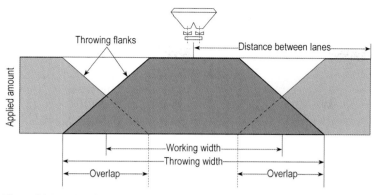

Figure 36. Broadcasting pattern for spinner spreaders

Broadcasting fertilizer spreaders distribute the fertilizer over a semicircular to semi-elliptical area around the spreading device. The spreading width is greater than the effective working width. As the amount of fertilizer which is applied per unit area decreases towards the borders of the spreading swath, overlapping of adjacent swaths is necessary for uniform transverse distribution of the fertilizer (Fig. 36). Modern broadcasting spreaders have broadcasting patterns with extremely flat sides, so that the risk of distribution errors in the overlap zone is minimized.

Three types of broadcasting spreaders with different spreading devices exist: single-disk, twin-disk, and oscillating spout (pendulum) spreaders (Fig. 37). The twin-disk spreader is gaining importance over the single-disk and oscillating-spout spreaders, since its basic concept is better suited for further technical development. Consequently, the spreading quality of twin-disk spreaders has been raised to a level almost matching that of pneumatic spreaders.

The spreading device in oscillating-spout spreaders is a pendulum tube that moves to and fro. Since this spreading technique allows working widths of only 6 – 15 m, whereas the large fields of modern agriculture require increasingly greater swath widths, the importance of oscillating spout spreaders is also diminishing.

Fixed-Width Fertilizer Spreaders. With fixed-width spreaders the transverse distribution of the fertilizer is achieved by means of booms, either pneumatically or mechanically. Unlike spinner spreaders, only one working width is possible.

In pneumatic fertilizer spreaders, the fertilizer is metered at the hopper by cam wheels, fed into distributor tubes, and transported by blast air to discharge outlets along the spreader booms, which are provided with deflectors (Fig. 38). Depending on the boom length, working widths of up to 24 m are possible. When working with a pneumatic spreader, overlapping of adjacent swaths is required, as with a spinner spreader. However, due to the steep flanks of the broadcasting pattern, the required overlap is much smaller (Fig. 39). Pneumatic spreaders can be used for the application of granulated and

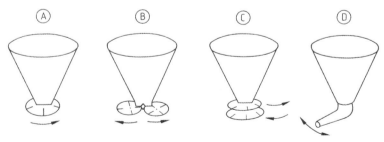

Figure 37. Broadcasting spreaders: single-disk centrifugal distributor (A), twin-disk centrifugal distributors with side-by-side (B) and over-and-under spinners (C), oscillating-spout distributor (D)

crystalline mineral fertilizers, if possible free of dust.

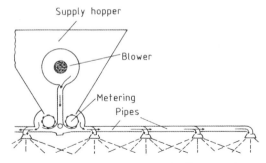

Figure 38. Schematic of a pneumatic fertilizer spreader

Spreaders with mechanical transverse distribution are generally equipped with a screw or auger (auger-type spreaders) Fertilizer flowing from the hopper is conveyed to the driven auger by various means and is then transported to the spreader tubes with adjustable discharge spouts (Fig. 40). Depending on the boom length, working widths of 6 – 8 m are possible. Auger-type spreaders are preferably used for the application of powdered fertilizers. Boom length and working width are identical. For uniform fertilizer distribution, exact driving in parallel runs is indispensable (Fig. 41).

Electronically Controlled Spreading Equipment. Electronic control units are increasingly being used to monitor and to improve the metering and distribution of mineral fertilizers. Thus, the correct amount of fertilizer can be automatically discharged when the traveling speed of the spreader is changed, and the amount of applied fertilizer can be displayed, based on permanent monitoring of the fertilizer discharge.

Site-Specific Fertilizer Application Guided by GPS. Until now farmers' fields generally received a uniform fertilizer rate. With the steadily increasing size of individual fields this practice is bound to change, as increasing field size also implies increasing heterogeneity of the soil and thus of the crop yield. For economical and environmental reasons, a future development towards the site-specific management of large-sized fields is imminent. Under this aspect, the global positioning system (GPS) is currently being tested under practical conditions for its suitability for crop management, including fertilizer application. GPS allows the exact position in the field to be determined and relocated, regardless of weather, location, and time. On this basis, it should be possible to apply fertilizers site-specifically in accordance with the existing soil and crop variations within a field. The complete technology, from yield recording by the harvester to nutrient application with the fertilizer spreader, has yet to be further developed. This will probably take another 5 to 10 years.

9. Environmental Aspects of Fertilizer Application

Fertilizers are used to enhance soil productivity or fertility (amelioration fertilization) or to maintain it at an economically and ecologically acceptable level (maintenance fertilization). The task of fertilizers is to replace the nutrients lost through the harvest and by other causes: irreversible fixation in the soil, leaching, escape in gas form, etc. Along with the positive effect of fertilizer application, there are also negative effects, especially with improper use.

Up to the beginning and middle of the 1970s, environmental problems triggered by fertilization practices were subject to local discussion only; examples were the buildup of heavy metals in soils and crops when sewage sludge was applied, the buildup of nitrates in the groundwater where vegetables were raised in the catchment areas, and the eutrophication of surface waters due to phosphates. Only later was the partly regional character of local environmental problems acknowledged.

From the early 1990s onwards, increasing emphasis was also placed on global aspects of the impact of fertilization on the environment, i.e., the influence of agricultural production in general and fertilization in particular on the increased emission of greenhouse gases. The awareness of the mutual relationship between surplus production of agricultural products and environmental problems within the EU has continuously grown over recent decades. In particular the interest in mineral emissions from agriculture has increased substantially, both in policy making and research and in practical farming.

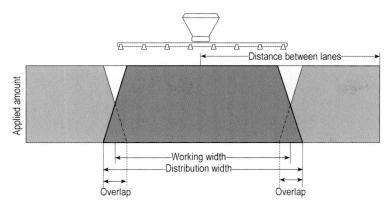

Figure 39. Broadcasting pattern for pneumatic spreaders

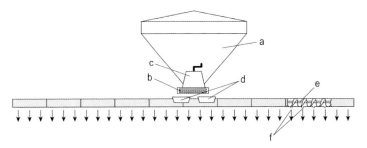

Figure 40. Schematic of a spreader with mechanical transverse distribution
a) Supply hopper; b) Elevator belt; c) Slide valve; d) Feed auger; r) Outlet

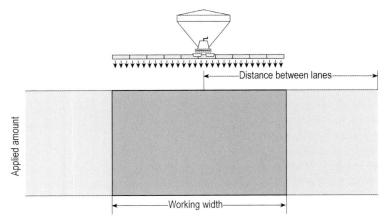

Figure 41. Broadcasting pattern of a spreader with mechanical transverse distribution

9.1. Nitrogen

Of all plant nutrients, nitrogen is the most effective in economic terms, but in ecological terms the most problematic. The plant absorbs nitrogen mainly as nitrate (NO_3^-) but partly as ammonium (NH_4^+). The plant's ability to assimilate larger molecules with organically bound nitrogen is limited.

Problems with nitrogen arise from a local or regional excess of nitrogen, regardless of its origin. Several cases, which are mostly of a regional character, are described first, then solutions are

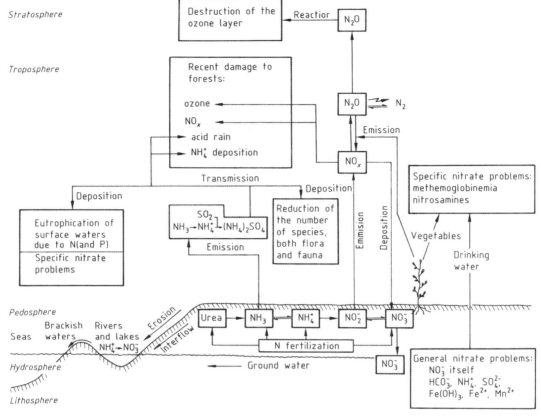

Figure 42. Agricultural nitrogen as an environmental factor

listed. Figure 42 summarizes the environmental aspects of nitrogen.

9.1.1. Ground Water

Various sectors of the economy — agriculture, forestry, water resources and wastewater management, transport, energy, and industry — are involved in ground water and surface water problems having to do with nitrogen. With regard to ground water, the quality of drinking water is the prime concern. A reduction in the maximum allowable nitrate content in drinking water within the EU, from 90 mg/L to 50 mg/L, has made the nitrogen problem more acute.

Until 1970–1975, virtually the only negative aspects considered with respect to nitrate were methemoglobinemia and the nitrate–nitrite–nitrosamine problem. Later, a number of other reactions triggered by nitrate and sulfate in the aquifer or during transport to it have come under discussion. Depending on the presence of aerobic or anaerobic conditions, the intensity of NO_3^- (and SO_4^{2-}) application, and the reserves of microbially available carbon sources, the reactions listed in Table 34 can yield the following compounds:

HCO_3^- or $Ca(HCO_3)_2$	water hardness
N_2O	stratospheric ozone breakdown
NH_4^+;	undesirable, toxic, or
H_2S;	foul-smelling
Fe^{2+}, Mn^{2+}	
$Fe(OH)_3$	ferric incrustation of wells

If available carbon sources are not present, or exhausted, nitrate (or sulfate) can break through into the ground water.

From around 1980 up to now increasing emphasis was placed on ground water as prime

Table 34. Sequence of environmental problems in the presence of excessive nitrate and sulfate in the pedosphere and hydrosphere

Oxygen and glucose in soil and ground water	Reduction processes	Energy released per mole of glucose, kJ	Problematic parameter Input	Output*
	Catabolic reduction of O_2, NO_3^-, and SO_4^{2-}			
	Respiration $C_6H_{12}O_6 + 6O_2 \longrightarrow 6CO_2 + 6H_2O$	2876		
	Denitrification $5C_6H_{12}O_6 + 24NO_3^- + 24H^+ \longrightarrow 30CO_2 + 42H_2O + 12N_2$ $C_6H_{12}O_6 + 6NO_3^- + 6H^+ \longrightarrow 6CO_2 + 9H_2O + 3N_2O$	2712 2578	NO_3^-	N_2O
	Nitrate respiration $C_6H_{12}O_6 + 12NO_3^- \longrightarrow 6CO_2 + 6H_2O + 12NO_2^-$, $NO_2^- \xrightarrow{H^+} NO_x$	1946	NO_3^-	NO_x
	Ammonification $C_6H_{12}O_6 + 3NO_3^- + 6H^+ \longrightarrow 6CO_2 + 3H_2O + 3NH_4^+$	1817	NO_3^-	NH_4^+
	Sulfate respiration (in the absence of NO_3^-) $C_6H_{12}O_6 + 3SO_4^{2-} + 6H^+ \longrightarrow 6CO_2 + 6H_2O + 3H_2S$	486	SO_4^{2-}	H_2S
	Inorganic reduction of NO_3^-			
	Dentrification with pyrite $5FeS_2 + 14NO_3^- + 4H^+ \longrightarrow 7N_2 + 10SO_4^{2-} + 5Fe^{2+} + 2H_2O$		NO_3^-	SO_4^{2-}, Fe^{2+}
	Dentrification with soluble iron $10Fe^{2+} + 2NO_3^- + 24H_2O \longrightarrow N_2 + 10Fe(OH)_3 + 18H^+ + N_2$		NO_3^-	$Fe(OH)_3$
	Dentrification with elemental or organic sulfur $5S + 6NO_3^- + 4H^+ \longrightarrow 3N_2 + 5SO_4^{2-} + 4CO_2 + 2H_2O$		NO_3^-	SO_4^{2-}

Left margin labels: Decreasing O_2 supply; Decreasing supply of available organic carbon; Supply of available organic carbon exhausted.

* Any reaction producing CO_2 is able to increase the concentration of $CaHCO_3$.

source of diffuse nitrogen pollution of surface waters [525, 526].

Causes of Nitrate (and Sulfate) Damage. The following cases have been described in the Federal Republic of Germany:

1) Ground water lowering, plowing of pastureland, intensive cropping with N fertilization [527]
2) Intensive cropping with vegetables and/or intensive stockbreeding with appropriate N fertilization (Fig. 44) [528 – 532]
3) Intensive agricultural land use — commercial gardening, animal husbandry — with appropriate N fertilization [533]
4) Commercial gardening, intensive white asparagus cultivation, overuse of mineral fertilizers (Fig. 45) [534, 535]
5) In the Netherlands, the ground water in regions with intensive stockbreeding (the provinces of Geldern and Brabant) displays not only elevated nitrate levels but often high ammonium values at depths between 10 and 25 m [536].

Recent surveys on the nitrate concentration of the ground water in Germany resulted in quite different frequency distributions for a representative overall set of measurement points and a special set of measurement points restricted to agricultural "loading areas" according to EU Directive 91/676. In the representative survey, only 10 % of the samples exceeded the EU threshold concentration for drinking water, whereas in the near-surface groundwater of so-called agricultural-loading areas this was the case in more than 60 % of the samples (Fig. 43).

A distinction should be made between acute and chronic nitrate problems. Extreme overfertilization or pastureland plowing can cause sudden, acute nitrate damage in soils threatened by leaching (sandy soils, high water table, low or declining denitrification capacity). For example, years of overapplication of N (mainly for asparagus, with more then ten times the N depletion by

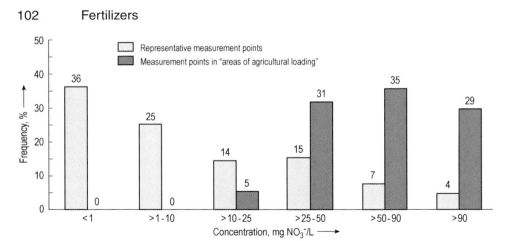

Figure 43. Frequency distribution of NO_3^- concentration in ground water in Germany [537, 538]

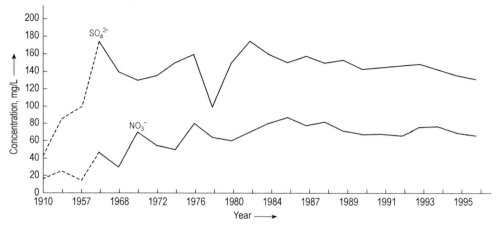

Figure 44. Nitrate and sulfate in the raw drinking water of Mussum, North Rhine-Westphalia, Federal Republic of Germany, 1910 – 1996 [532, 539].

the crop) in the drinking-water catchment area of the Bruchsal (Federal Republic of Germany) waterworks overloaded the denitrification capacity and caused a jump in nitrate in well waters (Table 35, Fig. 45) [535, 540].

Over a long period, an acute nitrate problem may become chronic. Even under economically optimal fertilization matched to base outputs, elevated ground water nitrate levels can come about [535, 541, 542]. The base outputs arise through leaching of water-soluble nitrogen-bearing substances. The unavoidable nitrogen base outputs, primarily in winter, are between 20 and 40 kg ha^{-1} a^{-1} [543 – 545]. A chronic nitrate problem is much more severe than an acute one with respect to detection and propagation, as well as the cost and the success of rehabilitation measures.

The causes of damage can be classified as follows:

1) Plowing of pastureland, forced in part by prior ground water lowering. In three to four years, this process robs light soils of 6 – 7 t of organic nitrogen through mineralization and leaching out of the root zone [546, 547]. In heavy soils, the process takes much longer. The nitrate formed by mineralization is first denitrified by microorganisms, with the simultaneous degradation of the microbially available carbohydrates. The denitrification of 6 – 7 t of nitrate N requires 16 – 19 t of such carbohydrates. In this way, the available

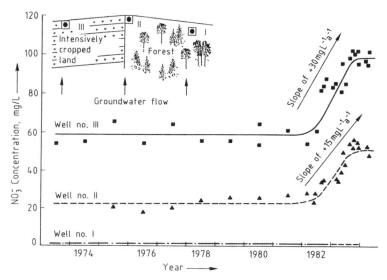

Figure 45. Nitrate concentration in water from three wells of Bruchsal Waterworks, 1973 – 1983 [540]
The insert shows the location of wells no. I – III.

Table 35. Denitrification potential and nitrate nitrogen concentration 2 – 3 m and 3 – 4 m under the surface of forest (pine or alder) and farmed land (cereals or asparagus) at Bruchsal (Federal Republic of Germany) in February, 1986

	Forest near well number I		Intensively cropped land near well number III	
	Pine	Alder	Cereals	Asparagus
	2 – 3 m under the surface			
Denitrification potential *, kg/ha	427	583	67	21
Nitrate nitrogen, kg/ha	6	18	23	56
	3 – 4 m under the surface			
Denitrification potential *, kg/ha	238	258	50	17
Nitrate nitrogen, kg/ha	4	3	50	60

* The denitrification potential is the concentration of water-soluble carbon, as glucose, multiplied by 10/27, which corresponds to the stoichiometry $5\ C_6H_{12}O_6 + 24\ NO_3^-$ (see Table 34).

organic matter soon becomes the limiting factor, and the denitrification capacity of the soil is exhausted. In soils with shallow foundation, there is thus a danger that the N output of a nearby intensive farming operation will appear almost quantitatively in the ground water. Depending on how close the wells are, the raw water may show elevated nitrate levels for decades [548].

2) Inappropriate crop rotation. Examples of poor crop rotation include high fractions of fallow or semifallow land (e.g., corn [maize] or sunflowers), summer grains without or with inadequate (winter) intercrops, and large fractions of legumes, especially if they are plowed in too soon [545, 549]. These practices usually result in high leaching losses, chiefly in winter.

3) Corn rotation with excessive liquid manure. In regions of large-scale stockbreeding, for example, in parts of the Netherlands, Belgium, Germany, and France, excessive amounts of liquid manure were formerly — and in many forms are still — applied.

4) Incorrect nitrogen fertilization. The problem is usually excessive fertilization, poor timing, and unsuitable fertilizer form, usually not an inappropriate application method.

The danger of overfertilization with mineral fertilizers is especially great in commercial gardening and in single crops with low N depletion (orchards, vineyards). Vegetables are usually harvested in the principal growth period of the vegetative phase. Leafy vegetables, above all, offer good market quality only if a good nitrogen supply is available at the time of harvesting. Thus high residual levels of readily soluble nitrogen remain in the soil. Here again, it often happens that too much fertilizer is applied. The same holds for vinegrowing, especially over permeable limestone. In sugar beet farming, fertilizer nitrogen was formerly applied far in excess of the depletion. However, from 1980 until 1994 there was already a continuous decrease from 250 to 125 kg N per hectare in German sugar beet growing, on average [550].

In general, the danger of overapplication is less for mineral fertilizers than for organic fertilizers, since the nitrogen content in the mineral fertilizer, unlike that in manure, is known and constant. What is more, the mineral fertilizer can be applied more precisely with respect to both quantity and schedule. Finally, cost is an important factor in determining the rate of mineral fertilization.

Overapplication of organic fertilizers is a problem chiefly in heavily stock-oriented farming with liquid manure fertilization and in pasturage with grazing. Despite overapplication of organic fertilizers by a factor of one or two, compared to the residual nitrogen loss, mineral fertilizer N is often supplied in addition to improve yields. This is demonstrated by the total nitrogen balances of different farming types in Germany. The mean nitrogen surpluses were estimated for the financial year 1995/96 at 19 kg N per hectare for arable farms, 107 kg N per hectare for cattle farms, and 166 kg N per hectare for special pig and poultry farms [551].

9.1.2. Surface Waters

In common with the groundwater situation, many Western European rivers have shown a strong tendency towards increasing levels of dissolved nitrogen. As an example, in the period between 1954 and 1995 the nitrogen load, mainly as nitrate, from diffuse sources of the Rhine at the Lobith control station increased from 95 000 to 185 000 t N per hectare per annum [552].

The present share of diffuse sources in the total nitrogen input into surface waters in Germany is estimated at about 60 % [526, 553, 554], because the N input from point sources was drastically reduced since the mid-1980s due to the introduction of nitrification and denitrification treatments in sewage plants. Of the diffuse sources, the pathway via groundwater is highly dominant, accounting for roughly two-thirds. Drain water was responsible for an average input share of about 10 % of the diffuse nitrogen pollution of surface water in Germany. Hence, about three-quarters of total diffuse nitrogen input was due to leaching of nitrogen from upper soil layers, and nitrogen fertilization with both mineral and farm-produced fertilizers has an important impact on the level of leaching losses.

In connection with the nitrate N (and phosphorus) loading of rivers and wetlands, special attention should be paid to the drainage system in the agricultural area of the northern European lowland plains, with their intensive stockbreeding and liquid manure disposal. The danger of eutrophication not only by increasing phosphate inputs but also by rising nitrogen inputs into estuaries, coastal zones, and especially the tidal mud flats of the North and Baltic Sea coasts has come under discussion [555 – 557]. Sediments in these areas since about 1950 – 1960 display increasing contents of both P and N [558, 559].

9.1.3. Atmosphere

Agriculture is one source of problematic N compounds in the atmosphere: N_2O from nitrification and denitrification processes in the soil and, chiefly in stockbreeding areas, ammonia (Fig. 42).

N_2O Emissions. N_2O [10024-97-2], with an atmospheric residence time of 100 – 200 years, serves as a source of NO [10102-43-9]. In the upper stratosphere NO, along with HO radicals and halogenated hydrocarbons, contributes to the breakdown of ozone, which is associated with the danger of increased short-wavelength radiation on the earth and a correspondingly increased susceptibility to skin cancer in humans [560].

$$O_3 \xrightarrow{h\nu} O + O_2$$
$$O + N_2O \longrightarrow 2NO$$
$$2NO + 2O_3 \longrightarrow 2NO_2 + 2O_2$$
$$\overline{3O_3 + N_2O \longrightarrow 3O_2 + 2NO_2}$$

Soil management is the most important agricultural N_2O source [561]; denitrification and nitrification are the responsible microbial processes [562, 563]. Intensive studies on N_2O emissions from agricultural soils carried out from the beginning of the 1990s gave average emission rates of 2 – 5 kg N_2O per hectare per annum for temperate zones [564 – 566], but much higher for tropical soils [567].

The total N_2O emission in Germany in 1990 was estimated by the Federal Environment Agency (UBA) at 143×10^3 t N per annum, with an agricultural contribution of 34 % [553].

The quantity of N_2O emitted from soils is determined by the content of available nitrogen, that is, by the level of N fertilization. Nitrogen fertilization was responsible for 50 – 80 % of the emissions, and between 1 and 3 % of the fertilized N was lost as N_2O [568 – 571].

A worldwide estimate of direct and indirect emissions of N_2O from nitrogen fertilization and nitrogen fixation is given in Table 36. The direct emissions of N_2O are estimated at 1.25 ± 1 % of the applied N. The indirect emissions are ascribed to N from past years' fertilization, crop residues, from subsurface aquifers, and from recent atmospheric depositions [571]. An additional 0.75 % of the applied N will eventually be evolved from these indirect sources [572]. With increasing nitrate levels in the soil, the denitrification ratio N_2O/N_2 shifts in favor of N_2O [573]. The average share of N_2O in the total denitrification loss was assessed at 8.5 % [574].

Ammonium-based fertilizers lead to higher N_2O emissions than nitrate-based fertilizers [569, 575]. Farm manures generally produce higher N_2O emissions, mainly due to the simultaneous application of quickly decomposable organic matter [576].

N_2O emissions due to N fertilization are mitigated by all measures to improve the efficiency of N use. The mitigation potential is estimated at up to 20 % [577].

NO_x (NO + NO_2) Emissions. NO_x, which has an average atmospheric residence time of 1.5 d, contributes to ozone synthesis in the troposphere:

$$NO_2 + O_2 \xrightarrow{h\nu} NO + O_3$$

The reverse reaction does not take place in the presence of carbon monoxide or reactive hydrocarbons, including methane, which is produced mainly by ruminants. Ozone can combine with unsaturated hydrocarbons from automobile exhausts to form ozonides, which react with NO_2 to yield peroxyalkyl nitrates

$$R-C\begin{smallmatrix}O-O-NO_2\\ \\O\end{smallmatrix}$$

Ozone and the peroxy compounds are strong oxidizing agents with high phytotoxicity. They have been blamed for damage to vegetation in both forestry and agriculture.

The volume of NO_x emissions due to fertilization (from N reactions in the soil, nitrification more than denitrification [578]) and of plant NO_x emissions [579 – 581] is hard to assess. The contribution of agriculture to global NO_x emissions was estimated at 22 % (ca. 11 Tg NO_x N per annum), which consists of biomass burning (14 %) and the influence of mineral fertilization and manuring (each ca. 4 %) [582]. When the photooxidation of emitted NH_3 in the troposphere is taken into account, the figure rises to 27 %. However, NO_x is reabsorbed and metabolized by the plant, so that gross and net processes must be differentiated [583 – 587].

NH_3 Emissions. Ammonia emitted into the atmosphere reacts fairly rapidly (residence time < 9 d) to NH_4^+ and, after reaction with, say, SO_4^{2-}, is precipitated as ammonium sulfate, $(NH_4)_2SO_4$. Atmospheric NH_3 or NH_4^+ promotes the long distance transport of SO_2 or SO_4^{2-}. Ammonia and ammonium nitrogen forms do have fertilizing action, but in Western Europe the negative effects outweigh the beneficial ones in close-to-natural (forests, heath) and natural ecosystems (nature preserves, surface waters). Aside from the phytotoxic action of NH_3 in the area close to the emission source, deposition of NH_4^+ causes long-range damage:

1) Leaching of nutrients (e.g., K^+, Ca^{2+}, Mg^{2+}) out of the phyllosphere (leaves, needles) and the pedosphere (soil) followed by acidulation of the soil and waters [588 – 592]
2) Buildup of nitrogen in the pedosphere and nutrient imbalances in vegetation [589]
3) Injuries to mycorrhizae, the roots in symbiotic association with fungi [593]
4) Changes in the flora and fauna of terrestrial and aquatic ecosystems due to the promotion

Table 36. Estimates of direct and indirect global emissions of N_2O from application of fertilizer N (synthetic or animal waste) to agricultural soils and from soils growing biological N-fixing crops (10^6 t/a N_2O N) [577]

Region	Mineral N	Animal waste	N fixation	Total	Range
Africa	0.04	0.21	0.05	0.30	0.15 – 0.45
North and Central America	0.26	0.11	0.11	0.48	0.24 – 0.72
South America	0.03	0.22	0.09	0.34	0.17 – 0.51
Asia	0.75	0.52	0.19	1.46	0.73 – 2.19
Europe	0.27	0.22	0.02	0.51	0.26 – 0.77
Oceania	0.01	0.03	0.01	0.05	0.03 – 0.08
Former Soviet Union	0.17	0.18	0.03	0.30	0.19 – 0.57
Total	1.53	1.49	0.50	3.50	1.80 – 5.30

of nitrophilic species and the inhibition of nitrophobic species [589, 594]
5) Increased loss of N with leach water, especially in damaged coniferous forests [595 – 597]
6) Increased N inputs to surface waters [598], corresponding N buildups in marine sediments, especially in the North Sea and Baltic Sea coastal regions [558, 559]

Around the turn of the century (in remote regions even today) the mean annual depositions of atmospheric NO_3^- N and NH_4^+ N were each ca. 1.3 kg/ha. Today, 10 – 15 kg/ha are deposited for each nitrogen form. On the basis of the filter capacity of the forest, the atmospheric N inputs may be increased by a factor of 1.2 – 2.0 (deciduous forest) or 3.0 (coniferous forest). In addition, a coniferous forest has a specific filtration capacity for NH_4^+ N but not for NH_3^- N [596, 599].

Recent measurements of the atmospheric nitrogen deposition into the soil – plant system with a new integral measurement system for wet, dry, and gaseous depositions even show annual net depositions of between 65 and 73 kg N per hectare at two locations in central Germany [600]. Clearly the earlier standard measurements underestimated the true magnitude of nitrogen deposition. Such nitrogen depositions exert a high pressure on natural or close-to-natural ecosystems because they greatly exceed the critical loads of these ecosystems.

The estimate of NH_3 emissions by agriculture is based on statistical figures of animal husbandry and fertilizer production and application, which all are multiplied by source-specific emission coefficients. Because of the different values of these specific coefficients given in the literature, all estimates of NH_3 emissions have a high range of uncertainty. An estimate for Germany (1992) resulted in the range from about 350 to 840 with a mean value of 577 Gg NH_3 N [601]. There are very large differences in the emission densities between and even within the federal states.

The estimates for Western Europe from different authors are in the range of 3.1 – 4.0 Tg/a NH_3 N [602]. As an example, Table 37 contains estimates of the anthropogenic NH_3 emissions from agriculture, industry, and other sources for 16 European countries. 90 % of the total emissions are due to agricultural activities, 74 % alone originates from animal husbandry (stabling, grazing, application of manure), and 12 % is attributed to the use of mineral fertilizers. The total emission is 4.0 Tg/a NH_3 N $\pm$ 30 %, that is, 2.2 – 5.2 Tg/a NH_3 N.

Worldwide, including natural emissions, mineral fertilizers account for 2 – 5 % and stockbreeding 12.5 % of the total emissions of about 120 Tg/a NH_3 N [560]. The high proportion due to stockbreeding in Western Europe is accounted for by the high population density of animals and by the changeover from solid manure to slurry manure, which is less laborious but higher in NH_3 emissions. In comparison with stockbreeding, NH_3 emissions are lower even when solid mineral fertilizers high in NH_4 or amide N (ammonium sulfate, diammonium phosphate, urea) are used. This is especially true for soils where the pH and the free lime content are high or where directly incorporating the fertilizer into the soil is not possible. Liquid fertilizers with NH_4^+, because they infiltrate the soil, emit less NH_3 than solid NH_4^+ fertilizers.

Based on the specific NH_3 emission coefficients of the various N fertilizers and their application, NH_3 emission rates from mineral fertilizers in Germany are calculated by different authors to be from 2.1 to 8.1 kg per hectare per annum NH_3 N [603].

Table 37. Anthropogenic ammonia emissions from 16 European countries [602]

Country	Animal husbandry	Mineral fertilizers	Cultivated plants	Industry	Other sources	Total
Belgium/Luxembourg	99	4.2	2.3	1.1	9.2	115
Germany	504	78.4	27.1	1.5	53.1	664
Denmark	97	7.5	4.2	0.4	9.5	119
Finland	36	2.7	3.9	0.7	3.8	47
France	564	127.6	46.1	2.0	64.3	804
Greece	57	18.6	13.8	0.5	7.8	98
Great Britain	366	55.9	26.9	1.4	39.1	489
Ireland	124	12.5	8.5	0.8	12.6	158
Italy	390	64.0	21.0	3.3	41.6	520
Netherlands	200	8.5	3.0	3.6	18.7	234
Norway	32	1.1	10.5	0.9	3.1	39
Austria	62	1.6	5.3	0.4	6.1	76
Portugal	59	6.2	6.8	0.3	6.3	78
Sweden	54	2.2	5.1	0.3	5.4	67
Switzerland	48	3.4	3.0	0.0	4.7	59
Spain	270	96.1	45.8	1.8	36.0	449
Total (abs.)	2961	490	224	19	321	4016
Total (rel.)	74	12	6	0.5	8	100

Apart from the impact of nitrogen fertilization on greenhouse gases, there is growing concern about its influence on the sink capacity of soils for methane, the most important greenhouse gas after CO_2. Oxidation by methanotropic bacteria in the soil is the only biological sink for methane. An adverse effect of N fertilization on methane oxidation was first reported in 1989 [604]. Later it was found that ammonium strongly inhibits methane oxidation, whereas nitrate has only a minor effect if at all [605 – 607]. This is due to the competition between NH_4^+ and CH_4 for the enzyme methane monooxygenase [608].

9.1.4. Biosphere

Nitrogen fertilizer may produce undesirable N forms in the crop and changes in the composition of flora and fauna in nearby close-to-natural or natural ecosystems.

Undesirable N Forms in Crop. Because of the potential formation of nitrosamines, crops should not contain too much nitrate. However, the nitrosation of substances such as vitamin C [50-81-7] and α-tocopherol [59-02-9] suppresses nitrosamine formation.

Fertilization with nitrogen alone increases the content of stored N forms (such as NO_3^- N, basic amino acids, amines, and amides, and lowers the content of vitamin C in vegetative plant parts. This is more true for NH_4^+ fertilization than for NO_3^- fertilization. The content of α-tocopherol is virtually unaffected [609]. A balanced NPK application, on the other hand, can raise the vitamin C content [610, 611]. This is important above all to vegetable farming, especially with regard to the NO_3^- content arising from fertilizer, since the nitrate intake of human beings is derived about 70 % from vegetables consumed if the NO_3^- content in drinking water is low (<10 mg/L). As a result, countries in Western Europe have adopted guidelines or maximum levels for nitrate in vegetables. Limits are imposed, for example, in the Netherlands, Austria, and Switzerland. The Federal Republic of Germany has established guidelines, e.g., for fresh lettuce and spinach, 3000 and 2000 mg/kg, respectively.

Biodiversity. The diversity of both faunal and floral communities is influenced by agriculture and fertilizer use. On a site-specific local scale, fertilizer use in general and nitrogen fertilization in particular promote the growth of crops more than that of accompanying flora. In grassland, heavy use of nitrogen inhibits the growth of many herbs and legumes.

High amounts of fertilizer salts or anhydrous ammonia may have adverse effects on earthworm population by direct contact. However, the detrimental effect on the total population is very low. The greater supply of fresh organic material that becomes available when soils are raised low to high fertility by fertilization has positive effects on both earthworms and soil microorganisms.

On a global scale the use of fertilizers has positive effects on biodiversity in reducing population pressure for cultivating unsuitable fragile soils, felling of rainforests, and overgrazing, that is, by reducing or preventing soil erosion and degradation.

9.1.5. Pedosphere (Soil)

Nitrogen in cultivated soil (root zone, 0 – 120 cm) is an asset for agriculture, because nitrogen reserves and availability in this region are an important criterion for soil fertility. However, in close-to-natural or largely natural ecosystems, atmospheric N inputs in the forest, heath, nature preserves, lakes, and rivers can lead, over a long period, to a buildup of nitrogen in the soils; the harmless removal of this N (possible only through denitrification) is of concern to ecologists.

9.1.6. Countermeasures

Measures aimed at preventing or overcoming nitrogen problems due to N fertilization in all environmental zones can be divided into direct measures and indirect measures. The aim of both is to minimize the residual nitrogen in the root zone at the end of the vegetation period and hence the leaching potential.

Direct Measures in Fertilization. For arable crops such as sugar beet [550], significant progress has been made in reducing the fertilizer-related nitrogen leaching potential. To provide an economically and an ecologically accurate nitrogen recommendation, the nitrogen supply from soil reserves is assessed by methods such as electroultrafiltration (EUF) or, if not measurable, by empirical estimation. Nitrogen fertilizer is added only to account for the difference between the demand of the crop and the supply from the soil. The guideline for the plant N demand over time is the nitrogen depletion with the harvest. Because of inevitable N losses (denitrification, leaching, etc.), roughly 20 – 30 % extra nitrogen must be supplied over and above the depletion. Optimal distributing systems apply the fertilizer uniformly. Most farmers follow these principles.

The only way to accurately find the actual nitrogen demand is a nitrogen balance for a single farm or field. The ecologically acceptable N level may be higher than, equal to, or lower than the optimal fertilization level: this fact represents a potential incompatibility between economics and ecology. It is very important in this concern, that the new German "Fertilizer Utilization Decree" contains a commitment of farmers to bookkeeping, thus calculating the balances of mineral nutrients at form level.

Farm-Produced Fertilizers. The most urgent nitrogen problem in Western Europe is that the livestock population in large regions has long been above acceptable levels. To comply with clean-air and clean-water requirements, the former liquid manure regulations of some states in Germany, that still permitted using liquid manure of three live-animal units per hectare of agricultural land, an amount still far beyond what is economically beneficial and compatible with the environment, were replaced in 1996 by a federal Fertilizer Utilization Decree with some special requirements for the application of farm-produced manures, including the following:

– Limiting gaseous NH_3 losses by appropriate technical measures, consideration of weather conditions, and the incorporation of liquid manure into unplanted soils immediately after application.
– Using liquid manure in autumn after the harvest of main crops only for subsequent catch or in combination with straw incorporation and only in limited amounts (equivalent to 40 kg of NH_4 N or 80 kg of total N per hectare).
– No liquid manure application in winter from November 15 to January 15.
– Limitation of farm manure application to 170 and 210 kg N per hectare per annum on arable land and grassland, respectively, including nitrogen in animal droppings on pastures. Up to 20 % of total N in manure may be subtracted to compensate unavoidable NH_3 losses during spreading.

Mineral Fertilizers. The matching of N fertilizer application to economic and ecological requirements is much easier with mineral fertilizers than with farm fertilizers, since the mineral products — in contrast to the farm-produced ones — are defined within narrow limits as to N content, N form, and availability.

The nitrogen form is of secondary importance, since NH_4^+ and amide N are usually converted in a few days, perhaps 1 – 2 weeks, to nitrate nitrogen in agricultural soils. Nitrification inhibitors, added to NH_4^+ fertilizers to retard the conversion of NH_4^+ to NO_3^-, may afford an extra measure of safety where there is no acute ground water nitrate problem; however, they are only a partial solution to the problem [612]. (Nitrification inhibitors may promote the liberation of NH_3 from NH_4^+ fertilizers, especially when the fertilizers are not incorporated in the soil [613].)

Controlled-release forms of N may represent an improvement from the ecological point of view when they make it possible to satisfy the crop N demand over time better than can be done with conventional nitrogen fertilizers, and when no substantial amount of nitrogen remains in the soil after the vegetation period.

Indirect Measures. All practices that lead to gains in yield without additional N alleviate environmental nitrogen problems due to fertilization.

Plant Cultivation. The selection of species and varieties and the planning of the entire rotation are crucial factors in reducing economically and ecologically undesired nitrogen losses. The *evergreen system,* that is, keeping fields covered with crop plants year-round, is a desirable goal; whenever possible, the intercrops should moderate N fertilization (nitrogen harvesting). This practice also protects against erosion. Fallow land should be avoided; semifallow area, such as under corn, sunflowers, or summer grains, should be minimized.

An extraordinary reduction of nitrate loss by leaching is achieved by overhead crop irrigation, and, the extreme case, by moderate spraying to avoid frost damage and possibly the loss of a crop in an early stage.

Plant Protection. Plant protection measures, whether mechanical (especially to combat weeds and grass), biological, or chemical, generally increase yield and thus better utilize the nitrogen supply. In this way they indirectly reduce ecologically undesirable nitrogen losses.

Plant Breeding. Nitrogen-efficient plant species and varieties produce more dry matter per unit of nitrogen delivered than do inefficient species.

Soil Tillage. Tillage also has a decisive effect on ecologically undesirable N losses. Tilling as deep as possible, while still preserving soil structure, aids the availability of the nitrogen and reduces the leaching loss by creating the optimal pore volume in combination with an ordered humus regime, thus optimizing the water regime and the supply of water to the crop.

9.2. Phosphorus

Environmental problems connected with phosphate fertilizers are eutrophication and the buildup of heavy metals.

9.2.1. Eutrophication

The pollution of nonflowing or low-circulation surface waters (lakes, estuaries, and coastal zones, especially tidal mud flats) with plant nutrients is to be avoided on account of eutrophication. This term refers to the excessive growth of algae and aquatic plants due to too great a supply of nutrients. Because the putrefaction of these organisms requires a great deal of oxygen, the water becomes depleted in oxygen. Fish die, and the biological purification of the water ceases to function. The lake has *turned over.*

The primary limiting factor for eutrophication is phosphate. Surface waters should contain ≤ 50 µg P per liter. Nitrogen can also become a factor for eutrophication when increased biomass growth takes place.

Figure 46 compares the amounts and main sources of phosphates in surface waters of Germany from 1975 to 1995 [554]. The P input by point sources was markedly decreased by improved sewage treatment practices and by the transition to phosphate-free detergents. Consequently, the share of diffuse sources in the total P input increased from nearly 20 % to about 50 %, although the absolute diffuse input also began to fall from ca. 1990 onwards. Soil erosion and surface runoff from agricultural fields are the most important diffuse sources (Fig. 47).

In contrast to nitrogen, phosphorus pollution via groundwater and drain water is not yet important, although vertical transport of this nutrient may create a pollution problem where the P

110   Fertilizers

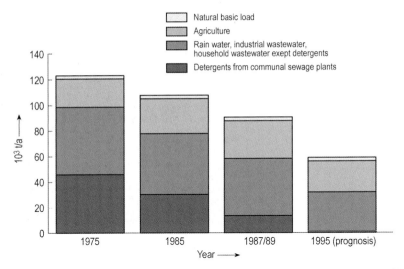

Figure 46. Changes in phosphorus input into surface waters in Germany from 1975 to 1995 [525, 526, 552, 554, 614, 615]

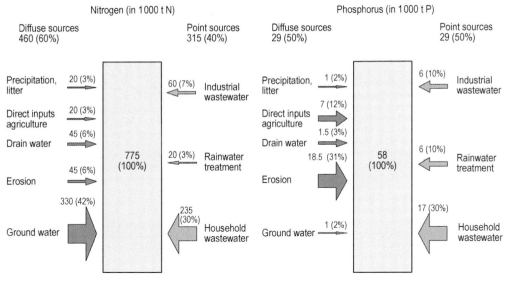

Figure 47. Nitrogen and phosphorus inputs into running waters by diffuse and point sources in Germany: prognosis for 1995 [553]

sorption capacity of deeper soil layers is already highly saturated. This has been reported for areas with high livestock densities in the Netherlands [616] and in north-western Germany [617].

The agricultural contribution to phosphorus in surface waters on a regional scale makes the implementation of efficient countermeasures in agriculture necessary [618, 619]:

– Prevention of erosion by field arrangement (e.g., reducing the slope length), crop management (mulched seeding, underseeding, changes in crop rotation), and soil management (tilling across the slope, improvement of soil structure and infiltration rate).
– Prevention of surface runoff from slopes by omitting the application of mineral or organic fertilizers in autumn and winter.
– Prevention of direct inputs of eroded soil and runoff water by planted margins along the banks of surface waters.
– Consequent adjustment of phosphorus fertilization to plant requirements: supply and re-

moval must be kept in balance on soils with optimum P supply status. On P-accumulated soils, fertilization can temporarily be reduced or even omitted.

In Germany, the implementation of countermeasures is accelerated by the German "Fertilizer Utilization Decree" of 1995.

Besides economic reasons, environmental concerns were also responsible for changes in mineral fertilizer consumption since 1985. As far as phosphorus is concerned, the balance surplus showed a mean annual decrease of 7 % for the period of 1990 – 1995 [551], in particular due to reduced application of mineral P fertilizer (mean application 11.0 kg P/ha in 1995 compared to 25.5 kg P/ha in 1985).

9.2.2. Heavy Metals Buildup

With regard to the impact of fertilizer use on heavy metals buildup in soils, the element cadmium is of primary concern. Fertilizer phosphates produced from phosphate rock contain cadmium [7440-43-9]. The amount depends on the origin of the rock and the digestion process employed. The cadmium content of phosphate rock varies from almost zero to about 700 mg/kg P. The acidification of phosphate rock partitions the Cd between the fertilizer and byproducts of phosphoric acid production such as phosphogypsum.

The Cd input into agricultural land by phosphate fertilizers in Germany was significantly decreased during the last two decades, due to both the preference of low-cadmium phosphate rock for fertilizer production and the dramatic decrease in mean phosphorus application rates. The mean Cd content is phosphorus fertilizers is now reported with about 100 mg/kg P [620]. The mean application rate in 1995/1996 was 11 kg P/ha. Based on these figures, the average annual Cd input is 1.1 g/ha, which compares to about 3.0 g/ha in 1980.

The quantities removed by the crops and by leaching are calculated at about 2.0 g Cd/ha. The background Cd contents in agricultural fields therefore, could not be augmented by mineral fertilizers alone, even if current application rates were doubled [621 – 623]. However, additional cadmium sources must be taken into account.

Atmospheric input from industrial air pollution is the most important source in all industrial countries (3 – 6 g Cd/ha). Manures, sewage sludges, and biowaste composts may be locally or regionally important heavy metal sources [624]. Legal regulations in Germany allow annual maximum heavy metal inputs into soils that are much higher than the realistic inputs via mineral fertilizers. For cadmium these threshold values are 18 g/ha by atmospheric depositions, 10 g/ha by biowaste composts or 16.7 g/ha by sewage sludges.

10. Legal Aspects

Since the beginning of the mineral fertilizer industry, many countries have issued regulations to protect fertilizer users. These rules deal, in particular, with nutrient contents and with safety for plants, human beings, animals, and the environment.

Fertilizer regulations are based on special fertilizer legislation or else form part of a more comprehensive regulatory scheme that governs areas such as animal feeds or pesticides. As a rule, the government department of agriculture has jurisdiction over fertilizers.

The rules on fertilizers extend only to commerce in fertilizers (offering, selling, trading); in only a few cases do they cover other areas (Table 38).

Definition. Laws on fertilizers either contain a definition or simply refer to a list of fertilizers. A definition may be scientifically or pragmatically oriented. While mineral fertilizers present fewer difficulties, the diverse agents used for soil amendment are harder to define.

Examples:

Material, the main function of which is to provide plant food. (ISO)

The term "fertilizers" as used within this Law shall be defined as being any substance applied to the soil for the purpose of supplying nutrients to plants; for producing a chemical change in the soil which will contribute to the cultivation of plants; or which, when applied to the plant, will supply plant nutrients. (Japan)

Fertilizers are substances that are intended to be supplied directly or indirectly to crop plants in order to promote their growth, increase their yield, or improve their quality; excepted are . . . (Federal Republic of Germany)

Table 38. National and international regulations concerning fertilizers

Federal Republic of Germany	Düngemittelgesetz vom 15. November 1977 (*Bundesgesetzblatt I*, p. 2134) Düngemittelverordnung vom 19. November 1977 (*Bundesgesetzblatt I*, p. 2845)
France	*Journal officiel* du 29 juin 1980 (Loi No. 79–595 du 13 juillet 1979 relative à l'organisation du contrôle des matières fertilisantes et des supports de culture ; *Journal officiel* du 14 juillet 1979)
Great Britain	Statuory Instruments 1977 No. 1489 *The Fertilisers Regulations*
Italy	Legge 19 ottobre 1984, n. 748. Nouve norme per la disciplina dei fertilizzanti *Gazetta Ufficiale* 6. Novembre 1984
European Economic Community	Council Directive of 18 December 1975 on the approximation of the laws of the member states relating to fertilizers (76/116 EEC) Council Directive of 22 June 1977 on the approximation of the laws of the member states relating to sampling and methods of analysis of fertilizers (77/535 EEC)
Austria	Bundesgesetz vom 7. November 1985 über den Verkehr mit Düngemitteln, Bodenhilfsstoffen, Kultursubstraten und Pflanzenhilfsmitteln (Düngemittelgesetz – DMG)
Japan	Fertilizer Control Law of Japan. MAFF Ordinance No. 87, Juli 5, 1978
United States	Association of Official Analytical Chemists (AOAC), Washington, DC
International	International Organization for Standardization (ISO), Genève, Switzerland

In the regulations and annexes, fertilizers are systematically classified, for example, into straight and multinutrient fertilizers. The physical form, for example, in liquid fertilizers, may also figure in the classification.

Approval, Registration, Type Lists, Standards. In principle there are two ways of regulating the trade in fertilizers: type approval and individual approval of fertilizers. The European Economic Community (EEC) has chosen the following approach in the interest of harmonization of laws: A type list describes the fertilizers covered (type designation, minimum nutrient contents, expression of nutrients, method of production, essential ingredients, nutrient forms, and nutrient solubilities). Fertilizers that meet these requirements may be marketed provided the labeling rules in all member states are complied with.

Such a system exists in similar form in many of the developed countries. Allowance is made for the fact that many fertilizers are products well known in international trade, such as urea, triple superphosphate, and NPK fertilizers. The administrative cost is low. The type list must be supplemented from time to time to take account of technical development. As a rule, effectiveness must first be proved (plant tests), as must safety for human beings, animals, the soil, and the environment.

Another approach is individual licensing for each fertilizer offered by a manufacturer or importer. An administrative action must precede the sale of any product. The government thus has a complete picture of the registered fertilizers and of the suppliers. Often, approval is granted only for a certain time. Provisional approvals are usually possible.

Because both systems have advantages and drawbacks, many countries prefer a mixed approach: type approval and registration. Standards are defined for fertilizers. Manufacturers and importers must seek registration, which requires stated tests. The certificates granted often have only a limited term.

The choice of approach is dictated, above all, by the system of laws and government in each country, in particular by the degree of supervision desired by the state.

Labeling, Terminology, Packaging. Next to approval, labeling of the product on the package or accompanying documents is the most important rule. The prescribed manner of labeling varies greatly from country to country. Before doing business, it is therefore vital to have an exact knowledge of these regulations. In general, the following information is required as a minimum:

1) type designation or name of the fertilizer
2) guaranteed content of each nutrient
3) name and address of the person responsible for sale
4) weight of product

Special rules apply to trademarks and trade names. There are special rules for packaging and sealing; most of these are restricted to certain fertilizers.

Official Inspections, Sampling, Analysis. As part of the supervision of fertilizer marketing, the sampling and analysis methods to be employed in official inspections are set forth. The variety of methods and their continual development make a selection for official tests necessary. The official checking of the guaranteed nutrient contents must be governed by rules that state whether and to what extent, unavoidable deviations are tolerated. For example, the EEC has adopted definite unified tolerances for each type of fertilizer.

Fertilizer legislation also includes the action to be taken when violations occur (fines, seizure).

Finally, besides the special laws pertaining to fertilizers there are also provisions in other areas of legislation that must be observed in the fertilizer market. The regulations on transportation and storage and the rules aimed at preventing epidemics, important for organic fertilizers, should be kept in mind.

11. Economic Aspects

Consumption, production, and international trade of fertilizers are determined by technical, economic and legal factors, i.e., by output – input efficiencies, factor and product prices, and legal constraints. This applies to the single plot of an individual farm up to the agricultural area of entire nations and continents. While in developing countries consumption of fertilizers is still growing, in developed countries environmental concerns are making themselves felt, leading to legal and economic disincentives against unduly high fertilization rates. However, because of sustained world population growth, world consumption of mineral fertilizers will further increase within the next decades.

11.1. Economics of Fertilization

11.1.1. Input – Output Relationships: The Yield Function

Assessing economic benefits of fertilization is a relatively complex task, since the outputs of crop production processes are determined by numerous input factors, affecting the output level and being interdependent among each other. The precondition for economic considerations, therefore, is quantitative knowledge about (crop) yield functions. Although at present this knowledge is still not complete, it can safely be stated that plant growth obeys basic laws of nature, especially chemical laws. The relevant yield function for a crop — as JUSTUS LIEBIG first showed with his "law of minimum" — can thus be written as:

$$y = \min \{b_1 x_1; \ldots b_j x_j; \ldots b_m x_m\} \quad (1)$$

y = amount of crop output
x_j = supply of necessary inputs
b_j = partial output – input coefficients

The x_j comprise inputs which are controllable by the farmer (seed, fertilizers, pesticides, etc.), as well as those which are not controllable (solar energy, genetic yield potential, water in rainfed agriculture, etc.). The output level y is determined by the input whose supply is minimal compared to all other inputs. Given certain supplies of all but one particular input, the output increases proportionally with increasing supply of this input, until some other input becomes the minimum factor. Further augmentation of the variable input does not lead to any further output; it would be wasted. When a fertilizer is the variable input, the relationship is called a linear response and plateau function (LRP function; Fig. 48) [627].

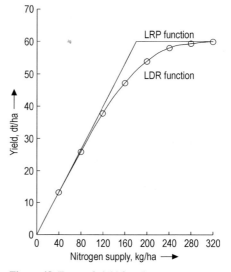

Figure 48. Types of yield function

The LRP function, however, is not compatible with results of fertilizer field experiments, investigated by means of regression analysis [628 – 630, 640]. Usually, these empirical investigations lead to curves shaped according to the "law of diminishing returns" (LDR function; Fig. 48). The difference between the two types of response function can be explained by the fact that available supplies of one or more of the other inputs vary from plant to plant in a plot (e.g., genetic potential, water) [641 – 643].

To simplify further discussion, it shall be assumed that the output (e.g., wheat) is only restricted by two inputs: the variable input x_v (e.g., nitrogen) and a given input x_g, whose amount varies randomly from plant to plant (e.g., genetic potential). Equation (1) then simplifies to:

$$y = \min \{b_g x_g; b_v x_v\} \quad (2)$$

Assuming a discrete distribution for the amounts of x_g, Equation (2) can be expanded to:

$$y = \sum_{i=1}^{n} p_i \min \{b_g x_{gi}; b_v x_v\} \text{ with } \sum_{i=1}^{n} p_i = 1 \quad (3)$$

y = wheat output in dt/ha
i = number of classes in the discrete distribution of the genetic potential
p_i = probabilities of the different classes of the discrete distribution
x_{gi} = genetic potentials of the plants in the different classes of the discrete distribution, measured in dt/ha
x_v = variable supply of nitrogen, measured in kg/ha
b_g, b_v = partial input – output coefficients

If, e.g., $i = 5$, $p_i = [0.1; 0.2; 0.4; 0.,2; 0.1]$, $x_{gi} = [20; 40; 60; 80; 100]$, $b_g = 1$ and $b_v = 0.33$, increased amounts of nitrogen (in steps of 40 kg/ha) result in the plot yields shown as dots in Figure 48. Applying regression analysis leads to the LDR function also depicted. Although each plant grows according to a LRP function, the analysis of this small "experiment" suggests a law of diminishing returns.

The shape of the LDR function depends upon the variance of the spatial distribution of the input x_g within a plot. A smaller variance produces LDR curves that more closely resemble the LRP function. When the variance approaches 0, the LDR function transforms into a LRP function. This relation is especially important because modern plant breeding has reduced genetic variance of varieties (sometimes to zero), and modern land cultivation has homogenized soil conditions.

11.1.2. Factors Controlling the Optimal Nitrogen Fertilization Level

The above relation is of essential importance in determining the economically optimal supply of the variable input nitrogen. Assuming the LDR function of Figure 48, a wheat price of 11 €/dt, a nitrogen price of 0,6 €/kg, and a nitrogen delivery from the soil of 80 kg/ha, the optimal nitrogen fertilization rate can be derived as shown in Figure 49. The farmer tries to maximize the difference between the monetary return (LDR_1) and the nitrogen costs (NC_1). This maximum is obtained at the nitrogen input level (Nop_1) for which the slope of the monetary return curve equals the (constant) slope of the nitrogen cost line, that is, when marginal monetary returns equal marginal nitrogen costs [631].

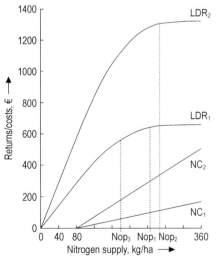

Figure 49. Economically optimal nitrogen supply in crop production, assuming the law of diminishing returns (LDR function)

Changes of output and/or input prices lead to different optimal nitrogen input levels and profitabilities. For example, doubling the wheat price to 22 €/dt results in the monetary return curve LDR$_2$. Tripling the nitrogen price to 1,8 €/kg results in the cost line NC$_2$. The first variation leads to the increased optimal nitrogen supply level Nop$_2$, the second variation to the decreased optimal nitrogen supply Nop$_3$. Generally speaking, increasing crop prices and/or decreasing factor prices result in higher fertilizer consumption, and vice versa.

If, however, the crop yield responds to variable nitrogen input levels according to an LRP function, then the farmer who tries to maximize the profitability would always provide the same nitrogen input level, regardless of price situations. Figure 50 shows monetary return functions for wheat prices of 11 €/dt (LRP$_1$), and 22 €/dt (LRP$_2$). In addition, the nitrogen cost lines, repeated from Figure 49, are depicted. Clearly, for all price situations the farmer would try to ensure the same optimal nitrogen input level Nop$_1$.

In modern agriculture, the yield functions are approaching LRP functions, so that price variations have less and less impact on levels of fertilization.

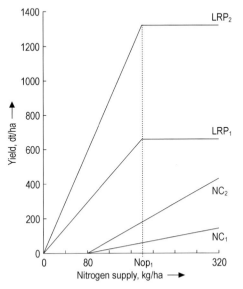

Figure 50. Economically optimal nitrogen supply in crop production, assuming linear response and plateau functions (LRP functions)

11.1.3. Factors Influencing the Optimal Nitrogen Fertilization Level

As can easily be derived from Equation (1), other factors, especially the levels of noncontrollable inputs for plant growth, exert a substantial influence on the optimal nitrogen input level. Firstly, the soil and climate conditions of a particular piece of land play a major role. Increased availabilities of solar energy and/or plant-available water allow for more productive fertilizer consumption, returning higher yields. As an example, Figure 51 shows the nitrogen response functions LRP$_1$ and LRP$_2$ for two parcels of land with different water conditions. The higher optimal yield y_2 of the "better" parcel can be obtained only if the nitrogen supply is increased from Nop$_1$ to Nop$_2$. Water and nitrogen are complementary inputs with respect to crop yield.

Secondly, modern plant breeding successively produces varieties with higher yield potentials, enabling the plants to consume water and/or solar energy more effectively. This, of course, leads to higher yields, provided the fertilizer input level is increased appropriately. Figure 51 may be also interpreted in terms two varieties of different productivity.

Thirdly, modern crop-protection agents keep the plants healthy during the vegetation period, enabling them to consume more solar energy and water. This again leads to higher yields, provided the fertilizer input level is adjusted accordingly. The situation with and without the application of modern pesticides can also be represented by the LRP functions of Figure 51.

Under a wide range of price situations, the farmer would make full use of favorable soil and climatic conditions, as well as of the advantages of improved seeds and crop-protection agents. As a rule, the additional monetary returns of the increased yields are much higher than the additional costs for fertilizers, seeds, and pesticides.

11.1.4. Environmental Aspects of Fertilization

Ecological concerns with respect to fertilization are becoming more and more important. They arise from the fact that farmers apply more fertilizers than may be consumed productively by the crops. Depending on soil conditions, some part of the surplus will not be stored in the soil to be used by the following crop, but will be leached into the underground, eventually contaminating the ground water.

However, even state-of-the-art fertilization causes some nitrogen surplus. This is due mainly to two facts: First, as discussed above, the levels of certain noncontrollable plant growth inputs are distributed randomly over a plot of land. Since the farmer does not know the exact locations, he will fertilize the whole field evenly or, in case of precision agriculture methods, at least larger parts of it. He will try to provide a nitrogen level such that the average marginal costs for the plot are equal to the average marginal monetary returns. Unavoidably, there will always be some low-yielding spots which receive too much fertilizer.

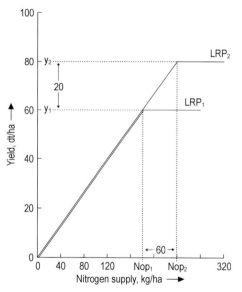

Figure 51. Effects of increased yield potentials on nitrogen requirement

Second, under most climatic conditions for rainfed agriculture, plant-usable water supply varies from year to year. Since the farmer does not know the water supply of a particular year in advance, he will always "shoot" for a "good" year. If, for example, in a good year the higher yield y_2 in Figure 51 is achievable, and in a "dry" year only the lower yield y_1, the farmer will nevertheless always provide for the higher nitrogen level Nop_2 instead of the lower level Nop_1.

The reason is simple: the expected profitability value of strategy Nop_2 is under realistic price and weather conditions much higher than that of the strategy Nop_1.

If the farmer provided only for the low fertilization level Nop_1, in a good year he would lose the monetary return of 20 dt/ha (see Fig. 51), because the insufficient nitrogen supply would limit the yield. At a wheat price of 11 €/dt, this corresponds to a loss of 220 €/ha. At the higher nitrogen level Nop_2, in a dry year 60 kg/ha nitrogen would be wasted (Fig. 51). This corresponds to a loss of 36 €/ha.

These economically induced fertilizer surpluses can be reduced by further improvements of inputs, production techniques, and prognostic information. Seed varieties with less genetic variance and homogenization of soil conditions lead to less variance of the noncontrollable growth factors within a field. Splitting nitrogen fertilization into several applications as the vegetation period develops and new information about the water supply becomes available also reduces waste of nutrients. Finally, improved long-term weather forecasts would be most valuable in reducing the uncertainty involved in fertilizing decisions.

Strict legal constraints for nitrogen application rates, however, would not be the strategy of choice. It would lead to a waste of valuable yield potential and in addition to economic losses for farmers.

11.2. World Consumption, Production, and Trade

Consumption. World consumption of the three primary nutrients — nitrogen, phosphate and potash — reached a total of 135.0×10^6 t in 1996/97. The previous record level of 145.6×10^6 t in 1988/89 was once again not reached. The consumption figures broke down as follows [625] in 10^6 t:

Nitrogen	82.9 N
Phosphate	31.1 P_2O_5
Potash	21.0 K_2O

Over the past 10 years, agricultural consumption of fertilizers experienced a deep recession, mainly due to developments in Central and Eastern Europe, as well as in the former Soviet

Union. World consumption fell by 17% from 1988/89 to 1993/94 and then recovered by 12% up to 1996/97. Over the ten year period, nitrogen increased by 16%, phosphate fell by 10%, and potash fell by 19%.

The consumption ratio $N : P_2O_5 : K_2O$ was 1 : 0.6 : 0.5 in 1976/77; it changed in favor of nitrogen to 1 : 0.5 : 0.4 in 1986/87, and in 1996/97 it reached 1 : 0.4 : 0.25. While in North America and Western Europe the ratio is 1 : 0.4 : 0.4, it is less well balanced in other regions of the world (Table 39). Fertilizer consumption and production are given for fertilizer years, running from 1 July to 30 June. For example, 1996/97 is the period from 1 July 1996 to 30 June 1997.

In case only calendar year figures are available, e.g., 1996 is summarized under 1996/97. A regional analysis of fertilizer consumption shows that over 60% is used in developing countries (in Asia, Latin America, Africa). The Asian markets are particularly important. About 50% of all fertilizers consumed are applied in Asia. North America takes second place with 17%, followed by Western Europe. Three decades ago, Western Europe was the world's largest fertilizer consumer; after 10 years, the Eastern Bloc had overtaken Western Europe; and another 10 years later, the developing countries have taken the lead — having increased their consumption tenfold in 30 years. This change corresponds to a mean annual growth rate of 8%.

The intensity of fertilizer use, however, is still greatest in Western Europe. In terms of nutrient per hectare of agricultural area, Western Europe averaged 121 kg applied in 1996/97 (Table 40). The corresponding figure for the developing countries is 24 kg. In these countries, however, virtually no fertilizer is applied to pastureland. If the figures are referred to agricultural area minus pastureland, the rate in the developing countries was 88 kg of nutrients per hectare, 56 kg of this being N. Fertilizer application is extremely low in the former Soviet Union (8 kg/ha).

Three countries account for over half of world fertilizer consumption: China with 36.6×10^6 t, the United States with 21.2×10^6 t, and India with 14.3×10^6 t. Next on the list are India with 14.3×10^6 t, France with 5.1×10^6 and Brazil with 4.8×10^6.

In years to come, a further worldwide increase is expected. This growth will take place chiefly in Asia, Latin America and Africa. The recovery of fertilizer use in Central/Eastern Europe and the former Soviet Union is anticipated but will take time.

Production. A survey of production in 1996/97 showed that Asia ranks first as world fertilizer producer, mainly for nitrogen, but also for phosphate fertilizers, followed by North America, the leading potash producer (Table 41).

Over the past ten years the production pattern by region shows a considerable decrease in Western and Central/Eastern Europe (-27% and -37%, respectively) but above all in the former Soviet Union (-50%). In North America ($+27\%$) and especially Asia ($+61\%$) production increased markedly. While overall world production remained nearly stagnant ($+3.5\%$), nitrogen production increased by 18%, but phosphate and potash dropped by 8% and 16%, respectively.

The leaders in production of fertilizers are China, the United States, and India. They are followed by Canada, the world's largest potash producer, and Russia (Table 42).

The rise in fertilizer production and consumption in recent decades could not have taken place without an enormous technical advance. Significant cost savings in ammonia production, the intermediate for nitrogen fertilizer production, resulted from the changeover from multi-train to single-train plants and the conversion to more economical feedstocks, mainly natural gas.

Due to improved product quality and well-balanced particle size spectrum, rational bulk transportation and easier application become possible. Higher analysis fertilizer grades brought savings in shipping and storage per unit of nutrient.

The fertilizer processes were optimized to decrease energy consumption. Emission reduction with modern techniques improved the environmental performance of the processes and at the same time increased the nutrient yield from the raw materials. These improvements are of great importance: the 149×10^6 t of nutrients produced worldwide in 1996/97 meant 430×10^6 t of fertilizers that had to be stored, transported, and applied (authors estimate).

Table 39. World nitrogen, phosphate and potassium consumption in millions of tonnes over the period 1976/77 – 1996/97 broken down by region/country [625] *

Region/country	Fertilizer season					
	1976/77		1986/87		1996/97	
	10^6 t	%	10^6 t	%	10^6 t	%
Nitrogen (N) fertilizer consumption						
Western Europe	8.8	19.0	11.4	15.9	10.1	12.2
Central/Eastern Europe	3.7	8.0	4.4	6.2	2.4	2.9
Former Soviet Union	7.3	15.7	11.5	16.1	2.8	3.4
North America	10.3	22.2	10.4	14.5	12.9	15.6
Latin America	2.2	4.7	3.8	5.3	4.1	4.9
Asia **	12.4	26.7	27.7	38.7	47.4	57.2
Africa	1.5	3.2	1.9	2.7	2.3	2.8
Other	0.2	0.4	0.4	0.6	0.9	1.1
World	46.4	100.0	71.5	100.0	82.9	100.0
Phosphorus (P_2O_5) fertilizer consumption						
Western Europe	5.8	21.6	5.3	15.3	3.7	11.9
Central/Eastern Europe	2.7	10.1	2.0	5.8	0.7	2.3
Former Soviet Union	4.9	18.3	9.1	26.2	0.9	2.9
North America	5.6	20.9	4.3	12.4	4.8	15.4
Latin America	2.0	7.5	2.8	8.1	2.9	9.3
Asia **	3.8	14.2	9.2	26.5	15.7	50.5
Africa	0.9	3.4	1.0	2.9	1.0	3.2
Other	1.1	4.1	1.0	2.9	1.4	4.5
World	26.8	100.0	34.7	100.0	31.1	100.0
Potassium (K_2O) fertilizer consumption						
Western Europe	5.5	5.7	5.9	22.6	4.3	20.5
Central/Eastern Europe	2.9	3.0	2.8	10.7	0.7	3.3
Former Soviet Union	5.6	5.8	6.7	25.7	1.2	5.7
North America	5.5	5.7	4.8	18.4	5.2	24.8
Latin America	1.2	1.2	1.9	7.3	2.6	12.4
Asia **	1.8	1.9	3.4	13.0	6.2	29.5
Africa	0.3	0.3	0.4	1.5	0.4	1.9
Other	0.2	0.2	0.2	0.8	0.4	1.9
World	23.0	23.9	26.1	100.0	21.0	100.0
Total fertilizer consumption ($N + P_2O_5 + K_2O$)						
World	96.2		132.3		135.0	

* FAO Fertilizer Yearbook 1980, 1990 and 1997.
** Excluding Asian republics of former Soviet Union.

Table 40. Nutrient use per hectare of agricultural area, [625, 632]

Region	Nutrient application, kg/ha			
	N	P_2O_5	K_2O	Total
Western Europe	67.6	24.7	28.8	121.2
Central/Eastern Europe	36.1	10.6	9.9	56.6
Former Soviet Union	4.8	1.5	2.1	8.4
North America	26.0	9.8	10.5	46.4
Other	15.2	5.8	2.7	23.6
World average	16.9	6.3	4.3	27.5

a Consumption: FAO Fertilizer Yearbook, Agricultural Area: FAOSTAT Database results

World Trade. Recent decades have seen a sharper rise in world trade than in consumption. Because of the limited number of potash deposits, the ratio of trade to consumption is highest for potash, with 91 %. Some 40 % of phosphate and 29 % of nitrogen are sold internationally (Table 43).

In nitrogen, Russia and the USA have the largest exports, but Canada, Ukraine, and the Netherlands are also important in the world market. The most important importing regions are China and the USA.

American exporters dominate the phosphate market, with sales in all regions, but primarily in the developing countries in Asia, Latin America

Table 41. World fertilizer production by nutrient and region 1996/97 [625]

Region	Nutrient production, 10^6 t							
	N		P_2O_5		K_2O		Total	
	1986/87	1996/97	1986/87	1996/97	1986/87	1996/97	1986/87	1996/97
Western Europe	11.8	9.5	4.4	2.9	8.3	5.4	24.5	17.8
Central/Eastern Europe	6.1	4.8	2.8	0.9	0.0	0.0	8.9	5.6
Former Soviet Union	15.0	8.7	8.5	2.5	10.2	5.5	33.7	16.7
North America	13.6	19.3	9.4	11.3	8.2	8.9	31.2	39.5
Latin America	3.0	3.2	1.9	1.8	0.1	0.4	5.0	5.4
Asia *	26.0	42.6	7.7	11.5	2.0	3.2	35.7	57.4
Africa	1.7	2.7	1.9	2.5	0.0	0.0	3.6	5.2
Oceania	0.2	0.3	0.8	0.7	0.0	0.0	1.0	1.0
World	77.4	91.1	37.4	34.1	28.8	23.4	143.6	148.6
World supply **	72.6	85.8	34.9	32.2	26.1	21.8	133.6	139.7

* Excluding Asian republics of former Soviet Union.
** Available world supply was arrived at by deducting from production estimated amounts for technical uses, further processing, transport, storage, and handling losses.

Table 42. Fertilizer production by country, 1996/97 [625]

Country	Fertilizer production, 10^6 t			
	N	P_2O_5	K_2O	Total
China	20.1	5.8	0.2	26.1
United States	12.5 *	10.9	0.8	24.2
India	8.6	2.6	0.0	11.2
Canada	2.7 *	0.4	8.1	11.2
Russia	4.9	1.6	2.6	9.1

* Author's estimate (available supply)

Table 43. World fertilizer trade in millions of tonnes (exports) (10^6 t) [625] *.

Nutrient	Fertilizer season		
	1976/77	1986/87	1996/97
N	8.5	17.4	24.2
P_2O_5	4.4	9.5	12.4
K_2O	13.7	17.6	19.1
Total	26.6	44.5	55.7

* FAO Fertilizer Yearbooks

and Africa. The largest exporters of potash are Canada and Germany, followed by Belarus and Russia.

The largest exporters and importers in 1996/97 are listed in Table 44.

Types of Fertilizers. As the quantities produced and applied have increased, the importance of fertilizer grades has changed. Low-analysis grades have decreased, while highly concentrated grades such as urea, triple superphosphate, and NPK fertilizers have gained (Fig. 52). Sulfur-containing fertilizers have gained importance to cope with sulfur deficiencies.

Urea has become the preferred nitrogen fertilizer, especially in the developing countries. In Western Europe, for example, ca. 80 % of nitrogen consumption is in the form of ammonium nitrate, calcium ammonium nitrate, and ammonium nitrate-based multinutrient fertilizers. Phosphate consumption is dominated by multinutrient fertilizers. In the case of potash, straight potassium chloride is the most important grade, but potassium sulfate is of growing importance due to its sulfur contents.

11.3. Future Outlook

The decisive factors for the development of fertilizer demand are population growth and changes in the available income per capita. A growing world population needs more food and fiber. In addition, a wealthier population has a

Table 44. Major exporting and importing countries for nitrogen, phosphate, and potash fertilizers, 1996/97 [625]

Country	Exports, 10^6 t	Country	Imports, 10^6 t
Nitrogen (N)			
Russia	3.6	China	4.7
United States	3.0	United States *	2.0
Ukraine	1.2	France	1.3
Netherlands	1.2	India	1.2
Canada *	1.3	Germany	1.2
Phosphate (P_2O_5)			
United States	5.7	China	2.8
Russia	1.2	India	0.7
Morocco	0.8	Australia	0.7
Tunisia	0.7	France	0.6
Netherlands	0.4	Thailand	0.5
Potassium (K_2O)			
Canada	7.9	United States	5.2
Germany	2.6	Brazil	1.8
Belarus	2.0	France	1.1
Russia	1.9	India	0.7
Israel	1.2	Malaysia	0.6

* Author's estimates.

higher daily energy intake per capita and it generally consumes a higher share of animal products (Table 45). Compared to direct human consumption, three to four times more plant products such as cereals are necessary to produce the same amount of energy in the form of animal products like milk, eggs, and meat.

11.3.1. Food Situation

Since the 1950s world food production per capita has continuously increased, although world population has more than doubled (from 2.5 to almost 6.0 billion). New crop varieties and improved production techniques enabled a substantial growth of average yields. A significant advance was the broader use of effective cereal fungicides in industrialized countries and the introduction of high-yielding varieties of rice, wheat, and maize in several developing countries, known in the 1970s as the "green revolution". These varieties, bred for tropical and subtropical regions, already offer a higher yield potential under traditional growing conditions. But their full potential can only be exploited when crop husbandry and plant nutrition are adapted to their needs, and agrochemicals provide protection against insects, diseases, and competition from weeds. In many cases they rendered the use of fertilizers and agrochemicals economic for the first time. The average yields of cereals, the most important staple food of humans, increased substantially (Table 46).

According to the FAO, on average each kilogram of fertilizer nitrogen applied to the soil produces about 12 kg of cereal units (1 kg of cereal unit is equal to the nutritional value of 1 kg of barley); therefore, it can be calculated that the food for half of the world's population can only be provided due to the use of $(75-80) \times 10^6$ t nitrogen in mineral fertilizers annually. Nevertheless, 840 million individuals still suffer from undernutrition and an estimated 20 million die annually of hunger and undernutrition-related diseases. Civil wars, natural disasters, and poverty, often related to unemployment, are the main reasons for starvation.

Growth in Food Demand. The 1996 update of the UN World Population Prospect shows that the rate of world population growth has declined to 1.48 % per annum, but this still means an additional 80 million humans to be fed every year. By the year 2020 world population is predicted to reach 7.67 billions. Since economic growth is expected to continue, a larger part of the population will be able to spend more money on food. Thus an increased demand per capita for plant products to be used as animal feed must be taken into account. The demand for cereals is expected to increase by 41 % from 1993 to 2020, that for meat by 63 %, and the consumption of tubers and root is estimated to rise by 40 % [634].

Table 45. Dietary energy supply per capita by economic group 1969 – 1971 and 1990 – 1992 [633]

Country Group	Total kcal/d		% animal	
	1969 – 71	1990 – 92	1969 – 71	1990 – 92
Industrialized	3120	3410	30.4	29.8
Transition	3330	3230	24.4	27.6
Low-income	2060	2430	6.1	9.3
Least developed	2060	2040	6.6	6.0
World	2440	2720	15.6	15.7

Table 46. World production and average yields of cereals 1972 – 1974 and 1996/97

	Production, 10^6 t		Yield, t/ha	
	1972 – 74	1996 – 97	1972 – 74	1996 – 97
Wheat	359	581	1.66	2.53
Maize	300	562	2.75	4.02
Barley	164	155	1.88	2.33
Rye, Oats	81	54	1.68	1.85
Rice	325	558	3.23	3.76
Total	1229	1910	2.19	3.11

Agricultural Land Use. More than 70 % of the world's surface is covered by water. Of the 13.4 billion hectares of dry land, 4.7 billion hectares are used by agriculture. The major portion is covered by extensively used grasslands, often in low-rainfall areas. Most of the food is supplied by the 1.4 billion hectares of arable land — a mere 10 % of the dry land. Much of the remaining area is not suitable for intensive agricultural production since it is either too dry, too steep, too cold, or infertile. The majority of soils usable for arable crops are covered by biotopes such as rain forests. There is no agreement to which extent they should be used for agriculture and how much must be preserved. FAO [635] calculated that Sub-Saharan Africa and Latin America have the most land reserves with crop-production potential.

At the same time 7 million hectares agricultural land are lost annually due to erosion and salination. The agricultural area available per capita declined from 0.44 ha (1960) to 0.27 ha (1990) and is expected to shrink further to 0.17 ha by 2020. Therefore FAO estimates that although newly cleared land may supply 21 % of the necessary increase in food production, and more frequent use of existing arable land 13 %, the most important contribution (66 %) must come from higher yields on soils already under cultivation. Although neither the quality of soils nor the agroclimatic conditions will allow the same high yield to be achieved everywhere, the wide variation in actual yields offers a vast potential for further improvement (Figure 53).

Intensity of Agricultural Production. The intensity of agricultural production is highest in industrialized and transition economies, where the purchasing power for food is high. Attractive markets create an efficient production system, whereas the subsistence farmer cannot invest in yield-raising inputs. Annual nutrient application rates therefore vary widely (Figure 53). Western European countries, Egypt, Saudi Arabia, and Mexico for example apply on average more than 200 kg N + P_2O_5 + K_2O per hectare wheat and harvest 4 – 7 t of grain, whilst countries using less than 50 kg plant nutrients obtain barely more than 2 t per hectare [636]. This of course can also be the result of unfavorable growing conditions, primarily insufficient supply of water, but for most countries it reflects the socio-economic situation.

Wheat is the crop which receives most (22 %) of the fertilizer nutrients, followed by rice and maize (17 % each). Cereals in total count for 65 % of the nutrient consumption, while 9 % are used on oil crops, 5 % each on vegetables and sugar beet/cane). Thus, fertilizers are predominately used for food production. Fiber crops, tobacco, and stimulants receive 4 and 2 %, respectively [636].

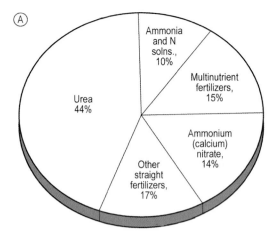
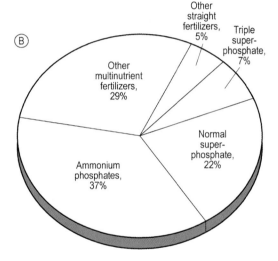
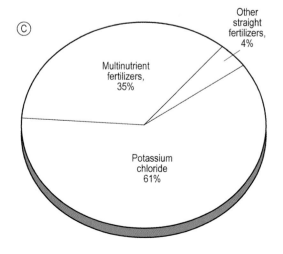

Figure 52. World fertilizer consumption by type, 1996/97 preliminary [632]

A) Nitrogen fertilizer; B) Phosphate fertilizer; C) Potassium fertilizer

In the past countries with economic and population growth have increased their fertilizer consumption (Table 47), whereas those with traditionally high consumption only show moderate growth (e.g., United States, Canada, Spain) or a decrease in consumption. This is the case for most EU countries, where nutrient levels in the soil have mostly been built up to the desirable level, environmental concern has led to strict regulations, and changes in agricultural policy, such as leaving arable land fallow, have been introduced. Changes of the economic system in the former centrally planned economies resulted in a dramatic decline in agricultural production and consequently in fertilizer consumption.

Table 47. Development of fertilizer use in the 20 countries with the highest actual nutrient consumption 1985/86 – 1995/96 in 10^6 t N + P_2O_5 + K_2O [637]

	1985/86	1995/96
China	16 852	35 527
USA	17 831	20 113
India	8504	13 876
France	5695	4915
Brazil	3197	4309
Germany	4823	2818
Indonesia	1972	2512
Pakistan	1511	2443
Canada	2325	2436
UK	2524	2264
Italy	2162	1883
Spain	1734	1817
Australia	1155	1735
Russian Fed.		1700
Turkey	1427	1700
Japan	2034	1642
Poland	3413	1511
Vietnam	386	1448
Thailand	434	1443
Malaysia	611	1247
World	128 613	130 865

11.3.2. Development of Fertilizer Consumption

World fertilizer consumption will continue to rise by more than 2 % per annum from 1996/97 to 2002/03 [638, 639]. The rate of growth will be highest in the densely populated countries of South Asia (+ 5.2 %), in Latin America (+

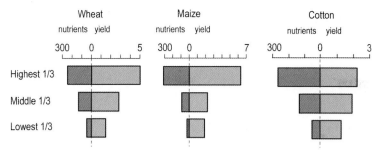

Figure 53. Average nutrient application rates (kg N + P$_2$O$_5$ + K$_2$O per hectare) and average yields (t/ha) of countries with high, medium, and low fertilizer application rates in 1995 [636]

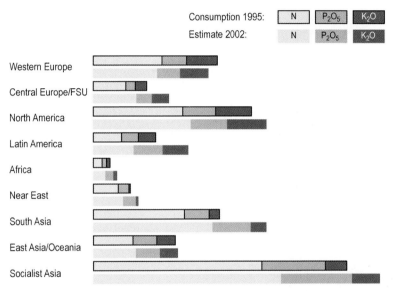

Figure 54. Fertilizer consumption forecast by region 1995 – 2002 [638]

5.0 %), Africa (+ 4.9 %) and Central Europe (+ 4.4 %) (Figure 54). The estimated increase in Central Europe signals a recovery from economic depression, while the growth rate in Africa simply reflects the low base. For Western Europe, a further decline in fertilizer consumption is forecast since the agricultural area in this densely populated, industrialized region will continue to decline, and more efficient use of organic manures (farmyard manure, sewage sludge, etc.) will be necessary to achieve the goals set for the protection of the environment.

12. References

1. K. Scharrer in, W. Ruhland (ed.): *Handbuch der Pflanzenphysiologie*, vol **IV**, Springer Verlag, Berlin-Göttingen-Heidelberg 1958, pp. 851 – 866.

Specific References

2. J. N. Burnell in R. D. Graham et al. (eds.): *Manganese in Soil and Plants*, Academic Press, London 1988, pp. 125 – 137.
3. H. Marschner: *The Mineral Nutrition of Higher Plants*, Academic Press, London 1993.

4. K. H. Falchuk, L. Ulpino, B. Mazus, B. L. Valee, *Biochem. Biophys. Res. Commun.* **74** (1977) 1206 – 1212.
5. M. G. Guerrero, J. M. Vega, M. Losada, *Ann. Rev. Plant Physiol.* **32** (1981) 169 – 204.
6. S. J. Brodrick, K. E. Giller, *J. Exp. Bot.* **42** (1991) 1339 – 1343.
7. J. v. Liebig: *Die organische Chemie in ihrer Anwendung auf Agrikultur and Physiologie*, Vieweg, Braunschweig 1841.
8. G. Hehl, K. Mengel, *Landwirtsch. Forsch. Sonderh.* **27** (1972) no. 2, 117 – 129.
9. A. K. Goswami, J. S. Willcox, *J. Sci. Food Agric.* **20** (1969) 592 – 595.
10. S. Haneklaus, E. Evans, E. Schnug, *Sulphur in Agriculture* **16** (1992) 31 – 34.
11. H. W. Scherer, A. Lange, *Biol Fertil Soils* **23** (1996) 449 – 453.
12. E. Saalbach, *Angew. Bot.* **58** (1984) 147 – 156.
13. E. H. Birnbaum, W. M. Dugger, B. C. A. Beasley, *Plant Physiol.* **59** (1977) 1034 – 1038.
14. H. E. Goldbach, *J. Trace Microprobe Techn.* **15** (1997) 51 – 91.
15. W. D. Loomis, R. W. Durst, *Biofactors* **3** (1992) 229 – 239.
16. G. K. Judel, *Landwirtsch. Forsch. Sonderh.* **34** (1977) no. 2, 103 – 108.
17. D. Werner, R. Roth, *Inorganic Plant Nutrition*. Encycl. Plant Physiol., New Ser. 15 B (1983) 682 – 694.
18. Y. Miyake, E. Takahashi, *Soil Sci. Plant Nutr. (Tokyo)* **29** (1983) 463 – 471.
19. A. D. M. Glass, M. Y. Siddiqi in P. B. Tinker, A. Läuchli (eds): *Advances in Plant Nutrition*, Praeger, New York 1984, vol. 1, pp. 103 – 147.
20. R. G. Wyn Jones, A. Pollard, *Inorganic Plant Nutrition*. Encycl. Plant Physiol. New Ser. 15 B, (1983) 528 – 562.
21. K. Mengel in R. D. Munson (ed): *Potassium in Agriculture*, American Society of Agronomy, Madison, Wis. 1985, pp. 397 – 411.
22. T. R. Peoples, D. W. Koch, *Plant Physiol,* **63** (1979) 878 – 881.
23. D. T. Clarkson, J. B. Hanson, *Annu. Rev. Plant Physiol.* **31** (1980) 239 – 298.
24. R. F. Hüttl, H. W. Zöttl, *Forest Ecol. Management* **61** (1993) 325 – 338.
25. C. R. Caldwell, A. Haug, *Physiol. Plant.* **54** (1982) 112 – 118.
26. J. B. Hanson in P. B. Tinker, A. Läuchli (eds): *Advances in Plant Nutrition*, Praeger, New York 1984, vol. 1, pp. 149 – 208.
27. C. B. Shear, *Hort. Science* **10** (1975) 361 – 365.
28. H. Marschner, *Neth. J. Agric. Sci.* **22** (1974) 275 – 282.
29. L. C. Campell, R. O. Nable in R. D. Graham, R. J. Hannan, N. C. Uren (eds.): *Manganese in Soils and Plants*, Kluwer Academic Press, Dordrecht 1988.
30. S. Spiller, N. Terry, *Plant Physiol.* **65** (1980) 121 – 125.
31. K. Mengel, G. Geurtzen, *J. Plant Nutr.* **9** (1986) 161 – 173.
32. J. C. G. Ottow, G. Benckiser, I. Watanabe, S. Santiago, *Trop. Agric. (Trinidad)* **60** (1983) 102 – 106.
33. K. Mengel, E. A. Kirkby: *Principles of Plant Nutrition*, 4th ed., International Potash Institute, Berne 1987.
34. D. Schroeder: *Soils – facts and concepts*, International Potash Institute, Berne 1984.
35. N. C. Brady: *The Nature and Properties of Soils*, Macmillan Publishing Co., New York 1984.
36. F. Scheffer, P. Schachtschabel: *Lehrbuch der Bodenkunde*, 14th ed., Enke Verlag, Stuttgart 1998.
37. R. Dudal in *SEFMIA Proc. Intern. Seminar on Soil Environment and Fertility Management in Intensive Agriculture*, Society of Science of Soil and Manure of Japan, Tokyo 1977, pp. 78 – 88.
38. Soil Taxonomy of the Soil Conservation Service, U.S. Department of Agriculture, Handbook no. 436.
39. E. N. Gapon, *J. Gen. Chem. USSR* **3** (1933) 144 – 152.
40. R. J. Haynes, *Adv. Agron.* **37** (1984) 249 – 315.
41. R. L. Parfitt, *Adv. Agron.* **30** (1978) 1 – 50.
42. C. D. Foy, R. L. Chaney, M. C. White, *Annu. Rev. Plant Physiol.* **29** (1978) 511 – 566.
43. C. E. Kallsen, T. W. Sammis, E. J. Gregory, *Agron. J.* **76** (1984) 59 – 64.
44. R. L Tate, *Soil Microbiology*, Wiley & Sons, New York 1995.
45. K. Mengel, *Adv. Soil Sci.* **2** (1985) 65 – 131.
46. G. L. Terman, *Adv. Agron.* **31** (1979) 189 – 223.
47. R. Vandré, J. Clemens, *Nutrient Cycling in Agrosystems.* **47** (1997) 157 – 165.
48. R. J. Buresh, S. K. DeDatta, *Adv. Argon.* **45** (1991) 1 – 59.
49. S. K. DeDatta, R. J. Buresh, *Adv. Soil Sci.* **10** (1989) 143 – 169.
50. N. K. Savant, S. K. DeDatta, *Adv. Agron.* **35** (1982) 241 – 302.
51. J. H. G. Slangen, P. Kerkhoff, *Fert. Res.* **5** (1984) 1 – 76.

52. L. Hendriks, N. Claassen, A. Jungk, *Z. Pflanzenernaehr. Bodenkd.* **144** (1981) 486 – 499.
53. P. H. Nye, *Adv. Agron.* **31** (1979) 225 – 272.
54. M. Silberbush, S. A. Barber, *Agron. J.* **75** (1983) 851 – 854.
55. C. A. Neyra, J. Döbereiner, *Adv. Agron.* **29** (1977) 1 – 38.
56. V. Römheld, H. Marschner, *Adv*, Plant Nutr. 2 (1986) 155 – 204.
57. R. Perrin, *Soil Use Manag.* **6** (1990) 189 – 195.
58. H. Marschner, B. Dell in A. D. Robson et al. (eds.): *Management of Mycorrhizas in Agriculture, Horticulture and Forestry*, 1994 , 89 – 102.
59. A. Finck: *Dünger und Düngung - Grundlagen und Anleitung zur Düngung der Kulturpflanzen*, VCH Verlagsgemeinschaft, Weinheim, Germany 1992.
60. A. L. Page, R. H. Miller, D. R. Keeney: *Methods of Soil Analysis - Chemical and microbiological properties*, 2nd ed., **part 2**, American Society of Agronomy, Inc., Madison, WI 1982.
61. E. Sibbesen, *J. Sci. Food Agric.* **34** (1983) 1368 – 1374.
62. G. Hoffmann, *Methodenbuch*, vol. **1, Untersuchung von Böden**, VDLUFA-Verlag, Darmstadt 1991/1997.
63. K. Németh, *Plant and Soil* **83** (1985) 1 – 19.
64. K. Németh, J. Maier, K. Mengel, *Z. Pflanzenernaehr. Bodenk.* **150** (1987) 369 – 374.
65. A. Link, Ph. D. Thesis, University of Hannover, Grauer Verlag Stuttgart 1997.
66. R. J. Poole, *Annu. Rev. Plant Physiol.* **29** (1978) 437 – 460.
67. E. A. Mitscherlich: *Bodenkunde für Landwirte, Forstwirte und Gärtner*, 7th ed., Verlag Paul Parey, Berlin – Hamburg 1954, p. 168.
68. S. P. Ceccotti, D. L. Messick, *Agrofood Industry hi-tech* **5** (1994) 9 – 14.
69. A. Vömel, *Z. Acker Pflanzenbau* **123** (1965/1966) 155 – 188.
70. U. Köpke: "Optimized Rotation and Nutrient Management in Organic Agriculture," in H. vanKeulen, E. A. Lantinga, H. H. van Laar (eds.): *Proc. Workshop Mixed Farming Systems in Europe*, Dronten, The Netherlands 1998.

General References

71. *Guidelines of the Council of the European Communities for the Harmonization of Legal Provisions in the Member States Concerned with Fertilizers*, 1975.

Specific References

72. J. Norden, *Raiffeisen* 1979, 50 – 55.
73. F. P. Achorn, W. C. Scott, J. A. Willbanks, *Fert. Solution* **14** (1970) 40.
74. V. Buhlmann, *Vereinigte Landwesenkaufleute* 1980, no. 3, 20, 22, 24.
75. J. Norden, *DIZ* (1978) 256 – 260, 512 – 515.
76. *Farm Chem.* **137** (1974) no. 10, 10.
77. T. P. Hignett, *Chem. Technol.* **2** (1972) 627 – 637.
78. T. P. Hignett (ed.): *Fertilizer Manual*, Martinus Nijhoff, Dr. W. Junk Publishers, Dordrecht – Boston – Lancaster 1985, p. 129.
79. D. M. Sloan, R. W. Veales, *J. AOAC* **60** (1977) 876.
80. R. Kümmel, *Chem. Techn. (Leipzig)* **25** (1973) 732.
81. W. R. Grace, NL-A 6 414 487, 1964.
82. *Derakane-Vinylesterharze*, DOW Chemical Europe (1977) Nov., p. 6.
83. *Handbuch der chemischen Beständigkeit der Dynamit Nobel AG*, 1982, p. 60.
84. W. von Beackmann, D. Funk, *3R, Rohre Rohrleitungsbau Rohrleitungstransp.* **17** (1978) 443.
85. L. G. Croysdale, W. E. Samuels, J. A. Wagner: "Urea – Ammonium Nitrate: Are They Safe?," *Chem. Eng. Prog.* **61** (1965) no. 1, 72 – 77.
   J. F. Anderson: "Explosions and Safety in Handling Urea – Ammonium Nitrate Solutions, *Saf. Air Ammonia Plants* **9** (1967) 70 – 71.
   Van Dolah et al.: "Explosion Hazards of Ammonium Nitrate under Fire Exposure," *Rep. Invest. U.S. Bur. Mines* no. 6773 (1966).
   Th. M. Groothiuzen: *Danger Aspects of Liquid Ammonium Nitrate. Part I. Detonation Properties*, report no. M3038–1 (1979).
   Th. M. Groothuizen: *Thermische stabiliteit van een ureum-ammonium-nitraatoplossing bij 60 °C*, report no. 484/485/70, Opdracht NSM-NV no. 9139 (1970).
86. J. M. Potts (ed.): "Fluid Fertilizers," Bulletin Y 185, Tennessee Valley Industry, Muscle thoals, Alabama 1984.
87. R. P. Rilo, *Zh. Prikl. Khim. (Leningrad) USSR* **50** (1977) 1676;
   J. Appl. Chem. Russ. (Engl. Transl.) **50** (1977) 1610.
88. A. L. Huhti, P. A. Gartaganis, *Can. J. Chem.* **34** (1956) 785.
89. W. P. Moore, "Liquid Fertilizers Have Long Storage Life," *Chem. Eng. News* **49** (1971) no. 44, 37.

90. E. U. Huffman, *New Developments in Fertilizer Technology*, Tennessee Valley Authority, Muscle Shoals, Ala. 1974.
91. A. Alon: "Recent Developments in Phosphoric Acid Manufacture," *ISMA Technical Conference*, Edinburgh, 14th to 16th Sept. 1965, Chap. V.
92. M. S. Casper, *Liquid Fertilizers*, Noyes Data Corp., Park Ridge, N.J., 1973.
93. A. V. Slack: *Fertilizer Developments and Trends*, Noyes Development Corp., Park Ridge, N. J., 1968, p. 344.
94. *Nitrogen* 1971, no. 4, 30.
95. R. D. Young: "Advances in Fertilizer Technology That Will Have Significant Impact," *Agric. Chem.* **26** (1971) 12 – 16, 29 – 31.
96. R. S. Meline, C. H. Davis, R. G. Lee, *Farm Chem.* **133** (1970) no. 11, 26, 28, 30.
97. National Fertilizer Solutions Association: Fluid Fertilizer Manual, Peoria, IL 1985.
98. Ugine Kuhlmann, DE-OS 2 100 413, 1971.
99. G. V. Kostyukhova, N. N. Malakhova, V. M. Lembrikoc, L. M. Soleveva, *Zh. Prikl. Khim. (Leningrad)* **51** (1978) 3; *J. Appl. Chem. USSR (Engl. Transl.)* **51** (1978) 1.
100. J. Silverberg, Assoc. Am. Fert. Control Off., Off. Publ., no. 21 168.
101. M. F. Abd El-Hameed, "Experimentation on the Use of Urea – Ammonium Sulphate (UAS) Liquid Fertilizer in Kuwait." Preprint J. F. Techn. Conf. (1984).
102. *Nitrogen*, 1985, no. 153, 20 – 31.
103. "Phosphoric Acid Sludge: An Inept Phosphatic Material?" *Phosphorus Potassium* 1985, no. 135, 26.
104. "Suspension Fertilizers without Polyphosphates," *Phosphorus Potassium* 1977, no. 89, 31.
105. J. A. Wilbanks: "Suspensions and Slurries," *Agric. Chem.* **22** (1967) 37.
106. "New Developments in Fertilizer Technology, 9th Demonstration," Tennessee Valley Authority, Muscle Shoals, Ala., 1972, pp. 53 – 55.
107. H. E. Mills, *Fert. Solutions* **6** (1962) no. 6, 38.
108. A. V. Slag, *Farm Chem.* **128** (1965) no. 5, 24.
109. M. N. Nabiev, A. F. Glagoleva, *Uzb. Khim, Zh.* 1960 no. 4, 3.
110. BASF, DE-AS 1 592 567, 1965.
111. W. C. Scott, J. A. Wilbanks, L. C. Faulkner: "Fluid Fertilizer," *Proc. Annu. Meet. Fert. Ind. Round Table* **25** (1975) 87.
112. "Urea-Ammonium Nitrate Solutions," *Nitrogen*, 1971, no. 69, 30 – 31.
113. S. H. Wittwer, F. G. Teubner, *Annu. Rev. Plant Physiol.* **10** (1959) 13 – 32.
114. W. Franke, *Planta* **55** (1960) 533 – 541; *Planta* **61** (1964) 1 – 16; *Z. Pflanzenkr. (Pflanzenpath.) Pflanzenschutz, Sonderh.* **8** (1977).
115. H.-P. Pissarek, *Dtsch. Landw. Presse* **22** (1972) no. 10.
116. G. Jürgens, H. Ramstetter, *Dtsch. Weinbau* **35** (1980) no. 19, 715 – 717.
117. A. Buchner, H. Sturm, *Gezielter düngen*, DLG-Verlag, Frankfurt 1985, pp. 148 – 155.
118. W. Bergmann, A. Hennig, *Chem. Tech. (Leipzig)* **25** (1973) 391 ff.
119. *Farm Chemicals Handbook '86*, Meister Publishing Co., Willoughby, Ohio, 1986.
120. BASF, DE 2 313 921, 1973 (J. Jung, B. Leutner, H. Sturm). C. A. Whitcomb, EP 0 023 364, 1980, prior. US 61 842, 1979.
121. BASF, EP 0 053 246, 1981 (B. Leutner et al.), DE 3 044 903, prior., 1980; EP-A 0 173 089 =DE-OS 3 427 980 1984 (A. Jungbauer, B. Leutner, G. F. Jürgens)
122. G. Jürgens: "Spurennährstoffe in die ähre," *Bauernblatt für Schleswig-Holstein* **33** (1979) no. 129, 24.
123. G. Jürgens, *Chemie und Technik in der Landwirtschaft* 1983, no. 4, 128 – 129.
124. A. Amberger: *Pflanzenernährung*, 4t ed., UTB 846, Eugen Ulmer Stuttgart, 1996.
125. A. Finck: "Fertilizers and their efficient use," in D. J. Halliday, M. E. Trenkel, W. Wichmann (eds.): *IFA (1992): World Fertilizer Use Manual*, International Fertilizer Industry Association, Paris 1992.
126. Association of American Plant Food Control Officials (AAPFCO) (ed.): *Official Publication No. 48*, Association of American Plant Food Control Officials, Inc., West Lafayette IN, 1995.
127. M. E. Trenkel: *Controlled-Release and Stabilized Fertilizers in Agriculture*, (FAO) IFA, Paris 1997.
128. IFA (ed.): *World Fertilizer Consumption*, International Fertilizer Industry Association, Paris 1996.
129. R. Powell: *Controlled Release Fertilizer*, Noyes Development Corp., Park Ridge, 1968.
130. J. Araten: *New Fertilizer Materials*, Noyes Development Corp., Park Ridge, N. J., 1968.

131. H. Petersen: "Synthesis of Cyclic Ureas by Ureidoalkylation," *Synthesis* 1973, 243 – 244, 262, 264.
132. A. V. Slack: *Fertilizer Development and Trends*, Noyes Development Corp., Park Ridge, N. J., 1968, p. 153 – 156.
133. H. Petersen: "Grundzüge der Aminoplastchemie," *Kunststoff-Jahrbuch*, 10th ed., Wilhelm Parnsegau Verlag, Berlin, 1968, pp. 30 – 107.
134. R. J. Church: "Chemistry and Processing of Urea and Ureaform," in V. Sauchelli (ed.): *Fertilizer Nitrogen Its Chemistry and Technology*, Reinhold Publ., New York 1964.
135. BASF (ed.): *Die Landwirtschaftliche Versuchsstation Limburgerhof 1914 – 1964. 50 Jahre landwirtschaftliche Forschung in der BASF*, BASF Aktiengesellschaft, Limburgerhof 1965.
136. R. Hähndel, "Langsamwirkende Stickstoffdünger — ihre Eigenschaften und Vorteile," *Mitt Landbau* (1986) 4.
137. DuPont, US 2 766 283, 1951 (E. T. Darden) = DE 1 002 765, 1952.
138. DuPont, CA 550 618, 1954 (H. M. Kvalnes, E. T. Darden), US prior., 1953/1954.
139. Allied Chemical, US 2 644 806, 1950 (M. A. Kise), US 2 845 401, 1955 (C. E. Waters).
140. J. M. O'Donnell, US 2 916 371, 1956.
141. Hercules Powder, US 3 227 543, 1965 (J. M. O'Donnell).
142. Borden Co., US 2 810 710, 1955 (D. R. Long).
143. Fisons, GB 789 075, 1954 (A. L. Whynes).
144. O. M. Scott & Sons, US 3 076 700 ,1957 (V. A. Renner); DE-OS 1 592 751, 1963 (V. A. Renner, R. H. Czurak).
145. Unpublished results of the BASF Ammonia Laboratory, 1962.
146. T. Ushioda, *Jpn. Chem. Q* **5** (1969) no. 4, 27.
147. E. Rother: "Depotdünger," 1975, and "Arbeiten zur Entwicklung eines Herstellverfahrens für Mehrnährstoffdünger mit Langzeitstickstoff," 1985, both internal BASF reports.
148. H. Schneider, L. Veegens: "Practical Experience with Ureaform Slow Release Nitrogen Fertilizer During the Past 20 Years and Outlook for the Future', *Proc. Fert. Soc.* **180** (1979) 4 – 39.
149. Aglukon company broschure: *"Nutralene, Nitroform, AZOLON — Biologically Controlled Nitrogen Release,"* Aglukon Spezialdünger GmbH, Düsseldorf 1993.
150. J. Jung, J. Dressel, *Landwirtsch. Forsch. Sonderh.* **26/I** (1971) 131.
151. Mitsubishi, DE 1 146 080, 1961 (M. Hamamoto, J. Sakaki). Mitsubishi, US 3 322 528, 1961 (M. Hamamoto).
152. Mitsubishi, DE 1 543 201, 1965 (S. Kamo, N. Yanai, N. Osako), JA prior (1964); GB 1 099 643, 1965, JA prior (1964). Hoechst, DE 1 303 018, 1963 (H. Schäfer, P. Krause); DE 1 244 207, 1962 (U. Schwenk, F. Kalk, H. Schäfer). BASF, DE-OS 1 618 128, 1967 (G. Rössler); DE 2 355 212, 1973 (O. Grabowsky et al.).
153. BASF, DE 1 081 482, 1959 (J. Jung et al.)
154. BASF, DE 1 176 39, 1963 (H. Kindler et al.); DE 1 244 ;207, 1962 (H. Brandeis, H. Petersen, A. Fikentscher).
155. Chisso Corp., JP 492 348, 1967; DE-OS 1 670 385, 1966 (K. Fukatsu et al.); US 3 488 351, 1966. (K. Fukatsu et al.).
156. C. J. Pratt, R. Noyes: *Ureaform in Nitrogen Fertilizer Chemical Process*, Noyes Development Corp., Pearl River, N. Y., 1965, pp. 232 – 234.
157. BASF: *Produktinformation für Erwerbsgartenbau, Landschaftsbau und Landwirtschaft*, COMPO GmbH, Münster 1995.
158. J. Jung, Z. *Pflanzenernaehr. Dueng. Bodenkd.* **94** (1961) 39; J. Jung, C. Pfaff, *Die Landwirtschaftliche Versuchsstation Limburgerhof 1914 bis 1964*, BASF, Ludwigshafen 1964.
159. A. J. Patel, G. C. Sharma, *J. Am. Soc. Hortic. Sci.* **102** (1977) no. 3, 364 – 367.
160. BASF, DE-OS 1 467 377, 1982 (J. Jung et al.).
161. *Chem. Ing. Tech.* **46** (1974) A 100; *Chem. Age (London)* **108** (1975) no. 2852, 17 – 18.
162. Hoechst, DE-OS 2 308 941, 1973.
163. W. Riemenschneider: "A New Simple Process for Manufacturing Oxamide," in *Prod. Tech. Plant Nutr. Effic. Br. Sulphur Corp. Int. Conf. Fert., 2nd, 1978*, (1978 – 1979) 1 – 17.
164. *Eur. Chem. News* **24** (1973) no. 604; **24** (1973) no. 609. T. Okada, *J. Ammonium Industry* **1** (1975) no. 2, 28; *Nitrogen* 1979, no. 117, 43.
165. T. Fujita, personal communication, 1996.
166. G. L. Bridger et al., *J. Agric. Food Chem.* **10** (1962) 181.

167. M. L. Salutzky, R. P. Steiger, *J. Agric. Food Chem.* **12** (1962) 486 – 491.
168. Mitsui Toatsu Chem., JP-Kokai 067 089/82, 1980.
    Nugata Sulphuric Acid Co, JP 7 035 210, (Prior) 1966;
    Chem. Abstr. **75** (1971) 34 643.
169. Wintershall AG, DE-OS 1 792 725, 1963 (W. Jahn-Held);
    DE 1 252 210, 1962 (W. Jahn-Held);
    DE 1 265 726, 1964 (W. Jahn-Held),
    DE 1 592 810, 1963 (W. Jahn-Held),
    DE 1 592 811, 1963 (W. Jahn-Held).
    F. Müller, DE 1 924 284, 1969.
    Kali & Salz, DE 3 204 238, 1982;
    DE 3 216 973, 1982.
    Knapsack, DE 1 260 450, 1966 (W. Kern, J. Cremer, H. Harnisch).
170. Monsanto, GB 1 226 256, 1968.
171. BFG Glass Group, BE 878 884, 1978.
172. J. Jung. *Z. Pflanzenernähr. Düng Bodenkd.* **91** (1960) 122 – 130.
    Archer Daniels Midland, US 3 223 518, 1961 (L. J. Hansen),
    DE 1 254 162, 1960 (L. J. Hansen).
173. M. J. Brown, R. E. Luebs, P. F. Pratt, *Agron. J.* **58** (1966) no. 2, 175 – 178.
    J. J. Oertli, O. R. Lunt, *Soil Sci. Soc. Am. Proc.* **26** (1962) 579 – 587.
174. Aglukon company brochure: *"Plantacote — The System of Crop-Specific Plant Nutrition,"* Aglukon Spezialdünger GmbH, Düsseldorf, 1992.
175. T. Fujita, "Invention of fertilizer coating technology using polyolefin resin and manufacturing of polyolefin coated urea," *Jap. J. Soil Sci. Plant Nutr.* **67** (1996) 3.
176. T. Fujita et al.: *MEISTER with controlled availability — Properties and efficient utilization,* Chisso-Asahi Fertilizer Co., Ltd., Tokyo.
177. T. Fujita: *"Technical Development, Properties and Availability of Polyolefin Coated Fertilizers,"* Proceedings Dahlia Greidinger Memorial International Workshop on Controlled/Slow Release Fertilizers, Technion – Israel Institute of Technology, Haifa, 7 – 12 March 1993.
178. Chisso Corp., CA 1 295 849 ,1992 (T. Fujita, S. Yoshida, K. Yamahira).
179. Chisso Corp., CA 1 295 848 ,1992 (T. Fujita, Y. Yamashita, S. Yoshida, K. Yamahira).
180. T. Fujita, S. Maeda, M. Shibata, C. Takahashi: *"Research & development of coated fertilizer,"* Proceedings Fertilizer, Present and Future, Japan. Society of Soil Science and Plant Nutrition, Tokyo 1989, pp. 78 – 100.
181. T. Fujita, S. Maeda, M. Shibata, C. Takahashi: *"Research & development of coated fertilizers,"* Japanese Society of Soil Science & Plant Nutrition, Proceedings of the Symposium on Fertilizer, Present and Future, Sept. 25 – 26, Tokyo 1989.
182. H. M. Goertz: *"Technology Developments in Coated Fertilizers,"* Proceedings Dahlia Greidinger Memorial International Workshop on Controlled/Slow Release Fertilizers, Technion – Israel Institute of Technology, Haifa, 7 – 12 March 1993.
183. S. P. Landels: *Controlled-Release Fertilizers: Supply and Demand Trends in U.S. Nonfarm Markets*, SRI International, Menlo Park, CA 1994.
184. W. P. Moore: *"Reacted Layer Technology for Controlled Release Fertilizers,"* Proceedings Dahlia Greidinger Memorial International Workshop on Controlled/Slow Release Fertilizers, Technion – Israel Institute of Technology, Haifa, 7 – 12 March 1993.
185. S. Shoji, A. T. Gandeza: *Controlled Release Fertilizers with Polyolefin Resin Coating*, Kanno Printing Co. Ltd., Sendai 1992.
186. The O.M. Scott & Sons Co., US 5 089 041 ,1992 (H. E. Thompson, R. A. Kelch).
187. Archer Daniels Midland, DE-AS 1 242 573, 1964 (Le Roy B. Sahtin), US prior., (1963).
    Archer Daniels Midland, US 3 264 088, 1965 (L. J. Hansen).
    Sierra Chemical Europe, EP-A 184 869, 1985; NL prior., 1984 (J. M. H. Lembie).
188. Chisso Asahi Fertilizer Co., JP 634 150, 1972, 650 860, 1972, DE-AS 2 461 668, 1974 (T. Fujita et al.), JP prior., 1973; US 4 019 890, 1974 (T. Fujita et al.), JP prior., 1973.
189. Chisso Asahi Fertilizer Co., DE 2 834 513, 1978 (T. Fujita et al.), JP prior., 1978.
190. RLC Technologies, L.L.C., US 5 599 374, 1997 (J. H. Detrick);
    US 5 547 486, 1996 (J. H. Detrick, F. T. Carney, Jr. US 5 374 292, 1993 (J. H. Detrick, F. T. Carney, Jr.
    US 4 969 947, 1990 (W. P. Moore); US 4 804 403, 1989 (W. P. Moore);
    US 4 711 659, 1987 (W. P. Moore).
191. Pursell Inc. company brochure: *"POLYON Polymer Coatings and the RLC Process,"* Pursell Industries, Inc., Sylacauga, Alabama 1995.
192. H. J. Detrick, PURSELL Technologies Inc., personal communication, 1997.

193. D. W. Rindt, G. M Bloin, J. G. Getsinger, *J. Agric. Food Chem.* **16** (1968) 773 – 778.
194. *Ag Chem Commer. Fert.* **28** (1973) no. 2, 14 – 15;
Chem. Eng. (N. Y.) **81** (1974) Dec. 23, 32 – 33.
Sulphur Inst. J. **10** (1974) no. 3/4, 7.
195. S. F. Allen et al., *Crops and Soils* **21** (1968) no. 3;
Agron. J. **63** (1971) 529.
196. D. W. Rindt, G. M. Bloin, O. E. Moore, *J. Agric. Food Chem.* **19** (1971) no. 5, 801 – 808.
197. Tennessee Valley Authority, US 3 295 950, 1965 (G. M. Bloin, D. W. Rindt);
US 3 342 577, 1966 (G. M. Bloin, D. W. Rindt);
US 3 903 333, 1972; (A. R. Shirley, R. S. Meline).
O. M. Scott & Sons, US 4 042 366, 1976 (K. E. Fersch, W. E. Stearus).
198. RLC Technologies, L.L.C., US 5 599 374 ,1997 (J. H. Detrick).
199. Bayer, DE-AS 2 240 047, 1972; DE-OS 3 020 422, 1980; EP 46 896, 1981; EP-A 95 624, 1983 (H. Heller, D. Schäpel).
200. Bayer, EP 90 992, 1983, DE 3 212 537, 1982 (H. Brunn et al.).
201. B. Crawford: *Florida Commercial Fertilizer Law. Rules and Regulations*, State of Florida, Department of Agriculture and Consumer Services, Tallahassee FL, 1995.
202. Bundesminister für Ernährung, Landwirtschaft und Forsten: *Verordnung zur Änderung düngemittelrechtlicher Vorschriften*, 1995, BGBl. I.
203. G. Kluge, G. Embert: *Das Düngemittelrecht mit fachlichen Erläuterungen*. Bonn 1996.
204. T. Fujita, personal communication, 1997.
205. A. Amberger: *Pflanzenernährung*, 4th ed., **Taschenbücher 846**, Verlag Eugen Ulmer, Stuttgart 1996, pp. 156, 157.
206. K. Mengel: *Ernährung und Stoffwechsel der Pflanze*, VEB Gustav Fischer Verlag, Jena 1968, pp. 287, 289.
207. Bundesrat, Drucksache 239/96 *Stellungnahme der Bundesregierung zu der Entschließung des Bundesrates zur Verordnung zur Änderung düngemittelrechtlicher Vorschriften*, Bundesanzeiger Verlagsgesellschaft mbH, Bonn 1996.
208. A. Amberger: "Potential of the Nitrification Inhibitor Dicandiamide to Control Nitrogen Management and Environmental Pollution," *Proceedings, Dahlia Greidinger Memorial International Workshop on Controlled/Slow Release Fertilizers, Technion*, Israel Institute of Technology Haifa, March 7 – 12 1993.
209. B. Scheffer, "Application of nitrogen fertilizers with nitrification inhibitors in water drainage areas," *GWF, Gas Wasserfach: Wasser/Abwasser* **135** (1994) 15 – 19.
210. B. Scheffer: "Nitratgehalte im Dränwasser bei Einsatz stabilisierter Stickstoffdünger," Stabilisierte Stickstoffdünger — ein Beitrag zur Verminderung des Nitratproblems, Proccedings: 15./16. Oktober 1991, Würzburg, Publisher; BASF Aktiengesellschaft, Limburgerhof/SKW Trostberg AG, 1991, pp. 83 – 94.
211. P. Schweiger: "Wege zur Minimierung des Nitratproblems," Proceedings: Stabilisierte Stickstoffdünger — ein Beitrag zur Verminderung des Nitratproblems. Fachtagung: 15./16. Oktober 1991, Würzburg. Publisher; BASF Aktiengesellschaft, Limburgerhof/SKW Trostberg AG, 1991, pp. 59 – 69.
212. W. Zerulla: "N_{min} — Gehalte im Boden nach der Düngung bzw. nach der Ernte," Proceedings: Stabilisierte Stickstoffdünger — ein Beitrag zur Verminderung des Nitratproblems. Fachtagung: 15./16. Oktober 1991, Würzburg. Publisher; BASF Aktiengesellschaft, Limburgerhof/SKW Trostberg AG, 1991, pp. 111 – 119.
213. H. Sturm, A. Buchner, W. Zerulla: *Gezielter düngen. Integriert, wirtschaftlich, umweltgerecht*, VerlagsUnion Agrar, DLG-Verlags-GmbH, Frankfurt a.M. 1994, pp. 90 – 95.
214. K. F. Bronson, A. R. Mosier, "Suppression of methane oxidation in aerobic soil by nitrogen fertilizers, nitrification inhibitors, and urease inhibitors," *Biol. Fertil. Soils* **17** (1994) 4, 263 – 280.
215. K. F. Bronson, A. R. Mosier: "Effect of Nitrogen Fertilizer and Nitrification Inhibitors on Methane and Nitrous Oxide in Irrigated Corn," *Biochemistry of Global Change. Radiatively Active Trace Gases*, Chapman & Hall, New York – London 1993.
216. K. F. Bronson, A. R. Mosier, S. R. Bishnoi, "Nitrous Oxide Emissions in Irrigated Corn as Affected by Nitrification Inhibitors," *Soil Sci. Soc. Am. J.* **56** (1992) 1, 161 – 165.
217. H.-J. Klasse: "Versuchsergebnisse zur Wirkung stabilisierter Stickstoffdünger auf die Nitratverlagerung bzw. -auswaschung. N_{min} — Gehalte im Boden nach der Düngung bzw. nach der Ernte," Proceedings: Stabilisierte

Stickstoffdünger — ein Beitrag zur Verminderung des Nitratproblems. Fachtagung: 15./16. Oktober 1991, Würzburg. Publisher; BASF Aktiengesellschaft, Limburgerhof/SKW Trostberg AG, 1991, pp. 103 – 110.
218. M. Koshino: *The Environmental Protection Framework Concerning Fertilizer Use in Japan*, National Institute of Agro-Environmental Sciences, Department of Farm Chemicals, Tsukuba 1993.
219. C. A. Grant, S. Jia, K. R. Brown, L. D. Baily: "Volatile losses of NH_3 surface-applied urea and urea ammonium nitrate with and without the urease inhibitors NBPT or ammonium thiosulphate," *Canadian Journal of Soil Science* **76** (1996) 3.
220. C. J. Watson et al., "Soil properties and the ability of the urease inhibitor N-(n-butyl) thiophosphoric triamide (nBTPT) to reduce ammonia volatilization from surface-applied urea," *Soil Biol. Biochem.* **26** (1994) 9, 1165 – 1171.
221. C. A. Grant, R. Ferguson, R. Lamond, A. Schlegel, W. Thomas: "Use of Urease Inhibitors in the Great Plains," *Proceedings Great Plains Soil Fertility Conference,* Denver, Co 1996.
222. M. E. Trenkel: *Controlled-Release and Stabilized Fertilizers in Agriculture*, (FAO) IFA, Paris 1997, p. 12.
223. J. A. Delgado, A. R. Mosier: "Mitigation alternatives to decrease nitrous oxides emissions and urea – nitrogen loss and their effect on methane flux," *J. Environ. Qual.* **25** (1996) 5.
224. A. Amberger: "Mobilization of rock phosphates in soil," *2nd African Soil Science Society Conference,* Cairo 1991.
225. F. Bayrakli, S. Gezgin: "Controlling ammonia volatilization from urea surface applied to sugar beet on a calcareous soil," *Commun. Soil. Sci. Plant Anal.* **27** (1996) 9, 10.
226. Z. P. Wang, O. van Cleemput, L. Baert: "Movement of urea and its hydrolysis products as influenced by moisture content and urease inhibitors," *Biol. Fertil. Soils* **22**, (1994) 1/2.
227. Z. P. Wang, O. van Cleemput, P. Demeyer, L. Baert: "Effect of urease on urea hydrolysis and ammonia volatilization," *Biol. Fertil. Soils* **11** (1991) 1, 43 – 47.
228. Z. P. Wang, L. T. Li, O. van Cleemput, L. Baert: "Effect of urease inhibitors on denitrification in soil," *Soil Use Manage.* **7** (1991) 4, 230 – 233.
229. B. Zacherl, A. Amberger: "Effect of the nitrification inhibitors dicyandiamide, nitrapyrin and thiourea on Nitrosomonas europea," *Fert. res.* **22** (1990) 37 – 44.
230. J. K. R. Gasser: "Nitrification Inhibitors — Their Occurence, Production and Effects of their Use on Crop Yields and Composition," *Soils Fert.* **33** (1970) 547 – 554.
231. K. Sommer, *Nitrifizide,* Landw. Forschung Sond. 27/II, 1972, pp. 64 – 82.
232. US 3 135 594, 1964 (Cleve A. I. Goring), Serial No. 653, 065 1957 expired in 1981; continuation-in-part of copending application.
233. US 3 424 754, 1969 (William H. Taplin III.), expired in 1986.
US 3 420 833, 1969 (William H. Taplin), expired in 1983.
234. DowElanco(ed.): *Quick Guide to Nitrogen Management,* DowElanco, Champaign, IL.
235. DowElanco(ed.): *References on N-Serve® Nitrogen Stabilizer — Bibliography,* DowElanco, Champaign, IL 1989.
236. J. Huffman, DowElanco, personal communication, 1997.
237. W. Zerulla, BASF Aktiengesellschaft, personal communication, 1996.
238. US 3 135 594, 1964 (Cl. A. I. Goring).
239. US 5 106 984, 1992 (M. E. Halpern, J. A. Orvik, J. Dietsche); EP 0 306 547 B1, 1992 (M. E. Halpern, J. Dietsche, J. A. Orvik, B. J. Barron).
240. FR 1 232 366, 1960 (L. Soubies, R. Gadet).
241. S. Hallinger: *"Untersuchungen zur biologischen Metabolisierung von Dicyandiamid,* Technische Universität München," Isnstitut für Bodenkunde, Pflanzenernährung und Phytopathologie, Lehrstuhl für Pflanzenernährung, Dissertation, 1992.
242. M. Hauser, K. Haselwandter: "Degradation of Dicyandiamide by Soil Bacteria," *Soil Biol. Biochem.*, **22** (1990) 113 – 114.
243. K. Vilsmeier: "Fate of ammonium-N in pot studies as affected by DCD addition," *Fert. res.* **29** (1991) 3, 187 – 189.
244. K. Vilsmeier: "Turnover of ^{15}N-Ammonium Sulfate with Dicyandiamide under Aerobic and Anaerobic Soil Conditions," *Fert. Res.* **29** (1991) 191 – 196.
245. R. Roll: *"Zur Toxikologie von Dicyandiamid,"* Proceedings: Stabilisierte Stickstoffdünger — ein Beitrag zur Verminderung des Nitratproblems. Fachtagung: 15./16. Oktober 1991, Würzburg. BASF Aktiengesellschaft, Limburgerhof/SKW Trostberg AG, 1991, pp. 77 – 82.

246. US 4 764 200, 1988 (O. Meiss, R. E. Nitzschmann); DE 4 128 828 A1, 1993 (K. Engelhardt, H. Dittmar, S. Luther).
247. US 4 994 100, 1991 (A. R. Sutton, Ch. W. Weston, R. L. Balser).
248. US 5 352 265, 1994 (Ch. W. Weston, L. A. Peacock, W. L. Thornsberry, Jr., A. L. Sutton).
249. DD 222 471, 1985 (H. J. Herbrich et al.).
250. DD 247 894, 1987 (K. Gunther et al.).
251. US 4 975 107, 1990 (K. E. Arndt et al.), US 4 969 946, 1990 (K. E. Arndt et al.), US 4 545 803, 1985 (K. E. Arndt et al.).
252. DE 4 446 194, 1996 (J. Dressel et al.), DE 4 137 011, 1993 (R. Kästner et al.), DE 3 820 739, 1990 (U. Baus et al.), DE 3 820 738, 1990 (U. Baus et al.), DE 3 409 317, 1985 (H. Bohm et al.).
253. DE 4 405 393, 1995 (M. Grabarse et al.), DE 4 405 392, 1995 (M. Grabarse et al.), DE 4 237 688, 1994 (I. Berg et al.), DE 4 237 687, 1994 (M. Grabarse et al.), DE 4 219 661, 1994 (M. Grabarse et al.), DE 4 218 580, 1993 (H. J. Hartbrich et al.), DE 4 211 808, 1993 (H. Böhland et al.), DE 4 018 395, 1991 (M. Grabarse et al.), DD 273 829, 1990 (H. J. Hartbrich et al.), DD 279 470, 1990 (H. J. Halbenz et al.).
254. DD 260 486, 1989 (H. Carlson et al.).
255. RU 2 046 116, 1996 (L. A. Gorelik et al.).
256. SU 1 680 683, 1992 (O. V. Kuznetsova et al.).
257. US 4 523 940, 1985 (K. E. Arndt et al.), US 4 522 642, 1985 (K. E. Arndt et al.).
258. DE 4 128 828 A1, 1993 (H. Dittmar et al.).
259. WO 98/05607, 1998 (Th Barth et al.)
260. S. Kincheloe: "The manufacture, agronomics and marketing of Agrotain," IFA Agro-Economics Committee Conference: 'Plant Nutrition in 2000', Tours, June 1997.
261. S. Kincheloe, A. R. Sutton: "The Manufacture, Agronomics and Marketing of agrotain," Book of Abstracts, 212th ACS National Meeting, American Chemical Society, Washington, D.C. 1996.
262. F. J. Dentener, P. C. Crutzen: "A Three-Dimensional Model of the Global Ammonia Cycle," *Atmos. Chem.* **19** (1994) 331 – 369.
263. W. H. Schlesinger, A. E. Hartley: "A Global Budget for NH$_3$," *Biogeochem.* **15** (1992) 191 – 211.
264. A. Finck: *Dünger und Düngung,* Verlag Chemie, Weinheim – New York 1979, pp. 52, 59.
265. US 4 530 714, 1985 (J. F. Kolc, M. D. Swerdloff, M. M. Rogic, L. L. Hendrickson, M. van der Puy).
266. Ch. F. Wilkinson: *N-(n-butyl) thiophosphoric triamide (NBPT, Agrotain®): A Summary of its Chemistry, Toxicology, Ecotoxicology and Environmental Fate,* IMC-Agrico Company, Mundelein, IL 1996, p. 6.
267. IMC Agrico (ed.): *Material Safety Data Sheet, AGROTAIN® Urease Inhibitor,* IMC-Agrico Company, Mundelein, IL 1995.
268. US 5 352 265, 1994 (C. W. Weston et al.), US 5 364 438, 1994 (C. W. Weston et al.).
269. US 5 698 003, 1997 (B. A. Omilinsky et al.).
270. Ministére de L'Environment de la République Français: "Bonne pratiques d'épandage et de stockage des fertlisants," Journal Officiel de La Republique Française, Jan. 5, 1994.
271. Bundesminister für Ernährung, Landwirtschaft und Forsten: *Verordnung zur Änderung düngemitelrechtlicher Vorschriften,* Bundesgesetzblatt Jahrgang 1995, Teil I., Bonn 1995.
272. G. Kluge, G. Embert: *Das Düngemittelrecht mit fachlichen Erläuterungen 1992,* Landwirtschaftsverlag GmbH, Münster – Hiltrup 1992.
273. KrW/AbfG: Kreislaufwirtschafts- und Abfallgesetz, Gesetz zur Vermeidung, Verwertung und Beseitigung von Abfällen, BGBl. I (1994) p. 2705.
274. G. Kluge, G. Embert: *Das Düngemittelrecht mit fachlichen Erläuterungen, (inkl. 1. Supplement 1997),* Landwirtschaftsverlag GmbH, Münster – Hiltrup 1996.
275. R. Gutser, *VDLUFA Schriftenreihe* **44** (1996), 29 – 43.
276. AbfKlärV: Klärschlammverordnung, BGBl. I (1992)
277. K. Severin, *VDLUFA Schriftenreihe* **44** (1996) 85 – 101.
278. R. Gutser, K. Teicher, P. Fischer, *Acta Horticulturae* **150** (1984) 175 – 184.
279. A. Fink: *Dünger und Düngung,* VCH Verlagsgesellschaft, Weinheim, Germany 1992, pp. 164 – 174.
280. E. Knickmann: "Handelshumusdüngemittel", in K. Scharrer, Linser (eds.): *Handbuch der Pflanzenernährung und Düngung,* Springer-Verlag, Wien 1968, pp. 1400 – 1431.
281. "Abfallstoffe als Dünger", *VDLUFA-Schriftenreihe* **23** (1988).
282. "Sekundärrohstoffe im Stoffkreislauf der Landwirtschaft", *VDLUFA-Schriftenreihe* **44** (1996).

283. R. Gutser, *Z. Branntweinwirtschaft* **138** (1998) 2 – 6.
284. U. Bayer, Diss. Techn. Univ. München, 1987.
285. T. v. Tucher, Diss. Techn. Univ. München, 1990.
286. Bundesministerium für Umwelt, Naturschutz- und Reaktorsicherheit: Entwicklung der Schwermetallgehalte von Klärschlämmen in der BRD von 1974 – 1994, Informationsblatt 1996.
287. A. Gronauer et al.: Bioabfallkompostierung — Verfahren und Verwertung, Bayerisches Landesamt für Umweltschutz, Schriftenreihe Heft 139 (1997).
288. P. Fischer, M. Jauch, *VDLUFA Schriftenreihe* **33** (1991) 751 – 756.
289. M. Bertoldi, P. Sequi, B. Lemmes, T. Papi: "The Science of Composting," *European Comm. Int. Symp.*, Udine 1995, Blackie Academic & Professional, an imprint of Chapman – Hall, 1996.
290. J. Köhnlein, H. Vetter: "Die Stalldüngerrotte bei steigender Stroheinstreu," *Z. Pflanzenernähr. Düng. Bodenkd.* **63** (1953) no. 108, 119 – 141.
291. G. J. Kohlenbrander, L. C. N. de la Lande Cremer: *Stahlmest en Gier*, H. Veenman & Zonen, Wageningen 1967.
292. H. Koriath und Kollektiv: *Güllewirtschaft, Gülledüngung*, VEB Deutscher Landwirtschaftsverlag, Berlin 1975.
293. K. Rauhe: "Wirtschaftseigene Düngemittel" in *Handbuch der Pflanzenernährung und Düngung*, vol. **2**, Springerverlag, Wien-New York, 1968 pp. 907 – 943.
294. D. Strauch, W. Baader, C. Tietjen: *Abfälle aus der Tierhaltung*, Verlag Eugen Ulmer, Stuttgart 1977.
295. H. Vetter, G. Steffens: *Wirtschaftseigene Düngung*, DLG-Verlag, Frankfurt 1986.
296. *Faustzahlen für die Landwirtschaft*, 10th ed., Landwirtschaftsverlag, Münster-Hiltrup 1983.
297. A. Amberger: "Gülle – ein schlechtgenutzter Dünger," *DLG-Mitteilungen* **2**, 1982 pp. 78 – 80.
298. J. C. Brogan (ed): *Nitrogen Losses and Surface Run-off from Landspreading of Manures*, Martinus Nijhoff/Dr. W. Junk Publishers, Den Hague-Boston-London 1981.
299. K. Früchtenicht, H. Vetter: "Phosphatzufuhr und Phosphatwirkung aus organischer Düngung," in *Phosphatdüngung aus heutiger Sicht*, Arbeitskreis Phosphat der deutschen Düngerindustrie, Hamburg 1983.
300. J. K. R. Gasser: *Effluents from Livestock*, Applied Science Publishers, London 1980.
301. T. W. G. Hucker, G. Catroux: *Phosphorus in Sewage Sludge and Animal Waste Slurriers*, Reichel, Dordrecht 1981.
302. G. Schechtner: "Nährstoffwirkung und Sonderwirkungen der Gülle auf dem Grünland," in *Bericht über die 7. Arbeitstagung "Fragen der Güllerei"*, Bundesversuchsanstalt für alpenländische Landwirtschaft, 1981, pp. 135 – 196.
303. J. H. Voorburg: *Utilization of Manure by Land Spreading*, Janssen Services, London 1977.
304. L. Diekmann: "Schadgase aus Flüssigmist" in *Rationalisierungskuratorium für Landwirtschaft* (*RKL*), 1983, pp. 839 – 885.
305. P. Förster: "Einfluß hoher Güllegaben und üblicher Mineraldünger auf die Stoffbelastung (NO_3, NH_4, P und SO_4) im Boden und Grundwasser in Sandböden Nordwestdeutschlands," *Z. Acker Pflanzenbau*, **137** (1973) 270 – 286.
306. W. R. Kelly: *Animal and Human Health Hazards Associated with the Utilization of Animal Effluents*, Commission of the European Community, Luxembourg 1978.
307. H. Kuntze: "Düngung und Gewässergüte," *Landwirtsch. Forsch. Sonderh.* **35** (1979) 14 – 24.

General References
308. *Fertilizer Manual, Development and Transfer of Technology*, Series no. 113, UNIDO, United Nations, New York 1980.
309. A. Schmidt: *Chemie und Technologie der Düngemittelherstellung*, Hüthig-Verlag, Heidelberg 1972.
310. A. J. More (ed.): *Granular Fertilizers and Their Production*, The British Sulphur Corp., London 1977.
311. *Fertilizers: Ammonium Phosphates*, **Process Economics Program Report no. 127**, SRI International, Menlo Park, Ca., 1979.
312. F. T. Nielsson: *Manual of Fertilizer Processing*, vol. **5 of the Fertilizer Science and Technology Series**, 1987.

Specific References
313. *Kirk-Othmer*, **10** (1980) 31 – 125.
314. T. P. Hignett, *Farm Chem.* 1963, no. 1, 34 – 35.
315. *Winnacker-Küchler*, 4th ed., **2** pp. 334 – 378.
316. J. Jung, *Proc. Int. Congr. Phosphorus Compd.*, 3rd, 1983, 463 – 495.
317. "Granular Fertilizers," supplement to *Phosphorus Potassium* **1985**.
318. V. Sauchelli, *Chemistry and Technology of Fertilizers, ACS Monograph Series*, Reinhold

Publishing Corporation, New York; Chapman & Hall Ltd., London 1960, p. 280.
319. A. T. Brook, "Developments in Granulation Techniques," paper read before the Fertiliser Society of London, 1957.
320. O. Bognati, L. Buriani, I. Innamorati, Proc. ISMA Tech./Agric. Conf., 1967.
321. T. E. Powell, *Proc. Technol. Intern.* **18** (1973) 271 –278.
322. *Nitrogen* 1981, no 131, 39 – 41.
323. *Nitrogen* 1975, no. 95, 31 – 36.
324. *World Guide to Fertilizer Process and Constructors*, 6th ed., The British Sulphur Corp., London 1979, pp. 138 – 139.
325. *Nitrogen* **175** (1988) 32 – 40.
326. O. E. Skauli: "Pan Granulation of Ammonium Nitrate and Urea," (Symposium on New Fertilizer Technology) ACS National Meeting, 168th, 1974.
327. J. P. Bruynseels: "Fluid Bed Granulation of Ammonium Nitrate and Calcium Ammonium Nitrate," *Proc. of the Fertiliser Society* 1985, no. 235.
328. J. P. Bruynseels, *Hydrocarbon Process.* **60** (1981) no. 9, 203 – 208.
329. H. B. Ries, *Aufbereit. Tech.* **18** (1977) 633 – 640.
330. *Nitrogen* **193** (1990) 22 – 26.
331. Private communication.
332. R. S. Meline, I. W. McCamy, J. L. Graham, T. S. Sloan *J. Agr. Food Chem.* **16** (1968) 235 – 240.
333. *Nitrogen* **177** (1989) 21 – 24.
334. J. W. McCamy, M. M. Norton, *Farm Chem.* **140** (1977) no. 2, 61 – 68, 70.
335. *New Developments in Fertilizer Technology, 11th Demonstration*, Tennessee Valley Authority, Muscle Shoals, Ala., 1976, pp. 33 – 35.
336. *Nitrogen* 1975 no. 98, 37 – 39.
337. L. M. Nunnely, F. T. Carney: "Melt Granulation of Urea by the Falling-Curtain Process," ACS National Meeting, 182th, 1981.
338. W. Pietsch, *Aufbereit. Tech.* **12** (1971) 684 – 690.
339. C. Debayeux in [340] , p. 8.1 – 8.5.
340. V. Bizzotto, IFA Technical Conference 5 – 8 Nov. 1984, Paris, p. 7.1 – 7.14 (Reprints).
341. *New Developments in Fertilizer Technology, 8th Demonstration*, Tennessee Valley Authority, Muscle Shoals, Ala., 1970, pp. 41 – 47.
342. G. C. Hicks, J. M. Stinson, *Ind. Eng. Chem. Proc. Des. Dev.* **14** (1975) 269 – 276.
343. A. S. Maartensdijk: "Direct Production of Granulated Superphosphates and PK-Compounds," *New Dev. Phosphate Fert. Technol. Proc. Tech. Conf. ISMA Ltd., 1976* (1977).
344. G. H. Wesenberg: "Concentrated Superphosphate: Manufacturing Process" in [312]
345. F. P. Achorn, D. G. Salladay: "Granulation Using the Pipe-Cross Reactor" in [312]
346. J. A. Benes, A. Hemm: "Modern Process Design in Compound Fertilizer Plants", *Seminar of Trends and Developments in the Fertiliser Industry,* 1985.
347. G. C. Hicks: "Review of the Production of Monoammonium Phosphate" in [312]
348. G. H. Wesenberg: "Diammonium Phosphate Plants and Process" in [312]
349. J. Steen, E. A. Aasum, T. Heggeboe: "The Norsk Hydro Nitrophosphate Process" in [312]
350. *Phosphorus Potassium* 1975, no. 76, 48 – 54.
351. I. W. McCamy, G. C. Hicks, *Proc. Annu. Meet. Fert. Ind. Round Table*, 26th, 1976, 168 – 176.
352. G. C. Hicks, I. W. McCamy, B. R. Parker, M. M. Norton; "Basis for Selection of Granulators at TVA," paper presented at the *Proc. Annu. Meet. Fert. Ind. Round Table*, 28th, 1978.
353. H. Stahl, *Aufbereit. Tech.* **21** (1980) 525 – 533.
354. G. Heiseler, D. Baier, D. Juling, W. Kretschmer, *Chem. Tech. (Leipzig)* **25** (1973) 410 – 414.
355. L. Diehl, K. F. Kummer, H. Oertel: "Nitrophosphates with Variable Water Solubility: Preparation and Properties," paper read before the Fertiliser Society of London, 1986.
356. M. Autti, M. Loikkanen, P. Suppanen: "Economic and Technical Aspects in some Differential NPK-Processes," *New Dev. Phosphate Fert. Technol. Proc. Tech. Conf. ISMA Ltd*, 1976 (1977).
357. H. B. Ries, *Aufbereit. Tech.* **16** (1975) 17 – 26.
358. I. S. E. Martin, F. J. Harris in [310], pp. 253 – 270.
359. J. Kotlarevsky, J. Cariou in [310], pp. 245 – 252.
360. M. Kuwabara, S. Hayamizu, A. Hatakeyama in [310] pp. 125 – 147.
361. J. L. Medbery, F. T. Nielsson, *Ind. Eng. Chem. Prod. Res. Dev.* **22** (1983) 291 – 296.
362. B. Dencs, Z. Ormós, *Int. Symp. Agglomeration, 3rd, 1981*, G 2 – G 15.
363. A. Maroglou: "Fluidized Bed Granulation," Ph. D. Thesis, University of Birmingham 1985.

364. S. Mortensen, S. Hoomand, *Chem. Eng. Prog.* **79** (1983) no. 4, pp. 37 – 42.
365. S. Mortensen, A. Kristiansen, *Int. Symp. Agglomeration, 3rd, 1981*, F 77 – F 90.
366. H. B. Ries, *Aufbereit., Tech.* **11** (1970) 262 – 280.
367. I. S. Mangat in [310] pp. 148 – 161.
368. J. L Hawksley in [310] pp. 110 – 124.
369. P. J. Sherrington, *Can. J. Chem. Eng.* **47** (1969) June, 308 – 316.
370. A. I. More (ed): *Fert. Nitrogen: Proc. Br. Sulphur Corp. Int. Conf. Fert. Technol., 4th, 1981*.
371. Y F. Berguin in [310] pp. 296 – 303.
372. P. J. Sherrington, R. Oliver: *Granulation*, Hlyden& Sons Ltd., London p. 172.
373. O. Uemaki, K. B. Mathur, *Ind. Eng. Chem. Process Des. Dev.* **15** (1976) 504 – 508.
374. F. E. Steenwinkel, J. W. Hoogendonk: "The Prilling of Compound Fertilisers," paper read before The Fertiliser Society (1969).
375. *Nitrogen* 1980, no. 127, 45.
376. P. J. van den Berg, G. Hallie: "New Developments in Granulation Techniques," *Proc. Fert. Soc.* 1960, no. 59.
377. *"Supra Today,"* 1977, p. 23.
378. R. P. Ruskan, *Chem. Eng. (N. Y.)* **83** (1976) June 7, 114 – 118.
379. I. R. Schoemaker, I. A. C. M. Smit, "Criteria for a Choice between various Granulation and Prilling Technologies for Fertilisers," paper read before the Fertiliser Society of London (1981).
380. R. G. Lee, M. M. Norton, H. G. Graham, *Proc. Annu. Meet. Fert. Ind. Round Table*, 24th, 1974, 79.
381. R. G. Lee, R. S. Meline, R. D. Young, *Ind. Eng. Chem. Process. Des. Dev.* **11** (1972) 90 – 94.
382. C. J. Pratt, *Br. Chem. Eng.* **14** (1969) 185 – 188.
383. H. B. Ries, *Int. Symp. Agglomeration, 3rd,* 1981, F 34 – F 59.
384. H. Rumpf, *Chem. Ing. Tech.* **30** (1958) 329 – 336.
385. G. Nöltner, *Int. Symp. Agglomeration, 3rd, 1981*, F 16 – F 33xs.
386. H. Schubert, E. Heidenreich, F. Liepe, Th. Neeße, *Mechanische Verfahrenstechnik II*, VEB Deutscher Verlag für Grundstoffindustrie, Leipzig 1979, p. 204.
387. G. Heinze: "Verfahrenstechniken des Agglomerierens," Seminar of the VDI-Bildungswerk, Stuttgart-Vaihingen 1984.
388. J. Eimers: "Granuliertechniken bei der Düngemittelherstellung," paper presented at the congress AGRICHEM 81, Bratislava 1981.
389. H. B. Ries, *Aufbereit. Tech.* **11** (1970) 147 – 153.
390. H. B. Ries, *Aufbereit. Tech.* **11** (1970) 615 – 621.
391. H. B. Ries, *Aufbereit. Tech.* **11** (1970) 744 – 753.
392. *New Developments in Fertilizer Technology, 13th Demonstration*, Tennessee Valley Autority, Muscle Shoals, Ala., 1980, pp. 75 – 79.
393. R. J. Milborne, D. W. Philip, "Adapting a Pipe Reactor to a Blunger for NPK Production," paper read before the Fertiliser Society of London, 1986.
394. W. Pietsch, *Aufbereit. Tech.* **7** (1966) no. 4, 177 – 191.
395. G. Heinze, *Chem. Tech. Heidelberg* **15** (1986) no. 6, 16, 18, 21.
396. R. H. Perry, C. H. Chilton: *Chemical Engineers' Handbook*, 5th ed., McGraw-Hill, New York 1973 pp. 57 – 65.
397. J. O. Hardesty, *Chem. Eng. Prog.* **51** (1955) 291 – 295.
398. F. T. Nielsson: "Granulation" in [312]
399. Ch. Davis, R. S. Meline, H. G. Graham, *Chem. Eng. Prog.* **64** (1968) no. 5, 75 – 82.
400. *Farm Chemicals Handbook*, 1986, p. B 35.
401. R. D. Young, G. C. Hicks, C. H. Davis, *J. Agric. Food Chem.* **10** (1962) , no. 1, 68 – 72.
402. F. G. Membrillera, J. L. Toral, F. Codina in [310], pp. 162 – 173.
403. "The Cros Fertilizer Granulation Process", *Phosphorus Potassium* **1977,** no. 87, 33 – 36.
404. *Phosphorus Potassium* 1986, no. 144, 27 – 33.
405. J. W. Baynham: "The SAI-R Process for the Manufacture of Compound Fertilisers containing Ammonium Nitrate and Diammonium Phosphate", *Proc. ISMA Tech. Conf. 1965.*
406. *Nitrogen* 1970, no. 64, 25 – 26.
407. E. Pelitti, J. C. Reynolds in [310], pp. 95 – 109.
408. R. M. Reed, J. C. Reynolds, *Chem. Eng. Prog.* **69** (1973) no. 2, 62 – 66.
409. E. Vogel, *Nitrogen*, 1986, no. 161, 28 – 31.
410. *Nitrogen* 1987, no. 166, 39.
411. K. Sommer, W. Herrmann, *Chem. Ing. Tech.* **50** (1978) 518 – 524.
412. F. D. Young, T. W. McCamy, *Can. J. Chem. Eng.* **45** (1967) 50 – 56.
413. H. Klatt, *Zem. Kalk Gips* **11** (1958) no. 4, 144 – 154.

414. Tennessee Corp., US 769 058, 1985.
415. J. W. McCamy, M. M. Norton in [310], pp. 68 – 94.
416. St. Ruberg, *Aufbereit.-Tech.* **28** (1987) 75 – 81.
417. H. B. Ries, *Tech. Mitt.* **77** (1984) 583 – 589.
418. H. B. Ries, *Maschinenmarkt* **92** (1986) 25 – 28.
419. H. Stahl: "Press-Agglomeration," seminar of the VDI-Bildungswerk, Stuttgart-Vaihingen 1984.
420. S. Maier in [310], pp. 283 – 295.
421. A. Seixas, J. D. Ribeiro in [340] p. 11.1 – 11.21.
422. H. Rieschel, *Aufber. Tech.* **11** (1970) 133 – 146.
423. A. Tassin, Chr. Fayard, *Inform. Chem.* **269** (1986) no. 1, 127 – 131.
424. P. Heroien, C. Fayard, *Proc. Inst. Briquet. Agglom. Bienn. Conf.* **22** (1991) 173 – 185.
425. W. Herrmann: *Das Verdichten von Pulvern zwischen zwei Walzen*, Verlag Chemie, Weinheim, Germany 1973.
426. W. Herrmann, K. Sommer, *Int. Symp. Agglomeration, 3rd*, 1981, F2 – F15.
427. W. Herrmann, *Chem. Ing. Tech.* **51** (1979) 277 – 282.
428. G. J. Thorne, J. D. C. Hemsley, H. Hudson, *Chem. Process. (London)* **19** (1973) no. 1, 17 – 18.
429. I. Podilchuck, W. T. Charlton, M. D. Pask: *Modern Approach to the Design of Granular Fertilizer Plants*, Fisons Ltd., Fertilizer Division, 1975.
430. H. Vernède, *Nitrogen* 1969, no. 60, 29 – 33.
431. Sahut-Conreur, *Verfahrenstechn.* 1987, no. 5, 8, 10, 12.
432. G. Hoffmeister: *"Physical Properities of Fertilizers & Methods for Measuring Them"*, TVA Bulletin Y 147, 1979.
433. L. Jäger, P. Hegner, *Fert. Technol.* **1** (1985) 191 – 221.
434. K. Niendorf, W. Teige, *Chem. Tech. (Leipzig)* **26** (1974) 771 – 774.
435. M. Brübach, A. Göhlich, *Verfahrenstechnik* **7** (1973) 268 – 272.
436. P. Runge, *Chem. Tech. (Leipzig)* **28** (1976) 163 – 165.
437. J. Silverberg, *Proc. Ann. Meet. Fert. Ind. Round Table 16th* 1966, 81.
438. D. C. Thompson: "Fertiliser Caking and its Prevention," *Proc. Fert. Soc.* 1972, no. 125.
439. O. Kjöhl, A. Munthe-Kaas: "Study of the Mechanism of Caking in some Nitrophosphate Fertilizers," Proc. ISMA Tech. Conf., 1970.
440. G. C. Lowrison: *Fertilizer Technology*, Ellis Horwood Limited, Chichester, 1989, p. 406.
441. P. A. Mackay, K. S. Sharples: "The Use of Special Oils and Coatings to Prevent Caking of Fertilizers," *Proc. Fert. Soc.* 1985, no. 239.
442. *Nitrogen* 1986, no. 164, 33 – 34.
443. *Nitrogen* 1990, no. 187, 20 – 27.
444. *Nitrogen* 1987, no. 166, 19.
445. E. Alto, P. Suppanen: "Dust Control in NPK Production," *Proc*, ISMA Tech. Conf. 1974.
446. J. Baretincic: "Environmental Regulations," in [312]
447. J. E. Seymour, in [310], pp. 174 – 180.
448. *Bulk Blend Quality Control Manual*, The Fertilizer Institute, Washington, D.C., 1975.
449. L. G. Croysdale, W. E. Samuels, J. A. Wagner, *Chem. Eng. Prog.* **61** (1965) 72 – 77.
450. G. Hoffmeister, "Bulk Blending," in [312]
451. L. Taylor, "Latest Technology and Methods Used for Bulk Blending and Fertilizer Storage in the United States," paper read before the Fertiliser Society of London, 1987.
452. H. W. Frizen: "Verfahrensketten für feste und flüssige Mehrnährstoffdünger in den USA," Dissertation, Universität Bonn, 1974.
453. K. H. Ullrich, *Vereinigte Landwarenkaufleute* 1979, no. 3.
454. Official Bulletin of the European Community, 20th ed., No. L213 from 22 August 1977, 77/535/EEC: Guidelines of the commission from 22 June 1977 on the approximation of the laws of the member states concerning sampling and methods of analysis of fertilizers. The Guidelines are extended and modified in: Bulletin L265 from 12.09.89, 89/519/EEC for Ca., Mg, Na and S; and Bulletin L113 from 07.05.93, 93/1/EEC for trace elements B, Co, Cu, Fe, Mn, Mo, Zn.
455. Methodenbuch, Band II: Die Untersuchung von Düngemitteln, [Book of Methods. Vol. II: The Analysis of Fertilizers] Verband Deutscher Landwirtschaftlicher Untersuchungs- und Forschungsanstalten (LUFA) [Association of German Agricultural Research Institutes] LUFA-Verlag, Darmstadt, 1995.
456. ISO (International Organization for Standardization).
457. Official Methods of Analysis of AOAC International, 16th ed, 3rd rev., AOAC International, Gaithersburg, Maryland, USA, 1997.

General References

458. B. Wermingshausen: *Kunststoffe im Landbau*, Kohlhammer-Verlag, Stuttgart-Berlin-Köln-Mainz 1985.

459. H. Baumann: *Plastoponik, Schaumkunststoffe in der Agrarwirtschaft*, Dr. Alfred Hüthig Verlag, Heidelberg 1967.
460. A. Kullmann: *Synthetische Bodenverbesserungsmittel*, VEB Deutscher Landwirtschaftsverlag, Berlin 1972.
461. M. G. Voronkov, G. I. Zelchan, E. Lukevitz in K. Ruhlmann (ed.): *Silizium und Leben, Biochemie, Toxikologie und Pharmakologie des Si*, Akademie-Verlag, Berlin 1975.

Specific References

462. H. Prün, *Rasen Turf Gazon* **17** (1986) no. 1, 120 – 127.
463. B. Werminghausen, *Chemie und Technik in der Landwirtschaft* 1962, 359 – 360.
464. H. Knobloch, *Wasser und Boden* **12** (1964) 403 – 405.
465. F. Penningsfeld, *Das Gartenamt* **28** (1979) 1 – 4.
466. J. Günther, *Gb + Gw*, Gärtnerbörse und Gartenwelt, 81 (1981) no. 31, 714 – 716.
467. B. Werminghausen, private communication, 1987.
468. B. Werminghausen, *Mitt. Landbau* 1964, no. 4, 1 – 8.
469. BASF, DE 1 245 205, 1962 (F. Stastny, J. Jung)
470. H. Will, paper presented at the meeting of the Gesellschaft für industriellen Pflanzenbau, Wien 1975.
471. H. Knobloch, *Wasser Boden* **18** (1960) 415 – 418.
472. H. Knobloch, *Z. Kulturtech. Flurbereinig.* **8** (1967) 54 – 59.
473. E. Zinck, R. Krötz, *Umweltmagazin* 1980, no. 3, 38 – 40.
474. E. Zinck, DE 2 742 606, 1977.
475. H. Rasp, *Z. Pflanzenernähr. Bodenk.* **133** (1972) no. 1/2, 111 – 123.
476. H. Rasp, *Mitteilungen Deutsche Bodenkundl. Gesellschaft* **15** (1972) 181 – 183.
477. W. Tepe, private communication based on an internal report (1976).
478. S. Maier, *Landwirtsch. Forsch. Sonderh.* **18** (1970) 14 – 20.
479. B. Werminghausen, *Acta Hortic.* **26** (1972) 159 – 164.
480. A. Buchner, J. Jung, P. Weisser, H. Will, *Landwirtsch. Forsch.* **22** (1969) no. 2, 94 – 99.
481. J. F. Kitchens, R. E. Casner, G. S. Edwards, W. E. Harward III, B. J. Macri, EPA 560/2–76–009, ARC 49–5681, Springfield, Va., 1976, pp. 1 – 200.
482. W. Skirde, *Z. Vegetationstechnik* **5** (1982) 1 – 6.
483. L. Unterstenhöfer, *Kunststoffe* **57** (1967) no. 11, 850 – 855.
484. F. Doehler, DE 839 944, 1948.
485. P. di Dio, paper presented at the Congress Chemistry in Agriculture, 2nd., Pressburg/Bratislawa 1972.
486. *Guide to Evaluating Sands and Amendments Used for High Trafficked Turfgrass*, University of California, Agricultural Extension, AXT no. 113 1970.
487. W. Ganssmann, *Phosphorsäure* **22** (1962) 223 – 241.
488. E. Seifert, Mitteilungen Leichtweiss-Institut Technische Universität Braunschweig, **1970,** nos. 25 and 27.
489. T. Niedermaier, *Landwirtsch. Forsch.* **32** (1979) no. 1/2, 129 – 137.
490. H. Gebhardt, *Mitteilungen Deutsche Bodenkundl. Gesellschaft* **15** (1972) 225 – 245.
491. W. Büring, *Das Gartenamt* **23** (1974) no. 5, 258 – 281.
492. K. Wiede, Dissertation, Bonn 1976.
493. W. Büring, *Rasen und Rasengräser.* 1969, no. 6, 78 – 83.
494. B. Walter, E. Petermann, *Weinberg und Keller* **21** (1974) no. 12, 513 – 532.
495. A. H. Sayegh, A. M. Osman, B. Hattar, M. Barasiri, M. Chaudri, M. Nimeh, *Agrochimica*, **26** (1982) no. 1, 22 – 30.
496. H. H. Becher, R. Woytek, *Z. Kulturtech. Flurbereinig.* **25** (1984) 46 – 53.
497. E. Schulze, unpublished reports of the Institut für Pflanzenbau, Universität Bonn, 1970 – 1977.
498. W. Büring, *Das Gartenamt* **31** (1982) 659 – 667.
499. A. Okuda, E. Takahashi, *Nippon Dojo Hiryogaku Zasshi* **32** (1961) 533.
500. L. H. P. Jones, K. A. Handreck, *Adv. Agron.* **19** (1967) no. 19, 107 – 149.
501. J. Vlamis, D. E. Williams, *Plant Soil* **27** (1967) 131.
502. H. Rasp, *Landwirtsch. Forsch. Sonderh.* **35** (1978) 394 – 411.
503. W. Friedmeann, W. Pesch, private communication (1979).
504. H. O. Leh, *Gesunde Pflanze* **38** (1986) no. 6, 246 – 256.
505. Henkel, DE 2 208 917, 1972 (E. Seifert, H. von Freyhold).
506. Guano-Werke, DE 2 612 888, 1976 (J. Bochem, G. Eisenhauer).
507. J. Cronjaeger, A. Hardt, H. Bernard, *Gesunde Pflanze* **45** (1993) 7.

508. M. Kretschmer, K. Schaller, K. Wiegand, *Landwirtsch. Forsch.* **35** (1982), no. 1 – 2, 66 – 76.
509. H. Prün, Landwirtsch. Forsch. *Sonderh.* **30** (1973) 288 – 296.
510. B. Meyer, *Mitteilungen Deutsche Bodenkundl. Gesellschaft* **15** (1972) 253 – 258.
511. F. Callebaut, D. Gabriels, M. D. Boodt, *Med. Fac. Landbouwwet., Rijksuniv. Gent* **49** (1984) no. 1, 53 – 61.
512. E. McGuire, R. N. Carrow, J. Troll, *Agron. J.* **70** (1978) 317 – 348.
513. M. J. Carrol, A. M. Petrovic, *Proceed. Int. Turfgrass Research Conference, 5th*, Avignon 1985, pp. 381 –389.
514. L. M. Rivière, S. Charpentier, B. Valat, *Rev. Horic.* **267** (1986) 43 – 49.
515. H. Prün, paper presented before the Deutsche Rasengesellschaft, Heidelberg 1984.
516. F. Masuda, *CEER Chem. Econ. Eng. Rev.* **15** (1983) no. 11, 19 – 22.
517. R. L. Flannery, W. J. Busscher, *Comm. Soil Sci. Plant Anal.* **13** (1982) no. 2, 103 – 111.
518. *Kirk-Othmer* 4th ed., vol. **22.**
519. C. A. York, N. A. Baldwin, *J. Sports Turf Res. Inst.* **68.**
520. C. A. York, N. A. Baldwin, *International Turfgrass Society Research Journal* **7** (1993).
521. Gefahrstoffverordnung vom 26.08.1986 (letzte Anpassung 06.98) ergänzt durch technische Regeln (TRGS511), Carl Heymann-Verlag, Köln.
522. M. Brübach: "Der Einfluß der Korngröße, der Granulatfestigkeit und der Reibung auf die Verteilung von Dünger- und Pflanzenschutzgranulaten," Dissertation, Berlin 1973.
523. A. Buchner, H. Sturm: *Gezielter düngen*, DLG-Verlag, Frankfurt 1985.
524. K. Rühle, "Die Verteilgenauigkeit pneumatischer Mineraldüngerstreuer," *KTBL-Schrift* no. 198, Darmstadt-Kranichstein 1975.
525. W. Werner, H. W. Olfs, K. Auerswald, K. Isermann: "Stickstoff- und Phosphoreintrag in Oberflächengewässer über diffuse Quellen," in A. Hamm (ed.): *Studie über die Wirkungen und Qualitätsziele von Nährstoffen in Fließgewässern*, Academia Verl., St. Augustin 1991, pp. 665 – 764.
526. W. Werner, H. P. Wodsack: "Stickstoff und Phosphoreintrag in die Fließgewässer Deutschlands unter besonderer Berücksichtigung des Eintragsgeschehens im Lockergesteinsbereich der ehemaligen DDR," *Agrarspektrum*, vol. **22**, DLG-Verl., Frankfurt 1994.
527. O. Strebel: "Untersuchungen zur Grundwasserbelastung bei land- und forstwirtschaftlicher Bodennutzung von Sandböden im Raum Hannover," *Z. Kulturtech. Flurbereinig.* **18** (1977) 310 – 312.
528. P. Obermann: "Hydrochemische/hydromechanische Untersuchungen zum Stoffgehalt von Grundwasser bei landwirtschaftlicher Nutzung," *Bot. Mitt. Dtsch. Gewässerkd.* **42** (1981) 1 – 217.
529. P. Obermann: "Die Belastung des Grundwassers aus landwirtschaftlicher Nutzung nach heutigem Kenntnisstand," in H. Nieder (ed.): *Nitrat im Grundwasser. Herkunft, Wirkungen, Vermeidung,* Verlag Chemie, Weinheim 1985, pp. 53 – 64.
530. P. Obermann, G. Einars, H. Zakosek, G. Schollmayer: *Hydrogeologische/bodenkundliche Untersuchungen im Bereich des Schüttensteiner Waldes bei Bocholt/Westfalen, 1st interim report to the Bundesministerium für Ernährung, Landwirtschaft und Forsten (Ministry of Nutrition, Agriculture, and Forestry): Beitrag der Landwirtschaft zur Verminderung der Nitratbelastung des Grundwassers*, **parts B2 und E**, 1985.
531. P. Obermann, W. Leuchs, G. Einars, H. Zakosek, G. Schollmayer, S. Schulte-Kellinghaus, H. Zepp: *Hydrogeologische/bodenkundliche Untersuchungen im Bereich des Schüttensteiner Waldes bei Bocholt/Westfalen*, final report, 1986.
532. P. Obermann, W. Leuchs: *Chemische Genese eines pleistozären Porengrundwasserleiters bei Bocholt/Westfalen, final report*, 1987.
533. *LWA Nordrhein-Westfalen: Grundwasserbericht 85,* Landesamt für Wasser- und Abfallwirtschaft Nordrhein-Westfalen, Düsseldorf 1985.
534. U. Rohmann: "Landwirtschaftliche und aufbereitungstechnische Maßnahmen zur Lösung des Nitratproblems," DVGW-Schriftenr. no. 46 (1985) 105 – 123.
535. K. Isermann: "Vertikale Verteilung von Nitrat- und Denitrifikationsparametern bis in 10 m Bodentiefe bei unterschiedlicher Landbewirtschaftung," *Herbsttagung Limburgerhof 1986.*
K. Isermann: "Vertical Distribution of NO_3-N and Denitrification Parameters in the Soil

535. [cont.] Down to Greater Depths (about 10 m) under Various Soil Management/Nitrogen Fertilization," International Symposium of CIEC Agricultural Waste Management and Environmental Protection, 4th, Braunschweig 1987.
536. I. F. I. Gast, J. Taat, W. van Duijvenbooden: "Landelijk meetnet — grondwaterkwaliteit," Rijksinstituut voor volksgezondheid en milieuhygiene 2, DEEL Rapport no. 840 382002, April 1985.
537. LAWA – Länderarbeitsgemeinschaft Wasser: "Nitrat," *Bericht zur Grundwasserbeschaffenheit*, Umweltministerium Baden-Württemberg, Stuttgart 1995.
LAWA – Länderarbeitsgemeinschaft Wasser: Ergebnisse der Überwachung der Gewässer gemäß Artikel 5 Abs. 6 Satz 2 der Richtlinie 91/676/EWG. Anlage 3 der Mitteilung der Bundesrepublik Deutschland (BMU) vom 25. Oktober 1996 über die Umsetzung der Richtlinie 91/676/EWG.
538. EU Nitrate Directive 19/676.
539. Bocholter Energie und Wasserversorgung GmbH, personal communication, August 17, 1998.
540. U. Rohmann, H. Sontheimer: *Nitrat im Grundwasser. Ursachen, Bedeutung, Lösungswege*, DVGW Forschungsstelle am Engler-Bunte-Institut, Karlsruhe 1985.
541. Der Rat von Sachverständigen für Umweltfragen: *Umweltprobleme der Landwirtschaft, Sondergutachten März 1985*, Verlag W. Kohlhammer, Stuttgart-Mainz 1985.
542. D. Sauerbeck: *Funktionen, Güte und Belastbarkeit des Bodens aus agrikulturchemischer Sicht. Materialien zur Umweltforschung*, published under the auspices of the Rat von Sachverständigen für Umweltfragen, Verlag W. Kohlhammer, Stuttgart-Mainz, 1985.
543. J. Jung, J. Dressel: "Zum Stand des Nitratproblems," *Herbsttagung Limburgerhof 1981*, 1 – 7.
544. A. Vömel: "Stickstoffdüngung und Stickstoffverluste auf ackerbaulich genutzten Böden," *Landwirtsch*. no. 197 Sonderh. (1981) 205 – 222.
545. O. J. Furrer, 1986: "Einfluß von Fruchtfolge und Düngung auf den Nitrateintrag ins Grundwasser," vol 10 C, Pro Aqua-Pro Vita, Internationale Fachmesse und Fachtagungen für Umweltschutztechnik und ökologie, 10th, Basel 1986.
546. W. Burghardt, cited in H. Kuntze: "Möglichkeiten und Grenzen des Bodenschutzes durch den Landwirt — aus ackerbaulich-bodenkundlicher Sicht," *Bodenschutz — mit der Landwirtschaft*, Arbeiten der DLG **185** (1986) 33 – 35.
547. O. Strebel, J. Böttcher, W. H. M. Duynisfeld: "Einfluß von Standortbedingungen und Bodennutzung auf Nitratauswaschung und Nitratkonzentration des Grundwassers," *Landwirtsch. Forsch., Sonderh.* **37** (1984) 34 – 44.
548. O. Strebel, J. Böttcher: "Ermittlung von Ursachen und Prognosen der mittleren Nitratkonzentration im Grundwasser von Wassereinzugsgebieten," *Mitt. Dtsch. Bodenkd. Ges.* **53** (1987) 305 – 307.
549. F. Hess: "Acker- und pflanzenbauliche Strategien zum verlustfreien Stickstofftransfer beim Anbau im Organischen Landbau," *Mitt. Ges. Pflanzenbauwiss.* **3** (1990) 241 – 244.
550. I. Günther, Farm Economic Data Working Group, Assoc. of Sugar Beet Growers, personal communication, 1995.
551. M. Bach, H. G. Frede, G. Lang: "Entwicklung der Stickstoff-, Phosphor- und Kaliumbilanz der Landwirtschaft in der Bundesrepublik Deutschland," Studie der Gesellschaft f. Boden- und Gewässerschutz e.V., Wettenberg 1997.
552. H. Behrendt: "Detection of anthropogenic Trends in Time series of river rhine lead using Windows of Discharge and long-term Means," Report of the ICES/OSPAR-Workshop "Identification of statistical methods for temporal trends" Kopenhagen, 25 – 26. Sept. 1997, pp. 20 – 24.
553. Umweltbundesamt Berlin (UBA): Forschungsbericht 1994, Berlin, 1995.
554. H. G. Frede, S. Dabbert (eds.): *Handbuch zum Gewässerschutz*, Landwirtschaft. Verlagsges. ECOMED, Landberg. 1998
555. H. Carlson: "Quality Status of the North Sea," *Dtsch. Hydrogr. Z.* (1986) suppl. vol. B, no. 16.
556. E. Schwedhelm, G. Irion: "Schwermetalle und Nährelemente in den Sedimenten der deutschen Nordseewatten," CFS Cour. Forschungsinst. Senkkenberg no. 73 (1985).
557. K. G. Malle: "Wie schmutzig ist die Nordsee?" *Chem. Unserer Zeit* **21** (1987) no. 1, 9 – 16.
558. G. Müller, J. Dominik, R. Reuther, R. Malisch, E. Schulte, L. Acker: "Sedimentary Record of Environmental Pollution in the Western Baltic Sea," *Naturwissenschaften* **67** (1980) 595 – 600.

559. G. Müller: "Zur Chronologie des Schadstoffeintrages in Gewässer," *Geowiss. Unserer Zeit* **1** (1983) 2 – 11.
560. P. J. Crutzen: "Atmospheric Interactions — Homogeneous Gas Reactions of C, N, and S Containing Compounds," in B. Bolin, R. B. Cook (eds.): *The Major Biochemical Cycles and their Interactions,* J. Wiley & Sons, Chichester, England 1983, pp. 65 –114.
561. T. Granli, O. C. Bøckman: "Nitrous Oxide from Agriculture," *Norwegian J. Agric. Sci.* (1994) Suppl. 12.
562. R. F. Weiss: "The Temporal and Spatial Distribution of Tropospheric Nitrous Oxide," *J. Geophys. Res. C* **86** (1981) 7185 – 7195.
563. R. J. Stevens, R. J. Laughlin, L. C. Burns, J. R. M. Arah, R. C. Hood: "Measuring the Contributions of Nitrification and Denitrification to the Flux of Nitrous Oxide from Soil," *Soil. Biol. Biochem.* **29** (1996) 139 – 151.
564. E. A. Kaiser, O. Heinemeyer: "Temporal Changes in N_2O-Losses from two arable Soils," *Plant and Soil* **181** (1996) 57 – 63.
565. F. C. Thornton, R. J. Valente: "Soil Emissions of Nitric Oxide and Nitrous Oxide from no-till Corn," *Soil Sci. Soc. Am. J.* **60** (1996) 1127 – 1133.
566. E. A. Kaiser et al., "What predicts Nitrous Oxide Emission and Denitrification N-Loss from European Soils?" *Z. Pflanzenernähr. Bodenk.* **159** (1996) 541 – 547.
567. E. A. Davidson, P. A. Matson, P. D. Brooks: "Nitrous Oxide Emissions controls and inorganic Nitrogen Dynamics in fertilized tropical agricultural Soils," *Soil Sci. Soc. Am. J.* **60** (1996) 1145 – 1152.
568. K. A. Smith, I. P. McTaggart, H. Tsuruta, K. Smith: "Emissions of N_2O and NO associated with Nitrogen Fertilization in intensive Agriculture, and the Potential for Mitigation," *Soil Use Manage.* **13** (1997) 296 – 304.
569. M. J. Eichner: "Nitrous Oxide Emissions from fertilized Soils: Summary of available Data," *J. Environ. Qual.* **19** (1990) 272 – 280.
570. A. F. Bouwman (ed.): *Soils and the Greenhouse Effect,* Wiley and Sons, New York 1990.
571. C. Henault, X. Devis, J. L. Lucas, J. C. Germon: "Influence of different agriculture Practices (type of Crop, form of N-Fertilizer) on Soil Nitrous Oxide Emission," *Biol. Fertil. Soils* **27** (1998) 299 – 306.
572. K. Isermann: "Agriculture's share in the emission of trace gases affecting the climate and some source-oriented proposals for sufficiently reducing this share", *Environm. Pollution* **83** (1994) 95 – 111.
573. K. L. Weier, J. W. Doran, J. F. Power, D. T. Walters: "Denitrification and the Dinitrogen/Nitrous Oxide Ratio as affected by Soil Water, available Carbon, and Nitrate," *Soil Sci. Soc. Am. J.* **57** (1993) 66 – 72.
574. J. C. G. Ottow, G. Benckiser, M. Kapp, G. Schwarz: "Denitrifikation — die unbekannte Größe," *DLG-Mitteilungen* **105** (1990) 8 – 10.
575. H. Flessa, W. Pfau, P. Dörsch, F. Beese: "The influence of Nitrate and Ammonium Fertilization on N_2O Release and CH_4 Uptake of a well-drained Topsoil demonstrated by a Soil Microcosm Experiment," *Z. Pflanzenernähr. Bodenk.* **159** (1996) 499 – 503.
576. H. Clayton, I. P. McTaggart, J. Parker, L. Swan: "Nitrous Oxide Emission from fertilized Grassland: a 2-Year Study of the Effects of N Fertilizer Form and environmental Conditions," *Biol. Fertil. Soil* **25** (1997) 252 – 260.
577. V. Cole et al.: "Agricultural Options for Mitigation of Greenhouse Gas Emissions," in R. T. Watson, M. C. Zinyowera, R. H. Moss (eds.): *Climate Change 1995*, Chap. 23, Cambridge University Press, Cambridge 1996, pp. 745 – 771.
578. R. Lipschultz, O. C. Zafiriou, S. C. Wofsy, M. B. McElroy, F. W. Valois, S. W. Watson: "Production of NO and N_2O by Soil Nitrifying Bacteria," *Nature* **294** (1981) 641 – 643.
579. C. A. Stutte, R. T. Weiland: "Gaseous Nitrogen Loss and Transpiration of Several Crops and Weed Species," *Crop. Sci.* **18** (1978) 887 – 889.
580. C. A. Stutte, R. T. Weiland, A. R. Blem: "Gaseous Nitrogen Loss from Soybean Foliage," *Agron. J.* **71** (1979) 95 – 97.
581. J. V. Dean, J. E. Harper: "Nitric Oxide and Nitrous Oxide Production by Soybean and Winged Bean During the In Vivo Nitrate Reductase Assay," *Plant Physiol.* **82** (1986) 718 – 723.
582. I. E. Galbally: "Factors controlling NO_x emissions from soils", *Exchange of trace gases between terrestrial ecosystems and the atmosphere,* John Wiley and Sons, New York 1989 pp. 23 – 27.
583. T. Yoneyama, A. Hashimoto, T. Totsuka: "Absorption of Atmospheric NO_2 by Plants and Soils. 4. Two Routes of Nitrogen Uptake

by Plants from Atmospheric NO$_2$: Direct Incorporation into Aerial Plant Parts and Uptake by Roots after Absorption into the Soil," *Soil Sci. Plant Nutr. (Tokyo)* **26** (1980) no. 1, 1 – 7.
584. R. Lensi, A. Chalamet: "Absorption de l'oxyde nitreux par les parties aeriennes du mais," *Plant Soil* **59** (1981) 91 – 98.
585. J. C. Ryden: "N$_2$O Exchange between a Grassland Soil and the Atmosphere," *Nature* **292** (1981) 235 –237.
586. H. H. Rogers, V. P. Aneja: "Uptake of Atmospheric Ammonia by Selected Plant Species," *Environ. Exp. Bot.* **20** (1980) 251 – 257.
587. M. J. Koziol, R. R. Whatley: *Gaseous Air Pollutants and Plant Metabolism*, Butterworths, Borough Green, Kent, England, 1984.
588. F. G. M. Roelofs, A. J. Kempers, A. L. F. M. Houdijk, J. Jansen: "The Effect of Air Borne Ammonium Sulphate on Pinus Nigra var. maritima in the Netherlands," *Plant Soil* **84** (1985) 45 – 56.
589. J. G. M. Roelofs, A. W. Boxman, H. E. G. Van Dijk: "Effects of Airbone Ammonium on Natural Vegetation and Forests," *Proceedings of the EEC Symposium*, Grenoble 1987.
590. K. Isermann: "Bodenversauerung beim natürlichen Wald- und Grasland-ökosystem von Rothamsted (England) unter dem Einfluß atmosphärischer und pedogener Protonen-Anlieferung von 1880 bis heute," a paper presented at the Kongreß der Internationalen Bodenkundlichen Gesellschaft, 13th, Hamburg 1986 and published in *AFZ, Allg. Forstzeitschr.* **1987**, pp. 379 – 381.
591. K. Isermann: "Die Rolle des Stickstoffs bei den "neuartigen" Waldschäden," paper presented at the Symposium Waldschäden am Biologischen Institut II der Universität Freiburg, 3rd, 1986.
592. L. De Temmermann, A. Ronse, K. Van den Cruys, K. Meeus-Verdinne: "Ammonia and Pine Tree Dieback in Belgium," *Proceedings of the EEC Symposium*, Grenoble 1987.
593. W. M. J. Den Boer: "Ammonia, Not only a Nutrient but also a Cause of Forest Damages," Wiss. Symposium zum Thema Waldschäden, Neue Ursachenhypothesen, Text 19, Umweltbundesamt, Berlin 1986.
594. H. Ellenberg: "Veränderungen der Flora unter dem Einfluß von Düngung und Immissionen," *Schweiz. Z. Forstwes.* **136** (1985) 19 – 39.
595. G. Büttner, N. Lamersdorf, R. Schultz, B: Ulrich: *Deposition und Verteilung chemischer Elemente in küstennahen Waldstandorten — Fallstudie Wingst; Abschlußbericht*, report of the Forschungszentrum Waldökosysteme/Waldsterben, series B, vol. 1, 1986.
596. K. Kreutzer, J. Bittersohl: "Stoffauswaschung aus Fichtenkronen (Picea abies (L.) Karst.) durch saure Beregnung," *Forstwiss. Centralbl.* **105** (1986) no. 4, 357 – 363.
597. M. Hauhs: "Die regionale Bedeutung des Nitrataustrages unter Wald für die Gewässerversauerung," DFG-Workshop Modellierung des bodeninternen Stickstoffumsatzes, Hohenheim 1987.
598. G. L. Hutchinson, F. G. Viets: "Nitrogen Enrichment of Surface Water by Adsorption of Ammonia Volatilized from Cattle Feedlots," *Science (Washington, D.C.)* **166** (1969) 514 – 515.
599. J. Gehrmann: "Derzeitiger Stand der Belastung von Waldökosystemen in Nordrhein-Westfalen durch Deposition von Luftverunreinigungen," *Forst-Holzwirt.* **42** (1987) no. 6, 141 – 145.
600. S. Mehlert: "Untersuchungen zur atmogenen Stickstoffdeposition und zur Nitratverlagerung", UFZ-Bericht 22, UFZ-Umweltforschungszentrum Leipzig-Halle GmbH, 1996.
601. L. Grünhage, H. J. Jäger, K. Freitag, K. Hanewald: "Emissionskataster Hessen — Landesweite Abschätzung der Emissionen aus biogenen und nicht gefaßten Quellen," Umweltplanung, Arbeits- und Umweltschutz, no. 184, Hessische Landesanstalt für Umwelt, Wiesbaden 1996.
602. ECETOC (ed.): "Ammonia Emissions to Air in Western Europe" Tech. Rep. No. 62, European Centre for Ecotoxicology and Toxicology of Chemicals, Brussels 1996.
603. "Gasförmige Stickstoffverbindungen aus der Landwirtschaft", Bericht für die DFG-Senatskommission zur Beurteilung von Stoffen in der Landwirtschaft, Schlußentwurf, Bonn April 1996.
604. P. A. Steudler, R. D. Bowden, J. M. Melillo, J. D. Aber: "Influence of nitrogen fertilization on methane uptake in temperate forest soils", *Nature* **341** (1989) 314 – 316.
605. S. Hansen, J. E. Maehlum, L. R. Bakken: "N$_2$O and CH$_4$ fluxes influenced by fertilization and tractor traffic", *Soil Biol. Biochem.* **259** (1993) 621 – 630.

606. B. W. Hütsch, C. P. Webster, D. S. Powlson: "Long-term effects of nitrogen fertilization on methane oxidation in soil of the Broadbalk Wheat Experiment", *Soil Biol. Biochem.* **25** (1993) 1307 – 1315.
607. K. F. Bronson, A. R. Mosier: "Suppression of methane oxidation in aerobic soil nitrogen inhibitors and urease inhibitors", *Biol. Fertil. Soils* **17** (1994) 68 – 84.
608. C. Bédard, R. Knowles: "Physiology, biochemistry and specific inhibitors of CH_4, NH_4^+ and CO oxidation by methanotrophs and nitrifier," *Microbiol. Rev.* **53** (1989) 68 – 94.
609. K. Mengel, E. A. Kirkby: *Principles of Plant Nutrition*, International Potash Institute, Worblaufen-Bern 1982.
610. G. Pfützer, C. Pfaff: "Einfluß der Düngung und Belichtung auf die Vitaminbildung in der Pflanze," *Umschau (1897)* **39** (1935) 917.
611. C. Pfaff, G. Pfützer: "über den Einfluß der Ernährung auf den Carotin- und Ascorbinsäuregehalt verschiedener Futterpflanzen," *Z. Angew. Chem.* **50** (1937) 179.
612. G. A. Rodgers, A. Penny, F. V. Widdowson, M. V. Hewitt: "Tests of Nitrification and of Urease Inhibitors, when Applied with Either Solid or Aqueous Urea, on Grass Grown on a Light Sandy Soil," *J. Agric. Sci.* **108** (1987) 109 – 117.
613. R. Prakasa, K. Puttanna: "Nitrification and Ammonia Volatilisation Losses from Urea and Dicyandiamide-treated Urea in a Sandy Loam Soil," *Plant Soil* **97** (1987) 201 – 206.
614. H. Bernhard: *Phosphor: Wege und Verbleib in der Bundesrepublik Deutschland. Probleme des Umweltschutzes und der Rohstoffversorgung*, Verlag Chemie, Weinheim-New York 1978.
615. A. Hamm: "Wie und woher kommen die Nährstoffe in die Flüsse?" in J. L. Lozan, H. Kausch (eds.): *Warnsignale aus Flüssen und Ästuaren, Wissenschaftliche Fakten*, Verl. Paul Parey, Berlin – Hamburg 1996 pp. 105 – 110.
616. A. Breeuwsma, J. G. A. Reijerink, O. F. Schoumans: Fosfaatverzadigde gronden in het oostelijk, central en zuidelijk zandgebied, Report 68, Staring Centrum, Wageningen 1990.
617. U. Pihl, W. Werner: "Erhebungsuntersuchungen zu Phosphatgehalten, Phosphatsorptionskapazität und relativer Phosphatsättigung der Böden in veredlungsstarken Gebieten Nordrhein-Westfalens als Prognosekriterien des potentiellen Phosphataustrags in Drain- und Grundwasser," Forschungsber. 64, Lehr- und Forschungsschwerpunkt "Umweltverträgliche und standortgerechte Landwirtschaft", Landwirtschaftliche Fakultät, Universität Bonn, Bonn 1998.
618. Th. Diez, K. Bucksteeg: "P-fluxes in a Watershead Area with Intensive Agriculture and Eutrophication of a Lake," paper presented at the Int. Symposium of CIEC Agricultural Waste Management and Environmental Protection, 4th, Braunschweig-Voelkenrode 1987.
619. W. Werner: "Implementation and efficiency of countermeasures against diffuse nitrogen and phosphorus inputs into surface waters from agriculture", in E. Romstad, J. Simonsen, A. Vatn (eds.): *Controlling mineral emissions in European agriculture*, CAB International, Oxon – New York 1997 , pp. 73 – 88.
620. Industrieverband Agrar e.V., personal communication, 1997.
621. K. Isermann: "Einfluß der Phosphatdüngung auf den Cadmiumgehalt des Bodens, ermittelt anhand zahlreicher Dauerversuche in Westeuropa," *Landwirtsch. Forsch. Sonderh.* **39** (1982) 283 – 301.
622. B. Machelett, W. Podlesak, J. Garz: "Zur Wirkung des Cadmiumgehalts im Düngerphosphat auf die Cadmiumkonzentration in Boden und Pflanze in einem langjährigen Phosphatdüngungsversuch," *Arch. Acker-Pflanzenbau Bodenk.* **28** (1984) no. 4, 247 – 251.
623. J. J. Mortvedt: "Cadmium Levels in Soils and Plant Tissues from Long Term Soil Fertility Experiments in the United States," *Congr. Int. Soc. Soil Science, 13th*, vol. **3**, Hamburg 1986.
624. W. Werner, C. Brenk: "Entwicklung eines integrierten Nährstoffkonzeptes als Basis eines umweltverträglichen, flächendeckenden Recyclings kommunaler Abfälle in Nordrhein-Westfalen und regionalisierte Bilanzierung der Schwermetallflüsse", Forschungsber. 48, Landwirtschaftliche Fakultät Universität Bonn, Bonn 1997.

General References

625. *FAO Fertilizer Yearbook* , Food and Agricultural Organization of the United Nations, Rome.
626. *FAO Production Yearbook* , Food and Agricultural Organization of the United Nations, Rome.

Specific References

627. Q. Paris: "The von Liebig-Hypothesis," *Am. J. Agric. Econ*, **74** (1992) 4, 1019 – 1028.
628. E. von Boguslawski: *Ackerbau, Grundlagen der Pflanzenproduktion*, Frankfurt/M. 1981.
629. K. Baeumer: "Düngung", in R. Diercks, R. Heitefuss (eds.): *Integrierter Landbau*, 2nd ed., München 1994 pp. 88 – 110.
630. E. O. Heady, J. L. Dillon: *Agricultural Production Functions*, Ames, Iowa 1961.
631. F. Kuhlmann: *Betriebslehre der Agrar- und Ernährungswirtschaft*, 2nd ed., Frankfurt/M. 2003.
632. *World Fertilizer Consumption Statistics April 1998* IFA International Fertilizer Industry Association, 1998. FAOSTAT Database results Internet http://www.FAO.org
633. FAO: *The Sixth World Food Survey*, Food and Agricultural Organisation of the United Nations, Rome 1996.
634. P. Prinstrup-Anderson et al.: *The World Food Situation: Recent Developments*, Emerging Issues and Long Term Prospects, IFPRI, Washington 1997.
635. FAO: *Dimension of Need, An Atlas of Food and Agriculture*, Food and Agricultural Organisation of the United Nations, Rome 1995.
636. G. Harris: *"An Analysis of Global Fertilizer Application Rates for Major Crops,"* International Fertilizer Development Centre (IFDC), Muscle Shoals 1998.
637. FAO: *Fertilizer Yearbook*, vol. **46**, Food and Agricultural Organisation of the United Nations, Rome 1996.
638. IFA: *"Fertilizer Demand Forecast,"* International Fertilizer Industry Association, Paris 1998.
639. FAO: *Fertilizer Outlook*, Food and Agricultural Organisation of the United Nations, Rome 1996
640. J. Debruck, G. Fischbeck, W. Kampe: *Getreidebau aktuell*, 7th ed., Frankfurt/M. 1983.
641. F. Kuhlmann: "Zum 50. Todestag von Friedrich Aereboe: Einige Gedanken zu seiner Intensitätslehre," *Agrarwirtschaft* **41** (1992) 222 – 230.
642. F. Kuhlmann, K. F. Frick: "Produktionstheoretische Betrachtungen über das "Gesetz des Minimums", und das "Gesetz vom abnehmenden Ertragszuwachs"," *Berichte über Landwirtschaft* **73** (1995) 591 – 623.
643. K. F. Frick: "Analyse und Entwicklung von Standardproduktionsfunktionstypen in der pflanzlichen Produktion," *Sonderheft der Agrarwirtschaft*, vol. **153**, Holm 1997.

Ammonia

Max Appl, Dannstadt-Schauernheim, Federal Republic of Germany

1.	Occurrence and History	144
2.	Physical Properties	144
3.	Chemical Properties	150
4.	Production	153
4.1.	Historical Development	153
4.2.	Thermodynamic Data	155
4.3.	Ammonia Synthesis Reaction	156
4.3.1.	General Aspects	156
4.3.2.	Catalyst Surface and Reaction Mechanism	158
4.3.3.	Kinetics	163
4.4.	Catalysts	167
4.4.1.	Classical Iron Catalysts	168
4.4.1.1.	Composition	169
4.4.1.2.	Particle Size and Shape	175
4.4.1.3.	Catalyst-Precursor Manufacture	177
4.4.1.4.	Catalyst Reduction	178
4.4.1.5.	Catalyst Poisons	181
4.4.2.	Other Catalysts	183
4.4.2.1.	General Aspects	183
4.4.2.2.	Metals with Catalytic Potential	184
4.4.2.3.	Commercial Ruthenium Catalysts	185
4.5.	Process Steps of Ammonia Production	186
4.5.1.	Synthesis Gas Production	187
4.5.1.1.	Feedstock Pretreatment and Raw Gas Production	187
4.5.1.2.	Carbon Monoxide Shift Conversion	193
4.5.1.3.	Gas Purification	196
4.5.2.	Compression	200
4.5.3.	Ammonia Synthesis	204
4.5.3.1.	Synthesis Loop Configurations	204
4.5.3.2.	Formation of Ammonia in the Converter	205
4.5.3.3.	Waste-Heat Utilization and Cooling	219
4.5.3.4.	Ammonia Recovery from the Ammonia Synthesis Loop	220
4.5.3.5.	Inert-Gas and Purge-Gas Management	221
4.5.3.6.	Influence of Pressure and Other Variables of the Synthesis Loop	224
4.5.3.7.	Example of an Industrial Synthesis Loop	225
4.6.	Complete Ammonia Production Plants	225
4.6.1.	Steam Reforming Ammonia Plants	225
4.6.1.1.	The Basic Concept of Single-Train Plants	225
4.6.1.2.	Further Developments	228
4.6.1.3.	Minimum Energy Requirement for Steam Reforming Processes	230
4.6.1.4.	Commercial Steam Reforming Ammonia Processes	232
4.6.1.4.1.	Advanced Conventional Processes	233
4.6.1.4.2.	Processes with Reduced Primary Reformer Firing	236
4.6.1.4.3.	Processes Without a Fired Primary Reformer (Exchanger Reformer)	239
4.6.1.4.4.	Processes Without a Secondary Reformer (Nitrogen from Air Separation)	241
4.6.2.	Ammonia Plants based on Partial Oxidation	242
4.6.2.1.	Ammonia Plants Based on Heavy Hydrocarbons	242
4.6.2.2.	Ammonia Plants Using Coal as Feedstock	245
4.6.3.	Waste-Heat Boilers for High-Pressure Steam Generation	247
4.6.4.	Single-train Capacity Limitations – Mega-Ammonia Plants	248
4.7.	Modernization of Older Plants (Revamping)	249
4.8.	Material Considerations for Equipment Fabrication	251
5.	Storage and Shipping	253
5.1.	Storage	253
5.1.1.	Pressure Storage	254
5.1.2.	Low-Temperature Storage	255
5.1.3.	Underground Storage	256
5.1.4.	Storage of Aqueous Ammonia	257
5.2.	Transportation	257
6.	Quality Specifications and Analysis	258
7.	Environmental, Safety, and Health Aspects	258

7.1.	Environmental Aspects of Ammonia Production and Handling	258
7.2.	Safety Features	260
7.3.	Health Aspects and Toxicity of Ammonia	262
8.	Uses	263
9.	Economic Aspects	263
10.	Future Perspectives	266
11.	References	270

1. Occurrence and History

Occurrence. Ammonia, NH_3, occurs in nature almost exclusively in the form of ammonium salts. Natural formation of ammonia is primarily by decomposition of nitrogen-containing organic materials or through volcanic activity. Ammonium chloride can deposit at the edges of smoldering, exposed coal beds, as observed in Persia before 900 A.D. Similar deposits can be found at volcanoes, for example, Vesuvius or Etna, in Italy. Ammonia and its oxidation products, which combine to form ammonium nitrate and nitrite, are produced from nitrogen and water vapor through electrical discharges in the atmosphere. These ammonium salts, as well as those arising from industrial and automotive exhausts, supply significant quantities of the nitrogen needed by growing plants when eventually deposited on the earth's surface. Ammonia and its salts are also byproducts of commercial processing (gasification, coking) of fossil fuels such as coal, lignite, and peat (see Fig. 1).

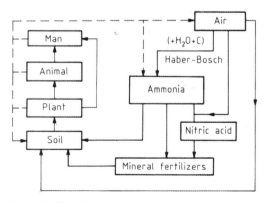

Figure 1. The nitrogen cycle

History. The name ammonia is derived from "sal ammoniacum" (Oasis Ammon in Egypt, today Siwa). Sal ammoniac was known to the ancient Egyptians; the Arabs were aware of ammonium carbonate. Free ammonia was prepared for the first time in 1774 by J. B. PRIESTLEY. In 1784, C. L. BERTHOLLET recognized that ammonia was composed of the elements nitrogen and hydrogen. W. HENRY, in 1809, determined the volumetric ratio of the elements as 1 : 3, corresponding to the chemical formula NH_3.

Following the discovery of the nature and value of mineral fertilization by LIEBIG in 1840, nitrogen compounds were used in increasing quantities as an ingredient of mineral fertilizers. At the end of the last century ammonia was recovered in coke oven plants and gas works as a byproduct of the destructive distillation of coal. The produced ammonium sulfate was used as fertilizer. Another nitrogen fertilizer was calcium cyanamide which was manufactured by the Frank – Caro process from 1898 onwards. Since both sources of nitrogen were limited in quantity they did not suffice for fertilization. Therefore, it was necessary to use saltpeter from natural deposits in Chile. In 1913, the first Haber – Bosch plant went on stream, representing the first commercial synthesis of ammonia from the elements. Subsequently, other ammonia plants were started up, and synthetic ammonia and/or its derivatives soon replaced Chile saltpeter. Today ammonia is a commodity product of the chemical industry. World production capacity in 2005 exceeded 168×10^6 t ammonia.

2. Physical Properties

Molecular Properties. Corresponding to its nuclear charge number, the nitrogen atom possesses seven shell electrons. One electron pair is in the ground state $1\,s$ (K shell), and five electrons are distributed over the four orbitals with the principal quantum number 2 (L shell). Of these, one electron pair occupies the $2\,s$ level and three unpaired electrons, respectively, a half of the remaining three levels, $2\,p_x$, $2\,p_y$, $2\,p_z$.

The unpaired electrons can enter into electron-pair bonds with the 1s electron of three hydrogen atoms. Thus, the three half-occupied orbitals of the L shell become about fully occupied (formation of an octet of the neon type in accordance with the octet theory of Lewis – Langmuir).

$$:\!\overset{..}{N}\!\cdot\; +\; 3\,H\cdot\; \longrightarrow\; :\!\overset{H}{\underset{H}{N}}\!:\!H$$

The nitrogen atom is at the apex of a pyramid above the plane of the three hydrogen atoms, which are arranged in an equilateral triangle. The H-N-H bond angle is about 107° [1]. Although covalent, the nitrogen – hydrogen bonds have a polar contribution because of the stronger electronegativity of nitrogen relative to hydrogen. Because of polarization of the bonds and the unsymmetrical molecular arrangement, the ammonia molecule has a considerable dipole moment, 1.5 D [2].

As the ammonia molecule possesses the same electron configuration as water (isosterism) and similar bond angles (water vapor bond angle 105°, dipole moment 1.84 D), ammonia and water behave similarly in many reactions. Ammonia and water are diamagnetic. The dielectric constant of liquid ammonia is about 15 and greater than that of most condensed gases; therefore, liquid ammonia has a considerable ability to dissolve many substances. The ammonia molecule, with its free electron pair, can combine with a proton.

$$H\!:\!\overset{H}{\underset{H}{N}}\!:\; +\; H\!:\!\overset{..}{\underset{..}{O}}\!:\!H \;\rightleftharpoons\; \left[H\!:\!\overset{H}{\underset{H}{N}}\!:\!H\right]^{+} + \left[:\!\overset{..}{\underset{..}{O}}\!:\!H\right]^{-}$$

In the resulting ammonium ion, the nitrogen atom is situated in the middle of a tetrahedron whose corners are occupied by hydrogen atoms. The four hydrogen atoms are equivalent in their behavior. According to LINUS PAULING, the positive excess charge is distributed about equally over all five atoms.

Compilations of Physical Data. The results of comprehensive investigations of the physical properties of ammonia have been published in [3] and [4]. Both papers provide numerous equations for physical properties derived from published data, the laws of thermodynamics, and statistical evaluation. These equations are supplemented by lists and tables of thermodynamic quantities and an extensive collection of literature references.

Moreover, data on physical properties and the complex systems important in synthesis may be found in [5 – 8], and of course, in the well-known tabulations *Landolt-Börnstein* [9] and *Handbuch der Kältetechnik* [10], among others. The most important physical data are compiled in Table 1.

$p - V - T$ Data. The $p - V - T$ data in Table 2 are calculated from the equations in [4] and further data in [3, 5 – 10]. Measured $p - V - T$ values for liquid ammonia in the pressure range from 7 to 180 MPa (70 – 1800 bar) and at temperatures from $-20\,°C$ to $40\,°C$ may be found in [11]. Detailed information on compressibility gained from experimental data can be found in [12 – 17]. An equation of state for liquid ammonia is given in [18]; for a more simple form, see [8]. Further data may be found in references [19 – 21]. Properties of liquid ammonia from -50 to $65\,°C$ and from saturation pressure up to 370 bar are reported in [22].

Caloric Data. Enthalpy and entropy of solid ammonia are given in [5]. Enthalpy and entropy may be calculated by the equations in [4]. Further data are given in [3, 23, 24] and [5 – 10]. An enthalpy log p diagram can be found in [10].

Specific Heat. In the range of $-45\,°C$ to $50\,°C$ the specific heat (in kJ kg^{-1} K^{-1}) of liquid ammonia can be calculated using the equation:

$$c_p = -3.787 + 0.0949T - 0.3734 \times 10^{-3}T^2 + 0.5064 \times 10^{-6}T^3 \tag{1}$$

where $T = \vartheta + 273.15$.

Assuming ammonia to be an ideal gas in the range 300 – 2000 K the following equation represents the specific heat:

Table 1. Properties of ammonia

M_r	17.0312
Molecular volume	
(at 0 °C, 101.3 kPa)	22.08 L/mol
Specific gas constant R	0.48818 kPa m^3 kg^{-1} K^{-1}
Liquid density	
(at 0 °C, 101.3 kPa)	0.6386 g/cm^3
Gas density	
(at 0 °C, 101.3 kPa)	0.7714 g/L
Liquid density	
(at -33.43 °C, 101.3 kPa)	0.682 g/cm^3
Gas density	
(at -33.43 °C, 101.3 kPa)	0.888 g/L
Critical pressure	11.28 MPa
Critical temperature	132.4 °C
Critical density	0.235 g/cm^3
Critical volume	4.225 cm^3/g
Critical compressibility	0.242
Critical thermal conductivity	0.522 kJ K^{-1} h^{-1} m^{-1}
Critical viscosity	23.90×10^{-3} mPa · s
Melting point (triple point)	-77.71 °C
Heat of fusion (at 101.3 kPa)	332.3 kJ/kg
Vapor pressure (triple point)	6.077 kPa
Boiling point (at 101.3 kPa)	-33.43 °C
Heat of vaporization	
(at 101.3 kPa)	1370 kJ/kg
Standard enthalpy of	
formation (gas at 25 °C)	-46.22 kJ/mol
Standard entropy	
(gas at 25 °C, 101.3 kPa)	192.731 J mol^{-1} K^{-1}
Free enthalpy of formation	
(gas at 25 °C, 101.3 kPa)	-16.391 kJ/mol
Net heating value, LHV	18.577 kJ/g
Gross heating value, HHV	22.543 kJ/g
Electrical conductivity	
(at -35 °C), very pure	1×10^{-11} Ω^{-1} cm^{-1}
Commercial	3×10^{-5} Ω^{-1} cm^{-1}
Ignition temperature acc.	
to DIN 51794	651 °C
Explosive limits	
NH$_3$ – O$_2$ mixture	
(at 20 °C, 101.3 kPa)	15 – 79 vol % NH$_3$
NH$_3$ –air mixture	
(at 0 °C, 101.3 kPa)	16 – 27 vol % NH$_3$
(at 100 °C, 101.3 kPa)	15.5 – 28 vol % NH$_3$

$$c_p = 1.4780 + 2.09307 \times 10^{-3} T - 2.0019 \times 10^{-7} T^2 \\ - 8.07923 \times 10^{-11} T^3 \quad (2)$$

The values of the specific heat at constant pressure in Table 3 have been calculated according to [4] and have been extrapolated to 500 °C using data from [3]. Further data can be found in [5 – 9, 25] and [26].

Properties of Saturated Ammonia Liquid and Vapor. Table 4 is derived from [4]. Further data may be found in [3, 7, 9], and [10]. For calculations from reduced data, among others, see [27, 28].

Vapor – Liquid Equilibria for the Ammonia, Hydrogen, Nitrogen, Argon, Methane System (Fig. 2). Because of the great importance of absorption processes in synthesis loop engineering (see Section 4.5.3.5), these binary and multicomponent systems have come under experimental and theoretical reexamination. The new efforts are characterized above all by consideration of thermodynamic relationships in combination with equations of state.

In [32] various publications are compared with one another and equilibrium methods for the ranges 273 – 323 K and 4.9 – 49 MPa (49 to 490 bar) are reported. Investigations of binary systems are reported in [30, 33 – 40] and equations and compilations in [31, 41], and [42], among others. The basis for the saturation concentration (Table 5) and solubilities in liquid ammonia (Table 6) are unpublished BASF calculations produced with the help of the Lee – Kesler equation of state [43] and publications [30, 31, 34 – 41]. The ammonia content of the gas may be calculated with the formulas from [37].

A mutual interaction of solubilities exists in all multicomponent systems. The interaction with methane is pronounced. This interdependence is treated in detail in [30] and [44 – 46].

Surface Tension. The following equation, from [47], represents the surface tension of ammonia (σ in Nm^{-1}) over the range -50 to 50 °C.

$$\sigma = 26.55 \times 10^{-3} - 2.3 \times 10^{-4} \vartheta \quad (3)$$

Experimental data are given in [48]. The surface tension of aqueous ammonia solutions over the ranges 0.1 – 1.2 MPa (1 – 12 bar) and 20 – 100 °C can be found in [49].

Dynamic Viscosity. A survey on dynamic viscosity data for the ranges 30 – 250 °C and 0.1 – 80 MPa (1 – 800 bar) appears in [7]; further data occur in [50] and [51]. Experimental viscosity data and correlating equations for the range 448 – 598 K and for pressures up to 12.1 MPa (121 bar) can be found in [52]. Generalized equations for viscosity calculations on refrigerants are presented in [53].

The following, from [47], is a formula for interpolation along the vapor pressure curve for liquid ammonia (viscosity η in N s m^{-2}):

$$\eta = 1.949 \times 10^{-4} - 1.72 \times 10^{-6} \vartheta + 7 \times 10^{-10} \vartheta^2 \quad (4)$$

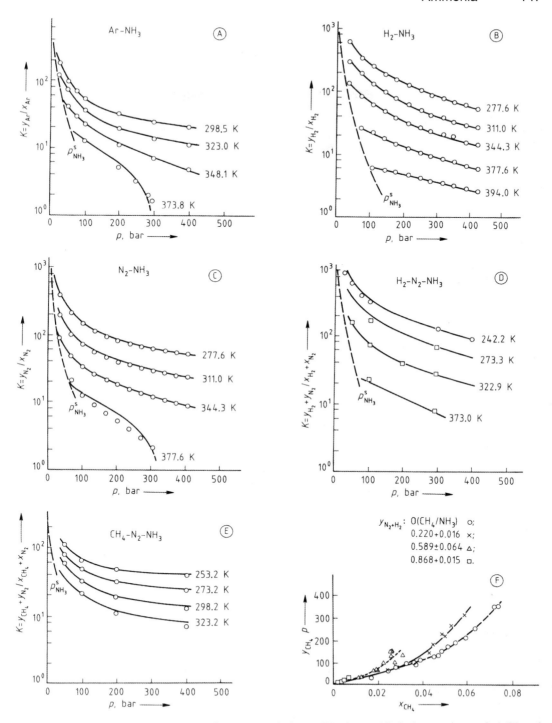

Figure 2. Equilibrium solubilities (K factors) for argon (A), hydrogen (B), nitrogen (C), hydrogen : nitrogen 3 : 1 (D), and methane : nitrogen in liquid ammonia (E). F shows the dependence of the methane partial pressure ($p \cdot y\text{CH}_4$) on the concentration of methane dissolved in the liquid, $x\text{CH}_4$, for three different gas mixtures
A – C and E according to [29], F according to [30], D was developed from data reported in [31]

Table 2. Specific volume of ammonia in L/kg

p, MPa	Temperature, °C											
	−33	−20	−10	0	10	20	50	100	150	200	250	300
0.1	1172.001	1235.492	1284.303	1333.094	1381.870	1430.634	1576.882	1820.589	2064.339	2308.163	2552.061	2796.020
0.5	1.467	1.503	1.533	1.566	257.613	269.524	303.259	356.206	407.353	457.602	507.299	556.610
1	1.466	1.503	1.533	1.565	1.600	1.638	144.981	173.883	200.728	226.667	252.117	277.272
2	1.465	1.501	1.531	1.563	1.598	1.636	64.713	82.476	97.322	111.121	124.391	137.352
3	1.464	1.500	1.530	1.562	1.596	1.634	1.772	51.740	62.785	72.589	81.818	90.721
4	1.463	1.499	1.528	1.560	1.594	1.631	1.768	36.088	45.448	53.305	60.532	67.414
5	1.462	1.498	1.527	1.558	1.592	1.629	1.764	26.361	34.981	41.713	47.754	53.432
6	1.461	1.497	1.526	1.557	1.590	1.627	1.760	19.379	27.944	33.968	39.229	44.109
7	1.460	1.495	1.524	1.555	1.589	1.625	1.756	2.171	22.859	28.422	33.137	37.449
8	1.459	1.494	1.523	1.554	1.587	1.623	1.752	2.151	18.986	24.251	28.565	32.455
10	1.458	1.492	1.520	1.551	1.583	1.618	1.745	2.115	13.375	18.385	22.163	25.467
15	1.453	1.486	1.514	1.543	1.574	1.608	1.727	2.048	4.321	10.478	13.632	16.176
20	1.448	1.481	1.508	1.536	1.566	1.599	1.712	1.999	2.780	6.534	9.408	11.574
25	1.443	1.475	1.501	1.529	1.558	1.590	1.697	1.959	2.516	4.474	6.957	8.863
30	1.438	1.470	1.496	1.522	1.551	1.581	1.684	1.926	2.376	3.540	5.443	7.113
35	1.434	1.465	1.490	1.516	1.544	1.573	1.672	1.898	2.282	3.094	4.494	5.922
40	1.429	1.460	1.484	1.510	1.537	1.565	1.660	1.873	2.211	2.839	3.893	5.088
45	1.424	1.454	1.479	1.504	1.530	1.557	1.650	1.850	2.155	2.670	3.497	4.491
50	1.419	1.449	1.473	1.498	1.523	1.550	1.639	1.830	2.108	2.547	3.221	4.054
55	1.414	1.444	1.468	1.492	1.517	1.543	1.630	1.811	2.068	2.453	3.020	3.727
60	1.409	1.439	1.462	1.486	1.511	1.536	1.620	1.794	2.033	2.377	2.865	3.476

Table 3. Specific heat of ammonia at constant pressure in kJ kg^{-1} K^{-1}

p, MPa	Temperature, °C												
	−33	−20	−10	0	10	20	50	100	150	200	250	300	400	500
0.1	2.324	2.236	2.197	2.175	2.163	2.159	2.174	2.240	2.328	2.422	2.518	2.612	2.797	2.966
0.5	4.468	4.528	4.572	4.618	2.673	2.553	2.383	2.339	2.385	2.459	2.542	2.628	2.815	2.981
1	4.466	4.526	4.569	4.615	4.668	4.736	2.707	2.470	2.456	2.504	2.574	2.653	2.838	2.998
2	4.462	4.521	4.563	4.607	4.659	4.725	3.785	2.787	2.611	2.597	2.636	2.696	2.886	3.028
3	4.458	4.516	4.557	4.600	4.651	4.714	5.039	3.216	2.788	2.697	2.701	2.742	2.934	3.060
4	4.454	4.511	4.552	4.594	4.642	4.703	5.014	3.832	2.994	2.805	2.769	2.788	2.981	3.091
5	4.450	4.506	4.546	4.587	4.634	4.693	4.991	4.826	3.236	2.923	2.840	2.837	3.028	3.121
6	4.446	4.501	4.541	4.580	4.626	4.683	4.969	6.854	3.526	3.050	2.916	2.887	3.067	3.168
7	4.442	4.497	4.535	4.574	4.618	4.673	4.947	6.776	3.882	3.190	2.994	2.939	3.069	3.203
8	4.439	4.492	4.530	4.567	4.610	4.664	4.927	6.551	4.330	3.342	3.077	2.992	3.155	3.238
10	4.432	4.484	4.520	4.555	4.595	4.645	4.889	6.215	5.724	3.695	3.255	3.103	3.242	3.304
15	4.415	4.463	4.496	4.527	4.561	4.603	4.806	5.701	20.410	4.965	3.776	3.407	3.470	3.463
20	4.400	4.444	4.474	4.501	4.530	4.566	4.736	5.396	8.827	6.899	4.399	3.739	3.644	3.607
25	4.387	4.427	4.454	4.477	4.502	4.532	4.676	5.187	6.826	8.074	5.046	4.080	3.812	3.734
30	4.376	4.412	4.435	4.456	4.477	4.502	4.624	5.032	6.048	7.426	5.543	4.397	3.980	3.861
35	4.367	4.399	4.419	4.436	4.454	4.475	4.578	4.912	5.614	6.605	5.741	4.652	4.104	3.967
40	4.361	4.387	4.404	4.418	4.432	4.450	4.538	4.814	5.330	6.033	5.693	4.818	4.229	4.074
45	4.358	4.377	4.391	4.402	4.413	4.427	4.502	4.734	5.128	5.643	5.539	4.894	4.310	4.159
50	4.359	4.369	4.379	4.387	4.395	4.407	4.469	4.665	4.975	5.365	5.364	4.903	4.392	4.244
55	4.367	4.363	4.369	4.374	4.379	4.388	4,440	4.607	4.855	5.157	5.202	4.873	4.442	4.301
60	4.385	4.360	4.361	4.362	4.365	4.370	4.413	4.556	4.757	4.996	5.060	4.824	4.492	4.359

An interpolation formula for ammonia vapor for the temperature range −20 to 150 °C at 0.098 MPa is given in [47]:

$$\eta = 9.83\times10^{-6}+2.75\times10^{-8}\vartheta+2.8\times10^{-11}\vartheta^2 \quad (5)$$

Reference [54] reports viscosity measurements for the hydrogen – ammonia system; [55] for nitrogen – ammonia, oxygen – ammonia, methane – ammonia, and ethylene – ammonia; and [56] for argon – ammonia.

Thermal Conductivity. In [57], experimental data and calculation methods for ammonia liquid and vapor for the ranges 0.1 – 49 MPa (1 – 490 bar) and 20 – 177 °C are reported. The anomaly in the thermal conductivity at the critical point is discussed in full detail. A nomogram for these data appears in [58]. A third-degree equation for the range 358 – 925 K can be found in [59]. For the liquid phase the following formula may be used to interpolate in the range

Table 4. Properties of saturated ammonia liquid and vapor

t, °C	p, kPa	Specific volume		Enthalpy		Heat of vaporization, kJ/kg	Entropy	
		Liquid, L/kg	Vapor, L/kg	Liquid, kJ/kg	Vapor, kJ/kg		Liquid, kJ kg^{-1} K^{-1}	Vapor, kJ kg^{-1} K^{-1}
−40	71.72	1.4490	1551.6	180.53	1568.7	1388.1	0.8479	6.8017
−35	93.14	1.4620	1215.4	202.80	1576.5	1373.1	0.9423	6.7105
−30	119.49	1.4754	962.9	225.19	1584.1	1358.9	1.0352	6.6241
−25	151.54	1.4892	770.95	247.69	1591.5	1343.8	1.1266	6.5419
−20	190.16	1.5035	623.31	270.31	1598.6	1328.3	1.2166	6.4637
−15	236.24	1.5184	508.49	293.05	1605.4	1312.4	1.3053	6.3890
−10	290.77	1.5337	418.26	315.91	1611.9	1296.0	1.3927	6.3175
−5	354.77	1.5496	346.68	338.87	1618.0	1279.1	1.4787	6.2488
0	429.35	1.5660	289.39	361.96	1623.6	1261.7	1.5636	6.1826
5	515.65	1.5831	243.16	385.17	1628.9	1243.8	1.6473	6.1188
10	614.86	1.6009	205.55	408.51	1633.7	1225.2	1.7299	6.0571
15	728.24	1.6194	174.74	432.01	1638.1	1206.1	1.8116	5.9972
20	857.08	1.6387	149.31	455.67	1642.0	1186.3	1.8922	5.9390
25	1002.7	1.6590	128.19	479.52	1645.3	1165.8	1.9721	5.8822
30	1166.6	1.6801	110.54	503.57	1648.1	1144.5	2.0512	5.8267
35	1350.0	1.7023	95.699	527.86	1650.3	1122.4	2.1295	5.7722
40	1554.6	1.7257	83.150	522.40	1651.9	1099.5	2.2075	5.7185
45	1781.7	1.7505	72.484	577.22	1652.8	1075.6	2.2849	5.6656
50	2033.0	1.7766	63.373	602.36	1653.0	1050.6	2.3619	5.6131

Table 5. Ammonia concentration at saturation in 1 : 3 mixtures of nitrogen and hydrogen, in vol %

t, °C	p, MPa							
	5	10	15	20	30	40	50	100
−30	2.9	1.7	1.34	1.15	0.95	0.9	0.87	0.8
−20	4.6	2.70	2.05	1.58	1.5	1.38	1.3	1.2
−10	6.8	4.06	3.10	2.63	2.22	2.03	1.94	1.75
0	10.0	5.80	4.50	3.80	3.20	2.90	2.75	2.5
10	15.0	8.25	6.40	5.36	4.50	4.05	3.84	3.55
20	19.6	11.4	8.6	7.35	6.05	5.55	5.25	4.9
30	20.0	15.2	11.5	9.80	8.20	7.50	7.10	6.70
40	26.5	20.0	15.0	14.0	10.75	9.80	9.40	9.0

Table 6. Solubility of hydrogen and nitrogen in liquid ammonia for 1 : 3 mixtures of nitrogen and hydrogen in cm^3 gas per gram NH$_3$ (STP)

t, °C		Total pressure, MPa							
		5	10	15	20	30	40	50	100
−30	N$_2$	0.632	1.161	1.637	2.036	2.743	3.291	3.754	5.317
	H$_2$	1.291	2.598	3.896	5.157	7.554	8.625	12.380	22.287
−20	N$_2$	0.864	1.438	2.036	2.568	3.452	4.195	4.195	8.852
	H$_2$	1.581	3.167	4.759	6.248	9.294	12.518	15.219	33.945
−10	N$_2$	0.922	1.730	2.488	3.129	4.258	5.207	6.027	8.853
	H$_2$	1.845	3.869	5.756	7.623	11.444	14.954	18.349	33.945
0	N$_2$	1.081	2.088	2.994	3.811	5.269	6.457	7.532	11.218
	H$_2$	2.215	4.560	6.956	9.163	13.740	18.220	22.461	41.863
10	N$_2$	1.226	2.461	3.583	4.564	6.342	7.800	9.190	14.085
	H$_2$	2.585	5.490	8.360	11.178	16.590	22.056	27.300	51.783
20	N$_2$	1.372	2.874	4.414	5.415	7.654	8.303	11.285	17.733
	H$_2$	2.969	6.423	9.970	13.338	20.142	26.731	33.310	63.530
30	N$_2$	1.505	3.315	4.954	6.418	9.169	11.619	13.968	22.528
	H$_2$	3.300	7.557	11.719	15.711	24.272	32.343	40.396	79.345
40	N$_2$	1.584	3.717	5.767	7.647	10.967	13.954	16.852	28.527
	H$_2$	3.631	8.761	13.745	18.916	28.028	39.127	49.451	97.253

from -10 to $100\ °C$ [47] (thermal conductivity λ in $W\ m^{-1}\ K^{-1}$):

$$\lambda = 0.528 - 1.669 \times 10^{-2} \vartheta - 6.2 \times 10^{-6} \vartheta^2 \qquad (6)$$

Also from [47] is the formula for ammonia vapor at 101.3 kPa and -20 to $150\ °C$:

$$\lambda = 0.0217 + 1.17 \times 10^{-4} \vartheta + 1.87 \times 10^{-2} \vartheta^2 \qquad (7)$$

The thermal conductivities of gas mixtures of ammonia with argon, neon, hydrogen, and methane are reported in [55] and [60 – 62].

Solubility in Water. Tables 7 and 8 and Figures 3 and 4 show miscellaneous information. General physicochemical and chemical engineering handbooks can be consulted for further data.

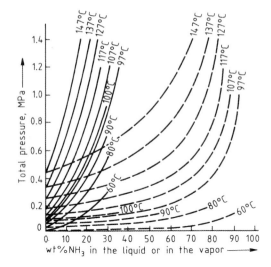

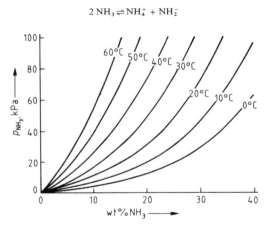

Figure 4. Total vapor pressure and composition of the liquid (——) and vapor (– – –) in the system NH_3-H_2O Isotherms

Inorganic salts such as nitrates, nitrites, iodides, cyanides, thiocyanides, and acetates are very soluble in liquid ammonia. The hydroxides, fluorides, and salts with di- and trivalent anions, such as oxides and sulfides, in general are insoluble. Ammonium salts are especially soluble.

A compilation of the solubilities of organic compounds in liquid ammonia shows notable solubility of saccharoses [64].

Liquid ammonia in combination with water as antisolvent and methylamine as prosolvent is also an excellent extraction medium for all types of petroleum fractions [65].

Figure 3. Ammonia partial pressure in aqueous ammonia solutions

Reference [63] gives $p - V - T$ values for ammonia containing up to 0.5 wt % water at pressures in the range 113 – 221 kPa.

Solubility in Liquid Ammonia. Liquid ammonia is a good solvent for many salts as well as for some metals and nonmetals. Because of its dipole moment, the ammonia molecule interacts with these ions to form solvates in a manner analogous to the water molecule in aqueous solutions. Solutions in liquid ammonia show significant electrical conductance. Pure ammonia, like water, itself has a conductivity that, although limited, is based on dissociation according to:

$$2\ NH_3 \rightleftharpoons NH_4^+ + NH_2^-$$

3. Chemical Properties

The ionic complex NH_4^+ results from addition of a proton to the ammonia molecule. This ammonium ion is similar to the alkali metal ions in its alkaline and salt-forming attributes.

The free radical NH_4 has been isolated only in the form of its amalgam (DAVY, 1811) which precipitates the more noble metals, such as copper, cadmium, and zinc, from their solutions. Solutions of ammonium in liquid ammonia are blue and usually exhibit properties similar to liquid ammonia solutions of sodium and potassium.

Table 7. Table for determining the percentage ammonia content of aqueous solutions from the density at 15 °C

Density, g/cm^3	Ammonia content	Density, g/cm^3	Ammonia content
1.000	0.00	0.935	17.12
0.995	1.14	0.930	18.64
0.990	2.31	0.925	20.18
0.985	3.55	0.920	21.75
0.980	4.80	0.915	23.35
0.975	6.05	0.910	24.99
0.970	7.31	0.905	26.64
0.965	8.59	0.900	28.33
0.960	9.91	0.895	30.03
0.955	11.32	0.890	31.73
0.950	12.74	0.885	33.67
0.945	14.17	0.880	35.60
0.940	15.63		

Table 8. Heat of solution of ammonia in water, in kJ/mol NH$_3$

t, °C	Mixture ratio (moles of water per mole ammonia)						
	1	2.33	4	9	19	49	99
0	30.69	34.25	35.17	35.80	36.22	36.47	36.09
20	27.38	32.87	33.66	34.50	34.67	34.83	34.92
40	24.53	31.99	32.91	33.62	33.87	33.95	34

Aqueous solutions of ammonia react as weak bases because hydroxyl ions result from the addition of a proton or H$_3$O$^+$ from the water to the ammonia molecule. However, the equilibrium is very strongly shifted to the side of free ammonia, as may be seen from the equilibrium constant at 18 °C:

$$\frac{[NH_4^+][OH^-]}{[NH_3]} = K[H_2O] = K'_{NH_3} = 1.75 \times 10^{-5} \quad (8)$$

A 1 N solution of ammonium hydroxide has a pH of 11.77 – at 18 °C.

Ammonium salts result from reaction with acids in aqueous solution or in the gas phase (→ Ammonium Compounds).

Ammonia is oxidized by *oxygen* or air, depending on the reaction conditions, to NO, NO$_2$, or N$_2$O or to nitrogen and water. On alkaline surfaces, such as quicklime or soda lime, ammonia – air mixtures are oxidized readily at 350 °C to nitrite and further to nitrate. The reaction rate can be increased by adding traces of nickel and cobalt oxides to the catalyst surfaces [66]. Gaseous ammonia reacts very violently or even explosively with nitrogen oxides to form nitrogen, water, ammonium nitrate or nitrite. The reaction with N$_2$O requires ignition.

At higher temperatures, especially in the presence of surface-active materials, ammonia forms *hydrogen cyanide* with carbon monoxide, methane, or charcoal. The catalytic oxidation of methane in the presence of ammonia is employed for the industrial production of hydrogen cyanide. With CO$_2$, SO$_3$, and P$_2$O$_5$, NH$_3$ forms amides of carbonic acid (→ Urea), sulfuric acid, and phosphoric acid.

Metals may replace one or all of the hydrogen atoms of ammonia. The *amide*, for example, sodium amide, NaNH$_2$, is obtained with evolution of hydrogen by passing ammonia vapor over metallic sodium. Sodium amide reacts with N$_2$O to form highly explosive sodium *azide*. An ammonium azide, NH$_4$N$_3$, may be prepared also. The *imide*, for example, lithium imide, Li$_2$NH, results from replacing a further hydrogen atom by metals. Finally, there are the *nitrides*, such as lithium or magnesium nitride, Li$_3$N or Mg$_3$N$_2$. The nitrides of reactive metals, such as lithium, magnesium, calcium, aluminum, and titanium, are formed directly from the elements at elevated temperature. In many cases, it is better to obtain nitrides through the action of ammonia on the metals or metal compounds at higher temperature. A nitride also may result from thermal decomposition of an amide.

Sulfur, phosphorus, and halogens also can take the place of the hydrogen atoms in ammonia. At relatively low temperatures, halogens react with ammonia to form *nitrogen – halogen compounds* and *ammonium halides*. At higher temperatures the products are nitrogen and hydrogen halides. With sodium hypochlorite, ammonia forms *hydrazine hydrate* (Raschig synthesis) with chloramine, H_2NCl, as an intermediate.

Ammonia undergoes numerous industrially important reactions with organic compounds. Reaction with alkyl halides or alcohols serves to manufacture *amines* or *imines*. For example, methanol forms mono- through trimethylamine; dichloromethane yields ethylene imine in the presence of calcium oxide. With organic acid halides, the reaction leads to the *acid amide*, with elimination of hydrogen chloride. Likewise, acid amides result from acylating ammonia with esters, acid anhydrides, or the acids themselves (above 100 °C); for example, formamide from methyl formate. Catalytic hydrogenation converts the acid amides to *primary amines*. Adding ammonia to aldehydes and ketones and splitting off water leads through unstable, intermediate amino compounds to three different forms of stable end products: for example, on reaction with formaldehyde, *hexamethylenetetramine* (urotropine) is obtained; with acetaldehyde, *ammono acetaldehyde*, which can again be dissociated hydrolytically; and in the case of benzaldehyde, *hydrobenzamide*. With ethylene and propylene oxides, aqueous ammonia reacts to form *ethanol-* or *propanolamine*. By reaction of olefins with ammonia, especially employing corona electrical discharges, *alkylamines* can be obtained, e.g., up to 10 % yield of ethylamine, or 15 % of 1-aminopropane. Otherwise, these compounds may be prepared by decomposing the acid amide or by reacting ammonia with alkyl halides in multistage processes [67]. The catalytic gas-phase oxidation of olefins in the presence of ammonia on vanadium- or molybdenum-containing catalysts, so-called *ammonoxidation*, results in economic yields of the commercially important *acid nitriles*, for example, acrylonitrile from propylene [68]. Ammonoxidation of *o*-xylene yields phthalodinitrile in a single reaction step.

Reactions in Liquid Ammonia. Reactions in liquid ammonia [69] sometimes proceed differently from those in water because of the solubility difference of many salts between water and ammonia. For example, the reaction

$$2\,AgNO_3 + BaCl_2 \underset{NH_3}{\overset{H_2O}{\rightleftharpoons}} 2\,AgCl + Ba(NO_3)_2$$

leads in water to precipitation of silver chloride, and in ammonia, to precipitation of barium chloride. In liquid ammonia, ammonium salts have an "acid" character. For example, alkali and alkaline-earth metals, cerium, lanthanum, manganese, cobalt, nickel, and iron and its alloys are soluble in solutions of ammonium bromide, iodide, cyanide, and thiocyanide, with formation of the corresponding metal amine salt and evolution of hydrogen:

$$2\,NH_4^+ + Mg \rightarrow Mg^{2+} + 2\,NH_3 + H_2$$

Correspondingly, metal amides in liquid ammonia have a "basic" character. The reaction of ammonium salt with metal amide in liquid ammonia is analogous to the neutralization of acid and base in water. The heats of neutralization in ammonia are even larger than in water. The process of hydrolysis corresponds to ammonolysis in ammonia. This results in ammonobasic compounds, for example, in the infusible precipitate $HgNH_2Cl$ from $HgCl_2$.

Liquid ammonia is a good solvent for white phosphorus. Rhombic sulfur is dissolved by liquid ammonia at $-11.5\,°C$. Once dissolved, at lower temperature a blue solution results in which partial ammonolysis of sulfur to sulfur nitride and ammonium sulfides or polysulfides takes place.

Liquid ammonia's ability to dissolve alkali and alkaline-earth metals has been well known for a long time. In concentrated solutions, the metals largely remain in the metallic state. The magnetic properties and the electrical conductivity, which is comparable to that of mercury, confirm this. In the more dilute blue solutions, the metals are completely dissociated to positive metal ions and solvated electrons [70]. The ammoniacal solutions allow preparation of many compounds otherwise unobtainable or obtainable only with difficulty. Examples are the sodium selenides, Na_2Se through Na_2Se_6, tellurides up to Na_2Te_4, antimonides, arsenides, polystannides, and polyplumbides. Introducing oxygen into the solution of alkali metals forms peroxides, Na_2O_2, K_2O_2, as well as dioxides, KO_2, RbO_2, and CsO_2. White explosive salts of the hypothetical acetylenediols, such as $KO-C\equiv C-OK$, result with carbon monoxide [71]; with acetylene, $NaC\equiv CH$; and with phosphine, KPH_2. Reactions of ammoniacal metal solutions with halogen-containing organic compounds can be used for the synthesis of higher hydrocarbons, amines, and free radicals. For example, reacting triphenylchloromethane with dissolved sodium produces the triphenylmethyl radical $Ph_3C^\bullet$.

A survey of organic reactions in liquid ammonia appears in [72].

4. Production

Ammonia is the second largest synthetic chemical product; more than 90 % of world consumption is manufactured from the elements nitrogen and hydrogen in a catalytic process originally developed by FRITZ HABER and CARL BOSCH using a promoted iron catalyst discovered by ALWIN MITTASCH. Since the early days there has been no fundamental change in this process. Even today the synthesis section of practically every ammonia plant has the same basic configuration as the first plants. A hydrogen – nitrogen mixture reacts over the iron catalyst (today's formulation differs little from the original) at elevated temperature in the range of 400 – 500 °C (originally up to 600 °C) and pressures above 100 bar with recycle of the unconverted part of the synthesis gas and separation of the ammonia product under high pressure. End of the 1990s a ruthenium-based catalyst was introduced which allows slightly lower synthesis pressure and is used in a few world-size plants (KAAP process of Kellogg – Braun – Root [73 – 75]).

BOSCH was already well aware that the production of a pure hydrogen – nitrogen mixture is the largest single contributor to the total production cost of ammonia [76]. So, in contrast to the synthesis reaction, dramatic changes happened over the years in the technology of synthesis-gas generation, and technical ammonia processes differ today mainly with respect to synthesis-gas preparation and purification. The elements nitrogen and hydrogen are abundantly available in the form of air and water, from which they can be separated by physical methods and/or chemical reactions using almost exclusively fossil energy. The predominant fossil fuels are natural gas, liquefied petroleum gas (LPG), naphtha, and higher petroleum fractions; coal or coke is used only under special economic and geographical conditions (China, India, South Africa). Recovery of ammonia as byproduct of other production processes, e.g., coke ovens, is no longer of great importance.

Of course, some of the hydrogen comes also from the hydrocarbons themselves (methane has the highest content), and the carbon acts as a reducing agent for water and in some processes may also facilitate the separation of oxygen from nitrogen by formation of carbon dioxide which can be removed by various operations.

4.1. Historical Development

The catalytic synthesis of ammonia from its elements is one of the greatest achievements of industrial chemistry. This process not only solved a fundamental problem in securing our food supply by production of fertilizers but also opened a new phase of industrial chemistry by laying the foundations for subsequent high-pressure processes like methanol synthesis, oxo synthesis, Fischer – Tropsch Process, coal liquefaction, and Reppe reactions. Continuous ammonia production with high space yields on large scale combined with the ammonia oxidation process for nitric acid, which was developed immediately after ammonia synthesis, enabled the chemical industry for the first time to compete successfully against a cheap natural bulk product, namely, sodium nitrate imported from Chile.

The driving force in the search for methods of nitrogen fixation, of course, was to produce fertilizers. In principle there are three ways of breaking the bond of the nitrogen molecule and fixing the element in a compound:

1) To combine the atmospheric elements nitrogen and oxygen directly to form nitric oxides
2) To combine nitrogen and hydrogen to give ammonia

3) To use compounds capable of fixing nitrogen in their structure under certain reaction conditions.

A vast amount of research in all three directions led to commercial processes for each of them: the electric arc process, the cyanamid process, and ammonia synthesis, which finally displaced the other two and rendered them obsolete.

The availability of cheap hydrolelectric power in Norway and the United States stimulated the development of the *electric arc process*. Air was passed through an electric arc which raised its temperature to 3000 °C, where nitrogen and oxygen combine to give nitric oxide. In 1904 CHRISTIAN BIRKELAND performed successful experiments and, together with SAM EYDE, an industrial process was developed and a commercial plant was built, which by 1908 was already producing 7000 t of fixed nitrogen. Working in parallel, SCHOENHERR at BASF developed a different electric arc furnace in 1905. The Norwegians and BASF combined forces in 1912 to build a new commercial plant in Norway. However, since at this time pilot-plant operation of ammonia synthesis was already successful, BASF withdrew from this joint venture soon after. Nevertheless, the Norwegian plants operated throughout World War I and had total production of 28 000 t/a of fixed nitrogen with a power consumption of 210 000 kW [27]. The specific energy consumption was tremendous: 60 000 kW per tonne of fixed nitrogen. Had this electricity been generated from fossil fuels this figure would correspond to about 600 GJ per tonne nitrogen, which is about 17 times the consumption of an advanced steam-reforming ammonia plant in 1996.

The cyanamide process, [27, 77 – 79], developed by FRANK and CARO in 1898, was commercially established by 1910. Calcium carbide, formed from coke and lime in a carbide furnace reacts with nitrogen to give calcium cyanamide, which can be decomposed with water to yield ammonia. The process was energetically very inefficient, consuming 190 GJ per tonne of ammonia. Some other routes via barium cyanide produced from barytes, coke and nitrogen, or using the formation of titanium nitride were investigated in Ludwigshafen by BOSCH and MITTASCH but did not appear promising. In 1934, 11 % of world nitrogen production (about 2×10^6 t/a) [80] was still based on the cyanamide process, and some plants even continued to operate after World War II.

After BERTHOLLET proved in 1784 that ammonia consists of nitrogen and hydrogen and was also able to establish the approximate ratio between these elements, many experiments in the 1800s were aimed at its *direct synthesis*, but remained unsuccessful [81 – 83]. One of the reasons for the lack of success was the limited knowledge of thermodynamics and the incomplete understanding of the law of mass action and chemical equilibrium. It was the new science of physical chemistry, which developed rapidly in the late 1800s, that enabled chemists to investigate ammonia formation more systematically.

Around 1900 FRITZ HABER began to investigate the ammonia equilibrium [84] at atmospheric pressure and found minimal ammonia concentrations at around 1000 °C (0.012 %). Apart from HABER, OSTWALD and NERNST were also closely involved in the ammonia synthesis problem, but a series of mistakes and misunderstandings occurred during the research. For example, OSTWALD withdrew a patent application for an iron ammonia synthesis catalyst because of an erroneous experiment, while NERNST concluded that commercial ammonia synthesis was not feasible in view of the low conversion he found when he first measured the equilibrium at 50 – 70 bar [85, 86].

After a controversy with NERNST, HABER repeated his measurements at atmospheric pressure and subsequently at higher pressures [87 – 90], overcoming his colleague's preoccupation with the unfavorable equilibrium concentrations. HABER concluded that much higher pressures had to be employed and that, perhaps more importantly, a recycle process was necessary.

The amount of ammonia formed in a single pass of the synthesis gas over the catalyst is much too small to be of interest for an economic production. HABER therefore recycled the unconverted synthesis gas. After separating the ammonia by condensation under synthesis pressure and supplementing with fresh synthesis gas to make-up for the portion converted to ammonia, the gas was recirculated by means of a circulation compressor to the catalyst-containing reactor. This process, described in the patent DRP 235 421 (1908), became the basis for the industrial manufacture of ammonia and since then the same principle has found widespread application in numerous high-pressure processes. HABER also anticipated the preheating of the synthesis gas to reaction temperature (at that time 600 °C) by heat exchange with the hot exhaust gas from the reactor, the temperature of which would be raised by the exothermic ammonia formation reaction sufficiently (about 18 °C temperature rise for a 1 % increase of the ammonia concentration in converted synthesis gas).

In 1908 HABER approached BASF (Badische Anilin & Soda Fabrik at that time) to seek support for his work and to discuss the possibilities for the realization of an industrial process. His successful demonstration in April 1909 of a small laboratory scale ammonia plant having all the features described above finally convinced the BASF representatives, and the company's board decided to pursue the technical development of the process with all available resources. In an unprecedented effort, CARL BOSCH, together with a team of dedicated and highly skilled co-workers, succeeded in developing a commercial process in less than five years [76, 91 – 97]. The first plant started production at Oppau in September 1913 and had a daily capacity of 30 t of ammonia. Expansions increased the capacity to about 250 t/d in 1916/17 and a second plant with a capacity of 36 000 t/a went on stream in 1917 in Leuna. Further stepwise expansions, finally reaching 240 000 t/a, already decided in 1916, came in full production only after World War I [77]. After World War I ammonia plants were built in England, France, Italy, and many other countries based on a BASF license or own process developments, with modified process parameters, but using the same catalyst.

Up to the end of World War II, plant capacities were expanded by installing parallel lines of 70 – 120-t/d units, and synthesis-gas generation continued to be based on coal until the 1950s. With growing availability of cheap petrochemical feedstocks and novel cost-saving gasification processes (steam reforming and partial oxidation) a new age dawned in the ammonia industry. The development started in the USA, where steam reforming of natural gas was used for synthesis gas production. This process was originally developed by BASF and greatly improved by ICI, who extended it to naphtha feedstocks. Before natural gas became available in large quantities in Europe, partial oxidation of heavy oil fractions was used in several plants. The next revolution in the ammonia industry was the advent of the single-train steam reforming ammonia plants, pioneered by M. W. KELLOGG and others. The design philosophy was to use a single-train for large capacities (no parallel lines) and to be as far as possible energetically self-sufficient (no energy import) by having a high degree of energy integration (process steps in surplus supplying those in deficit). Only through this innovative plant concept with its drastic reduction in feedstock consumption and investment costs, could the enormous increase in world capacity in the following years became possible. Increasing competition and rising feedstock prices in the 1970s and 1980s forced industry and engineering companies to improve the processes further.

The LCA process of ICI (page 239) and the KRES/KAAP process, which is the first process since 1913 to use a non-iron synthesis catalyst, are recent advances that make a radical breakaway from established technology.

A short survey of the history of ammonia process technology can be found in [94, 95, 98 – 103].

4.2. Thermodynamic Data

Ammonia synthesis proceeds according to the following reaction:

$$0.5\,N_2 + 1.5\,H_2 \rightleftharpoons NH_3 \quad \Delta H_{298} = -46.22\,\text{kJ/mol}$$

To fix a kilogram of nitrogen in ammonia requires reacting 2.4 m^3 (STP) of hydrogen and 0.8 m^3 (STP) of nitrogen. About 3.27 MJ of heat is evolved. Table 9 is a compilation of the most important thermodynamic data for the reaction at atmospheric pressure.

Chemical Equilibrium. The reaction equilibrium has been investigated experimentally and theoretically many times. Values for the equilibrium constant are available for pressures up to 350 MPa (3500 bar).

GILLESPIE and BEATTIE [104] (see also [6]) were by far the most successful experimentally in establishing a firm basis for an analytical expression of the equilibrium constant in the range of industrial interest. The values in Tables 10

Table 9. Thermodynamic data for the reaction $0.5\ N_2 + 1.5\ H_2 \rightleftharpoons NH_3$ at atmospheric pressure

	t, K							
	300	400	500	600	700	800	900	1000
ΔH, kJ	−46.35	−48.48	−50.58	−52.04	−53.26	−54.28	−55.06	−55.68
ΔS, J K^{-1}	−99.35	−105.63	−110.03	−112.71	−114.55	−115.89	−116.77	−117.48
$\Delta G/T$, J K^{-1}	−55.22	−15.66	8.88	26.25	38.48	48.02	55.56	61.81
c_p, NH$_3$, J mol^{-1} K^{-1}	35.52	38.80	41.97	45.04	47.98	50.80	53.49	56.03
c_p, H$_2$, J mol^{-1} K^{-1}	28.868	29.199	29.278	29.341	29.454	29.634	29.89	30.216
c_p, N$_2$, J mol^{-1} K^{-1}	29.144	29.270	29.601	30.132	30.777	31.451	32.117	32.724

and 11 were calculated using their equation. A detailed description, with literature data and many tables, appears in [6]. A description of the equilibrium using the REDLICH – KWONG equation of state is given in [105].

Heat of Reaction. HABER investigated the heat of reaction at atmospheric pressure [106]. Numerous authors have estimated the pressure dependence under various assumptions. Today, most people use the Gillespie – Beattie equation [107]. This equation was used in calculating the values in Table 12. For further data see, for example, [6]. Reference [108] contains test results for the range 120 – 200 MPa (1200 – 2000 bar) and 450 – 525 °C. Additional literature can be found in [109].

Physical Properties of the Reactants. Various authors have measured the $p - V - T$ behavior of nitrogen, hydrogen, and 3 : 1 hydrogen – nitrogen mixtures. Reference [6] contains a survey. For the specific heat, thermal conductivity, and viscosity of the reactants, see [110, 111]. It is to be noted that heats of mixing must be considered for synthesis gas. This applies especially for high pressures and low temperatures. Reference [112] gives viscosities of hydrogen – nitrogen and hydrogen – ammonia mixtures.

4.3. Ammonia Synthesis Reaction

4.3.1. General Aspects

Usually, a system having an exothermic heat of reaction under operating conditions should react spontaneously. However, to form ammonia from hydrogen and nitrogen molecules, significant energy input is required for the nitrogen molecule to achieve the activated state. This is because of its high dissociation energy of 941 kJ/mol, which is considerably higher than that of hydrogen. According to estimates [113], initiation of ammonia synthesis homogeneously in the gas phase requires an activation energy of 230 – 420 kJ/mol. For purely thermal energy supply with a favorable collision yield, this activation barrier requires temperatures well above 800 – 1200 K to achieve measurable reaction rates. However, at such high temperatures and industrially reasonable pressures, the theoretically achievable ammonia yield is extremely small because of the unfavorable position of the thermodynamic equilibrium. In fact, all older attempts to combine molecular nitrogen purely thermally with atomic or molecular hydrogen failed. On the other hand, both ammonia and hydrazine result from reacting atomic nitrogen with atomic hydrogen (see for example [113, p. 186]).

At pressures above 200 MPa (2000 bar), the synthesis of ammonia proceeds even in the absence of specific catalysts. At such extreme pressures the vessel walls appear to catalyze the formation of ammonia.

In the homogeneous phase under thermodynamically favorable temperature conditions, the formation of ammonia may be forced by employing other forms of energy, such as electrical energy or ionizing radiation. The principal difficulty with these so-called plasma processes, which also impedes their economic use, is that the energy supplied is useful only in part for ammonia formation. A greater part is transformed in primary collision and exothermic secondary

Table 10. Ammonia content (in mol %) in equilibrium synthesis gas; $N_2 : H_2 = 1 : 3$

t, °C	p_{abs}, MPa									
	5	10	20	30	40	50	60	70	80	90	100
300	39.38	52.79	66.43	74.20	79.49	83.38	86.37	88.72	90.61	92.14	93.39
310	36.21	49.63	63.63	71.75	77.35	81.51	84.73	87.29	89.35	91.03	92.42
320	33.19	46.51	60.79	69.23	75.12	79.53	82.98	85.74	87.98	89.83	91.35
330	30.33	43.45	57.92	66.64	72.79	77.46	81.13	84.09	86.52	88.52	90.20
340	27.64	40.48	55.04	63.99	70.39	75.29	79.18	82.34	84.95	87.12	88.94
350	25.12	37.60	52.17	61.31	67.93	73.04	77.14	80.49	83.28	85.62	87.59
360	22.79	34.84	49.33	58.61	65.41	70.72	75.01	78.55	81.52	84.02	86.15
370	20.64	32.21	46.53	55.89	62.85	68.33	72.80	76.52	79.66	82.33	84.61
380	18.67	29.71	43.79	53.19	60.26	65.89	70.53	74.42	77.72	80.54	82.97
390	16.87	27.36	41.12	50.50	57.66	63.41	68.19	72.23	75.69	78.67	81.25
400	15.23	25.15	38.53	47.86	55.06	60.91	65.81	69.99	73.59	76.71	79.43
410	13.74	23.08	36.04	45.26	52.47	58.39	63.40	67.69	71.42	74.68	77.54
420	12.40	21.16	33.65	42.72	49.91	55.87	60.96	65.36	69.20	72.58	75.57
430	11.19	19.38	31.37	40.26	47.39	53.37	58.50	62.99	66.93	70.43	73.53
440	10.10	17.74	29.20	37.87	44.92	50.88	56.05	60.60	64.63	68.22	71.43
450	9.12	16.23	27.15	35.57	42.50	48.43	53.61	58.20	62.29	65.97	69.28
460	8.24	14.84	25.21	33.36	40.16	46.03	51.19	55.80	59.95	63.69	67.08
470	7.46	13.57	23.39	31.26	37.89	43.67	48.81	53.42	57.60	61.39	64.85
480	6.75	12.41	21.69	29.55	35.71	41.38	46.46	51.06	55.25	59.09	62.60
490	6.12	11.36	20.10	27.34	33.61	39.16	44.17	48.74	52.92	56.78	60.33
500	5.56	10.39	18.61	25.54	31.60	37.02	41.94	46.46	50.62	54.48	58.06
510	5.05	9.52	17.24	23.84	29.68	34.95	39.77	44.22	48.36	52.20	55.80
520	4.59	8.72	15.96	22.24	27.86	32.97	37.68	42.05	46.13	49.96	53.55
530	4.19	8.00	14.77	20.74	26.13	31.07	35.65	39.94	43.96	47.75	51.32
540	3.82	7.34	13.68	19.34	24.49	29.26	33.71	37.89	41.84	45.58	49.13
550	3.49	6.74	12.67	18.02	22.95	27.54	31.85	35.92	39.79	43.47	46.97
560	3.20	6.20	11.74	16.80	21.49	25.90	30.06	34.02	37.80	41.41	44.86
570	2.93	5.70	10.88	15.65	20.13	24.35	28.37	32.20	35.88	39.41	42.81
580	2.69	5.26	10.09	14.59	18.84	22.88	26.75	30.46	34.04	37.48	40.81
590	2.47	4.85	9.36	13.60	17.64	21.50	25.22	28.80	32.26	35.62	38.87
600	2.28	4.48	8.69	12.69	16.52	20.20	23.76	27.22	30.57	33.83	37.00

processes into undesirable heat or unusable incidental radiation.

In the catalytic combination of nitrogen and hydrogen, the molecules lose their translational degrees of freedom by fixation on the catalyst surface. This drastically reduces the required energy of activation, for example, to 103 kJ/mol on iron [114]. The reaction may then proceed in the temperature range 250 – 400 °C. In 1972, it was discovered that electron donor – acceptor (EDA) complexes permit making ammonia with measurable reaction rate at room temperature.

This section concentrates mainly on the ammonia synthesis reaction over iron catalysts and refers only briefly to reactions with non-iron catalysts. Iron catalysts which are generally used until today in commercial production units are composed in unreduced form of iron oxides (mainly magnetite) and a few percent of Al, Ca, and K; other elements such as Mg and Si may also be present in small amounts. Activation is usually accomplished in situ by reduction with synthesis gas. Prereduced catalysts are also commercially available.

Numerous investigations have been performed to elucidate the mechanism of catalytic reaction of nitrogen and hydrogen to form ammonia. References [115 – 118] give reviews of the older and some of the newer literature. Since the 1980s a large variety of surface science techniques involving Auger electron spectroscopy, X-ray photoelectron spectroscopy, work-function measurements, temperature-programmed adsorption and desorption, scanning tunnelling microscopy, and others have been developed [119, 120]. Many of these methods are based on interaction of slow electrons, ions, or neutral particles and exhibit high sensitivity to surface structures. With these powerful tools the kinetics of nitrogen and hydrogen adsorption and desorption could be investigated, and it was also possible to identify adsorbed intermediates. The results of these experiments allow the mechanism of ammonia synthesis in the

Table 11. Equilibrium ammonia content (in mol %) in the presence of inert gases (2.5 mol % Ar; 7.5 mol % CH_4); $N_2 : H_2 = 1 : 3$

t, °C	p_{abs}, MPa									
	5	10	20	30	40	50	60	70	80	90	100
300	32.11	43.10	54.27	60.62	64.95	68.12	70.56	72.49	74.03	75.28	76.31
310	29.52	40.51	51.98	58.63	63.20	66.59	69.22	71.31	72.99	74.37	75.51
320	27.05	37.96	49.66	56.57	61.38	64.98	67.79	70.04	71.87	73.38	74.63
330	24.71	35.46	47.32	54.45	59.48	63.29	66.28	68.70	70.67	72.31	73.68
340	22.51	33.03	44.96	52.30	57.52	61.52	64.69	67.27	69.39	71.16	72.65
350	20.46	30.68	42.62	50.11	55.51	59.68	63.02	65.76	68.03	69.93	71.54
360	18.56	28.42	40.30	47.90	53.46	57.79	61.29	64.18	66.59	68.63	70.36
370	16.80	26.27	38.01	45.69	51.37	55.85	59.49	62.52	65.08	67.24	69.10
380	15.19	24.33	35.77	43.48	49.26	53.86	57.64	60.81	63.49	65.79	67.77
390	13.72	22.31	33.59	41.29	47.14	51.85	55.74	59.03	61.84	64.27	66.36
400	12.39	20.50	31.48	39.13	45.02	49.81	53.81	57.21	60.14	62.68	64.89
410	11.18	18.81	29.44	37.01	42.91	47.76	51.84	55.34	58.38	61.03	63.35
420	10.08	17.24	27.49	34.93	40.83	45.71	49.86	53.44	56.57	59.32	61.75
430	9.10	15.79	25.62	32.92	38.77	43.67	47.86	51.52	54.73	57.57	60.09
440	8.21	14.45	23.85	30.97	36.75	41.64	45.87	49.58	52.86	55.78	58.39
450	7.41	13.22	22.17	29.09	34.78	39.65	43.88	47.63	50.97	53.96	56.65
460	6.70	12.09	20.59	27.29	32.87	37.68	41.92	45.68	49.06	52.11	54.87
470	6.06	11.05	19.10	25.56	31.02	35.77	39.97	43.75	47.16	50.25	53.06
480	5.49	10.11	17.71	23.92	29.23	33.90	38.06	41.83	45.25	48.38	51.24
490	4.97	9.24	16.41	22.36	27.51	32.08	36.20	39.94	43.36	46.51	49.40
500	4.51	8.46	15.19	20.89	25.87	30.33	34.37	38.08	41.49	44.64	47.56
510	4.10	7.74	14.07	19.50	24.30	28.64	32.61	36.26	39.65	42.79	45.72
520	3.73	7.09	13.02	18.19	22.81	27.02	30.89	34.49	37.84	40.97	43.90
530	3.40	6.50	12.05	16.96	21.39	25.46	29.24	32.76	36.07	39.17	42.09
540	3.10	5.97	11.16	15.81	20.05	23.98	27.65	31.09	34.34	37.41	40.31
550	2.83	5.48	10.33	14.73	18.79	22.57	26.12	29.48	32.66	35.68	38.56
560	2.59	5.04	9.57	13.73	17.60	21.23	24.67	27.93	31.04	34.01	36.84
570	2.38	4.64	8.87	12.79	16.48	19.96	23.28	26.44	29.47	32.38	35.17
580	2.18	4.27	8.22	11.92	15.43	18.76	21.95	25.01	27.96	30.80	33.54
590	2.01	3.94	7.63	11.12	14.44	17.63	20.69	23.65	26.51	29.28	31.96
600	1.85	3.64	7.08	10.37	13.52	16.56	19.50	22.35	25.12	27.81	30.43

pressure range of industrial interest to be elucidated [121 – 123].

As with every catalytic gas-phase reaction, the course of ammonia synthesis by the Haber – Bosch process can be divided into the following steps:

1) Transport of the reactants by diffusion and convection out of the bulk gas stream, through a laminar boundary layer, to the outer surface of the catalyst particles, and further through the pore system to the inner surface (pore walls)
2) Adsorption of the reactants (and catalyst poisons) on the inner surface
3) Reaction of the adsorbed species, if need be with participation by hydrogen from the gas phase, to form activated intermediate compounds
4) Desorption of the ammonia formed into the gas phase
5) Transport of the ammonia through the pore system and the laminar boundary layer into the bulk gas stream

Only the portion of the sequence that occurs on the catalyst surface is significant for the intrinsic catalytic reaction. Of special importance is the adsorption of nitrogen. This assumption is decisive in representing the synthesis reaction kinetics. The transport processes occurring in the pores of the catalyst in accordance with the classical laws of diffusion are of importance in industrial synthesis (see also Sections 4.4.1.2 and 4.3.3).

4.3.2. Catalyst Surface and Reaction Mechanism

Earlier studies [124 – 133] had already suggested that on iron catalysts nitrogen adsorption and dissociation can be regarded as the rate-determining step for the intrinsic reaction.

Table 12. Heat of reaction ΔH (in kJ) for the reaction $0.5\ N_2 + 1.5\ H_2 \rightleftharpoons NH_3$, from [104]

t, °C	p_{abs}, MPa										
	0 (ideal)	10	20	30	40	50	60	70	80	90	100
0	44.39	55.20	66.02	76.84	87.66	98.47	109.29	120.11	130.92	141.74	152.56
25	44.92	53.48	62.03	70.59	79.15	87.70	96.26	104.82	113.38	121.93	130.49
50	45.44	52.37	59.30	66.23	73.16	80.09	87.02	93.95	100.88	107.80	114.73
75	45.96	51.69	57.41	63.14	68.86	74.58	80.31	86.03	91.76	97.48	103.20
100	46.47	51.28	56.10	60.91	65.72	70.53	75.34	80.16	84.97	89.78	94.59
125	46.97	51.08	55.19	59.29	63.40	67.51	71.62	75.72	79.83	83.94	88.04
150	47.46	51.02	54.57	58.12	61.68	65.23	68.78	72.34	75.89	79.44	83.00
175	47.95	51.06	54.17	57.28	60.39	63.50	66.61	69.72	72.83	75.94	79.05
200	48.42	51.17	53.92	56.68	59.43	62.18	64.93	67.69	70.44	73.19	75.94
225	48.88	51.34	53.80	56.26	58.71	61.17	63.63	66.09	68.55	71.01	73.47
250	49.33	51.54	53.76	55.97	58.19	60.40	62.62	64.84	67.05	69.27	71.48
275	49.76	51.77	53.79	55.80	57.81	59.82	61.84	63.85	65.86	67.87	69.88
300	50.18	52.02	53.86	55.70	57.54	59.38	61.22	63.06	64.90	66.74	68.59
325	50.59	52.28	53.98	55.67	57.37	59.06	60.75	62.45	64.14	65.83	67.53
350	50.99	52.55	54.12	55.69	57.26	58.82	60.39	61.96	63.53	65.10	66.66
375	51.36	52.82	54.28	55.74	57.20	58.66	60.12	61.58	63.04	64.49	65.95
400	51.73	53.09	54.46	55.82	57.18	58.55	59.91	61.28	62.64	64.00	65.37
425	52.07	53.35	54.64	55.92	57.20	58.48	59.76	61.04	62.32	63.60	64.88
450	52.40	53.61	54.82	56.03	57.23	58.44	59.65	60.86	62.06	63.27	64.48
475	52.71	53.86	55.00	56.14	57.29	58.43	59.57	60.72	61.86	63.00	64.14
500	53.01	54.09	55.18	56.26	57.35	58.43	59.52	60.60	61.69	62.78	63.86
525	53.28	54.31	55.35	56.38	57.42	58.45	59.48	60.52	61.55	62.58	63.62
550	53.53	54.52	55.51	56.50	57.48	58.47	59.46	60.45	61.44	62.42	63.41
575	53.76	54.71	55.66	56.60	57.55	58.50	59.44	60.39	61.33	62.28	63.23
600	53.98	54.88	55.79	56.70	57.61	58.52	59.43	60.34	61.24	62.15	63.06

This has now been fully confirmed by microkinetic simulations based on the results of surface science studies [121, 134, 135]. In single-crystal experiments it was found that the activation energy for dissociative nitrogen adsorption is not constant; it increases as the surface becomes increasingly covered with adsorbed nitrogen species. Under reaction conditions it can attain values of 63 – 84 kJ/mol [123, 125, 126, 136] which are typical for ammonia synthesis. In the case of nitrogen the first step seems to be an adsorption in molecular state followed by chemisorption in the atomic state. From experimental results it was concluded that for hydrogen a direct transition from the gaseous H_2 molecule into the chemisorbed H_{ad} is most likely, and evidence for the stepwise hydrogenation of surface nitrogen atomic species was found. In IR spectroscopic investigations, NH and NH_2 species were identified [137], and secondary ion emissions detecting NH^+ also confirmed the presence of NH species on the surface [138]. It was further shown that at lower temperatures nitrogen becomes adsorbed only in the molecular state, but subsequently dissociates when the temperature is raised. Isotopic experiments with $^{30}N_2$ and $^{29}N_2$ showed that the surface species resulting from low-temperature adsorption was molecular, whereas that from high-temperature adsorption was atomic [139]. Ammonia synthesis is highly sensitive to the orientation of the different crystal planes of iron in the catalyst [140 – 144]. Measurements on defined single crystal surfaces of pure iron performed under ultrahigh vacuum [145 – 147] clearly showed that Fe(111) is the most active surface. This was also demonstrated by the rate of ammonia formation on five different crystallographic planes for unpromoted iron at 20 bar, as shown in Figure 5.

Activation energy, reactivity, adhesion coefficient (the probability that a nitrogen molecule striking the surface will be adsorbed dissociatively) and work function show a clear dependence on surface orientation [148, 149]:

Ammonia synthesis activity:
(111) > (211) > (100) > (210) > (110)
Activation energy of nitrogen adsorption:
(111) > (100) > (110)
Work function:
(210) > (111) > (211) > (100) > (110)

160   Ammonia

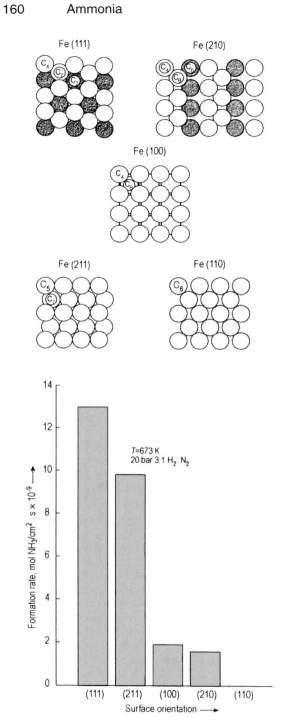

Figure 5. Ammonia formation rate on different iron surfaces [135, 148]

Surface roughness:
(210) > (111) > (211) > (100) > (110)

Ultrahigh vacuum (UHV) experiments with single crystals show that the activation energy of the nitrogen adsorption at zero coverage increases from about zero for Fe(111) to 21 kJ/mol for Fe(110) [121, 125, 150 – 152]. These values increase significantly for higher coverage [151, 152]. Adsorption and desorption under higher pressure on finely dispersed catalyst indicate that the reaction is highly activated under these conditions with high coverage, about 100 kJ/mol [124, 153].

According to [150, 151] adsorption on the planes Fe(111) and Fe(110) is associated with a regrouping of the surface atoms.

A possible explanation for the high activity of faces (111) and (211) is that these are the only surfaces which expose C_7 sites (iron with seven nearest neighbors) to the reactant gases. There are theoretical arguments [154] that highly coordinated surface atoms should show increased catalytic activity due to low-energy charge fluctuations in the d-bands of these highly coordinated atoms. This argument might probably be the key for the special role of C_7 sites. Other reasons discussed are based on charge transfer and interaction of iron d-bands with antibonding 2 π^* orbitals of nitrogen [155].

Promotion with potassium of single iron crystals enhances the sticking probability for nitrogen dissociation much more on the Fe(100) and (110) than on the Fe(111) (factors 280, ca. 1000 and 8, respectively) to the effect that the differences in surface orientation disappear [156]. A similar effect was not found for the ammonia synthesis at 20 bar and catalyst temperature of 400 °C: only a two-fold increase of the ammonia formation rate was measured for Fe(111) and Fe(100), and the face (110) was found to be inactive with and without potassium [148]. Other experiments [157] show that even the least active face Fe(110) becomes as active for the synthesis as Fe(111) after addition of alumina with subsequent annealing with oxygen and water vapor. A proposed mechanism for these findings — backed by X-ray photoelectron spectroscopy, temperature programmed desorption and electron microscopy — assumes that first alumina forms an iron aluminate $FeAl_2O_4$ on the surface. This new surface then may serve as a template on which iron grows with (111) and (211) orientation upon exposure to the synthesis-gas mixture in the reaction [158].

Based on these experimental results a reaction scheme for the ammonia synthesis may be formulated comprising the following sequence of individual steps [121]:

$$H_2 + * \rightleftharpoons 2\,H_{ad} \qquad (9)$$

$$N_2 + * \rightleftharpoons N_{2,ad} \qquad (10)$$

$$N_{2,ad} \rightleftharpoons 2\,N_{ad} \qquad (11)$$

$$N_{ad} + H_{ad} \rightleftharpoons NH_{ad} \qquad (12)$$

$$NH_{ad} + H_{ad} \rightleftharpoons NH_{2,ad} \qquad (13)$$

$$NH_{2,ad} + H_{ad} \rightleftharpoons NH_{3,ad} \qquad (14)$$

$$NH_{3,ad} \rightleftharpoons NH_3 + * \qquad (15)$$

The progress of the reaction may be described in the form of an energy profile, as shown in Figure 6. Industrial ammonia synthesis in the homogeneous gas phase is not feasible because of the high dissociation energies for the initial steps. The reaction over a catalyst avoids this problem since the energy gain associated with the surface atom bonds overcompensates these dissociation energies and the first steps have actually become exothermic.

Dissociative nitrogen adsorption remains nevertheless the rate-determining step, not so much on account of its activation barrier but rather because of the very unfavorable preexponential factor in its rate constant. The subsequent hydrogenation steps are energetically uphill, but the energy differences involved can easily be overcome at the temperatures (ca. 700 K) used in industrial ammonia synthesis. It is, however, quite apparent that the rate-controlling step would switch from dissociative nitrogen adsorption to hydrogenation of adsorbed atomic nitrogen species if the temperature were lowered sufficiently because of these differences in activation energy.

Some critics [159] of the above energy diagram question the low net activation barrier from the gas phase. The arguments are based on an analysis of activation energies from early measurements of the nitrogen adsorption kinetics on

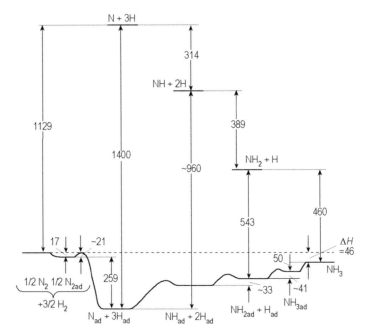

Figure 6. Schematic energy profile of the progress of ammonia synthesis on Fe (energies in kJ/mol) [121]

singly (Al_2O_3) promoted catalysts and on the results of experiments with supersonic molecular beams [160, 161].

An attempt to explain these differences is given in [162]. In more recent investigations of the adsorbed nitrogen species [163 – 166] a second molecularly adsorbed species was detected. This so-called α state was interpreted as a bridge-bonded species with electron donation from the surface to the antibonding π levels of N_2, whereas the γ state is regarded as a terminally bound species. Thus the following picture for the nitrogen adsorption emerges:

$N_2 \rightarrow S*-N_2$ (γ state)

$\rightarrow S*-N_2-S*$ (α state) $\rightarrow 2 S*-N$ (β state)

where S* denotes a surface atom.

For industrial catalysts made by careful reduction of magnetite fused with nonreducible oxide promoters the important role of the (111) face seems to be confirmed [167]. However, the question whether the active industrial catalyst exposes mostly (111) faces remains unresolved. If not, further improvements of the catalyst are at least theoretically possible [168]. This was indeed the case with the new AmoMax-10 catalyst [169, 170] of Südchemie. This catalyst in its oxidic form is based on wustite instead of magnetite and has a significantly better activity at lower temperature than magnetite. Wustite is a nonstoichiometric iron oxide($Fe_{1-x}O$) with a cubic crystal structure, x ranging from 0.03 to 0.15. With still (111) faces present in the reduced state the catalyst has a higher specific surface area and an improved pore structure.

A critical evaluation of the present knowledge of the mechanism of the synthesis reaction was made by SCHLÖGL [152].

Other reaction mechanisms have been debated for reaction temperatures below 330 °C [138, 139, 143, 171 – 181]. These propose participation of diatomic nitrogen, or of adsorption complexes containing diatomic nitrogen, in the rate-determining step (see 115, 116 for further literature).

NIELSEN et al. investigated ammonia synthesis on a commercial Topsøe catalyst, KM IR, over a wide temperature range. They found evidence that a different reaction mechanism predominates below and above 330 °C [182]. Also, at low temperatures, chemisorbed hydrogen blocks the catalyst surface [183]. The latter finding is in agreement with the observations of ERTL's group [136].

Reaction Mechanism on Non-Iron Catalysts. Non-iron systems which exhibit some potential to catalyze ammonia synthesis can be divided into the following groups [184]:

- Platinum group metals such as Ru, Os Ir, Pt (no nitrides)
- Mn, Co, Ni, Tc, Rh and their alloys (no nitride formation under synthesis conditions)
- Mn, Mo, V (present as nitrides under the reaction conditions)

In the non-iron systems the rate-determining step is also dissociative adsorption of nitrogen and the catalyst effectivity seems to be primarily dictated by the activation energy of the dissociation reaction [184]. This is somewhat surprising in view of the marked differences in the heats of adsorption of nitrogen and the adsorption activation energy. This even holds for tungsten, which has no significant activation energy and a high adsorption enthalpy for nitrogen, so that hydrogenation of adsorbed atomic nitrogen could be expected to be the rate-determining step. The factor common with the iron catalyst is the structure sensitivity.

The only system which seems to be promising for industrial application is ruthenium promoted with rubidium on graphite as carrier (see Section 4.4.2.3). Further information on structure, activity and reaction mechanism of non-iron catalysts is given in [116, 184 – 187]. Specific references: vanadium [188], uranium [189], molybdenum [190 – 192], tungsten [193].

4.3.3. Kinetics

Knowledge of the reaction kinetics is important for designing industrial ammonia synthesis reactors, for determining the optimal operating conditions, and for computer control of ammonia plants. This means predicting the technical dependence on operating variables of the rate of formation of ammonia in an integral catalyst volume element of a converter.

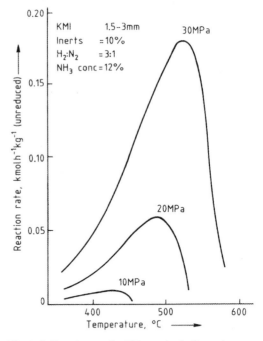

Figure 8. Reaction rate for NH_3 synthesis. Dependence on the temperature at various pressures.

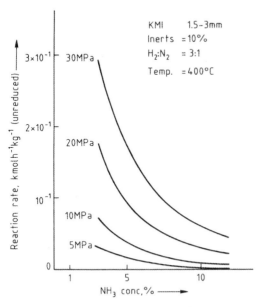

Figure 7. Reaction rate for NH_3 synthesis. Dependence on the ammonia concentration at various pressures.

High pressure promotes a high rate of ammonia formation; high ammonia concentration in the synthesis gas (recycle gas) restricts it (Fig. 7). High temperatures accelerate ammonia formation but imply a lower value of the equilibrium ammonia concentration and so a lower driving force. Therefore, the rate of formation at first increases with rising temperature but then goes through a maximum as the system approaches thermodynamic equilibrium (Fig. 8). A similar situation exists for the dependence of the reaction rate on the ratio of the hydrogen and nitrogen partial pressures; with lower temperature, the maximum rate shifts to a lower hydrogen – nitrogen ratio (Fig. 9). Figure 9 presents data obtained using a commercial iron catalyst, Topsøe KMIR. The data show a sharp drop in reaction rate with declining temperature at $H_2/N_2 = 3:1$ ratio in contrast to a $H_2/N_2 = 1:1$ ratio. This may be attributed to a hindering effect of adsorbed hydrogen at low temperature [183].

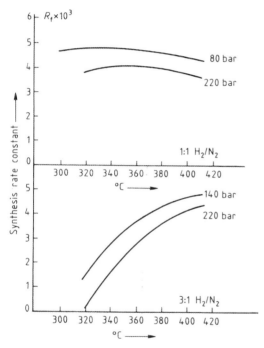

Figure 9. Ammonia synthesis rate constant dependence on hydrogen – nitrogen ratio

Equations for describing ammonia synthesis under industrial operating conditions must represent the influence of the temperature, the pressure, the gas composition, and the equilibrium composition. Moreover, they must also take into consideration the dependence of the ammonia formation rate on the concentration of catalyst poisons and the influence of mass-transfer resistances, which are significant in industrial ammonia synthesis.

Since the beginning of commercial ammonia synthesis, a large number of different kinetic equations have been suggested, emanating in each case from a proposed reaction mechanism or from empirical evaluations. A critical review of the data and equations published up to 1959 appears in [194]. A discussion of kinetics proposed up to 1970, insofar as they have been based on measurements in the operating range of commercial interest, can be found in [171]. An evaluation of present knowledge is given in [121, 134, 195, 196].

Contradictory data on the kinetics of ammonia synthesis, especially in the earlier literature, in some circumstances may reflect a lack of attention to the influence of impurities in the gas. If oxygen compounds are present in the synthesis gas, reversible poisoning of the adsorbing areas, in accordance with an equilibrium depending on the temperature and the water vapor – hydrogen partial pressure ratio, must be taken into account when developing rate equations (see also Section 4.4.1.4).

Experimental Measurements of Reaction Kinetics. The reaction expressions discussed in the following model the intrinsic reaction on the catalyst surface, free of mass-transfer restrictions. Experimental measurements, usually made with very fine particles, are described by theoretically deduced formulas, the validity of which is tested experimentally by their possibility for extrapolation to other reaction conditions. Commonly the isothermal integral reactor is used with catalyst crushed to a size of 0.5 – 1.5 mm to avoid pore diffusion restriction and heat-transfer resistance in the catalyst particles. To exclude maldistribution effects and back mixing, a high ratio of bed length to bed diameter is chosen. Sometimes the catalyst is also diluted with inert material. In some investigations, differential reactors were used. To exclude any poisoning by the synthesis gas, very pure reactants were prepared by decomposition of anhydrous ammonia [195].

Classical Expressions for Ammonia Synthesis Kinetics. The first expression useful for engineering purposes was the Temkin – Pyzhev Equation (1) proposed in 1940 [197, 198]. It is

(1) $v = k_1 p_{N_2} \left(\dfrac{p_{H_2}^3}{p_{NH_3}^2}\right)^\alpha - k_{-1} \left(\dfrac{p_{NH_3}^2}{p_{H_2}^3}\right)^{1-\alpha}$

$\alpha = 0.5 - 0.75$

(2) $v = k' p_{H_2}^a p_{N_2}^{1-\alpha}$

Near equilibrium (fugacities instead of partial pressures in Eq. 1) ← Higher pressure, some ammonia in the synthesis gas → Far from equilibrium

(3) $v = \dfrac{k_{-1}^0 \left(a_{N_2} K_a^2 - \dfrac{a_{NH_3}^2}{a_{H_2}^3}\right)}{\left(1 + K_3 \dfrac{a_{NH_3}}{a_{H_2}^w}\right)^{2\alpha}}$

$w = 1.5;\ \alpha = 0.75$

(4) $v = \dfrac{k p_{N_2}^{1-\alpha} \left(1 - \dfrac{1}{K}\dfrac{p_{NH_3}^2}{p_{N_2} p_{H_2}^3}\right)}{\left(\dfrac{1}{p_{H_2}} + \dfrac{1}{K}\dfrac{p_{NH_3}^2}{p_{N_2} p_{H_2}^3}\right)^\alpha \left(1 + \dfrac{1}{p_{H_2}}\right)^{1-\alpha}}$

based on the assumption that dissociative adsorption is the rate-determining step, that hydrogen and ammonia have no significant influence on nitrogen adsorption, and that the kinetics of nitrogen adsorption and desorption can be described adequately by Elovich-type adsorption on an energetically inhomogeneous surface. For many years this kinetic expression was the basic design equation for ammonia converters. Values for the factors α between 0.5 [199–201] and 0.75 [6, 202, 203] were used. A problem with this equation was that the α values (reaction order) were dependent on temperature, and the rate constants on pressure [182, 202–206]. More serious (not so much for industrial purposes, where the converter feed has always a certain ammonia content) was the fact that for zero ammonia content, as in some laboratory measurements, the equation gives an infinitely high reaction rate. To avoid this, a simpler expression (Eq. 2) was often used [204, 205].

An important modification was made by TEMKIN [207] who incorporated hydrogen addition to the adsorbed nitrogen as a second rate-determining step (Eq. 4). ICI demonstrated that this equation gives a better fit with experimental data [208]. It was also shown later that the original Equation (1) is a simplified form of a more general model which can be derived from the concept of energetically homogeneous (Langmuir–Hinshelwood adsorption isotherm) as well as for heterogeneous surfaces (Elovich-type isotherm). The applicability of a particular equation resulting from this concept also depends on the state of reduction of the catalyst [206, 209] and the type of promoter [210]. Equation (3), used by NIELSEN et al. [211], is a combination of these model equations, developed by OZAKI et al. [212], that uses fugacities instead of partial pressures. A similar equation is found in [213]. Additionally, a number of modified equations were proposed and tested with existing experimental data and industrial plant results [214–217]. Near the thermodynamical equilibrium, Equation (3) transforms into Equation (1) [171, 182].

Surface science based ammonia kinetics [121, 122, 195, 196] are presently still viewed as an academic exercise rather than as a practical tool for engineering. The large amount of available data on nitrogen and hydrogen adsorption from ultrahigh-vacuum studies on clean iron surfaces, acquired with all the modern spectroscopic techniques, has prompted some research groups, such as BOWKER et al. [218, 219] and STOLTZE and NØRSKOW [220, 221], to attempt the generation of a kinetic expression for ammonia synthesis from a detailed microscopic model of the reaction mechanism consisting of a number of discrete steps at molecular and atomic level. Potential energy diagrams for the various intermediate steps and species, were set up and Arrhenius expressions for each single step with known or estimated values for all pre-factors and activation energies were formulated. The best results have been achieved so far by calculating the overall rate from the rate of dissociation of the adsorbed nitrogen and equilibrium constants for all other reaction steps. The adsorption–desorption equilibria were treated with approximation of competitive Langmuir-type adsorption and by evaluation of the partition functions for the gaseous and adsorbed species. The data from single-crystal experiments for potassium-promoted Fe(111) surface were used for the rate of the dissociative nitrogen adsorption. Comparison of the calculated ammonia yields with those determined experimentally on a commercial Topsøe KM1 catalyst showed an agreement within a factor better than 2. Figure 10 demonstrates these encouraging results. These conclusions have been confirmed by calculations results of another independent group [134, 222].

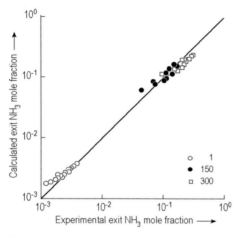

Figure 10. Comparison of ammonia concentrations calculated from surface science kinetics with experimentally measured values [221, 223]

To compete with the empirical models (Temkin and improved expressions) for the best fit to experimental data cannot be the prime objective of the microkinetic approaches. Rather, they are means of checking whether our knowledge and understanding of the elementary steps correspond to the reality of catalysts under industrial synthesis conditions.

Transport Phenomena. For practical application, the above kinetic equations have to be modified to make allowance for mass and energy transfer since the reaction rates actually observed in a commercial converter are lower. One aspect is interparticle mass transfer and heat transfer through the stagnant film which surrounds the catalyst particles. The high velocity of the gas passing through the converter creates sufficient turbulence to keep the film thickness rather small in relation to the catalyst grain size. For this reason the largest concentration gradient (with respect to the concentration in the bulk gas stream) is within the catalyst particles. Since the thermal conductivity of the iron catalysts is much higher than that of the synthesis gas, the major temperature difference is in the external gas film, while the catalyst particles themselves operate under approximately isothermal conditions. As can be seen from Table 13 the differences in temperature and ammonia concentration between the bulk gas stream and the external catalyst surface are small. It also appears that the effects are oppositely directed and will partly compensate each other. So it can be concluded that their combined influence on the reaction rate is negligible compared to inaccuracies of the experimental data for the intrinsic catalyst activity [195].

For the particle sizes used in industrial reactors (≤ 1.5 mm), intra particle transport of the reactants and ammonia to and from the active inner catalyst surface may be slower than the intrinsic reaction rate and therefore cannot be neglected. The overall reaction can in this way be considerably limited by ammonia diffusion through the pores within the catalysts [224]. The ratio of the actual reaction rate to the intrinsic reaction rate (absence of mass transport restriction) has been termed as pore effectiveness factor E. This is often used as a correction factor for the rate equation constants in the engineering design of ammonia converters.

For pore diffusion resistances in reactions having moderate heat evolution, the following phenomena characteristically hold true in industrial ammonia synthesis [225]: in the temperature range in which transport limitation is operative, the apparent energy of activation falls to about half its value at low temperatures; the apparent activation energy and reaction order, as well as the ammonia production per unit volume of catalyst, decrease with increasing catalyst particle size [224, 226–228]. For example at the gas inlet to a TVA converter, the effective rate of formation of ammonia on 5.7-mm particles is only about a quarter of the rate measured on very much smaller grains (Fig. 11) [171].

Mathematical models for calculating these effectiveness factors involve simultaneous differential equations, which on account of the complex kinetics of ammonia synthesis cannot be solved analytically. Exact numerical integration procedures, as adopted by various research groups [171, 229–231], are rather troublesome and time consuming even for a fast computer. A simplification [232] can be used which can be integrated analytically when the ammonia kinetics

Table 13. Mass and heat transfer effects at the external surface of catalyst particle

Position in catalyst bed, vol% from inlet	NH_3 concentration in bulk gas, mol%	NH_3 concentration at catalyst surface, mol%	Temperature in bulk gas, °C	Temperature at catalyst surface, °C
0	2.500	2.592	400.0	401.4
20	4.500	4.288	428.1	429.5
40	5.960	6.045	455.1	456.5
60	7.000	7.778	481.9	483.2
80	9.300	9.361	505.5	506.6
100	10.500	10.536	522.7	523.3

are approximated by a pseudo-first-order reaction [227, 228, 233], according to the equation:

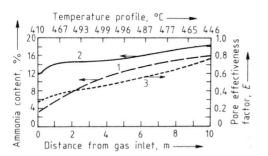

Figure 11. Ammonia content in the bulk stream (1) and in the catalyst pores (at $r = 0.5 R$) (2) and pore effectiveness factor, E (3)
21.4 MPa; 12 % inerts; SV = 15000 h^{-1}; particle diameter, $2 R = 5.7$ mm

$$r = k_v \left(c_{NH_3, equilibrium} - c_{NH_3} \right) \quad (16)$$

For this case, the pore effectiveness factor E is a function of the so-called Thiele modulus m [234]:

$$E = \frac{3}{m} \left[\frac{1}{\tanh m} - \frac{1}{m} \right] \quad (17)$$

The Thiele modulus m is defined by

$$m = \frac{d_{eff}}{2} \sqrt{\frac{k_v}{D_{eff}}} \quad (18)$$

where

d_{eff} = effective particle diameter
D_{eff} = effective diffusion coefficient of ammonia in the catalyst particle
k_v = reaction rate constant referred to a unit of particle volume
tnh = hyperbolic tangent

The practical application of kinetic equations to the mathematical calculation of ammonia synthesis converters is described in [208, 229, 235 – 241].

4.4. Catalysts

The ammonia synthesis catalyst may be viewed as the heart of an ammonia plant. For a given operating pressure and desired production, it determines the operating temperature range, recycle gas flow, and refrigeration requirement. As a result, it directly fixes vessel and exchanger design in the synthesis loop. It also indirectly influences the make-up gas purity requirement, and so the operating pressure, and capital cost, and energy consumption for synthesis gas production and purification. Although the proportionate cost of catalysts compared to the total cost of a modern ammonia synthesis plant is negligible, the economics of the total process are determined considerably by the performance of the ammonia catalyst [242].

Industrial catalysts for ammonia synthesis must satisfy the following requirements:

1) High *catalyst activity* at the lowest possible reaction temperatures in order to take advantage of the favorable thermodynamic equilibrium situation at low temperatures. Average commercial catalysts yield about 25 vol % ammonia when operating at 40 MPa (400 bar) and 480 °C catalyst end temperature, which corresponds to a 535 °C equilibrium temperature. With catalysts that would function at a reaction temperature about 100 K lower, a yield of 45 vol % ammonia can be expected with the same approach to equilibrium, or the pressure may be reduced to 15 MPa (compare Tables 9 and 10).

2) The highest possible *insensitivity to oxygen- and chlorine-containing catalyst poisons*, which may be present in even the very effectively purified synthesis gas of a modern process (see Sections 4.4.1.5, and 4.5.1.3). In assessing the newly developed catalyst systems recommended for operation at very low temperatures (see Section 4.4.2), it must be kept

in mind that the effect of poisons, for example, oxygen compounds, may become more severe as temperature declines (see Fig. 23).
3) *Long life*, which is determined essentially by resistance to thermal degradation and to irreversible poisoning (see Section 4.4.1.4). In older high-pressure plants (60 – 100 MPa), catalyst life was a big issue; because the catalysts in these plants showed a markedly reduced life owing to the severe operating conditions, the necessary downtime for removing, replacing, and reducing the catalyst had a considerable effect on the ammonia manufacturing cost. In modern single-train ammonia plants, conventional iron catalysts achieve service lifetimes up to 14 years.
4) *Mechanical Strength*. Insufficient pressure and abrasion resistance may lead to an excessive increase in converter pressure drop, and so to a premature plant shutdown. For example, mechanical disintegration during operation along with oxygen sensitivity thwarted the industrial application of uranium carbide catalysts [243].
5) Because of the high and increasing world demand for ammonia, a *reliable primary raw material source*. For example, osmium, which was planned as the first industrial catalyst, is so scarce that, in 1910, as a precautionary measure for this option, BASF had secured almost the total world supply [243].

The ammonia synthesis catalyst problem has been more intensively studied than the catalysis of any other industrial reaction. At BASF A. MITTASCH et al. started a tremendous program, in which up to 1911 more than 2500 different formulations were testet in more than 6500 runs. The experiments were finally brought to an end in 1922 after a total of 22 000 tests. They tested almost all elements of the periodic table for their suitability as ammonia catalysts [81, 244]. From these experiments came a series of technical findings, for example, concerning the relationships between catalytic effectiveness and the strength of the nitrogen bond and reducibility, or relating to the mechanism of opposing activation or inactivation in doubly promoted systems.

In principle, metals or metal alloys are suitable as ammonia catalysts, above all those from the transition-metal group [245] (Table 14). Metals or metal compounds for which the chemisorption energy of nitrogen is neither too high nor too low show the greatest effectiveness (Figs. 12, 13), [246, 247], so the magnetite-based catalyst proved suitable for industrial use.

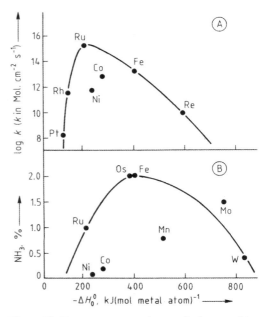

Figure 12. The rate constants of ammonia decomposition (A) on and the ammonia synthesis capacities (B) of metals as a function of $-\Delta H_0^0$. (Mol. denotes molecule, mol denotes mole)

4.4.1. Classical Iron Catalysts

From the early days of ammonia production to the present, the only catalysts that have been used have been iron catalysts promoted with nonreducible oxides. Recently, a ruthenium-based catalyst promoted with rubidium has found industrial application. The basic composition of iron catalysts is still very similar to that of the first catalyst developed by BASF.

The catalytic activity of iron was already known well before the advent of industrial ammonia synthesis. RAMSAY and YOUNG used metallic iron for decomposing ammonia. PERMAN [248], as well as HABER and OØRDT [249], conducted the first catalytic synthesis experiments with iron at atmospheric pressure. NERNST [85] used elevated pressures of 5 – 7 MPa. Pure iron showed noticeable initial activity which, however, could be maintained for longer operating periods only with extremely pure synthesis gas.

Table 14. Effectiveness of various elements as catalysts, promoters, or catalyst poisons

	Catalysts	Promoters	Poisons
I		Li, Na, K, Rb, Cs	
II		Be, Mg, Ca, Ba, Sr	Cd, Zn
III	Ce and rare earths	Al, Y, La, Ce and rare earths	B, Tl
IV	(Ti), (Zr)	Si, Ti, Zr, Th	Sn, Pb, C
V	(V)	Nb, Ta	P, As, Bi
VI	(Cr), Mo, W, U	Cr, Mo, W, U	O, S, Se, Te
VII	(Mn), Re		F, Cl, Br, I
VIII	Fe, Ni, Co, Ru, (Rh), Os, (Ir)		

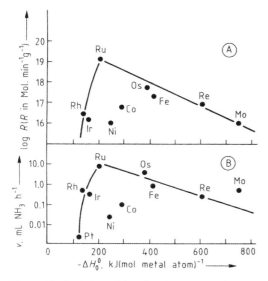

Figure 13. Catalytic activity of carbon-supported metals promoted by metallic potassium as a function of $-\Delta H_0^0$.
A) The rate of isotopic equilibration of N_2 at 623 K, 20 kPa of pN_2 (Mol. denotes molecule, mol denotes mole);
B) The rate of ammonia synthesis at 523 K, 80 kPa of total pressure

Table 15. Effect of various elements or their oxides on the activity of iron catalysts in ammonia synthesis

a)	positive: Al, Ba, Be, Ca, Ce, Cr, Er, K, La, Li, Mg, Mn, Mo, Na, Nb, Nd, Rb, Sm, Sr, Ta, Th, Ti, U, V, W, Y, Zr
b)	negative: As, B, Bi, Br, C, Cd, Cl, F, J, P, Pb, S, Sb, Sn, Te, Tl, Zn
c)	doubtful: Au, Co, Cu, Hg, Ir, Ni, Os, Pd, Pt, Si

The ammonia synthesis catalyst problem could be considered solved when the catalytic effectiveness of iron in conversion and its onstream life were successfully and substantially improved by adding reduction-resistant metal oxides [244] (Table 15). The iron catalysts promoted with aluminum and potassium oxides proved to be most serviceable [250]. Later, calcium was added as the third activator. Development work in the United States from 1922 can be found in [251].

Modern catalysts additionally contain other promoters that were present in the older catalysts only as natural impurities from the raw materials. Onstream life and performance were enhanced considerably by optimizing the component ratios (Section 4.4.1.1), conditions of preparation (Section 4.4.1.3), and catalyst particle size and form (Section 4.4.1.2). The high-purity gas of modern processes and the trend to lower synthesis pressures especially favor the development of more active and easily reducible types of catalysts, at some sacrifice in temperature stability and resistance to poisons. To some extent even today, ammonia plant operating conditions and types of converters (Section 4.5.3.2) can differ greatly one from another. Thus, individual catalyst manufacturers now offer several catalyst types in various particle size distributions, in oxidic and prereduced states.

4.4.1.1. Composition

Table 16 gives a composition survey of commercial ammonia catalysts in the years 1964 – 1966. The principal component of oxidic catalysts is more or less stoichiometric magnetite, Fe_3O_4, which transforms after reduction into the catalytically active form of α-iron.

The degree of oxidation of industrial catalysts has a considerable influence on their catalytic properties. MITTASCH in 1909 established that catalysts manufactured by reducing a magnetite phase were superior to those prepared from other oxides. For industrial catalysts, the highest ammonia yields are observed with an Fe(II) – Fe(III) ratio of 0.5 – 0.6, about the degree of oxidation of stoichiometrically com-

Table 16. Examples of commercial ammonia catalysts from the years 1964 – 1966. Values for the composition in weight %. The numbers at the beginning of the lines are keyed to the catalyst origin.

Origin, type	Fe total	FeO	Fe$_2$O$_3$	Al$_2$O$_3$	MgO	SiO$_2$	CaO	K$_2$O	Other	Particle size, mm	Bulk density, kg/L
1	68.6	36.07	57.85	3.30	0.09	0.75	2.13	1.13	–	2 – 4	2.37
2 normal	60.0	32.91	60.18	2.90	0.37	0.35	2.80	0.54	–	4 – 10	2.94
2	68.2	31.30	62.53	2.90	0.30	0.35	1.65	0.97	–	6 – 10	2.80
2 HT	66.9	32.47	59.18	2.95	1.55	0.40	2.95	0.50	–	6 – 10	2.80
2 prereduced	88.1			3.70	0.43	0.45	3.60	0.70	–	6 – 10	2.30
3	71.3	39.22	58.2	1.80	0.18	0.27	1.43	0.89	–	2 – 4	2.86
4	66.3	22.27	49.0	0.59	4.47	0.77	0.65	0.50	0.7 Cr$_2$O$_3$		
4 prereduced	90.6			0.10	6.08	1.23	0.10	0.86	1.05 Cr$_2$O$_3$		
5	71.5	33.0	65.5	2.96	1.55	–	0.20	0.01	–		
6 (1964)	69.5	23.85		3.15	0.26	0.40	1.85	1.10	–	3 – 9	2.71
6 (1966)	66.9			2.73	0.29	0.43	1.84	1.15	–	5 – 10	2.73
7 prereduced	90.4			3.12	1.00	0.46	0.25	0.58	0.4 MnO	5/5	2.55
8	68.4	35.35		3.16	0.56	0.50	3.54	0.58	–	2 – 4	2.61
9	70.0	32.14		3.17	0.28	0.10	2.40	0.32	–	2 – 4	2.81
10	70.8	33.62		1.58	0.28	1.14	0.67	1.57	–	2 – 4	2.66
11 normal (1964)	66.7	35.95	56.97	3.27	0.67	0.55	3.00	0.65	–		
11 (1966)	68.2	38.70	54.60	2.42	0.35	0.64	2.85	0.58	–		
11 normal (1964)				3.0	0.3	0.5	2.0	1.0	–		
11 (1966)	69.5	38.20	56.70	2.34	0.35	0.57	1.85	0.57	–		
11 HT	66.3	38.22	52.38	2.94	3.56	0.30	2.66	0.63	–		
11 prereduced	84.9			3.62	0.43	0.94	4.70	0.68	–	5 – 11	2.11
12				3.9	4.0	0.8	2.3	1.8	–		
13		23.15		2.9	0.1	0.42	3.12	0.52	–		
14				4		0.7	3	1			

posed magnetite [252 – 254] (Fig. 14). To obtain optimal catalyst composition, careful control of the manufacturing process, especially the melting conditions, which determine the oxygen content, is necessary.

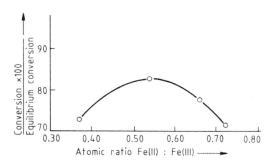

Figure 14. Dependence of the ammonia yield on the degree of oxidation of the iron in the unreduced catalyst

In general, the catalysts contain varying quantities of the oxides of aluminum, potassium, calcium, magnesium, and silicon as promoters. Patents recommend adding sodium [255], beryllium [256], vanadium [257], uranium [258], or platinum [259]. Reference [260] describes cesium-containing catalysts. Catalysts patented by Lummus [261] and Ammonia Casale [262] contain cerium as additional promoter. ICI [263] has developed a cobalt-containing catalyst, as has Grande Paroisse [264].

Nature of the Surface of Commercial Iron Catalysts. Freshly reduced commercial iron catalysts that contain aluminum, potassium, and calcium oxides as basic promoters consist of approximately 30-nm primary crystallites; the spaces between them form an interconnecting system of pores. Besides a maximum at a pore radius of about 10 nm that originates on reduction of the Fe$_3$O$_4$ (magnetite) phase of the nonporous oxidic catalyst, the pore distribution curve (Fig. 15) generally shows a peak at 25 – 50 nm that is formed on reduction of the wustite phase [171, 265]. The pore volume measures 0.09 – 0.1 cm^3/g; with an apparent density of 4.8 – 4.9 g/cm^3, accordingly, the pores represent 44 – 46 % of the volume of a catalyst granule [6]. The surface of the walls of the pores, the so-called inner surface, amounts to about 15 m^2/g.

The novel AmoMax catalyst [169, 170] of Südchemie is iron-based but uses wustite instead of magnetite—which previously was considered undispensible for the production of industrial ammonia catalysts—has an improved pore structure and higher specific surface area.

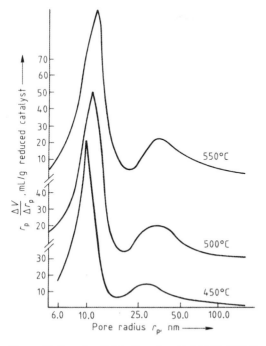

Figure 15. Pore size distribution of a commercial catalyst after reduction at various temperatures [265]

The composition of the outermost atomic layers of the pore walls deviates considerably from the overall average concentrations. Auger electron spectroscopic (AES) measurements on an industrial catalyst (BASF S 6-10) have shown that a significant enrichment of the promoters into the surface results using the unreduced as well as the reduced catalyst [123] (see Table 17). The free iron surface of the reduced BASF catalyst [123] and Topsøe catalyst KM-I [266] comprises only a fraction of the total surface, as could be deduced from the results of prior investigations [171, 267 – 273].

The aluminum oxide promoter exists partly in the form of larger crystallites and, moreover, is relatively homogeneously distributed over the iron area of the surface, although with low concentration [123, 266]. After reduction, about 1 wt % of the alumina also remains statistically distributed in the form of $FeAl_2O_4$ molecular groups built into the α-iron lattice of the reduced catalyst [274, 275]. According to [123] the potassium, in the form of a K + O adsorbed layer, covers about 20 – 50 % of the iron surface. According to [265, 266], a correlation exists between the distribution of the potassium and that of aluminum and/or silicon. Calcium oxide segregates, essentially at the grain boundaries, into separate regions, probably as a mixture of the silicate and ferrite [276]. Auger spectroscopic investigations on reduced BASF and Topsøe catalysts reveal large local differences in composition [123, 265]. Large, apparently homogeneous regions that have originated from reduction of Fe_3O_4 crystallites alternate with nonhomogeneous regions that are formed by the reduction of FeO crystals or consist of amorphous phases [265].

Extensive studies in the last decade have provided a more refined picture of the morphology of the active catalyst (reduced state) and its precursor (oxidic state). A review is given in [277]. With methods such as scanning transmission electron microscopy (TEM) and electron microdiffraction a textural hierarchy has been modeled. Macroscopic particles in the reduced catalyst are confined by fracture lines running through a system of blocks consisting of stacks of slabs in parallel orientation. This structure is already preformed in the preparation of the catalyst precursor, and in the reduction process a further subdivision of the slabs into even smaller platelets might occur. This texture is stabilized by structural promoters, which act as spacers and "glue," separating neighboring platelets and thus providing voids for the interconnection of the pore system. There is also evidence that the basal plane of many platelets has the Fe(111) orientation [167, 277].

Influence of the Promoters. Promoters can be arranged in different groups according to the specific action of the metal oxides:

Structural stabilizers, such as Al_2O_3, produce a high inner surface during reduction and stabilize it under thermal stress by restraining iron crystallite growth [157, 254, 278]. The ability of the various metal oxides to create a high specific surface decreases in the following order [279]:

$Al_2O_3 > TiO_2 > Cr_2O_3 > MgO > MnO$
$\quad = CaO > SiO_2 > BeO$

Table 17. Composition by volume of an industrial ammonia catalyst in comparison to the surface composition before and after reduction (an approximately 10^{-4} cm^2 size typical surface). Numerical values in atomic % [123]

	Fe	K	Al	Ca	O
Volume composition	40.5	0.35	2.0	1.7	53.2
Surface composition before reduction	8.6	36.1	10.7	4.7	40.0
Surface composition after reduction	11.0	27.0	17.0	4.0	41.0

So-called *electronic promoters*, such as the alkali oxides, enhance the specific activity (based on a unit surface) of iron – alumina catalysts. However, they reduce the inner surface or lower the temperature stability and the resistance to oxygen-containing catalyst poisons [280, 281]. In the alkali-metal series, the promoter effect increases with increasing atomic radius, and the destructive effect with decreasing atomic radius [282]. In striving to improve the activity or stability of iron catalysts, a multitude of structural and electronic promoters has been investigated, among them rare-earth oxides [283, 284], such as Sm_2O_3 [285], Ho_2O_3, Dy_2O_3, and Er_2O_3 [286].

Promoter oxides that are reduced to the metal during the activation process and form an alloy with the iron (see also Section 4.4.2) are a special group. Among those in use industrially, cobalt is of special interest [287, 288].

The effect of a given promoter depends on concentration and on the type of promoter combination and the operating conditions, especially the reaction temperature and the synthesis gas purity [257, 280, 282, 289 – 292].

A graphic picture is conveyed in [282] of how the activity of a quadruply promoted (4 % Al_2O_3, 1 % K_2O, 1 % CaO, 1 % SiO_2) catalyst changes with varying promoter concentration and operating conditions (Fig. 16). Under normal operating conditions [14 – 45 MPa, 380 – 550 °C, 10 000 – 20 000 m^3 m^{-3} h^{-1} (STP)], the optimal activity corresponds to a composition of 2.5 – 3.5 % CaO, 2.3 – 5.0 % Al_2O_3, 0.8 – 1.2 % K_2O, and 0 – 1.2 % SiO_2 (Fig. 16 B). Raising or lowering the concentration of a particular oxide causes a reduction in activity. Changes in the potassium and aluminum oxide concentrations have an especially strong influence. Adding magnesium oxide decreases the catalyst performance. If before the test the catalyst is overheated at 700 °C for 72 h (Fig. 16 C) or poisoned with 2000 cm^3 water vapor per cubic meter of gas at 550 °C (Fig. 16 D), the optimum composition shifts to higher Al_2O_3 and SiO_2 and lower K_2O and CaO concentrations. Magnesium oxide addition now shows a favorable effect.

For ammonia plants operating at pressures up to 35 MPa (350 bar), catalyst end temperatures of 520 – 530 °C maximum, and with highly purified synthesis gas, the preferred catalysts contain 2.5 – 4.0 % Al_2O_3, 0.5 – 1.2 % K_2O, 2.0 – 3.5 % CaO, 0 – 1.0 % MgO, and a natural content of about 0.2 – 0.5 % SiO_2 [254]. Less active but more poison- and temperature-resistant catalysts containing up to 3.6 % magnesium oxide were recommended for older plants, for example, classical Casale plants, which operated at up to 80 MPa (800 bar) pressure and in which catalyst end temperature reached 650 °C (see for example [6]). An industrial catalyst for operating temperatures up to 550 °C is stabilized against deterioration by 2 – 5 % V_2O_5 besides 3.5 – 4.0 % Al_2O_3, 2.0 – 2.5 % CaO, and 0.7 % K_2O [257]. For higher operating temperatures, still higher V_2O_5 contents are recommended. Silicon dioxide additions shift the optimum potassium oxide concentration to higher values [280]. For example, the Bulgarian catalyst K-31 contains 3.9 % Al_2O_3, 2.3 % CaO, 0.4 % MgO, 0.8 % SiO_2, and 1.8 % K_2O. An older Norsk Hydro catalyst, besides the usual additives, contained 1.14 % SiO_2 and 1.57 % K_2O. The ICI catalyst with composition 5.2 % CoO, 1.9 % CaO, 0.8 % K_2O, 2.5 % Al_2O_3, 0.2 % MgO, 0.5 % SiO_2, remainder Fe_3O_4, is substantially more active than the conventional cobalt-free catalysts [263, 293]. Reducing the synthesis pressure and/or the synthesis temperature should enable application of the Lummus [261, 294] and Ammonia Casale [262] cerium-containing catalysts.

The effect of the promoters on the rate of reduction and the temperature required for reducing the iron oxide phase is also significant in industrial practice. The structural promoters,

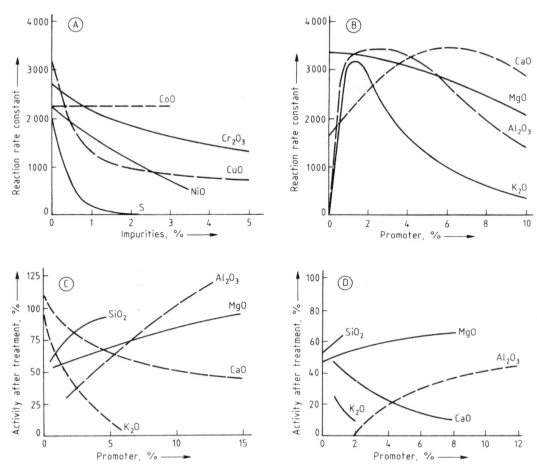

Figure 16. Dependence of the catalyst activity on various factors
A) Concentration of impurities; B) Concentration of promoters; C) Overheating to 700 °C with increasing promoter concentrations; D) Poisoning with water at increasing promoter concentrations

such as Al_2O_3, lower the rate of reduction [277, 295, 296]. Magnesium oxide-activated iron, said to be thermally stable up to 650 °C, needs a higher reduction temperature than aluminum oxide-promoted iron catalyst [297]. Greater differences in reducibility have also been observed in commercial catalysts with similar chemical composition [298] and in connection with particular oxide phases, such as $CaFe_3O_5$ and FeO [298 – 300]. Early work [301] referred to the rough parallels between the reducibility and the thermal stability of catalysts. All published experience appears to demonstrate that it is not possible to combine in a catalyst high thermal stability with easy reducibility and high activity at low temperatures. Hence it may be advantageous to use a combination of active and thermally-resistant catalysts in the same converter.

Mechanism of the Promoter Effect. The action of the so-called structural promoters (stabilizers), such as Al_2O_3, is closely associated with their solubilities in the iron oxide matrix of the unreduced catalyst or with the capability of the regular crystallizing magnetite to form solid solutions with iron – aluminum spinels [6, 302 – 304]. The solid solutions of Fe_3O_4 and the spinel $FeAl_2O_4$ have a miscibility gap below 850 °C [305]; at 500 °C, the solubility limit in the magnetite mixed crystal is a maximum of 7.5 % $FeAl_2O_4$, i.e., 3 % Al_2O_3, referred to iron. Higher alumina contents lead to separation into two phases, whereby only the portion dissolved in the magnetite phase appears to be responsible for the specific promoter action of alumina [267, 268, 306]. According to [307], there exists a close connection with the mechanism of the reduction which consists schematically of the following partial steps [308]:

1) Phase-boundary reaction of hydrogen with oxygen ions of the magnetite lattice, with formation of water vapor and release of electrons
2) Formation of metallic iron nuclei by combination the electrons with Fe^{2+} ions
3) Diffusion of Fe^{2+} ions and electrons to the nuclei and growth of the nuclei to iron crystals of various size

A more detailed discussion of the reduction mechanism is given in Section 4.4.1.4.

In the presence of dissolved aluminum ions, at not too high a temperature, the diffusion rate of the iron ions in the magnetite lattice is low. Hence, nucleus formation proceeds rapidly relative to crystal growth. Therefore, small iron crystallites, about 30 nm, form with correspondingly large specific surfaces. The aluminum probably remains partly in the iron crystallite in the form of very small $FeAl_2O_4$ areas statistically distributed over the lattice [274, 275, 307, 309], where an $FeAl_2O_4$ molecule occupies seven α Fe lattice positions [116].

According to this concept, the stabilizer function of alumina reduces to paracrystalline lattice defects; an analogous effect is to be expected with Cr_2O_3, Sc_2O_3, etc. [274, 309]. Another theory is based on the observation that during reduction part of the alumina precipitates with other promoters into the surface of the iron crystallite in a molecularly dispersed distribution [270, 310] or in small islands [266]. This "patchy" monolayer of alumina acts like a "spacer" between iron atoms of neighboring crystallites and prevents sintering by means of a "skin" effect [275]. (See also [116, 265, 276].)

Insofar as small crystals of nonreducible oxides dispersed on the internal interfaces of the basic structural units (platelets) will stabilize the active catalyst surface Fe(111), the paracrystallinity hypothesis will probably hold true. But the assumption that this will happen on a molecular level on each basic structural unit is not true. The unique texture and anisotropy of the ammonia catalyst is a thermodynamically metastable state. Impurity stabilization (structural promotion) kinetically prevents the transformation of platelet iron into isotropic crystals by Ostwald ripening [167]. Thus the primary function of alumina is to prevent sintering by acting as a spacer, and in part it may also contribute to stabilizing the Fe(111) faces [168, 311].

Calcium oxide, which also acts as a structural promoter [265] has a limited solubility in magnetite. It tends to stretch the magnetite lattice [116]. In the main, on cooling the magnetite melt, it separates at the grain boundaries as $CaFe_3O_5$ (at very rapid cooling rates) [298] and, in the presence of SiO_2, forms poorly reducible intermediate layers of calcium ferrite and silicates [265, 299]. In the reduced catalysts, it segregates between the iron crystallites [272, 312] and so possibly prevents sintering together at high operating temperatures. One may also presume that by partial neutralization of the "acid" components by calcium, more potassium is made available for activating the iron [208].

Potassium is likewise scarcely soluble in magnetite because of its ionic size [116]. In the unreduced catalysts, separate potassium- and iron-rich regions were found [123]. The appearance of a $K_2Fe_{22}O_{34}$ phase besides an unidentified phase, however, has been proved [313]. According to [314], during reduction, the emerging K_2O migrates to the iron crystallite surface. While doing this it reacts with the more or less homogeneously distributed aluminum (silicon) compounds. In this way, it is distributed over the iron phase. In the reduced catalysts, the potassium exists as a K + O adsorption layer that covers about 20–50% of the iron surface [123]. According to [271, 315], potassium associates partly with the alumina in the surface, partly as KOH with iron [316]. It was found that the enhancement of the catalysts' specific activity by potassium oxide is accompanied by a decrease in the electron work function [315, 317, 318]. The promoting effect of potassium seems to be based on two factors which probably act simultaneously. One mechanism is the lowering of the activation energy for the dissociative adsorption of nitrogen [115, 123]. The explanation is based on an electrostatic model [121, 155, 319]. As indicated by the strong decrease in the work function upon potassium adsorption, there is a considerable electronic charge transfer to the substrate, which creates a $M^{\delta-} - K^{\delta+}$ dipole. A nitrogen molecule adsorbed near such a site will experience a more pronounced back-bonding effect from the metal to its antibonding π orbitals. This will increase the bond strength to the metal and further weaken the N – N bond, as can be seen from a further reduction of the N – N stretching frequency [320]. The other effect consists of lowering the adsorption energy of ammonia, which avoids hindering of nitrogen adsorption by blocking (poisoning) of the catalyst surface by adsorbed ammonia molecules [113, p. 234], [123]; hence, potassium oxide ought to improve the catalyst performance less at low than at higher operating pressures [202, p. 316], [321].

The negative effect of K_2O concentrations higher than about 0.58 % has not been explained unequivocally [157]. With increasing potassium concentration, this manifests itself by the increasing size of the average iron crystallite or the decreasing specific surface in the reduced catalyst [280, 313]. Since potassium oxide prevents the formation of solid solutions between alumina and magnetite to a certain extent [322], the recrystallization-promoting effect of higher K_2O concentrations may be attributed to a lowering of the portion of alumina dissolved in the magnetite phase [313]. Remarks in [314] and [323] reveal another possible interpretation: the K_2O located at the phase boundary surface and not bound to acid or amphoteric oxides may be converted by water vapor concentrations over 10^{-2} ppm in the synthesis gas into potassium hydroxide or by hydrogen into potassium and potassium hydroxide, which would exist in molten form at operating conditions [314, 316].

An extensive review on promoters can be found in [122].

4.4.1.2. Particle Size and Shape

The choice of particle size and shape of commercial ammonia catalysts is determined mainly by two factors:

1) Catalyst performance
2) Pressure drop

From the standpoint of space – time yield, it is desirable to use the finest possible particle, which, practically speaking, is about 1 – 2 mm (Fig. 17); however, with decreasing particle size, the pressure drop and the risk of destructive fluidization of the catalyst increase (Fig. 18).

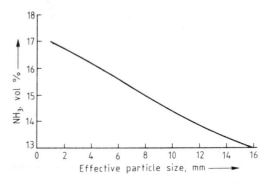

Figure 17. Influence of the particle size on the ammonia production (BASF catalyst).
25 MPa; SV = 12 000 m^3 m^{-3} h^{-1} (STP); gas composition (vol %): N_2 20.59, H_2 60.06, NH_3 2.55, Ar 5, CH_4 10

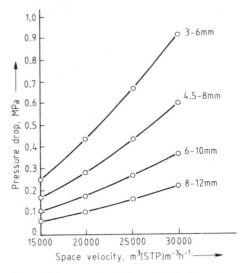

Figure 18. Pressure drop in the catalyst bed for various catalyst particle size ranges, from [6].
Depth of the catalyst bed, 7 m; reaction pressure 27.1 MPa; reaction temperature 450 °C

For processes operating at pressures of 25 – 45 MPa (250 – 450 bar) and at space velocities of 8000 – 20 000 m^3 m^{-3} h^{-1} (STP) a grain size of 6 – 10 mm is preferred. Larger granulations, for example, 8 – 15 mm or 14 – 20 mm, are used only in plants where the lowest possible pressure drop is essential because of very high gas velocities. In catalyst zones in which the ammonia formation rate is so high that the allowable temperature limits are exceeded, it may be advantageous to use coarse particles for suppressing the reaction. Radial-flow converters and the horizontal crossflow Kellog converter (Fig. 52), which operate at comparatively low gas velocities [324], allow the use of small granulations (1.5 – 3 or 2 – 4 mm) with optimal use of the converter volume. Fluidized-bed processes, which were explored especially in the Soviet Union, have so far been unsuccessful [325, 326].

Two effects cause the low production capacity of coarse-grained catalyst: first, large grain size retards transport of the ammonia from the particle interior into the bulk gas stream, because this proceeds only by slow diffusion through the pore system. Slow ammonia diffusion inhibits the rate of reaction. At the high reaction rate typical for the converter inlet layer, only a surface layer of the catalyst grains, about 1 – 2 mm thick, participates in the reaction.

The second effect is a consequence of the fact that a single catalyst grain in the oxidic state is reduced from the outside to the interior of the particle [327]: the water vapor produced in the grain interior by reduction comes into contact with already reduced catalyst on its way to the particle outer surface; this induces a severe recrystallization [6]. The effect is very significant. As an example, if the particle size increases from about 1 to 8 mm, the inner surface decreases from $11 - 16 \, m^2/g$ to $3 - 8 \, m^2/g$ [328].

To allow for the influence of various particle shapes and size distributions within a defined sieve fraction, in lay-out calculations it is customary to employ an effective particle diameter, d_{eff}, as nominal size. The diameter d_{eff} is defined as the ratio of equivalent diameter Λ and a form factor ψ. Λ is equal to the diameter of a sphere with a volume equal to the (average) volume of the particles, and ψ is the average ratio of the particle surface to the surface of a sphere of equal volume.

Table 18 shows the relation between the catalyst size classification, the equivalent particle diameter, and the percentage saving in catalyst or converter volume relative to the 6 – 10 mm standard size [293].

Table 18. Effect of catalyst size on catalyst volume

Catalyst size classification, mm	Approximate equivalent particle diameter, mm	Relative catalyst volume, %
6 – 10	7.5	100
4.5 – 8	5.5 – 6.5	92 – 95
3 – 6	4.5 – 4.7	88 – 90
1.5 – 3	2.0 – 2.2	80 – 82
1 – 1.5	1.2 – 1.3	77 – 79

An irregular grain shape, for example with a shape factor of 1.5, has a more favorable effective activity for the individual particle and for radial intermixing of mass and heat in an industrial converter [329] than a more cubic or spherical shape, with a shape factor close to one. According to a patent by Chemie Linz AG [330], the catalyst particle ought to be 2 – 20 times as long as it is broad, preferably 5 – 10 mm long and 1 – 2 mm thick (broad). On the other hand, regular shapes have the advantages of greater abrasion resistance and lower pressure drop (see Fig. 19) [331].

The advantages of regular catalyst shapes and the need to compensate for the above-described negative effects of larger grain size by a system of macropores in the oxidic and reduced catalyst stimulated various attempts to manufacture shaped, macroporous catalysts. Various manufacturing techniques have been proposed [332 – 341]. As an example, magnetite is melted with the additives at high temperature (> 1600 °C) and the melt is cooled, broken, and ground to powder. After water is added and, if required, a binding agent, such as bentonite [340], or a promoter salt, such as cerium nitrate [262], the powder is pelletized. The pellets subsequently are dried and sintered in an inert atmosphere at higher temperatures (about 1350 °C).

The application of macroporous catalysts ought to be especially useful for very low synthesis pressures and in plants in which large catalyst particles must be used for reasons of low pressure drop. For example, the performance of the macroporous Topsøe catalyst KMG 6 mm at 5 MPa synthesis pressure is said to be at least equivalent to KM I 1.5 – 3 mm and distinctly superior to KM I 6 – 10 mm [342].

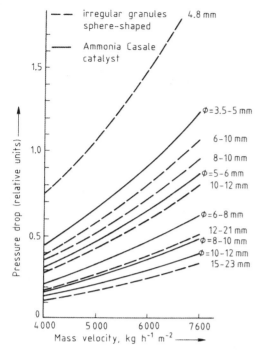

Figure 19. Comparison of the Ammonia Casale spherical catalyst and irregularly shaped catalyst [293]

4.4.1.3. Catalyst-Precursor Manufacture

The term "ammonia catalyst" commonly refers to the oxidic form consisting of magnetite and oxidic promoters. In fact this is only the catalyst precursor which is transformed into the active catalyst composed of α-iron and promoters by reduction with synthesis gas, usually in situ. The reduction step is very important for catalyst performance.

The first effective catalysts were made by the oxygen-melt process. The purest possible iron (e.g., Swedish charcoal iron), together with the additives, was burned to Fe_3O_4 in a stream of oxygen. This process was largely replaced by melt processes in which natural or, less frequently, synthetic magnetite, together with the activators, was melted electrically or in electric arc furnaces [253, 343]. The cooled melt is ground to the proper granulation and reduced with hydrogen – nitrogen mixtures.

A process developed by Farbenfabriken (formerly Friedrich Bayer) had only local significance. Complex iron cyanides were decomposed thermally in the presence of hydrogen.

The hydrodecomposition proceeds via the carbide and nitride, finally to α-iron. This is activated in the well-known manner by reduction-resistant metal oxides. The so-called Mont Cenis process employed such catalysts [344]. Repeatedly described as a means of manufacture, although only in the scientific literature, is coprecipitation of the catalyst components, for example, from aqueous solutions of the metal salts, with subsequent calcining and reduction [345 – 348]. With magnesium oxide as support, very small (under 10 nm) iron particles are obtained with high specific iron surfaces [348], similar to those obtained by exchange of magnesium ions by iron ions in the surfaces of magnesium hydroxy carbonate crystals [348, 349]. Granulation and sintering techniques have been used for the preparation of shaped macroporous iron catalysts, which, however, have not gained industrial importance.

Impregnating the pore surface of prereduced passivated catalysts is a possibility for incorporating promoters into iron ammonia catalysts. A United Kingdom patent [350], by way of example, claims catalysts manufactured by impregnating the reduced catalysts with cerium salts. Improving the performance and life of industrial catalysts by radioactivity and X-rays [351], treatment with ultrasound [352, 353], mechanical treatment [354], or high-frequency, alternating-field heating [355] has been attempted also.

The superiority of the catalyst manufacturing processes that use a molten iron oxide stage is mainly due to the fact that above 1000 °C in air, magnetite, Fe_3O_4, is the thermodynamically stable oxide phase of iron [81, 356]. Magnetite leads to especially efficient catalysts, and its electrical conductivity allows the use of economical electrical melting processes.

In 1996 the prices of commercial ammonia catalysts were about 2 $/lb (3.58 €/kg) for oxidic and about 5.5 $/lb (10.23 €/kg) for prereduced. Therefore, they are among the least expensive catalysts.

Operating conditions in the individual manufacturing steps — proportioning and mixing the raw materials, melting, cooling, crushing or if necessary grinding and preforming, and reduction — influence the quality of the finished catalyst (Fig. 20) [242].

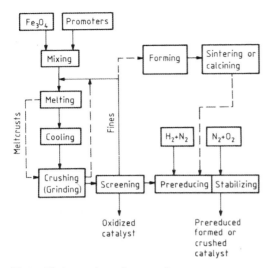

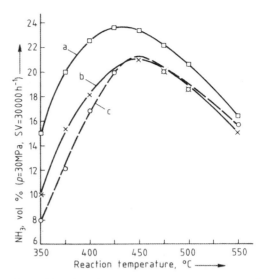

Figure 20. Ammonia catalyst manufacture

Figure 21. Effect of the melting temperature and rate of cooling of the melt on the activity of ammonia catalysts a) Melt overheated to 3500 °C (rapid cooling); b) Melt temperature 1800 – 2000 °C (rapid cooling); c) Melt temperature 1800 – 2000 °C (slow cooling)

The raw materials — usually, natural magnetite, lime, potash, and alumina — must, as far as possible, be free of catalyst poisons (see Section 4.4.1.5). Many ores have too high a content of free or bound silica, which can be lowered with magnetic separators [357]. Melting is accomplished in electrical resistance or induction furnaces (arc furnaces in the past) operating at 1600 – 2000 °C. The walls of these furnaces should consist of a weakly basic tamping material indifferent to the melt, such as magnesium oxide, which is also an activating component of the catalyst [358]. A homogeneous distribution of the promoters in the magnetite melt and a degree of oxidation at or under that of stoichiometric magnetite should be obtained. This is said to be promoted by initially overheating to temperatures up to 3500 °C [359] (Fig. 21). It is claimed to be advantageous to bring the promoters into the melt as common chemical compounds that are isomorphous with magnetite [360].

Induction furnaces are optimal for the melting operation. Their good temperature control permits accurate adjustment of the degree of oxidation. Since the melt is held in constant turbulent motion by the magnetic field produced in the primary coil, it is well mixed, even for short melt times. In comparison to the most frequently used resistance furnaces, plant cost and power consumption are higher.

Another important factor in catalyst manufacture is the melt cooling rate, which is affected by the design and dimensions of the ingot molds. Quenching or fast cooling in thin sheets leads to a less abrasion-resistant, sharp-edged chip after crushing. Very slow cooling results in a more cubic chip, but with inferior catalyst quality (Fig. 21). In practice, slow cooling is avoided.

With falling temperature, both the solubility of the activator oxides (see page 173) in magnetite and the rate of adjustment to the new phase equilibrium decline. Therefore, by rapid cooling of the melt the activator oxide distribution can be frozen in a condition corresponding to that of a higher temperature [361]. According to [298], there may exist relationships between the melt cooling rate, the appearance of certain phases, and the reducibility of the catalysts.

4.4.1.4. Catalyst Reduction

The reduction of oxidic catalyst is generally effected with synthesis gas. The magnetite is converted into a highly porous, high surface area, highly catalytically active form of α-iron. The promoters, with the exception of cobalt, are not reduced [6].

To ensure maximum effectiveness of the catalyst, a defined reduction procedure must be followed. Above all, it is important to hold the partial pressure of the resulting water vapor as low as possible and to insure that the water vapor does not come into contact with regions that have already been reduced. High temperature and high water vapor partial pressure markedly accelerate premature catalyst aging by recrystallization. Therefore, the reduction should be carried out at high gas velocities [about 5000 – 15 000 m^3 m^{-3} h^{-1} (STP)], at the lowest temperatures sufficient for complete reduction, and at not too high pressures (7 – 12 MPa in low-pressure, 25 – 30 MPa in high-pressure plants) to hold the exothermic formation of ammonia under better control during the reduction. When the reduction of the oxidic catalyst is carried out in the production plants, long reduction times are needed at low temperatures and low pressures with a consequential loss of production.

In practice, the reduction temperature is raised stepwise by using the exothermic heat of ammonia formation. The progress of the reduction is controlled according to the catalyst temperature and the water concentration by means of the synthesis gas flow. As a rough guideline, the water content of the gas effluent from the catalyst should not exceed 2 – 3 g/m^3 (STP). Under these conditions, depending on its size and operating pressure, a synthesis converter with a fresh load of oxidic catalyst attains its full production capacity in 4 – 10 d.

The minimum temperatures necessary for reduction are somewhat different for the various catalyst types. Catalysts conventionally employed in medium-pressure plants may be reduced from about 340 to 390 °C, although a slow induction period starts somewhat lower. Generally, temperatures above 440 °C are required to complete the reduction.

The reducibility of industrial catalysts is dependent on both the combination of promoters and the degree of oxidation. The FeO (wustite) phase is reduced faster and at lower temperatures than the Fe$_3$O$_4$ (magnetite) phase [298]. According to [298], the rather considerable differences in the reduction rates of commercial catalysts with similar compositions may be attributed to differences in manufacturing methods or operating conditions. Commonly, the manufacturers hold these in strict secrecy.

Some older publications deal with the influence of catalyst granulation on the optimum reducing conditions [362]. Directions for reduction in multibed converters that combine fine- and coarse-grained catalyst appear in [363]. The influence of the hydrogen/nitrogen ratio during reduction on the catalyst performance after reduction is discussed in [364].

The influence of the reduction conditions (gas flow rate, temperature in the range 300 – 600 °C, nitrogen content in the hydrogen in the range 0 – 100 %) on the production capacity of an ammonia catalyst has been investigated [365].

The gas/solid reaction between magnetite and hydrogen has been studied in great detail by rate measurements, microscopy, and X-ray diffraction [366 – 369]; a summary is given in [277]. On the atomic scale the reaction is controlled by two processes:

1) Metallic iron is formed from wustite by direct chemical reaction controlled in the initial phase by the reaction rate (activation energy ca. 65 kJ/mol) and in the final stage by diffusion processes involving hydrogen and water on the reaction site:

$$FeO + H_2 \rightarrow Fe + H_2O$$

2) The chemical reaction creates an iron(II) ion concentration gradient in the solid. This gradient leads to a rapid diffusion of iron(II) ions from magnetite through wustite to the chemical reaction interface, where they are reduced and precipitated as iron nuclei. This is made possible by the structural defects of the wustite. The precipitation of further wustite nuclei on the magnetite/wustite reaction interface seems to be effected by ion/electron diffusion processes rather than by direct contact of magnetite with hydrogen [370]:

$$O^{2-} + H_2 \rightarrow H_2O + 2e^-$$

$$Fe_3O_4 + Fe^{2+} + 2e^- \rightarrow 4 FeO$$

The topotactic reduction process leads to a core and shell structure which is visible under the optical microscope and is shown schematically in Figure 22 (a). Figure 22 (b) shows the wustite/magnetite interface as observed by electron microscopy [277]. This concept was developed for single-crystal magnetite grains, but the

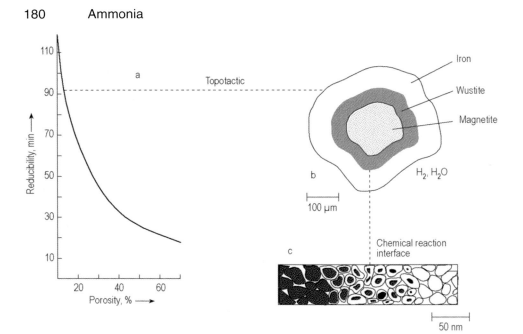

Figure 22. Mechanism of catalyst reduction [119]
a) Reducibility of catalyst under standard conditions as a function of its porosity; b) Core and shell structure of catalyst; c) Reaction interface

shell and core model is also valid for commercial polycrystalline catalysts [277, 371]. Information on the reduction kinetics, including industrial catalysts, can be found in [167, 327, 372 – 375]. Newer findings seem to question the shell and core model with the topotactic reaction interface [152]. On account of the simplifications involved this model should be regarded as a formalistic approach to describe the reaction kinetics rather than a mechanistic approximation. A more detailed atomistic picture leads to the assumption that there is no homogeneous topotactic reaction interface. According to this concept the active catalyst should not grow as a shell around a precursor core but as a core within an oxidic matrix.

Prereduced, stabilized catalyst types, introduced on the market some years ago, have gained a considerable market share. Prereduced catalysts have the full pore structure of active catalysts, although the pore surface has been oxidized to a depth of a few atomic layers to make these catalysts nonpyrophoric.

Reactivating such catalysts usually takes only 30 – 40 h. Ammonia formation begins at substantially lower temperatures, so that altogether the downtime of a production unit is reduced markedly. This and further advantages, such as reducing the risk of damaging the catalyst during activation by too high a local concentration of water, the quantitatively lower yield of aqueous ammonia solution (which can be added to the production), as well as the roughly 20 % lower bulk weight (which reduces the design loadings and costs of the internals of large converter units), make using prereduced catalysts increasingly attractive, especially in single-train plants, in spite of higher prices. The somewhat inferior mechanical strength is a disadvantage that requires special care when a charge is being loaded into the converter. When removed from the hydrogen – nitrogen atmosphere at low to moderate pressure in the producing furnace, the prereduced catalysts are pyrophoric and must be stabilized before transport and installation by passivating the surface. Usually, the method recommended by BURNETT is used. The reduced charge is treated with nitrogen containing 100 – 1000 ppm oxygen at 50 – 70 °C (maximum 95 °C) and a pressure of 0.1 – 0.2 MPa (1 – 2 bar) and up [376]. The reducible oxygen content of the prereduced catalyst ranges between

2 and 7 %. Part of this is only loosely bound and is removed in reactivation even below 200 – 300 °C.

Detailed data on the manufacturing steps most important to the catalyst performance, reduction, prereduction, passivating, and reactivation, appear in [6], including a discussion of the most important literature in this field.

4.4.1.5. Catalyst Poisons

The activity of an ammonia synthesis catalyst may be lowered by certain substances, commonly referred to as poisons. These substances can be minor gaseous constituents of the synthesis gas or solids introduced into the catalysts during the manufacturing procedure, derived from impurities in the natural magnetite from which the catalyst is made. These latter should not play a major role with catalysts from manufacturers of repute and are not discussed in detail in this section because of the proprietary nature of the production processes. General measures to avoid this sort of contamination include selecting a rather pure magnetite, the application of pretreatment processes, and the use of high-purity promoters. The melting process itself may also contribute to minimizing the content of some minor impurities. For gaseous poisons in the synthesis gas, a distinction can be made between permanent poisons that cause irreversible damage to the catalyst and temporary poisons which lower the activity while present in the synthesis gas. In contrast to temporary poisons, permanent poisons can be detected by chemical analysis. Oxygen-containing compounds such as H_2O, CO, CO_2, and O_2 are the most common temporary poisons encountered in ammonia synthesis.

Oxygen compounds have a reversible effect on iron catalysts at not too high temperatures. That is, the activity of a damaged catalyst may be practically completely restored by reduction with clean synthesis gas. Equivalent concentrations of oxygen compounds, for example, 100 ppm of O_2 or CO_2 and 200 ppm CO or H_2O, lead to the same degree of poisoning, presumably because as soon as they enter the catalyst bed they rapidly and completely transform into H_2O [6]. The damage depends approximately linearly on the quantity of adsorbed water taken up by the catalyst [377], which is proportional to $\sqrt{pH_2O}/\sqrt{pH_2}$ [378]. Corresponding to the adsorption equilibrium, the degree of poisoning therefore rises with growing partial pressure ratio, pH_2O/pH_2 and falls with increasing temperature.

Under the assumption of a displacement equilibrium in accordance with

$$(N_2)_{ads} + H_2O_{gas} + 2\,H_2{}_{gas} \rightleftharpoons (O)_{ads} + 2\,NH_3$$

I. A. SMIRNOV et al. set up a rate equation for ammonia synthesis [379, 380] that takes the effect of water vapor into consideration over a wide range of temperature and pressure:

$$v = \frac{k_1 p_{N_2} - k_2 p_{NH_3}^2 / p_{H_2}^3}{\left(p_{NH_3}^2/p_{H_2}^2 + C p_{H_2O}/p_{H_2}\right)^{1/2}} \quad (19)$$

and found the values of C listed in Table 19.

A more recent investigation [381 – 383] proposed multiplying the rate equation by a correction factor $1 - \Theta$, where $\Theta = a + bT + cT \ln X_{H_2O}$ and X_{H_2O} is the molar fraction of H_2O. Some authors assume a different route for the formation of adsorbed atomic oxygen [384, 385]:

$$H_2O\,(g) + 3* \rightleftharpoons 2\,H* + O*$$

where * denotes a surface site.

The equivalence of H_2O, CO, CO_2 and O_2 [386] with respect to their poisoning effect has been confirmed.

The experimentally determined effect of water vapor concentrations up to about 30 ppm on the activity of a commercial catalyst (BASF S 6–10) at 30 MPa is evident from Figure 23.

With continuing exposure oxygen compounds also cause irreversible damage to the catalyst activity that is causally linked with growth of the iron primary crystallite [383, 387]. This is probably one of the main causes of the decline in converter performance over the course of the catalyst operating life. This damage depends on the water-vapor partial pressure and is especially serious, in contrast to reversible poisoning, at high temperatures. In a pilot plant, Österreichische Stickstoffwerke (ÖSW), Linz, established that at 30 MPa and a water-vapor content of 250 ppm, for example, the production declined by about 15 % per month. For a carbon monoxide content of 5 ppm, they determined about a 4 – 5 % decrease in activity per year. The

Table 19. C values for the ammonia synthesis rate equation

Catalyst	Temperature, °C				
	400	425	450	475	500
Fe + Al$_2$O$_3$	0.63	0.39	0.24	0.17	0.12
Fe + Al$_2$O$_3$ + K$_2$O	0.74	0.46	0.28	0.20	0.14

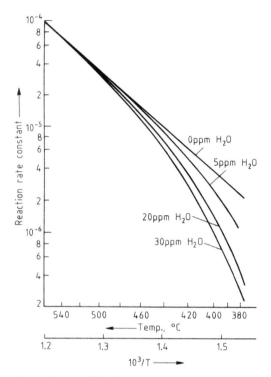

Figure 23. Reversible effect of increasing water vapor concentrations in the synthesis gas on the activity of industrial ammonia catalysts

mance is observed after a month of operation. With the highly purified synthesis gas of modern synthesis processes (with final gas purification by methanation the carbon monoxide level is lowered to below 5 ppm), operating periods of up to 14 years can be achieved for a converter charge without significant loss of activity (Fig. 24).

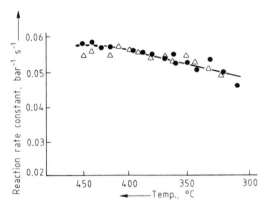

Figure 24. Activity of BASF ammonia catalyst S 6–10 in a commercial converter after 14 years of operation
● fresh catalyst; △ after 14 years operation

influence of the operating temperature is evident from the data. The performance of a catalyst operated at 570 °C is about 35 – 40 % under that of a catalyst charge operated at 520 °C. That is, the higher the temperature, the greater is the harmful effect of oxygen compounds [388]. Correspondingly to the data, they established that above all, temperatures above 500 – 520 °C must be avoided in order to achieve a converter charge operating period of more than 2 – 3 years. If the temperature is kept at this level, generally recommended for modern plants, then even at CO concentrations of 20 ppm in the recycle loop gas, no serious deterioration of the catalyst perfor-

As already mentioned in Section 4.4.1.1, the concentration and combination of promoters affect the degree of irreversible damage. This must be considered in the choice of catalyst for a particular plant.

Sulfur, Phosphorus, and Arsenic Compounds. Sulfur, occasionally present in synthesis gases from coal or heavy fuel oil, is more tightly bound on iron catalysts than oxygen. For example, catalysts partially poisoned with hydrogen sulfide cannot be regenerated under the conditions of industrial ammonia synthesis. Compounds of phosphorus and arsenic are poisons but are not generally present in industrial syngas. There are indications that these permanent poisons exert the most detrimental effect when present as hydrogen compounds and are less harmful in higher oxidation states [389].

With regard to the sulfur bound on the catalyst surface, differences exist between the various types of ammonia catalysts, especially between those containing or free of alkali and alkaline earths. Pure iron and catalysts activated simply with alumina chemisorb S_2N_2 or thiophene when treated with concentrations of H_2S that lie below the equilibrium for the FeS bond, a maximum of 0.5 mg of sulfur per m^2 of inner surface or free iron surface; this corresponds to monomolecular coverage [390, 391]. The monolayer is also preserved on reduction with hydrogen at 620 °C, whereas FeS formed by treatment above 300 °C with high H_2S concentrations is reducible as far as the monolayer. For total poisoning, $0.16 - 0.25$ mg S/m^2 is sufficient. Like oxygen, sulfur promotes recrystallization of the primary iron particle.

Under similar poisoning conditions, alkali- and alkaline earth-containing industrial catalysts adsorb more H_2S. In spite of this, however, in terms of activity, they are more stable toward the action of sulfur and are partially regenerable [391]. In a catalyst bed, most of the sulfur already has been taken up in the gas inlet layer. A catalyst sulfur content of several 100 ppm suffices to impair its activity [6].

In industrial plants, sulfur may reach the ammonia converter in various forms. In some plants, traces of H_2S and COS may not be removed in the upstream purification steps and so may enter the converter with the make-up gas. However, the sulfur contained in the compressor oil constitutes the main danger. On cracking the oil to lower molecular mass hydrocarbons, sulfur is freed as H_2S. It is therefore very important to use an oil with low sulfur content in ammonia plants, especially those still using reciprocating compressors. If after mixing with the recycle gas the make-up gas first runs through an ammonia condensation stage in which the H_2S and also to a certain extent COS are very effectively washed out by condensing ammonia, a sulfur content of the oil of $0.1 - 0.2$ wt % ought to be sufficiently low. Otherwise, a value under 0.1 % is recommended [392]. In modern plants designed with centrifugal compressors and in which the sulfur content of the synthesis gas is extremely low because of very effective purification (about $0.5 - 1$ μg S/m^3 (STP)), sulfur poisoning is of lesser importance than carbon monoxide and chlorine poisoning.

Chlorine compounds. The permanent poisoning effect of chlorine compounds is two orders of magnitude worse than that of oxygen compounds. Concentrations of about 0.1 ppm are viewed as the uppermost allowable limit in order not to affect adversely the life of ammonia catalysts [392]. The deactivation effect is based at least in part on the formation of alkali chlorides that are volatile at the upper synthesis temperatures.

Further information on catalyst poisoning is given in [382, 383].

4.4.2. Other Catalysts

4.4.2.1. General Aspects

For a long time efforts to improve the efficiency of industrial ammonia production concentrated on synthesis gas production, and major progress was achieved over the years. In ammonia synthesis itself considerable progress was made in converter design and recovery of the reaction energy at high temperature, but there has been no substantial improvement in the catalyst since the 1920s. The standard commercial iron catalyst still requires relatively high pressures (usually in excess of 130 bar), high temperatures ($400 - 500$ °C) and large reactor volumes (more than 60 m^3 for a capacity of 1500 t/d) to achieve good economics, although in a few cases a pressure as low as 80 bar has been used. From the vast amount of experimental and theoretical studies of the iron catalyst one can conclude that there is only limited potential for further improvement. Substantial energy savings would require lowering the synthesis pressure considerably, down to the synthesis gas production level, say. However, to compensate for the less favorable equilibrium situation much lower operating temperatures would be necessary, because otherwise too low an ammonia concentration would result, and additional energy would be needed for recovery, thus cancelling the energy saving from synthesis gas compression. To reach this goal a synthesis catalyst with a volumetric efficiency some two orders of magnitude greater than magnetite would be necessary. Process studies show an energy saving potential of about 1 GJ per tonne NH_3 [393].

In the search for an alternative catalyst, most metals have been tested, either as primary components or as promoters. Much of this work was performed in the early, pioneering studies in the BASF laboratories [243, 244]. Most of the studies in the following years concentrated on the magnetite system in the sense of more fundamental and general catalytic research. Rising energy costs since the mid-1960s have given a new incentive to the search for other catalyst systems with improved performance. The first development which found commercial application was a cobalt-modified magnetite catalyst introduced in 1984 by ICI. With similar kinetic characteristics its volumetric activity is about twice that of the standard iron catalyst. The only other catalyst system which exhibits a promising potential for industrial application is based on ruthenium [184]. These new efforts to find improved catalysts could use methods and knowledge of modern surface science as developed on the example of the magnetite catalyst. Structural sensitivity and nature of the nitrogen adsorption and dissociation steps could serve as guidelines [185].

For the overall performance of potential catalysts in practical application additional factors, such as number of active sites, physical form, and porosity must also be taken into account. The classical commercial iron catalyst is an unsupported catalyst. First of all iron is a cheap material and secondly by the incorporation of alumina a surface area similar to that attained in highly dispersed supported catalysts can be obtained. Of course, for an expensive material such as the platinum group metals, the use of a support material is the only viable option. The properties of the supported catalyst will be influenced by several factors [184]

– Adequate surface of the carrier to achieve a reasonable metal loading.
– Dispersion stability by using a more active support with strong interaction between the support phase and the metal precursor. Too strong an interaction may cause difficulties in the metal reducing.
– Promoter localization with respect to metal sites and support sites.
– Gas transport effects will be governed by pore size, pore distribution, and tortuosity.
– Anion retention capability, which plays a role in the catalyst preparation by impregnation of a support with metal impregnation of the support.

4.4.2.2. Metals with Catalytic Potential

Materials that show significant ammonia synthesis activity can be divided into three categories according to their ability to form nitrides:

1) Platinum group metals: no stable nitrides (Ru, Os, Ir, Pt)
2) Metals forming nitrides unstable under reaction conditions (Mn, Fe, Co, Ni, Tc, Re)
3) Metals likely to be present as nitrides under synthesis conditions (groups 3 – 6 of the periodic table)

Of the platinum group metals only ruthenium and osmium show an activity superior to iron, though only in presence of alkali metal promoters, as may be seen from Table 20[394].

Although osmium was the first active catalyst used by Haber in 1909 [395] in his laboratory-scale unit to demonstrate the technical viability of the high-pressure recycle process, this metal never became an industrial catalyst because of its limited availability and its dangerous properties. The interest therefore shifted to ruthenium. The most active ruthenium catalysts use a graphite support and alkali metal promotion, preferentially with rubidium or cesium [396, 397]. It is unlikely that the promoter is in metallic state as its high vapor pressure would probably lead to substantial losses under synthesis conditions. It is assumed that a charge transfer complex $M^+ \cdots C^-$ is formed between the metal and the graphite. A major advantage of graphite is its ability to stabilize high loadings of alkali metals. A special limitation until recently was the chemical reactivity of carbon supports. In a typical ammonia synthesis environment with high hydrogen partial pressure, the catalyst may also catalyze the methanization of carbon, which would lead to destruction of the support. Indeed, this phenomenon has been observed [398], but it can be avoided by careful heat treatment of the support above 1500 °C [399]. With this modified carbon material lifetimes of at least six years are expected [394].

The group of metals forming low-stability or unstable nitrides includes Mn, Fe, Co, Ni, Tc, and Re. As in the case of iron a clear structural sensitivity was found for rhenium but the role

Table 20. Ammonia synthesis activity of metals supported on carbon with potassium metal promotion (mL NH_3/ mL catalyst, 573 K, 128.313 kPa, H:N = 3:1)

		Fe 0.72	Co 0.4	Ni 0.04
Mo 0.6		Ru 22.2	Rh 0.52	Pd 0
	Re 0.36	Os 5.6	Ir 0.68	Pt 0.008

of promoters remains the subject of discussion. There are also indications of structure sensitivity for cobalt and nickel. It was attempted to improve the activity of the classical magnetite catalyst by alloying with nickel or cobalt. The only commercial catalyst is a cobalt containing magnetite [400].

Of the group of metals forming stable nitrides, only molybdenum is of some interest. Under synthesis conditions it is present as a nitride with some ammonia formation activity and structural sensitivity [401]. Molybdenum also seems to exhibit activity in biological nitrogen fixation [402] and is synthetically active at ambient conditions in the air-sensitive Glemser compounds [402, 403].

The results of the intensive research in this field over the last decades demonstrate that, irrespective of the catalyst, the rate-determining step in the ammonia synthesis reaction is the dissociation of the nitrogen molecule and the catalyst effectiveness is determined in the first instance by the activation energy of the dissociation reaction. The other common factor for the ammonia catalysis is the structure sensitivity of molecular nitrogen adsorption. Only if both conditions are favorable, and other factors such as hydrogen and ammonia inhibition do not play a major role, can a sufficient overall reaction rate be expected. The available data show that these conditions are fulfilled only for a limited number of metals: iron, ruthenium, and osmium. On account of its very strong ammonia inhibition rhenium is not an option. Extensive literature on non-iron catalysts is given in [184, 185]; kinetic investigations are reported in [185].

4.4.2.3. Commercial Ruthenium Catalysts

Since the early days of industrial ammonia synthesis only minor improvements have been achieved for the magnetite system: optimization of manufacturing procedures, promoter concentrations, and particle size to give somewhat higher activity and longer service life.

A notable development for the magnetite system was the introduction of cobalt as an additional component by ICI in 1984 [404, 405]. The cobalt-enhanced catalyst formula was first used in an ammonia plant in Canada using ICI Catalco's AMV process (later also in other AMV license plants) and is also successfully applied in ICI's LCA plants in Severnside. Commercially successful is a new type of iron-based catalyst, the AmonMax-10 of Südchemie which uses wustite instead of magnetite. This catalyst is already used in seven ammonia plants with a total capacity of 5600 t/d [169, 170, 406, 407]. Production of synthetic wustite is described in [408], its application for ammonia synthesis catalyst in [409]. For a literature report on promoted iron catalysts see [410].

In 1979 BP disclosed to M.W. Kellogg a new catalyst composed of ruthenium on a graphite support [404, 411]. In October 1990, after a ten-year test program, Kellogg started the commercialization of the Kellogg Advanced Ammonia Process (KAAP) using this catalyst [411], which is claimed to be 10 – 20 times as active as the traditional iron catalyst: The KAAP catalyst. According to the patent [400] the new catalyst is prepared by subliming ruthenium-carbonyl $[Ru_3(CO)_{12}]$ onto a carbon-containing support which is impregnated with rubidium nitrate. The catalyst has a considerably higher surface area than the conventional catalyst and, according to the patent example, it should contain 5 wt % Ru and 10 wt % Rb. Besides having a substantially higher volumetric activity, the promoted ruthenium catalyst works best at a lower than stoichiometric H/N ratio of the feed gas as shown in Figure 25. It is also less susceptible to self-inhibition by NH_3 (Figure 26) and has excellent low-pressure activity.

Three ammonia plants in Trinidad, each with a nameplate capacity of 1850 t/d, use the KAAP ruthenium catalyst [412]. The first layer of the four-bed reactor [100, 102] is loaded with conventional magnetite catalyst, the others with the ruthenium–graphite catalyst. For the

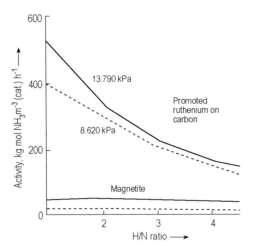

Figure 25. Effect of H/N ratio on activity of Ru and Fe_3O_4 catalysts [184]

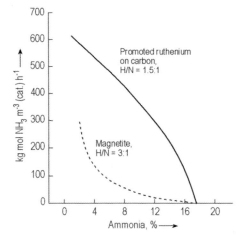

Figure 26. Ammonia inhibition of Ru and Fe_3O_4 catalysts [184]

special properties of ruthenium, which have to be considered in handling these catalysts see [100]. Topsøe has developed a ruthenum catalyst which uses boron nitride instead of graphite as support. Boron nitride, occasionally called "white graphite", because of its structure similar to graphite, is completely stable towards hydrogenation under all conditions relevant to technical ammonia synthesis. A Ba-Ru/BN catalyst proved completely stable in a 5000 h test at 100 bar and 550 °C using a 3:1 $H_2 – N_2$ mixture

[413]. There is no industrial application so far. A discussion on ruthenium as ammonia synthesis catalyst with an extensive literature review is found in [414].

The potential for ruthenium to displace in the long run iron in new plants will depend on whether the benefits of its use are sufficient to compensate the higher costs. Ruthenium prices increased by a factor of four in 2000 to 170 $/oz. due to new applications in the electronic industry. Ruthenium price development is occasionaly published in FINDS, a Stokes Engineering Publication.

In common with the iron catalyst ruthenium will also be poisoned by oxygen compounds. Even with some further potential improvements it seems unlikely to reach an activity level which is sufficiently high at low temperature to allow operation of the ammonia synthesis loop at the pressure level of the syngas generation. An overview over ammonia synthesis catalyst development history and newer research is given in [413].

4.5. Process Steps of Ammonia Production

The term "ammonia synthesis" is increasingly used when referring to the total ammonia production process. Synthesis conditions are no longer viewed in isolation. Of course, they are an important consideration in the total process but can be determined properly only in relation to the total plant integration (see Section 4.6). A literature summary of ammonia production is contained in [415] and [416]; the United States patent literature in the field from 1972 to 1980 is covered in [417]. More modern and comprehensive reviews of ammonia production technology can be found in [100, 102, 418 – 420]. The journal *Nitrogen*, now named *Nitrogen + Syngas*, published by British Sulphur, presents an update of the state of the art from time to time.

The complete process of industrial ammonia production may be subdivided into the following sections:

A) Synthesis gas production
  1) Feedstock pretreatment and gas generation
  2) Carbon monoxide conversion
  3) Gas purification

B) Compression
C) Synthesis and purge gas management

The most fundamental changes over the years have occurred in synthesis gas production and gas compression. In the synthesis section itself, some progress has been made in converter design and optimization of heat recovery.

4.5.1. Synthesis Gas Production

The goal is preparing a pure mixture of nitrogen and hydrogen in the stoichiometric ratio of 1 : 3. The raw materials are water, air, and a carbon-containing reducing medium, that, for its part, may contain hydrogen (natural gas, CH_4; naphtha, $\approx CH_2$; petroleum, $\approx CH$) and nitrogen; for example, natural gas from the Slochteren field in the Netherlands contains 14 % nitrogen.

Usually only the carbon-containing materials and hydrogen from other sources are regarded as raw materials in the narrow sense because of the abundance of air, which provides all of the nitrogen, and water, which generally supplies most of the hydrogen. The term feedstock is often applied to the total consumption of fossil fuel, although strictly speaking a distinction should be made between gasification feed and fuel for energy generation.

Certain raw materials for synthesis gas production that were once of primary importance currently are used only under special economic and geographical circumstances (e.g., China, where 66 % of production is based on coal). These include solid fuels, coke oven gas, and hydrogen produced by electrolysis. Reference [421] covers coke oven gas as a feedstock for ammonia synthesis and references [422 – 425] describe producing hydrogen by water electrolysis for ammonia production.

Table 21 provides an overview of the raw material sources (apart from water and air) for world ammonia capacity.

Table 21 indicates that new ammonia plants are based almost exclusively on natural gas and naphtha. This trend is also expected to continue in the near future. Naturally, the regional distribution is diverse. In North America, for example, natural gas dominates, with 95 % of capacity. In the EU, 86 % of capacity is based on natural gas and 8 % on naphtha [426].

The capital cost and the specific energy requirement (i.e., feed and fuel, and so the manufacturing cost) largely depend on the raw material employed [427, 428]. Table 22 shows the relative capital cost and the relative energy requirement for a plant with a capacity of 1800 t/d ammonia. For the natural gas based plant the current best value of 28 GJ per tonne NH_3 is used. If water electrolysis (4.5 kWh/Nm^3 H_2) is used together with an air separation unit for the nitrogen supply the energy requirement amounts to 34 GJ per tonne NH_3 when the electric energy is valuated just with the caloric equivalent, which would be only justified when electric power is generated from water power. In case of electricity generation from fossil energy with 40 % efficiency the consumption figure is 85 GJ per tonne NH3, which is 300 % of the consumption of a modern steam reforming plant [429]. The investment could according to a rather rough estimate be about three times the investment for a steam reforming plant with natural gas.

4.5.1.1. Feedstock Pretreatment and Raw Gas Production

The chemical reaction of hydrocarbons with water, oxygen, air, or any combination of these is generally referred to as gasification. It yields a gas mixture made up of CO and H_2 in various proportions along with carbon dioxide and, where air is used, some nitrogen. Any carbon containing feedstock will undergo a reaction according to Equation (5) or (6) or both simultaneously.

$$[CH_x] + H_2O \rightleftharpoons CO + H_2 + x/2\, H_2 \quad \Delta H > 0 \qquad (20)$$

$$[CH_x] + 1/2\, O_2 \rightleftharpoons CO + x/2\, H_2 \quad \Delta H < 0 \qquad (21)$$

Light hydrocarbons ranging from natural gas (methane) to naphtha (max. C_{11}) undergo reaction with steam over a catalyst according to Equation (5) which is usually called steam reforming. Corresponding to Equation (6), commonly known as partial oxidation, all carbon-containing feedstocks can be processed in a noncatalytic reaction with oxygen (together with a minor amount of steam for process reasons,

Table 21. Feedstock distribution of world ammonia production capacity

	1962		1972		1983		1998	
	10^3 t N	%	10^3 t N	%	10^3 t N	%	10^3 t N	%
Coke oven gas and coal	2800	18	4600	9	7200	8	16500	14
Natural gas	7800	50	32100	63	66850	74	94300	77
Naphtha	2050	13	10700	21	9050	10	7300	6
Other petroleum products	2950	19	3600	7	7200	8	4400	3
Total	15600	100	51000	100	90300	100	122500	100

Table 22. Relative ammonia plant investment and relative energy requirement for 1800 t/d NH_3

	Natural gas	Naphtha	Fuel oil	Coal
Relative investment	1.0	1.15	1.5	2.5
Relative specific energy requirement (based on lower heating values)	1.0	1.1	1.3	1.6

which gives rise to a simultaneous reaction according to Eq. 5). An additional equilibrium reaction involved in any gasification process is the water gas shift reaction (7).

$$CO + H_2O \rightleftharpoons CO_2 + H_2 \quad \Delta H^0_{298} = -41 \text{ kJ/mol} \quad (22)$$

Although reaction in the right hand direction is favored by lower temperatures it is responsible for the initial carbon dioxide content of the raw synthesis gas. To maximize the hydrogen yield, this reaction is carried out in a separate step over a different catalyst at a lower temperature than the preceding gasification step (Section 4.5.1.2).

From Equation (5) it can be seen that in the steam reforming variant the proportion of hydrogen supplied by the feedstock itself increases with its hydrogen content. It attains the theoretical maximum of 66% with methane. The hydrogen – oxygen bond energy in water is higher than the hydrogen – carbon bond energy in the hydrocarbon. The positive enthalpy per mole of hydrogen therefore decreases as the proportion of hydrogen contributed from the feedstock itself increases. Natural gas consists predominantly of methane and is therefore the most hydrogen-rich and energetically the best raw material for the steam-reforming route. In the partial-oxidation route less hydrogen is produced in the primary gasification step and the raw synthesis gas has a rather high CO content.

The raw gas composition is thus strongly influenced by the feedstock and the technology applied. But for the different feedstocks there are some constraints on the applicability of the various gas generation processes. The catalytic steam reforming technology can only be applied to light hydrocarbon feedstock (up to naphtha) but not for heavy hydrocarbons such as fuel oil or vacuum residue. These raw materials contain a substantial amount of sulfur and also minor quantities of heavy metals, which would poison the sensitive reforming catalyst. In addition cracking reactions will occur on the catalyst, depositing carbon, which not only blocks the catalyst pores but also restricts interparticle flow. Thus for heavy feedstocks the only choice is noncatalytic partial oxidation, which, however, is capable of processing any type of hydrocarbon feedstock. The various commercial coal gasification processes may also be classified as partial oxidations.

Steam Reforming Processes. As the nickel-containing catalysts are sensitive to poisons, any sulfur compounds present in the hydrocarbon feedstock have to be removed by hydrodesulfurization, generally with a combination of cobalt – molybdenum and zinc oxide catalysts [430 – 432].

Adsorption on activated carbon is an alternative when the feed is natural gas with a rather low sulfur content.

$$R-SH + H_2 \rightarrow RH + H_2S \quad (23)$$

$$H_2S + ZnO \rightarrow ZnS + H_2O \quad (24)$$

The general overall reaction can be formulated as

$$C_nH_{(2n+2)} + n\,H_2O \rightleftharpoons nCO + (2n+1)\,H_2 \qquad (25)$$

or more specifically for methane, usually the major constituent of natural gas, as:

$$CH_4 + H_2O \rightleftharpoons CO + 3\,H_2 \quad \Delta H = +206\,\text{kJ/mol} \qquad (26)$$

Simultaneously to this equilibrium the water gas shift reaction (Eq. 7) proceeds.

To introduce nitrogen to achieve the required stoichiometric hydrogen/nitrogen ratio for ammonia synthesis, the reforming reaction is split into two sections. In the first section, the primary reformer, the reaction proceeds in indirectly heated tubes filled with nickel-containing reforming catalyst and is controlled to achieve a partial conversion only [in conventional plants 65 % based on methane feed, leaving around 14 mol % methane (dry basis) in the effluent gas]. In the following secondary reformer — a refractory-lined vessel filled with nickel catalyst — the gas is mixed with a controlled amount of air introduced through a nozzle (burner). By combustion of a quantity of the gas the temperature is raised sufficiently (to about 1200 °C) that the endothermic reforming reaction is completed with the gas adiabatically passing the catalyst layer. In this way the outlet temperature is lowered to around 1000 °C, and a residual methane content of 0.5 % or lower (dry basis) is attained in conventional plants [100, 418]. Nitrogen already present in the natural gas tends to cause a reduced specific air ratio in the secondary reformer and a reduced secondary reformer temperature rise. Therefore, to maintain the same methane leak, the primary reformer exit temperature must be increased.

The primary reformer consists of a multitude of reformer tubes loaded with the nickel catalyst (15 – 25 % NiO on α-aluminum oxide, calcium aluminate, or magnesium aluminum spinel support) [102] in a furnace box in which the heat needed for the reaction is transferred to the tubes by radiation. The heat is generated in burners, generally gas-fired, in the furnace box.

A special consideration is the lifetime of the expensive reformer tubes, made of highly alloyed chromium-nickel steel by centrifugal casting, because under the severe reaction conditions the material exhibits creep which finally leads to rupture [433, 434]. The time to rupture for a specific material depends on the tube-wall temperature and on the internal pressure. This limits the reforming pressure, which to save energy in synthesis-gas compression should be as high as possible. As the reforming reaction is endothermic and proceeds with volume increase the negative effect of a pressure increase (lower conversion) has to be compensated by a higher reaction temperature and hence higher wall temperatures, but this is limited by the material. Another possibility to compensate is a higher steam surplus (steam/carbon ratio), but this is economically unfavorable. The furnace box usually accommodates 200 – 400 tubes (depending on plant capacity), 10 – 13 m long, with an inner diameter of 75 – 140 mm and a wall thickness of 11 – 18 mm. The standard material for a long time was HK 40 (20 Ni/25 Cr) but for replacements and new plants, HP modified (32 – 35 Ni/23 – 27 Cr stabilized with about 1.5 % Nb) is being increasingly used on account of its superior high-temperature properties [435]. With this latter tube material a reforming pressure of 40 bar is possible at outer tube-wall temperatures of around 900 °C. A further improvement are *microalloys* that additionally contain Ti and Zr [100, 102, 436, 437].

According to the disposition of the burners primary reformers can be classified as top-fired, side-fired, terraced-wall, or, less common, bottom-fired reformers [100, 102]. The steam/carbon ratio used in modern commercial primary reformers for natural gas is between 2.8 and 3.5, and markedly higher for naphtha, for which an alkalized nickel catalyst has to be used to prevent carbon deposition, which causes catalyst deactivation and local overheating of tubes (hot bands and hot spots). A special requirement for the catalyst is a low content of silica, which could be volatile under the hydrothermal conditions and deposit downstream of the secondary reformer at lower temperature, causing fouling of the waste-heat recovery surfaces.

When higher hydrocarbons (naphtha) are used, a so-called "rich gas stage" can be utilized upstream of the primary reformer. In this, the higher hydrocarbon – steam mixture is transformed at relatively low temperature (400 – 500 °C) and steam – carbon ratios less than two into a methane-rich gas. This is then converted in the primary reformer under normal reforming conditions. For this rich-gas stage, three processes are commercially available: the British Gas Process [438], the BASF – Lurgi Process [439 – 441], and the Japan Gasoline Process [442].

This process has recently become the subject of renewed interest for increasing the capacity of existing natural-gas-based ammonia plants in which the primary reformer has been identified as bottle-neck. Under the name pre-reforming, it is installed up-stream of an existing tubular reformer [100, 102, 443 – 450]. The natural gas enters such a pre-reformer with a temperature of 530 °C instead of being fed to the primary reformer tubes with the same temperature. A temperature drop of about 60 – 70 °C occurs in the catalyst bed due to the overall endothermic reaction. Medium-grade heat is used to reheat the exit gas to the correct primary reformer entrance temperature. This compensation heat may be derived from a variety of sources, including flue gas, process gas, or gas turbine exhaust.

Another possibility for reducing the load on the primary reformer is to transfer part of the conversion duty to the secondary reformer with application of an super-stoichiometric amount of air. This requires removal of the surplus of nitrogen from the synthesis gas either ahead of the synthesis loop (by cryogenic methods or pressure swing adsorption) or by purge gas and hydrogen recovery [451, 452]. In the extreme case the whole reforming reaction could be performed without a tubular reformer by autothermal catalytic reforming in a design similar to a secondary reformer. In this case it would be necessary to use oxygen or oxygen-enriched air instead of air [102, 418, 453 – 457]. Unlike to a secondary reformer, which is fed with partially reformed gas, having a substantial concentration of hydrogen, the autothermal reformer (ATR) is fed directly with the hydrocarbon feedstock. Because of the higher heat of reaction in the internal combustion (temperature >2000 C), the flow conditions, heat release characteristics and the risk of soot formation are very different from the situation of a normal secondary reformer. Therefore special considerations in the design of burner and reactor are necessary. See also → Gas Production.

A recent development which avoids a fired primary reformer is the exchanger reformer, which with some simplification may be viewed as tubular heat exchanger with the catalyst inside the tubes, which are heated by the hot secondary reformer effluent flowing on the shell-side. In some designs the tubes may be open at the lower end, in which case the gas flow on the shell side consists of a mixture of the off-gases from the secondary reformer and from the reformer tubes. Commercially operating designs are the GHR of ICI (Figure 27) [100, 458 – 461] and the KRES of M. W. Kellogg [102, 406, 462 – 468] (Fig. 28). Similar concepts are offered by other licensors and contractors e.g., Braun & Root, or Topsøe [100].

ICI has come out with a modified design, the AGHR, with "A" standing for advanced. As shown in Figure 27 B, the bayonet tubes are replaced by normal tubes attached to a bottom tubesheet using a special seal to allow some expansion. In this way the delicate double tubesheet of the GHR is avoided. The seal which prevents leakage of methane-rich gas to the secondary reformer effluent flowing on the shell side has a unique design which is subject to patent applications of ICI. The AGHR will allow a single-line concept for worldscale plants whereas with the GHR several parallel units for large plants would be necessary [100].

Because of the smaller size compared to a conventional fired reformer considerable investment savings can be achieved. To close the heat balance between the exchanger reformer and secondary (autothermal) reformer, the latter has to take on a higher reforming duty, which may be achieved by using an over-stoichiometric amount of air or oxygen-enriched air. In some configurations the exchanger reformer is partially by-passed, part of the feed being fed directly into the autothermal reformer.

Haldor Topsøe has developed an exchanger reformer called HTER which may be used in various process configurations as described in [469]. The design may be also used as a convective reformer, called HTCR with a dedicated burner.

A concept developed by Uhde goes a step further in this direction: exchanger reforming and subsequent noncatalytic partial oxidation, which provides the reaction heat, are accommodated in a single vessel. This combined autothermal reformer (CAR) design, shown in Figure 29 was operated in a demonstration unit producing 13 000 m^3/h of synthesis gas [470 – 473].

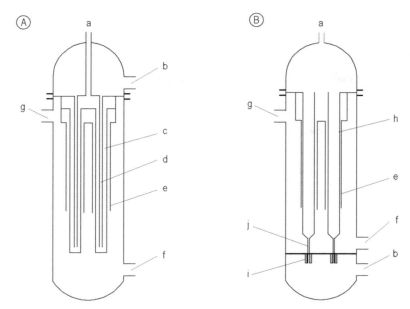

Figure 27. ICI gas heated reformer
A) GHR; B) AGHR
a) Tubeside inlet; b) Tubeside outlet; c) Scabbard tube; d) Bayonet tube; e) Sheath tube; f) Shellside inlet; g) Shellside outlet; h) Catalyst tube; i) Seal; j) Tail pipe; k) Catalyst; l) Refractory lining

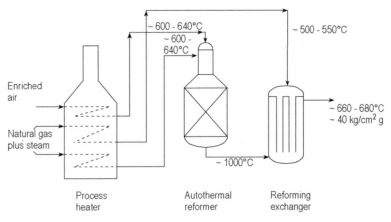

Figure 28. Kellogg reforming exchanger System (KRES) [100]

A unique steam reforming process [474 – 476] has been developed in Japan to the pilot-plant stage. It reportedly can operate without upstream desulfurization and should be able to gasify naphtha, crude oil, and atmospheric or vacuum residues.

For literature on steam reforming see [100, 418, 419, 453, 477 – 485]; reaction kinetics and thermodynamics [102]; for steam reforming of naphtha, [486 – 488]; for steam reforming catalysts, [100, 102, 489 – 492] for reformer design, [100, 102, 493, 494]. An overview of the steam reforming technology, its historical development, present state of the art, newer developments and future perspectives together with quotation of newest literature is given in [433]. Detailed discussion on exchanger reforming and autothermal reforming is found in [469].

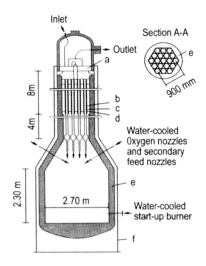

Figure 29. Uhde combined autothermal reformer (CAR)
a) Sandwich type tubesheet; b) Enveloping tube; c) Reformer tubes; d) Tubesheet; e) Refractory lining; f) Water jacket

Partial Oxidation. Hydrocarbons or coal will react with an amount of oxygen insufficient for total combustion to CO_2 according to:

$$C_nH_m + n/2\, O_2 \rightarrow n\, CO + m/2\, H_2 \quad (27)$$

$$C + 1/2\, O_2 \rightarrow CO \quad \Delta H = -123 \text{ kJ/mol} \quad (28)$$

In practical operation some steam must always be added, the quantity depending on feedstock and process configuration, so that the following reactions proceed in parallel:

$$C_nH_m + n\, H_2O \rightarrow n\, CO + (n+m/2)H_2 \quad (29)$$

$$C + H_2O \rightarrow CO + H_2 \quad \Delta H = +119 \text{ kJ/mol} \quad (30)$$

As the overall reaction is exothermic no external heat supply is necessary. Since in some processes with coal feedstock (e.g., the Lurgi Process) the reaction according to Equation (15) may proceed to a considerable extent, they are more often referred to as coal gasification rather than as partial oxidation, but this is just a matter of definition.

Certain processes have achieved particular significance. The Texaco and Shell processes can handle fuel oil, vacuum residues, and coal. The Lurgi process handles lump coal, operates at 2.5 MPa, the Koppers – Totzek process operates on coal dust at atmospheric pressure.

Partial Oxidation of Hydrocarbons. The two dominant processes are the Texaco Syngas Generation Process (TSGP) and the Shell Gasification Process (SGP). In both the reaction is performed in an empty pressure vessel lined with alumina. The reactants (oil and oxygen, along with a small amount of steam) are introduced through a nozzle at the top of the generator vessel. The nozzle consists of concentric pipes so that the reactants are fed separately and react only after mixing at the burner tip or in the space below. The temperature in the generator is between 1200 and 1400 °C. Owing to insufficient mixing with oxygen, about 2 % of the hydrocarbon feed is transformed into soot, which is removed from the gas by water scrubbing. Depending on the process configuration the gas is either cooled by quenching or in a waste-heat boiler. The TSGP and SGP are rather similar, they mainly differ in the nozzle design, soot removal and recirculation, and process gas cooling. The reaction pressures may be as high as 80 bar; there are no mechanical or material limitations to raising the gasification pressure further, but with respect to the overall ammonia process this might be beyond the energy optimum because of the increasing energy demand for nitrogen and oxygen compression. Maximum raw gas generation capacity of a single generator corresponds to about 1000 t/d ammonia. World-scale ammonia plants based on partial oxidation exist in Germany, India, China, and other countries [495]. Further information is found for the TSGP in [496 – 504], for the SGP in [505 – 515].

Partial Oxidation of Coal. So far the Koppers – Totzek, Texaco and Lurgi gasifiers, and probably the Winkler process in some smaller installations, have been used in ammonia plants, but the successful demonstration of the Shell process in other applications makes it a potential candidate for ammonia production, too. Additional processes in different stages of technical development are the HTW and the Dow process. Information on the status and the development in the gasification of coal can be obtained from [516 – 524]. A special case is China, where a number of small plants with a capacity of 30 – 80 t/d still produce their synthesis gas with the today largely outdated water-gas process which operates at atmospheric pressure.

Coke or anthracite is reacted intermittently with air and steam in a fixed bed. The heat produced by the exothermic reaction of coal and air in the "blow phase" is stored in the fixed bed and provides the heat needed for reaction of coal with steam in the "run phase" [98, 525]. The Koppers – Totzek Process [526 – 531] used in several ammonia plants in China, India, and South Africa, operates practically at atmospheric pressure. Dry coal dust is fed to the two (sometimes four) burners of the gasifier. Oxygen, together with a small amount of steam, is introduced immediately at the head of the burners, and the mixture enters the reaction zone with high velocity. The residence time is less than 1 s and the temperature is 1500 – 1600 °C. The gas leaving the top of the reactor vessel is cooled in a waste-heat boiler, followed by a water scrubber to remove carbon and ash traces from the raw gas and to effect further cooling. Liquid slag is withdrawn from the reactor bottom. The process can handle bituminous coal and lignite. To overcome the disadvantage of atmospheric pressure operation, a version was developed capable of operating at 25 – 30 bar. This Prenflow (pressurized entrained flow) process [420, 532] is being tested in a 48 t/d pilot plant. The Texaco Coal Gasification Process [518, 533 – 540] is rather similar to the Texaco partial oxidation process for heavy hydrocarbons. An aqueous slurry containing 60 – 70 % coal is fed by reciprocating pumps to the generator at a pressure of 20 – 40 bar. Waste-heat boiler, quencher, and carbon scrubber are especially adapted to deal with the ash and slag introduced with the bituminous coal feed.

In its coal gasification process [518, 541 – 546]. Shell has completely departed from the concept of its process for partial oxidation of heavy hydrocarbons. With reversed flow pattern in the gasifier (from bottom to top), dry coal dust is introduced via lock hoppers into the reactor vessel operating at 20 – 40 bar.

The Lurgi Dry Gasifier [518, 547 – 553] performs the reaction in a moving bed, usually operating at 25 – 30 bar. Crushed coal with a particle size of 4 – 40 mm enters the top of the gasifier through a lock hopper and is evenly distributed over the cross-section of the coal bed surface by a distribution disk equipped with scraper arms. Ash with only 1 % of residual carbon is removed at the bottom of the gasifier by a revolving grid with slots through which steam and oxygen are introduced. The temperature in the lower section of the bed is around 1000 °C, and at the top where the raw gas exits about 600 °C. As a result of this lower temperature the raw gas has an increased content of impurities such as tars, phenols, and some higher hydrocarbons. In addition the methane content is relatively high (up to 10 – 15 %), so that purification and conditioning of the raw gas is a rather elaborate task. The process actually can use any sort of coal and can handle ash contents higher than 30 %. The British Gas/Lurgi Slagging Gasifier [518, 549, 554] operates without a grate with withdrawal of the liquid slag.

The classic Winkler gasifier, which operates at atmospheric pressure, is today still in use in some smaller plants (e.g., in China). In a further development (HTW process) by Rheinbraun it has been tested in a demonstration plant with lignite at 10 bar [555, 556].

The Advanced Coal Gasification Process (ACGP) [557], a new concept developed by AECI Engineering, Kynoch Ltd and Babcock Wilson, is an entrained gasification process at atmospheric pressure. Pulverized coal and oxygen are injected near the base of the gasifier through eight burners using a proprietary feed technique which employs a special pump. The gasifier is a slender column with a square cross-section. The combustion area is lined with heat resistant ceramic material. The upper part is made of finned tubes which are welded together and are circulated with water by thermosyphon action. So far no commercial installation exists but it is claimed that one unit can produce the synthesis gas for 750 t/d ammonia.

A block diagram showing steam reforming and partial oxidation together with the further steps needed to transform the raw gas into pure make-up gas for the synthesis is shown in Figure 30.

4.5.1.2. Carbon Monoxide Shift Conversion

As ammonia synthesis needs only nitrogen and hydrogen, all carbon oxides must be removed from the raw synthesis gas of the gasification process. Depending on feedstock and process technology, this gas contains 10 – 50 % carbon monoxide and also varying amounts of carbon dioxide. In the water gas shift reaction, tradi-

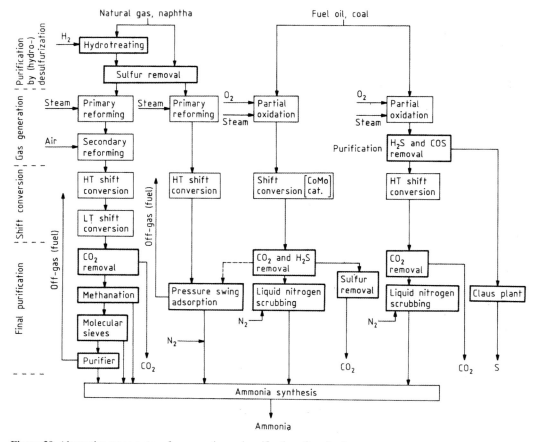

Figure 30. Alternative process steps for generation and purification of synthesis gas

tionally known as carbon monoxide shift conversion (Eq. 7), the carbon monoxide serves as reducing agent for water to yield hydrogen and carbon dioxide. In this way not only is the carbon monoxide converted to readily removable carbon dioxide but also additional hydrogen is produced:

$$CO + H_2O \rightleftharpoons CO_2 + H_2 \quad \Delta H = -41.2 \text{ kJ/mol} \qquad (31)$$

As no volume change is associated with this reaction, it is practically independent of pressure, but as an exothermic process, it is favored by lower temperatures, which shift the equilibrium to the right-hand side. Even with a low excess of steam in the gas, the equilibrium concentrations of CO are low; for example, 0.2 vol % at 220 °C and 0.12 vol % at 200 °C for a steam/gas ratio of 0.4.

To keep the temperature low the heat of reaction must be removed in appropriate way and to achieve a sufficient reaction rate effective catalysts have to be applied. The process is therefore performed in steps, with intermediate heat removal between the individual catalyst beds in which the reaction runs adiabatically. Quasi-isothermal reactors have been developed in which cooling tubes run though the catalyst layers. As the process configuration and catalysts are to some extent different for steam reforming and partial oxidation, they are treated separately here.

Shift Conversion in Steam Reforming Plants. In the traditional plant concept, the gas from the secondary reformer, cooled by recovering the waste-heat for raising and superheating steam, enters the high-temperature shift (HTS) reactor loaded with an iron–chromium catalyst at 320–350 °C. After a temperature increase of around 50–70 °C (depending on initial CO concentration) and with a residual CO content of around 3 % the gas is then cooled to 200 – 210 °C for the low temperature shift (LTS), which is carried out on a copper – zinc – alumina catalyst in a downstream reaction vessel and achieves a carbon monoxide concentration of 0.1 – 0.3 vol %.

In the unreduced state the HTS catalyst is iron(III) oxide (Fe_2O_3) containing additionally 5 – 10 % chromic oxide (Cr_2O_3). During operation, it is reduced more or less stoichiometrically to the composition of magnetite (Fe_3O_4). This catalyst is active in the temperature rage of 300 – 500 °C. Steam surplus is not only necessary for thermodynamic reasons but also to suppress undesirable side reactions. Decreasing the steam surplus lowers the oxygen to carbon ratio in the HTS to such an extent that the atmosphere can reduce magnetite partially to metallic iron. In addition the Boudouard reaction can occur under these conditions. The resulting carbon is deposited within the catalyst particles causing their disintegration, and iron carbides will be formed, which are effective Fischer – Tropsch catalysts that lead to the formation of some methane and higher hydrocarbons [100, 558, 559]. Newly introduced HTS catalysts with additional copper promotion suppress this side reaction [560] and are therefore less sensitive to lower steam-to-gas ratios. The classical HTS iron catalyst is resistant against sulfur compounds, but this is of greater importance in partial oxidation processes and less for the practically sulfur-free steam reforming gas.

In some ammonia process schemes operating without a secondary reformer and applying pressure swing adsorption (PSA) for further purification (KTI PARC) only a HTS is used.

The LTS catalyst, supplied in pellets like the HTS catalyst, consists of 40 – 55 % copper oxide, 20 – 30 % zinc oxide, the balance being alumina. The catalyst properties are influenced far more by the formulation and manufacturing procedure [561] than by its chemical composition. The copper oxide is reduced in situ with hydrogen and a carrier gas (usually nitrogen) to form fine copper crystallites on which the activity depends. Sulfur, usually present as H_2S, has to be below 0.1 ppm, but even with such low concentrations, the catalyst is slowly poisoned. The ZnO adsorbs the sulfur and it finally transforms into bulk ZnS. When the ZnO is exhausted in a given layer of the catalyst, the H_2S causes deactivation of the copper by sintering. The poisoning process moves through the catalyst as a relatively sharp front and can be seen in the change of the catalyst temperature profile over time [562, 563]. The LTS catalyst is protected by a guard bed, formerly loaded with ZnO, but nowadays usually with LTS catalyst [564]. Changing the guard bed more frequently prolongs the service life of the main LTS catalyst bed. Traces of chlorine compounds [565], which may be introduced with the natural gas or more often with the process air to the secondary reformer, may also deactivate the LTS catalyst by accelerating the sintering of the copper particles. Unlike sulfur poisoning, chlorine is more diffusely distributed over the whole catalyst bed by migration as volatile zinc and copper chlorides.

As chemical composition and formulation of the LTS catalyst are very similar to methanol production catalysts, small quantities of methanol are formed and are found in the process condensate after cooling the LTS effluent. In a consecutive reaction, amines (mainly methylamine) are formed from the methanol and traces of ammonia originating from the secondary reformer and the HTS. These pollutants are removed from the process condensate by steam stripping and ion exchange. Byproduct formation is higher with fresh catalyst and declines with operating time. New catalyst types with increased activity and higher selectivity have reduced the problem. The tendency for methanol formation increases with decreasing steam/gas ratio [566].

A relatively new process concept is the intermediate temperature shift (ITS) [567] which performs the shift conversion in a single step. The catalyst is based on a copper – zinc – alumina formulation and optimized for operating in a wider temperature range than the standard LTS catalyst. The reaction heat can be removed by use of a tube-cooled reactor raising steam or heating water for gas saturation to supply process steam in the reforming section (Linde LAC, ICI Catalco LCA). In a new plant using the spiral-wound Linde reactor [568], a methane slip of only 0.7 mol % (dry basis) is achieved. Further purification is performed by PSA. Generally the shift conversion reactors have an axial gas flow pattern, but recently radial gas flow configurations have been chosen in some instances.

Additional literature on shift conversion can be found in [100, 102, 418, 477, 478, 569 – 577].

Shift Conversion in Partial Oxidation Plants. The raw synthesis gases from partial oxidation of heavy hydrocarbons and coal differ mainly in two aspects from that produced from light hydrocarbons by steam reforming. First, depending on the feedstock composition, the gas may contain a rather high amount of sulfur compounds (mainly H_2S with smaller quantities of COS); second, the CO content is much higher, in some cases in excess of 50 %. The sulfur compounds (see Section 4.5.1.3) can be removed ahead of the shift conversion to give a sulfur-free gas suitable for the classical iron HTS catalyst. In another process variant the sulfur compounds are removed after shift conversion, which thus has to deal with a high-sulfur gas. As the standard iron catalyst can tolerate only a limited amount of sulfur compounds, the so-called dirty shift catalyst is used in this case. This cobalt – molybdenum – alumina catalyst [569, 578 – 582] is present under reaction conditions in sulfidized form and requires for its performance a sulfur content in the gas in excess of 1 g S/m^3. Reaction temperatures are between 230 and 500 °C.

Irrespective of the catalyst type used, the high initial carbon monoxide concentration means that the reaction must generally be performed in steps, with intermediate cooling. It has been reported that the CO content can be reduced from 50 to 0.8 % in a single step in a large hydrogen plant by using a quasi-isothermal reactor (e.g., the Linde spiral-wound reactor).

4.5.1.3. Gas Purification

In further purification, carbon dioxide, residual carbon monoxide, and sulfur compounds (only present in the synthesis gas from partial oxidation) have to be removed as they are not only a useless ballast but above all poisons for the ammonia synthesis catalyst.

The total sulfur contained in the coal and hydrocarbon feedstock is converted in gasification to H_2S and a smaller amount of COS, which are removed as described below. In contrast steam reforming requires removal of sulfur from the natural gas and light hydrocarbon feedstocks upstream of gasification to avoid poisoning of the sensitive reforming catalysts. This is usually performed by hydrodesulfurization and adsorption of the H_2S by ZnO. As this is an essential part of the steam reforming process it is treated in Section 4.5.1.1.

The classical method for CO_2 removal is to scrub the CO_2 containing synthesis gas under pressure with a solvent capable of dissolving carbon dioxide in sufficient quantity and at sufficient rate, usually in countercurrent in a column equipped with trays or packings. The CO_2-laden solvent is flashed, often in steps, to around atmospheric pressure, and the spent scrubbing liquid is subsequently heated and regenerated in a stripping column before being recycled to the pressurized absorption column. In the early days of ammonia production water, often river water, served as solvent in a once-through process without regeneration and recycling.

Today a variety of solvents are used and they can be categorized as physical or chemical solvents. In the physical solvents the carbon dioxide dissolves without forming a chemical compound, and this allows recovery simply by flashing. In the chemical solvents the carbon dioxide becomes fixed as a chemical compound, which needs heat for its decomposition. At low carbon dioxide partial pressures, the chemical solvents absorb substantially more carbon dioxide than the physical solvents; at higher partial pressures the physical solvents (according to Henry's Law, the loading is approximately proportional to the CO_2 partial pressure) have a higher loading capacity than the chemical solvents, for which the solubility approaches a saturation value. Figure 31 shows the loading characteristic for various solvents.

As both sour gases, CO_2 and H_2S, have good solubility in the applied solvents, special process configurations are required for partial oxidation gases to recover separately a pure CO_2 fraction and an H_2S-rich fraction suitable for sulfur disposal.

Very suitable for the partial pressure range of CO_2 in steam reforming plants (4 – 7 bar) are chemical solvents based on aqueous solutions of potassium carbonate or alkanolamines containing additional activators to enhance mass transfer and, in some cases, inhibitors to limit or prevent corrosion processes. Primary and secondary amines, for example, monoethanolamine (MEA) and diethanolamine (DEA) exhibit a high mass-transfer rate for carbon dioxide but have a high energy demand for regeneration. For this reason tertiary amines are commonly used today, for example, methyldiethanolamine together with an activator. Triethanolamine does not achieve the required final CO_2 concentration, and in the few cases where it was used it was followed by additional scrubbing with monoethanolamine (MEA).

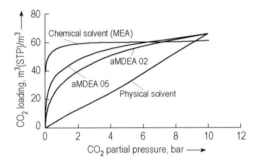

Figure 31. CO_2 loading characteristics of various solvents

The potassium carbonate processes from the various licensors differ with respect to the activator, corrosion inhibitor and to some extent in the process configuration.

Over the years considerable progress has been achieved in improving the efficiency of the carbon dioxide removal systems. The first generation of single-train steam reforming ammonia plants used MEA and consumed about 5.8 GJ per tonne NH_3, which was about 14 % of the total energy consumption. Table 23 demonstrates the progress made in energy consumption.

Table 23. Heat requirements for regeneration in CO_2 removal systems [100, 583]

Process	Heat requirement, kJ/mol CO_2
MEA (without Amine Guard)	209
MEA (with Amine Guard III)	116
Benfield hot potash (1-stage)[a]	88
BASF aMDEA process (1-stage)[a]	73
Benfield LoHeat	28 – 35
BASF aMDEA process (2-stage)[b]	28 – 30

[a] Single-stage regeneration.
[b] Two-stage regeneration.

The first progress was made by addition of corrosion inhibitors (e.g., Amine Guard introduced by Union Carbide) [583 – 587], which allow a higher loading of the solvent. The energy consumption of the modern systems, as shown in the last two lines of Table 23 depends largely on process configuration and required final purity which may range from 50 to 1000 ppm in the purified gas. There is also a trade-off between energy consumption and investment costs.

The hot potash systems, for example, the Benfield process licensed by UOP, differ in the type of activator used to increase the reaction rate between the CO_2 and the solvent. The activators enhance mass transfer and thus influence not only the regeneration energy demand (circulation rate of the solvent) but also the equipment dimensions. The following activators are used: in Benfield [588 – 591] and Carsol [592, 593] process ethanolamines; in the Giammarco Vetrocoke process glycine (originally arsenic oxide) [594 – 597]; in the Catacarb system [598, 599] amine borate; in the Exxon Flexsorb HP process [600, 601] a sterically hindered amine; in the Carbosolvan process [602, 603] sulfosolvan B. New activators named ACT-1 (UOP) [604, 605] and LRS-10 [606, 607] (British Gas) have been introduced. All hot potash systems need corrosion inhibitors the concentration of which must carefully to be monitored.

A typical example of an efficient modern hot potash system is the Benfield LoHeat [608] system of UOP. The high thermal efficiency of this two-stage adsorption process using lean and semi-lean solvent is achieved by recompression of the flash steam with an injector or a mechanical vapor compressor. UOP has also developed a number of other process configurations [600, 603, 609 – 611].

BASF's aMDEA [612 – 618] process uses an aqueous solution of the tertiary amine methyldiethanolamine (MDEA) with a special amine activator. No corrosion inhibitors are necessary, and unlike MEA no solvent degradation is observed, so recovery installations are not required. Owing to the low vapor pressure of MDEA and the activator, there are no losses of the active solvent components. The carbon dioxide binds much less strongly to MDEA than to MEA, and the solvent character is more like a hybrid between a strong chemical and a purely physical solvent. On account of the relative weak binding forces, a substantial amount of carbon dioxide can be recovered simply by flashing to low pressure, and only a small amount has to be recovered by stripping. The process is very versatile: increasing the activator concentration shifts the character of the solvent more to the chemical side and vice versa, and process configurations include two- and single-stage designs, the latter allowing old MEA units to be revamped [619 – 621] simply by swapping the solvent without changing the process equipment.

Physical solvent based processes can also be applied in steam reforming ammonia plants. The Selexol process (UOP) [598, 622 – 630] uses polyethylene glycol dimethyl ether, the Sepasolv MPE process (BASF) [598, 613] polyethylene glycol methyl isopropyl ether, and the Fluor Solvent Process [631 – 634] polypropylene carbonate. These solvents are stable, noncorrosive, nontoxic, and not very volatile, but they have a rather high capacity for adsorbing water. For this reason the raw gas has to be dry, which is usually achieved by chilling and operating at low temperatures. The Selexol Process for example operates at 5 °C. The carbon dioxide is recovered simply by flashing; no heat is necessary. Before the solvent is recycled it has to be vacuum treated and/or air stripped to remove traces of carbon dioxide. When comparing the energy consumption of physical and chemical solvent based processes for a given application not only the heat for solvent regeneration but also the mechanical energy for solvent circulation has to be taken into account.

Raw synthesis gas produced by partial oxidation has a carbon dioxide partial pressure between 10 and 30 bar, depending mainly on feedstock and gasification pressure. In the lower half of this partial pressure range chemical solvents based on tertiary amines (e.g., MDEA) might be suitable, but at higher values physical solvents become increasingly preferable. Examples of physical solvent processes are Selexol Process, Sepasolv MPE process, Lurgi Purisol Process (N-methylpyrrolidone) [635 – 637], Sulfinol process (tetrahydrothiophene-1,1-dioxide + diisopropanol) [638, 639], and, probably the most important in this connection, the Rectisol process (Linde and Lurgi) [598, 640 – 645], with methanol as solvent operating at -15 to -40 °C.

The presence of sulfur compounds, mainly H_2S along with a minor quantity of COS, introduces some complications for CO_2 removal from partial oxidation gases. Since the sour gases are both soluble in the solvents and a separate recovery in the regeneration stage is only partially possible (only pure CO_2 can be obtained along with a CO_2 fraction more or less rich in H_2S) two principal plant flowsheets are possible. In one, a first scrubbing stage, which removes the sulfur compounds, is positioned upstream of the shift conversion. The second stage, removing pure CO_2, is located downstream of the shift reactors. This concept reduces the effort required to receive a highly concentrated H_2S fraction suitable for further processing in a Claus plant or a sulfuric acid plant. In the other arrangement, the sour gases are removed after shift conversion, which in this case has to operate over dirty shift catalyst (see Section 4.5.1.2) with the consequence that the H_2S is diluted by a large amount of CO_2. In this case, too, methanol regeneration in the Rectisol process can be tailored to achieve a H_2S-rich fraction for downstream processing and pure CO_2, a task difficult to fulfil with the other physical solvents mentioned.

Extensive surveys and additional literature of the CO_2 removal processes are found in [583, 645 – 660] a theoretical analysis is given in [661].

Final Purification. After bulk removal of the carbon oxides has been accomplished by shift reaction and CO_2 removal, the typical synthesis gas still contains 0.2 – 0.5 vol % CO and 0.005 – 0.2 vol % CO_2. These compounds and any water present have to be removed down to a very low ppm level, as all oxygen-containing substances are poisons for the ammonia synthesis catalyst [662].

Methanation. Methanation is the simplest method to reduce the concentrations of the carbon oxides well below 10 ppm and is widely used in steam reforming plants. It is actually the reverse reaction of steam reforming of methane:

$$CO + 3H_2 \rightleftharpoons CH_4 + H_2O \quad \Delta H = -206 \text{ kJ/mol} \quad (32)$$

and

$$CO_2 + 4H_2 \rightleftharpoons CH_4 + 2H_2O \quad \Delta H = -165 \text{ kJ/mol} \quad (33)$$

The advantages of simplicity and low cost more than outweigh the disadvantages of hydrogen consumption and production of additional inerts in the make-up gas to the synthesis loop.

The reaction is carried out over a supported nickel catalyst at a pressure of 25 – 35 bar and a temperature of 250 – 350 °C. The required catalyst volume is relatively small. If a breakthrough of carbon monoxide from the low-temperature shift or carbon dioxide from the absorption system occurs, the intensely exothermic methanation reaction can reach temperatures exceeding 500 °C very quickly [663]. For example, 1 % CO_2 breakthrough leads to an adiabatic temperature rise of 60 °C. Controls should be installed and other security measures taken to avoid these high temperatures because the catalyst may be damaged or the maximum allowable operating temperature of the pressure vessel wall exceeded.

Methanation as final purification for the raw gas from partial oxidation was proposed by Topsøe [664]. In this case the shift conversion is carried out in two stages with a special sulfur-tolerant shift catalyst followed by removal of hydrogen sulfide and carbon dioxide in an acid gas removal unit. Because of the potential danger of a sulfur break-through causing poisoning, the normal copper – zinc – alumina catalyst is usually not applied, which is surprising as the same risk exists in partial oxidation based methanol plants for the similarly composed methanol catalyst.

Selectoxo Process. The Selectoxo process (Engelhard) reduces the hydrogen consumption of the methanation system, as well as the inert gas content of the purified synthesis gas fed to the synthesis loop. After low-temperature shift conversion, the cooled raw gas is mixed with the stoichiometric quantity of air or oxygen needed to convert the carbon monoxide to carbon dioxide. The mixture is then passed through a precious-metal catalyst at 40 – 135 °C to accomplish this selective oxidation [665 – 668]. The carbon dioxide formed by the Selectoxo reaction adds only slightly to the load on the downstream carbon dioxide absorption system.

Methanolation [669, 670] has been proposed for partially replacing methanation. It converts the residual carbon oxides to methanol, preferably at higher pressure in an intermediate stage of synthesis gas compression. Methanol is removed from the gas by water scrubbing. The methanol may be recycled to the steam reformer feed or recovered as product. As full conversion of the carbon oxides is not achieved, a clean up methanation unit must follow the methanolation section.

Dryers. It is energetically advantageous to add the purified synthesis gas at a point in the synthesis loop where it can flow directly to the synthesis converter (see Section 4.5.3.1). For this reason water and traces of carbon dioxide must be removed from the make-up gas downstream of methanation. This is accomplished by passing the make-up gas through molecular sieve adsorbers.

Cryogenic methods are usually used for final purification of partial oxidation gases, but may be also incorporated in steam reforming plants. A prominent example is the Braun Purifier process [451, 671 – 677].

The purifier is a cryogenic unit placed downstream of the methanator and its duty is to remove the nitrogen surplus introduced by the excess of air used in the secondary reformer of the Braun ammonia process (4.6.1.4.2). Additionally the inert level in the synthesis loop is reduced through this unit because methane is completely and argon is partially removed from the make-up gas. Another advantage of the process is that it separates the front-end and the synthesis loop, permitting the H/N ratio in the synthesis loop to be set independent of the secondary reformer. The purifier is a relatively simple unit composed of feed/effluent exchanger, a rectifier column with an integrated condenser and turbo-expander. At −185 °C methane and argon are washed out. The cooling energy is supplied by expansion of the raw gas over the turbo-expander (pressure loss about 2 bar) and expanding the removed waste gas to the pressure level of the reformer fuel.

Liquid Nitrogen Wash [678 – 680]. Normally, for the partial oxidation processes, only a high-temperature shift conversion is used. This results in a carbon monoxide content of the gas after shift conversion in the range 3 – 5 vol %. Copper liquor scrubbing for carbon monoxide removal, commonly employed in early plants, has become obsolete and is now operated in only a few installations. Not only does it have a high energy demand, but it is also environmentally undesirable because of copper-containing wastewater. Liquid nitrogen wash delivers a gas to the synthesis loop that is free of all impurities, including inert gases and is also the means for adding some or all of the nitrogen required for synthesis.

The nitrogen is liquefied in a refrigeration cycle by compression, cooling, and expansion. It flows to the top of a wash column, where it countercurrently contacts precooled synthesis gas from which most of the methane and hydrocarbons have been condensed. All of the cold equipment is installed in an insulated "cold box." The wash column temperature is about $-190\,°C$. Liquid nitrogen wash systems are in operation at pressures up to 8 MPa corresponding to the highest gasification pressures. Careful surveillance of the inlet gases is required. Water and carbon dioxide in the inlet gas will freeze, causing operating difficulties. Traces of nitric oxide (NO) may react with olefinic hydrocarbons, causing explosions [681, 682].

Normally, an air separation plant is installed in conjunction with liquid nitrogen wash for economy in operation. In modern plants, the air separation and the nitrogen wash frequently are closely integrated with one another so that economies can be realized in the refrigeration system.

Pressure Swing Adsorption. This process can be used to replace the LT shift conversion, carbon dioxide removal, methanation, and even the secondary reformer as well [100, 652, 683 – 686]. It uses molecular sieves as adsorbents in a series of vessels operated in a staggered cyclic mode changing between an adsorption phase and various stages of regeneration. The regeneration of the loaded adsorbent is achieved by stepwise depressurization and by using the gas from this operation to flush other adsorbers at a different pressure level in the regeneration cycle. The hydrogen recovery may be as high as 90 % depending on the number of adsorbers in one line, which may be as high as 10. Very high purity can be obtained, with about 50 ppm argon and less than 10 ppm of other impurities.

The process scheme for the ammonia plant may consist of production of pure hydrogen followed by separate addition of pure nitrogen from an air separation unit [687 – 689]. In a special version the nitrogen can be added in the PSA unit itself to increase hydrogen recovery [690 – 693]. In some processes it may also remove the excess nitrogen introduced with the process air fed to the secondary reformer, e.g. the LCA Process of ICI (page 239). Since this technology has proven its reliability in rather large hydrogen plants for refineries it is now also used for world-scale ammonia plants, e.g., the Linde LAC process (page 241).

4.5.2. Compression

Up to the mid-1960s reciprocating compressors were used to compress the synthesis gas to the level of the synthesis loop, which was around 300 bar in the majority of the plants at that time. Higher pressures were used in a few installations, for example, in Claude and Casale units. Prior to about 1950 gas generation processes and shift conversion operated at essentially atmospheric pressure. The gas was first compressed to the level of the CO_2 removal section (usually 25 bar) and afterwards to around 300 bar for final purification (at that time usually copper liquor scrubbing) and synthesis. Reciprocating compressors with as many as seven stages in linear arrangement with intermediate cooling were used, whereby the CO_2 removal section was usually installed between the 3rd and 4th stages. Machines with a suction volume up to 15 000 m^3 (STP) for the first stage were not uncommon. Huge flywheels were designed as the rotors of synchronous motors (ca. 125 rpm) with two crankshafts on both sides connected over crossheads with the piston rod for the horizontally arranged stages. In some instances gas engines were used as drivers.

The rapid technical progress in the hydrocarbon based technologies of steam reforming and partial oxidation made it possible to generate the synthesis gas at a pressure level sufficient for the CO_2 removal operation. As gasification proceeds with a considerable volume increase and feedstocks such as natural gas are usually already available under pressure at battery limits, considerable savings in compression energy are achieved in this way.

Along with the introduction of pressure gasification, horizontally balanced compressors in which the cylinders are in parallel configuration on both sides of a common crankshaft became the preferred design. In these machines a good dynamic balance can readily be achieved, higher speeds are possible and also the use of asynchronous motors is possible. The low height of the arrangement has less severe requirements for foundations, allows simpler piping connections and facilitates maintenance. When gas engine drivers (two-stroke type) were used instead of electric motors, some designs applied a common crankshaft for the piston rods of the gas machine cylinders and compressor cylinders. In a very few cases steam turbines with special speed reduction gears have been used. In smaller plants, the various compression services, e.g., natural gas, process air, and synthesis gas compression, were apportioned among the crankshaft throws in such a manner that a single compressor can perform all compression duties [694]. Further information on reciprocating compressors is given in [695 – 698].

One of the most important features of the energy integrated single-stream steam reforming ammonia plant pioneered by M. W. Kellogg in 1963 was the use of centrifugal compressors for the compression duties of synthesis gas and recycle, process air, and refrigeration. From this time onwards application of centrifugal compressors became standard practice in most ammonia plants irrespective of the synthesis gas generation technology. The fundamental advantage of these machines are low investment (single machines even for very large capacities) and maintenance cost, less frequent shutdowns for preventive maintenance, and high reliability (low failure rate) [699]. In most cases the centrifugal compressors in ammonia plants are directly driven by steam turbines. This avoids the losses associated with generation and transmission of electric power. For this reason the overall efficiency of a plant with steam-driven centrifugal compressors is superior, although the centrifugal compressors are inherently less efficient than reciprocating units. A further advantage is that centrifugal compressors require only a fraction of the space needed for reciprocating compressors.

Manufacturing capabilities limit the minimum possible passage width (today about 2.8 mm) at the outer circumference of a centrifugal compressor impeller and this imposes a limit on the minimum effective gas volume leaving the last impeller. Unless the total volumetric gas flow has a reasonable relationship to the passage width of the last impeller and the pressure ratio, excessive pressure losses would occur within the passage and in the diffusers between the impellers, rendering the machine extremely ineffective. The first single-train ammonia plants with a capacity of 550 – 600 t/d had to lower the synthesis pressure to 145 – 150 bar to meet the required minimum gas flow condition. Today, with improved manufacturing techniques, the minimum gas flow from the last wheel is 350 m^3 for synthesis gas with a molecular mass of about 9 and an efficiency of around 75 %. This corresponds to a capacity of 400 t/d at 145 bar. As newer synthesis catalysts allow a pressure of 80 bar in the synthesis loop (ICI's LCA Process, Kellogg's KAAP) a centrifugal compressor could be used down to 220 t/d. Of course, for today's world-scale capacities of 1200 – 2000 t/d these technical limitations have no influence on the synthesis pressure, which even for plants with 1800 t/d is between 155 and 190 bar [100, 418, 700 – 702].

The tensile strength of the steels normally used to manufacture the compressor impellers allow a maximum wheel tip speed of about 330 m/s, which limits the pressure increase attainable by each impeller. A pressure increase, for example, from 25 to 200 bar would require 18 – 20 impellers. However, a compressor shaft must have sufficient rigidity to avoid excessive vibration, and this limits the possible length such that a compressor shaft cannot accommodate more than eight or nine impellers. It is therefore necessary to arrange several compressor casings in series, with compression ratios from 1.8 to 3.2.

To overcome the pressure drop (5 – 20 bar) in the synthesis loop re-compression of the recycle gas is required. In practically all modern ammonia plants, the shaft of the final casing also bears the impeller for the compression of the recycle gas. Depending on synthesis configuration, mixing of make-up gas and recycle can be performed inside the casing or outside (three or four-nozzle arrangement; Fig. 34).

In older plants which used a reciprocating compressor for the recycle, a recycle cylinder was often mounted together with the other cylinders on the reciprocating frame. Sometimes special rotary compressors, so-called mole pumps were also used, with the unique feature that compressor and electric driver were completely enclosed in a common high-pressure shell. In old Casale plants, the make-up gas was introduced into the high pressure recycle loop and acted as

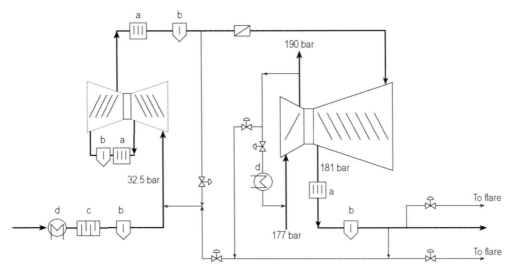

Figure 32. Centrifugal compressor for make-up and recycle gas compression of an ammonia plant (courtesy of Uhde)
a) Air cooler; b) Separator; c) Silencer; d) Water cooler

the driving fluid of an injector which compressed the recycle gas.

Today modern plant concepts for world-scale capacity plants tend to limit the number of compressor casings to two. Figure 32 shows an example of the synthesis gas compressor of a large ammonia plant.

Geared or metal diaphragm couplings are used to connect the shafts of the individual casings (two in Figure 32). These flexible couplings prevent possible compressor damage resulting from slight misalignment and shaft displacement.

Sealing of the rotating shaft against the atmosphere is an important and demanding task. The high pressures and the high rotational speeds involved do not allow mechanical contact shaft seals. Usually, liquid-film shaft seals with cylindrical bushings (floating rings) are applied [703]. In this concept, an oil film between the shaft and a floating ring, capable of rotation, provides the actual sealing. The floating ring is usually sealed to the compressor casing by O-rings. Seal oil flows between both halves of the floating ring. Part of the oil returns to the reservoir, while the remainder flows against the gas pressure into a small chamber from which, together with a small quantity of gas, it is withdrawn through a reduction valve. Figure 33 is a schematic diagram of a liquid film shaft seal.

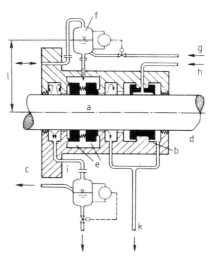

Figure 33. Liquid film shaft seal with cylindrical bushing for a high-pressure centrifugal compressor
a) Shaft; b) Bearing; c) Pressure side; d) Ambient side; e) Floating seal rings; f) Seal oil surge vessel; g) Seal oil; h) Lube oil; i) Drain to gas – oil separator; k) Drain to lube oil tank; l) Elevation for seal oil head

The seal oil pressure in the floating ring cavity must always be slightly higher than the gas pressure within the casing which is provided by static height difference of the oil level in the elevated oil buffer vessel. In this way normally no oil should enter into the synthesis gas. As the labyrinth section on the pressure side is connected with the suction side through an equalizing line, it is necessary to seal the compressor shaft towards the atmosphere only at suction pressure level.

Often, seal oil supply is combined with the lubricating oil system, with oil reservoir, filters and (in part) pumps in common.

The high rotational speeds and the relatively large masses of the compressor rotors place high demands on the performance of the bearings. This is especially true for the thrust bearings, which must withstand high thrust forces. The minimum clearances necessary at the labyrinths and the impellers allow practically zero wear. Any wear resulting in friction could lead instantly to severe damage. For these reasons, in addition to measuring bearing temperatures, the axial position of the rotor and the radial vibrational deflections are continuously monitored by sensors. An increased vibrational amplitude is often an early indication of mechanical faults, such as rotor imbalance, bearing damage, occurrence of friction, or misalignment.

An interesting concept is dry (oil-less) gas seals for centrifugal compressors. Development dates back to 1969 and the first commercial application was in a natural-gas compressor in 1973. Since then they have been widely used in off-shore service [704 – 706]. Only recently have they gained acceptance in the ammonia industry, where several compressors for synthesis gas and refrigeration duty equipped with dry seals have been successfully placed in service. Nitrogen is used as an inert fluid for the seal, which is achieved at the radial interface of rotating and stationary rings. During operation the seal is not completely tight; some of the seal gas flows back to the suction side to be recompressed, and a small amount from the suction side may go to the atmospheric side and is sent to the flare on account of its content of combustibles. During stops, when the shaft is not rotating, the seal ring is pressed tight against the seat by means of a spring and the differential gas pressure. Dry gas seals in combination with oil-lubricated bearings (dry/wet) have the advantage that a much smaller oil system is required and that there is no contact between oil and gas, which eliminates an emission source.

A new development, already in commercial service, but so far not used in ammonia plants, is magnetic bearings [704, 707]. Magnetic bearings promise a wider temperature range, are less prone to wear, are less prone to developing vibrations due to imbalance and require less maintenance. A combination of magnetic bearings and dry seals (dry/dry) could totally eliminate the oil system.

Integrated geared centrifugal compressors developed by DEMAG and GHH are a new development which might become of interest for plants of smaller capacity operating at lower pressure. The driver (e.g., a steam turbine) drives a common gear to which the individual compression stages are connected. Each stage has a single impeller which runs with very high speed, for example, 25 000 rpm or higher. Compressors with three or four stages are in operation, for example, a methanol synthesis compressor for 75 000 m^3/h and a pressure of 75 bar. It is likely that this concept can be extended to a final pressure of 120 bar with ammonia synthesis gas.

Compressor control is achieved basically by controlling the rotational speed of the driver, in modern plants often with the help of a distributed control system (DCS). If the volumetric flows through the machine at start-up or during reduced load operation deviate too far from the design values, it is necessary to re-circulate gas through individual stages or through the whole machine. Otherwise the compressor can enter a state of pulsating flow, called surge, which could cause damage. Anti-surge control (minimum bypass control, kickback control) is designed to prevent this condition, as well as to minimize the incidence and degree of uneconomical re-circulation. A point of discussion is sometimes the minimum load at which the compressor can run without kickback, which means without loss of efficiency. Usually a load of 75 % is possible, and with special impeller design, the value may be lowered to 70 % but with a slight sacrifice of full-load efficiency.

Other compression duties in the plants, such as process air in steam reforming plants and air, and oxygen, and nitrogen compression in partial oxidation plants, are also performed by centrifugal compressors. Also for the ammonia compression in the refrigeration section centrifugal compressors are normally in service. In some cases screw compressors have been used for this duty on account of their good efficiency and load flexibility, which is of interest in plants where the ammonia product is split between direct users at the site and cold storage in changing ratios.

Criteria for compressor selection and economic comparisons are discussed in [708 – 716]. Additional information is given in [717 – 727].

In modern plants the synthesis gas compressors, including recycle, are almost exclusively driven by a steam turbines. These are generally extraction turbines with a condensing section. Steam is extracted at suitable pressure levels (e.g. 45 – 55 bar) to provide, for example, the process steam in steam reforming plants, and for other drivers, e.g., air compressor, ammonia compressor, boiler feed water pumps, and blowers.

As failures and breakdowns of these large rotary machines could lead to long and expensive repairs and to a corresponding loss of production it is advisable to keep the essential spare parts, for example, spare rotors, in stock. In older steam turbines, sometimes blade failures occurred, but this is no longer a problem due to improved blade design and shroud bands, which are standard today.

Gas turbines have also been used as drivers for compressors in ammonia plants. The exhaust may be used for steam production, for preheating duties, or as combustion air in the primary reformer [728 – 732].

4.5.3. Ammonia Synthesis

Under the conditions practical for an industrial process ammonia formation from hydrogen and nitrogen

$$N_2 + 3\,H_2 \rightleftharpoons 2\,NH_3 \quad \Delta H = -92.44\,\text{kJ/mol}$$

is limited by the unfavorable position of the thermodynamic equilibrium, so that only partial conversion of the synthesis gas (25 – 35 %) can be attained on its passage through the catalyst. Ammonia is separated from the unreacted gas by condensation, which requires relatively low temperatures for reasonable efficiency. The unconverted gas is supplemented with fresh synthesis gas and recycled to the converter. The concentration of the inert gases (methane and argon) in the synthesis loop is controlled by withdrawing a small continuous purge gas stream. These basic features together with the properties of the synthesis catalyst and mechanical restrictions govern the design of the ammonia synthesis converter and the layout of the synthesis loop. Evaluation criteria are energy consumption, investment and reliability.

4.5.3.1. Synthesis Loop Configurations

A number of different configurations are possible for the synthesis loop. They can be classified according to the location of ammonia condensation and the point at which the make-up gas is introduced. Figure 34 shows the principal possibilities.

If the make-up gas is absolutely free of catalyst poisons, such as water and carbon dioxide (for example, after molecular sieve dehydration or liquid nitrogen wash), it can be fed directly to the synthesis converter (Fig. 34 A). After the gas leaves the synthesis converter, ammonia is condensed by cooling and the recycle gas is referred to the recycle compressor. This represents the most favorable arrangement from a minimum energy point of view. It results in the lowest ammonia content at the entrance to the converter and the highest ammonia concentration for condensation.

When the make-up gas contains water or carbon dioxide, advantage is taken of the fact that these materials are absorbed completely by condensing ammonia. This requires that the condensation stage be located partially or wholly between the make-up gas supply point and the converter. This arrangement has the disadvantage that the ammonia concentration for condensation is reduced by dilution with the make-up gas. Also, at equal condensing temperature, a higher ammonia concentration exists at the inlet to the converter. Figure 34 B shows the simplest such configuration. An additional drawback of this arrangement is that all the ammonia produced must be compressed with the recycle gas in the recycle compressor.

The scheme shown in Figure 34 C, the frequently used "four-nozzle compressor", avoids this waste of energy. With this arrangement, recycle compression follows directly after condensing and separating the ammonia. In this configuration, it is possible to cool the recycle gas using cooling water or air immediately before

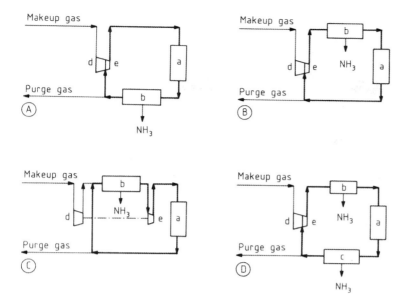

Figure 34. Schematic flow diagrams of typical ammonia synthesis loops
A) Synthesis loop for pure and dry make-up gas; B) Product recovery after recycle compression; C) Product recovery before recycle compression (four-nozzle compressor design); D) Two stages of product condensation
a) Ammonia converter with heat exchangers; b) Ammonia recovery by chilling and condensation; c) Ammonia recovery by condensation at ambient temperature; d) Synthesis gas compressor; e) Recycle compressor

admixing the make-up gas (i.e., before diluting the recycle gas) and thereby to reduce the energy expenditure for refrigerated cooling.

Splitting the cooling step for ammonia condensation also offers advantages when the recycle gas is compressed together with the make-up gas. This applies especially at synthesis pressures above about 25 MPa (250 bar). At these pressures, a greater portion of the ammonia formed can be liquefied by cooling with cooling water or air (see Fig. 34D).

When ammonia-containing recycle gas and carbon dioxide containing make-up gas mix together under certain conditions of concentration and temperature, precipitation of solid ammonium carbamate can result.

In recent years, also as a retrofit in existing plants, molecular sieve drying of make-up gas has increasingly been applied in order to realize the energy-saving arrangement of the synthesis loop corresponding to Figure 34 A.

4.5.3.2. Formation of Ammonia in the Converter

The central part of the synthesis system is the converter, in which the conversion of synthesis gas to ammonia takes place. Converter performance is determined by the reaction rate, which depends on the operating variables (cf. Section 4.3.3). The effect of these parameters is discussed briefly in the following.

With increasing pressure, ammonia formation increases (Fig. 35). This results not only from the more favorable equilibrium situation for the reaction, but also from the effect on the reaction rate itself. In industrial practice, there are plants that operate at about 8 MPa (80 bar), but there are also those that operate at more than 40 MPa (400 bar). Today, plants are built mainly for synthesis pressures of 150 – 250 bar. Typical operating parameters for modern synthesis loops with different pressures are listed in Table 24.

Table 24. Typical operating parameters for modern synthesis loops at 140 and 220 bar (1000 t/d NH$_3$) [418, p. 226]

Parameters	Inlet pressure, bar	
	140	220
Inlet flow, Nm^3/h	500 000	407 000
Inlet NH$_3$ conc., mol %	4.1	3.8
Outlet NH$_3$ conc., mol %	17.1	19.9
Inlet inert conc., mol %	8.0	12.0
NH$_3$ separator temperature, °C	−5	−5
Relative catalyst volume	1	0.6

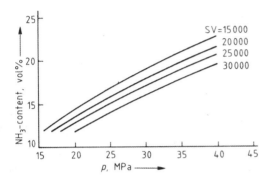

Figure 35. Performance for a four-bed quench converter as a function of operating pressure with space velocity (per hour) as parameter; 10 % inerts in the inlet synthesis gas

Converter performance decreases with increasing inert gas content (Fig. 36). The usual range is 0 – 15 vol %. For a secondary loop based on purge gas, it can be 30 % or more (see Section 4.5.3.5).

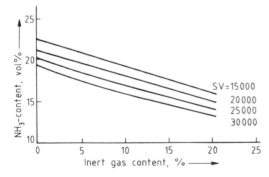

Figure 36. Performance of a converter as a function of inlet inert gas (CH$_4$ and Ar) content with space velocity (per hour) as parameter, inlet NH$_3$ content is 3.5 %; 30 MPa pressure; catalyst particle size is 6 – 10 mm

Converter performance also diminishes (Fig. 37) with increasing oxygen content of the synthesis gas. Today, a level of 10 ppm in the make-up gas, corresponding to about 3 ppm in the converter inlet gas, is usually, not exceeded (cf. Section 4.4.1.5).

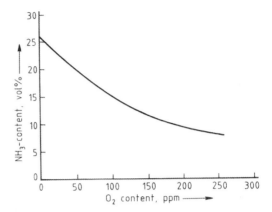

Figure 37. Performance of a converter as a function of oxygen content (all oxygen-containing impurities) in the inlet synthesis gas

In contrast to the above-mentioned variables, the dependence of the converter performance on the H$_2$/N$_2$ ratio shows a true maximum (Fig. 38). The optimum conversion at high space velocity [SV = m^3 (STP) gas h^{-1} · m^{-3} catalyst] lies close to an H$_2$/N$_2$ ratio of 2 and approaches 3 at low space velocities. The reason is that equilibrium plays a greater role at low space velocities and has a maximum at a ratio of 3, except for small corrections [6] with regard to the behavior of real gases. Usually, the ratio is adjusted to 3, because in most plants, conversions near equilibrium are attained.

In practice, space velocities vary from about 12 000 h^{-1} at about 15 MPa (150 bar) to about 35 000 h^{-1} at about 80 MPa (800 bar). Usually, with increasing space velocity, the ammonia concentration in the effluent synthesis gas from a given converter does indeed go down (Figs. 35, 36, 38). However, the operating point normally chosen means that the increase in gas flow rate more than compensates for the reduced ammonia concentration. Thus, a still higher ammonia production rate is achieved. Plant operation often takes advantage of these phenomena. For example, this characteristic can be used to maintain ammonia production rate when the synthesis catalyst ages and its activity declines. Increasing converter flow rate and declining synthesis catalyst activity can reach a point, even with careful control, where the reaction "blows out" and production ceases. This occurs when the heat of reaction is no longer sufficient to provide the temperatures necessary for operation of the feed – effluent heat exchanger. The heat exchanger then fails to heat the cold converter feed gas to the required reaction "ignition" temperature. Achieving maximum ammonia production requires operating in the neighborhood of this "blow out" point, in turn requiring very careful control [734 – 736]. If the synthesis converter is to be operated in this region, then it is advisable to oversize the converter feed – effluent heat exchanger system to attain a higher degree of control stability.

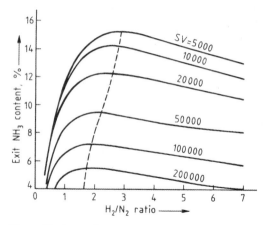

Figure 38. Ammonia conversion as a function of hydrogen/nitrogen ratios in the inlet synthesis gas with space velocity (per hour) as parameter; 9.7 MPa pressure [733]

Converter Design. Design of ammonia synthesis reactors is not just the calculation of the required catalyst volume; other parameters have to be considered, too, and for some of them optimum values have to be found. This raises the question of the definition of optimum. In the early days with more strict material and fabrication-related limitations the converters were usually designed for minimum high-pressure volume, and this meant maximum use of the catalyst. Today the objective is to optimize the heat recovery (at the highest possible level) and to minimize the investment for the total synthesis loop.

The design of an ammonia converter is a demanding engineering and chemical engineering task. To calculate the parameters for the design, including dimensions and number of catalyst beds, temperature profiles, gas compositions, and pressure drop, a suitable mathematical model is required.

Two differential equations describe mathematically the steady-state behavior of the reactor section of a converter. The first models the concentration – position relationship for transformation of the reactants to products, i.e., the reaction kinetic equation (cf. Section 4.3.3). The second handles the temperature – position behavior of the reacting synthesis gas, the catalyst, and the vessel internals. The form of the latter is characteristic of the type of converter. The temperature profile depends not only on the rate of reaction heat evolution but also on the method and nature of the system for removing heat from the catalyst bed or beds. Additional equations describe the behavior of the separate feed – effluent heat exchanger system [737 – 741]. General information on converter calculations are given in [195, 740 – 742]. Computer programs and applications can be found in [743 – 746]. For a discussion of modeling of different converter types, see [737 – 741, 747, 748]. Models for multibed quench converters are described in [747 – 754]. Tubular reactors are treated in [755 – 762]. The individual effects of the operational parameters are evaluated in [6, 265, 763].

The reaction temperature profile is of particular importance because the reaction rate responds vigorously to temperature changes. Figure 39 plots lines of constant reaction rate illustrating its dependence on temperature and ammonia concentration in the reacting synthesis gas. The line for zero reaction rate corresponds to the temperature – concentration dependence of the chemical equilibrium. From Figure 39 it is apparent that there is a definite temperature at which the rate of reaction reaches a maximum for any given ammonia concentration. Curve (a) represents the temperature – concentration locus of maximum reaction rates. To maintain maximum reaction rate, the temperature must decrease as ammonia concentration increases.

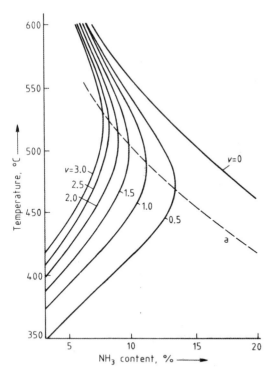

Figure 39. Nitrogen reaction rate v in m³ NH₃/(m³ catalyst · s) as a function of temperature and ammonia concentration at 20 MPa pressure and 11 vol % inerts in the inlet synthesis gas
a) Locus of temperatures resulting in maximum reaction rate at a given ammonia concentration

If the objective in design or operation were optimizing catalyst utilization, then Figure 39 shows that the converter temperature – composition profile should follow curve (a), which corresponds to maximum reaction rate at all points. It is also obvious that in reality this "ideal" temperature – concentration profile cannot be achieved. For example, a synthesis gas with about 3 % ammonia concentration entering the converter cannot be heated to the "ideal" temperature by heat exchange because the very high temperature required does not exist in the converter system. To reach the "ideal" temperature, the first portion of the catalyst must initially operate adiabatically. Consideration of the service life of the catalyst requires that this maximum initial temperature not exceed that recommended by the manufacturer, usually 530 °C (cf. Section 4.4.1.4). Following this initial adiabatic temperature rise, it is possible to minimize the required catalyst volume by cooling the reacting synthesis gas such that, as ammonia formation progresses, the temperature follows curve (a). In the days when converters were designed to operate at very high pressures and temperatures and before the advent of improved construction materials, the converter design represented a real limitation on plant capacity. To maximize converter output and plant capacity to achieve the most favorable overall manufacturing cost, it was necessary to optimize catalyst utilization. A converter temperature – concentration profile was often compared to the "ideal" for optimum usage of high-pressure vessel and catalyst volumes [763, 764].

Commercial Ammonia Converters.
Principal Reactor Configurations. Commercial converters can be classified into two main groups:

1) Internally cooled with cooling tubes running through the catalyst bed or with catalyst inside the tubes and the cooling medium on the shell side. The cooling medium is mostly the reactor feed gas, which can flow counter- or cocurrently to the gas flow in the synthesis catalyst volume (*tube-cooled converters*).
2) The catalyst volume is divided into several beds in which the reaction proceeds adiabatically. Between the individual catalyst beds heat is removed by injection of colder synthesis gas (*quench converters*) or by indirect cooling with synthesis gas or via boiler feed water heating or raising steam (*indirectly cooled multibed converters*).

The gas flow can have an axial, cross-flow or radial flow pattern. The different cooling methods can be combined in the same converter.

The severe conditions of high pressure, high temperature, and high hydrogen partial pressures place strict requirements on the construction materials and design for both groups. For example, almost all converters consist of an outer pressure vessel containing a separate inner vessel housing the catalyst and other internals, such as gas distributors and heat exchangers. Relatively cool converter feed gas flows through the annular space between the outer pressure shell and the internal "basket". This shields the outer shell from the high-temperature "basket", permitting use of comparatively low-alloy chromium – molybdenum steels for its construction. Often, part of the converter feed – effluent heat-exchange system surface is placed within the converter pressure shell. By this means, the nozzle penetration through the pressure shell for the converter effluent gas is also maintained at relatively low temperature. Today, this latter feature is not always necessary; the state of the art in converter construction materials now allows design of exit nozzles for the maximum anticipated temperatures, i.e., up to about 530 °C. References [765] and [766] review some literature on ammonia converters.

Tube-Cooled Converters. To remove the heat evolved in the synthesis reaction, converters have been designed in which cooling tubes run through the catalyst bed. With these tubes, the heat is transferred to the converter feed gas to heat it to the reaction ignition temperature or to an external cooling medium. The known designs for such converters are suited only for small production capacities and therefore currently of limited interest. When designed to utilize the heat of reaction for heating the converter feed gas, such converters have the further disadvantage that temperature control is sluggish and temperature oscillations dampen out very slowly, if at all, a phenomenon called "hunting".

Typical of such converters is the countercurrent design of the Tennessee Valley Authority (TVA) [734, 760, 764, 767 – 772]. Part of the feed gas to this unit enters the converter at the top and flows down through the annular space between the pressure shell and the basket. The main gas flow enters the converter at the bottom and joins the shell cooling gas. The mixture is heated to about 200 °C in an internal exchanger located beneath the catalyst bed. The gas then enters the cooling tubes that run through the catalyst bed (Fig. 40 A). There it absorbs the heat released in the reaction and reaches the required reaction ignition temperature of about 400 °C. The reaction begins almost adiabatically in the catalyst bed. As the reacting gas temperature rises, the temperature difference between the reacting gas and the cooling tubes increases, resulting in increasing heat removal (Fig. 40 B). As the reacting gas reaches the bottom of the catalyst bed, the rate of reaction begins to decrease sufficiently, because of high ammonia concentration, that cooling predominates and the temperature of the reacting gas begins to fall. Figure 40 C shows the converter temperature – concentration profile. The reaction temperature at first somewhat exceeds that for maximum reaction rate but eventually falls below curve (a). Cold converter feed gas can be admitted at a

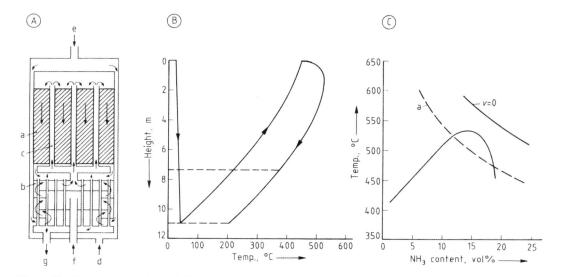

Figure 40. Countercurrent tube-cooled converter (TVA converter) [764]
A) Converter layout; a) Catalyst; b) Heat exchanger; c) Cooling tubes; d) Main gas inlet; e) Vessel wall cooling gas inlet; f) Temperature control gas (cold-shot) inlet; g) Gas exit
B) Gas temperature profile through the converter; C) Ammonia concentration versus temperature (cf. Fig. 39)

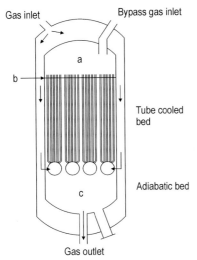

Figure 41. ICI tube-cooled ammonia converter
a) Top of catalyst bed; b) Cooling tubes; c) Catalyst;

point within the internal heat exchanger to control both the total system and reaction temperature profiles through bypassing. Casale has employed a similar converter design concept [773].

Other tube-cooled converters with countercurrent flow are the Mont Cenis reactor [774 – 777], the original Haber – Bosch reactors [774], the Claude converter [778, 779], and the older Fauser design [778]. These converters were all used in relatively small plants and are now obsolete.

An interesting rebirth of the countercurrent principle is the new ICI tube cooled converter used in the LCA process (Fig. 41) [780].

The Nitrogen Engineering Corporation (NEC) converter applies cocurrent flow by means of bayonet tubes [764, 767, 769 – 771, 781 – 785]. This design places maximum heat-exchange temperature difference — therefore, maximum cooling performance — at the catalyst bed inlet, the point where maximum reaction rate — therefore maximum rate of heat evolution — is taking place. The intent is to obtain closer approach to curve (a) (cf. Fig. 42 C). Figure 42 A shows the general arrangement of the converter, catalyst bed, and heat exchanger; Figure 42 B, the temperature profile of the system. Chemico, a derivative of NEC, continued to apply such converters with only slight changes [773].

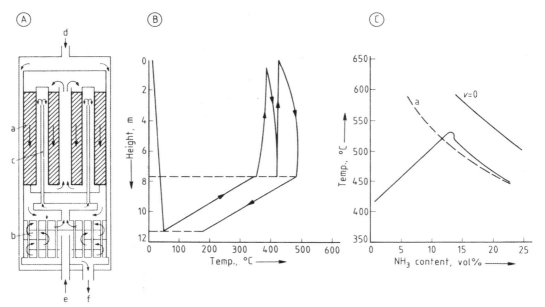

Figure 42. Cocurrent-flow tube-cooled converter [764]
A) Converter; a) Catalyst; b) Heat exchanger; c) Cooling tubes; d) Gas inlet; e) Temperature control gas (cold-shot) inlet; f) Gas exit
B) Gas temperature profile through the converter;
C) Ammonia concentration versus temperature (cf. Fig. 39)

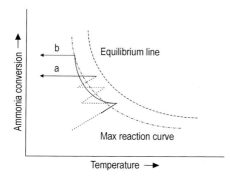

Figure 43. Expected thermal profile of catalyst along the converter and comparison with plant data [786]

Ammonia Casale developed a new converter concept which uses no tubes but arranged cooling plates. The plates are radially immersed in the axial–radial catalyst bed to remove the reaction heat while it is formed. Figure 43 [786] shows the temperature profile of this pseudo-isothermal reactor. The profile follows the line of the maximum reaction rate to obtain the highest possible conversion per pass from a given catalyst volume. The use of plates allows a design without tubesheets and simplifies catalyst loading and unloading and the use of small size high active catalyst. The new Casale IAC ammonia converter is suited for very high capacities up to 4500 t/d for the the so-called mega-ammonia plants presently in discussion. The first commercial use of a converter of this type is operating successfully in a 1050 stpd (953 mtpd) plant in Trinidad [787]. In a recent Casale company brochure the plate arrangement is shown taking a methanol converter as an example (Fig. 44).

Today, the great majority of synthesis converters are designed with catalyst distributed in several beds, within one or more reactor vessels. In each bed, the synthesis gas reacts adiabatically, and direct or indirect cooling is provided between the catalyst beds for cooling the reacting mixture from a temperature above to a value below the "ideal" (curve (a), Fig. 45).

Multibed Reactors with Direct Cooling (Quench Converters). In quench converters cooling is effected by injection of cooler, unconverted synthesis gas (cold shot) between the catalyst beds. The catalyst beds may be separated by grids designed as mixing devices for main gas flow and cold shot, or be just defined by the location of cold gas injection tubes as for example in the ICI lozenge converter.

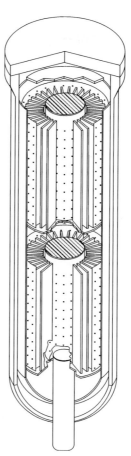

Figure 44. Internals of a Casale isothermal converter (courtesy of Ammonia Casale)

In this type of converter only a fraction of the recycle gas enters the first catalyst layer at about 400 °C. The catalyst volume of the bed is chosen so that the gas leaves it at ca. 500 °C (catalyst suppliers specify a maximum catalyst temperature of 530 °C). Before it enters the next catalyst bed, the gas is "quenched" by injection of cooler (125 – 200 °C) recycle gas. The same is done in subsequent beds. In this way the reaction profile describes a zig-zag path around the maximum reaction rate line. A schematic drawing of a quench converter together with its temperature/location and temperature/ammonia concentration profiles is presented in Figure 45.

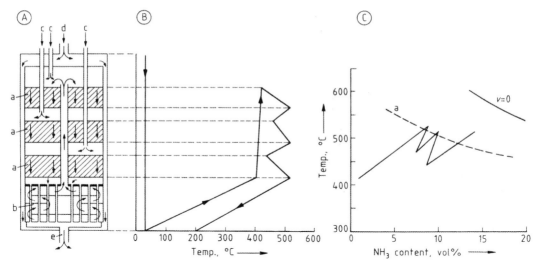

Figure 45. Multibed converter with quench cooling [788]
A) Converter; a) Catalyst; b) Heat exchanger; c) Quench gas inlets; d) Gas inlet; e) Gas exit
B) Gas temperature profile through the converter;
C) Ammonia concentration versus temperature (cf. Fig. 39)

A disadvantage is that not all of the recycle gas passes over the entire catalyst volume so that considerable ammonia formation occurs at higher ammonia concentration and therefore at lower reaction rate. Therefore a higher catalyst volume is needed compared to an indirectly cooled multibed converter. However, no extra space is required for interbed heat exchangers, so that the total volume of the high-pressure vessel will remain about the same as that for the indirectly cooled variant [789].

As the quench concept was well suited to large capacity converters it had a triumphant success in the early generation of large single-train ammonia plants constructed in the 1960s and 1970s. Mechanical simplicity and very good temperature control contributed to the widespread acceptance. For example, M. W. Kellogg alone has installed more than 100 of its quench converters. Though being increasingly replaced by the indirect-cooling concept by revamp or substitution they are still extensively used. Descriptions of earlier designs of Kellogg, BASF, and Uhde can be found in [766, 770, 790, 791].

The most important example is the M. W. Kellogg three- or four-bed converter [766, 792 – 794] (Figure 46). In this design, the catalyst "basket" is not easily removable from the pressure vessel. The catalyst can be changed by draining it at the bottom of the converter through "downcomers" that connect all catalyst beds with one another. The converter feed – effluent exchanger, attached to the top head, is designed for disassembly.

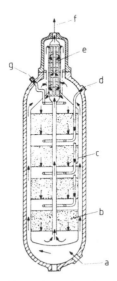

Figure 46. Kellogg four-bed vertical quench converter
a) Gas inlet; b) Catalyst bed; c) Basket; d) Quench;
e) Interchanger; f) Gas outlet; g) Bypass

Other designs were used by BASF [766], Casale [766], Chemico, Grand Paroisse [765], ICI [766], Uhde [766], and others.

An interesting variant in this group is the ICI lozenge converter [293, 766, 773, 774, 792]. This design (Fig. 47) has a single catalyst bed that is divided into several zones (usually four) by quench gas distributors, through which colder recycle gas is injected evenly across the whole cross section of the catalyst bed. For this reason it is justifiable to classify this converter as a multibed type. The distributors consist of banks of transverse sparge pipes which deliver gas at regular intervals along their length. The spargers are in a void space within horizontal mesh-covered structures, whose cross section is lozenge shaped so that the catalyst particles can flow freely past them during loading and unloading. A special version of this reactor concept is the opposed flow design [795, 796], suggested for very large capacities. In this configuration the converted gas is collected and withdrawn from the middle of the catalyst bed, with down-flow in the upper half and up-flow in the lower half of the catalyst bed. The uninterrupted catalyst bed is maintained in the opposed-flow converter. A design similar to that of ICI with direct injection of the quench gas into the single catalyst bed has been proposed by Chemico [797].

Converters with axial flow face a general problem: with increasing capacity the depth of the catalyst beds must be increased because for technical and economic reasons it is not possible to increase the bed diameter, and thus the pressure vessel diameter, above a certain limit. To compensate for the increasing pressure drop conventional axial flow converters have to use relatively large catalyst particles, which have the disadvantage of lower activity compared to smaller particles mainly on account of diffusion restriction. *Radial gas flow* in the converter avoids this dilemma, and with this concept it is possible to design converters for very large capacities without excessive diameters and with low pressure drop, even with small catalyst particle size. The advantages of radial flow are discussed in [798 – 801]. Radial flow has been also applied in tube-cooled converters [802]. The first radial flow converter introduced commercially and then widely used was the Topsøe S 100 converter [766, 774, 792, 803 – 808], which has two catalyst layers with a catalyst particle size of 1.5 – 3 mm. Figure 48 shows a schematic of the converter [418].

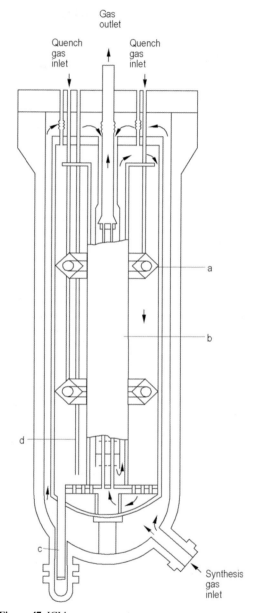

Figure 47. ICI lozenge converter
a) Quench gas distributors; b) Heat exchanger; c) Catalyst discharge nozzle; d) Tube for thermocouples

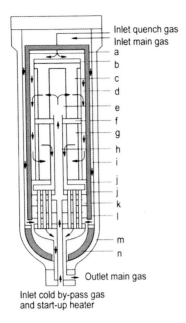

Figure 48. Haldor Topsøe S100 converter
a) Outer internal lid; b) Inner internal lid; c) First catalyst chamber; d) Inner annular space; e) Perforated center tube; f) Catalyst support plate 1; g) Second catalyst chamber; h) Transfer tube; i) Outer annular space; j) Catalyst support plate 2; k) Heat exchanger; l) Refractory fiber; m) Pressure shell; n) Refractory cement

The major part of the gas enters the vessel at the top and flows down as shell cooling gas. It then passes through the feed – effluent heat exchanger and flows upwards through a central pipe to the first catalyst bed, which is traversed from the inside to the outside. After the effluent from the first bed has been quenched with cooler recycle gas, it enters the second bed and passes through it in the inward direction. The cold gas enters through the bottom of the vessel and is mixed with the inlet gas to the first bed for temperature control.

Radial flow quench converters have also been used by Chemoproject [809], Österreichische Stickstoffwerke [810], and Lummus [811].

Axial – radial flow pattern was introduced by Ammonia Casale. Converters with strictly radial gas flow require mechanical sealing of the top of each catalyst bed and dead catalyst volume with little or no flow to avoid bypassing of the catalyst. In the Casale concept there is no need for a dead catalyst zone as the annular catalyst bed is left open at the top to permit a portion of the gas to flow axially through the catalyst. The remainder of the gas flows radially through the bulk of the catalyst bed. As shown in Figure 49 this is achieved by leaving the upper part of the catalyst cartridge at outlet side unperforated so that gas entering from the top is forced to undergo partially axial flow.

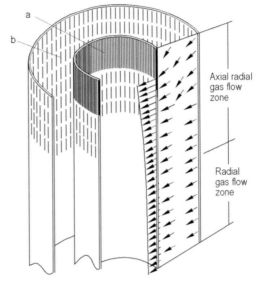

Figure 49. Ammonia Casale axial – radial flow pattern
a) Unperforated wall; b) Perforated wall

Crossflow was chosen as a different approach by M. W. Kellogg in their horizontal quench converter to obtain low pressure drop even with small catalyst particles [765, 804, 805, 812 – 815]. The catalyst beds are arranged side by side in a removable cartridge which can be removed for catalyst loading and unloading through a full-bore closure of the horizontal pressure shell. As the cartridge is equipped with wheels it can be moved in and out on tracks, thus needing no crane. The gas flows vertically from the top to the bottom. The temperature difference between the top and the bottom requires special design measures to prevent uneven circumferential warming of the pressure shell and to avoid bending.

Multibed Converters with Indirect Cooling. In converters of this type cooling between the individual beds is effected by indirect heat exchange with a cooling medium, which may be

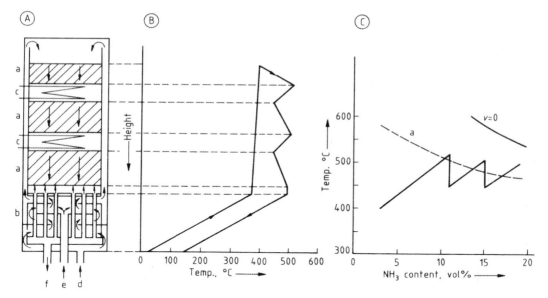

Figure 50. Multibed converter with indirect cooling
A) Converter; a) Catalyst; b) Heat exchanger; c) Cooling section; d) Gas inlet; e) Temperature control gas (cold-shot) inlet; f) Gas exit
B) Gas temperature profile through the converter;
C) Ammonia concentration versus temperature (cf. Fig. 39)

cooler synthesis gas and/or boiler feed water heating and steam raising. The heat exchanger may be installed together with the catalyst beds inside one pressure shell, but an attractive alternative, especially for large capacities, is to accommodate the individual catalyst beds in separate vessels and use separate heat exchangers. This approach is especially suitable when using the reaction heat for raising high-pressure steam. The indirect cooling principle is applied today in most large new ammonia plants, and also in revamps an increasing number of quench converters are modified to the indirect cooling mode. Figure 50 shows a schematic of the principle together with temperature/location and temperature/ammonia concentration profile.

Converters with indirect cooling have been known since the early days of ammonia production, for example, the Fauser – Montecatini reactor [765, 766, 770, 774, 816 – 818]. In this converter, tube coils between catalyst beds transfer the reaction heat to a closed hot water cycle under pressure, operating by natural draft. The hot water releases the absorbed heat in an external steam boiler generating about 0.8 t of steam per tonne of ammonia at about 45 bar (ca. 250 °C).

In the well-known Uhde – Chemie Linz converter with three catalyst beds — described in various versions [763, 765, 766, 792, 819 – 822] — the indirect cooling is provided by converter feed gas. Feed gas enters at the top, passes down the annulus between basket and shell to cool the pressure wall, flows through the shell side of the lower feed – effluent heat exchanger, and then via the center pipe and interbed exchangers to the top of the first catalyst bed. The gas passes downwards through the catalyst beds and the tube side of the interbed exchangers and the lower heat exchanger to leave the reactor vessel. For trimming purposes a quench is foreseen.

Further development of the radial flow concept used in the quench converter Topsøe Series 100 has led to the successful launch of the Topsøe Series 200 converter [566, 798, 823 – 829], designed for indirect cooling. Two versions are shown in Figure 51, with and without a lower internal heat exchanger. A "cold shot" ahead of the first catalyst bed is installed for temperature adjustment.

In the converter without a lower exchanger the feed gas enters at the bottom and flows as pressure wall cooling gas to the top of the converter. After passing the centrally installed

Ammonia

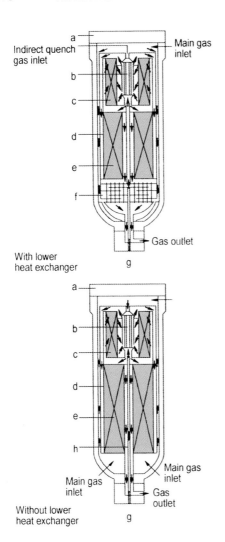

Figure 51. Topsøe Series 200 converter
a) Pressure shell; b) Interbed heat exchanger; c) 1st Catalyst bed; d) Annulus around catalyst bed; e) 2nd Catalyst bed; f) Lower heat exchanger; g) Cold bypass; h) Cold bypass pipe

interbed exchanger on the tube side, the gas is mixed with cold gas for temperature adjustment and passes through the first catalyst bed radially from the outside to the inside. The exit gas flows through the shell side of the interbed exchanger before it enters the second bed, which is crossed in the same direction as the first one. The Topsøe Series 300 converter contains three catalyst beds and two central interbed exchangers in a single pressure shell.

Casale [830 – 835] has also successfully commercialized converters based on the axial – radial flow concept with indirect cooling.

Kellogg has re-engineered its horizontal cross-flow quench converter for indirect cooling [765, 812, 836 – 838]. (Fig. 52) As in the quench version the pressure shell has a full-bore closure to remove the catalyst cartridge for loading and unloading. The reactor contains two catalyst beds, with the second one split into two parallel sections. Reactor feed gas passing between cartridge and shell is used to keep the pressure wall cool, and an inlet – outlet heat exchanger is located between first and second bed. A cold shot is installed for adjusting the inlet temperature of the first catalyst bed. A horizontal design with indirect cooling is also proposed in [839].

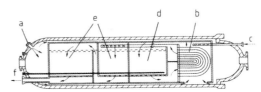

Figure 52. Kellogg horizontal intercooled ammonia converter
a) Inlet; b) Interbed heat exchanger; c) Bypass; d) Bed 1; e) Bed 2; f) Outlet

In the C. F Braun ammonia synthesis concept [840 – 845], separate vessels were used for the two catalyst beds in the original version. The feed – effluent heat exchanger was located between the first and second reactor vessels, and the second catalyst vessel was followed by a waste-heat boiler for high pressure steam. This basic concept was already introduced in the heyday of the quench converters and prior to the first energy crisis. In a later version three catalyst vessels [845 – 849] with boilers after the second and the third were used. The gas flow is axial and the mechanical design rather simple. Contrary to most converter designs, in which the pressure shell is kept below 250 °C by means of insulation or by flushing with cooler gas, the pressure vessel wall of the C. F. Braun reactors is at 400 °C, which is possible with modern steels in compliance with the Nelson diagram (see Sec-

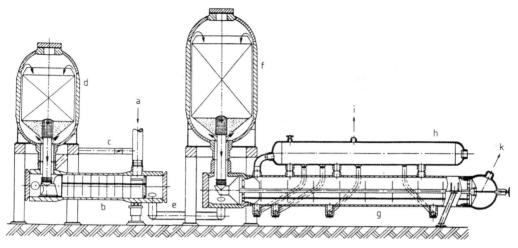

Figure 53. C. F. Braun converters with interbed heat exchanger and waste-heat boiler
a) Gas inlet; b) Feed – first bed effluent exchanger; c) Bypass for temperature control; d) First bed; e) Line to second bed; f) Second bed; g) waste-heat boiler (Borsig); h) Steam drum; i) Steam outlet; k) Gas outlet

tion 4.8). Figure 53 shows the two-bed version of the C. F. Braun converter.

To withstand the high outlet temperature level (450 – 500 °C) needed for the high-pressure boilers, a special design is employed to keep the outlet nozzle cool. C. F. Braun has devised a good solution by direct coupling the boilers and the exchanger to the converter.

Uhde has used three catalyst beds for capacities up to 2300 t/d (Fig. 54). The first two are accommodated in a single pressure vessel together with an inlet – outlet exchanger. Then a waste-heat boiler generating high-pressure steam cools the gas before it enters the second vessel containing the third bed, which discharges through a second external high-pressure boiler. The gas flow is in radial mode in all three catalyst beds.

A new development is the Uhde dual pressure process [850 – 855], suited for very large capacities. The syngas with 110 bar from the low-pressure casing of the compressor is fed to a three-bed, intercooled, once-through converter which can produce a third of the total ammonia yield. From the effluent about 85 % of the ammonia produced is removed from the gas before it is compressed in the high presure casing to 210 bar to enter the well-known Uhde two-vessel-three-bed combination as described above. Figure 55 shows the Uhde dual pressure concept. The first plant based on this concept with a capacity of 3300 t/d will come on stream 2006 in Al-Jubail, Saudi Arabia (Safco IV). This is so far worldwide the largest capacity in a single synthesis (see also 4.6.4).

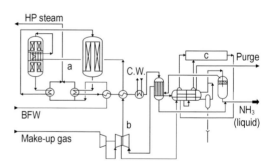

Figure 54. Uhdes standard synloop configuration
a) Ammonia converter; b) Syngas compressor; c) Refrigeration

Topsøe has now introduced a hot-wall converter with only one catalyst bed and no internal heat-exchange equipment, similar to C.F. Braun, but with radial flow. Three of these converters can be combined with an external heat exchanger and two high-pressure boilers to give an arrangement as described for Braun. This Topsøe Series 50 converter, can also be combined with the Series 200 reactor and two external high-pressure boilers to attain the configuration as described for Uhde (Topsøe Series 250).

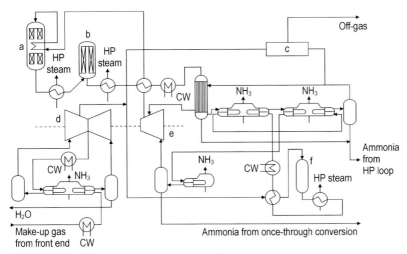

Figure 55. Uhde dual pressure ammonia synthesis concept
a) First ammonia converter; b) Second ammnia converter; c) Purge gas recovery unit; d) LP casing; e) HP casing; f) Once-through ammonia converter

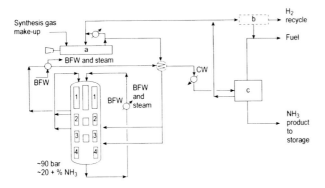

Figure 56. Kellogg advanced ammonia process (KAAP)
a) Compressor; b) Purge gas recovery unit; c) Refrigeration

Kellogg has developed for its ruthenium catalyst based KAAP ammonia process [100, 465] a special converter design. Four radial flow beds are accommodated in a single pressure shell with intermediate heat exchangers after the first, second and third bed. The first bed is loaded with conventional iron catalyst, the following ones with the new ruthenium catalyst. Figure 56 is a simplified sketch of the synthesis loop of the KAAP Figure 57 shows the internals of the KAAP converter [856].

For revamps Kellogg has also proposed a two-bed version completely loaded with ruthenium catalyst to be placed downstream of a conventional converter [466].

Optimizing the Temperature Profile and Bed Dimensions in Multibed Converters. With closer approach to equilibrium, the volume of catalyst required in the individual beds increases, requiring greater total catalyst volume and appropriate changes in the catalyst bed dimensions. The literature has frequently treated the problem of optimizing both the distribution of catalyst volume between a given number of beds and the bed inlet temperatures [737, 739, 764, 857 – 860]. Reference [861] examines optimizing the temperature profile in a given four-bed quench converter. For additional literature on this important issue in converter layout and design, see also page 207.

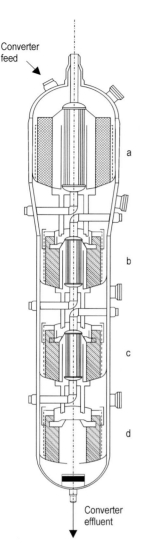

Figure 57. KAAP four bed ammonia converter
a) Bed no. 1 magnetite catalyst; b) Bed no. 2 KAAP catalyst; c) Bed no. 3 KAAP catalyst; d) Bed no. 4 KAAP catalyst

4.5.3.3. Waste-Heat Utilization and Cooling

The reaction heat of ammonia synthesis is 46.22 kJ/mol at STP, which corresponds to 2.72 GJ per tonne NH_3. Utilization of this heat at the highest possible temperature for generating high-pressure steam is a major contribution to the high total energy efficiency of modern ammonia plants [862]. Early converters, operating at about 300 bar, equipped with a lower heat exchanger for raising the inlet temperature for the (first) catalyst bed to the ignition temperature (ca. 400 °C), received the converter feed at about ambient temperature and therefore had outlet temperatures of ca. 250 °C (ca. 15.5 °C per mol % of ammonia formed).

Initially there was practically no heat recovery, and nearly the total heat content of the gas down to ambient temperature and thus the reaction heat was transferred to cooling water. Subsequently plants were modified to use this heat to some extent, but the low temperature level allowed only boiler feed water heating and generation of low-pressure steam (ca. 20 bar). In some instances, water circulation was installed to use this heat in other plants or production steps. As for any type of converter the outlet temperature rises with increasing inlet temperature (ΔT is determined by the degree of conversion), in further developments an additional heat exchanger for converter feed versus the converted gas was installed, downstream of the above-mentioned heat recovery. In this way the temperature level at which heat could be recovered was increased, ultimately to the point where the inlet temperature to waste-heat recovery is equal to the outlet temperature of the last catalyst bed. In practice this corresponds to moving the lower heat exchanger (which in multibed converters exchanges feed to the first catalyst bed against effluent from the last bed) partially or completely to a position outside of the converter and downstream of the waste-heat recovery installation. In this way the waste-heat downstream of the synthesis converter in modern plants is available in the temperature range around 480 to 290 °C. The steam pressure, formerly 100 bar, has now been raised to 125 bar, which means that the gas can be cooled in the boiler to ca. 350 °C; the remaining recoverable heat is used for boiler feed water heating.

The trend followed in newer plants is to increase conversion per pass with the result of higher ammonia outlet concentrations and lower outlet temperatures from the last bed. However, as optimum energy efficiency of the whole ammonia plant requires maximum high-pressure steam generation, part of the heat must be recovered before the reaction is completed in the reactor system. This can be accomplished [825, 826, 862, 863] by using three catalyst beds in separate pressure vessels with boilers after the second and the third vessel and an inlet – outlet heat exchanger for the first catalyst bed.

Advanced ammonia concepts produce as much as 1.5 t of high-pressure steam per tonne of ammonia, which correspond roughly to 90 % of the standard reaction enthalpy. Figure 58 is a temperature – enthalpy diagram for a converter system corresponding to the original C. F. Braun arrangement. High-pressure steam (113 bar, 320 °C) is generated after the second catalyst bed.

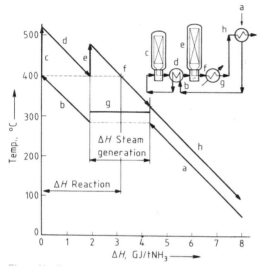

Figure 58. Temperature versus enthalpy diagram for a two-bed system with steam generation
a) Heating in main feed effluent exchanger; b) Further heating in feed – first bed effluent exchanger; c) Temperature rise in first bed; d) Cooling in feed – first bed effluent exchanger; e) Temperature rise in second bed; f) Cooling by steam generation; g) Temperature level of waste-heat boiler; h) Cooling in main feed effluent exchanger

Appropriate designs of waste-heat boilers are described in 4.6.3.

4.5.3.4. Ammonia Recovery from the Ammonia Synthesis Loop

In all commercial plants ammonia is recovered from the synthesis loop by cooling the synthesis gas to condense the ammonia under synthesis pressure. The liquid ammonia product is separated from the gas, which is recycled. Arrangement and location of the ammonia separator(s), recirculation compression, addition of make-up gas and extraction of purge gas are discussed in Section 4.5.3.1; see also Figure 34).

In older high-pressure synthesis loops (> 450 bar) cooling by water and or air is sufficient to obtain the required low residual ammonia concentration in the gas. In modern plants, which operate at moderate pressures, the cooling has to be supplemented by refrigeration, for which a mechanical ammonia refrigeration cycle, with one or more temperature levels is generally used. Refrigeration down to − 25 °C is used, which with inclusion of the necessary temperature difference in the chiller requires ammonia evaporation at about atmospheric pressure. The amount of ammonia vaporized (and consequently reliquefied by compression and water or air cooling) can be substantial. At a loop pressure in the range 100 to 150 bar the quantity of reliquefied ammonia may be twice the ammonia product flow.

The liquid ammonia of the high-pressure separator is flashed to about 20 bar, whereby the majority of the dissolved gases are released in the let-down vessel. This gas is normally used as a fuel, preferably after removal of ammonia vapor to avoid NO_x formation in the combustion furnace.

The ammonia from the let-down vessel may be sent directly to downstream users or flashed further to atmospheric pressure for storage in a cold tank. All ammonia vapors removed from flash gases and from purge gas by water scrubbing can be recovered as pure liquid product by distillation if there is no direct use for the aqueous ammonia.

Absorption refrigeration with aqueous ammonia instead of a mechanical refrigeration system [864 – 869] did not find widespread application.

Although ammonia condensation was already used in HABER's lab-scale ammonia plant and in early pilot plants of BOSCH, the first commercial units of BASF used ammonia absorption in water to remove the product from the cool synthesis loop gas, because various technical problems were encountered with refrigeration at that time. It was only in 1926 that ammonia condensation was introduced in the Haber – Bosch plants. In the early 1920s LUIGI CASALE successfully used condensation for his first plant. Water cooling was sufficient on account of the very high synthesis pressure.

Recovery of ammonia by water scrubbing offers the advantage of achieving a very low residual ammonia content, but the drawback is that the whole recycle gas has to be dried afterwards and in addition distillation of aqueous ammonia is necessary to yield liquid ammonia. Nevertheless the scrubbing route was again proposed for a synthesis loop to be operated under extremely low pressure (around 40 bar) [870]. Snam Progetti [293, 871, 872] has proposed removing the ammonia from the loop gas at ambient temperature down to 0.5 mol % by absorption in dilute aqueous ammonia.

The extent to which the ammonia concentration in the gas can deviate from that expected for ideal behavior can be seen from Tables 4 and 5. For example, at 30 bar the ammonia vapor pressure according to Table 4 is 1.167 bar, corresponding to 5.84 mol % at the total pressure of 20 bar. In contrast, Table 5 gives the ammonia concentration in a 1/3 nitrogen/hydrogen mixture at 20 bar total pressure as 9.8 mol %.

4.5.3.5. Inert-Gas and Purge-Gas Management

Apart from nitrogen and hydrogen, the fresh make-up gas supplied to the synthesis loop usually contains small quantities of inert gases. These include methane (from gas generation), argon (from the process air), and helium (from the natural gas). Because they are inert, they tend to concentrate in the synthesis loop and must be removed to maintain the loop material balance. A portion of the inert gases dissolves in the liquid produced in the ammonia separator. Table 6 gives a rough approximation of the extent to which these inerts are dissolved. Figure 2 gives vapor – liquid equilibrium ratios for use in making precise calculations of the dissolved quantities of inerts. If the synthesis gas pressure is high, for example, 300 bar), and the inert gas concentration in the synthesis loop make-up gas low enough, for example, under 0.2 vol % [873], then dissolution in the product ammonia suffices to remove the inerts from the synthesis loop.

With a higher inert gas content in the make-up gas this method is not applicable, because the required partial pressure of the inert gas at equilibrium in the loop gas would become so high that a synthesis under moderate pressure would be virtually impossible. So, in addition to removal as dissolved gases (flash gas), inerts have also to be removed from the gas phase by withdrawing a small purge-gas stream from the loop. At the same time expensive synthesis gas is also lost from the loop, which lowers the ammonia yield. Therefore, determining the appropriate inert gas concentration requires a precise economic calculation.

A high inert gas level has various drawbacks. It decreases the specific converter performance by reducing the hydrogen and nitrogen partial pressures. The gas recycle flow is increased by the amount of inert gas. Piping and equipment must correspondingly be increased in size, and the associated power consumption for recycle gas increases. Moreover, there is an unfavorable effect on condensation of the ammonia product. Because of the dilution, less ammonia can be condensed from the recycle synthesis gas by less expensive air or water cooling or higher temperature level refrigeration.

There are several possibilities for reducing the losses associated with the purge gas. The most capital-intensive method consists of feeding the purge gas to a second synthesis loop operating at a slightly lower pressure [874, 875]. As this loop can operate at a very high inert level (40 % or more), only a very small final purge stream is necessary and no recompression is needed. Up to 75 % of the hydrogen from the first-loop purge stream can be recovered. This system has the advantage that nitrogen is also recovered, but it is too expensive for use in modern plants and revamps. For this reason other methods have been developed.

Hydrogen Recovery by Cryogenic Units [876]. Ammonia is first removed from the purge gas by cooling or in a water wash operating at 7.5 MPa (75 bar). Molecular sieve adsorbers then eliminate moisture and traces of ammonia (Fig. 59). The dry, ammonia-free purge gas from the adsorbers next enters the cold box. Heat exchange with cold product hydrogen fraction and with gas rejected to fuel cools the purge gas to a temperature of about $-188\,°C$ (85 K). Partial condensation liquefies methane and argon as well as some of the nitrogen and helium. These are removed in a separator, leaving a hydrogen-rich gas.

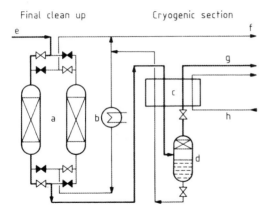

Figure 59. Simplified flow diagram of cryogenic hydrogen recovery unit
a) Molecular sieve adsorbers; b) Heater; c) Plate-fin exchanger; d) Separator; e) Ammonia-free purge gas; f) Fuel gas to reformer burners; g) Hydrogen product to syngas compressor; h) NH$_3$ refrigerant

The liquid flows through a control valve, reducing its pressure, and into a brazed aluminum (plate-fin or core-type) heat exchanger. The hydrogen-rich gas also flows into the same exchanger (in separate passages) where the vaporizing liquid and the hydrogen are warmed by cooling the entering purge gas. Liquid ammonia from the ammonia plant may be used to provide additional refrigeration, especially at plant startup.

The warm hydrogen-rich gas flows back to the suction side of the second stage of the synthesis gas compressor (6.5 – 7 MPa). About 90 – 95 % of the hydrogen and 30 % of the nitrogen in the purge gas can be recovered.

The remaining gas, with a high concentration of inerts, serves as fuel for the primary reformer. After heating in a preheater, a portion serves to regenerate the molecular sieves and then likewise flows to reformer fuel.

Cryogenic hydrogen recovery units are supplied by firms such as Costain Engineering (formerly Petrocarbon Development) [877 – 887], Linde [888], and L'Air Liquide, among others. Reference [889] reports on the changes in operating conditions of an ammonia plant resulting from the operation of a cryogenic hydrogen recovery unit.

Hydrogen Recovery by Membrane Separation. The Monsanto Prism membrane separator system uses selective gas permeation through membranes to separate gases. This principle has been applied to separating hydrogen from other gases [890 – 893] (→ Membranes and Membrane Separation Processes). The membranes are hollow fibers with diameters of about 0.5 mm. The fiber is a composite membrane consisting of an asymmetric polymer substrate and a polymer coating. The design of a single separator module (length, 3 – 6 m; diameter, 0.1 – 0.2 m) resembles a shell and tube heat exchanger. A bundle with many thousands of hollow fibers is sealed at one end and embedded in a tubesheet at the other. The entire bundle is encased in a vertical shell (Fig. 60).

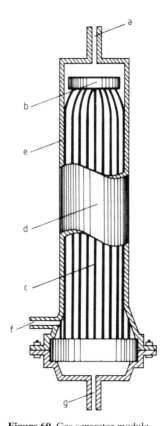

Figure 60. Gas separator module
a) Nonpermeate gas outlet; b) Fiber bundle plug; c) Hollow fiber; d) Separator (length, diameter, and number of separators determined by ammonia process); e) Coated carbon steel shell; f) Feed stream of mixed gases; g) Permeate gas outlet

The purge gas is water scrubbed at 135 – 145 bar, reducing the ammonia concentration to less than 200 ppm. The scrubbed purge gas is heated to 35 °C and sent directly to the Prism separators. Trace concentrations of ammonia and water vapor in the gas stream pose no problem to the membrane. Therefore, a dryer system is not required.

The gas stream enters the separator on the shell side, i.e., the outside of the hollow fibers. Hydrogen permeates through the wall of the fibers. Hydrogen-rich permeate gas flows down the bore of the fiber and through the tubesheet and is delivered at the bottom of the separator. The remaining (nonpermeating) gases, nitrogen, methane, and argon, are concentrated on the shell side, recovered through the top and pass to the next separator module. Several separators operate in series. The rate of permeation decreases across a bank of separators as the hydrogen partial pressure differential across the membrane approaches zero. Therefore, a second bank of separators with lower pressure on the tube side is used to increase the hydrogen recovery. Of the recovered hydrogen, 40 – 70 % leaves the first bank of separators at 7 MPa (70 bar) and is returned to the second-stage suction of the syngas compressor. The second bank permeate hydrogen is recovered at 2.5 – 2.8 MPa (25 – 28 bar) and returned to the first-stage suction of the syngas compressor. Overall hydrogen recovery is 90 – 95 %. The remaining nonpermeate gas stream normally flows to primary reformer fuel.

The main advantages of the Prism separator system are simplicity, ease of operation, and low maintenance. Reference [894] compares membrane and cryogenic separation units for a large ammonia plant.

For further literature on gas separation by membranes, see [895 – 901].

Membrane technology is also offered by other licensors; an example is the Polysep Membrane System of UOP [902]. In addition to the systems based on hollow fibers, membrane modules have been developed in which the membrane is in the form of a sheet wrapped around a perforated center tube using spacers to separate the layers. The raw gas flows in axial direction in the high pressure spacer and the permeate is withdrawn in the low pressure spacer. Linde, for example, offers such a module under the name Serarex [903, 904].

Hydrogen Recovery by Pressure Swing Adsorption. Pressure swing adsorption on zeolite molecular sieves (PSA) (see Section 4.5.1.3) may be also used for hydrogen recovery from purge gas [685]. The process, originally developed by Union Carbide under the name HYSIV, is now marketed as Polybed PSA by UOP [902, 905 – 907]. PSA technology is also offered by Linde and other companies. If required, the process also offers the possibility to supply pure hydrogen from the purge gas for other uses. PSA units usually operate at adsorption pressures of 20 – 30 bar and achieve recovery rates higher than 82 % for hydrogen and 20 % for nitrogen. Carbon-based adsorbents for pressure swing adsorption have also been investigated [905, 906] and a process developed by Bergbau-Forschung [907] is offered by Costain.

Hydrogen Recovery Using Mixed Metal Hydrides. A proprietary hydrogen separation process utilizing the reversible and selective adsorption capability of mixed metal hydrides has been proposed. The hydride, such as $LaNi_5$, FeTi, or Mg_2Cu, is in the form of ballasted pellets. The ballast material serves as a heat sink to store the heat of adsorption. Subsequently, this is used to supply the heat of desorption. The ballast also is the binder for the pellets, preventing attrition. Each type of metal hydride is susceptible to certain contaminants. Therefore, selection of the metal hydride must be based on the analysis of the gas to be treated. No ammonia removal step is required upstream of the unit. The system yields 99 mol % hydrogen product at a recovery efficiency of 90 – 93 % [904, 908, 909]. No commercial installations are known to be in operation in ammonia production processes.

Argon Recovery from Ammonia Purge Gas. The waste gas from hydrogen recovery plants is more highly enriched in argon than the purge gas. If potential markets for argon exist, then it may be possible to supplement the hydrogen recovery plant with one for recovering argon. Cryogenic argon recovery from ammonia purge gas is discussed in [910 – 912]. Typical argon recoveries are in excess of 90 %, with a purity of 99.999 %.

4.5.3.6. Influence of Pressure and Other Variables of the Synthesis Loop

The influences of individual parameters can be summarized as given in [913, p. 231]:

- Pressure: increasing pressure will increase conversion due to higher reaction rate and more favorable ammonia equilibrium.
- Inlet temperature: there are two opposed effects as increasing temperature enhances reaction rate but decreases the adiabatic equilibrium concentration.
- Space velocity: increasing the space velocity normally lowers the outlet ammonia concentration, but increases total ammonia production.
- Inert level: increasing the inert level lowers the reaction rate for kinetic and thermodynamic reasons (Section 4.3.3).
- Hydrogen/nitrogen ratio: a true maximum reaction rate for a certain H/N ratio; at lower temperatures the maximum lies at lower H/N ratios (Section 4.3.3). Position of the maximum also depends on the space velocity values (Section 4.5.3.2).
- Recycle rate: at constant pressure and production rate, the consequence of higher recycle rate is a lower ammonia concentration. In this case the difference to the equilibrium concentration and thus the reaction rate increases with the result that less catalyst is required. However, the temperature level for waste-heat recovery decreases and with lower temperature differences, larger heat exchanger surface areas become necessary and the cross sections of piping and equipment have to be enlarged on account of the higher gas flow.
- Separator temperature: together with pressure and location of make-up gas addition, the temperature of the ammonia separator determines the ammonia concentration at the converter inlet. A lower temperature means lower ammonia concentration, which translates into either a lower catalyst volume or a higher conversion.
- Catalyst particle size (Section 4.4.1.2): smaller catalyst particles give higher conversion because of lower diffusion restrictions (higher pore efficiency).

The question of the best synthesis pressure is rather a difficult one and the answer is extremely dependent on optimization parameters such as feedstock price, required return on investment, and site requirements. In principle it is possible to calculate the minimum amount of mechanical work needed in the synthesis loop. If plots of kilowatt hours versus synthesis pressure for make-up gas compression, recycle, and refrigeration are superimposed the result will be a minimum, for which a value of 155 bar is reported in [914, 915]. The result is strongly dependent on assumed boundary conditions, and other studies came to higher values in the range 180 – 220 bar. However, such diagrams should be interpreted with care. First, this type of plot is strongly influenced by the catalyst activity, and thus by any factors that affect it, such as grain size and minimum possible temperature for the first bed, and especially the equilibrium temperature of the last bed [916]. Second, this approach considers only the mechanical energy and ignores completely the recovered energy of reaction, its energy level, and its impact on the energy balance of the complete ammonia plant. Third, the temperature attainable at a given site by air- or water cooling will affect the refrigeration duty. And fourth, the type of front end can profoundly alter the result. The front-end pressure determines the suction pressure of the synthesis gas machine. In a plant based on partial oxidation with an operating pressure of 80 bar, less than half as much energy is needed to compress the make-up gas to 180 bar, for example, as in a steam reforming plant with a suction pressure of only 25 bar.

Thus the problem of choosing the best synthesis pressure is complex because not only does the entire energy balance have to be examined, but also the mechanical design and the associated investment costs. For an actual project, the costs for energy (i.e., the feedstock) have to be weighed against investment. With the exception of some older processes which used extremely high pressures (Casale, Claude) a pressure of around 300 bar was common for the old multi-stream plants operated with reciprocating compressors. The centrifugal compressor is one of the most important features of single-train ammonia plants. The first of these plants, with a capacity of 600 t/d, were built in the mid-1960s and the maximum attainable synthesis pressure was restricted to ca. 150 bar by technical limitations of the centrifugal compressor, which needed a minimum gas flow (see also Section 4.5.2). Of course, at current plant sizes of 1200 – 2000 t/d this constraint is no longer of importance. These plants usually operate in the pressure range of 170 – 190 bar and it is a confirmation of the argumentation presented above that the different pressures are not reflected in the overall energy consumption of the complete ammonia plant. Analysis of the effects of various parameters on the energy consumption of the synthesis loop are also reported in [825, 826]. At constant equilibrium temperature for the effluent from the last bed (constant ammonia concentration) energy consumption of the loop was found to be almost independent of pressure in the range of 80 – 220 bar [916].

4.5.3.7. Example of an Industrial Synthesis Loop

Figure 61 is an example of modern ammonia synthesis loop (Krupp – Uhde) with two converter vessels and three indirectly cooled catalyst beds producing 1500 t/d NH_3 at 188 bar.

The gas enters the converter (a) at 295 °C and is subsequently heated in the internal heat exchanger to 390 °C before it enters the first catalyst layer. The outlet gas from the first layer then passes through the aforementioned heat exchanger and enters the second bed, after which the gas leaves the converter with 469 °C and passes a waste-heat boiler generating 125 bar steam. The inlet gas of the second vessel, which accommodates the third catalyst bed, has a temperature of 401 °C and the outlet enters a further waste-heat boiler generating 125 bar steam.

4.6. Complete Ammonia Production Plants

The previous sections mainly considered the individual process steps involved in the production of ammonia and the progress made in recent years. The way in which these process components are combined with respect to mass and energy flow has a major influence on efficiency and reliability. Apart from the feedstock, many of the differences between various commercial ammonia processes lie in the way in which the process elements are integrated. Formerly the term ammonia technology referred mostly to "ammonia synthesis technology" (catalyst, converters, and synthesis loop), whereas today it is interpreted as the complete series of industrial operations leading from the primary feedstock to the final product ammonia.

The major determinant for process configuration is the type of feedstock, which largely governs the mode of gas generation and purification. The other important factor is the plant capacity, which, together with consumption and costs of feedstock and energy, is decisive for the production economics. An important development was the concept of single-train plants, first introduced for steam reforming based production by M.W. Kellogg in 1963 with a capacity of 600 t/d. Before then maximum capacities had mostly been about 400 t/d, with several parallel trains in the synthesis gas preparation stage and the synthesis loop. Today world-scale plants have capacities of 1200 – 2000 t/d. The lowest capital cost and energy consumption result when steam reforming of natural gas is used. In addition site requirements can influence the layout considerably. In contrast to a stand-alone plant, ammonia production at a developed industrial site may import and/or export steam and power, which affects the total energy consumption.

With the exception of the Koppers – Totzek coal gasification process, which operates at near atmospheric pressure, all modern gasification processes operate at elevated pressure. Steam reforming of light hydrocarbons at 30 – 40 bar and partial oxidation of heavy hydrocarbons at 40 – 90 bar are generally used.

4.6.1. Steam Reforming Ammonia Plants

4.6.1.1. The Basic Concept of Single-Train Plants

The innovative single-train concept, introduced in 1963 by Kellogg, was a technical and an economical breakthrough and triggered a tremendous increase in world ammonia capacity. No parallel lines, even for high capacity, and a highly efficient use of energy, with process steps in surplus supplying those in deficit, were the main features. Figure 62 shows a block flow diagram and the gas temperature profile for a steam reforming ammonia plant [917].

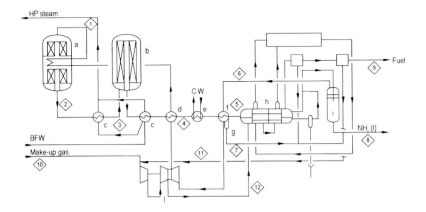

Figure 61. Example of a synthesis loop for 1500 t/d NH$_3$ (Krupp – Uhde)
a) Converter with two radial beds and internal heat exchange; b) Converter with one radial bed; c) Waste-heat recovery; d) Heat exchanger; e) Water cooling; f) Heat exchanger; g) No. 1 separator; h) Refrigerated cooling; i) No. 2 separator; j) NH$_3$ recovery; k) H$_2$ recovery

No.	Flow rate, kmol/h	T, °C	p, MPa	Gas composition, mol %					Production of NH$_3$ liquid, t/h
				N$_2$	H$_2$	CH$_4$	Ar	NH$_3$	
1	28344	295	18.80	23.53	61.20	8.20	2.89	4.18	
2	25368	469	18.48	20.43	50.79	9.15	3.23	16.40	
3	25368	401	18.38	20.43	50.79	9.15	3.23	16.40	
4	24672	336	18.26	19.60	47.99	9.43	3.31	19.67	
5	21613	21	18.02	22.33	54.71	10.67	3.78	8.51	
6	28344	0	17.88	23.54	61.20	8.20	2.88	4.18	
7	3060	21	18.02	0.30	0.55	0.50	0.09	98.55	51.36
8	3678	20	2.50	0.02	0.03	0.12	0.01	99.82	62.51
9	320	38	0.25	46.55	21.72	24.34	7.21	0.18	0.01
10	7667	35	3.23	25.89	72.73	1.07	0.31	0.00	
11	374	38	3.25	5.22	90.20	2.86	1.72	0.00	
12	8041	35	18.20	24.93	73.54	1.16	0.37	0.00	

Table 25. Main energy sources and sinks in the steam reforming ammonia process

Process section	Origin	Contribution
Reforming	primary reforming duty	demand
	flue gas	surplus
	process gas	surplus
Shift conversion	heat of reaction	surplus
CO$_2$ removal	heat of solvent regeneration	demand
Methanation	heat of reaction	surplus
Synthesis	heat of reaction	surplus
Machinery	drivers	demand
Unavoidable loss	stack and general	demand
Balance	auxiliary boiler or import	deficit
	export	surplus

High level surplus energy is available from the flue gas of the reformer and the process gas streams of various sections, while heat is needed, for example, for the process steam for the reforming reaction and in the solvent regenerator of the carbon dioxide removal system. Because considerable mechanical energy is needed to drive compressors, pumps, and blowers, it was most appropriate to use steam turbine drives, since sufficient steam could be generated from the waste-heat. As the level was high enough to raise high-pressure steam (100 bar) it was possible to use the process steam first to generate mechanical energy in the synthesis gas compressor turbine before extracting it at the pressure level of the primary reformer. Table 25 lists all significant energy sources and sinks within the process.

The earlier plants operated at deficit, and needed an auxiliary boiler, which was integrated in the flue gas duct. This situation was partially caused by inadequate waste-heat recovery and low efficiency in some energy consumers. Typically, the furnace flue gas was discharged in the stack at rather high temperature because there was no air preheating and too much of the reaction heat in the synthesis loop was rejected to the cooling media (water or air). In addition, efficiency of the mechanical drivers was low and the heat demand for regenerating the solvent from the CO$_2$ removal unit (at that time aque-

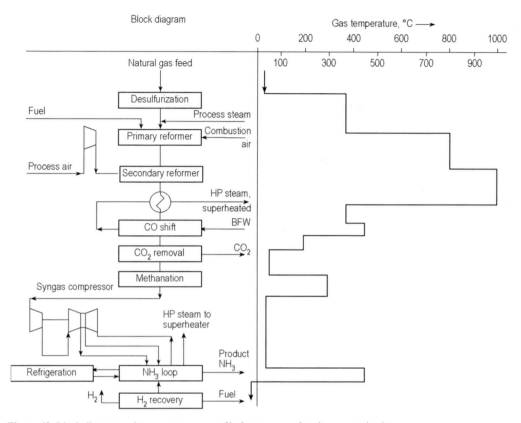

Figure 62. Block diagram and gas temperature profile for a steam reforming ammonia plant

ous MEA) was high. Maximum use was made of direct steam turbine drive, not only for the major machines such as synthesis gas, process air, and refrigeration, but even for relatively small pumps and blowers. The outcome was a rather complex steam system. Even after substitution of the smaller turbines by electric motors, the steam system in the modern plant is still a complex system as shown in Figure 64. Ammonia plant steam systems are described in [100, 102, 418, 918 – 921].

The first generation of the single-train steam reforming plants is discussed in [479, 922 – 927], and the required catalysts are reviewed in [573, 928, 929]. A survey of the development of the steam reforming concept through 1972 can be found in [930]. Other references which cover the development of the steam reforming before the introduction of the single-train concept (1940 to 1960) can be found in [418 p. 276].

The new plant concept had a triumphant success story. By 1969, 30 new Kellogg large single-train plants with capacities of 1000 t/d or more were in operation, and other contractors were offering similar concepts.

The decrease in energy consumption compared to the older technology was dramatic, and with the low gas prices at that time it is understandable that greater emphasis was placed on low investment cost, although there was a considerable potential for further reducing the energy consumption.

With the advent of the single-train steam reforming plants, it became standard for licensors and engineering contractors to express the total net energy consumption per tonne of ammonia in terms of the lower heating value of the feedstock used. The total net energy consumption is the difference between all energy imports (mainly feedstock) and all energy exports (mostly steam and/or electric power) expressed as lower heating value of the consumed feedstock, whereby electric power is converted with an efficiency of 25 – 30 % and steam is accounted for with its caloric value.

4.6.1.2. Further Developments

The significant changes in energy prices from 1973 onwards were a strong challenge to process licensors, engineering contractors and plant owners to obtain better energy efficiency. The overall energy consumption was reduced from around 45 GJ per tonne NH_3 for the first large single-train units to less than 29 GJ per tonne NH_3 (Table 26).

Energy saving modifications are described in [293, 872, 931 – 946]. For catalyst improvements see [947]. Some of the most important improvements compared to the first generation of plants are discussed below.

Feedstock Purification. In feedstock purification, mainly desulfurization, adsorption on active carbon was replaced by catalytic hydrogenation over cobalt – molybdenum or nickel – molybdenum catalyst, followed by absorption of the H_2S on ZnO pellets with formation of ZnS. By itself this measure has no direct influence on the energy consumption but is a prerequisite for other energy saving measures, especially in reforming and shift conversion.

Reforming Section. In the reforming section energy savings were achieved by several, often interrelated, measures, of which the most important are the following:

– Reduction of the flue gas stack temperature to reduce heat losses to the atmosphere [566]
– Avoiding excessive heat loss by better insulation of the reformer furnace
– Introduction of combustion air preheating [943]
– Preheating the reformer fuel
– Increased preheat temperatures for feed, process steam and process air
– Increased operating pressure (made possible by using improved alloys for the reformer tubes and improved catalysts)
– Lowering of the steam to carbon ratio [948]
– Shifting some reformer duty from primary to secondary reformer with the use of excess air [451, 452] or oxygen-enriched air [949] in the secondary reformer, including the possibility of partially bypassing the primary reformer [950, 951]
– Installing a prereformer or rich-gas step is another possibility to reduce primary reformer duty and stack temperature of the flue gas [443 – 450], especially in LPG- and naphtha-based plants

A more recent development that breaks away from the usual plant configuration is to replace the traditional fired primary reformer with an exchanger reformer which uses the heat of the effluent of the secondary reformer [458 – 465, 951, 952]. Also other applications have been reported in which flue gas [952] heat from the fired reformer is used to perform a part of the reforming.

Shift Conversion. Improved LT shift catalysts can operate at lower temperatures to achieve a very low residual CO content and low byproduct formation. A new generation of HT shift catalysts largely avoids hydrocarbon formation by Fischer – Tropsch reaction at low vapor partial pressure, thus allowing lower steam to carbon ratio in the reforming section (see Section 4.5.1.2).

Carbon Dioxide Removal Section. In the carbon dioxide removal section the first generation of single-train plants often used MEA with a rather high demand of low-grade heat for solvent regeneration. With additives such as UCAR Amine Guard [583, 587], solvent circulation could be reduced, saving heat and mechanical energy. Much greater reduction of energy consumption was achieved with new solvents and processes, for example BASF aMDEA or Benfield LoHeat. Other hot potash systems (Giammarco Vetrocoke, Catacarb) and physical solvents (Selexol) were introduced (Section 4.5.1.3).

Final make-up gas purification was improved by removing the water and carbon dioxide traces to a very low level by using molecular sieves. Some concepts included cryogenic processes with the benefit of additional removal of methane and argon.

Table 26. Development of the net energy consumption of natural gas based steam reforming ammonia plants (real plant data) in GJ per tonne NH_3

Year	1966	1973	1977	1980	1991
Plant	A	B	C	D	E
Feed	23.90	23.32	23.48	23.36	22.65
Fuel, reformer	13.00	9.21	7.35	5.56	5.90
Fuel, auxiliary boiler	2.60	5.02	3.06	1.17	
Export					0.55
Total	39.50	37.55	33.89	30.18	28.00

Ammonia Synthesis Section. In the ammonia synthesis section conversion was increased by improved converter designs (see Section 4.5.3.2), larger catalyst volumes, and to some extent with improved catalysts. The main advances in converter design were the use of indirect cooling instead of quenching, which allowed the recovery of reaction heat high pressure steam. Radial or cross-flow pattern for the synthesis gas instead of axial flow was introduced. All modern plants include installations for hydrogen recovery (cryogenic, membrane, or PSA technology; see Section 4.5.3.5).

Machinery. Developments in compressor and turbine manufacturing have led to higher efficiencies.

– Gas turbine drive for a compressor and/or an electric generator combined with the use of the hot exhaust as combustion air for the primary reformer or raising steam (combined cycle) [728 – 732, 953].
– Employing electric motors instead of condensation turbines [954].
– Application of liquid and gas expansion turbines can recover mechanical work (e.g., letdown of the CO_2-laden solvent, liquid ammonia, purge and fuel gas).

Steam system and waste-heat recovery were improved by the following measures: increased pressure and superheating temperature of high-pressure steam; providing a part of the process steam by natural gas feed saturation [452, 583, 955 – 957]; inclusion of a steam superheater downstream of the secondary reformer [921, 958].

Process Control and Process Optimization. Progress in instrumentation and computer technology has led to increased use of advanced control systems and computerized plant optimization. Advanced control systems [959 – 968] allow operating parameters to be kept constant in spite of variations in external factors such as feedstock composition or ambient or cooling water temperatures. These systems may be operated in open loop fashion (set values changed manually by the operator) or in closed fashion (set points automatically adjusted to optimum values by using a computer model with input of operational and economic data). Also plant simulation [969 – 972] is possible by using extensive computer models of complete plants. These models can simulate in real time the dynamic response to changes in operating parameters, plant upsets, etc. Such systems are used for off-line optimization studies and for operator training [973, 974].

The above list, by no means complete, is also a survey of options for plant revamps (Section 4.7). Quite a number of options can be found in papers presented at the "AIChE Annual Symposium Ammonia Plants and Related Facilities" giving practical experience and presenting case stories [975].

Many of these elements are strongly interrelated with each other and may affect different sections of the plant concept. It is thus a demanding engineering task to arrive at an optimum plant concept, which can only defined by the conditions set by the feedstock price, the site influences, and the economic premises of the customer. An evaluation of the individual merits of the described measures in terms of investment and operational cost in a generalized form is not possible and can be done only from case to case in real project studies.

To illustrate the forgoing discussion of the concept of the single-train steam reforming plant, Figure 63 presents a modern low-energy ammonia plant with flow sheet and process streams (UHDE process).

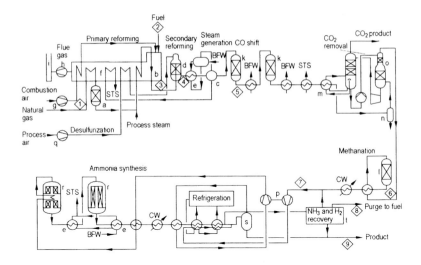

Figure 63. Modern integrated single-train ammonia plant based on steam reforming of natural gas (Uhde process)
a) Sulfur removal; b) Primary reformer; c) Steam superheater; d) Secondary reformer; e) waste-heat boiler; f) Convection section; g) Forced draft fan; h) Induced draft fan; i) Stack; k) HT and LT shift converters; l) Methanator; m) CO_2 removal solvent boiler; n) Process condensate separator; o) CO_2 absorber; p) Synthesis gas compressor; q) Process air compressor; r) Ammonia converter; s) High-pressure ammonia separator; t) Ammonia and hydrogen recovery from purge and flash gas

	1	2	3	4	5	6	7	8	9
CH_4, mol %	91.24	91.24	14.13	0.60	0.53	0.65	1.16	24.34	0.12
C_nH_m, mol %	5.80	5.80							
CO_2, mol %	1.94	1.94	10.11	7.38	18.14	0.01			
CO, mol %			9.91	13.53	0.33	0.40			
Ar, mol %				0.28	0.25	0.30	0.37	7.21	0.01
H_2, mol %			65.52	54.57	59.85	73.08	73.54	21.72	0.03
N_2, mol %	1.02	1.02	0.33	23.64	20.90	25.56	24.93	46.55	0.02
NH_3, mol %								0.18	99.82
Drygas, kmol/h		1713.7	534.43	5296.4	8414.2	9520.7	7764.0	8041.4	319.9	3676.6
H_2O, kmol/h			3520.6	4086.1	2979.6	22.8	13.3	0.2	0.9
Total, kg/h	30213	9422	121183	199555	199555	70084	71004	6496	62626
p, MPa	5.00	0.25	3.95	3.90	3.61	3.43	3.23	0.25	2.50
T, °C	25	25	808	976	229	50	35	38	20

Figure 64 shows a simplified diagram of the steam system. Even in such an advanced plant the quantity of steam generated from waste-heat is as much as 3.4 times the weight of ammonia produced.

4.6.1.3. Minimum Energy Requirement for Steam Reforming Processes

The energy saving measures described in Section 4.6.1.2 have considerably reduced the demand side (e.g., CO_2 removal, higher reforming pressure, lower steam to carbon ratio, etc.). On the supply side, the available energy has been increased by greater heat recovery. The combined effects on both sides have pushed the energy balance into surplus. Because there is no longer an auxiliary boiler which can be turned down to bring the energy situation into perfect balance, the overall savings usually could not be translated into further actual reduction of the gross energy input to the plant (mainly natural gas). In some cases this situation can be used advantageously. If the possibility exists to export a substantial amount of steam, it can be economically favorable (depending on feedstock price

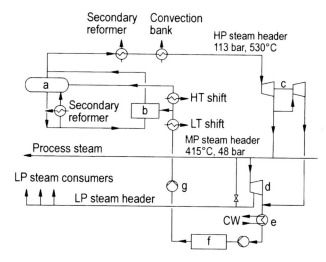

Figure 64. Steam system of a modern steam reforming
a) Steam drum, 125 bar; b) NH_3 loop; c) Turbine for syngas compressor; d) Turbine for process air compressor and alternator; e) Surface condenser; f) Condensate treatment; g) BFW pump

Table 27. Increase of plant efficiency by steam export (GJ per tonne NH_3)

	Plant		Difference
	A	B	
Natural gas	27.1	32.6	+ 5.5
Electric power	1.1	1.1	
Steam export		− 6.4	− 6.4
Total energy	28.2	27.3	− 0.9

and value assigned to the steam) to deliberately increase the steam export by using additional fuel, because the net energy consumption of the plant is simultaneously reduced (Table 27).

A reduction in gross energy demand, that is, a lower natural gas input to the plant, can only be achieved by reducing fuel consumption, because the actual feedstock requirement is determined by the stoichiometry. So the only way is to decrease the firing in the primary reformer, which means the extent of reaction there is reduced. This can be done by shifting some of the reforming duty to the secondary reformer with surplus air or oxygen-enriched air, although this makes an additional step for the removal of surplus nitrogen necessary. A more radical step in this direction is total elimination of the fired primary reformer by using exchanger reformers like the ICI GHR and the Kellogg KRES.

Based on pure methane, it is possible to formulate a stoichiometric equation for ammonia production by steam reforming:

$$CH_4 + \underbrace{0.3035\,O_2 + 1.131\,N_2}_{1.4345\text{ air}} + 1.393\,H_2O \longrightarrow CO_2 + 2.262\,NH_3$$

$\Delta H = -86$ kJ/mol and $\Delta F = -101$ kJ/mol at 25 °C

From a mere thermodynamic point of view, in an ideal engine or fuel cell, heat and power should be obtainable from this reaction. Since real processes show a high degree of irreversibility, a considerable amount of energy is necessary to produce ammonia from methane, air and water. The stoichiometric quantity of methane derived from the above reaction is 583 m³ per tonne of ammonia, corresponding to 20.9 GJ per tonne NH_3 (LHV), which with some justification could be taken as minimum value. If full recovery of the reaction heat is assumed, then the minimum would be the lower heating value of ammonia, which is 18.6 GJ per tonne NH_3. Table 28 compares the specific energy requirement for ammonia production by steam reforming with the theoretical minimum.

Table 28. Specific energy requirement for ammonia production compared to the theoretical minimum

	GJ per tonne NH$_3$ (LHV)	% theory
Classical Haber – Bosch (coke)	80 – 90	(338 – 431)
Reforming, 0.5 – 10 bar (1953 – 55)	47 – 53	225 – 254
Reforming, 30 – 35 bar (1965 – 75)	33 – 45	139 – 215
Low energy concepts (1975 – 84)	29 – 33	139 – 158
Modern concepts (since 1991)	< 28	134
Stoichiometric CH$_4$ demand	20.9	= 100

Comparison of energy consumption figures without precise knowledge of design and evaluation criteria can be misleading. First of all the state of the ammonia product should be noted. Relative to the delivery of liquid ammonia to battery limits at ambient temperature, the production of 3 bar ammonia vapor at the same temperature would save 0.6 GJ per tonne NH$_3$, while delivery as liquid at − 33 °C would need an extra 0.3 GJ per tonne NH$_3$. The temperature of the available cooling medium has a considerable influence. Increasing the cooling water temperature from 20 to 30 °C increases energy consumption by 0.7 GJ per tonne NH$_3$. A detailed energy balance, listing all imports and exports, together with the caloric conversion factors used for steam and power is needed for a fair comparison of plants. The beneficial effect of energy export to the net energy consumption is discussed above. Gas composition is also of some importance. Nitrogen content is marginally beneficial: 10 mol % nitrogen leads to a saving of about 0.1 GJ per tonne NH$_3$, whereas a content of 10 mol % carbon dioxide would add 0.2 GJ per tonne NH$_3$ to the total consumption value [976 p. 263].

Energy requirements and energy saving possibilities are also discussed in [293, 872, 940 – 946].

The energy consumption figures discussed so far represent a thermodynamic analysis based on the first law of thermodynamics. The combination of the first and second laws of thermodynamics leads to the concept of ideal work, also called exergy. This concept can also be used to evaluate the efficiency of ammonia plants. Excellent studies using this approach are presented in [977 p. 258], [978]. Table 29 [977] compares the two methods. The analysis in Table 29 was based on pure methane, cooling water at 30 °C (both with required pressure at battery limits), steam/carbon ratio 2.5, synthesis at 140 bar in an indirectly cooled radial converter.

Almost 70 % of the exergy loss in the process occurs in the reforming section and in steam generation. From conventional first law analysis it can be seen that almost all of the losses are transferred to the cooling water. As the analysis assumes water in liquid state, the LHV analyses in Table 29 is not completely balanced. For a perfect balance the heat of evaporation of water (as a fictive heating value) would have to be included.

4.6.1.4. Commercial Steam Reforming Ammonia Processes

Especially with ammonia processes based on steam reforming technology it has become a habit to differentiate between processes from various licensors and engineering contractors. This is not so much the case for partial oxidation plants (see Section 4.6.2). Strong competition together with increased plant size and the associated financial commitment has reduced the number of licensors and engineering contractors to a few companies capable of offering world-scale plants, often on a lump-sum turnkey basis. In some cases these companies sub-license their processes and special engineering know-how to competent engineering companies possessing no knowledge of their own in the ammonia field. There are also several smaller companies with specific and sometimes proprietary know-how which specialize in revamps of existing plants or small plant concepts.

In the following, each of the commercially most important processes is discussed in some detail and a shorter description of economically less important processes is given. The process configuration offered and finally constructed by a given contractor may vary considerably from case to case, depending on economic and site conditions and the clients' wishes. Thus plants from the same contractor and vintage often differ considerably. It is possible to categorize steam reforming plants according to their configuration in the reforming section:

1) Advanced conventional processes with high duty primary reforming and stoichiometric process air in the secondary reformer

Table 29. Energy analysis of a low energy ammonia plant (GJ per tonne NH_3)

	HHV	LHV	Exergy
Input			
Natural gas consumption			
Reformer feed	24.66	22.27	23.28
Reformer fuel	7.49	6.78	7.08
Auxiliary boiler fuel	0.34	0.29	0.33
Total consumption	32.49	29.34	30.69
Losses			
Reforming	0,38	0,38	4,94
Steam generation	0.33	0.33	2.39
Shift, CO_2 removal, methanation	1.30	1.30	0.67
Synthesis	1.70	1.70	1.55
Turbines and compressors	6.50	6.50	0.54
Others (including stack)	1.30	0.68	0.46
Total losses	11.51	10.89	10.55
NH_3 product	20.98	17,12	20.14
Efficiency, %	64,60	58,40	65,60

2) Processes with reduced primary reformer firing and surplus process air
3) Processes without a fired primary reformer (exchanger reformer)
4) Processes without a secondary reformer using nitrogen from an air separation plant

In principle the amount of flue gas emitted should be related to the extent of fired primary reforming, but generalizations are questionable, because sometimes the plant layout, as dictated by site requirements, may considerably change the picture for the specific flue gas value.

4.6.1.4.1. Advanced Conventional Processes

Kellogg Low-Energy Ammonia Process [101, 946, 979 – 985]. The Kellogg process is along traditional lines, operating with steam/carbon ratio of about 3.3 and stoichiometric amount of process air and low methane slip from the secondary reformer. The synthesis pressure depends on plant size and is between 140 and 180 bar. Temperatures of the mixed feed entering the primary reformer and of the process air entering the secondary reformer are raised to the maximum extent possible with today's metallurgy. This allows reformer firing to be reduced and, conversely, the reforming pressure to be increased to some extent to save compression costs. An important contribution comes from Kellogg's proprietary cross-flow horizontal converter, which operates with catalyst of small particle size, low inlet ammonia concentration, and high conversion. Low-energy carbon removal systems (Benfield LoHeat, aMDEA, Selexol) contribute to the energy optimization.

When possibilities to export steam or power are limited, part of the secondary reformer waste-heat is used, in addition to steam generation, for steam superheating, a feature in common with other modern concepts. Proprietary items in addition to the horizontal converter are the traditional Kellogg reformer, transfer line and secondary reformer arrangement, waste-heat boiler, and unitized chiller in the refrigeration section.

According to Kellogg 27.9 GJ per tonne NH_3 can be achieved in a natural gas based plant with minimum energy export, but with export of larger quantities of steam this value could probably be brought down to about 27 GJ per tonne NH_3. Figure 65 shows a simplified flowsheet of the process [101] with Selexol CO_2 removal systems (other options are, e.g., Benfield or BASF aMDEA).

Haldor Topsøe Process. In addition to technology supply, Haldor Topsøe also produces the full catalyst range needed in ammonia plants. The energy consumption of a basically classic plant configuration has been reduced considerably by applying systematic analysis and processes engineering. Descriptions and operational experience are given in [445, 566, 943, 986 – 997].

Topsøe offers two process versions. The first operates at steam/carbon ratio of 3.3 and with rather high residual methane content from the secondary reformer. Shift conversion is conventional, the Benfield or Vetrokoke process is used

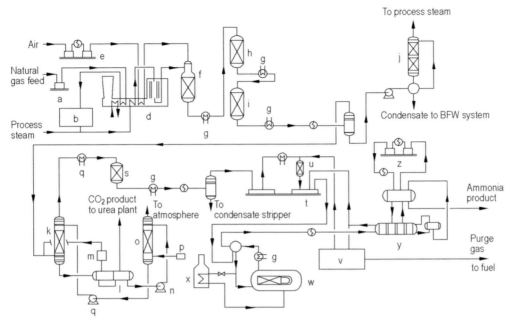

Figure 65. M.W. Kellogg's low energy process
a) Feed gas compressor; b) Desulfurization; d) Primary reformer; e) Air compressor; f) Secondary reformer; g) Heat recovery; h) High temperature shift converter; i) Low temperature shift converter; j) Condensate stripper; k) CO_2 absorber; l) CO_2 flash drum; m) Recycle compressor; n) Semi-lean Pump; o) Stripper (other options are, e.g., Benfield or BASF aMDEA); p) Stripper air blower; q) CO_2 lean pump; r) Methanator feed preheater; s) Methanator; t) Synthesis gas compressor; u) Dryer; v) Purge gas H_2 recovery; w) Ammonia converter; x) Start-up heater; y) Refrigeration exchanger; z) Refrigeration compressor

for carbon dioxide removal, and the synthesis pressure depends on plant size ranging between 140 and 220 bar when the proprietary Topsøe two-bed radial converter S 200 is used. A simplified flowsheet is presented in Figure 66.

An actual plant has reported a consumption of 29.2 GJ per tonne NH_3 [995].

The second version has a S/C ratio of 2.5 and shift conversion with medium- and low-temperature catalysts, both copper-based. For CO_2 removal Selexol or aMDEA is chosen. The synthesis is performed at 140 bar with a Topsøe two-bed S 200 radial converter, followed by a single-bed radial S 50 converter (S 250 configuration). After the converters, high-pressure steam is generated. An additional proprietary item is the side-fired reformer.

For this most energy-efficient concept a figure of 27.9 GJ per tonne NH_3 is claimed [994].

Uhde Process. Uhde, in the ammonia business since 1928, markets a low-energy ammonia plant with classic process sequence and catalysts [850, 998 – 1005]. High plant reliability at competitive overall costs was a major objective. A process flow diagram together with the main process stream is presented in Figure 63.

Key features are the high reforming pressure (up to 43 bar) to save compression energy, use of Uhde's proprietary reformer design [999] with rigid connection of the reformer tubes to the outlet header, also well proven in many installations for hydrogen and methanol service. Steam to carbon ratio is around 3 and methane slip from the secondary reformer is about 0.6 mol % (dry basis). The temperature of the mixed feed was raised to 580 °C and that of the process air to 600 °C. Shift conversion and methanation have a standard configuration, and for CO_2 removal BASF's aMDEA process (1340 kJ/Nm^3 CO_2) is preferred. Synthesis is performed at about 180 bar in Uhde's proprietary converter concept with two catalyst beds in the first pressure vessel and the third catalyst bed in the second vessel.

After each converter vessel high pressure steam (125 – 130 bar, up to 1.5 t per tonne NH_3)

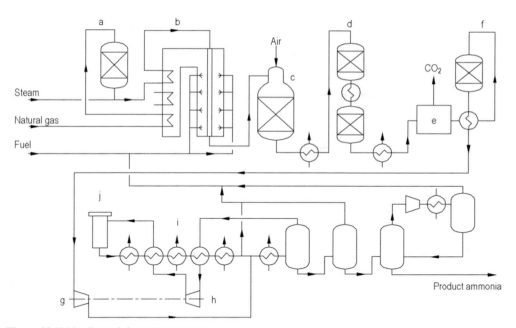

Figure 66. Haldor Topsøe's low energy process
a) Desulfurization; b) Primary reformer; c) Secondary reformer; d) Shift conversion; e) CO_2 removal; f) Methanation; g) Main compressor; h) Recycle compressor; i) Heat recovery; j) Converter

is generated (Uhde also offers its own boiler design in cooperation with an experienced boiler maker). Heat exchange between inlet and outlet of the first bed is performed in the first vessel, and gas flow in all beds is radial. When only a minimum of energy export (steam or power) is possible, the process heat from the secondary reformer outlet is partly used to raise high-pressure steam, and partly for superheating high-pressure steam. Refrigeration uses screw compressors with high operational flexibility and efficiency. Instead of the standard synloop Uhde offers the Dual Pressure Process for very large capacities (Fig. 55) (see Section 4.5.3.2). In this variant a once-through converter at lower pressure (100 bar), which produces about one third of the capacity, is followed by Uhde's standard loop (Fig. 54) at around 200 bar [850 – 855].

Achieved net energy consumption is about 28 GJ per tonne NH_3 and Uhde's engineers expect values of below 27 GJ per tonne NH_3 when a gas turbine and large steam export is included [1003].

LEAD Process (Humphreys & Glasgow, now Jacobs) [872, 1006]. The LEAD process is a highly optimized conventional approach with synthesis at 125 bar and two converter vessels, the first of which contains two catalyst beds with axial-flow quenching, while the second has a third bed with small particle size catalyst and radial flow. A consumption of 29.3 GJ per tonne NH_3 is claimed.

Exxon Chemical Process. The Exxon Chemical process [1007, 1008], was specifically designed for the company's own site in Canada and so far not built for third parties. It uses a proprietary bottom-fired primary reformer furnace and a proprietary hot potash carbon dioxide removal system with a sterically hindered amine activator. Synthesis loop and converter are licensed by Haldor Topsøe A/S. Synthesis is carried out at 140 bar in a Topsøe S-200 converter and total energy consumption is reported to be 29 GJ per tonne NH_3.

Fluor Process. The Fluor process [293, 866, 872, 1009] uses the proprietary propylene carbonate based CO_2 removal system with adsorption refrigeration using low level heat downstream of the low-temperature shift. Methanation and CO_2 removal are placed between the compression stages and thus operate at higher pressure. With a value of 32 GJ per tonne NH_3 [866] this is not really a low-energy concept.

Lummus Process. For the Lummus Process schemes [293, 872, 1010 – 1012] a consumption of 29.6 [872] to 33.5 GJ per tonne NH_3 [1013] is quoted. In the synthesis section either an axial flow quench converter or a radial flow converter with indirect cooling is used. CO_2 removal is performed with a physical solvent, and there are no special features compared to other conventional process configurations.

Integrating the ammonia and urea process into a single train was proposed by Snam Progetti to reduce investment and operating costs [1014].

4.6.1.4.2. Processes with Reduced Primary Reformer Firing

Braun Purifier Process [451, 673, 676, 677, 872, 1015 – 1025]. Characteristic of the low-energy Braun purifier process (Fig. 67) is the reduced primary reformer duty, which is achieved by shifting a proportion of the reforming reaction to the secondary reformer by using about 150 % of the stoichiometric air flow. The excess nitrogen introduced in this way is removed after the methanation step in a cryogenic unit known as a purifier [1020], which also removes the methane and part of the argon. The result is a purer synthesis gas compared to conventional processes, and only minimal purge from the loop is required. A typical flow diagram of this process is shown in Figure 67.

Synthesis is carried out in the proprietary Braun adiabatic hot-wall converter vessels (Fig. 53). Each catalyst bed (of which three are now used in newer plants [1019]) is accommodated in a separate vessel with an inlet – outlet heat exchanger after the first and high-pressure steam boilers after the following. The smaller furnace produces less flue gas and consequently less waste-heat, which makes it easier to design a balanced plant with no energy export. The lower reforming temperature allows a reduction of the steam/carbon ratio to about 2.75 without adverse effects on the HT shift, because of the less reductive character of the raw gas on account of its higher CO_2 content. In energy balanced plants, the use of waste-heat in the secondary reformer effluent is split between steam raising and steam superheating.

The concept shows great flexibility [1021] for design options. It is possible, for example, to aim for minimal natural gas consumption, even at the cost of importing some electric power. On the other hand, it is possible to improve the overall efficiency further by exporting a greater amount of energy. In this latter case it is advantageous to incorporate a gas turbine to drive the process air compressor. The hot exhaust (about 500 °C) of the turbine contains 16 – 17 mol % of oxygen and can serve as preheated combustion air of the primary reformer. In addition it is possible to include an electric generator to cover the plant demand and export the surplus. The C. F. Braun process can attain 28 GJ per tonne NH_3 in a balanced plant, but with steam export and realization of the available improvement possibilities a value of 27 GJ per tonne NH_3 seems feasible.

ICI AMV Process. The ICI AMV process [452, 780, 997, 1026 – 1038] also operates with reduced primary reforming (steam/carbon ratio 2.8) and a surplus of process air in the secondary reformer, which has a methane leakage of around 1 %. The nitrogen surplus is allowed to enter the synthesis loop, which operates at the very low pressure of 90 bar with an unusually large catalyst volume, the catalyst being a cobalt-enhanced version of the classical iron catalyst. The prototype was commissioned 1985 at Nitrogen Products (formerly CIL) in Canada, followed by additional plants in China. A flow sheet is shown in Figure 68.

In the Canadian plant, only the air compressor is driven by a steam turbine, which receives the total steam generated in the plant and has an electric generator on the same shaft. All other consumers, including synthesis gas compressor, are driven by electric motors. Separate machines are used for make-up gas and recycle compression. The make-up gas compressor is located upstream of the methanator to make use of the compression heat to warm up the cold gas coming from the Selexol carbon dioxide scrubber.

A further key feature is that about half of the process steam is supplied by feed gas saturation. The synthesis converter is a three-bed design with quench between the first two beds and an exchanger after the second bed to raise the

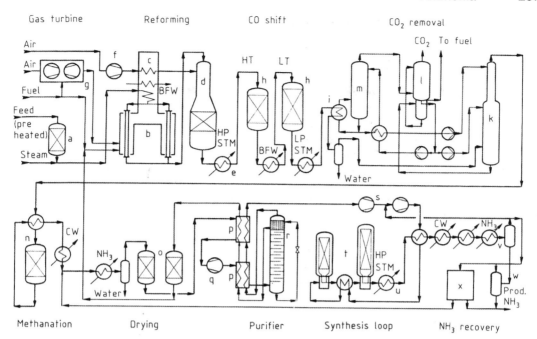

Figure 67. The Braun purifier ammonia process
a) Sulfur removal; b) Primary reformer; c) Convection section; d) Secondary reformer; e) waste-heat boiler; f) Process air compressor; g) Gas turbine; h) High- and low-temperature shift converters; i) CO_2 removal solvent reboiler; k) CO_2 absorber; l) CO_2 desorber; m) CO_2 stripper; n) Methanator; o) Driers; p) Purifier heat exchanger; q) Expansion turbine; r) Purifier column; s) Synthesis gas compressor; t) Synthesis converters; u) waste-heat boiler; v) High-pressure ammonia separator; w) Ammonia letdown vessel; x) Ammonia recovery from purge gas

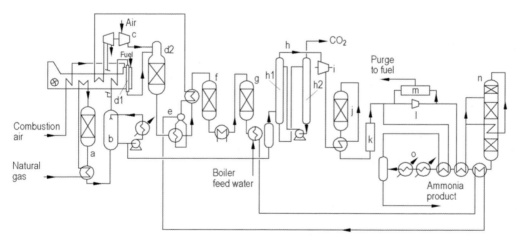

Figure 68. ICI AMV process
a) Desulfurization; b) Natural gas saturation; c) Process air compression; d1) Primary reformer; d2) Secondary Reformer; e) Boiler; f) High temperature shift; g) Low temperature shift; h) Selexol CO_2 removal; h1) CO_2 absorber; h2) Regenerator; i) Single stage compression; j) Methanation; k) Cooling and drying; l) Circulator; m) Hydrogen recovery; n) Ammonia converter; o) Refrigeration system

gas temperature of the feed to the first bed. Excess nitrogen and inerts (methane and argon) are removed by taking a purge gas stream from the circulator delivery and treating it in a cryogenic unit operating at loop pressure. The recovered hydrogen is returned to the circulator suction. Demonstrated efficiency is 28.5 GJ per tonne NH_3.

Foster Wheeler AM2 Process. The Foster Wheeler AM2 process [950, 1039], also belongs to the group of processes that shift load from the primary to the secondary reformer, but differs from the preceding concepts in that only 20 – 50 % of the total feedstock is treated in the tubular primary reformer. The remaining feed is directly sent to the secondary (autothermal) reformer which operates with a high surplus of process air (up to 200 %) and a rather high methane slip of 2.75 % (dry basis). After conventional shift, further purification is performed by Selexol CO_2 removal, methanation, and molecular sieve drying. A cryogenic unit operating at synthesis pressure rejects the nitrogen surplus from the loop. An energy consumption of 29.3 GJ per tonne NH_3 is claimed.

Humphreys & Glasgow BYAS Process. The Humphreys & Glasgow (now Jacobs) BYAS process [1036, 1040, 1041] resembles the above-described processes in its principal process features: a considerable proportion of the feed is sent directly to the secondary reformer, bypassing the fired primary reformer; use of excess air in the secondary reformer; installation of a cryogenic unit as last step of make-up gas production to remove excess nitrogen, residual methane, and the majority of the argon. As a consequence the inert level in the loop can be kept rather low, with only a small withdrawal of purge gas. An energy consumption as low as 28.7 GJ per tonne NH_3 is claimed [1042]. The process is especially suited for revamps, where it allows plant capacity to be increased.

Jacobs Plus Ammonia Technology [1043] is especially tailored for small capacities in the 300 to 450 t/d range, with a load shift from primary to secondary reformer and use of excess process air. To produce a stoichiometric synthesis gas the surplus nitrogen has to be rejected in the final purification. This is done in a PSA unit, which receives the purge gas and part of the synthesis gas taken ahead of the methanation step. All nitrogen, methane, residual carbon oxides, and argon are adsorbed to give a stream of pure hydrogen. Hydrogen and the remainder of the synthesis gas downstream of methanation are mixed to achieve a 3:1 H_2:N_2 gas composition, with a lower inerts content than the synthesis gas after methanation. The consumption figure reported for a totally energy-balanced plant is 28.8 GJ per tonne NH_3, and with substantial steam export a value of 26.8 GJ per tonne NH_3 is claimed.

Montedison Low-Pressure Process. The Montedison low-pressure process [872, 951, 1044, 1045] involves a split flow to two primary reformers. About 65 % of the feed – steam mixture flows conventionally through the radiant tubes of a fired primary reformer followed by a secondary reformer. The balance of the feed – steam mixture passes through the tubes of a vertical exchanger reformer. This exchanger reformer has a tubesheet for the catalyst tubes at the mixed feed inlet. There is no tubesheet at the bottom of the tubes, where the reformed gas mixes directly with the secondary reformer effluent. The combined streams flow on the shell side to heat the reformer tubes in a manner similar to that described for the M. W. Kellogg KRES reformer, see Section 4.6.1.4.3 and 4.5.1.1). The process air flow is stoichiometric. Synthesis is performed at 60 bar in a proprietary three-bed indirectly cooled converter with ammonia separation by water, from which ammonia is then recovered by distillation using low-grade heat. Other process steps are conventional. As driver of the process air compressor the installation of a gas turbine is suggested with use of the hot exhaust as preheated combustion air for the fired primary reformer. For this process, which has been tested in a 50 bar pilot plant, an energy consumption of 28.1 GJ per tonne NH_3 is claimed [951].

Kellogg's LEAP Process. In the late 1970s Kellogg [293, 870, 872, 1046] proposed a process which extends the basic idea of the concept described above even further. The flow of the preheated gas stream mixture is split into three streams, with 47 % through catalyst tubes in the radiant section of the fired primary reformer, 12 % through catalyst tubes in the convection section, and 41 % through the tubes of an exchanger reformer heated by the effluent of a secondary reformer. It was intended to operate the ammonia synthesis at the pressure of the front end by using no synthesis gas compression or only a small booster. An enormous quantity of the classical ammonia synthesis catalyst would have been necessary, and for recovery of the ammonia from the loop a water wash with subsequent distillation was suggested, using low-level heat in an integrated absorption refrigerator. A consumption below 28 GJ per tonne NH_3 was calculated.

4.6.1.4.3. Processes Without a Fired Primary Reformer (Exchanger Reformer)

ICI LCA Process. The ICI LCA process [1047 – 1051] is a radical breakaway from the design philosophy of the highly integrated large plant used successfully for the last 25 years. Figure 69 shows a diagram of the so-called core unit which includes only the sections essential for ammonia production (up to 450 t/d). A separate utility section, shown in Figure 70, supplies refrigeration, steam, electricity, includes cooling and water-demineralization system, and, if needed, recovers pure carbon dioxide. Both figures show the configuration of the first two plants built at Severnside (UK) with a capacity of 450 t/d each.

Feed gas is purified in a hydrodesulfurization unit operating at lower than usual temperatures and passes through a saturator to supply a part of the process steam, while the balance is injected as steam. Heated in an inlet – outlet exchanger to 425 °C the mixed feed enters the ICI gas heated reformer (GHR) [458 – 460] at 41 bar, passing to the secondary reformer at 715 °C. The shell side entrance temperature of the GHR (secondary reformer exit) is 970 °C, falling to 540 °C at the exit of the GHR. Methane levels at the GHR exit and the secondary reformer are 25 % and 0.67 % respectively (dry basis). Overall steam to carbon ratio is 2.5 – 2.7. The gas, cooled to 265 °C in the inlet/outlet exchanger, enters a single-stage shift conversion reactor with a special copper – zinc – alumina-based catalyst that operates in quasi-isothermal fashion and is equipped with cooling tubes in which hot water circulates, whereby the absorbed heat is used for feed gas saturation, as described above. CO_2 removal and further purification is effected by a PSA system, followed by methanation and drying.

Synthesis operates at 82 bar [780] in a proprietary tubular converter loaded with a cobalt-enhanced formulation of the classical iron catalyst. Purge gas is recycled to the PSA unit, and pure CO_2 is recovered from the PSA waste gas by an aMDEA wash. Very little steam is generated in the synthesis loop and from waste gases and some natural gas in the utility boiler in the utility section (60 bar), and all drivers are electric.

The original intention was to design an ammonia plant which can compete with modern large capacity plants in consumption and specific investment, and, by means of lower energy integration, to achieve greater flexibility for start-up and reduced-load operation, needing minimum staffing. The basic plant features (GHR, isothermal shift, and synthesis) can in principal be applied to larger capacities. The flow sheet energy consumption is 29.3 GJ per tonne NH_3.

Kellogg, today Kellogg Brown and Root (KBR), offers several process schemes with use of its KRES exchanger reformer and its KAAP ammonia technology [463 – 466]. With including the Purifier (of C.F. Braun, later Braun and Root) KBR offers now the KAAPplusProcess [1052 – 1056] shown in Figure 71.

Desulfurized gas is mixed with steam and then split into two streams in approximate ratio 2:1. These are separately heated in a fired heater. The smaller of the two enters the exchanger reformer at 550 – 550 °C, while the remainder is passed directly to the autothermal reformer at 600 – 640 °C. The exchanger reformer and the autothermal reformer use conventional nickel-based primary and secondary reforming catalysts, respectively. To satisfy the heat balance, the autothermal reformer operates with surplus of air. The required heat for the endothermic reaction in the tubes of the exchanger reformer comes from the gases on the shell side, comprising a mixture of the effluent from the autothermal reformer and the the gas emerging from the tubes. The shell side gas leaves the vessel at 40 bar. The purifier removes the nitrogen surplus together with residual methane and part of the argon.

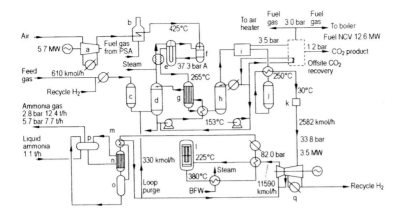

Figure 69. ICI LCA process (core unit)
a) Process air compressor; b) Start-up air heater; c) Hydrodesulfurization; d) Saturator; e) GHR; f) Secondary reformer; g) Shift converter; h) Desaturator; i) PSA system; j) Methanator; k) Gas dryer; l) Ammonia converter; m) Two-stage flash cooling (one stage shown); n) Chiller; o) Catchpot; p) Flash vessel q) Syngas compressor

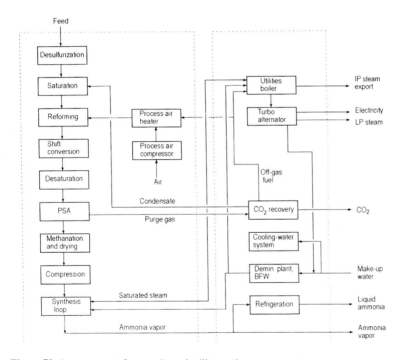

Figure 70. Arrangement of core units and utility section

Synthesis proceeds at about 90 bar in a four-bed radial-flow converter (Fig. 57) (hot-wall design) with interbed exchangers. The first bed is charged with conventional iron-based catalyst for bulk conversion and the others with Kellogg's high activity ruthenium-based catalyst, allowing an exit ammonia concentration in excess of 20 % to be obtained. The other process steps are more along traditional lines. The overall energy consumption claimed for this process can be as low as 27.2 GJ per tonne NH_3.

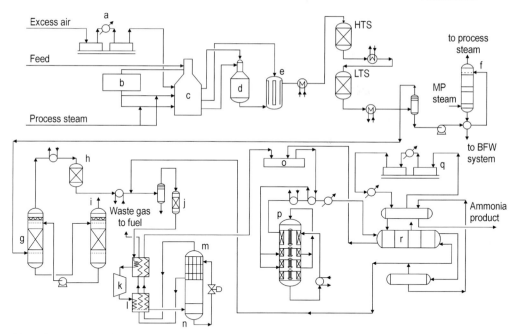

Figure 71. KBR KAAPplus Process
a) Air compressor; b) Sulfur removal; c) Process heater; d) Automatic reformer (ATR); e) Reforming exchanger (KRES); f) Condensate stripper; g) CO_2 absorber; h) Methanator; i) CO_2 stripper; j) Dryer; k) Expander; l) Feed/effluent exchanger; m) Condenser; n) Rectifier column; o) Synthesis gas compressor; p) KAAP ammonia converter; q) Refrigeration compressor; r) Refrigeration exchanger

The LCA and KAAPplusProcess are environmentally favorable because atmospheric emissions of both nitrogen oxides and carbon dioxide are dramatically reduced as there is no reforming furnace.

Chiyoda Process [1057]. In this process the traditional fired primary reformer is also replaced by an exchanger reformer and the heat balance requires excess air in the secondary reformer with the consequence of a cryogenic unit as final step in the make-up gas preparation to remove the surplus of nitrogen. Additionally, gas turbines are proposed as drivers for the process air compressor and synthesis gas compressor with the hot exhaust being used for steam generation and feed gas preheating.

Topsøe has described the lay-out for an ammonia plant based on fully autothemic reforming [469].

4.6.1.4.4. Processes Without a Secondary Reformer (Nitrogen from Air Separation)

KTI PARC Process. The KTI PARC ammonia process [583, 1036, 1058 – 1064] uses the following process elements: air separation unit, classical primary reformer at 29 bar, standard HT shift, power generation in a Rankine cycle with CFC to generate electric power (optional), carbon dioxide removal (optional, only when pure CO_2 product is required), PSA, nitrogen addition, synthesis loop. In this concept four sections of the classical process sequence (secondary reforming, LT shift, CO_2 removal, methanation) can be replaced by the fully automatic high-efficiency PSA system, which has a proprietary configuration (UOP) in which nitrogen flushing enhances hydrogen recovery. The overall efficiency ranges from 29.3 to 31.8 GJ per tonne NH_3.

Linde LAC Process. The Linde LAC process [1065 – 1069] consists essentially of a hydrogen plant with only a PSA unit to purify the synthesis gas, a standard cryogenic nitrogen

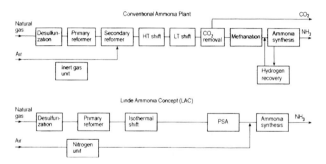

Figure 72. Comparison of Linde LAC process with a conventional process

unit, and an ammonia synthesis loop. In principle it is similar to the PARC process, but designed for world-scale capacities. First application was for a 1350 t/d plant in India. Figure 72 compares the LAC process to a conventional one. If pure CO_2 is needed, it can be recovered by scrubbing the off-gas from the PSA unit at low pressure or, probably with better energy efficiency, by installing the CO_2 removal unit directly in the synthesis gas train ahead of the PSA system. The synthesis converter and loop are based on ICI and Casale know-how. According to Linde the process should consume about 28.5 GJ per tonne NH_3 or, with inclusion of pure CO_2 recovery, 29.3 GJ per tonne NH_3.

Humphreys & Glasgow MDF Process (now Jacobs) [293, 687, 688, 1012, 1013, 1070 – 1073]. This concept has a configuration similar to the Linde LAC process. Energy consumption with inclusion of pure CO_2 recovery (which is optional) is 32.8 GJ per tonne NH_3.

4.6.2. Ammonia Plants based on Partial Oxidation

4.6.2.1. Ammonia Plants Based on Heavy Hydrocarbons

Although partial oxidation processes can gasify any hydrocarbon feedstock, on account of its higher energy consumption and investment costs, commercial use of this technology is restricted to the processing of higher hydrocarbons, often containing as much as 7 % sulfur. Where natural gas is unavailable or the heavy feedstock can be obtained at a competitive price, this gasification technology can be an economic choice.

There are two commercially proven partial oxidation routes for heavy feedstocks: the Shell process and the Texaco process. In contrast to the steam reforming, for which most contractors have their own proprietary technology for the individual process steps, the engineering firms which offer ammonia plants based on heavy hydrocarbons have often to rely for the individual process stages on different licensors. Lurgi, for example, has built very large capacity ammonia plants that use Shell's gasification process, its own proprietary version of the Rectisol process [645], Linde's air separation and liquid nitrogen wash, and Topsøe's technology for synthesis converter and loop.

Independent, experienced engineering companies, not directly active in ammonia plant design may be used as general contractors to coordinate a number of subcontractors supplying the different technologies required. This is in line with the fact that the degree of energy integration is usually lower than in steam reforming technology, because in absence of a large fired furnace, there is no large amount of flue gas and consequently less waste-heat is available. Therefore, a separate auxiliary boiler is normally necessary to provide steam for mechanical energy and power generation. Nevertheless, some optimization has successfully reduced the overall energy consumption, for which in older installations values of around 38 GJ per tonne NH_3 were reported. More recent commercial bids quote values as low as 33.5 GJ per tonne NH_3.

The arguments presented above suggest describing the two principal routes, Shell and Texaco, which differ in the gasification process,

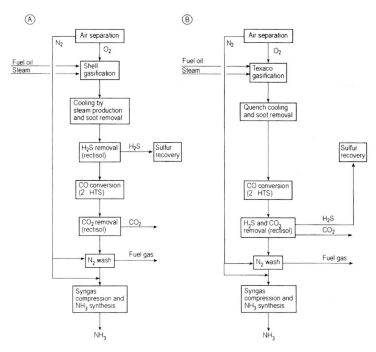

Figure 73. Flow diagrams of ammonia production from fuel oil or heavy residues by the Shell (A) and the Texaco (B) process (standard configuration)

rather than listing all individual contractor design approaches. Figure 73 shows the classical sequence of process steps for both cases.

Processes Using Shell Gasification (e.g., Lurgi) [505 – 515]. A cryogenic air separation plant provides oxygen for the gasification and the nitrogen for the liquid nitrogen wash and for supplying the stoichiometric amount for the synthesis of ammonia. Oil enters the alumina-lined gasification vessel through a central jet in the burner nozzle. A substantial pressure drop is needed to ensure atomization of the oil and proper mixing with oxygen, fed through the annulus between the jet and the outer case of the burner nozzle. The temperature in the gasification vessel (generator) is between 1200 and 1400 °C, and the pressure between 35 and 65 bar.

The hot gas contains soot, formed because of insufficient mixing of the reactants, and fly ash. A waste-heat boiler, a proprietary item of the Shell process, raises 100 bar steam and cools the gas to 340 °C. Soot is removed from the raw gas in a two-stage water wash. Older installations used an elaborate technique to remove the soot from the water by extraction with naphtha and light oil to form soot pellets which could be burnt or recycled to the feed oil. In newer installations the carbon – water slurry is filtered off in automatic filters, and the moist filter cake is subjected to a controlled oxidation in a multiple-hearth furnace.

A selective Rectisol unit with methanol of about −30 °C as solvent is used to remove H_2S and COS (together with some CO_2) to less than 0.1 %. The removed sulfur-rich fraction is sent to a Claus plant for recovery of elemental sulfur or converted to sulfuric acid. The gas is heated subsequently by heat exchange, supplied with steam in a saturator, and then fed to shift conversion, which proceeds stepwise with intermediate heat removal. The gas is cooled by a direct water cooler, and the hot water is recycled to the saturator.

A second Rectisol wash stage follows to remove CO_2 by absorption at −65 °C in methanol, which is regenerated by flashing and stripping. Molecular sieve adsorption then removes residual traces of methanol and CO_2. To remove residual CO a liquid nitrogen wash is applied for final purification with the advantage of also lowering the argon content in the make-up gas, which is adjusted by nitrogen addition to the stoichiometric ratio $N_2 : H_2 = 1 : 3$. Converter and synthesis loop configuration depend on the licensor chosen. Plant descriptions are given in [1074 – 1077].

Processes Using Texaco Gasification (e.g., Foster Wheeler, Linde, Uhde). Temperatures in the generator are similar to those in the Shell process; units with operating pressures up to 90 bar are in operation [497, 500 – 504, 1078]. Some modern installations (e.g., Linde) use pumps for liquid oxygen instead of oxygen compressors. In contrast to the Shell arrangement, oxygen enters the gasifier through the central nozzle of the burner, and oil is fed through the annular space between central nozzle and outer burner tube. Instead of a waste-heat boiler a direct water quench is applied for cooling the raw synthesis gas, which is subsequently scrubbed first in a Venturi scrubber and then in a packed tower to remove the soot. Texaco also offers a version operating with a waste-heat boiler instead of a water quench. Although this is preferable when producing CO-rich synthesis gases (e.g., methanol or oxogas), quench is thought to be more economical when hydrogen-rich gases are manufactured. Soot recovery from the water is performed by extraction with naphtha. The soot – naphtha suspension is mixed with feed oil, and the naphtha is distilled off and recycled to the extraction stage. The shift reaction uses a cobalt – molybdenum – alumina catalyst [569, 578 – 582] which is not only sulfur-tolerant but also requires a minimum sulfur content in the gas for proper performance. The conversion is subdivided into stages with intermediate cooling by raising steam. The following Rectisol process has a somewhat more elaborate configuration than the version used in the Shell route. The large amount of carbon dioxide formed in shift conversion lowers the H_2S concentration in the sour gas, and for this reason a special concentration step is required for methanol regeneration to obtain pure CO_2 and a fraction sufficiently rich in H_2S for a Claus plant or a sulfuric acid plant. The remaining process steps are identical with the Shell route. Figure 74 gives an example of a Linde flow diagram. Descriptions of plants using the Texaco process can be found in [569, 1079].

Topsøe Process. A concept using enriched air instead of pure oxygen and methanation instead of a liquid nitrogen wash was proposed by Topsøe [1080, 1081]. A Shell gasifier with a waste-heat boiler or a Texaco generator with a quench are equally well suited to this process. After soot removal, shift conversion is performed on a sulfur-tolerant catalyst in several beds with intermediate cooling, leaving a residual CO content of 0.6 mol %. An appropriate process (Rectisol or amine based) removes the sour gases H_2S, COS, and CO_2, and this is followed by methanation. Make-up gas drying, compression, and synthesis loop have no special features. The anticipated energy consumption is 34.8 GJ per tonne NH_3. A basically similar synthesis gas preparation, but based on gasification with pure oxygen, is already used in large commercial plant in Japan [1082].

Foster Wheeler Air Partial Oxidation Process [950, 1083] is a proposed modification of the Texaco gasification process. It is intended to operate at 70 bar with highly preheated air (815 °C) instead of pure oxygen. The synthesis gas purification train comprises soot scrubbing followed by shift conversion, acid gas removal (for example Selexol), and methanation. The gas is dried and finally fed to a cryogenic unit, which removes the surplus nitrogen by condensation together with methane, argon, and residual carbon monoxide. The rejected nitrogen is heated and expanded in a turbine, which helps to drive the air compressor.

A major aspect in the design concept is the separation of nitrogen and oxygen. A conventional air separation plant is based on the fractional distillation of oxygen and nitrogen, which differ in boiling point by only 13 °C. In the cryogenic unit for the Foster Wheeler process a lesser quantity of nitrogen is separated from hydrogen with a much higher boiling point difference (57 °C). According to Foster Wheeler this leads to considerable saving in capital investment and energy consumption compared to the traditional approach using pure oxygen from an air separation plant and a liquid nitrogen wash

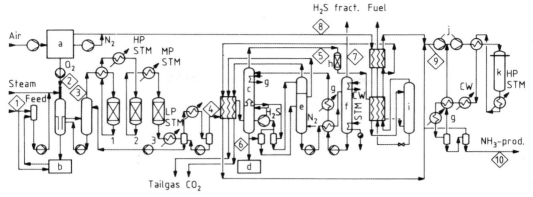

Figure 74. Ammonia production based on heavy fuel oil (Linde flow scheme with Texaco gasification)
a) Air separation unit; b) Soot extraction; c) CO_2 absorption; d) Methanol/H_2O distillation; e) Stripper; f) Hot regenerator; g) Refrigerant; h) Dryer; i) Liquid N_2 scrubber; j) Syngas compressor; k) NH_3 reactor

Material Balance

① Feed: heavy fuel oil	② Oxygen	⑩ Ammonia
$m = 44$ t/h	$m = 49.6$ t/h	$m = 56.25$ t/h
$p = 0.4$ MPa	$p = 9.0$ MPa	$p = 2.5$ MPa
$t = 90\,°C$	$t = 150\,°C$	$t = 11\,°C$
Composition, wt%	Purity	
C 84.6	$O_2 = 98.5$ vol%	
H 10.0	balance: Ar and N_2	
O 0.4		
N 0.4		
S 4.5		
Ash 0.1		

		③	④	⑤	⑥	⑦	⑧	⑨
H_2	vol%	43.0	60.8	96.3	0.9	0.1	–	75.00
CO	vol%	47.7	1.5	2.3	0.1	–	–	–
CO_2	vol%	6.9	35.8	–	98.8	67.4	–	–
N_2	vol%	0.1	0.1	0.2	0.1	2.5	100.00	25.00
CH_4	vol%	0.8	0.5	0.8	0.1	–	–	–
Ar	vol%	0.4	0.3	0.4	0.0	–	–	–
H_2S + COS	vol%	1.1	0.8	–	–	30.0	–	–
H_2O	vol%		0.2	–	–	–	–	–
	kmol/h	5680	8250	5200	1800	200	1750	6650
p	MPa	8.37	7.65	7.30	0.15	0.25	7.4	7.1
t	°C	255	45	−63	−15	15	15	10

for gas purification. A figure of 35.6 – 37.6 GJ per tonne NH_3 is claimed for heavy oil feedstock and 31.4 – 32.7 for natural gas as feedstock. A similar variant also using air instead of pure oxygen is offered by Humphreys & Glasgow (now Jacobs) [1084].

4.6.2.2. Ammonia Plants Using Coal as Feedstock

In the early days the entire ammonia industry was based on coal feedstock. Today coal or coke (including coke oven gas) are used as feed for only a smaller part of world-wide ammonia production. In 1990, for example, only 13.5 % of the world ammonia capacity used this raw material [426]. A newer statistic estimates a figure of 19 % for 2001 [1085]. But with the enormous increase of the natural gas prices in the USA the share of coal in the feedstock pattern of ammonia might become larger in the future (see Chapter 9).

Apart from a few plants operating in India and South Africa, the majority of coal-based ammonia plants are found in China. Commercially proven coal gasification processes are Lurgi (dry gasifier), British Gas/Lurgi (slagging gasifier), Winkler/HTW, Koppers – Totzek, Shell, Texaco, and Dow, [516 – 518, 1086]. So far only the Koppers – Totzek, Texaco, and Lurgi processes have been used commercially for ammonia production [492, 516, 520 – 524, 659, 1085, 1087, 1088]. The Shell process, demonstrated commercially in another application with a capacity equivalent to a world-scale ammonia plant, is also a potential candidate for ammonia production processes.

In recent years little development work has been done on complete ammonia plant concepts based on coal. The traditional leading ammonia contractors have to rely on proprietary processes licensed from different companies, which similarly tend not to have specific ammonia technology of their own. Again, compared to a steam reforming plant, the degree of integration is considerably lower; power generation facilities are usually separate. Thus it is difficult to identify specific ammonia processes for the individual contractors and the following descriptions serve as examples, without striving for completeness.

The Koppers – Totzek Process gasifies coal dust with oxygen in the temperature range 1500 – 1600 °C at about atmospheric pressure. For a more detailed description of the gasification, refer to Section 4.5.1.1 and [526 – 531, 1086]. The cooled gas, free of coal dust and fly ash, contains about 60 % of CO. The next step is compression to about 30 bar, followed by sulfur removal at -38 °C with chilled methanol (Rectisol process). Steam is added for the shift conversion, carried out stepwise with intermediate heat removal and with a standard HTS catalyst. A second Rectisol stage, operating at -58 °C and 50 bar, removes the CO_2, and the final purification step is a liquid nitrogen wash. Any of the well known converter and synthesis loop concepts may be used, with no purge or minimal purge, due to the practically inert-free make-up gas. Several plants are operating in South Africa [1089, 1090] and India [1091].

The atmospheric-pressure gasification is a considerable disadvantage of this process route, which substantially increases equipment dimensions and costs, as well as the power required for synthesis gas compression. An energy input of 51.5 GJ per tonne NH_3 (LHV) has been reported. According to [557], the atmospheric ACGP gasifier could lower the consumption to 44 GJ per tonne NH_3 (HHV).

Lurgi Process. Lurgi [517, 518, 547 – 551, 1092] offers a concept using its proprietary Lurgi dry bottom gasifier, described in Section 4.5.1.1. The moving-bed generator, which can handle any sort of coal (ash content may exceed 30 %), operates at 30 bar, and the product gas contains up to 15 % CH_4, higher hydrocarbons, troublesome phenolic material, and tars. After washing with process condensates to remove ash and dust, the gas is cooled further with recovery of waste-heat. Several process steps treat the separated gas liquor to recover tar, phenols, and some ammonia. Shift conversion, Rectisol process, and liquid nitrogen wash are the further operations in the production of make-up gas. The liquid nitrogen wash produces a methane-rich fraction, which is separately processed in a steam reformer, and the reformed gas rejoins the main stream at the Rectisol unit for purification. The gasification has a power consumption of 32 – 34 GJ per tonne NH_3, and steam generation consumes 18 – 22 GJ per tonne NH_3, resulting in a total energy consumption of 50 – 56 GJ per tonne NH_3.

The Texaco Coal Gasification Process [518, 533 – 540, 1093 – 1098] (see Section 4.5.1.1) originates from Texaco's partial oxidation process for heavy oil fractions and processes a coal – water slurry containing 60 – 70 % coal. A lock hopper system removes ash and glassy slag as a suspension from the quench compartment of the generator. The process can handle bituminous and sub-bituminous coal but not lignite. The further gas purification steps used to arrive at pure make-up gas correspond to those described for an ammonia plant using the Texaco partial oxidation of heavy oil fractions.

Ube Industries commissioned a 1000 t/d ammonia plant in 1984 using Texaco's coal gasification process [1082, 1099]. An energy consumption of 44.3 GJ per tonne NH_3 is stated, which is lower than the 48.5 GJ per tonne NH_3 quoted for another Texaco coal gasification-based ammonia plant [517].

4.6.3. Waste-Heat Boilers for High-Pressure Steam Generation

Because of its great influence on reliability and efficiency of all ammonia plants a special section for the generation of high-pressure steam seems to be appropriate. The operating conditions for the boilers are more severe than those normally encountered in power plants; on account of the high pressure on both sides the heat transfer rates and thus the thermal stresses induced are much higher.

In steam reforming plants, for example, the temperature of the gas from the secondary reformer has to be reduced from 1000 °C to 350 °C before entering the HT shift vessel. In earlier plant generations two boilers were usually installed in series, with a bypass around the second to control the inlet temperature for the HT shift. Common practice for a long time was to use a boiler with water tube design. A famous example is the Kellogg bayonet-tube boiler, applied in more than 100 plants. Besasuse of size limitations two parallel units were installed. For sufficient natural water circulation these boilers needed a steam drum at a rather high elevation and a considerable number of downcomers (feed water) and risers (steam/water mixture). An alternative tube-bundle design which can directly substitute the bayonet-tube internals was recently developed. This concept uses twisted tubes [1100].

In contrast fire-tube boilers are much better suited for natural circulation and the steam drum can sit in a piggyback fashion right on top of the boiler. This makes it possible to provide each boiler with its own separate steam drum, which allows a greater flexibility in the plot plan. In a fire-tube boiler, the inlet tubesheet and the tubesheet welds are exposed to the extreme temperature of the reformed gas, which creates rather large temperature gradients and therefore high expansive stress. A positive feature of the design, however, is that debris in feed water (mainly magnetite particles spalling from the water side of the tubes) can collect at the bottom of the horizontally mounted vessel without creating difficulties and are removed easily by blowdown. Water-tube boilers, especially bayonet-types, are very sensitive in this respect, because the deposits may collect in the lowest and most intensively heated part of the tube. In an extreme case of scaling, this may restrict the water flow to the point where boiling occurs irregularly (film boiling). The risk is overheating and tube failure.

The key factor which allowed the use of fire-tube boilers after the secondary reformer was the development of thin-tubesheet designs. Thick tubesheets in this kind of service are too rigid and have too high a temperature gradient, and the resultant stress on the tube-to-tubesheet welds can lead to cracks. The inherent flexibility of thin tubesheets assists in dispersing stresses and reduces the risk of fatigue failure of the tube-to-tubesheet welds and tubesheet-to-shell welds. In all the various designs of this concept, the tubesheet is only 20 – 30 mm thick. The hot inlet channel and the tubesheet are shielded by a refractory layer, and the tube inlets are protected by ferrules. In one concept (Uhde [1101, 1102], Steinmüller [1103]) the flexible tubesheet is anchored to and supported by the tubes to withstand the differential pressure, which poses some restriction on the tube length. The tubesheet of Babcock-Borsig (today Borsig) [1104 – 1106] is reinforced by stiffening plates on the back side (Fig. 75). Both solutions have full penetration tube-to-tubesheet welds. Steinmüller's waste-heat boiler product line is now owned by Borsig. A critical analysis of the two concepts is given in [1107].

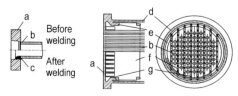

Figure 75. Reinforced tubesheet of the Borsig boiler
a) Tubesheeet; b) Tube; c) TIG welded root pass; d) Shell; e) Supporting ring; f) Stiffener plate; g) Anchor

In the synthesis loop boilers at the exit of the converter up to 50 % of the total steam is generated. As much as 1.5 t of steam per tonne ammonia, equivalent to 90 % of the reaction heat can be generated. For this service also fire-tube versions have been used, including Borsigs's thin-tubesheet design. But compared to the secondary reformer service, where the gas pressure is lower than the steam pressure, the conditions and stress pattern are different. In the synthesis loop boiler the opposite is the case with the result that the tubes are subjected to longitudinal compression instead of beeing under tension. Several failures in this application have been been reported [1108, 1109].

The thick-tubesheet concepts in various configurations are more generally accepted now. Proven U-type designs (Fig. 76) are available from Uhde [853, 1110], Balcke – Dürr [1111 – 1113], KBR [1114], and Borsig [1115, 1116]. A horizontal synthesis waste-heat boiler was developed by Balcke – Dürr with straight tubes and thick tubesheets on both ends [1117, 1118].

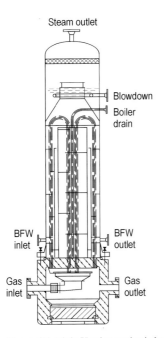

Figure 76. Uhde U-tube synthesis loop boiler

A special design is Borsig's hot/cold tubesheet. The hot and cold end of the tube are arranged alternately, so that a hot shank is always to a cold shank and vice versa. The advantage is that the tubesheet can kept below 380 °C [102, 1119].

Waste-heat boilers in partial oxidation plants, which cool the exit gas of the generator from 1400 °C to around 350 °C, face additional difficulties. The gas contains soot and probably some fly ash particles. Very high gas velocities and appropriate design are necessary to prevent any deposition on the heat-exchange surfaces and to reduce the danger of attrition as well. A special design for Texaco gasifications (orginally by Schmidt'sche Heißdampf GmbH and Steinmüller GmbH) is offered now in an improved version (Fig. 77) by Borsig [1120]. Forced water circulation around the entrance nozzles helps to stand the high heat flux (700 kJ m^{-2}s^{-1}) at this location. The Shell process uses propretary designs (Fasel – Lentjes) [1121].

Additional information on waste-heat boilers is found in [100, 418, 820, 1053, 1122]. Many papers on this subject were presented at the "AIChE Annual Symposia for Safety In Ammonia Plants and Related Facilities" [975].

4.6.4. Single-train Capacity Limitations – Mega-Ammonia Plants

The principle of the economy of scale is well known in production processes. For ammonia plants a scale exponent 0.7 is widely used as a rule of thumb:

$$\text{cost}_2 = \text{cost}_1 \times (\text{capacity}_2/\text{capacity}_1)^{0.7}$$

The advantage of larger plants is not only the investment but also the reduction of fixed costs, such as labor, maintenance, and overhead expenses. But the simple formula should be handled with care because it is only valid as long as a given process configuration is scaled up and the dimensions of the equipment for very large capacities do not lead to disproportionate price increases. For example there may be fewer vendors or special fabrication procedures might become necessary. So the scale factor could increase for very large single-train capacities. Nevertheless the specific investment will decrease with the increase of capacity. On the other hand certain process developments like autothermal reforming and exchanger reforming could reduce the investment for the higher capacities.

very large capacities. Several studies [412, 855, 1056, 1123 – 1127] have been made by various contractors to investigate the feasibility and the size limitations for steam reforming plants. According to most of these studies a capacity of 5000 t/d seems to be possible. With a conventional process concept for the front end there is no restriction for producing syngas for 5000 t/d of ammonia. Steam reforming furnaces of the required size have already been built for other applications (methanol) and neccessary vessel diameters seem also to be fabricable. A limitation could be the synthesis compressor which can probably be designed for 4500 t/d only. The KBR KAAP process has there an advantage because of its low synthesis pressure. This also holds true for the Dual Pressure Process of Uhde, where just the loop which produces only two thirds of the ammonia is at high pressure whereas the once-through converter is at lower pressure. Another capacity limiting factor is pipe diameter in the synthesis section. The largest plant so far was built by Uhde in Saudi-Arabia. It has a capacity of 3300 t/d and came on-stream in 2006.

4.7. Modernization of Older Plants (Revamping)

With rising feedstock prices and hard competition in the market, many producers have looked for possibilities to "revamp" or modernize their older, less efficient plants so that they can stay competitive. Most revamp projects have been combined with a moderate capacity increase because some of the original equipment was oversized and only specific bottlenecks had to be eliminated, not entailing excessive cost. As the market possibilities for a company do not increase in steps of 1000 or 1500 t/d but slowly and continuously, such a moderate capacity addition will involve less risk and will be more economical than building a new plant.

For a revamp project first an updated baseline flow sheet of the existing plant should be prepared from which the proposed improvement can be measured [1128 – 1130]. Depending on the objective (energy saving and/or capacity increase) the following guidelines should be kept in mind: maximum use of capacity reserves in existing equipment; shifting duties from over-

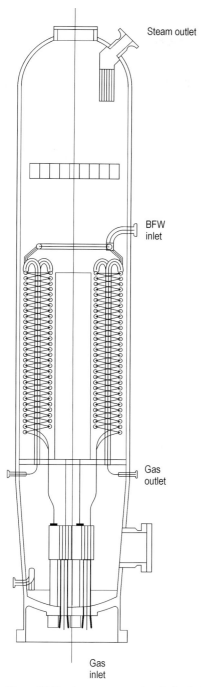

Figure 77. Borsig syngas waste-heat boiler for Texaco gasification

For export-orientated plants at locations with cheap natural gas there is growing interest in

taxed units to oversized ones; if possible, simple modifications of existing equipment are preferable to replacement; the amount of additional equipment should be kept to a minimum [1129].

To give an exhaustive list or description of the individual modification options is beyond the scope of this article, but reviews on this subject and useful information are given in [942, 946, 1131 – 1148]. Section 4.6.1.2 describes modifications that lower the energy consumption in newer plant generations compared to the first generation of single-train ammonia plants, and this also represents an overview of the revamp options for existing steam reforming plants.

Just a few of the frequently used revamp possibilities should be mentioned here. In steam reforming plants it is often possible to lower the steam/carbon ratio by using improved reforming catalysts and copper-promoted HT shift catalysts. More active LT shift catalysts lower the residual CO content, which will reduce H_2 loss (methanation) and inert content in the make-up gas. Drying of the make-up gas, addition of hydrogen recovery from purge gas, and installing a more effective CO_2 removal are other options. With the *aMDEA* system, which can be flexibly tailored to fit into existing process configurations, it is, for example, possible to simply replace the MEA solvent with the aMDEA solution, adjusting the activator concentration accordingly to achieve zero or only minor equipment modification. Also hot potash processes have been converted in this way to aMDEA.

Other measures, involving more additional hardware and engineering work, are introduction of combustion air preheating and reducing the primary reformer load. This latter option is used when the revamp objective is capacity increase and the primary reformer is identified as a bottleneck. One possibility is to increase the duty of the secondary reformer and use air in excess of the stoichiometric demand. Elegant variants of this principle are the *Jacobs BYAS process* [1036, 1040, 1041] and the *Foster Wheeler AM2 process* [950, 1039, 1149]. Another method is to perform a part of the primary reforming in a pre-reformer [443 – 450] that uses low-level heat. Alternative methods to enlarge the reforming capacity make use of the process heat of the secondary reformer in an exchanger reformer such as *ICI's GHR* [1150] or *Kellogg's KRES* [465, 468]. If oxygen is available, installation of a parallel autothermal reformer or a parallel *Uhde CAR unit* [470 – 473, 1151], (see also Section 4.5.1.1) could be considered. Description of executed revamp projects are given in [1042, 1152 – 1157].

Similarly, numerous modernization possibilities exist for partial oxidation plants, and they may even outnumber those for steam reforming plants. Common to both plant types is the potential for improvement of the synthesis loop and converter. Application of indirect cooling and smaller catalyst particles are frequently chosen to reduce energy consumption through lower pressure drop, reduced synthesis pressure, higher conversion, or a combination thereof. Apart from replacing existing ammonia converters, in situ modification of the internals of installed converters is a very economic approach. *Topsøe* [1136, 1158] mostly uses its Series 200 configuration; and *Ammonia Casale* [830 – 833, 1122, 1159 – 1162] its proprietary ACAR technology, with which more than 90 plants have been revamped so far. With a "split-flow" configuration Kellogg proposed an in situ revamp option to change its four-bed quench converter into a two-stage intercooled converter (with two parallel beds for the second stage) using smaller catalyst particles of 1.5 – 3.5 mm and 3 – 6 mm [1163, 1164]. A promising loop modernization option for the future will be the *Kellogg KAAP process* [466, 468, 1165].

The fact that more than 45 % of world ammonia plants are older than 30 – 35 years suggests that there is a major potential for revamp projects, even in plants which have already made modifications. Revamp histories are a constant topic at the "AIChE Annual Symposium Ammonia Plant and Related Facilities" [975] and a wealth of information and practical experience can be found there.

A special revamping option is the integration of other processes into an ammonia plant [102]. For example a CO production from a side stream upstream of the HT shift is possible [1166]. Hydrogen can also be produced from a side stream by using PSA. An example of coproduction of methanol is shown in Figure 78.

In the less integrated partial oxidation plants coproduction schemes are easier to be incorporated and there are several large installations which were directly designed to produce ammonia, methanol and hydrogen [1167, 1168].

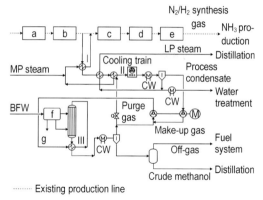

Figure 78. Methanol coproduction (side-stream loop)
a) Secondary reformer; b) Waste-heat boiler; c) HT shift; d) LT shift; e) Purification; f) Steam drum; g) Methanol converter; h) Catchpot; I) Syngas compressor; j) Let-down vessel

4.8. Material Considerations for Equipment Fabrication

Hydrogen Attack. In several steps of the ammonia production process, especially in the synthesis section, the pressure shells of reaction vessels as well as the connecting pipes are in contact with hydrogen at elevated pressure and temperature with a potential risk of material deterioration [1169 – 1171].

Chemical Hydrogen Attack. Under certain conditions chemical hydrogen attack [1172 – 1175] can occur. Hydrogen diffuses into the steel and reacts with the carbon that is responsible for the strength of the material to form methane, which on account of its higher molecular volume cannot escape. The resulting pressure causes cavity growth along the grain boundaries, transforming the steel from a ductile to a brittle state. This may finally reach a point where the affected vessel or pipe ruptures, in most cases without any significant prior deformation. This phenomenon was already recognized and principally understood by BOSCH et al. [76] when they developed the first ammonia process. The resistance of steel against this sort of attack can be enhanced by alloy components which react with the carbon to form stable carbides (e.g., molybdenum, chromium, tungsten, and others). The rate of deterioration of the material depends on the pressure of the trapped methane, the creep rate of the material, and its grain structure. Areas highly susceptible to attack are those which have the greatest probability of containing unstable carbides, such as welding seams [1176]. The type of carbides and their activity are strongly influenced by the quality of post-weld heat treatment (PWHT). The risk of attack may exist at quite moderate temperatures (ca. 200 °C) and a hydrogen partial pressure as low as 7 bar.

Numerous studies, experiments and careful investigations of failures have made it possible to largely prevent hydrogen attack in modern ammonia plants by proper selection of hydrogen-tolerant alloys with the appropriate content of metals that form stable alloys. Of great importance in this field was the work of NELSON [1174, 1177, 1178], who summarized available experimental and operational experience in graphical form. These Nelson diagrams give the stability limits for various steels as a function of temperature and hydrogen partial pressure. In [1179, 1180], CLASS gives an extensive survey, still valid today, on this subject. Newer experience gained in industrial applications required several revisions of the original Nelson diagram. For example, 0.25 and 0.5 Mo steels are now regarded as ordinary nonalloyed steels with respect to their hydrogen resistance [1172].

Physical Hydrogen Attack. A related phenomenon is physical hydrogen attack, which may happen simultaneously with chemical attack. It occurs when adsorbed molecular hydrogen dissociates at higher temperatures into atomic hydrogen, which can diffuse through the material structure. Wherever hydrogen atoms recombine to molecules in the material structure (at second-phase particles or material defects such as dislocations) internal stress becomes established within the material. The result is a progressive deterioration of the material that lowers its toughness until the affected piece of equipment cracks and ultimately ruptures.

The phenomenon is also referred to as hydrogen embrittlement. It is most likely to occur in welds that not received proper PWHT. Holding a weld and the heat-affected zone for a prolonged period at elevated temperature (an operation known as soaking) allows the majority of included hydrogen to diffuse out of the material. But this may not be sufficient if moisture was present during the original welding operation (for example if wet electrodes or hygroscopic fluxes were used), because traces of atomic hydrogen are formed by thermal decomposition of water under the intense heat of the welding procedure. Highly critical in this respect are dissimilar welds [1181], such as those between ferritic and austenitic steels [1182], where the formation of martensite, which is sensitive to hydrogen attack, may increase the risk of brittle fracture.

At higher temperature and partial pressure, hydrogen is always soluble to a minor extent in construction steels. For this reason it is advisable not to cool vessels too rapidly when taking them out of service, and to hold them at atmospheric pressure for some hours at 300 °C so that the hydrogen can largely diffuse out (soaking). In contrast to the hydrogen attack described above this phenomenon is reversible.

It has been reported [1183, 1184] that hot-wall converters in which the pressure shell is in contact with ammonia-containing synthesis gas at 400 °C have developed cracks in circumferential welding seams to a depth nearly approaching the wall thickness in places. This was surprising because the operating conditions for the material (2.25 Cr/1 Mo) were well below the Nelson curve. One investigation [1185] concluded that hydrogen attack had occurred by a special mechanism at temperatures lower than predicted by the Nelson diagram. Nitriding proceeding along microcracks could transform carbides normally stable against hydrogen into nitrides and carbonitrides to give free active carbon, which is hydrogenated to methane. High residual welding stress and internal pressure are considered to be essential for the propagation of the cracks. A rival theory [1184, 1186] attributes the damage to physical hydrogen attack resulting from the use of agglomerated hygroscopic flux in combination with insufficient soaking. However, for a longer time no problems were experienced for the same converter design in other cases where non-hygroscopic flux was used for welding and a more conservative vessel code was adopted, leading to less stress on account of thicker walls. But in the meantime also these converters developed cracks. So there are still questions left regarding the two theories.

Nitriding is a problem specific to the ammonia converter. It occurs in the presence of ammonia on the surface of steel at temperatures above 300 °C [1187 – 1190]. With unalloyed and low-alloy steels, the nitride layer grows with time to a thickness of several millimeters. Austenitic steel, used for the converter basket, develops very thin but hard and brittle nitride layers, which tend to flake off. In the nitrided areas, the risk of formation of brittle surface cracks exists.

Temper Embrittlement. For heat-resistant steels long-term service at temperatures above 400 °C (e.g., high-pressure steam pipes) can lead to a decline in impact strength [1191]. Normally, transition temperatures (precipitous decline of notched bar impact values) of below 0 °C are encountered, but this can increase to 60 °C and more (temper embrittlement).

The susceptibility to temper embrittlement can be reduced by controlling the level of trace elements (Si, Mn, P, Sn) in steels [1192, 1193]. Vessels for which temper embrittlement is anticipated should not be pressurized below a certain temperature.

Metal Dusting. Metal dusting [418, 1194 – 1200] [102, 1201] is a corrosion phenomenon which has come into focus again in the last few years with the introduction of exchanger reformer technology and the operation of steam superheaters in the hot process gas downstream of the secondary reformer waste-heat boiler. Conventional carburization is a familiar problem with high-temperature alloys in steam reforming furnaces caused by inward migration of carbon leading to formation of carbides in the metal matrix. It happens at high temperatures, typically above 800 °C, and the carbon originates from cracking of hydrocarbons. In contrast, metal dusting occurs at 500 – 800 °C on iron – nickel and iron – cobalt based alloys with gases containing carbon monoxide. The Boudouard reaction, strongly catalyzed by iron, nickel, and cobalt, is generally regarded as the source of the carbon in this case. It is assumed that thermodynamically favored sites exist for these elements at the surface and enhance the carbon deposition if the gas composition corresponds to a carbon activity >1 [1198, 1199].

As the name implies, the affected material disintegrates into fine metal and metal oxide particles mixed with carbon. Depending on the defects in a protective oxide film on the metal surface and the ability of the material to sustain this film, an induction period may be observed until metal dusting manifests itself as pitting or general attack. A possible mechanism was proposed by GRABKE [1202] and HOCHMANN [1203].

At least from a theoretical point of view, alloys formulated to form chromium, aluminum, or silicon oxide films should exhibit an increased resistance. Efforts to find solutions for this problem are continuing, but at present the following situation must be accepted: virtually all high-temperature alloys are vulnerable to metal dusting; higher steam/carbon ratios reduces this sort of corrosion; improvements may be expected by additional surface coating (for example with aluminum). With materials such as Inconel 601, 601H, 625 and similar alloys it is at least possible to reduce the attack to a level which is tolerable in practical operation. The active sites at the metal surface which catalyze the Boudourd reaction can be poisoned by H_2S, thus inhibiting the initiation of metal dusting [1199].

Hydrogen Sulfide Corrosion. Corrosion by hydrogen sulfide in partial oxidation plants can be controlled by the use of austenitic steels, but special care to ensure proper stress relief of welds is advisable to avoid stress corrosion cracking in these plants caused by traces of chlorine sometimes present in the feed oil.

Stress Corrosion Cracking. Stress corrosion cracking (SCC) of many steels in liquid ammonia is a peculiar phenomenon. It occurs at ambient temperature under pressure as well as in atmospheric storage tanks at -33 °C. Extensive studies [1204 – 1216] have defined the conditions under which SSC in liquid ammonia is likely to occur and how it may largely be prevented, but the mechanism is so far not fully understood. Preventive measures include maintaining a certain water content in the ammonia and excluding even traces of oxygen and carbon dioxide. Welds in pressurized vessels must be properly stress-relieved, and in atmospheric tanks it is important to select the appropriate welding electrodes, avoid excessive differences of the thickness of plates welded together and choose the correct geometry of welding seams.

5. Storage and Shipping

Producing and processing ammonia requires storage facilities to smooth out fluctuations in production and usage or shipments. If manufacture and use occur in separate locations, then appropriate transport must be arranged [1217]. This may be by ocean-going ships, river barges, rail or tank cars, or by pipeline. Ammonia is usually handled in liquid form, but in some cases delivery of ammonia vapor to downstream consumers on site may have some advantage due to savings of refrigeration energy in the ammonia plant. If there is an opportunity for on-site usage or marketing of aqueous ammonia, obtained as a byproduct of purge-gas scrubbing or, less frequently, deliberately produced, storage and handling facilities for this product will also be needed, but compared to liquid ammonia the demand is negligible.

Liquid ammonia is a liquefied gas. Its storage and distribution technologies therefore have much in common with other liquefied gases. Reference [415, vol. IV] summarizes the literature on storage, handling, and transportation of ammonia.

5.1. Storage

Three methods exist for storing liquid ammonia [1218, 1219]:

1) Pressure storage at ambient temperature in spherical or cylindrical pressure vessels having capacities up to about 1500 t
2) Atmospheric storage at -33 °C in insulated cylindrical tanks for amounts to about 50 000 t per vessel
3) Reduced pressure storage at about 0 °C in insulated, usually spherical pressure vessels for quantities up to about 2500 t per sphere

The first two methods are preferred, and there is growing opinion that reduced-pressure storage is less attractive [415]. However, in many cases, the combination of atmospheric and pressure storage may be the most economic concept (Fig. 79). The determining factors for the type of storage — apart from the required size — are temperature and the quantity of ammonia flowing into and out of storage [1220].

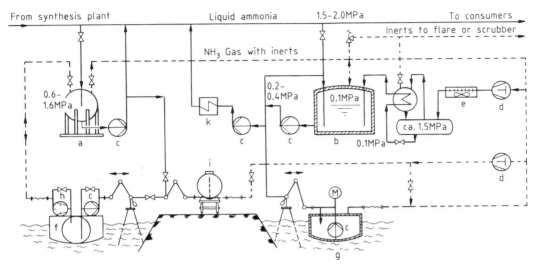

Figure 79. Ammonia terminal with loading and unloading facilities
a) Sphere at ambient temperature; b) Tank at ambient pressure (refrigerated); c) Pumps; d) Compressors; e) Air-cooled condenser; f) Barge with pressure tank; g) Barge for ammonia at ambient pressure; h) Booster; i) Rail car or truck; k) Heater

Table 30. Characteristic features of ammonia storage tanks

Type	Typical pressure, bar	Design temperature, °C	t ammonia per t steel	Capacity, t ammonia	Refrigeration compressor
Pressure storage*	16 – 18	ambient	2.8	< 270*	none
Semi-refrigerated storage	3 – 5	ca. 0	10	450 – 2700	single stage
Low-temperature storage	1.1 – 1.2	– 33	41 – 45	4500 – 45 000	two-stage

* Refers to cylindrical tanks ("bullets"); spherical vessels see Section 5.1.1.

Characteristic features of the three types of storage are summarized in Table 30[418]. For pressurized storage, spheres with a capacity up to 1500 t have been constructed, in which case the ratio tonnes ammonia per tonne of steel will become about 6.5.

5.1.1. Pressure Storage

This system is especially suitable for:

– Storing small quantities of ammonia
– Balancing production variations with downstream units processing pressurized ammonia
– Loading and unloading trucks, tank cars, and marine vessels carrying pressurized ammonia
– Entrance to or exit from pipeline systems

Usually, cylindrical pressure vessels are designed for about 2.5 MPa. The larger spherical vessels are designed only for about 1.6 MPa to avoid wall thicknesses above 30 mm. A coat of reflecting paint or, frequently in hot climates, an outer covering of insulation may be used on the vessels to avoid solar radiation heating. Spraying the vessel with water is very effective against intense solar radiation but does stain and damage the paint. As a rule, liquid ammonia fed to storage from a synthesis loop carries inert gases with it. Besides the prescribed safety relief valves, pressure control is provided for the storage drum by controlled bleeding of the inert gases through a pressure-reducing valve, for example, into a water wash system (Fig. 79).

Sometimes high-strength or fine-grained steels are used in making pressure vessels. These may be susceptible to stress corrosion cracking by ammonia. Safely avoiding this hazard requires careful thermal stress relief after completing all welding on the vessel.

The shape of the vessel depends above all on its capacity:

Cylindrical, usually horizontal for up to about 150 t

Spherical vessels resting on tangentially arranged support columns but also more recently, for static and safety reasons, in a suitably shaped shallow depression for about 250 – 1500 t.

Ancillary equipment, designed for at least 2.5 MPa (25 bar): meters and flow controls for pressurized ammonia feed and effluent streams; centrifugal pumps for discharging into liquid ammonia supply piping and for liquid ammonia loading; equipment for safe pressure relief for ammonia vapor and inerts (see Fig. 79 and [1219]). Design of pressure storage tanks and the related safety aspects are discussed in [1221].

Stress corrosion cracking in pressurized ammonia vessels and tanks is a problem which has been discussed in many papers [1205 – 1216, 1222 – 1229]. The mechanism of this phenomenon, the influence of water, and the role of oxygen are not yet completely understood, in spite of extensive research. A review is given in [1230]. As it is generally accepted that addition of water may inhibit stress corrosion [415, 1216, 1222] it has become a widely used practice to maintain a water content of 0.2 % in transport vessels [1222]. Protection may also be achieved by aluminum or zinc metal spray coating [1228, 1230]. More recent research [1226, 1230], however, has shown that water may not give complete protection.

The prevailing opinion was that stress corrosion should not occur in atmospheric storage tanks [1231]. Therefore, it was somewhat surprising when cases of stress corrosion in atmospheric ammonia tanks were reported [1232 – 1234]. Descriptions of further incidents, inspection techniques, and repair procedures can be found in [1229, 1235 – 1243].

In 1995, the capital investment, including ancillary equipment, for a 1000-t pressurized ammonia storage facility amounted to about $ 3.5 $\times 10^6$.

5.1.2. Low-Temperature Storage

Modern single-train plants need to have large-volume storage facilities available to compensate for ammonia production or consumption outages. Storage equivalent to about 20 days is customary for this. Moreover, transporting large amounts of ammonia by ship or pipeline and stockpiling to balance fluctuations in the ammonia market have great importance. Therefore, correspondingly large storage facilities are necessary at ship and pipeline loading and unloading points.

For comparable large storage volumes, the capital investment costs for atmospheric pressure storage are substantially lower than for pressure storage. In spite of higher energy costs for maintaining the pressure and for feed into and out of atmospheric storage, it is still more economical than pressure storage. This applies especially to storage of ammonia coming from the synthesis loop at low temperature and loading and unloading of refrigerated ships.

For atmospheric storage at $-33\,°C$ single tanks with a capacity up to 50 000 t of ammonia are available [1243]. Design pressure is usually 1.1 – 1.5 bar (plus the static pressure of the ammonia). The cylindrical tanks have a flat bottom and a domed roof and are completely insulated. Refrigeration is provided by recompression of the boil-off, usually with two-stage reciprocating compressors. Incoming ammonia of ambient temperature is flashed to $-33\,°C$ before entering the tank, and the vapors from the flash vessel are also fed to the refrigeration unit. Refrigeration units have at least one stand-by unit, mostly powered by a diesel engine. Single-wall and double-wall tanks are used in the industry.

Single-Wall Tanks. The single-wall tank has a single-shell designed for the full operating pressure. Mats or panels of rock wool or foamed organics (e.g., in-situ applied polyurethane foam) are used for insulation of single wall tanks. The outside insulation must be completely vapor tight to avoid icing and requires the highest standards of construction and maintenance to avoid hazardous deterioration by meteorological influences. A metal sheet covering is normally applied, and sometimes a bond wall of reinforced concrete or steel is added [1244, 1245].

Double-wall tanks are known in various designs. In the simplest version an inner tank designed for storage temperature and pressure is surrounded by a second tank. The annular space between the two walls is filled with insulation material, for example, Perlite. The main purpose of the outer shell is to contain and hold the insulation. Today's usual practice is to design the outer shell to the same standard as the inner shell. This so-called double-integrity tank concept provides an additional safety measure as the outer tank can hold the full content if the inner shell fails.

Further safety provisions include surrounding the tanks by dikes or placing them in concrete basins to contain the liquid ammonia in the event of a total failure. Discussions of such secondary containment are found in [881, 1244 – 1246].

Two principle "double integrity" designs are used as shown in Figure 80. In Figure 80 A and B the inner tank has a solid steel roof and is pressure-tight, whereas the gap between inner and outer shell, which can withstand the full hydrostatic pressure at operating temperature, is only covered by the water-tight insulation. The insulation for the inner shell may fill the total annular space [e.g., lightweight concrete [1247] or Perlite (B)] or may consist of an organic foam attached to the outside of the inner shell (A).

In the design according to Figure 80 C the outer tank is pressure-tight. The inner tank has a suspended roof and the insulation fills the annular gap, which contains an atmosphere of ammonia vapor.

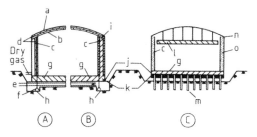

Figure 80. Arrangements of refrigerated ammonia tanks
a) Waterproof roofing; b) Steel roof; c) Inner steel shell; d) Foamed organic materials or rock wool; e) Heating coils; f) Stay bolts; g) Inner steel base; h) Concrete ring wall; i) Loose Perlite or lightweight concrete and rock wool; j) Outer steel shell and base; k) Cellular glass or lightweight concrete; l) Suspended deck with insulation; m) Pile foundation; n) Steel roof and shell or prestressed concrete lined for vapor containment; o) Loose Perlite or insulation on liner or in space

Because of the static loading, cellular glass or lightweight concrete [1247] is used for the floor insulation in all cold tank systems. A foundation as shown in Figure 80 A and B requires underground heating to prevent formation of a continuous ice sheet under the tank, which could lift it. If, as in Figure 80 C, the tank bottom is on piles above ground level, heating is unnecessary.

The tanks are fabricated by welding on site, from steels which retain their notch ductility strength at low temperature. Corresponding to the decreasing static pressure, steel plate thickness in the cylindrical shell is reduced stepwise from bottom to top. Technical details and design questions are treated in [415 Vol IV], [1246, 1248 – 1250]. Influence of climatic conditions on design and operation is considered in [1251]. Specific cases and foundation problems are discussed in [1249, 1252 – 1254]. Retrofitting of existing storage tanks is described in [1255]. Maintenance and inspection procedures are covered in [1087, 1256 – 1261]. For stress corrosion occurred in atmospheric tanks see Section 5.1.1.

A rough budget figure of the capital investment in 2005 for a double-integrity tank with a capacity of 45 000 t including cooling compressors is around $ 10×10^6. Inclusion of facilities to receive or deliver pressurized ammonia at ambient temperature would add a few millions more. Energy consumption would also increase considerably, to about 80 kWh/t ammonia throughput. In addition, substantial investment will be necessary for a complete terminal including, for example, a jetty and equipment for ship loading and facilities for loading rail and/or tank cars.

5.1.3. Underground Storage

For years, the liquefied petroleum gas (LPG) industry has used pressurized underground liquefied gas storage. This technique has been applied to ammonia also. DuPont has operated a rock cavern in the United States with a capacity of 20 000 t. Norsk Hydro has one in Norway at 50 000 t. Because of the contaminants occurring in liquid ammonia stored this way and the lack of suitable construction sites, no further storage facilities of this kind have been built for a long time.

5.1.4. Storage of Aqueous Ammonia

To allow storage and transportation at temperatures up to 35 °C, the concentration of aqueous ammonia should not exceed 25 %, because of its vapor pressure. A small facility uses mostly cylindrical vessels. The capacity of the transporting equipment determines the storage volume, e.g., at least 40 m^3 storage volume for 30 m^3 tank cars. For larger capacities, tanks are used. If it is necessary to avoid contaminating the aqueous ammonia by iron hydroxide, austenitic (stainless) steels may be used instead of the usual carbon steels.

5.2. Transportation

Transportation in Small Containers. The most common containers are:

- Cylindrical steel bottles and pressurized flasks for about 20 – 200 kg anhydrous ammonia to meet the requirements of laboratories, small refrigeration systems, and the like
- Polyethylene canisters, metal casks, and the like for 25 % aqueous ammonia.

Transportation in Trucks and Rail Cars with Capacities Normally up to 100 m^3; Jumbo Rail Cars Holding up to 150 m^3. The most widely used methods are:

– With pressure vessels for anhydrous ammonia (maximum allowable operating pressure about 2.5 MPa)
– With atmospheric pressure vessels for 25 % aqueous ammonia
– With vessels designed for elevated pressure for high-concentration aqueous ammonia (maximum allowable operating pressure in accordance with the ammonia content up to 1.6 MPa).

A modern rail car loading station for anhydrous and aqueous ammonia is described in [1262, 1263].

The distribution in rail cars and trucks primarily serves to supply smaller processing operations and wholesale merchants. However, rail transport of liquid ammonia may to some extent supplement large marine and pipeline shipments. Reference [1264] examines rail freight cost. Normally, shipping liquid ammonia by truck is used only where other means of transport are not available, e.g., in the agricultural practice of direct fertilization.

Shipping in Ocean-Going Vessels and River Barges. Regarding the transport volume, shipping of anhydrous ammonia is far more important than transport by rail. In 2004, for example, a total of 13×10^6 t of anhydrous ammonia was transported by ocean-going vessels (IFA statistics). Overseas shipping gained great momentum through exports from producers in countries with low natural gas prices or low-price policies [1265]. A comparison of shipping costs is given in [1266]; for more up-to-date information consult the journal *Nitrogen* (later *Nitrogen & Methanol*, today *Nitrogen and Syngas*), published by British Sulphur, which regularly reports shipping prices. Most river barges have loading capacities of 400 to 2500 t and mostly have refrigerated load, but a few are pressure vessels. Ocean-going vessels may transport as much as 50 000 t of fully refrigerated ammonia.

Pipelines. Transport of large volumes of ammonia by pipeline [1266, 1267] over great distances is far more economical than by river barge or rail. In the USA the MidAmerica Pipeline System transports ammonia to terminals in Kansas, Nebraska, and Iowa, which are intensive agricultural areas. The Gulf Central Pipeline connects the major producers along the Texas and Louisiana Gulf coast with terminals in Iowa, Illinois, Nebraska, and Missouri. [1268]The MidAmerica Pipeline, for example, has a peak delivery capacity of 8000 t/d for a number of destinations [1269], the hold up is about 20 000 t. Maintenance is described in [1270].

As ammonia is transported at a temperature of at least 2 °C it has to be warmed up at the supplier terminal and cooled down again to − 33 °C at the receiver terminal. Exact knowledge of the $p - V - T$ properties is important [1271]. Automatic lock valves are installed at intervals of 10 miles, so that the volume which can be released between two valves is limited to 400 t.

The world's longest ammonia pipeline has been in operation since 1983 in Russia, connecting the large production facilities

Togliatti/Gordlovka with the terminals Grigorowski/Odessa at the Black Sea over a distance of 2424 km [1272]. Apart from some short pipeline connections, most of them shorter than 50 km, there is no significant pipeline transport in Europe, where ammonia is predominantly further processed in downstream facilities on site and where no widespread direct application to agricultural crops exists.

6. Quality Specifications and Analysis

The quality of the ammonia product depends to some extent on the operating conditions of the production plant and storage. For example, water content from a synthesis loop receiving a dry make-up gas is about zero, whereas plants which receive the feedgas after a methanation without further drying may give a water content of 0.1 – 0.2 %. Oil may be introduced by the seal oil of the synthesis compressor, but on account of its low solubility in liquid ammonia it usually settles out on storage, so that only a minor concentration will remain in commercial deliveries. There are two commercial qualities of anhydrous ammonia: commercial (technical) grade (ammonia as received from production and storage) and refrigeration grade (technical product purified by distillation [1273] or molecular sieve adsorption). Table 31 lists the commercial specifications. The relevant standards are US Specification OA-445a, Supplement 5 (1963) and, in Germany, DIN 8960 (1972) for refrigeration-grade ammonia. To inhibit stress corrosion cracking a water content of at least 0.2 % for shipped and pipelined liquid ammonia is generally recommended and is mandatory in the USA.

Various concentrations and purities of *aqueous ammonia* are on the market. Mostly, the concentration is 25 – 30 % NH_3 and the iron content less than 10 ppm. Shipping in pressure vessels is necessary for ammonia contents above 25 % because of its elevated vapor pressure. For more stringent purity requirements for aqueous ammonia, the containers should be made of seawater-resistant aluminum (magnesium alloyed) or austenitic steels.

Analysis. Ammonia is readily detectable in air in the range of a few parts per million by its characteristic odor and alkaline reaction. Specific indicators, such as Nessler's reagent (HgI in KOH), can detect ammonia in a concentration of 1 ppm. For the quantitative determination of ammonia in air, synthesis gas, and aqueous solutions, individual (manual) and continuous (recorded) analyses can be made (for a measurement station for automatic determination of ammonium/ammonia, see [1274]). The methods used include, among others:

Acidimetry and volumetric analysis by absorption
Gas chromatography
Infrared absorption
Thermal conductivity measurement
Electrical conductivity measurement
Measurement of heat of neutralization
Density measurement (for aqueous ammonia)

Normally, the water content of liquid ammonia is determined volumetrically as the ammonia-containing residue on evaporation or gravimetrically by fully vaporizing the ammonia sample and absorbing the water on KOH.

The oil content of liquid ammonia can be tested gravimetrically by first evaporating the ammonia liquid and then concentrating the ether extract of the residue. Iron, aluminum, calcium (ammonia catalyst) and other impurities can then be determined in the ether-insoluble residue.

The inert gas content is analyzed volumetrically after the vaporized ammonia has been absorbed in water. Then the inert gas composition is analyzed chromatographically.

7. Environmental, Safety, and Health Aspects

7.1. Environmental Aspects of Ammonia Production and Handling

Measured by its overall environmental impact — air, water and soil pollution, materials and energy consumption — ammonia production is a rather clean technology, with low emissions, low energy consumption due to high-efficiency process design, and no severe cross-media dilemmas, in which improving one environmental effect worsens another. Typical emissions are

Table 31. Minimum quality requirements for ammonia

Quality		Commercial grade		Refrigeration grade	
		USA	Germany	USA	Germany
Purity	wt %, min	99.5	99.5	99.98	99.98 *
Water	wt %, max	0.5	0.2	0.015	0.2
Inerts **	mL/g, max	not spec'd	not spec'd	0.1	0.08
Oil	ppm by wt	5.0	5.0	3.0	not spec'd
Free of H_2S, pyridine, and naphthene					

* Allowable boiling point change on vaporization of 5 – 97 % of the test sample, 0.9 °C
** The noncondensable gases dissolved in ammonia are H_2, N_2, CH_4, and Ar. Their amounts depend on the methods of synthesis and storage. The inerts amount to about 50 mL/kg for atmospheric storage.

given in EFMA's publication Production of Ammonia in the series "Best Available Techniques and Control in the European Fertilizer Industry" [1275].

Steam Reforming Ammonia Plants. Table 32 summarizes the emission values for steam reforming ammonia plants.

The source of the NO_x emissions is the flue gas of the fired primary reformer, and in plants without a fired tubular reformer NO_x is emitted from fired heaters and auxiliary boilers, but in considerably smaller quantities. However, NO_x emissions from ammonia production, compared to the total amount from human activities, is in fact a marginal quantity. Only about 0.16 % of the anthropogenic NO_x emissions come from ammonia production. Emissions to water, generally originating from the condensate from the condensation of surplus process steam ahead of the CO_2 removal system, can largely be avoided. Minor concentrations of methanol and amines can be removed and recycled to the reformer feed by stripping with process steam, and the stripped condensate can be recycled to the boiler feed water after polishing with ion exchangers. Steam stripping without recycling to the reformer would produce a cross media effect, because in this case the pollutants would be transferred from water to air. Another possibility is use of the process condensate for feed gas saturation [452].

Partial oxidation ammonia plants have the same emission sources except for the primary reformer flue gas. The plants have an auxiliary boiler to generate steam for power production and fired heaters, which on account of the sulfur content of the fuel oil release a flue gas containing SO_2 (< 1500 mg/m^3). Other possible emissions are H_2S (< 0.3 mL/m^3), CO (30 mL/m^3) and traces of dust. The NO_x content of the flue gas depends on the configuration of the auxiliary boiler and on the extent electric power generation on the site as opposed to outside supply. The total NO_x emission per tonne of product may be somewhat lower than for steam reforming plants.

Noise. An increasing awareness has developed for noise generation and emission from ammonia plants especially in the neighborhood of residential areas [1276]. Many investigations have dealt with noise generation and noise abatement in ammonia plants [1277 – 1281]. The following major sources can be identified: depressurizing of large gas quantities for control or venting, steam blowing, burner noise, resonance vibrations in the flue gas ducts, and noise from compressors, blowers, and pumps. Measures for noise reduction include installing low-noise letdown valves, use of silencers, sound-reducing enclosures for compressors or housing them in closed buildings.

Emission Limits and Guideline Values for Ammonia Production. There are two categories of regulations:

1) Legally binding emission values for certain pollutants associated with ammonia production
2) Guideline values not legally binding but providing the background for requirements laid down in individual operating permits

In Europe such legally binding emission levels relevant to ammonia production exist only in Germany. In the Netherlands and Germany limits for emissions from boilers also apply for ammonia plants, for example for the reformer

Table 32. Emissions from steam reforming ammonia plants

			Existing plants	New plants
Emission to air	NO_x (as NO_2)*	ppmv (mL/m^3)	150	75
		mg/Nm^3	300	150
		kg/t NH_3	0.9	0.45
Emission to water	NH_3/NH_4 (as N)	kg/t NH_3	0.1	0.1
Waste material**		kg/t NH_3	< 0.2	< 0.2

* At 3 % O_2.
** Spent catalysts.

furnace. The present limit in Germany is 200 mg NO_x/m^3 for furnaces up to 300 MW$_{th}$. Specific emission guideline values are laid down in the United Kingdom. In the other European countries no national emission limits or guidelines exist; individual operation permits are negotiated, usually orientated on other cases and countries. In the United States, for example, NO_x emission level values are categorized in the Clean Air Act, which defines a significance level for the source according to the total emission (10 to 100 t/a) on the one hand, and assigns a threshold limit to the geographical area (100 t/a to 10 t/a) on the other.

For ammonia in air the authorities in Germany require that the achievable minimum level should correspond to the state of the art; originally projected values in a draft of TA Luft [1282], were not subsequently adopted; some MIK values (maximum imission concentrations) are given in VDI Richtlinie 2310 (Sept. 1974). In Germany there is no maximum ammonia concentration laid down for wastewater, it is only referred to the general interdiction to introduce poisons or harmful substances into rivers and lakes. A concentration of 1.25 mg NH_3 per liter, stated to be harmful to fish, could serve as a guideline. In the "Catalog of Water Endangering Substances" (Germany) anhydrous ammonia is not listed, but aqueous ammonia is classified in the Water Endangering Group 1 as a weakly endangering substance.

To harmonize European regulations concerning emission value permits for industrial production, the EU commission has proposed a Council Directive Concerning Integrated Pollution Prevention Control (IPPC Directive). A key element is the concept of Best Available Techniques (BAT). This directive requests an exchange between national environmental authorities, industry and nongovernmental organizations with the aim of preparing IPPC BAT Reference Documents (BREFs), which could provide a common accepted basis for emission limit value regulations of the individual EU member states. A BREF for ammonia is being prepared to serve as pilot project. A previous BAT document on ammonia was published in 1990 [1283]. For noise emissions usually local permits are negotiated, whereby in most instances the contribution to the sound level in the surrounding residential areas will be limited. For the limitation of sound levels at working places within the plant occupational safety regulations are valid which also require ear protection above a certain level. The European Fertilizer Manufacture's Associatiom (EFMA) has published a series: "Best Available Techniques for Pollution Prevention and Control in the European Fertilizer Industry" Booklet No. 1 is "Production of Ammonia" [1275]. For most up-to-date immission regulations contact the relevant authorities in the different countries.

7.2. Safety Features

In ammonia production three potential hazard events can be identified: fire/explosion hazard from the hydrocarbon feed system; fire/explosion hazard due to leaks in the synthesis gas purification, compression, or synthesis section (75 % hydrogen); and toxic hazard from release of liquid ammonia from the synthesis loop. In addition there is also a potential toxic hazard in handling and storing of liquid ammonia. The long history of ammonia production since 1913 has demonstrated this production technology is a very safe operation. Despite the complexity, size and partially severe operating conditions of vessels and piping remarkably few failures occur. It is largely thanks to such forums as the annual AIChE Safety in Ammonia Plants and Related Facilities Symposium that failures, incidents, and accidents are reported and discussed including the manner how the problems were handled and dealt with. Thus the industry as a whole has the opportunity to learn from the specific experiences of its individual members. A short desciptions of a number of accidents is found in [1284, 1285]. The severe impacts of rare events with explosions seem to be confined to a radius of around 60 m [1275]. The long industrial experience is summarized in a number of national codes and standards which have to be applied for design, material selection, fabrication, operation, and periodic technical inspections of the equipment used. This is especially important for equipment operating at high temperatures and/or pressures. Apart from proper design, skilled and well-trained operators and effective and timely maintenance are essential for safe plant operation.

General practice today is to make so-called HAZOP (hazards and operability) studies with an experienced team consisting of operating personal, process engineers (also from contractors), experts in process control and safety experts (often independent consultats from outside the company) in which, following a very meticulous procedure apparatus after apparatus is checked for potential failure and risk possibilities together with a proposal for the appropriate remedies. HAZOP studies introduced by ICI [1286 – 1289] for general application in chemical plants proved to be very helpful not only for existing plants but also in the planning phase of new ones. It is especially important to define and configure trip systems and trip strategies for safe plant shut downs in case of offset conditions. A similar approach, called Reliability, Availability, Maintainability (RAM) was developed by KBR [1290]. A special point is the assessment of remaining service life of equipment because many ammonia plants operate beyond there originally anticipated lifetime, which was usually defined by economic criteria [1291, 1292]. In this context the question of adequate maintenance is very important and to which extent preventive maintenance is provided. Usually single-train ammonia plants are shut down every two to five years for some weeks for changing of catalyst, equipment inspections, maintenance and, if they fit in the time schedule, also smaller revamps. The time frame of unscheduled shut downs because of problems which need urgent repairs can be used for routine maintenance operations, too.

The inspections are widely carried out with non-destructive testing [1293]. Further information is given in [975]. Statistics of plant on-stream factors, shut-down frequencies and time, as well as an investigation of the causes are found in [1294, 1295].

An area which deserves special attention with respect to safety is the storage of liquid ammonia. In contrast to some other liquefied gases (e.g., LPG, LNG), ammonia is toxic and even a short exposure to concentrations of 2500 ppm may be fatal. The explosion hazard from air/ammonia mixtures is rather low, as the flammability limits [1296 – 1300] of 15 – 27 % are rather narrow. The ignition temperature is 651 °C. Ammonia vapor at the boiling point of − 33 °C has vapor density of ca. 70 % of that of ambient air. However, ammonia and air, under certain conditions, can form mixtures which are denser than air, because the mixture is at lower temperature due to evaporation of ammonia. On accidental release. the resulting cloud can contain a mist of liquid ammonia, and the density of the cloud may be greater than that of air [1296, 1301]. This behavior has to be taken into account in dispersion models.

Regulations and Guidelines. For maintaining safety in ammonia storage, transfer, and handling many regulations and guidelines exist in the various countries. It is beyond the scope of this article to list them completely and to check for the most up-to-date status. It is recommended to consult the relevant governmental authorities of the individual countries and the relevant industrial codes, applying to fabrication and operation of equipment. An overview of such regulations and standards is given in [1291].

Ammonia Spills. In an ammonia spill a portion of the liquid ammonia released flashes as ammonia vapor. This instantaneous flash is followed by a period of slow evaporation of the remaining liquid ammonia [1301, 1302]. In a release from a refrigerated tank operating slightly above atmospheric pressure, the amount of the initial flash is only a few percent of the total, whereas in a release from a pressure vessel at about 9 bar and 24 °C approximately 20 % of the spilled ammonia would flash. Figures for the proportion flashed as a function of pressure and temperature of the spilled ammonia are given in [1302]. This reference also provides quantitative information on temperature development and evaporation rate of small and large ammonia pools at certain wind velocities and also takes radiation influences (sunny or overcast sky, day or night) into account.

Spill experiments on land and on water in various dimensions have been carried out by various companies and organizations. Underwater releases were also studied. Models have been developed to mathematically describe ammonia dispersion in such events [1303 – 1308].

A number of real incidents with tank cars, rail cars, tanks, loading/unloading of ships and barges, and pipeline transport in which major spills occurred is described in [1309 – 1318] and in an excellent review on safety in ammonia storage [1319]. Training of operating personal for ammonia handling with special regard to ammonia spills is important [1317, 1320, 1321].

7.3. Health Aspects and Toxicity of Ammonia

In ammonia production, storage, and handling the main potential health hazard is the toxicity of the product itself. For this reason this section concentrates on ammonia only. Other toxic substances such as carbon monoxide or traces of nickel carbonyl (which may be formed during shut down in the methanation stage) may be only a risk in maintenance operations and need appropriate protection provisions as well as blanketing or flushing with nitrogen.

Human Exposure. The threshold of perception of ammonia varies from person to person and may also be influenced by atmospheric conditions, values as low as $0.4 - 2$ mg/m^3 (0.5 – 3 ppm) are reported in [1322], but 50 ppm may easily detected by everybody [1323]. Surveys [1324] found concentrations from 9 – 45 ppm in various plant areas. Though initially irritated, exposed persons may quickly become accustomed to these concentrations. Another report [1325] gives concentration limits for short time exposures as follows: 100 ppm (10 min); 75 ppm (30 min) 50 ppm (60 min). The time-weighted average Threshhold Limit Value (TLV) of the ACGIH is 25 ppm (or 18 mg/m^3) [1326]. This recommendation was supplemented by a value for short-time exposure: 35 ppm for 15 min. The MAK value is 50 ppm.

Exposure to higher ammonia concentration has the following effects: 50 – 72 ppm does not disturb respiration significantly [1327]; 100 ppm irritates the nose and throat and causes a burning sensation in the eyes and tachypnoe [1327]; 200 ppm will cause headache and nausea, in addition to the above symptoms [1328]; at 250 – 500 ppm tachypnoe and tachycardia [1328]; at 700 ppm, immediate onset of burning sensations in the eyes [1329]; 1000 ppm causes immediate coughing [1330]. The symptomatology of various exposure levels is also described in [1326, 1331, 1332]. At 1700 ppm coughing with labored breathing, sometimes with momentary inability to breath (coughing of rescued persons may continue for hours). 2500 to 4500 ppm may be fatal after short exposure; 5000 ppm and higher causes death by respiratory arrest [1319].

The metabolism seems not to be significantly changed after exposure to 800 ppm [1330]. A discussion of metabolism and of acute and chronic health problems caused by ammonia can be found in [1333].

Toxicology. Ammonia is a strong local irritant. On mucous membranes alkaline ammonium hydroxide is formed, which dissolves cellular proteins and causes severe necrosis (corrosive effect).

The primary target organ is the pulmonary system, and the following symptoms can be observed: pharyngitis, laryngitis, tracheobronchitis, nausea, vomiting, increased salivation, reflectoric bradycardia, and life-threatening symptoms, such as edema of the glottis, laryngospasm, bronchospasm, and interstitial lung edema [1334].

Ammonia or ammonium hydroxide can penetrate the cornea rapidly, leading to keratitis, damage of the iris, cataract, and glaucoma [1335].

Oral ingestion of aqueous ammonia can corrode the mucous membranes of the oral cavity, pharynx, and esophagus and cause the shock syndrome, toxic hepatitis, and nephritis. Because of its corrosive action constrictions of the esophagus may result.

Ammonia is absorbed rapidly by the wet membranes of body surfaces as ammonium hydroxide, converted to urea, and excreted by the kidneys [1336]. The capacity of detoxification via urea is sufficient to eliminate the ammonium ion when ammonia is inhaled in nonirritating concentrations. The inhaled ammonia is partly neutralized by carbon dioxide present in the alveoli [1337]. Only a small fraction of the ammonia is exhaled unchanged by the lungs (12.3 % at an inhalation concentration of 230 ppm) [1338]. Repeated inhalation can cause a higher tolerance because the mucous membranes become increasingly resistant [1339]. Additional information on the toxicology of ammonia can be found in [1340 – 1356].

Carcinogenicity. Ammonia failed to produce an increase in the incidence of tumors in Sprague Dawley rats even when the protein ratio in the diet was increased or when urea was added [1357].

Lifetime ingestion of ammonium hydroxide in drinking water by mice was without any carcinogenic effects [1358].

Mutagenicity. Ammonia is not mutagenic in the Ames *Salmonella* system and in *Saccharomyces cerevisiae* [1359].

8. Uses

In 2003 about 83 % [1360] of ammonia production was consumed for fertilizers. Ammonia is either converted into solid fertilizers (urea; ammonium nitrate, phosphate, sulfate) or directly applied to arable soil.

The industrial use of ammonia is around 17 % [1360]. Actually every nitrogen atom in industrially produced chemicals compounds comes directly or indirectly from ammonia. An important use of the ammonia nitrogen, partly after conversion to nitric acid, is the production of plastics and fibers, such as polyamides, urea – formaldehyde – phenol resins, melamine-based resins, polyurethanes, and polyacrylonitrile. Another application is the manufacture of explosives, hydrazine, amines, amides, nitriles and other organic nitrogen compounds, which serve as intermediates for dyes and pharmaceuticals. Major inorganic products are nitric acid, sodium nitrate, sodium cyanide, ammonium chloride, and ammonium bicarbonate. Urea production consumed about 40 % of the ammonia produced in 1995.

In the environmental sector ammonia is used in various processes for removing SO_2 from flue gases of fossil-fuel power plants. The resulting ammonium sulfate is sold as fertilizer. In the selective catalytic reduction process (SCR) the NO_x in flue gases is reduced in a catalytic reaction of the nitrogen oxides with a stoichiometric amount of ammonia [1361 – 1364]. Also non-catalytic reduction is applied with ammonia or urea solutions.

Ammonia may also serve as a solvent in certain processes. Another application is the nitriding of steel. An old and still flourishing business is the use of ammonia as refrigerant, based on its low boiling point and its high heat of evaporation. For some time heavy competition came from chlorofluorocarbons (CFCs), but with increasing environmental concern regarding the application of CFCs ammonia's position is strengthening again. Ammonia is applied in a large number of industrial and commercial refrigeration units and air-conditioning installations [1365 – 1368]. In addition to the high specific refrigeration effect, the ammonia has the following advantages: it is noncorrosive; it tolerates moisture, dirt, and oil contaminants; it is cheap and there are many suppliers. A drawback is its toxicity.

The production of smaller volumes of hydrogen/nitrogen mixtures used as protective gases for chemical products [1369] and for metalworking processes [1370] by decomposition of ammonia over iron- or nickel-based catalysts at 800 – 900 °C may be an economic alternative where production or purchase of pure hydrogen is too expensive (see also → Hydrogen, Chap. 4.5.2), [1371]. Energy-related applications of ammonia are proposed in [1363].

9. Economic Aspects

About 1.4 % of the world consumption of fossil energy (not including combustion of wood) goes into the production of ammonia. In developing countries, ammonia is generally one of the first products of industrialization. As 83 % of world nitrogen consumption is for fertilizers it might be expected that ammonia production should develop approximately in proportion to the growth of world population. This was roughly the case in the mid-1980s as can be seen from Figure 81 [100]; since then the rate of increase in production has been markedly slower than that of world population. This has been mainly for economic reasons in developing countries and for ecological reasons in industrialized countries.

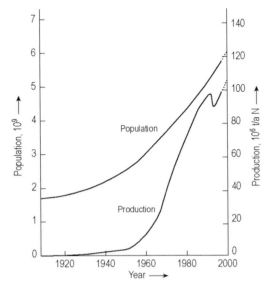

Figure 81. Development of ammonia production and world population

The political changes in the former Eastern Bloc caused a dramatic slump in 1991 – 1993. Table 33 shows the development from 1985 to 2006. Because of the surplus of capacity over demand from the mid-1980s to the mid-1990s the tendency to invest in new plants was rather small and revamping measures aiming at energy saving dominated. A retarding effect resulted also from the financial crises in Asia in the early 1990s associated with a retreat from goverment subsidizing.

From 2002/2003 onwards new ammonia projects materialized in increasing capacity numbers (see Fig. 82). The required plant availabilities to satisfiy the expected demand are between 80 and 85 %. However, since the industry is capable to reach plant availabilities of 91 – 93 %, around 40 000 t/d of capacity is idle now. Much of this idle capacity is in the United States (because of the extreme increase of the natural gas price) and in Eastern Europe. The shift is to plants in areas with low gas cost (Middle East, Trinidad, Venezuela).

The future growth of the nitrogen demand supplied by synthetic ammonia is expected to be larger than the growth of the global population. Economic growth with the tendency of higher meat consumption and improvement of nutrition in developing countries are driving forces for increased fertilizer production. Meat consumption is presently increasing with around 3 % per year [1360], the nitrogen required to produce the animal feed is in the range of 20 – 30 % of the total nitrogen demand for fertilizers. The nonfertilizer use of nitrogen is governed by the general economic development and by environmental legislation [1377].

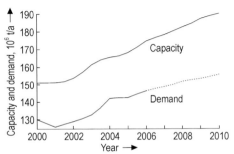

Figure 82. World ammonia capacity and demand

The geographical distribution of world ammonia capacity is shown in Figure 83. Europe and North America, which now together have a 26 % capacity share, lost their leading position (54 % in 1969) to Asia, which now accounts for 46 % (17 % in 1969), as may be seen from Figure 83.

Acccording to the figures for 2002 [1378], the P. R. China is the largest producer with 36.6×10^6 t/a NH_3, followed by North America (USA and Canada) with 16.6×10^6 t/a, the FSU countries with 15.9×10^6 t/a, India 11.9×10^6 t/a and Western Europe 11.4×10^6 t/a.

Because of the steadily increasing natural gas prices in the United States and Western Europe ammonia imports from low-gas-price locations are constantly growing and a number of ammonia plants have been temporarily or permanently closed.

Table 33. World ammonia supply/demand balance (10^6 t/a ammonia) [429, 1372 – 1377]

Year	1985	1990	1991	1992	1993	1994	1995	1998	2000	2001	2003	2004	2005	2006
Capacity	143.9	148,7	148.8	151.2	153.9	141.7	142.7	147.6	151.1	151.1	163.2	165.5	168.3	175.0
Demand	111.7	114.6	119.2	113.1	109.7	113,4	115.6	126.7	129.8	126.0	132.3	142.4	142.2	146.5
Capacity utilization, %	77.7	77.1	80.1	74.8	71.3	80.0	81.0	85.9	85.9	8.34	81.07	86.0	84.5	83.7

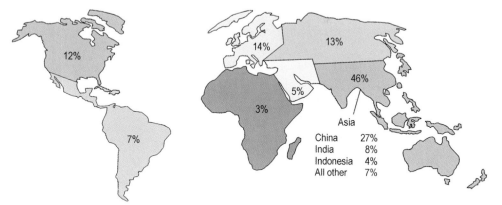

Figure 83. Geographical distribution of world ammonia capacity [429]

The major importers of ammonia are Western Europe (2002: 3.7×10^6 t) and the United States (2002: 5.6×10^6 t), whereas, on account of their enormous natural gas reserves, the FSU countries (4.5×10^6 t), Trinidad (3.5×10^6 t) and the countries in the Middle East, whose producers export predominantly to Asia, are dominant in ammonia export. Nearly three-quarters of world natural gas reserves (2003) are concentrated in two areas: in the former Soviet Union (32 %) and the Arabian Gulf (41 % including Iran). Iran, which owns 15 % of the world gas reserves, is so far not a major ammonia producer and not an ammonia exporter. An indication of the natural gas prices is given in Table 34.

Table 34. Natural gas prices at different locations around 2004 [1377, 1379, 1380]

Location	$/$10^6$ Btu*
USA	> 6.00
Canada	4.8
Western Europe	4.4
Trinidad	1.6
Venezuela	0.75
Argentina	1.25
Russia	0.9

* 1 Btu (British thermal unit) = 1055 J.

For for comparative production cost evaluations these figures should be taken with some care, because they are statistically compiled and might not always be suited for an individual case or a concrete situation. In 2005 gas prices dramatically increased in the USA, where they reached in some areas values above $ $9/10^6$ Btu, and also in Europe the $ 5 line was crossed.

The feedstock question is discussed in [1381]. Natural gas is by far the most economical feedstock for ammonia production, achieving the lowest energy consumption and requiring the lowest investment. This can be seen from Table 35, which compares ammonia production costs in North-West Europe for different feedstocks using today's best technological standards for each process.

Consumption figures for older plants can be considerably higher than those for modern ones as used in Table 35., so that the world-wide average figure for steam reforming plants is probably between 31 and 33 10^6 Btu per tonne NH_3. For the annual production 330 days of operation per year were assumed. For a specific plant the on-stream factor might be lower for market reasons or because of outages due to technical problems, which will increase the fixed costs per tonne.

With natural gas prices above $ $9/10^6$ Btu in the USA coal-based production would be competitive at full production costs [1384]. But the

Table 35. Ammonia production costs from various feedstocks (2004)[a]

Feedstock (Process)	Natural gas (steam reforming)	Vacuum residue (partial oxidation)	Coal (partial oxidation)
Feedstock price, $/10^6 Btu[b] (LHV)	4.6*	2.7*	2.5*
Consumption, 10^6 Btu[b] per tonne NH_3	27	36	45
Feed + energy (fuel), $ per tonne NH_3	124.2	97.2	112.5
Other cash costs, $ per tonne NH_3	38.0	51.0	75.0
Total cash costs, $ per tonne NH_3	162.2	148.2	187.5
Capital related costs[c], $ per tonne NH_3	65.1	96.3	162,8
Total production costs, $ per tonne NH_3	227.3	243.5	350.3
Investment[c], 10^6 $	250	370	625

[a] The 2004 Feedstock prices are taken from the BP Statistical Review of World Energy 2005 [1382]. The price of the vaccuum residue is the heavy fuel oil price minus a deduction of $ 50/t [1383].
[b] 1 Btu (British thermal unit) = 1055 J.
[c] The investment for steam reforming is a budget figure based on recent estimates for a 2000 t/d plant inside battery limits and excludes any offsites, a 45 000 t ammonia tank with refrigeration is included. It is assumed that the utilities are supplied at battery limit in required specification. Investment figures for the vacuum residue and coal-based plants are derived from the steam reforming figure by the well-known investment relation factors. Capital related costs depend highly on the financing condition and on ROI (return on investment) objectives of the investor. So the figures for the capital related costs in the table assuming debt/equity ratio 60:40, depreciation 6%, 8 % interest on debts, 16 % ROI on equity, are only an example to illustrate this substantial contribution to the total costs.

relative high investment cost for the coal-based plants create more difficulties for the project financing. Nevertheles there are already feasibility studies made in the USA for coal-based production using Illinois coal ($ 30/t, $ 1.30/10^6 Btu) [1385]. For coal as an alternative to natural gas see also [1386].

The total investment for a new ammonia production can vary considerably depending on the location and its specific conditions. In the example of Table 35. an industralized site is assumed where all the necessary off-site installations are already existing. For a remote site with heavy forest vegetation where all neccessary offsite facilities like workshops, fire fighting, ambulance, administration, and a few housing buildings and roads have to be included, the total investment for a 2000 t/d capacity could easily reach $ 450 for a natural gas steam reforming plant.

A special case is the P. R. China where as a consequence of its large coal reserves and limited gas and oil reserves more than 60 % of the ammonia production uses coal, 20 % uses natural gas and the rest is oil-based [1292]. In the rest of the world 90 % of the ammonia production uses natural gas. For the future coal has prospects [1384, 1387] as can be seen from world reserves of fossil energy and their present consumption rate (Table 36).

Only about 14 % of the ammonia is traded as such. Only for this proportion prices are regularly published in *Nitrogen + Syngas*, FINDS, by Ferticon, Green Market, IFA, EFMA, TFI statistics and other publications. Apart from some direct uses (3%) all ammonia is converteted into downstream products (Fig. 84).

Even with inclusion of the shipping costs from the low-gas-price areas Trinidad ($ 25/t) and from Russia/Ukraine ($ 65-70 /t) to the US Gulf Coast these imports can easily compete with the domestic production as to be seen from Table 37.

Because of the economy of scale there is the tendency to choose a capacity of 2000 – 2200 t/d for new projects at low-gas-price locations and some investors are looking for even larger sizes. The largest single stream plant with 3300 t/d was built by Uhde in Saudi Arabia and started up in 2006. Technical feasability studies for 5000 t/d plants are made by various contractors (see Section 4.6.4). But besides technical feasability there are questions about the economic risk and financing.

10. Future Perspectives

It is difficult to venture a prognosis for the future development of ammonia production technology. As about 85-87 % of the ammonia consumption goes into the manufacture of fertilizers, it is obvious that the future of the ammonia industry is very closely bound up with future fertilizer needs and the pattern of the world supply.

Table 36. World reserves and consumption rate of fossil feestocks in 2004 [1380]

	Coal, t	Oil, t	Natural gas, Nm^3
Reserves	456×10^9	162×10^9	180×10^{12}
Consumption per year	2778×10^6	3868×10^6	2692×10^9
Years of expected supply	164	42	67

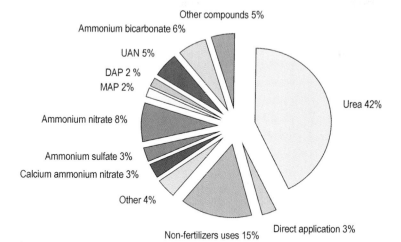

Figure 84. Applications of ammonia
UAN = Urea — ammonium nitrate; DAP = Diammonium phosphate; MAP = Magnesium — ammonium phosphate

Table 37. Ammonia cash costs 2005*

	USA	Trinidad	Russia	Arabian Gulf
Nat. gas price $/$10^6$ Btu	9–10	1.6–2.0	0.9	0.5–1.0
Feedstock costs, $ per tonne NH_3	310	50	30	16–31
Other costs, $ per tonne NH_3	30	30	30	30
Freight, $ per tonne NH_3	-	25 [1292] to USA	70 [1292] to USA	25–70 to Asia
Cash costs, $ per tonne NH_3	309–340	105–155	130	71–131

* For simplicity reasons equal gas consumption and other costs are assumed.

Nitrate pollution, which may become a problem in some countries, is sometimes used as an argument against the use of industrially produced "inorganic" or "synthetic" fertilizers and in favor of more "organic" or "natural" fertilizers. But this is a misconception, because degradation of biomass and manure also leads to nitrate run-off. In addition, the amount of this biological material is far to small to supply the increasing demand of the agriculture necessary to feed the growing world population. The solution to the run-off problem therefore can also not consist in cutting back fertilizer application, which would be disastrous in developing countries, but is likely to lie in a better management of the timing of fertilizer application, requiring a better understanding of the biological nitrogen cycle. Estimates of fixed nitrogen from natural sources and antropogenic activities on global basis are found in [1388 – 1392].

A fundamental question is, whether there are other options for the future than the present ammonia technology, which involves severe temperatures and pressures. The research for new approaches should strive for the following, admittedly rather ambitious and idealized objectives: ambient reaction conditions, less or no consumption of fossil energy, lower investment, simple application, even in remote developing areas, and wide capacity range. As the major demand is in agriculture, biological processes for in vivo conversion of molecular nitrogen into fixed nitrogen would probably be the first choice. Alternatively abiotic processes (in vitro) with homogeneous catalysis under mild reaction conditions could be discussed for ammonia production.

Biological Processes. Certain bacteria and blue-green algae are able to absorb atmospheric nitrogen, by themselves or in symbiosis with a host plant, and transform it into organic nitrogen compounds via intermediate ammonium ions, whereby the host plant indirectly supplies the large amount of energy needed. A well known example is the symbiotic relationship between legumes (e.g., peas, beans, soybeans, lupins) and Rhizobium bacteria, which settle in their root nodules. Ammonia synthesis is performed by the enzyme nitrogenase [1393 – 1395]. For the synthesis of the nitrogenase in the cell the so-called NIF gene is responsible. Apart from Rhizobium bacteria a number of other organisms have nitrogen fixation potential, living in symbiosis or providing the required energy from their own metabolism. Genetic engineering to extend biological nitrogen fixation to other plant groups pursues various routes [1396, 1397]:

1) Modifying Rhizobium to broaden its spectrum of host plants
2) Transferring the NIF gene to other bacteria which have a broader spectrum of host plants but have no natural nitrogen-fixing capability
3) Engineering soil bacteria to absorb and convert nitrogen to ammonia and release it to the soil [1398]
4) Inserting the NIF gene directly into plants

Especially the last option, to design nitrogen-fixing species of traditional crops such as wheat, rice, and corn, leads to the following questions: Would it be preferable to fix nitrogen in the leaves where energy is available or in the roots, to which carbohydrates have to be transported? If fixation is performed in the leaves, then how can the problem of the proximity of oxygen generated in the photosynthesis and oxygen-sensitive nitrogenase be dealt with? To what extent will the energy required by a plant to fix nitrogen affect overall productivity?

The last question has been discussed in some detail [1399]. For the conversion of molecular nitrogen into ammonia the following stoichiometric equation may be formulated:

$0.5\,N_2 + 1.5\,H_2O \rightarrow NH_3 + 0.75\,O_2$

$\Delta F^0_{298} = +340\,\text{kJ/mol}, \quad \Delta H^0_{298} = +383\,\text{kJ/mol}$

The oxidation of glucose generates 3140 kJ/mol and with 100 % efficiency only about 0.11 mol of glucose would be required to produce one mole of ammonia. This ideal can, of course, never be achieved. In the molybdenum-containing nitrogenase a transfer of eight electrons is involved in one catalytic cycle:

$N_2 + 8\,H^+ + 8\,e^- \rightarrow 2\,NH_3 + H_2$

As seen from this equation two electrons are "wasted" in forming molecular hydrogen, which also seems to be unavoidable in vivo. This stoichiometry suggests an efficiency of 75 %, but if additional functions associated with fixation are taken into account, a maximum overall efficiency of 10 – 15 % is estimated for the enzyme, and even less for the bacteria as a whole [1400]. In investigations on Rizobium [1401, 1402] it was shown for the nodulated roots of peas that 12 g of glucose are consumed to produce 1 g of fixed nitrogen. Using this efficiency for the case where all the fixed nitrogen in the plant is supplied only via this route, a yield loss of 18 % would result. Nitrate, for example from a fertilizer, absorbed by the plant is transformed to ammonia involving a free energy change $\Delta F^0_{298} = +331$, which is comparable to the free energy change for converting molecular nitrogen to ammonia, as shown in the stoichiometric equation above.

This is a somewhat surprising result, which suggests that nitrogen supply with nitrate has an energetic advantage over the biological fixation process [1403]. A definite answer to this question will only be possible when genetically engineered nitrogen-fixing plants are available in sufficient amounts for field tests. There is no doubt that some day in the future genetic engineering will succeed in designing nitrogen-fixing plants for agricultural crops.

But until this will become a reality, application of selected natural nitrogen-fixing microorganisms to crops would be an intermediate solution to reduce industrial fertilizer consumption. In the United States field tests with *Azospirillium* applied to corn could demonstrate a considerable increase in harvest yield [1397]. Careful directed application of blue-green algae, (*Anabaena azolla*) and water fern (*Azolla pinata*) in tropical rice fields is also very promising [1404,

1405]. Unlike most diazotropic bacteria, blue-green algae are self-sufficient in that they themselves produce the energy necessary for the nitrogen fixation by photosynthesis.

Abiotic Processes [1399]. Possibilities of converting molecular nitrogen into ammonia in homogeneous solution by using organometallic complexes have been studied from around 1966 onwards. Numerous investigations resulted in the laboratory-scale synthesis of ammonia, but with rather low yields. The prospects of this route are not judged to be very promising in terms of energy consumption and with respect to the costs of the rather sophisticated catalyst systems. Photochemical methods [1406 – 1408], to produce ammonia at ambient temperature and atmospheric pressure conditions in presence of a catalysts have up to now attained ammonia yields which are far too low to be economically attractive. For a detailed review of biological and abiotic nitrogen fixation, see → Nitrogen Fixation and [1399]; additional literature references can be found in [1390 – 1394].

Conclusions. Based on the foregoing discussion and the evaluation of present technology (see also [100]) and research efforts one may formulate the scenario for the future development of ammonia production as follows:

1) For the major part, ammonia production in the next 15 to 20 years will rely on the classic ammonia synthesis reaction combining nitrogen and hydrogen using a catalyst at elevated temperature and pressure in a recycle process.
2) It is very likely that genetic engineering will succeed in modifying some classical crops for biological nitrogen fixation and that large-scale application will occur predominantly in areas with strongly growing populations to secure the increasing food demand. This development may be pushed by the fact that compared to the classical fertilizer route less capital and less energy would be needed. This might happen in the future but an estimate regarding extent and time horizon is difficult. But even with the introduction of this new approach, traditional ammonia synthesis will continue to operate in parallel, because it might be necessary to supplement the biological nitrogen fixation with classical fertilizers. In addition, the existing ammonia plants represent a considerable capital investment and a great number of them may reliably operate for at least another 20 to 30 years from a mere technical point of view.
3) Natural gas will remain the preferred feedstock of present ammonia production technology in the medium term (15 – 20 years) as may be assumed from the world energy balance shown in Table 26. Partial oxidation of heavy hydrocarbon residues will be limited to special cases, but coal gasification could have a renaissance within this period.
4) Technical ammonia production based on non-fossil resources, e.g. electrolysis (power from nuclear, solar, water or geothermic energy) or gasification of biomass in this time frame will play no role.
5) Ammonia technology will not change fundamentally, at least not in the next 10 to 15 years. Even if there are radical, unforeseeable developments, they will need some time to mature to commercial applicability. With the traditional concepts, the margins for additional improvements have become rather small after years of continuous research and development. Thus further progress may be confined to minor improvements of individual process steps, catalysts and equipment design.
6) The best energy figure so far achieved in commercial production is around 28 GJ per tonne NH_3. With the ruthenium-based catalysts M.W. Kellogg claims a figure of 27.2 GJ per tonne NH_3. Further reduction would need an even more active catalyst to operate at a much lower synthesis pressure, for example, at around 30 bar, which is the usual pressure level of steam reforming. The probability of finding such a catalyst is rather small after all the years of intensive research. Figures between 27 and 28 GJ per tonne NH_3 already correspond to a rather high efficiency of around 75 % with respect to the theoretical minimum of 20.9 GJ per tonne NH_3 (calculated as stoichiometric methane demand, see Section 4.6.1.3).
7) The bulk of ammonia production in the near future will still be produced in world-scale plants of capacity 1000 – 2000 t NH_3 per day. Small capacities will be limited to locations where special logistical, financial or feedstock conditions favor them. More plants in high-

gas-cost areas will be closed and the gap will be filled by higher imports from low-price gas areas, where new plants, intended for export, probably also mega-plants, will be built.

8) Most newer technology developments will mainly reduce investment costs and increase operational reliability. Smaller integrated process units, such as exchanger reformers in various configurations, contribute to this reduction and will achieve additional savings by simplifying piping and instrumentation. Progress in instrumentation and computer control could reduce the effective overall energy consumption achieved over the course of a year by tuning plants continuously to optimal operating conditions.

11. References

1. L. Pauling: *The Nature of the Chemical Bond*, 3rd ed., Cornell University Press, Ithaca, N.Y., 1960.
   L. Pauling: *Die Natur der chemischen Bindung*, 3rd ed., Verlag Chemie, Weinheim 1968.
2. F. Hund, *Z. Phys.* **31** (1925) 95.
3. L. Haar, *J. Phys. Chem. Ref. Data* **7** (1978) no. 3, 635 – 792.
4. *VDI Forschungsh.* 1979, no. 596, VDI-Verlag Düsseldorf.
5. F. Din: *Thermodynamic Functions of Gases*, vol. 1, Butterworths, London 1956.
6. A. Nielsen: *An Investigation on Promoted Iron Catalysts for the Syntheses of Ammonia*, Gjellerup, Kopenhagen 1968.
7. M. P. Wukalowitsch et al., *Teploenergetika (Moscow)* **7** (1960) no. 1, 63 – 69;
   Chem. Ing. Tech. **33** (1961) no. 4, 5, 291, 292.
8. R. Döring: *Thermodynamische Eigenschaften von Ammoniak (R 717)*, Verlag C. F. Müller, Karlsruhe 1978;
   Ki Klima + Kälte Ing. extra no. 5.
9. *Landolt-Börnstein*, vol. 2, part 1 – 10, Springer Verlag, Berlin-Heidelberg-New York 1971.
10. *Handbuch der Kältetechnik*, vol. **IV**, Springer Verlag, Berlin – Heidelberg – New York 1956.
11. A. Kumagai, T. Toriumi, *J. Chem. Eng. Data* **16** (1971) no. 3, 293.
12. R. Wiebe, V. I. Gaddy, *J. Am. Chem. Soc.* **60** (1938) 2300.
13. E. P. Bartlett, H. L. Cupples, T. H. Tremearne, *J. Am. Chem. Soc.* **50** (1928) 1275.
14. E. P. Bartlett et al., *J. Am. Chem. Soc.* **52** (1930) 1363.
15. B. H. Sage, R. H. Olds, W. N. Lacey, *Ind. Eng. Chem.* **40** (1948) 1453.
16. A. Michels et al., *Appl. Sci. Res.* **A3** (1953) 1.
17. A. Michels et al., *Appl. Sci. Res.* **A4** (1954) 180.
18. R. Plank, *Kältetechnik* **12** (1960) 282 – 283.
19. W. K. Chung, *Can. J. Chem. Eng.* **55** (1977) no. 6, 707 – 711.
20. G. G. Fuller, *Ind. Eng. Chem. Fundam.* **15** (1976) no. 4, 254 – 257.
21. B. D. Djordjevic et al., *Chem. Eng. Sci.* **32** (1977) no. 9, 1103 – 1107.
22. M. Zander, W. Thomas, *J. Chem. Eng. Data* **24** (1979) 1.
23. L. J. Christiansen, J. Kjaer: *Enthalpy Tables of Ideal Gases*, Haldor Topsøe, Copenhagen 1982, p. 1.
24. Dow Chemical Corp. *JANAF Thermodynamical Tables* 2nd ed., 1971, NSRDS NBS-37.
25. L. Haar, *J. Res. Natl. Bur. Stand. Sect. A* **72** (1968) 2.
26. E. R. Grahl, *Pet. Process.* **8** (1953) 562.
27. I. Hay, G. D. Honti, in G. D. Honti (ed.): *The Nitrogen Industry*, part 1, Akademiai Kiado, Budapest 1976, pp. 68 – 73.
28. R. G. Barile, G. Thodos, *Can. J. Chem. Eng.* **43** (1965) 137 – 142.
29. C. G. Alesandrini et al., *Ind. Eng. Chem. Process Des. Dev.* **11** (1972) no. 2, 253.
30. H. Zeininger, *Chem. Ing. Tech.* **45** (1973) no. 17, 1067.
31. H. H. Reamer, B. H. Saga, *J. Chem. Eng. Data* **4** (1959) no. 4, 303.
32. K. V. Reddy, A. Husain, *Ind. Eng. Chem. Process Des. Dev.* **19** (1980) 580 – 586.
33. H. Masuoka et al., *J. Chem. Eng. Jpn.* **10** (1977) no. 3, 171 – 175.
34. A. T. Larson, C. A. Black, *J. Am. Chem. Soc.* **47** (1925) 1015.
35. A. Michels et al., *Physica (Utrecht)* **16** (1950) 831 – 838.
36. A. Michels et al., *Physica (Utrecht)* **25** (1959) 840.
37. G. Guerreri, *Hydrocarbon Process.* **49** (1970) no. 12, 74 – 76.
38. H. H. Reamer, B. H. Saga, *J. Chem. Eng. Data* **4** (1959) no. 2, 152.
39. R. Wiebe, T. H. Tremearne, *J. Am. Chem. Soc.* **55** (1933) 975.
40. F. Heise, *Ber. Bunsenges. Phys. Chem.* **76** (1972) 938.

41. B. Lefrancois, C. Vaniscotte, *Genie Chim.* **83** (1960) no. 5, 139.
42. H.-D. Müller, *Z. Phys. Chem. (Leipzig)* **255** (1974) no. 3, 486.
43. B. I. Lee, M. G. Kesler, *AIChE J.* **21** (1975) no. 3, 510.
44. H. Masuoka, *J. Chem. Eng. Jpn.* **10** (1977) no. 3, 171 – 175.
45. K. Konoki et al., *J. Chem. Eng. Jpn.* **5** (1972) no. 2, 103 – 107.
46. G. Sarashina, *J. Chem. Eng. Jpn.* **7** (1974) no. 6, 421 – 425.
47. R. Heide, *Luft Kältetech.* **7** (1971) no. 3, 128 – 130.
48. I. Durant, *J. Chem. Soc.* 1934, no. 135, 730.
49. Yu. V. Efremov, I. F. Golubev, *Russ. J. Phys. Chem. Eng. (Engl. Transl.)* **36** (1962) no. 5, 521 – 522.
50. J. A. Jossi et al., *AIChE J.* **8** (1962) no. 1, 59 – 63.
51. H. Shimotake, G. Thodos, *AIChE J.* **9** (1963) no. 1, 63.
52. B. K. Sun, T. S. Storwick, *J. Chem. Eng. Data* **24** (1979) no. 2, 88.
53. I. T. Perelsatein, E. B. Parushin, *Kholod. Tekh.*, 1980, no. 6, 34 – 37.
54. A. K. Pal, A. K. Barura, *J. Chem. Phys.* **47** (1967) no. 1, 216.
55. E. A. Mason, L. Mouchik, *J. Chem. Phys.* **36** (1962) no. 10, 2746.
56. A. B. Rakshit, C. S. Roy, A. K. Barua, *J. of Chemical Physics* **59** (1973) 3633 – 3688.
57. D. P. Needham, H. Ziebland, *Int. J. Heat Mass Transfer* **8** (1965) 1387 – 1414.
58. A. M. P. Tans, *Ind. Chem.* **39** (1963) no. 3, 141.
59. R. Afshar, S. Murad, S. C. Saxena, *Chem. Eng. Commun.* **10** (1981) 1 – 11.
60. D. A. Kouremenos, *Kältetechnik* **20** (1968) no. 10, 319.
61. A. O. S. Maczek, P. Gray, *Trans. Faraday Soc.* **66** (1970) 127.
62. P. Correia et al., *Ber. Bunsenges. Phys. Chem.* **72** (1968) 393.
63. W. Inkofer et al., *Ammonia Plant Saf.* **13** (1971) 67.
64. F. W. Bergstrom, *Ind. Eng. Chem.* **24** (1932) 57.
65. P. Barton et al., *Ind. Eng. Chem. Process Des. Dev.* **7** (1968) no. 3, 366.
66. K. A. Hofmann: *Anorganische Chemie*, 20th ed., Vieweg & Sohn, Braunschweig 1969.
67. U. Müller, A. Greiner, *Chem. Tech. (Leipzig)* **18** (1966) no. 6, 327.
68. *Chem. Zentralbl.* 1966, 14 – 0884.
69. J. Jander: *Chemie in wasserfreiem flüssigem Ammoniak*, Vieweg & Sohn, Braunschweig – Interscience, New York 1963.
70. J. P. Lenieur et al., *C. R. Acad. Sci. Ser. C* **268** (1969) no. 20, 1791.
71. E. Weiss, W. Buchnery cited in [66].
72. H. Smith: *Organic Reactions in Liquid Ammonia*, Vieweg & Sohn, Braunschweig – Interscience, New York–London 1963.
73. R. Strait, *Nitrogen & Methanol* **238** (1999) 37–43.
74. R. Strait, paper presented at Nitrogen '99, Caracas, Venezuela (March 1999) International Conference & Exhibition, British Sulphur Publishing.
75. "Maximising output from a single train", *Nitrogen & Methanol* **238** (1999) 44–48.
76. C. Bosch, Nobel prize address 1933.
77. F. Ritter, F. Walter in K. Winnacker, E. Weingaertner (eds.): *Chemische Technologie*, **vol. 2**, Carl Hanser, München 1950 pp. 368.
78. J. Hess, *Chem. Ind. (Berlin)* **45** (1922) 538.
79. *Chem. Ind. (Berlin)* **48** (1925) 718.
80. G. Plumpe: *Die I. G. Farbenindustrie*, — Wirtschaft, Technik und Politik Dunker & Humblot, Berlin 1990.
81. A. Mittasch: Chemische Grundlegung der industriellen Ammoniakkatalyse, Ludwigshafen 1953.
82. *Ullmann*, 3rd ed., **vol. 3**, p. 544 – 546.
83. A. Mittasch, W. Frankenburger, *Z. Elektrochem. Angew. Phys. Chem.* **35** (1929) 920.
84. F. Haber, G. van Oordt, *Z. Anorg. Allg. Chem.* **43** (1904) 111; **44** (1905) 341; **47** (1905) 42.
85. W. Nernst, *Z. Elektrochem. Angew. Phys. Chem.* **13** (1907) 521.
86. F. Jost, *Z. Anorg. Allg. Chem.* **57** (1908) 414.
87. F. Haber, R. Le Rossignol, *Ber. Dtsch. Chem. Ges.* **40** (1907) 2144.
88. P. Krassa, *Chem. Ztg. Chem. Appar.* **90** (1966) 104.
89. F. Haber, *Z. Elektrochem. Angew. Phys. Chem.* **14** (1908) 181; **20** (1914) 597.
90. R. Le Rossignol, *Naturwissenschaften* **16** (1928) 1070.
91. B. Timm, *Chem. Ing. Tech.* **35** (1963) no. 12, 817.
92. F. A. Henglein, *Chem. Ztg.* **75** (1951) 345 – 350, 389 – 392, 407 – 410.
93. K. Holdermann: *Im Banne der Chemie – Carl Bosch*, Econ Verlag, Düsseldorf 1953.

94. M. Appl in W. F. Furter (ed.): *A Century of Chemical Engineering*, Plenum Publ. Corp., 1982 p. 29.
95. M. Appl: *IIChE Conference*, Hyderabad, 1986;
Indian Chemical Engineer **XXIX** (1987) no. 1, 7 – 29.
96. S. A. Tophamy: *Catalysis, Science and Technology*, vol.7, Springer-Verlag, Berlin, Heidelberg, New York, Tokyo pp. 1 – 50.
97. A. Nagel et al.: *Stickstoff*, Schriftenreihe der BASF, 2nd ed., 1991.
98. M. Appl, *Nitrogen* **100** (1976) 47.
99. M. Appl: "The Haber-Bosch Heritage: The Ammonia Production Technology", *50th Anniversary of the IFA Technical Conference*, Sept. 1997, Sevilla.
100. M. Appl: *Ammonia, Methanol Hydrogen, Carbon Monoxide — Modern Production Technologies*, CRU, London 1997.
101. M. Appl: *Ammonia, Methanol, Hydrogen, Carbon Monoxide — Modern production Technologies* CRU Publ., London 1997, p. 44, 45.
102. M. Appl: *Ammonia, Principles and Industrial Practice*, WILEY-VCH, Weinheim-New York 1999.
103. G. P. Williams, paper presented at 50th AIChE Symposium on Safety in Ammonia Plants and Related Facilities, Toronto 2005.
104. L. J. Gillespie, J. A. Beattie, *Phys. Rev.* **36** (1930) 743.
105. S. Peter, H. Wenzel, *Chem. Ing. Tech.* **45** (1973) no. 9 + 10, 573.
106. F. Haber, S. Tamaru, *Z. Elektrochem. Angew. Phys. Chem.* **21** (1915) 191.
107. L. J. Gillespie, J. A. Beattie, *Phys. Rev.* **36** (1930) 1008.
108. V. G. Telegin, G. M. Lutskii, *J. Appl. Chem. USSR (Engl. Transl)* **37** (1964) 2271.
109. L. J. Christiansen in W. A. Nielsen (ed.): *Ammonia — Catalysis and Manufacture*, Springer, Heidelberg 1995 pp. 1 – 16.
110. I. Granet, *Pet. Refiner* **33** (1954) no. 5, 205.
111. I. Granet, P. Kass, *Pet. Refiner* **32** (1953) no. 3, 149.
112. A. K. Pal, A. K. Barua, *J. Chem. Phys.* **47** (1967) 116.
113. W. G. Frankenburger, *Catalysis 1954 – 1960* **3** (1955) 171.
114. R. Brill et al., *Ber. Bunsenges. Phys. Chem.* **73** (1969) 999.
115. A. Ozaki, K. Aika: "The Synthesis of Ammonia by Heterogeneous Catalysis," in R. W. F. Hardy et al. (eds.): *A Treatise on Dinitrogen Fixation, Sect. I and II*, Wiley, New York 1979, pp. 169 – 247.
116. A. Ozaki, K. Aika: "Catalytic Activation of Dinitrogen," in R. H. Anderson, M. Boudart (eds.): *Catalysis: Science and Technology*, **vol. 1**, Springer Verlag, Berlin–Heidelberg–New York 1981, pp. 87 – 158.
117. P. H. Emmett in E. Drauglis, R. I. Jaffee (eds.): *The Physical Basis for Heterogenous Catalysis*, Plenum Publ., New York 1975, p. 3.
118. K. Tamaru in J. Chatts et al. (eds.): *New Trends in the Chemistry of Nitrogen Fixation*, Academic Press, New York 1980, p. 13.
119. G. A. Somorjai: *Chemistry in Two Dimensions: Surfaces*, Cornell Press, Ithaca 1981.
120. G. Ertl, J. Küppers: *Low Electrons and Surface Chemistry*, 2nd ed., VCH Verlagsgesellschaft, Weinheim 1985.
121. G. Ertl in J. R. Jennings (ed.): *Catalytic Ammonia Synthesis*, Plenum Press, New York and London 1991, pp. 109 – 131.
122. P. Stolze in A. Nielson (ed.): *Ammonia — Catalysis and Manufacture*, Springer-Verlag, Heidelberg 1995, pp. 17 – 102.
123. G. Ertl: "Kinetics of Chemical Processes on Well-defined Surfaces," in R. H. Anderson, M. Boudart (eds.): *Catalysis: Science and Technology*, **vol. 4**, Springer-Verlag, Berlin–Heidelberg–New York 1983, pp. 257 – 282.
124. J. J. F. Scholten et al., *Trans. Faraday Soc.* **55** (1959) 2166.
125. J. J. F. Scholten, Dissertation, Techn. Hochschule Delft 1959.
126. P. Mars et al. in J. H. de Boer (ed.): *The Mechanism of Heterogenous Catalysis*, Elsevier, Amsterdam 1960, p. 66.
127. K. Tanaka, A. Matsumaya, *J. Res. Inst. Catal. Hokkaido Univ.* **2** (1971) 87.
128. M. I. Temkin, *Adv. Catal.* **28** (1979) 173.
129. P. H. Emmett, S. Brunauer, *J. Am. Chem. Soc.* **56** (1934) 35.
130. H. S. Taylor, *J. Am. Chem. Soc.* **53** (1931) 578.
131. W. Frankenburger, *Z. Elektrochem.* **39** (1933) 45, 97, 269.
132. K. T. Kozhenova, M. Y. Kagan, *Zh. Fiz. Khim.* **14** (1940) 1250.
133. P. H. Emmett, S. Braunauer, *J. Am. Chem. Soc.* **59** (1937) 1553.

134. L. M. Aparicio, J. A. Dumesic: *Topics in Catalysis*, **vol. 1**, Baltzer AG, Basel 1994, p. 233 – 252.
135. D. R. Strongin, G. A. Somorjai in J. R. Jennings (ed.): *Catalytic Ammonia Synthesis*, Plenum Press, New York 1991, pp. 133 – 178.
136. G. Ertl, M. Huber, S. B. Lee, Z. Paal, M. Weiss, *Appl. Surf. Sci.* **8** (1981) 373.
137. T. Nakata, S. Matsushita, *J. Phys. Chem.* **72** (1968) 458.
138. V. I. Shvachko et al., *Kinet. Katal.* **7** (1966) 635, 734.
139. Y. Morikawa, A. Ozaki, *J. Catal.* **12** (1968) 145.
140. R. Westrik, P. Zwieterring, *Proc. K. Ned. Akad. Wet. Ser. B: Palaeontol. Geol. Phys. Chem.* **56** (1953) 492.
141. R. Brill et al., *Angew. Chem.* **79** (1967) 905.
142. H. Taube, Dissertation, Freie Universität Berlin 1968.
143. R. Brill, *Ber. Bunsenges. Phys. Chem.* **75** (1971) 455.
144. J. A. Dumesic, H. Topsøe, S. Khammouma, M. Boudart, *J. Catal.* **37** (1975) 503; see also [365].
145. F. Boszo et al., *J. Cat.* **49** (1977) 103.
146. G. Ertl et al., *J. Vac. Sci. Technol.* **13** (1976) 314.
147. G. Ertl et al., *Surf. Sci.* **114** (1982) 515.
148. G. A. Somorjai et al.: *Topics in Catalysis*, **vol. 1**, Baltzer AG, Basel 1994, p. 215 – 231.
149. N. D. Spencer, R. C. Schoonmaker, G. A. Somorjai, *J. Catal.* **74** (1982) 129.
150. G. Ertl: "Plenary Lecture," in T. Seiyama, K. Tanabe (eds.): New Horizons in Catalysis, Proc. Int. Congr. Catal. 7th 1980, Elsevier 1981, p. 21.
151. F. Bozso, G. Ertl, *J. Catal.* **49** (1977) 18; **50** (1977) 519.
152. R. Schlögl: "Ammonia Synthsis," in G. Ertl, H. Knözinger, J. Weitcamp (eds.): *Handbook of Heterogeneous Catalysis*, **vol. 4**, Wiley-VCH, Weinheim 1997, pp. 1698 – 1748.
153. B. Fastrup: *Topics in Catalysis*, **vol. 1**, Baltzer AG, Basel 1994, p. 273 – 283.
154. L. Falicov, G. A. Somorjai, *Proc. Natl. Akad. Sci. USA* **82** (1985) 2207.
155. G. Ertl et al., *Surf. Sci.* **114** (1982) 527.
156. G. Ertl et al., *Surface Science* **114** (1982) 527.
157. S. R. Bare, D. R. Strongin, G. A. Somorjai, *J. Phys. Chem.* **90** (1987) 4726.
158. D. R. Strongin et al., *J. Catal.* **103** (1987) 289.
159. M. Bowker: *Topics in Catalysis*, **vol. 1**, Baltzer AG, Basel 1994, p. 265 – 271.
160. C. Rettner, H. Stein, *Phys. Rev. Let.* **56** (1987) 2768.
161. C. Rettner, H. Stein, *J. Chem. Phys.* **87** (1987) 770.
162. J. K. Norskov: *Topics in Catalysis*, **vol. 1**, Baltzer AG, Basel 1994, p. 385 – 403.
163. M. Gunze et al., *Phys. Rev. Lett.* **53** (1984) 850.
164. L. J. Whitman, C. E. Bartosch, W. Ho, G. Strasser, M. Grunze, *Phys: Rev. Lett.* **56** (1986) 1984.
165. L. J. Whitman, C. E. Bartosch, W. Ho, *J. Chem. Phys.* **85** (1986) 3688.
166. M. Grunze, G. Strasser, M. Golze, *Appl. Phys. A* **44** (1987) 19.
167. J. Schütze et al.: *Topics in Catalysis*, **vol. 1**, Baltzer AG, Basel 1994, p. 195 – 214.
168. M. Boudart: *Topics in Catalysis*, **vol. 1**, Baltzer AG, Basel 1994, p. 405 – 414.
169. N. Ringer, paper presented at Süd-Chemie Catalyst Symposium, Manama, Bahrain, June 2004.
170. N. Ringer, paper presented at 50th AIChE Symposium on Safety in Ammonia Plants and Related Facilities, Toronto 2005.
171. A. Nielsen, *Catal. Rev.* **4** (1970) no. 1, 1.
172. W. A. Schmidt, *Angew. Chem.* **80** (1968) 151.
173. R. Brill et al., *Z. Phys. Chem. (Wiesbaden)* **64** (1969) 215.
174. N. Takezawa, I. Toyoshima, *J. Catal.* **19** (1970) 271.
175. S. Carra, R. Ugo, *J. Catal.* **15** (1969) 435.
176. I. I. Tretyakov et al., *Dokl. Akad. Nauk SSSR* **175** (1967) 1332.
177. G. Parravano, *J. Catal.* **8** (1967) 29.
178. M. I. Temkin et al., *Kinet. Katal.* **4** (1963) 494.
179. S. R. Logan, J. Philp, *J. Catal.* **11** (1968) 1.
180. N. Takezawa, P. H. Emmett, *J. Catal.* **11** (1968) 131.
181. G. Schulz-Ekloff, *Ber. Bunsenges. Phys. Chem.* **75** (1971) 110.
182. A. Nielsen et al., *J. Catal.* **3** (1964) 68.
183. P. E. H. Nielsen, unpublished work, Haldor Topsøe A/S, cited in [265].
184. S. R. Tennison in J. R. Jennings (ed.): *Catalytic Ammonia Synthesis*, Plenum Press, New York and London 1991, pp. 303 – 364.
185. K. Aika, K. Tamaru in W. A. Nielsen (ed.): *Ammonia — Catalysis and Manufacture*, Springer-Verlag, Heidelberg 1995, pp. 103 – 148.
186. J. N. Nwalor, J. G. Goodwin, jr.: *Topics in Catalysis*, **vol. 1**, Baltzer AG, Basel 1994, p. 228 – 148.

187. K. Aika, A. Ozaki, *J. Catal.* **16** (1970) 97.
188. D. A. King, F. Sebba, *J. Catal.* **4** (1965) 253.
189. N. Segal, F. Sebba, *J. Catal.* **8** (1967) 105.
190. K. Aika, A. Ozaki, *J. Catal.* **14** (1969) 311.
191. G. E. Moore, F. C. Unterwald, *J. Chem. Phys.* **48** (1968) 5409, 5393.
192. M. R. Hillis et al., *Trans. Faraday Soc.* **62** (1966) 3570.
193. K. Azuma, *Nature (London)* **190** (1961) no. 4775, 530.
194. H. Kubota, M. Shindo, *J. Chem. Eng. Jpn.* **23** (1959) 242.
195. G. Gramatica, N. Pernicone in: J. R. Jennings (ed.): *Catalytic Ammonia Synthesis*, Plenum Press, New York and London 1991, pp. 211 – 252.
196. J. P. Hansen in A. Nielson (ed.): *Ammonia — Catalysis and Manufacture*, Springer-Verlag, Heidelberg 1995, pp. 149 – 190.
197. M. I. Temkin, V. Pyzhev, *Acta Physicochim. URSS* **12** (1940) 327.
198. M. I. Temkin, *Zh. Fiz. Khim.* **24** (1950) 1312.
199. R. M. Adams, E. W. Comings, *Chem. Eng. Prog.* **49** (1953) 359.
200. P. H. Emmett, J. T. Kummer, *Ind. Eng. Chem.* **35** (1943) 677.
201. D. Annable, *Chem. Eng. Sci (Genie Chimique)* **1** (1952) no. 4, 145.
202. C. Bokhoven et al., *Catalysis 1954 – 1960* **3** (1955) 265.
203. R. Brill, *J. Chem. Phys.* **19** (1951) 1047.
204. S. Kiperman, V. S. Granovskaya, *Zh. Fiz. Khim.* **26** (1952) 1615.
205. M. I. Temkin et al., *Kinet. Katal.* **4** (1963) 224.
206. A. Ozaki et al., *Proc. R. Soc. London Ser. A* **258** (1960) 47.
207. M. I. Temkin, N. M. Morozow, E. M. Shapatina, *Kinet. Catal. (Engl. Trans.)* **4** (1963) 565.
208. ICI: *Catalyst Handbook with Special Reference to Unit Processes in Ammonia and Hydrogen Manufacture*, Wolfe Scientific Books, London 1970.
209. V. D. Livshits, I. P. Sidorov, *Zh. Fiz. Khim.* **26** (1952) 538.
210. K. Aika, A. Ozaki, *J. Catal.* **13** (1969) 232.
211. A. Nielsen, J. Kjaer, B. Hansen, *J. Catal.* **3** (1964) 68.
212. A. Ozaki, H. S. Taylor, M. Boudart, *Proc. R. Soc. London Ser. A* **258** (1960) 47.
213. R. Brill, *J. Catal.* **16** (1970) 16.
214. A. Cappelli, A. Collina, *Inst. Chem. Eng. Symp. Ser.* **35** (1972) 10.
215. G. Buzzi-Ferraris et al., *Eng. Chem. Sci.* **29** (1974) 1621.
216. U. Guacci et al., *Ind. Eng. Chem.* **16** (1977) 166.
217. D. G. Dyson, J. M. Simson, *Ind. Eng. Chem. Fundam.* **7** (1968) 601.
218. M. Bowker, I. Parker, K. Waugh, *Appl. Catal.* **14** (1985) 101.
219. M. Bowker, I. Parker, K. Waugh, *Surface Science* **97** (1988) L 233.
220. P. Stolze, J. K. Norskov, *Phys Rev. Let.* **55** (1985) 2502.
221. P. Stolze, J. K. Norskov, *Surface Science* **189/190** (1987) 91; *Surface Science* **197** (1988) L 230; *J. Catal.* **110** (1988) 1.
222. J. A. Dumesic, A. A. Trevino, *J. Catal.* **116** (1989) 119.
223. P. Stolze, *Phys. Sci.* **36** (1987) 824.
224. E. Wicke, *Z. Elektrochem.* **60** (1956) 774.
225. A. Wheeler, *Adv. Catal.* **3** (1951) 249.
226. T. Akehata et al., *Can. J. Chem. Eng.* **39** (1961) 127.
227. C. Bokhoven, W. van Raayen, *J. Phys. Chem.* **58** (1954) 471.
228. J. Hoogschagen, *Ind. Eng. Chem.* **47** (1955) 906.
229. D. G. Dyson, J. M. Simon, *Ind. Eng. Chem. Fundam.* **7** (1968) 605.
230. C. Chu, O. A. Hougen, *Chem. Eng. Sci.* **17** (1962) 167.
231. C. P. P. Singh, N. D. Saraf, *Ind. Eng. Chem. Process Des. Dev.* **18** (1979) 364.
232. A. Cappelli, A. Collina, *Inst. Chem. Eng. Symp. Ser.* **35** (1972) 10.
233. C. Wagner, *Z. Phys. Chem. (Leipzig)* **193** (1943) 1.
234. E. W. Thiele, *Ind. Eng. Chem.* **31** (1939) 916.
235. J. Kjaer: *Calculation of Ammonia Converters on an Electronic Digital Computer*, Akademisk Forlag, Kopenhagen 1963.
236. J. J. Hag, I. M. Palla, *Br. Chem. Eng.* **8** (1963) no. 3, 171.
237. J. Kubec, J. Burianova, Z. Burianec, *Int. Chem. Eng.* **14** (1974) no. 4, 629.
238. R. F. Baddour, R. L. T. Brian, B. A. Logeais, J. P. Eymery, *Chem. Eng. Sci.* **20** (1965) 281. R. L. T. Brian, R. F. Baddour, J. P. Eymery, *Chem. Eng. Sci.* **20** (1965) 297.
239. M. J. Shah, *Ind. Eng. Chem.* **59** (1967) no. 1, 73.
240. A. D. Stephens, *Chem. Eng. Sci.* **30** (1975) 11.
241. L. D. Gaines, *Ind. Eng. Chem. Process. Des. Dev.* **16** (1977) no. 3, 381.

242. T. Huberich, R. Krabetz in A. V. Slack (ed.): *Ammonia*, part III, Marcel Dekker, New York – Basel 1977.
243. A. Mittasch: *Geschichte der Ammoniaksynthese*, Verlag Chemie, Weinheim 1951.
244. A. Mittasch, *Adv. Catal.* **2** (1950) 82.
245. C. Bosch, *Z. Elektrochem. Angew. Phys. Chem.* **24** (1918) 361.
246. K. Aika, J. Yamaguchi, A. Ozaki, *Chem. Lett.* 1973, 161; cited in [116].
247. S. R. Logan, C. Kemball, *Trans. Faraday Soc.* **56** (1960) 144.
248. E. P. Perman, *Proc. R. Soc. London* **76** (1905) 167.
249. F. Haber, G. von Oørdt, *Z. Anorg. Allg. Chem.* **43** (1905) 111.
250. BASF, DE 249447, 1910.
251. A. T. Larson, R. S. Tour, *Chem. Metall. Eng.* **26** (1922) 493, 555, 588, 647.
252. J. A. Almquist, E. D. Crittenden, *Ind. Eng. Chem.* **18** (1926) 1307.
253. G. L. Bridger et al., *Chem. Eng. Prog.* **43** (1947) 291.
254. H. Hinrichs, *Br. Chem. Eng.* **12** (1967) 1745.
255. Indianapolis Center for Advanced Research, US 4235749, 1980.
256. Topsøe, GB 989242, 1961.
257. P. D. Rabina et al., *Khim. Prom.* **5** (1969) 350.
258. The Gas Council London, DE-OS 1931758, 1969 (T. Nicklin, D. Oyden).
259. Chevron Res. Comp., US 3653831, 1972.
260. J. L. Carter, C. G. Savini, US 3472794, 1969; SU 173204 (749 751) 23-4, 1961.
261. The Lummus Comp., US 3992328, 1976.
262. Ammonia Casale, IT-A 47920 A, 1979.
263. ICI, EP-A 7830276, 1979.
264. The Lummus Comp., US 3951862, 1976. Société Chimique de La Grande Paroisse, Azote et Produits Chimiques, US 3839229, 1974.
265. A. Nielsen: "Ammonia Synthesis: Exploratory and Applied Research," *Catal. Rev. Sci. Eng.* **23** (1981) 17 – 51.
266. D. C. Silverman, M. Boudart, *J. Catal.* **77** (1982) 208.
267. C. Peters et al., *Z. Elektrochem.* **64** (1960) 1194.
268. K. Schäfer, *Z. Elektrochem.* **64** (1960) 1190.
269. P. H. Emmett, S. Brunauer, *J. Am. Chem. Soc.* **59** (1937) 310; *J. Catal.* **15** (1969) 90.
270. S. Brunauer, P. H. Emmett, *J. Am. Chem. Soc.* **62** (1940) 1732.
271. D. C. Silverman, Dissertation, Stanford University 1976.
272. H. C. Chen, R. B. Anderson, *J. Colloid Interface Sci.* **38** (1972) 535; *J. Catal.* **28** (1973) 161.
273. G. Ertl, N. Thiele, *Appl. Surf. Sci.* **3** (1979) 99.
274. R. Hosemann, A. Preisinger, W. Vogel, *Ber. Bunsenges. Phys. Chem.* **70** (1966) 786. H. Ludwiszek, A. Preisinger, A. Fischer, R. Hosemann, A. Schönfel, W. Vogel, *J. Catal.* **51** (1977) 326. R. Buhl, A. Preisinger, *Surf. Sci.* **47** (1975) 344.
275. W. S. Borghard, M. Boudart, *J. Catal.* **80** (1983) 194.
276. M. Boudart: "Kinetics and Mechanism of Ammonia Synthesis," *Catal. Rev. Sci. Eng.* **23** (1981) 1 – 15.
277. R. Schlögl in J. R. Jennings (ed.): *Catalytic Ammonia Synthesis*, Plenum Press, New York and London 1991 pp. 19 – 108.
278. C. Peters, K. Schäfer, R. Krabetz, *Z. Elektrochem.* **64** (1960) 1194.
279. M. E. Dry, J. A. K. du Plessis, B. M. Leuteritz, *J. Catal.* **6** (1966) 194.
280. R. Krabetz, C. Peters, *Angew. Chem. Int. Ed. Engl.* **4** (1965) 341.
281. A. Kazusaka, I. Toyoshima, *Z. Phys. Chem. (Wiesbaden)* **128** (1982) 111.
282. L. M. Dmitrenko et al., *Proc. Int. Congr. Catal. 4th 1968* **1** (1971) 404.
283. K. Karaslavova, M. D. Anastasov, *Geterog. Katal. Tr. Mezhdunar. Simp. 3rd 1975*, 1978, 297.
284. K. Z. Zakieva, P. D. Rabina, J. E. Zubova, I. E. Klaustova, N. Z. Pavlova, L. D. Kusnetzov, *Tr. Mosk. Khim. Tekhnol. Inst. im. D. I. Mendeleeva* 1973 no. 73, 120.
285. B. Aleksic, N. Jovanovic, A. Terlecki-Baricevic, *Geterog. Katal. Tr. Mezhdunar. Simp. 3rd 1975*, 1978, 147.
286. M. G. Berengarten et al., *Kinet. Katal.* **15** (1974) 250.
287. Yu. N. Artyukh, M. T. Rusov, N. A. Boldyreva, *Kinet. Katal.* **8** (1967) 1319.
288. P. J. Smith, D. W. Taylor, D. A. Dowden, C. Kemball, D. Tayler, *Appl. Catal.* **3** (1982) 303.
289. I. D. Kusnetsov, *Chem. Tech. (Leipzig)* **15** (1963) 211.
290. J. Herrmann et al., *Chem. Tech. (Leipzig)* **18** (1966) 472.
291. D. G. Ivanov, N. D. Anastasov, *Chem. Tech. (Leipzig)* **15** (1963) 229.

292. L. Sokol, *Chem. Tech. (Leipzig)* **15** (1963) 214.
293. U. Zardi, A. Antonini, *Nitrogen* **122** (1979) 33.
294. M. C. Sze, *Hydrocarbon Process.* **56** (1977) no. 12, 127.
295. N. Pernicone, G. Liberti, G. Servi in C. Eyraud, M. Escoubes (eds.): *Progress in Vacuum Mikrobalance Techniques*, **vol. 3**, Heyden, London 1975, p. 304.
296. A. Mittasch, E. Keunecke, *Z. Elektrochem. Angew. Phys. Chem.* **38** (1932) 666.
297. H. Uchida, I. Terao, K. Ogawa, *Bull. Chem. Soc. Jpn.* **37** (1964) 653.
298. N. Pernicone, F. Traina: *Preparation of Catalysts*, **vol. 2**, Elsevier, Amsterdam 1979, p. 321.
299. B. S. Clausen, S. Mørup, H. Topsøe, R. Candis, E. J. Jensen, A. Baranski, A. Pattek, *J. Phys. (Orsay, Fr.)* **37** (1976) C6-245.
300. P. D. Rabina, T. Y. Malysheva, L. D. Kusnetzov, V. A. Batyrev, *Kinet. Catal. (Engl. Transl.)* **11** (1970) 1030.
301. A. Mittasch, E. Keunecke, *Z. Elektrochem. Angew. Phys. Chem.* **38** (1932) 666.
302. R. Westrik, *J. Chem. Phys.* **21** (1953) 2049.
303. M. E. Dry, L. C. Ferreira, *J. Catal.* **7** (1967) 352.
304. F. Garbassi, G. Fagherazzi, M. C. Calcaterra, *J. Catal.* **26** (1972) 338.
305. A. C. Turnock, H. P. Eugster, *J. Petrol.* **3** (1962) 533.
306. R. Brill, *Allg. Prakt. Chem.* **17** (1966) 94.
307. R. Hosemann et al., *Ber. Bunsenges. Phys. Chem.* **70** (1966) 769.
308. L. v. Bogdandy et al., *Ber. Bunsenges. Phys. Chem.* **67** (1963) 958.
309. N. Pernicone et al., *Catal. Proc. Int. Congr. 5th 1972* 1973, 1241.
310. V. Solbakken, A. Solbakken, P. H. Emmett, *J. Catal.* **15** (1969) 90.
311. K. C. Wough et al.: *Topics in Catalysis*, **vol. 1**, Baltzer AG, Basel 1994, p. 295 – 301.
312. G. Ertl: "Zum Mechanismus der Ammoniaksynthese," *Nachr. Chem. Tech. Lab.* **31** (1983) no. 3, 178.
313. Y. Sasa, M. Uda, I. Toyoshima, *Chem. Lett.* 1982, no. 12, 2011.
314. S. K. Egeubaev et al., *Kinet. Katal.* **6** (1965) no. 4, 676; also see [171] and [282].
315. M. M. Ivanov et al., *Kinet. Katal.* **9** (1968) 1239; *Kinet. Katal.* **10** (1969) 349.
316. J. G. Ommen et al., *J. Catal.* **38** (1975) 120.
317. A. Ozaki et al., *Catal. Proc. Int. Congr. 5th 1972* **2** (1973) 1251.
318. E. K. Enikeev, A. V. Krylova, *Kinet. Katal.* **3** (1962) 116; *Chem. Abstr.* **57** (1962) 7965.
319. G. Ertl, M. Weiss, S. B. Lee, *Chem. Phys. Lett.* **60** (1979) 391.
320. M. C. Tsai et al., *Surf. Sci.* **155** (1985) 387.
321. J. A. Dumesic, Dissertation, Stanford University 1974.
322. T. Yoshioka, J. Koezuka, I. Toyoshima, *J. Catal.* **14** (1969) 281.
323. Y. E. Sinyak, Thesis, Inst. of Chemical Physics Akad. Sci. USSR, Moscow 1960; cited in [324].
324. Haldor Topsøe, ZA 645279, 1964; also see [328].
325. I. P. Nukhlenov et al., *J. Appl. Chem. USSR (Engl. Transl.)* **37** (1964) 248.
326. V. N. Anokhin et al., *J. Appl. Chem. USSR (Engl. Transl.)* **35** (1962) 29.
327. A. Baranski et al., *J. Catal.* **26** (1972) 286.
328. V. Věk, *Chem. Ing. Tech.* **45** (1973) 608.
329. K. Feind, *Chem. Ing. Techn.* **38** (1966) 1081.
330. Chemie Linz, US 3965246, 1976.
331. E. Comandini, A. Passariello, U. Zardi, *Ammonia Plant Saf.* **23** (1981) 44.
332. Wargons Aktiebolag, Haldor Topsøe, US 3243386, 1961.
333. Kuhlmann, GB 1080838, 1964.
334. Kuhlmann, CH 434218, 1964.
335. Haldor Topsøe, GB 989242, 1961.
336. ICI, BE 576059, 1959.
337. IG Farbenind., DE 748620, 1938.
338. I. V. Nicolescu et al., *Genie Chim.* **82** (1959) 33.
339. I. V. Nicolescu et al., *Chem. Tech. (Leipzig)* **15** (1963) 226.
340. S.I.R.I. Società Italiana Ricerche Industriali, US 4073749, 1978.
341. ICI, GB 1484864, 1977.
342. J. E. Jarvan, H. U. Larsen, unpublished work, (Haldor Topsøe A/S) cited in [265].
343. A. T. Larson et al., *Chem. Trade J. Chem. Eng.* **100** (1937) 403.
344. G. M. Schwab (ed.): *Handbuch der Katalyse*, **vol. 5: Heterogene Katalyse II**, Springer-Verlag, Wien 1957, p. 566.
345. A. M. Rubinštejn et al., *Izv. Akad. Nauk SSSR Ser. Khim.* 1966, 1707; *Chem. Zentralbl.* 1968, 8 – 0639.
346. A. M. Rubinštejn et al., *Kinet. Katal.* **6** (1965) 285; *Chem. Zentralbl.* 1965, 42 – 0622.

347. P. Bussiere, R. Dutartre, G. A. Martin, J. P. Mathieu, *C. R. Hebd. Seances Acad. Sci. Ser. C* **280** (1975) 1133.
R. Dutartre, M. Primet, G. A. Martin, *React. Kinet. Catal. Lett.* **3** (1975) 249.
348. H. Topsøe, J. A. Dumesic, E. C. Derouane, B. S. Clausen, S. Mørup, J. Villadsen, N. Topsøe in: *Preparation of Catalysts*, **vol. 2**, Elsevier, Amsterdam 1979, p. 365.
H. Topsøe et al. in T. Seiyama, K. Tanabe (eds.), New Horizons in Catalysis, Proc. Int. Congr. Catal. 7th 1980, Elsevier 1981.
349. M. Boudart, A. Delbouille, J. A. Dumesic, S. Khammouna, H. Topsøe, *J. Catal.* **37** (1975) 486.
J. A. Dumesic, H. Topsøe, S. Khammouna, M. Boudart, *J. Catal.* **37** (1975) 503.
J. A. Dumesic, H. Topsøe, M. Boudart, *J. Catal.* **37** (1975) 513.
350. The Lummus Comp., GB 1529823, 1978.
351. IG Farbenind., DE 554855, 1928; FR 611139, 1935.
352. N. W. Sseljakow, SU 48210, 1936; *Chem. Zentralbl.* **1937** II, 119.
353. N. N. Pschenitzyn et al., *Zh. Prikl. Khim. (Leningrad)* **13** (1940) 76; *Chem. Zentralbl.* 1940 II, 1932.
354. Österreichische Stickstoffwerke, AT 252957, 1964.
355. M. V. Tovlin et al., *Kinet. Katal.* **7** (1966) 749; *Chem. Zentralbl.* 1966, 42 – 0597.
356. H. Flood, D. G. Hill, *Z. Elektrochem.* **61** (1957) 18.
357. J. Valcha, *Chem. Tech. (Leipzig)* **15** (1963) 222.
358. Österreichische Stickstoffwerke, GB 833878, 1958.
359. D. B. Christozvonov et al., *Int. Chem. Eng.* **9** (1969) 387.
360. SU 169077, 1963.
361. H. Uchida, N. Todo, *Bull. Chem. Soc. Jpn.* **29** (1956) 20; also see [359].
362. Z. N. Bardik et al., *Khim. Promst. (Moscow)* **42** (1966) 351.
363. CS 117658, 1966 (L. Sokol).
364. P. A. Strelzov et al., *Khim. Promst. (Moscow)* **43** (1967) no. 1, 46.
365. H. Amariglio, G. Rambeau, in G. C. Bond et al. (eds.): Proc. Int. Congr. Catal. 6th 1976, Chemical Society London 1977, p. 1113.
366. T. L. Joseph, *Trans. AJME* **120** (1936) 72.
367. P. K. Stangway, H. U. Ross, *Trans. AJME* **242** (1968) 1981.
368. M. Moukassi et al., *Metall. Trans.* **B14** (1983) 125.
369. Y. K. Rao et al., *Metall. Trans.* **B10** (1979) 243.
370. J. O. Edström, *J. ISI* **175** (1953) 289.
371. A. Baranski, M. Lagan, A. Pattek, A. Reizer, L. J. Christiansen, H. Topsøe in: *Preparation of Catalysts*, **vol. 2**, Elsevier, Amsterdam 1979, p. 353.
372. A. Baranski et al., *Archivum Hutniktva* **25** (1980) 153.
373. A. Baranski et al., *Appl. Catal.* **3** (1982) 207.
374. A. Baranski et al., *Appl. Catal.* **19** (1985) 417.
375. A. Pattek-Janczyk et al., *Appl. Catal.* **6** (1983) 35.
376. J. A. Burnett et al., *Ind. Eng. Chem.* **45** (1953) 1678.
377. R. Royen, G. H. Langhans, *Z. Anorg. Allg. Chem.* **315** (1962) 1.
378. P. H. Emmett, S. Brunauer, *J. Am. Chem. Soc.* **52** (1930) 2682; *J. Am. Chem. Soc.* **62** (1940) 1732.
379. I. A. Smirnov et al., *Kinet. Katal.* **6** (1965) 351.
380. I. A. Smirnov, *Kinet. Katal.* **7** (1966) 107.
381. P. E. Højlund Nielsen, paper at the ACS meeting, Washington 1983.
382. P. E. Højlund Nielsen in W. A. Nielsen (ed.): *Ammonia — Catalysis and Manufacture*, Springer-Verlag, Heidelberg 1995, pp. 191 – 199.
383. P. E. Højlund Nielsen in J. R. Jennings (ed.): *Catalytic Ammonia Synthesis*, Plenum Press, New York 1991, pp. 285 – 301.
384. P. Stolze, *Phys. Sci.* **36** (1983) 824.
385. P. Stolze, J. K. Nørskov, *J. Vac. Sci. Technol.* **A5** (1987) 581.
386. T. Kirkerød, P. Skaugset, Abstracts of IV Nordic Symp. on Catalysts, Trondheim 1991.
387. I. P. Sidorov, K. E. Istomina, *Chem. Abstr.* **54** (1960) 6049.
388. H. Hinrichs, *Chem. Ing. Tech.* **40** (1968) 723.
389. L. N. Marakhovets et al., *Izv. Vyssh. Uchebn. Zaved. Khim. Khim. Tekhnol.* **15** (1972) no. 5, 735.
390. R. Brill et al., *Ber. Bunsenges. Phys. Chem.* **72** (1968) 1218.
391. J. D. Bulatnikova et al., *Zh. Fiz. Khim.* **32** (1958) 2717.
392. H. J. Hansen in V. Sauchelli (ed.): *Fertilizer Nitrogen*, its Chemistry and Technology, Reinhold Publ. Co., New York 1964.
393. I. Dybkjaer, E. A. Gam, AIChE Symposium on Safety in Ammonia Plants, San Francisco, Calif., 1984.
394. A. Ozaki, *Acc. Chem. Res.* **14** (1981) 16.
395. F. Haber, *Z. Elektrochem.* **16** (1910) 244.

396. A. Ozaki, K. Urabe, K. Shimazaki, S. Sumiya: *Preparation of Catalysts*, vol. 2: Scientific Basis for the Preparation of Heterogenous Catalysts, Elsevier, Amsterdam 1979, p. 381.
397. British Petroleum Co., US 4163775, 1979.
398. F. F. Gadallah et al. in G. Poncelet et al. (eds.): *Preparation of Catalysts*, **vol. 3**, Elsevier, Amsterdam 1983.
399. GB 1 468 441, 1977.
400. GB 174 079, 1985.
401. L. Volpe, M. Boudart, *J. Phys. Chem.* **90** (1986) 4874.
402. O. Glemser, *Naturwissenschaften* **37** (1950) 539.
403. D. J. Lowe, B. E. Smith, R. N. F. Thornely in P. M. Harrison (ed.): *Metalloproteins*, Part 1, Verlag Chemie, Weinheim 1985, p. 207.
404. *Nitrogen* **193** (1991) 17 – 21.
405. W. K. Taylor, S. A. Hall, D. G. Heath, ICI/CFDC Technical Symp., Shanghai 1989.
406. *Nitrogen & Methanol* **270** (2004) 37.
407. N. Ringer, M. Michel, R. Stockwell, paper presented at 50th AIChE Symposium on Safety in Ammonia Plants and Related Facilities, Toronto 2005.
408. J.-C. Park, D. Kim, C.-S. Lee, D.-K. Kim, *Bull. Korean Chem. Soc.* **20** (1999) no. 9, 1005–1009.
409. L. Huazhang et al., US 5,846 507 (1998) Zhejiang University of Technology, China.
410. S. Bencic: "Ammonia synthesis promoted by iron catalysts", Literature report, Michigan State University (April 2001).
411. J. P. Shirez, J. R. LeBlanc, Kellogg Ammonia Club Meeting, San Francisco 1989.
412. J. Abughazaleh, J. Gosnell, D. P. Mann, R. Strait: 2005 KBR Indonesian Ammonia Seminar, 20-23 February 2005, Jakarta, Indonesia.
413. C. J. H. Jacobson, S. E. Nielsen, paper presented at 46th AIChE Symposium on Safety in Ammonia Plants and Related Facilities, Toronto 2001.
414. "Is there any real competition for iron", *Nitrogen & Methanol* **257** (2002) 34–39.
415. A. V. Slack, G. Russel James (eds.): *Ammonia*, part I – IV, Marcel Dekker, New York 1973 /74.
416. S. Strelzoff: *Technology and Manufacture of Ammonia*, Wiley-Interscience, New York 1981.
417. F. J. Brykowski (ed.): *Ammonia and Synthesis Gas*, Noyes Data Corp., Park Ridge, N.J., 1981, p. 280.
418. I. Dybkjaer in W. A. Nielsen (ed.): *Ammonia — Catalysis and Manufacture*, Springer-Verlag, Heidelberg 1995, pp. 199 – 327.
419. C. W. Hooper in J. R. Jennings (ed.): *Catalytic Ammonia Synthesis*, Plenum Press, New York 1991, pp. 253 – 283.
420. H. Schmidt-Traub, H. C. Pohl, *Chem. Ing. Tech.* **55** (1983) no. 11, 850. *Hydrocarbon Process.* **63** (1984) no. 4, 95.
421. D. Wagner, J. F. Meckel, K. H. Laue in A. I. More (ed.), *Proc. Br. Sulphur Corp. Int. Conf. Fert. Technol. 4th 1981* 1982, 471.
422. *Nitrogen* 1975, no. 97, 35.
423. T. Grundt, K. Christiansen in A. I. More (ed.), *Proc. Br. Sulphur Corp. Int. Conf. Fert. Technol. 4th 1981* 1982, 73.
424. H. Wendt, *Chem. Ing. Tech.* **56** (1984) no. 4, 265.
425. G. A. Crawford, S. Benzimra, *Fertilizer '85 Conference (British Sulphur)*, London, Feb. 1985.
426. The Ultimate in Ammonia/Urea Studies, Fertecon 1991.
427. K. J. Mundo, *Chem. Ing. Tech.* **45** (1973) no. 10 a, 632.
428. T. A. Czuppon, L. J. Buividas, *Hydrocarbon Process.* **58** (1979) no. 9, 197.
429. J. Gosnell, paper presented at Hydrogen Conference, Argonne National Laboratory, 13 October 2005.
430. J. Y. Livingston, *Hydrocarbon Process.* **50** (Jan. 1971) 126.
431. H. W. Neukermans, J. P. Schurmans, *AIChE Symp. Saf. Ammonia Plants*, Montreal 1981.
432. B. W. Burklow, R. L. Coleman, *Ammonia Plant Saf.* **19** (1977) 21.
433. J. Brightling, P. Farnell, C. Forster, F. Beyer, paper presented at 50th AIChE Symposium on Safety in Ammonia Plants and Related Facilities, Toronto 2005.
434. J. Brightling, *Nitrogen & Methanol* **256** (2002) 29–39.
435. J. Thuillier, F. Pons, *Ammonia Plant Saf.* **20** (1978) 89.
436. Ch. J. Muhlenforth, *FINDS*, A Stokes Engineering Publication Volume XVII, Number 4, Fourth Quarter 2002.
437. Ch. J. Muhlenforth, *FINDS*, A Stokes Engineering Publication, Volume XV, Number 3, Third Quarter 2000.
438. R. G. Cockerham, G. Percival, *Trans. Inst. Gas. Eng.* **107** (1957/58) 390/433.
439. H. Jockel, B. E. Triebskorn, *Hydrocarbon Process.* **52** (1973) no. 1, 93.

440. W. Rall, *Erdöl Kohle Erdgas Petrochem.* **20** (1967) no. 5, 351.
441. H. Jockel, *GWF Gas Wasserfach* **110** (1969) no. 21, 561.
442. T. Ishiguro, *Hydrocarbon Process.* **47** (1968) no. 2, 87.
443. N N. Clark., W. G. S. Henson, 33rd AIChE Ammonia Safety Symp., Denver 1988.
444. B. J. Cromarty, *Nitrogen* **191** (1991) 30 – 34.
445. R. Vannby, S. Winter-Madsen, 36th AIChE Ammonia Safety Symp., Los Angeles 1991.
446. B. J. Cromarty, B. J. Crewdson, *The Application of Pre-reforming in the Ammonia and Hydrogen Industries*, ICI Catalco Tech. Paper, 59W/0590/0/CAT2.
447. K. Elkins, *Feedstock Flexibility Options*, ICI Catalco Tech. Paper, 313/035/0REF.
448. P. W. Farnell, *Pre-Reforming — a Retrofit Case Study*, ICI Catalco Tech. Paper, 291W/025/0/IMTOF.
449. W. D. Verduijn, 37th AIChE Ammonia Safety Symp., San Antonio 1992.
450. S. E. Nielsen, I. Dybkaer, 41th AIChE Ammonia Safety Symp., Boston 1996.
451. B. I. Grotz, *Hydrocarbon Process.* **46** (1967) no. 4, 197.
452. J. G. Livingstone, A. Pinto, *Chem. Eng. Prog.* **79** (1983) no. 5, 62.
453. I. Dybkær, *Fuel Processing Technology* **42** (1995) 85 – 107.
454. *Chem. Eng. (N.Y.)* **73** (1966) 24.
455. *Hydrocarbon Process.* **63** (1984) no. 4, 103.
456. I. Dybkjaer, *Fuel Processing Technology*, **42** (1995) 85–107.
457. S. Fritsch "Synthesis gas Processing Comparison of four synthesis gas routes", *Krupp-Unde fertilizer symposium*, Dortmund, June 11–13.
458. *Nitrogen* **178** (1989) 30 – 39.
459. K. J. Elkins, I. C. Jeffery, D. Kitchen, A. Pinto, "Nitrogen '91", *International Conference (British Sulphur)*, Copenhagen, June 4 – 6, 1991.
460. P. M. Armitage, J. Elkins, D. Kitchen, AIChE Ammonia Safety Symp., Los Angeles 1991.
461. D. Kitchen, K. Mansfield, Eurogas 92, Trondheim 1992.
462. R. Schneider, Kellogg Ammonia Club Meeting, San Diego 1990.
463. J. R. LeBlanc, Asia Nitrogen: British Sulphur Intern. Conf., Singapore 1996.
464. *Ammonia*, Kellogg brochure, HG 1/96.
465. J. R. LeBlanc, R. Schneider, K. W. Wright, 40th AIChE Ammonia Safety Symp., Tucson 1985.
466. T. A. Czuppon, S. A. Knez, 41th AIChE Ammonia Safety Symp., Boston 1996.
467. Syngas Technologies Supplement to Nitrogen & Methanol, 2005.
468. A. Malhotra, P. Kramer, S. Singh, paper presented at 48th AIChE Symposium on Safety in Ammonia Plants and Related Facilities, Orlando 2003.
469. I. Dybkjaer, paper presented at 50th AIChE Symposium on Safety in Ammonia Plants and Related Facilities, Toronto 2005.
470. H. D. Marsch, N. Thiagarajan, 33rd AIChE Ammonia Safety Symp., Denver 1988.
471. H. J. Herbort, Uhde Ammonia Symp., Dortmund 1992.
472. N. Thiagarajan, Uhde Ammonia Symp., Dortmund 1992.
473. H. D. Marsch, N. Thiagarajan, 37th AIChE Ammonia Safety Symp., San Antonio 1992.
474. T. Tomita, M. Kitugawa, *Chem. Ing. Tech.* **49** (1977) no. 6, 469.
475. *Nitrogen* **144** (1983) 33.
476. T. Tomita et al., Special Paper 9, 11th World Petroleum Congress, London 1983.
477. *Nitrogen* **168** (1987) 29 – 37.
478. *Nitrogen* **186** (1990) 21 – 30.
479. M. Appl, H. Gössling, *Chem. Ztg.* **96** (1972) 135.
480. O. J. Quartulli, *Hydrocarbon Process.* **44** (1965) no. 4, 151.
481. J. Davies, D. A. Lihou, *Chem. Process Eng. (London)* **52** (1971) no. 4, 71.
482. H. D. Marsch, H. J. Herbort, *Hydrocarbon Process.* **61** (1982) no. 6, 101.
483. *Nitrogen* **214** (1995) 38 – 56.
484. *Nitrogen* **195** (1992) 22 – 36.
485. A. I. Forster, B. J. Cromarty, *The Theory and Practice of Steam Reforming*, ICI/Catalco/KTI/UOP, 3rd Int. Seminar on Hydrogen Plant Operation, Chicago 1995.
486. G. W. Bridger, W. Wyrwas, *Chem. Process Eng. (London)* **48** (1967) no. 9, 101.
487. F. Marschner, H. J. Renner, *Hydrocarbon Process.* **61** (1982) no. 4, 176.
488. J. R. Rostrup-Nielsen, *Ammonia Plant Saf.* **15** (1973) 82.
489. G. W. Bridger, *Oil Gas J.* **74** (Feb. 16, 1976) 73;
Ammonia Plant Saf. **18** (1976) 24.
490. D. P. Rounthwaite, *Plant Oper. Prog.* **2** (1983) no. 2, 127.
491. *Nitrogen* **174** (1988) 23 – 24.
492. P. V. Broadhurst, P. K. Ingram, paper presented at Nitrogen 2005, International Conference and Exhibition, Bucharest, Romania 27 February – 2 March 2005.

493. *Nitrogen* **166** (1987) 24 – 31.
494. *Nitrogen* **167** (1987) 31 – 36.
495. G. E. Weismantel, L. Ricci, *Chem. Eng. (N.Y.)* **86** (1979) no. 21, 57.
496. H. Fricke, *Chem. Ztg.* **96** (1972) 123.
497. R. Lohmüller, *Chem. Ing. Tech.* **56** (1984) no. 3, 203.
498. C. I. Kuhre, C. I. Shearer, *Hydrocarbon Process.* **50** (1971) no. 12, 113.
499. L. J. Buividas, J. A. Finneran, O. J. Quartulli, *Chem. Eng. Prog.* **70** (1974) no. 10, 21; *Ammonia Plant Saf.* **17** (1975) 4.
500. *Nitrogen* **83** (1973) 40.
501. C. P. Marion, J. R. Muenger, *Energy Prog.* **1** (1981) no. 1 – 4, 27.
502. H. J. Madsack, *Hydrocarbon Process.* **61** (1982) no. 7, 169.
503. S. Strelzoff, *Hydrocarbon Process.* **53** (1974) no. 12, 79.
504. *Hydrocarbon Process.* **58** (1979) no. 4, 168.
505. G. J. van den Berg et al., *Hydrocarbon Process.* **45** (1966) no. 5, 193.
506. L. W. ter Haar, *Ind. Chim. Belge* **33** (1968) 655.
507. P. D. Becker et al., *Chem. Process Eng. (London)* **52** (1971) no. 11, 59.
508. C. L. Reed, C. J. Kuhre, *Hydrocarbon Process.* **58** (1979) no. 9, 191.
509. *Lurgi Gas Production Technology: The Shell Process* Lurgi comp. brochure 189e/6.92/2.20.
510. C. Higmann, *Perspectives and Experience with Partial Oxidation of Heavy Residues*, Lurgi company brochure.
511. *Integrated Gasification Combined Cycle Process*, Lurgi information paper, 1995.
512. W. Liebner, N. Hauser, ERPI Conference on Power Generation and the Environment, London 1990.
513. C. Higmann, G. Grünfelder, Conference on Gasification Power Plants, San Francisco 1984.
514. W. Soyez, 33rd AIChE Ammonia Safety Symp., Denver 1988.
515. *Hydrocarbon Process.* **63** (1984) no. 4, 90.
516. *Nitrogen* **126** (1980) 32.
517. *Nitrogen* **161** (1986) 23 – 27.
518. E. Supp: *How to Produce Methanol from Coal*, Springer-Verlag, Heidelberg 1989.
519. *Ammonia from Coal*, Tennessee Valley Authority, Muscle Shoals, Ala. 1979.
520. L. J. Buividas, *Chem. Eng. Prog.* **77** (1981) no. 5, 44; *Ammonia Plant Saf.* **23** (1981) 67.
521. *Ammonia from Coal*, Bull. Y-143, Tennessee Valley Authority, Muscle Shoals, Ala. 1979.
522. D. A. Waitzman, *Chem. Eng. (N.Y.)* **85** (1978) 69.
523. F. Brown, *Hydrocarbon Process.* **56** (1977) no. 11, 361.
524. H. Teggers, H. Jüntgen, *Erdöl Kohle Erdgas Petrochem.* **37** (1984) no. 4, 163.
525. *Nitrogen* **226** (1997) 47 – 56.
526. H. Staege, *Tech. Mitt. Krupp Werksber.* **40** (1982) no. 1, 1.
527. *Hydrocarbon Process.* **63** (1984) no. 4, 94.
528. H. J. Michaels, H. F. Leonard, *Chem. Eng. Prog.* **74** (1978) no. 8, 85.
529. H. Staege, *Hydrocarbon Process.* **61** (1982) no. 3, 92.
530. J. E. Franzen, E. K. Goeke, *Hydrocarbon Process.* **55** (1976) no. 11, 134.
531. A. D. Engelbrecht, Tennessee Valley Authority Symposium in Muscle Shoals, Alabama 1979.
532. *Nitrogen* 1985, no. 156, 35.
533. T. W. Nurse, *5th International Symposium Large Chemical Plants*, Antwerpen 1982.
534. E. T. Child, Tennessee Valley Authority Symposium in Muscle Shoals, Alabama 1979.
535. B. Cornils et al., *Hydrocarbon Process.* **60** (1981) no. 1, 149.
536. B. Cornils et al., *Chem. Ing. Tech.* **52** (1980) no. 1, 12.
537. *Hydrocarbon Process.* **63** (1984) no. 4, 97.
538. W. Konkol et al., *Hydrocarbon Process.* **61** (1982) no. 3, 97.
539. D. E. Nichols, P. C.Williamson in A. I. More (ed.), *Proc. Br. Sulphur Corp. Int. Conf. Fert. Technol. 4th 1981* 1982, 53.
540. B. Cornils et al., *CEER Chem. Econ. Eng. Rev.* **12** (1980) no. 6 – 7, 7.
541. *Hydrocarbon Process.* **63** (1984) no. 4, 96.
542. M. J. van der Burgt, *Hydrocarbon Process.* **58** (1979) no. 1, 161.
543. C. A. Bayens, 36th AIChE Ammonia Safety Symposium, Los Angeles 1991.
544. *Clean Coal Technology*, International Power Generation, 1990.
545. M. J. van der Burgt, J. E. Naber, *Ind. Chem. Bull.* (1983) 104 – 105.
546. J. N. Mahagaokar, J. N. Phillips, A. Kreweinghaus, 10th ERPI Conf. on Coal Gasification, 1986.
547. T. J. Pollaert, *Chem. Eng. Prog.* **74** (1978) no. 8, 95.
548. *Hydrocarbon Process.* **60** (1980) no. 11, 130.
549. M. K. Schad, C. F. Hafke, *Chem. Eng. Prog.* **79** (1983) no. 5, 45.

550. H. Hiller, *Erdöl Kohle Erdgas Petrochem.* **28** (1975) no. 2, 74.
551. P. D. Becker, Tennessee Valley Authority Symposium in Muscle Shoals, Alabama 1979.
552. W. Schäfer et al., *Erdöl Kohle Erdgas Petrochem.* **36** (1983) no. 12, 557.
553. *Hydrocarbon Process.* **61** (1982) no. 4, 151.
554. *Hydrocarbon Process.* **61** (1982) no. 4, 133.
555. K. A. Theis, E. Nitschke, *Hydrocarbon Process.* **61** (1982) no. 9, 233. *Hydrocarbon Process.* **63** (1984) no. 4, 93.
556. P. Speich, H. Teggers, *Erdöl Kohle Erdgas Petrochem.* **36** (1983) no. 8, 376.
557. W. L. E. Davey, E. L. Taylor, M. D. Newton, P. S. Larsen, P. S. Weitzel, "Asia Nitrogen," *Br. Sulphur Conf.*, Kuala Lumpur, Feb. 22 – 24, 1998, Proceedings of the Conf., vol. 2, p. 165.
558. P. S. Pedersen, J. H. Carstensen, J. Boghild-Hansen, 34th AIChE Ammonia Safety Symp., San Francisco 1989.
559. R. E. Stockwell, *FINDS V* (1990).
560. S. J. Catchpole et al., *Modern Catalyst Systems for Increased Ammonia Production and Efficiency*, ICI Catalco Tech. Paper.
561. H. Roos, H. Wanjek, 34th AIChE Ammonia Safety Symp., San Francisco 1989.
562. J. R. Rostrup-Nielsen et al., 37th AIChE Ammonia Safety Symp., San Antonio 1992.
563. J. B. Hansen, P. S. Pedersen, J. H. Carstensen, 33rd AIChE Ammonia Safety Symp., Denver 1988.
564. W. C. Lundberg, *Ammonia Plant Saf.* **21** (1979) 105.
565. P. W. Young, B. C. Clark, *Ammonia Plant Saf.* **15** (1973) 18.
566. I. Dybkjaer in A. I. More (ed.), *Proc. Br. Sulphur Corp. Int. Conf. Fert. Technol. 4th 1981* 1982, 503.
567. J. M. Stell et al., *The Operation of High, Low and Intermediate Temperature Catalysts*, ICI Catalco/KTI/UOP Hydrogen Plant Seminar, Chicago 1995.
568. J. Ilg, B. Kandziora, 41st AIChE Ammonia Safety Symp., Boston 1996.
569. D. F. Balz, H. F. Gettert, K. H. Gründler, *Plant Oper. Prog.* **2** (1983) no. 1, 47.
570. *Nitrogen* **226** (1997) 43 – 46.
571. J. M. Moe, *Chem. Eng. Prog.* **58** (1962) no. 3, 33.
572. *Nitrogen* **40** (1966) 28.
573. A. P. Ting, Shen-Wu-Wan, *Chem. Eng. (N.Y.)* **76** (1969) no. 11, 185.
574. J. F. Lombard, *Hydrocarbon Process.* **48** (Aug. 1969) 111.
575. L. Lloyd, M. V. Twigg, *Nitrogen* 1979, no. 118, 30.
576. *Nitrogen* 1981, no. 133, 27.
577. P. N. Hawker, *Hydrocarbon Process.* **61** (1982) no. 4, 183.
578. W. Auer, *Erdöl Kohle Erdgas Petrochem.* **24** (1971) 145.
579. *Hydrocarbon Process.* **61** (1982) no. 4, 154.
580. I. Dybkjaer, H. Bohlbro, *Ammonia Plant Saf.* **21** (1979) 145.
581. I. Dybkjaer, in: *Ammonia from Coal Symposium*, TVA Muscle Schoals Alabama, USA p. 133.
582. BASF, DE-AS 1250792, 1959.
583. K. J. Stokes, *Nitrogen* **131** (1981) 35. K. J. Stokes in A. I. More (ed.): *Proc. Br. Sulphur Corp. Int. Conf. Fert. Technol. 4th 1981* 1982, 525.
584. K. F. Butwell et al., *Chem. Eng. Prog.* **69** (1973) no. 2, 57 – 61.
585. *Nitrogen* **96** (1975) 33.
586. *Nitrogen* **102** (1976) 40.
587. K. F. Butwell, D. J. Kubek, P. W. Sigmund, *Ammonia Plant Saf.* **21** (1979) 156; *Chem. Eng. Prog.* **75** (1979) no. 2, 75. K. F. Butwell, D. J. Kubek, *Hydrocarbon Process.* **56** (1977) no. 10, 173.
588. H. E. Benson, R. W. Parrish, *Hydrocarbon Process.* **53** (1974) no. 4, 81.
589. H. E. Benson in A. V. Slack, G. J. James (eds.): *Ammonia*, **vol. 2**, Marcel Dekker, New York 1974, p. 159.
590. B. S. Grover, E. S. Holmes: "Nitrogen 86," *Br. Sulphur Conf.*, Amsterdam 1986, Proceedings of the Conf., p. 101.
591. R. K. Bartoo, S. J. Ruzicka: "Fertilizer '83," *Br. Sulphur Corp. 7th Int. Conf.*, London 1983, Proceedings of the Conf., p. 129.
592. US 4035 166, 1977 (F. C. van Hecke).
593. L. C. Crabs, R. Pouillard, F. C. van Hecke, 24th AIChE Ammonia Safety Symp., San Francisco 1989.
594. *Hydrocarbon Process.* **61** (1982) no. 4, 95 – 102.
595. R. N. Maddox, M. D. Burns, *Oil Gas J.* **66** (1968) 4 91.
596. G. Giammarco, in A. V. Slack (ed.): *Ammonia*, **vol. 12**, Marcel Decker New York 1974, p. 171.
597. L. Tomasi, *Nitrogen* **199** (1992) 35.
598. *Hydrocarbon Process.* **63** (1984) no. 4, 57 – 64.

599. A. G. Eickmeyer in A. V. Slack, G. J. James (eds.): *Ammonia*, **vol. 2**, Marcel Dekker, New York 1974.
600. *Nitrogen* **180** (1989) 20 – 30.
601. C. C. Song et al., 36th AIChE Ammonia Safety Symp., Los Angeles 1991.
602. S. Linsmayer, *Chem. Tech. (Leipzig)* **24** (1972) no. 2, 74.
603. K. Elberling, W. Gabriel, *Chem. Tech. (Leipzig)* **29** (1977) no. 1, 43.
604. T. M. Gemborys, 39th AIChE Ammonia Safety Symp., Vancouver 1994.
605. M. J. Mitariten, C. M. Wolf, T. J. DePaola, ICI Catalco/UOP Hydrogen Plant Seminar, Chicago 1995.
606. P. Clough: "Asia Nitrogen," *Br. Sulphur Conf.*, Bali 1994, Proceedings of the Conf., p. 191 – 193.
607. J. N. Iyengar, D. E. Keene, 38th AIChE Ammonia Safety Symp., Orlando 1993.
608. US 4702 898 (B. S. Grover).
609. R. K. Bartoo, T. M. Gemborys: "Nitrogen 91," *Br.Sulphur Conf.*, Copenhagen 1991, Proceedings of the Conf., p. 127 – 139.
610. R. K. Bartoo, *Removing Acid Gas by the Benfield Process*, UOP Comp brochure, GP 5132–1M-11/93.
611. R. K. Bartoo, *Chem. Eng. Prog.* **80** (1984) no. 10, 35
612. R. E. Meissner, U. Wagner, *Oil Gas J.* **81** (Feb. 7, 1983) 55.
613. K. Volkamer, E. Wagner, F. Schubert, *Plant Oper. Prog.* **1** (1982) no. 2, 134.
614. K. Volkamer, U. Wagner: "Fertilizer '83," *Br. Sulphur Corp. 7th Int. Conf.*, London 1983, Proceedings of the Conf., p. 139.
615. W. Gerhard, W. Hefner, 33rd AIChE Ammonia Safety Symp., Denver 1988.
616. D. W. Stanbridge, Y. Ide, W. Hefner, AIChE Ammonia Safety Symp., Anaheim 1984.
617. R. Welker, R. Hugo, R. Meissner, 41st AIChE Ammonia Safety Symp., Boston 1996.
618. R. Welker, R. Hugo, B. Büchele: Asia Nitrogen 96, Singapore 1996.
619. G. Ripperger, J. C. Stover, AIChE Spring National Meeting, Boston 1996.
620. J. E. Nobles, N. L. Shay: "Nitrogen 88," *Br. Sulphur Int. Conf.*, Geneva 1988.
621. W. Hefner, R. E. Meissner, 37th AIChE Ammonia Safety Symp., San Antonio 1992.
622. D. K. Judd, *Hydrocarbon Process.* **57** (1978) no. 4, 122.
623. C. G. Swanson, *Ammonia Plant Saf.* **21** (1979) 152.
624. C. G. Swanson, F. C. Burkhard, *Ammonia Plant Saf.* **24** (1984) 16.
625. V. A. Shah, *Energy Progress* **8** (1988) 67 – 70.
626. R. J. Hernandez, T. L. Huurdemann: "Nitrogen 88," *Br. Sulphur Int. Conf.*, Geneva 1988.
627. T. L. Huurdemann, V. A. Shah, 34th AIChE Ammonia Safety Symp., San Franzisco 1989.
628. V. A. Shah, *Hydrocarbon Process.* **67** (1988).
629. R. J. Hernandez, T. L. Huurdemann, *Chem. Eng.* **5** (1989) 154 – 156.
630. *Fertilizer Focos* **5** (1988) 27.
631. J. L. Lewis, H. A. Truby, M. B. Pascoo, *Oil Gas J.* **72** (June 24, 1974) 120.
632. J. L. Lewis, H. A. Truby, M. B. Pascoo, *Hydrocarbon Process.* **58** (1979) no. 4, 112.
633. R. W. Bucklin, R. L. Schendel, *Energy Progr.* **4** (1984) no. 3, 137.
634. J. P. Cook in A. V. Slack, G. J. James (eds.): *Ammonia*, **vol. 2**, Marcel Dekker, New York 1974, p. 171
635. G. Hochgesand, *Ind. Eng. Chem.* **62** (1970) 7 – 37.
636. L. Dailey, in A. V. Slack, G. J. James (eds.): *Ammonia*, Marcel Decker New York 1974.
637. W. V. Korf, K. Thormann, K. Braßler, *Erdöl Kohle Erdgas, Petroch.* **16** (1963) 2, 94.
638. K. E. Zarker in A. V. Slack, G. J. James (eds.): *Ammonia*, **vol. 2**, Marcel Dekker, New York 1974 p. 219.
639. J. P. Klein, *Erdöl Kohle Erdgas Petrochem.* **23** (1970) no. 2, 84.
640. B. Sehrt, P. Polster, *Chem. Tech. (Leipzig)* **32** (1980) no. 7, 345.
641. *The Rectisol Process for Gas Purification*, Lurgi brochure 1676 e/5.95/10.
642. *Rectisol for Gas Treating*, Lurgi Express Information O 1051/12.72.
643. G. Ranke, Linde Berichte aus Wissenschaft und Technik.
644. H. Haase, *Chem. Anlagen Verfahren* **10** (1970) 59.
645. "By no means a foregone conclusion", *Nitrogen & Methanol* **252** (2001) 33–51.
646. H. W. Schmidt, H. J. Henrici, *Chem. Ztg.* **96** (1972) no. 3, 154.
647. K. J. Stokes, *Ammonia Plant Saf.* **22** (1980) 178.
648. B. G. Goar, *Oil Gas J.* **69** (July 12, 1971) 75.
649. A. L. Kohl, F. C. Riesenfeld: *Gas Purification*, Gulf Publ. Co., Houston, Tex. 1979.
650. G. Hochgesand, *Chem. Ing. Techn.* **40** (1968) 43.

651. S. Strelzoff, *Chem. Eng. (N.Y.)* **82** (1975) no. 19, 115.
652. D. Werner, *Chem. Ing. Techn.* **53** (1981) no. 2, 73.
653. K. G.Christensen, *Hydrocarbon Process.* **57** (1978) no. 2, 125.
654. H. Thirkell in A. V. Slack G. J. James (eds.): *Ammonia*, **vol. 2**, Marcel Dekker, New York 1974, p. 117.
655. S. Strelzow: *Technology and Manufacture of Ammonia*, Wiley, New York 1981, p. 193.
656. F. C. Brown, C. L. Leci, *Proc. Fertilizer Soc.* **210** (1982) 1.
657. *Nitrogen* **180** (1989) 20.
658. *Nitrogen* **229** (1997) 37 – 51.
659. M. P. Sukumaran, *Nitrogen & Methanol* **264** (2003) 19–28.
660. *Nitrogen & Methanol* **275** (2005) 38–47.
661. H. W. Schmidt, *Chem. Ing. Tech.* **40** (1968) 425.
662. D. W. Allen, W. H. Yen, *Ammonia Plant Saf.* **15** (1973) 96.
663. A. J. M. Janssen, N. Siraa, J. M. Blanken, *Ammonia Plant Saf.* **23** (1981) 19.
664. *Chemical Economy and Engineering Review*, **12** (6 – 7), 33 – 36, (Jun/Jul 1980).
665. C. M. Buckthorp, *Nitrogen* **113** (1978) 34.
666. *Nitrogen* **123** (1980) 39.
667. J. C. Bonacci, T. G. Otchy, *Ammonia Plant Saf.* **20** (1978) 165.
668. J. H. Colby, G. A. White, P. N. Notwick, *Ammonia Plant Saf.* **21** (1979) 138.
669. *Nitrogen* **197** (1992) 18 – 25.
670. P. Soregaard-Andersen, O. Hansen, 36th AIChE Ammonia Safety Symp., Los Angeles 1991.
671. *Nitrogen* **217** (1995) 41 – 48.
672. US 3442 613, 1969 (B. J. Grotz).
673. *Nitrogen* **144** (1983) 30.
674. B. J. Grotz, *Nitrogen* **100** (1976) 30.
675. *Nitrogen* **182** (1989) 25.
676. B. J. Grotz, G. Good, (1980) *Chem. Age (London)* (November, 14), 18.
677. C. K. Wilson et al., *Nitrogen* **151** (1984) 31.
678. W. Scholz, *DECHEMA-Monogr.* **58** (1968) 31.
679. W. Förg, *Chem. Anlagen + Verfahren* 1970 (March), 33.
680. *Nitrogen* **95** (1975) 38.
681. S. Hahesa et al., *Saf. Air Ammonia Plants* **8** (1966) 14.
682. S. R. Krolikowski, *Chem. Eng. (London)* 1965 no. 186, 40.
683. F. Corr, F. Dropp, E. Rudelstorfer, *Hydrocarbon Process.* **58** (1979) no. 3, 119.
684. D. H. Werner, G. A. Schlichthärle, *Ammonia Plant Saf.* **22** (1980) 12.
685. *Nitrogen* **121** (1979) 37.
686. J. L. Heck, *Oil Gas J.* **78** (Feb. 11, 1980) 122.
687. P. Taffe, J. Joseph, *Chem. Age* 1978 (October) 14.
688. P. R. Savage, *Chem. Eng. (N.Y.)* **82** (1978) no. 25, 68D.
689. R. Rehder, P. Stead, FAI Seminar 1985, The Fertilizer Association of India, New Dehli, Tech II-2/1.
690. W. F. van Weenen, J. Tielroy, *Nitrogen* **127** (1980) 38.
691. *Oil Gas J.* **19** (1981) 270.
692. W. F. van Weenen, J. Tielroy, *Chem. Age India* **31** (1980) no. 12, Dev-2/1.
693. J. Voogd, FAI Seminar 1985, The Fertilizer Association of India, New Dehli, Tech 11–1/1.
694. D. J. Carra, R. A. McAllister, *Chem. Eng. (N.Y.)* **70** (1963) 62.
695. "Compressors" in *Encyclopedia of Chemical Engineering*, **Vol. 10**, Dekker, New York 1979, 157 – 409.
696. E. E. Ludwig, "Compressors" in *Applied Design for Chemical and Petrochemical Plants*, **Vol. 3**, Gulf, Houston, 1983, pp. 251 – 396.
697. C. A. Vancini: *La Sintesi dell Amoniaca*, Hoepli, Milano 1961, p. 497.
698. J. L. Kennedy, *Oil Gas J.* **65** 1967, no. 46, 105; no. 48, 95; no. 1, 72; no. 51, 76. *Oil Gas J.* **66** 1968, no. 4, 76.
699. G. P. Williams, W. W. Hoehing, *Chem. Eng. Prog.* **79** (1983) no. 3, 11.
700. G. A. J. Begg, *Chem. Process Eng. (London)* **49** (1968) no. 1, 58.
701. F. Fraschetti et al., *Quad. Pignone* 1967, no. 9, 5.
702. *Chem. Eng. (N.Y.)* **73** (1966) no. 9, 126.
703. P. Kerklo, *Hydrocarbon Process.* **61** (1982) no. 10, 112.
704. H. Strassmann, Uhde Ammonia Symp., Dortmund 1992.
705. C. Dickinson, *Why Dry Running Gas Seals?* J. Crane Inc. Comp. Publ.
706. W. Koch, *Dry Gas Seal — Principles — Capabilities — Application*, J. Crane Inc. Comp. Publ.
707. S. Gray, B. Baxter, G. Jones, "Magnetic Bearings can Increase Availability, Reduce Oil and Maintenance Costs," Power Engineering, 1990.
708. R. R. Poricha, D. G. Rao, *Technology (Sindri, India)* **3** (1966) 96.

709. H. Förster, *Chem. Tech. (Leipzig)* **23** (1971) 157.
710. *Hydrocarbon Process. Pet. Refiner* **45** (1966) no. 5, 179.
711. *Eur. Chem. News* **11** (1967) no. 270, 26.
712. W. Plötner, *Chem. Tech. (Leipzig)* **24** (1972) 324.
713. A. F. Wilck, *Energy Tech.* **23** (1971) no. 5, 161.
714. I. M. Kalnin, *Luft Kältetech.* **8** (1972) no. 3, 142.
715. W. L. Luther, *Ingenieur Digest* **10** (1971) no. 1, 60.
716. H. E. Gallus, *Brennst. Wärme Kraft* **23** (1971) no. 4, 172.
717. F. Fraschetti, U. Filippini, P. L. Ferrara, *Quad. Pignone* **9** (1967) 5.
718. G. A. J. Begg, *Chem. Proc. Eng. (London)* **49** (1968) no. 1, 58.
719. A. Vitti, *Het Ingenieursblad* **40** (1971) 619.
720. P. L. Ferrara, A. Tesei, *Quad. Pignone* **25** (1978) 131.
721. J. Salviani et al., in A. V. Slack, G. J. James (eds.): *Ammonia*, **vol. 2** , Marcel Dekker, New York 1974, p. 1.
722. *Nitrogen* **55** (1968) 37.
723. S. Labrow, *Chem. Proc. Eng. (London)* **49** (1968) no. 1, 55.
724. H. Foerster, *Chem. Technol.* **23** (1971) 93.
725. H. Foerster, *Chem. Technol.* **23** (1971) 157.
726. W. Plötner, *Chem. Technol.* **24** (1972) 324.
727. *Nitrogen* **181** (1981) 23.
728. S. D. Caplow, S. A. Bresler, *Chem. Eng. (New York)* **74** (1967) no. 7, 103.
729. K. J. Stokes, *Chem. Eng. Prog.* **75** (1979) no. 7, 88.
730. GB 1 134 621, 1968 (D. R. Twist, D. W. Stanbridge).
731. S. Ujii, M. Ikeda, *Hydrocarbon Process.* **60** (1981) no. 7, 94.
732. W. Rall, 35th AIChE Ammonia Safety Symp., San Diego 1990.
733. H. Uchida, M. Kuraishi, *Bull. Chem. Soc. Jpn.* **23** (1955) 106.
734. R. F. Baddour et al., *Chem. Eng. Sci.* **20** (1965) 281.
735. C. van Heerden, *Ind. Eng. Chem.* **45** (1953) 1242.
736. H. Inoue, T. Komiya, *Int. Chem. Eng.* **8** (1968) no. 4, 749.
737. H. Bakemeier et al., *Chem. Ing. Tech.* **37** (1965) 427.
738. I. Porubszky et al., *Br. Chem. Eng.* **14** (1969) no. 4, 495.
739. F. Horn, *Chem. Ing. Tech.* **42** (1970) no. 8, 561.
740. V. Hlavacek, *Ind. Eng. Chem.* **62** (1970) no. 7, 8.
741. J. Kjaer: *Measurement and Calculation of Temperature and Conversion in Fixed-bed Catalytic Reactors*, Gjellerup, Kopenhagen 1958.
742. G. F. Froment, *Chem. Ing. Tech.* **46** (1974) no. 9, 374.
743. J. E. Jarvan, *Ber. Bunsenges. Phys. Chemie* **74** (1970) no. 2, 74.
744. J. Kjaer, Computer Methods in Catalytic Reactor Calculations, Haldor Topsøe, Vedaek 1972.
745. J. E. Jarvan, *Oil Gas J.* **76** (1978) no. 5, 178.
746. J. E. Jarvan, *Oil Gas J.* **76** (1978) no. 3, 51.
747. C. P. P. Singh, D. N. Saraf, *Ind. Eng. Chem. Process Des. Dev.* **18** (1979) no. 3, 364.
748. J. Simiceanu, C. Petrila, A. Pop, *Chem. Tech. (Leipzig)* **35** (1983) no. 12, 628.
749. K. Lukas, D. Gelbin, *Chem. Eng. (London)* **7** (1962) 336.
750. M. J. Shah, *Ind. Eng. Chem.* **59** (1967) 72.
751. L. D. Gaines, *Ind. Eng. Chem. Process Des. Dev.* **18** (1979) no. 3, 381.
752. L. D. Gaines, *Chem. Eng. Sci.* **34** (1979) 37.
753. K. V. Reddy, A. Husain, *Ind. Eng. Chem. Process Des. Dev.* **21** (1982) no. 3, 359.
754. L. M. Shipman, J. B. Hickman, *Chem Eng. Progr.* **64** (1968) 59.
755. R. F. Baddour et al., *Chem. Eng. Sci.* **20** (1965) 281.
756. P. L. Brian, R. F. Baddour, J. P. Emery, *Chem. Eng. Sci.* **20** (1965) 297.
757. G. A. Almasy, P. Jedlovsky, *Chem. Eng. (London)* **12** (1967) 1219.
758. I. Porubsky, E. Simonyi, G. Ladanyi, *Chem. Eng. (London)* **14** (1969) 495.
759. A. Murase, H. L. Roberts, A. O. Converse, *Ind. Eng. Chem. Process Des. Dev.* **9** (1970) 503.
760. D. Balzer et al., *Chem. Technik (Leipzig)* **23** (1971) 513.
761. D. R. Levin, R. Lavie, *Ind. Chem. Symp. Ser.* **87** (1984) 393.
762. A. Nielsen: *Advances in Catalysis*, **vol. V** , Academic Press, New York 1953, p. 1.
763. H. Hinrichs, J. Niedetzky, *Chem. Ing. Tech.* **34** (1962) 88.
764. L. Fodor, *Chim. Ind. Genie Chim.* **104** (1971) 1002.
765. *Nitrogen* **140** (1982) 30.
766. U. Zardi, *Hydrocarbon Process.* **61** (1982) no. 8, 129.

767. J. J. Hay, I. M. Pallai, *Br. Chem. Eng.* **8** (1963) 171.
768. P. L. T. Brian et al., *Chem. Eng. Sci.* **20** (1965) 297.
769. W. Hennel, *Chem. Tech. (Leipzig)* **15** (1963) 293.
770. L. B. Hein, *Chem. Eng. Prog.* **48** (1952) no. 8, 412.
771. N. S. Zayarnyi, *Int. Chem. Eng.* **2** (1962) no. 3, 378.
772. A. Murase et al., *Ind. Eng. Chem. Process Des. Dev.* **9** (1970) no. 4, 503.
773. J. B. Allen, *Chem. Process Eng. (London)* **46** (1965) no. 9, 473.
774. J. B. Allen, *Chem. Proc. Eng. (London)* **46** (1965) 473.
775. L. B. Hein, *Chem. Eng. Progr.* **48** (1952) 412.
776. C. A. Vancini: *Synthesis of Ammonia*, Macmillan, London 1971.
777. I. Hay, G. D. Honti: *The Nitrogen Industry*, **vol. 1**, Akademiai Kiado, Budapest 1976.
778. *Nitrogen* **72** (1971) 34.
779. C. A. Vancini: *Synthesis of Ammonia*, Macmillian, London 1971, p. 249.
780. D. Hooper, paper presented at 47th AIChE Symposium on Safety in Ammonia Plants and Related Facilities, San Diego 2002.
781. US 2953 371, 1958 (A. Christensen, R. D. Rayfield).
782. US 3041 151, 1962 (A. Christensen).
783. US 3050 377, 1962 (A. Christensen).
784. US 2861 873, 1976 (G. A. Worn).
785. US 3032 139, 1976 (M. Vilceanu, C. Bors).
786. E. Filippi, F. Di Muzio, E. Rizzi. 2nd Casale Symposium for Customers and Licensees, Lugano 30 May – 2 June 2006.
787. E. West-Toolsee, E. Filippi, paper presented at 50th AIChE Symposium on Safety in Ammonia Plants and Related Facilities, Toronto 2005.
788. H. Hinrichs et al., *DECHEMA Monogr.* **68** (1971) 493 – 500.
789. P. Lesur, *Nitrogen* **108** (1977) 29.
790. C. A. Vancini: *Synthesis of Ammonia*, Macmillan, London 1971, p. 336, 240, 244.
791. A. V. Slack, G. J. James, *Ammonia*, Marcel Dekker, New York 1974, p. 310.
792. *Nitrogen* **75** (1972) no. 75, 33.
793. US 3475 136, 1969 (G. P. Eschenbrenner, C. A. Honigsberg).
794. S. Strelzow: *Technology and Manufacture of Ammonia*, Wiley, New York 1981, p. 34.
795. US 3694 169, 1972 (R. Fawcett, A. W. Smith, D. Westwood).
796. D. E. Riddler, 16th AIChE Ammonia Safety Symp., Atlantic City 1971.
797. US 3633 179, 1972 (D. D. Metha, E. J. Miller).
798. I. Dybkaer in H. I. Lasa (ed.): *Chemical Reactor Design and Technology*, Nasa ASI Series E: Applied Sciences, no. 110, Nijhoff Publ., Dordrecht 1986.
799. V. Vek, 7th Int. Symp. Large Chem. Plants, Brugge 1988, p. 77.
800. V. Vek, *Chem. Ing. Tech.* **45** (1973) 608.
801. V. Vek, *Ind Eng. Chem. Process Des. Dev.* **16** (1977) 412.
802. DE 3334 777, 1984 (K. Ohasaki, J. Zanma, H. Watanabe).
803. *Nitrogen* **31** (1964) 22.
804. T. Wett, *Oil Gas J.* **69** (1971) 70.
805. *Chem. Eng (New York)* **78** (1971) 90.
806. US 3372 988, 1968 (H. J. Hansen).
807. A. Nielsen, 16th AIChE Ammonia Safety Symp., Atlantic City 1971.
808. A. V. Slack in A. V. Slack, G. J. James (eds.): *Ammonia*, **vol. 3**, Marcel Dekker, New York 1974, p. 345.
809. A. V. Slack in A. V. Slack, G. J. James (eds.): *Ammonia*, **vol. 3**, Marcel Dekker, New York 1974, p. 354.
810. US 3754 078, 1975 (H. Hinrichs et al.).
811. US 3918 918, 1975 (H. B. Kohn, G. Friedman).
812. O. J. Quartulli, G. A. Wagner, *Hydrocarbon Process.* **57** (1978) no. 12, 115.
813. US 3567 404, 1971 (L. C. Axelrod et al.).
814. G. P. Eschenbrenner, G. A. Wagner, 16th AIChE Ammonia Safety Symp., Atlantic City 1971.
815. M. W. Kellogg, EP-A 106076, 1983 (C. van Dijk et al.)
816. C. A. Vancini: *Synthesis of Ammonia*, Macmillan, London 1971 p. 232.
817. US 2898 183, 1959 (G. Fauser).
818. S. Strelzow: *Technology and Manufacture of Ammonia*, Wiley, New York 1981, p. 27.
819. H. W. Graeve, *Chem. Eng. Prog.* **77** (1981) no. 10, 54;
Ammonia Plant Saf. **23** (1981) 78.
820. *Nitrogen* **81** (1973) 29, 37.
821. H. Hinrichs, H. Lehner, *Chem. Anlagen + Verfahren* 1972, no. 6, 55.
822. F. Förster, *Chem. Eng. (N.Y.)* **87** (1980) no. 9, 62.
823. *Nitrogen* **101** (1976) 42.
824. Topsøe Topics, June 1976.
825. I. Dybkjaer, E. A. Gam, 29th AIChE Ammonia Safety Symp., San Francisco 1994.

826. I. Dybkjaer, E. A. Gam, *CEER Chem. Econ. Eng. Rev.* **16** (1984) 29.
827. Haldor Topsøe, US 4181 701, 1980 (E. A. Gam).
828. Haldor Topsøe, US 2710 247, 1981.
829. I. Dybkjaer, J. E. Jarvan, Nitrogen 97, British Sulphur Conference, Geneva 1997.
830. *Nitrogen* **169** (1987) 33.
831. E. Commandini, U. Zardi, Fertilizer Latin America, Int. Conf. British Sulphur, Caracas 1989.
832. L. Sutherland, B. Wallace, 33rd AIChE Ammonia Safety Symp., Denver 1988.
833. G. Pagani, U. Zardi, FAI Seminar 1988, The Fertilizer Society of India, New Dehli, S II/1–19.
834. K. A. Clayton, N. Shannahan, B. Wallace, 35th AIChE Ammonia Safety Symp., San Diego 1990.
835. N. Shannahan, *Hydrocarbon Process* **68** (1989) 60.
836. S. E. Handman, J. R. LeBlanc, *Chem. Eng. Prog.* **79** (1983) no. 5, 56.
837. H. Stahl, *Symposium of the Fertilizer Society*, London 1982, Proceedings, p. 61.
838. US 4452760, 1984 (R. B. Peterson, R. Finello, G. A. Denavit).
839. US 3892 535, 1975 (W. Hennel, C. Sobolewsk).
840. W. A. Glover, J. P. Yoars, *Hydrocarbon Process.* **52** (1973) no. 4, 165; *Ammonia Plant Saf.* **15** (1973) 77.
841. *Nitrogen* **82** (1973) 34.
842. GB 1134 621, 1989 (D. R. Twist, D. W. Stanbridge).
843. W. A. Glover, J. P. Yoars, *Hydrocarbon Process* **52** (1973) 165.
844. US 3721 532, 1973 (L. E. Wright, A. E. Pickford).
845. US 3851 046, 1975 (L. E. Wright, A. E. Pickford).
846. US 4 744 966, 1988 (B. J. Grotz).
847. EP 268 469, 1988 (B. J. Grotz).
848. K. C. Wilson, B. J. Grotz, *J. Tech. Dev.* **14** (1990) 54.
849. B. J. Grotz, L. Grisolia, *Nitrogen* **199** (1992) 39.
850. Ammonia, Uhde company brochure F 110e/2000 09/2004 DÖ/Hi (09.08.2004).
851. *Nitrogen & Methanol*, **258** (2002) 47–48.
852. J. Larsen, D. Lippman, C. W. Hooper, *Nitrogen & Methanol*, **253** (2001) 41–46.
853. D. Lippmann: "Large-scale ammonia plant technology", paper presented at 4th Uhde Fertilizer Symposium, Dortmund June 2002.
854. J. Rüther, J. Larsen, D. Lippman, D. Claes, paper presented at 50th AIChE Symposium on Safety in Ammonia Plants and Related Facilities, Toronto 2005.
855. J. S. Larsen, D. Lippman, paper presented at 46th AIChE Symposium on Safety in Ammonia Plants and Related Facilities, San Diego, 2002.
856. K. Blanchard, St. Noe, E. Plaxco, *FINDS*, A Stokes Engineering Publication, Volume XV Number 4, Fourth Quarter 2000.
857. F. Horn, L. Küchler, *Chem. Ing. Tech.* **31** (1959) 1.
858. H. Bakemeier, R. Krabetz, *Chem. Ing. Tech.* **34** (1962) 1.
859. R. Jackson, *Chem. Eng. Sci.* **19** (1964) 19, 253.
860. D. C. Dyson et al., *Canadian J. Chem. Eng. Sci.* **45** (1967) 310.
861. L. D. Gaines, *Ind. Eng. Chem. Process Des. Dev.* **16** (1977) no. 3, 381.
862. I. Dybkjaer in A. Nielsen: *Ammonia — Catalysis and Manufacture*, Springer-Verlag, New York 1995, p. 251.
863. Ammonia — Energy Integration in Ammonia Plants, Uhde Engineering News 2–91, Hi 111 9 1000 91, 1991.
864. M. J. P. Bogart in A. I. More (ed.), *Proc. Br. Sulphur Corp. Int. Conf. Fert. Technol. 4th 1981* 1982, 141.
865. M. J. P. Bogart, *Plant Oper. Prog.* **1** (1982) no. 3, 147.
866. M. J. P. Bogart, *Hydrocarbon Process.* **57** (1978) no. 4, 145.
867. W. Malewski, *Chem. Ztg.* **95** (1971) 186.
868. K. Bohlscheid, *Chem. Prod.* **8** (1979) no. 3.
869. G. Holldorff, *Hydrocarbon Process.* **58** (1979) no. 7, 149.
870. Pullman Inc., DE-OS 2741851, 1978 (C. L. Becker).
871. G. Pagani, U. Zardi, *Hydrocarbon Process.* **51** (1972) no. 7, 106 – 110.
872. F. Saviano, V. Lagana, P. Bisi, *Hydrocarbon Process.* **60** (1981) no. 7, 99.
873. H. Hinrichs, *Chem. Ztg. Chem. Appar.* **86** (1962) 223.
874. H. Neth et al., *Chem. Eng. Prog.* **78** (1982) no. 7, 69.
875. *Chem. Week* **116** (Feb. 19, 1975) 29.
876. A. Finn, *Nitrogen* **175** (1988) 25 – 32.
877. R. Harmon, W. H. Isalski, *Ammonia Plant Saf.* **23** (1981) 39.
878. W. H. Isalski, *Nitrogen* **152** (1984) 100.
879. R. Harmon, W. H. Isalski, 25th AIChE Ammonia Safety Symp., Portland 1980.

880. C. A. Combs, 25th AIChE Ammonia Safety Symp., Portland 1980.
881. A. Haslam, P. Brook, H. Isalski, L. Lunde, *Hydrocarbon Process.* **55** (1976) no. 1, 103.
882. *Nitrogen* **102** (1976) 35.
883. R. Banks, *Chem. Eng. (N.Y.)* **84** (1977) no. 21, 90.
884. A. A. Haslam, W. H. Isalski, *Ammonia Plant Saf.* **17** (1975) 80.
885. R. Harmon in A. I. More (ed.), *Proc. Br. Sulphur Corp. Int. Conf. Fert. Technol. 4th 1981* 1982, 113.
886. R. Banks, *Ammonia Plant Saf.* **20** (1978) 79.
887. *Oil Gas J.* **77** (March 5, 1979) 182.
888. R. Fabian, D. Tilman, *Linde Ber. Techn. Wiss.* **59** (1986) 6.
889. C. A. Combs, *Ammonia Plant Saf.* **23** (1981) 32.
890. D. L. MacLean, C. E. Prince, Y. C. Chae, *Chem. Eng. Prog.* **76** (1980) no. 3, 98; *Ammonia Plant Saf.* **22** (1980) 1.
891. Y. C. Chae, G. S. Legendre, J. M. van Gelder in A. I. More (ed.), *Proc. Br. Sulphur Corp. Int. Conf. Fert. Technol. 4th 1981* 1982, 457.
892. *Nitrogen* **130** (1981) 40.
893. *Nitrogen* **136** (1982) 29.
894. R. L. Schendel, C. L. Mariz, J. Y. Mak, *Hydrocarbon Process.* **62** (1983) no. 8, 58.
895. A. K. Fritzsche, R. A. Narayan, *CEER Chem. Econ. Eng. Rev.* **19** (1987) 19.
896. W. A. Koros, R. A. Narayan, *Chem. Eng. Progr.* (1995) 68 – 81.
897. *Hyrocarbon Process.* **62** (1983) 43 – 62.
898. D. L. MacLean, D. J. Stockey, T. R. Metzger, *Hydrocarbon Process.* **62** (1983) no. 8, 47.
899. H. Knieriem, Jr., *Hydrocarbon Process.* **59** (1980) no. 7, 65.
900. M. D. Rosenzweig, *Chem. Eng. (N.Y.)* **88** (1981) no. 24, 62.
901. G. Schulz, H. Michele, U. Werner, *Chem. Ing. Tech.* **54** (1982) no. 4, 351.
902. G. Q. Miller, M. J. Mitariten, ICI Catalco/KTI/UOP Hydrogen Plant Seminar, Chicago 1995.
903. *Nitrogen* **121** (1979) 37 – 43.
904. G. Low, *Nitrogen* **147** (1984).
905. Union Carbide Corp., US 4077780, 1976.
906. *Chem. Eng. (London)* 1979, no. 345, 395.
907. K. Knoblauch et al., *Erdöl Kohle Erdgas Petrochem.* **32** (1979) 551.
908. J. G. Santangelo, G. T. Chen, *Chemtech* **13** (1983) no. 10, 621.
909. J. J. Sheridan, III, et al., *Less Common Met.* **89** (1983). 447.
910. R. J. Berry, *Chem. Eng. (N.Y.)* **86** (1979) no. 15, 62.
911. W. H. Isalski, G. J. Ashton in A. I. More (ed.), *Proc. Br. Sulphur Corp. Int. Conf. Fert. Technol. 4th 1981* 1982, 125.
912. S. Lynn, C. Alesandrini, *Ammonia Plant Saf.* **16** (1974) 80.
913. I. Dybkjaer in A. Nielsen: *Ammonia — Catalysis and Manufacture*, Springer-Verlag, New York 1995, p. 231.
914. O. J. Quartulli, J. B. Fleming, J. A. Finneran, *Hydrocarbon Process.* **47** (1968) 153.
915. O. J. Quartulli, J. B. Fleming, J. A. Finneran, *Nitrogen* **58** (1969) 25.
916. I. Dybkjaer in A. Nielsen (ed.): *Ammonia — Catalysis and Manufacture*, Springer-Verlag, New York 1995, p. 226, 227.
917. "A balancing act", *Nitrogen & Methanol* **273** (2005) 39–44.
918. L. Silberring, *Nitrogen* **120** (1979) 35.
919. P. Hinchley, *Chem. Eng. (N.Y.)* **86** (1979) no. 17, 120.
P. Hinchley, *Proc. Inst. Mech. Eng.* **193** (1979) no. 8.
920. H. Lachmann, *Ammonia Plant Saf.* **23** (1981) 51.
921. H. Weber et al., *Chem. Ing. Tech.* **56** (1984) no. 5, 356.
922. O. J. Quartulli, W. Turner, *Nitrogen* **80** (1972) 28.
923. O. J. Quartulli, W. Turner, *Nitrogen* **81** (1973) 32.
924. J. B. LeBlanc, M. N. Shah, L. J. Buividas, *Hydrocarbon Process.* **59** (1980) no. 4, 68-G.
925. *Hydrocarbon Process.* **62** (1983) no. 11, 81.
926. E. Futterer, E. Pattas, *Chem. Ztg.* **98** (1974) no. 9, 438.
927. K.-J. Mundo, *Chem. Anlagen + Verfahren* 1972, no. 6, 49.
928. J. D. Rankin, J. G. Livingstone, *Ammonia Plant Saf.* **23** (1981) 203.
929. J. S. Campbell, J. W. Marshall, *Nitrogen* 1976, no. 103, 33.
930. O. J. Quartulli, D. Wagener, *Erdöl Kohle Erdgas Petrochem.* **26** (1973) no. 4, 192.
931. U. Zardi, A. Antonini, *Nitrogen* **122** (1979) 33.
932. G. D. Honti, 4th Int. Conf on Fertilizer Technol., London 1991, Conf. Proc., p. 1.
933. F. Saviano, W. Lagana, P. Bisi, *Hydrocarbon Process.* **60** (1981) 99.
934. I. Dybkjaer, ECN Europ. Chemical News: Fertilizers 83 (suppl.), 1983, p.15.
935. G. R. James, K. J. Stokes, *Chem. Eng. Progr.* **62** (1984) 81.

936. J. M. Blanken, 33rd AIChE Ammonia Safety Symp., Denver 1988.
937. F. C. Brown, *Nitrogen* **100** (1976) 65.
938. K. J. Mundo, *Erdöl, Kohle, Erdgas Petrochemie* **31** (1978) 74.
939. *Nitrogen* **182** (1989) 25.
940. L. J. Buividas, *Hydrocarbon Process.* **58** (1979) no. 5, 257.
941. G. D. Honti in A. I. More (ed.), *Proc. Br. Sulphur Corp. Int. Conf. Fert. Technol. 4th 1981* 1982, 1.
942. G. R. James, 32nd AIChE Ammonia Safety Symp., Minneapolis 1987. USA 1981.
G. R. James, K. J. Stokes, *Chem. Eng. Prog.* **80** (1984) no. 6, 33.
943. I. Dybkjaer, *Eur. Chem. News* 1983, Feb. 21, 15 (Fertilizers '83 Suppl.).
944. A. Pinto, P. L. Rogerson, *Chem. Eng. Prog.* **73** (1977) no. 7, 95.
945. K. J. Mundo, *Erdöl Kohle Erdgas Petrochem.* **31** (1978) no. 2, 74.
946. J. R. LeBlanc, *Hydrocarbon Process.* **63** (1984) no. 7, 69.
947. "The chemical bearings of the ammonia process" *Nitrogen & Methanol* **244** (2000) 31–39.
948. A. Nielsen et al., *Plant Oper. Prog.* **1** (1982) no. 3, 186.
949. ICI, EP-A 93502, 1983 (A. Pinto).
950. P. H. Brook: "Fertilizer '83," *Br. Sulphur Corp. 7th Int. Conf.*, London 1983, Proceedings of the Conf., p. 159.
951. G. Pagani, G. Brusasco, G. Gramatica in A. I. More (ed.), *Proc. Br. Sulphur Corp. Int. Conf. Fert. Technol. 4th 1981* 1982, 195.
952. Pullman Inc., US 4079017 (D. B. Crawford, C. L. Becker, J. R. LeBlanc);
US 4162290, 1979 (D. B. Crawford, C. L. Becker, J. R. LeBlanc).
953. S. Uji, M. Ikeda, *Hydrocarbon Process.* **60** (1981) no. 7, 94.
954. E. Nobles, J. C. Stover, *Chem. Eng. Prog.* **80** (1984) no. 1, 81.
E. Nobles, J. C. Stover, *Ammonia Plant Saf.* **24** (1984) 41.
955. *Nitrogen* **162** (1986).
956. J. G. Livingstone, A. Pinto, 27th AIChE Ammonia Safety Symposium, Los Angeles 1982.
957. W. K. Taylor, A. Pinto, 31st AIChE Ammonia Safety Symposium, Boston 1986.
958. "Ammonia – Uhde's low energy technology", *Uhde engineering news 1 – 91 (1991).*
959. P. Tjissen, 21nd AIChE Ammonia Safety Symp., Atlantic City 1976.
960. G. R. Nieman, L. C. Daigre, III, 18th AIChE Ammonia Safety Symp., Vancouver 1973.
961. C. C. Yost, C. R. Curtis, C. J. Ryskamp, 24th AIChE Ammonia Safety Symp., San Francisco 1979.
962. *Nitrogen* **65** (1970) 32.
963. T. L. Huurdeman, 33rd AIChE Ammonia Safety Symp., Denver 1988.
964. R. L. Allen, Jr., G. A. Moser, 36th AIChE Ammonia Safety Symp., Los Angeles 1991.
965. D. Dekmush et al., 37th AIChE Ammonia Safety Symp., San Antonio 1992.
966. S. Weems, D. H. Ball, D. E. Griffin, *Chem. Eng. Prog.* **75** (1979) no. 5, 64; *Ammonia Plant Saf.* **21** (1979) 39.
967. G. Collier, J. D. Voelkers, D. E. Griffin, *Ammonia Plant Saf.* **22** (1980) 206.
968. F. Yazhari, *Hydrocarbon Process.* **61** (1982) no. 5, 187.
969. S. Madhaven, *Plant /Oper. Progr.* **3** (1984) no. 1, 14.
970. S. M. Solomon, 29th AIChE Ammonia Safety Symp., San Francisco 1984.
971. S C. Moore, T. M. Piper, C. C. Chen, 30th AIChE Ammonia Safety Symp., Seattle 1985.
972. *Nitrogen* **169** (1987) 35.
973. S. Mani, S. K. Shoor, H. S. Pedersen, 33rd AIChE Ammonia Safety Symp., Denver 1988;
Plant /Oper. Progr. **8** (1989) 33.
974. G. Grossman, J. Dejaeger, 37th AIChE Ammonia Safety Symp., Los Angeles 1991.
975. Ammonia Plants and Related Facilities Symposia – Proceedings from the last 50 Years, 3 CD-ROMs, American Institute of Chemical Engineers 2005, ISBN 0-8169-0995-4.
976. I. Dybkjaer in A. Nielsen (ed.): *Ammonia Catalysis and Manufacture*, Springer, New York 1995 p. 262, 263.
977. I. Dybkjaer in A. Nielsen (ed.): *Ammonia Catalysis and Manufacture*, Springer, New York 1995 p. 258 – 261; 269 – 271 .
978. P. Radke, *Nitrogen* **225** (1997) 27.
979. *Oilweek (Calgary, Alberta)* (May 23, 1983) 12.
980. *Chem. Week* **134** (March 21, 1984) 15.
981. *Nitrogen* **189** (1989) 25.
982. J. R. LeBlanc, *Chem. Econ. Eng. Rev.* **18** (1986) no. 5, 22.
983. *Fertilizer Focus* **4** (1987) no. 10, 40.
984. J. R. LeBlanc, *Energy Prog.* **5** (1985) no. 1, 4.
985. J. R. LeBlanc, AIChE Symposium on Safety in Ammonia Plants, San Francisco, Calif., 1984.

986. *CEER Chem. Econ. Eng. Rev.* **11** (1979) no. 5, 24.
987. I. Dybkjaer in A. Nielsen (ed.): *Ammonia Catalysis and Manufacture*, Springer, New York 1995 p. 262, 263.
988. I. Dybkjaer, Fertilizer Latin America Int. Conf., Caracas 1989; Brit. Sulphur Corp. Proc., vol. 1, p. 77.
989. I. Dybkjaer, FAI Seminar 1990, The Fertilizer Assoc. of India, New Delhi, p. SIII–1.
990. T. Bajpai, *Nitrogen 88, Brit. Sulphur 12th Int. Conf.*, Geneva 1988.
991. T. S. Hariharan, *J. Tech. Dev.* **7** (1987) no. 4, 42.
992. S. R. Sahore, T. S. Krishnan, *Fertilizer News* **15** (1989).
993. I. Dybkjaer, *Fertilizer Industry Ann. Rev.* **XIII** (1990) 42a.
994. I. Dybkjaer, IFA Technical Conf., The Hague 1992.
995. I. Dybkjaer, IFA – FADINAP Regional Conf. for Asia and Pacific, Bali 1992
996. *Hydrocarbon Process.* **62** (1983) no. 11, 79.
997. F. Brown, Proceedings no. 218, The Fertilizer Society, London 1983.
998. *Hydrocarbon Process.* **60** (1981) no. 11, 132.
999. *The Uhde Reformer: High Pressure, High Temperature Service*Uhde Brochure Hi 18 1500 11/1991.
1000. Energy Efficient Ammonia Production *Uhde Eng. News.* **3** (1991)
1001. Ammonia Plant – Energy Integration *Uhde Eng. News.* **2** (1991).
1002. *Uhde's Ammonia Technology* Uhde Brochure Hi 18 1500 11/1991; RRD, 1992.
1003. Ammonia, Uhde's Low Energy Technology *Uhde Eng. News.* **1** (1991).
1004. R. Hakmann, FAI Seminar 1990, The Fertilizer Assoc. of India, New Delhi, p. SIII–3.
1005. J. Dejaeger, E. Das, 38th AIChE Ammonia Safety Symp., Orlando 1993.
1006. F. C. Brown in A. I. More (ed.), *Proc. Br. Sulphur Corp. Int. Conf. Fert. Technol. 4th 1981* 1982, 39.
1007. P. A. Ruziska, C. C. Song, 29th AIChE Ammonia Safety Symp., San Francisco 1984.
1008. P. A. Ruziska, P. Dranze, C. C. Song, Nitrogen 86, Brit. Sulphur conf., Amsterdam 1986.
1009. *Chem. Eng. (N.Y.)* **86** (1979) no. 26, 88.
1010. L. J. Ricci, *Chem. Eng. (N.Y.)* **86** (1979) no. 3, 50, 54.
1011. *Oil Gas* J. 76 (1978) no. 49, 34.
1012. P. Taffe, *Chem. Age (London)* **117** (Dec. 15, 1978) 8.
1013. V. Pachaiyappan, *Fertilizer News* 1979, 41.
1014. V. Lagana, *Chem. Eng. (N.Y.)* **85** (1978) no. 1, 37.
1015. B. J. Grotz, *Nitrogen* **100** (1976) 71.
1016. B. Grotz, G. Good, *Chem. Age (London)* **121** (Nov. 14, 1980) S 18.
1017. *Nitrogen* **144** (1983) 30.
1018. K. C. Wilson et al., *Nitrogen* **151** (1984) 31.
1019. K. C. Wilson, B. J. Grotz, J. Richez, Nitrogen 86, Brit. Sulphur Conf., Amsterdam 1986.
1020. B. J. Grotz, *Nitrogen* **217** (1995) 41 – 48.
1021. K. G. Christensen, B. J. Grotz, K. G. Gosnell, *Nitrogen* **181** (1989) 31 – 36.
1022. W. Glover, J. P. Yoars, *Hydrocarbon Process.* **52** (1973) no. 4, 165.
1023. W. Glover, J. P. Yoars, 17th AIChE Ammonia Safety Symp., Minneapolis 1972.
1024. K. G. Christensen, B. J. Grotz, K. G. Gosnell, Fertilizer Latin America Int. Conf., Caracas 1989; Brit. Sulphur Corp. Proc., vol. 1.
1025. B. J. Grotz, L. Grisolia, *Nitrogen* **199** (1992) 39.
1026. ICI, EP-A 49967, 1982 (A. Pinto).
1027. J. G. Livingstone, A. Pinto, 27th AIChE Ammonia Safety Symp., Los Angeles 1982.
1028. Fertilizer Industry Annual Review, vol. XI, 1988
1029. S. A. Topham, S. A. Hall, D. G. Heath, ICI/CFDC Tech. Symp., Shanghai 1989.
1030. W. K. Taylor, A. Pinto, *Commisioning C–I–Ls Ammonia Plant*, 31st AIChE Ammonia Safety Symp., Boston 1986; *Plant/Operations Prog.* **6** (1987) 106 – 111.
1031. M. P. Robert, C. W. Hooper, ICI/CFDC Tech. Symp., Shanghai 1989.
1032. K. J. Elkins et al., Asia Nitrogen, Int. Conf., (Brit. Sulphur) Bali 1994.
1033. K. J. Elkins et al., *ICIs AMV Ammonia Technology* ICI Catalco tech. paper 246W/126/3/AMM.
1034. *Nitrogen* **162** (1989) 32 – 37.
1035. J. G. Livingstone, A. Pinto, Fertilizer 83, Int. Conf. (Brit. Sulphur), London 1988.
1036. *Nitrogen* **141** (1983) 37.
1037. *Chem. Week* **132** (Jan. 5, 1983) 23.
1038. *Hydrocarbon Process.* **62** (1983) no. 11, 80.
1039. D. L. Banquy, *Ammonia Plant Saf.* **24** (1984) 8.
1040. F. C. Brown, *Eur. Chem. News* 1982, no. 15, 10 (Process Review Suppl.).
1041. Humphreys & Glasgow Ltd., GB-A 2126208 A, 1983 (C. L. Winter).

1042. F. C. Brown, C. Topham, Proc. Europ. Conf. on Energy Efficient Prod. of Fertilizers, Bristol 1990, p. 1.
1043. *Jacobs Ammonia Technology*, Jacobs company brochure, 1996.
1044. *Chem. Eng. (N.Y.)* **88** (1981) no. 10, 33.
1045. P. Conolly, *Chem. Age (London)* **122** (Feb. 6, 1981) 12.
Eur. Chem. News **36** (1981) no. 967, 16.
1046. Pullman Inc., US 4148866, 1979 (C. L. Becker);
US 4153673, 1979 (C. L. Becker).
1047. *Nitrogen* **178** (1989) 30 – 39.
1048. J. M. Halstead, A. Pinto, FAI Seminar, Delhi 1988.
1049. A. Pinto, J. M. S. Moss, T. C. Hicks, 34th AIChE Ammonia Safety Symp., San Francisco 1989.
1050. K. Elkins, I. R. Barton, 39th AIChE Ammonia Safety Symp., Vancouver 1994; *Operational Performance of the ICI Leading Concept Ammonia (LCA) Process*, ICI Catalco tech. paper 274W/126/3/LCA.
1051. T. C. Hicks, A. Pinto, *Fertilizer News* (1989) 37.
1052. X. Shen, X. Xu, W. Feng, S. Singh, A. Malhotra, *Nitrogen & Methanol*, **274** (2005) 39.
1053. A. Malhotra, L. Hackemesser, paper presented at 46th AIChE Symposium on Safety in Ammonia Plants and Related Facilities, Montreal 2002.
1054. J. Gosnell, paper presented at Nitrogen 2000, Vienna (March 2000), International Conference & Exhibition, British Sulphur Publishing.
1055. J. Gosnell, paper presented at Gulf Nitrogen 2002, Doha (March 2002), International Conference & Exhibition, British Sulphur Publishing.
1056. "Mega-ammonia round-up", *Nitrogen & Methanol*, **258** (2002) 39–48.
1057. T. Miyasugi et al., *Chem. Eng. Prog.* **80** (1984) no. 7, 41.
T. Miyasugi et al., *Ammonia Plant Saf.* **24** (1984) 64.
1058. W. F. van Weenen, J. Tielroy, *Nitrogen* **127** (1980) 38.
1059. *Oil Gas J.* **79** (1981) no. 18, 270.
1060. S. Ratan, K. S. Jungerhans, KTI Symp., Scheveningen 1991.
1061. J. J. Westenbrink, J. Voogd, Nitrogen 86, Brit. Sulphur Int. Conf., Amsterdam 1986.
1062. J. J. Westenbrink, J. Voogd, FAI Seminar, New Delhi 1995.
1063. W. F. van Weenen, J. Tielroy, *Proc. Fertilizer Soc. (London)* (1980) no. 191, 1.
1064. W. F. van Weenen, J. Tielrooy, *32nd Ann. Meet., The Fertilizer Industry Round Table*, USA 1982, p. 268.
1065. *Nitrogen* **208** (1994) 44 – 49.
1066. M. Lembeck, Asia Nitrogen 96: Brit. Sulphur Int. Conf., Singapore 1996.
1067. M. Lembeck, *Fertilizer News* (1993) 15.
1068. *The Linde Ammonia Concept*, Linde Comp. publ. 1995.
1069. J. Ilg, B. Kandziora, 41st AIChE Ammonia Safety Symp., Boston 1996.
1070. M. Spear, *Chem. Eng. (London)* 1979, no. 340, 29.
1071. P. R. Savage, *Chem. Eng. (N.Y.)* **85** (1978) no. 25, 68 D.
1072. *ECN Euro Chem News* **20** (1978) 39.
1073. F. C. Brown, Proc. Dev. Symp. 1978, The Inst. of Engineers, p. 1.
1074. W. Armbruster, *Erdöl Kohle Erdgas Petrochem.* **33** (1980) no. 3, 118.
1075. G. Kammholz, G.-A. Müller, *Erdöl Kohle Erdgas Petrochem.* **26** (1973) no. 12, 695.
1076. H. E. Butzert, *Chem. Eng. Prog.* **72** (1976) no. 1, 56;
Ammonia Plant Saf. **18** (1976) 78.
1077. E. Supp, *Nitrogen* **109** (1977) 36.
1078. *Uhde's Design of the Texaco Oil Gasification*, Krupp – Uhde comp. brochure TOPG 2A/AB, 1995.
1079. J. Morrison, *Oil Gas J.* **67** (1969) Feb. 24, 76.
1080. *CEER Chem. Econ. Eng. Rev.* **12** (1980) no. 6 – 7, 33.
1081. J. Dybkjaer, J. Hansen *Chem. Age India* **31** (1980) (12): DEV. 3/1
1082. T. Sueyama, T. Tsujino, Fertilizer 85: Brit. Sulphur Int. Conf., London 1985.
1083. G. F. Skinner in A. I. More (ed.), *Proc. Br. Sulphur Corp. Int. Conf. Fert. Technol. 4th 1981* 1982, 491.
1084. Humphreys & Glasgow Ltd., GB-A 2126573 A, 1983 (F. C. Brown).
1085. "Coal as feedstock" Nitrogen 2003, Warsaw (February 2003), International Conference & Exhibition, British Sulphur Publishing.
1086. *Nitrogen* **126** (1980) 32 – 39.
1087. F. Prasek, 32nd AIChE Ammonia Safety Symp., Minneapolis 1987.
1088. *Nitrogen & Methanol* **264** (2003) 12–18.
1089. R. A. Sharpe, *Hydrocarbon Process.* **55** (1976) no. 11, 171.
1090. L. J. Partridge, *Chem. Eng. Prog.* **72** (1976) no. 8, 57;
Ammonia Plant Saf. **18** (1976) 73.

1091. S. McQueen, *Chem. Eng. (N.Y.)* **85** (1978) no. 25, 68 H.
1092. Lurgi Express Information Ö 1323/5.79, Ammonia Plant Based on Coal.
1093. *Chem. Eng. News* **57** (1977) no. 23, 27.
1094. G. W. Alves, D. A. Waitzmann, *Erdöl Kohle Erdgas Petrochem.* **35** (1982) no. 2, 70.
1095. D. Netzer, J. Moe, *Chem. Eng. (N.Y.)* **84** (1977) no. 23, 129.
1096. U. Buskies, F. Summers, *Erdöl Kohle Erdgas Petrochem.* **31** (1978) no. 10, 474.
1097. D. A. Waitzman et al., Fall Annual Meeting of the AIChE, Washington, D.C., Oct. 30, 1983.
1098. T. Matsunami, paper presented to the Internat. Fertilizer Industry Assoc., Johannesburg, South Africa, March 1983.
1099. *Ubes Texaco Process Coal Gasification Plant*, Ube comp. brochure 1984.
1100. "New ammonia WHB internals raise capacity, enhance reliability", *Nitrogen & Methanol* **266** (2003) 41–47.
1101. H. W. Graewe, 25th AIChE Ammonia Safety Symp., Portland 1980.
1102. H. W. Graewe, *Chem. Eng. Progr.* **77** (1981) no. 10, 54.
1103. Plant and process enineering equipment – Components and plant circuitry, Steimüller company brochure P 8604-06-0510 (1989).
1104. K. Nassauer, M. Fix, AFA/Abu — Qir Ammonia/Urea Technol. Symp., 1996.
1105. K. Nassauer: "Process gas waste heat boilers", Borsig AG, company brochure 290151 AC.
1106. Waste heat recovery for reformed gas and synthesis gas cooling in modern ammonia plants, Borsig AG company brochure.
1107. Process Gas Waste Heat recovery, Borsig AG company brochure.
1108. H. Lachmann, D. Fromm, 32nd AIChE Ammonia Safety Symp., Minneapolis 1987.
1109. *Nitrogen & Methanol* **268** (2004), 38 (references).
1110. Ammonia – Energy Integration in Ammonia Plants, Uhde Engineering News 2-91, Hi 111 9 1000 91, 1991.
1111. G. R. Prescott et al., 33rd AIChE Ammonia Safety Symp., Denver 1988.
1112. NH_3—Gas—Synthesis, Balke-Dürr comp. brochure, CBA 100o-7.91, 1991.
1113. Balcke-Dürr, Company Leaflet 31.07.1990.
1114. T. Timbres et al., 34th AIChE Ammonia Safety Symp, San Francisco 1989.
1115. Synloop Waste Heat Boiler in Ammonia Plants-Unique Hot/Cold Tube Sheet design, Babcock-Borsig AG company brochure.
1116. Synloop Waste Heat Boiler in Ammonia Plants, Borsig AG company brochure.
1117. M. Podhorsky et al., 40th AIChE Ammonia Safety Symp., Tucson 1995; T. Timbres et al., 34th AIChE Ammonia Safety Symp, San Francisco 1989.
1118. M. Podhorski: "Hydraulic expansion of tubes", paper presented at International Conference on Expanded and Rolled Joint Technology, Toronto 1993.
1119. Synloop Wast Heat Boiler in Ammonia Plants — Unique Hot/Cold Tubesheet Design, Babcock Borsig comp. brochure.
1120. Synthesis Gas Cooler Downstream of Partial Oxidation of Oil or Natural Gas, Borsig AG company brochure.
1121. Shell POX Waste heat boilers, Standard Fasel-Lentjes, Company brochure 1996.
1122. U. Zardi, E. Commandini, C. Gallazzi in A. I. More (ed.), *Proc. Br. Sulphur Corp. Int. Conf. Fert. Technol. 4th 1981* 1982, 173.
1123. J. Abughazaleh, J. Gosnell, R. Strait, paper presented at 47th AIChE Symposium on Safety in Ammonia Plants and Related Facilities, San Diego 2002.
1124. W. E. Davey, T. Wurzel, E. Filippi, *Nitrogen & Methanol* **262** (2003) 41–47.
1125. S. E. Nielsen, paper presented at 46th AIChE Symposium on Safety in Ammonia Plants and Related Facilities, Montreal 2002.
1126. J. Larsen, D. Lippmann, C. W. Hooper, *Nitrogen & Methanol* **253** (2001) 41–46.
1127. W. E. Davey, T. Wurzel, E. Filippi, paper presented at Nitrogen 2003, Warsaw (February 2003), International Conference & Exhibition, British Sulphur Publishing.
1128. D. C. Thomson, 4th Int. Conf. on Fertilizer Technol., London 1991, Conf. Proc., part 1, p. 1.
1129. J. A. Tonna, F. C. Brown, T. W. Nurse, *Nitrogen 91: Brit. Sulphur Conf.*, Copenhagen 1991, p. 127 – 139.
1130. S. Madhavan, B. Landry, *Nitrogen 91: Brit. Sulphur Conf.*, Copenhagen 1991, p. 127 – 139.
1131. R. W. Parrish, *Process Improvements*, Gulf Coast Ammonia Producers Meeting, Baton Rouge 1991.
1132. J. R. LeBlanc, *Hydrocarbon Process.* **65** (1986) no. 8, 39 – 44.
1133. M. Jung, *Nitrogen* **191** (1991) 42 – 52.
1134. *Nitrogen* **141** (1983) 38.
1135. F. C. Brown, *ECN Europ. Chem. News, Proc. Rev.* **8** (1982).

1136. I. Dybkjaer, Fertilizer Latin America, Int. Conf. Brit Sulphur, Caracas 1989.
1137. A. Nielsen et al., *Plant/Oper. Progr.* **1** (1982) no. 3, 186.
1138. H. Graewe, *Nitrogen 88: 12th Int. Conf. Brit. Sulphur*, Geneva 1988, p. 77.
1139. A. M. Dark, E. A. Stallworthy, *Nitrogen* **153** (1985) 25
1140. D. Singh, *Process Plant Eng.* **73** (1986) Jan — March.
1141. J. R. LeBlanc, D. O. Moore, R. V. Schneider, III, *Oil Gas J.* **80** (1982) 115.
1142. R. Darjat, J. R. LeBlanc, Nitrogen 88: 12th Int. Conf. Brit. Sulphur, Geneva 1988, p. 77.
1143. C. L. Becker in A. I. More (ed.), *Proc. Br. Sulphur Corp. Int. Conf. Fert. Technol. 4th 1981* 1982, 537.
1144. J. R. LeBlanc et al., *Oil Gas J.* **80** (1982) no. 38, 115.
1145. S. I. Wang, N. M. Patel, *Ammonia Plant Saf.* **24** (1984) 1.
S. I. Wang, N. M. Patel, *Plant Oper. Prog.* **3** (1984) no. 2, 101.
1146. G. Low, *Nitrogen* **147** (1984) 1 – 32 (Suppl. "Revamping Ammonia Plants").
1147. T. Evans, *Nitrogen & Methanol* **232** (1998) 41–52.
1148. S. E. Nielsen, P. Vaug, paper presented at 49th AIChE Symposium on Safety in Ammonia Plants and Related Facilities, Denver 2004.
1149. Foster Wheeler, US 4296 085, 1981 (D. L. Banquy).
1150. D. Kitchen, A. Pinto, 35th AIChE Ammonia Safety Symp., San Diego 1990.
1151. *Nitrogen* **214** (1996) 46.
1152. H. Bendix, L. Lenz, 33rd AIChE Ammonia Safety Symp., Denver 1988.
1153. D. L. MacLean, C. E. Prince, Y. C. Chae, 24th AIChE Ammonia Safety Symp., San Francisco 1979;
Chem. Eng. Progr. 1980, 98 – 104.
1154. R. G. Howerton, 24th AIChE Ammonia Safety Symp., San Francisco 1979.
1155. N. W. Patel, S. I. Wang, K. J. Kittelstad, 33rd AIChE Ammonia Safety Symp., Denver 1988.
1156. M. Tsujimoto et al., 34th AIChE Ammonia Safety Symp., San Francisco 1989.
1157. M. Tsujimoto et al., *Nitrogen 91: Brit. Sulphur Conf.*, Copenhagen 1991, p. 127 – 139.
1158. R. L. Newland, J. D. Pierce, D. M. Borzik, 32nd AIChE Ammonia Safety Symp., Minneapolis 1987.
1159. Ammonia Casale, DE-OS 3146778, 1981 (U. Zardi, E. Commandini).
1160. E. Commandini, U. Zardi: "Fertilizer '83", *Br. Sulphur Corp. 7th Int. Conf.*, London 1983, Proceedings of the Conf., p. 179.
1161. *Fertilizer News* (1987) (December) 19.
1162. F. C. Brown, U. Zardi, G. Pangani, *Nitrogen 88: 12th Brit. Sulphur Int. Conf.*, Geneva 1988, p. 1159 – 139.
1163. The M. W. Kellogg ammonia converter retrofit, Kellogg company brochure HG/2.5M/3–88 (1988)
1164. R. G. Howerton, S. A. Noe, *33rd AIChE Ammonia Safety Symposium*, Denver, Oct 1988.
1165. T. Czuppon et al., 38th AIChE Ammonia Safety Symp., Orlando 1993.
1166. Carbon monoxide production technologies, KTI Newsletter (Winter 1987).
1167. H. Jungfer: "Synthesegas aus Raffinerierückständen, Konzeption und Betriebsergebnisse von Lindeanlagen zur Partiellen Oxidation", *Linde-Bericht aus Technik und Wissenschaft* **57** (1985) 15–20 (ISBN 0024-3728).
1168. Lurgi References – Gas and Synthesis Gas Technology, Lurgi brochure 1569e/9.93/4.10.
1169. B. Szantay, E. Jahab in G. D. Honti (ed.): *The Nitrogen Industry*, Akademiai Kiado, Budapest 1976, p. 701 – 706.
1170. C. A. Vancini: *La Sintesi dell Amoniaca*, Hoepli, Milano 1961, p. 769.
1171. *Nitrogen* **16** (1962) 48.
1172. Steel for Hydrogen Service at Elevated Temperatures and Pressures in Petrochemical Refineries and Petrochemical Plants, API Publ. 941, Am. Petrol. Inst. 1990.
1173. W. Hausmann et al., *Stahl Eisen* **107** (1987) no. 12, 45 – 53.
1174. W. A. Bonner, *Hydrocarbon Process.* (1951) no. 5, 165.
1175. G. R. Prescott, B. Shannon, paper presented at 45th AIChE Symposium on Safety in Ammonia Plants and Related Facilities, Tuscon 2000.
1176. B. Granville, *Welding in the World* **31** (1993) no. 5, 308.
1177. API Publication 941, 2nd ed., American Petroleum Institute, Washington, D.C., June 1977.
1178. G. A. Nelson, *Transactions of ASME* **56** (1977) 205 – 213.
1179. I. Class, *Stahl Eisen* **80** (1960) 11, 17.
1180. I. Class, *Stahl Eisen* **85** (1965) 149.

1181. J. B. Sievert, paper presented at 46th AIChE Symposium on Safety in Ammonia Plants and Related Facilities, Montreal 2002.
1182. G. R. Prescott, *Plant Oper. Prog.* **1** (1982) no. 2, 94.
1183. A. Heuser, 37th AIChE Ammonia Safety Symp., Los Angeles 1991.
1184. G. R. Prescott, 37th AIChE Ammonia Safety Symp., Los Angeles 1991.
1185. G. H. Wagner, A. Heuser, G. Heinke, 37th AIChE Ammonia Safety Symp., Los Angeles 1991.
1186. G. R. Prescott, B. J. Grotz, 39th AIChE Ammonia Safety Symp., Vancouver 1994.
1187. H. D. Marsch, *Plant Oper. Prog.* **1** (1982) no. 3, 152.
1188. U. Jäkel, W. Schwenk, *Werkst. Korros.* **22** (1971) no. 1, 1.
1189. S. Y. Sathe, T. M. O'Connor 32nd AIChE Ammonia Safety Symp., Denver, 1988.
1190. C. A. van Grieken, 33rd AIChE Ammonia Safety Symp., Minneapolis, 1987.
1191. G. R. Precott, paper presented at 47th AIChE Symposium on Safety in Ammonia Plants and Related Facilities, San Diego 2002.
1192. Y. Murakami, T. Nomura, J. Watanabe, *MPC/ASTM Symposium on the Application of 2 1/4 Cr-1 Mo Steel for Thick Wall Pressure Vessels*, Denver, Col. 1980.
J. Watanabe et al., 29th Petroleum Mechanical Engineering Conf., Dallas, Tex. 1974.
G. Grote, *Chem. Ing. Tech.* **55** (1983) no. 2, 93.
1193. R. Bruscato, *Weld. J. (Miami)* **49** (1970) no. 4, 148.
1194. J. A. Richardson, *Nitrogen* **205** (1993) 49 – 52.
1195. H. Stahl, S. G. Thomson, 40th AIChE Ammonia Safety Symp., Tucson 1995.
1196. T. Shibasaki et al., 40th AIChE Ammonia Safety Symp., Tucson 1995.
1197. R. J. Gommans, T. L. Huurdeman, 39th AIChE Ammonia Safety Symp., Vancouver 1994.
1198. J. DeJaeger, L. Guns, J. Korkhaus, 39th AIChE Ammonia Safety Symp., Vancouver 1994.
1199. J. A. Richardson, Boudouard Carbon and Metal Dusting, ICI Catalco Tech. paper.
1200. O. J. Dunmore, *Proc. of UK Corrosion Conf.*, 1982.
1201. "GHR – out of the wood yet" *Nitrogen & Methanol* **266** (2003) 33–40.
1202. H. Grabke, R. Krajak, J. C. Nava Paz, *Corrosion Science* **35** (1993) 1141.
1203. R. F. Hochmann, 4th Int. Congr. on Metal Corrosion, Nat. Corr. Engineers, 1972.
1204. A. W. Loginow, E. H. Phelps, *Corrosion-NACE* **31** (1975) no. 11, 404.
1205. A. W. Loginow, E. H. Phelps, *Corrosion (Houston)* **18** (1962) 299.
1206. L. Lunde, *Ammonia Plant Saf.* **24** (1984) 154.
J. M. Blanken, *Ammonia Plant Saf.* **24** (1984) 140.
1207. *Plant Oper. Prog.* **2** (1983) no. 3, 247.
1208. Dechema-Werkstoff-Tabelle/Chem. Beständigkeit, keyword "Ammoniak," sheet 4, Nov. 1978.
1209. I. Class, K. Gering, *Werkst. Korros.* **25** (1974) no. 5, 314.
1210. A. W. Loginow, *Mater. Perform.* **15** (1976) no. 6, 33.
1211. B. E. Wilde, *Corrosion-NACE* **37** (1981) no. 3, 131.
1212. R. S. Brown, *Plant Oper. Prog.* **1** (1982) no. 2, 97.
1213. T. Kawamoto et al., *IHI Eng. Rev.* **10** (1977) no. 4, 17.
1214. N. K. Roy, *Fert. Technol.* **18** (1981) no. 1+2, 1.
1215. P. B. Ludwigsen, H. Arup, *Corrosion (Houston)* **32** (1976) no. 11, 430.
1216. E. H. Phelps, *Ammonia Plant Saf.* **16** (1974) 32.
1217. W. v. d. Heuvel, G. v. d. Lindenbergh, *Ingenieursblad* **43** (1974) no. 18, 540 – 546.
1218. C. C. Hale, *Nitrogen* **119** (1979) 30 – 36.
1219. C. C. Hale, *Nitrogen* **125** (1980) 5.
1220. T. Huberich, *Plant Oper. Prog.* **1** (1982) no. 2, 117 – 122.
1221. I. K. Suri, R. K. Bohla, *Fertilizer News* (1986) (May), 52
1222. E. H. Phelps, 18th AIChE Ammonia Safety Symp., Vancouver 1973.
1223. E. A. Olsen, 13th AIChE Ammonia Safety Symp., Montreal 1968.
1224. E. H. Phelps, 16th AIChE Ammonia Safety Symp., Atlantic City 1971.
1225. H. Arup, 21st AIChE Ammonia Safety Symp., Atlantic City 1976.
1226. L. Lunde, R. Nyborg, 27th AIChE Ammonia Safety Symp., Boston 1986; *Plant/Oper. Progr.* **6** (1967) 11 – 16.
1227. J. D. Stephens, F. Vidalin, 32nd AIChE Ammonia Safety Symp., Minneapolis 1987.
1228. L. Lunde, R. Nyborg, 34th AIChE Ammonia Safety Symp., San Francisco 1989.
1229. J. Blanken, 22nd AIChE Ammonia Safety Symp., Denver 1983;

1230. L. Lunde, R. Nyborg, The Fertilizer Society, London 1991, Proc. no. 307.
1231. H. Arup, *Ammonia Plant Saf.* **19** (1977) 73.
1232. J. R. Byrne, F. E. Moir, R. D. Williams, 33rd AIChE Ammonia Safety Symp., Denver 1988.
1233. M. Appl et al., 34th AIChE Ammonia Safety Symp., San Francisco 1989.
1234. R. A. Selva, A. H. Heuser, 21st AIChE Ammonia Safety Symp.,
1235. K. A. van Krieken, 16th AIChE Ammonia Safety Symp., Atlantic City 1976.
1236. A. Cracknell, 24th AIChE Ammonia Safety Symp., San Francisco 1979.
1237. R. S. Brown, *Plant/Oper. Progr.* **1** (1982) 97.
1238. D. C. Guth, D. A. Clark, 29th AIChE Ammonia Safety Symp., San Francisco 1994.
1239. D. C. Guth, D. A. Clark, *Plant/Oper. Progr.* **4** (1985) 16.
1240. S. Hewerdine, The Fertilizer Society, London 1991, Conf. Proc. no. 308, p. 1.
1241. B. G. Burke, D. E. Moore, 21st AIChE Ammonia Safety Symp., San Francisco 1989.
1242. M. J. Conley, S. Angelsen, D. Williams, 35th AIChE Ammonia Safety Symp., San Diego 1990.
1243. C. C. Hale, *Ammonia Plant Saf.* **16** (1974) 23; **21** (1979) 61; **24** (1984) 181.
C. C. Hale, *Nitrogen* **150** (1984) 27.
1244. J. J. Aarts, D. M. Morrison, *Ammonia Plant Saf.* **23** (1981) 124.
1245. N. A. Hendricks, *Ammonia Plant Saf.* **21** (1979) 69.
1246. J. M. Shah, *Plant Oper. Prog.* **1** (1982) no. 2, 90.
1247. K. Feind, *Ammonia Plant Saf.* **20** (1978) 46.
1248. Code of Practice for the Large Scale Storage of Fully Refrigerated Anhydrous Ammonia in the UK, Chemical Industries Association, London 1975.
1249. J. R. Thomson, 34th AIChE Ammonia Safety Symp., San Francisco 1989.
1250. J. R. Thomson, R. N. Carnegie, 33rd AIChE Ammonia Safety Symp., Denver 1988.
1251. C. C. Hale, *Plant Oper. Prog.* **1** (1982) no. 2, 107.
1252. E. T. Comeau, M. L. Weber, 21st AIChE Ammonia Safety Symp., Atlantic City 1976.
1253. K. A. Wick, J. B. Withaus, H. C. Mayo, 21st AIChE Ammonia Safety Symp., Atlantic City 1976.
1254. O. A. Martinez, S. Madhavan, D. J. Kellett, 27th AIChE Ammonia Safety Symp., Boston 1986;
Plant Oper. Prog. **6** (1967) 129
1255. C. C. Hale, J. A. Josefson, D. E. Mattick, 29th AIChE Ammonia Safety Symp., San Francisco 1984.
1256. P. N. Arunachalam, *Fertilizer News* **57** (1986).
1257. I. Dayasagan, *Fertilizer News* **60** (1986).
1258. P. P. Briggs, J. M. Richards, III, E. G. Fiesinger, 30th AIChE Ammonia Safety Symp., Seattle 1985.
1259. R. C. A. Wiltzen, 35th AIChE Ammonia Safety Symp., San Diego 1990.
1260. R. H. Squire, 35th AIChE Ammonia Safety Symp., San Diego 1990.
1261. S. B. Ali, R. E. Smallwood, 35th AIChE Ammonia Safety Symp., San Diego 1990.
1262. G. Schlichthärle, T. Huberich, *Plant Oper. Prog.* **2** (1983), no. 3, 165 – 167.
1263. A. A. Arseneaux, 29th Ammonia Safety Symp., San Francisco 1984.
1264. W. R. Southard et al., Interstate Commerce Comission, U.S. Govt., Informal Study of Freight Rate Structure Fertilizer and Fertilizer Products, vol. II.
1265. R. F. Schrader, *Nitrogen* **117** (1979) 26.
1266. T. P. Hignett, *Transportation and Storage of Ammonia*, Fertilizer Industry Round Table, Washington 1979.
1267. C. C. Hale, *Nitrogen 88: 12th Brit. Sulphur Int. Conf.*, Geneva 1988.
1268. *Nitrogen* **140** (1982) 20.
1269. G. V. Rohleder, 13th Ammonia Safety Symp., Montreal 1968.
1270. W. A. Inkofer, 13th Ammonia Safety Symp., Montreal 1968.
1271. W. A. Inkofer, G. M. Wilson, J. E. Adams, 15th Ammonia Safety Symp., Denver 1970.
1272. V. V. Kharlamov, Yu. M. Tsymbal, *Zh. Vses. Khim. Ova.* **28** (1983) no. 4, 433 – 438.
1273. T. F. Kohlmeyer, *Gas Wärme Int.* **19** (1970) no. 1, 15 – 20.
1274. R. Leschbaer, H. Schumann, *DECHEMA-Monogr.* **80** (1976) 1639 – 1669, part 2, 747 – 748.
1275. *Best Available Techniques for Pollution Prevention and Control in the European Fertilizer Industry*, Booklet 1: Production of Ammonia, European Fertilizer Manufacturers Assoc., 1995
1276. D. Eckhold et al., *Chem. Tech. (Leipzig)* **24** (1972) no. 2, 92.
1277. L. V. Caserta, *Chem. Eng. Prog.* **68** (1972) no. 5, 41.
1278. T. Dear, *Ammonia Plant Saf.* **16** (1974) 108.

1279. R. Werchan, *Ammonia Plant Saf.* **18** (1976) 87.
1280. J. G. Seebold, *Hydrocarbon Process.* **51** (1972) no. 3, 97.
1281. E. E. Allen, *Gas World* **172** (1970) no. 4501, 431.
1282. TA, Technische Anleitung zur Reinhaltung der Luft vom Dezember 1983.
1283. *Environmental Quality of Life*, Technical Note on the best Available Technology not Entailing Excessive Costs for Ammonia Production, EC Report EUR 13002 EN, 1990.
1284. V. Pattabathula, *FINDS*, A Stokes Engineering Publication, Volume XX, Number 3, Third Quarter 2005.
1285. V. Pattabathula, B. Rani, D. H. Timbres, paper presented at 50th AIChE Symposium on Safety in Ammonia Plants and Related Facilities, Toronto 2005.
1286. C. D. Swann, M. L Preston, *Loss Prev. Proc. Ind.* **8** (1995) 6, 349–353.
1287. R. D. Turney, *TransChem.* **68** B (1990) 12–16.
1288. R. C. McConnel, paper presented at 36th AIChE Symposium on Safety in Ammonia Plants and Related Facilities, Los Angeles 1991,
Ammonia Plant Saf. **32** (1992) 104.
1289. A. L. Ormond, *J. Chem. A. Loss Prev. Bulletin* **126** (1995).
1290. D. Bourgois, M. Felscher, W. Moore, paper presented at 48th AIChE Symposium on Safety in Ammonia Plants and Related Facilities, Orlando 2003.
1291. *Nitrogen & Methanol* **264** (2003) 37–45.
1292. C. E. Jaske, paper presented at 47th AIChE Symposium on Safety in Ammonia Plants and Related Facilities, San Diego 2002.
1293. M. Dressel, M. Heinke, U. Steinhoff, paper presented at 35th AIChE Symposium on Safety in Ammonia Plants and Related Facilities, San Diego 1990.
1294. G. P. Williams: "Safety Performance in Ammonia, Methanol and Urea Plants 1999-2001 Surveys", paper presented at 48th AIChE Symposium on Safety in Ammonia Plants and Related Facilities, Orlando 2003.
1295. G. P. Williams: "Safety Performance in Ammonia Plants, 1997–1998 Surveys", Plant Survey International, Inc., paper presented at 45th AIChE Symposium on Safety in Ammonia Plants and Related Facilities, Tuscon 2000.
1296. R. F. Griffiths, G. D. Kaiser, *J. Hazard Mater* **6** (1982) 197.
1297. C. Charp, Agricult. Anhydrous Ammonia Technol. Use Proc. Symp., San Louis 1966, p. 21.
1298. G. F. P. Harris, P. E. MacDermott, *Inst. Chem. Eng. Symp. Ser.* **49** (1977) 29.
1299. W. J. De Coursey et al., *Can. J. Chem. Eng.* **40** (1962) 203.
1300. E. Banik, *Explosivstoffe* **5** (1957) no. 2, 29.
1301. J. M. Blanken, 24th Ammonia Safety Symp., San Francisco 1979.
Chem. Eng. Progr. **76** (1980) 89 – 104.
1302. W. L. Ball, 14th AIChE Ammonia Safety Symp., Portland 1969.
1303. A. Resplandy, *Chim. Ind. Genie Chim.* **102** (1969) 691.
1304. H. C. Goldwire, *Chem. Eng. Progr.* **82** (1986) 35.
1305. P. K. Raj, J. H. Hagopian, A. S. Kalekar, 19th AIChE Ammonia Safety Symp., Salt Lake City 1969.
1306. J. C. Statharas et al., *Process Safety Progr.* **10** (1993) 118.
1307. K. P. Raj, R. C. Reid, *Environ. Sci. Technol.* **12** (1978) 1422.
1308. National Research Council Panel on Response to Causalities Involving Shipborne Hazards, AD-A075203, 1979, p. 1.
1309. H. v. Bell, *Chem. Eng. Prog.* **78** (1982) no. 2, 74 – 77.
1310. K. A. Vick, J. B. Witthaus et al., *Chem. Eng. Prog. Techn. Man.* **22** (1980) 54 – 62.
1311. B. H. Winegar, *Chem. Eng. Prog. Tech. Man.* **22** (1980) 226 – 230.
1312. F. J. Heller, *Chem. Eng. Prog. Tech. Man.* **23** (1981) 132 – 145.
1313. R. J. Eiber, *Chem. Eng. Prog. Tech. Man.* **23** (1981) 146 – 156.
1314. J. J. O'Driscol, *Ammonia Plant Saf.* **17** (1975) 119 – 122.
1315. R. F. Griffiths, G. D. Kaiser, *J. Hazard. Mater.* **6** (1982) no. 1/2, 197 – 212.
1316. P. P. K. Raj, R. C. Reid, *Environ. Sci. Technol.* **12** (1978) no. 13, 1422 – 1425.
1317. M. L. Greiner, *Plant Oper. Prog.* **3** (1984) no. 2, 66;
Ammonia Plant Saf. **24** (1984) 109.
1318. J. E. Lessenger, *Plant Oper. Prog.* **4** (1985) no. 1, 20.
1319. A. Nielsen: *Ammonia — Catalysis and Manufacture*, Springer, New York 1995.
1320. C. C. Hale, W. H. Lichtenberg, *Ammonia Plant Saf.* **22** (1980) 35.
1321. P. J. Baldock, *Chem. Eng. Prog. Loss Prev.* **13** (1980) 35 – 42.

1322. M. Y. Nuttonson, PB–209478, 1 (1972)
1323. M. B. Jacobs: *The analytical Toxicology of Industrial Inorganic poisons*, **509** Interscience, Wiley N.Y. (1967).
1324. National Inst. for Occupational Safety and Health, Rockville 1974, PB-246669, p. 1.
1325. National Research Council, Committee on Toxicology, Washington, PB-244336, p. 1.
1326. L. Legters, (1980), AD-A094501 p. 1
1327. I. M. Alpator, *Prom. Toksikol. i. Klinika Prof. Zabol. Khim. Etiol.* **200** (1962) .
1328. L. Matt, *Med. Inaugural Diss*, Würzburg (1889).
1329. A. C. S. Fieldner, S. H. Katz, S. P. Kinney, in *Noxious Gases*, **125** 2nd rev. ed. Reinhold, N.Y. (1943).
1330. L. Silvermaur, L. Whittenberger, I. Muller, *I. Industr. Hyg.* **31** (1949) 74.
1331. J. E. Lessenger, *Plant/Oper. Progr.* **4** (1985) 20.
1332. J. C. Barber, 22nd AIChE Ammonia Safety Symp., Denver 1977.
1333. J. E. Ryer-Powder, 35rd AIChE Ammonia Safety Symposium, San Diego 1990.
1334. M. Coplin, *Lancet* **241** (1949) 95.
1335. W. M. Grant, *Arch. Ophthal.* **44** (1950) 399.
1336. T. Sollmann, *A Manual of Pharmacology*, 6th ed., Saunders, Philadelphia 1944.
1337. R. C. Wands, in G. D. Clayton, F. E. Clayton: *Patty's Industrial Hygiene and Toxicology*, **vol. 2 B**, Wiley Interscience, New York 1981, p. 3045 – 3052.
1338. K. B. Lehmann, *Arch. Hyg.* **17** (1893) 329.
1339. P. Trendelenburg in A. Heffter: *Handbuch der experimentellen Pharmakologie*, Vol. I, p. 470, Springer, Berlin 1923.
1340. I. M. Alpator, *Gig. Tr. Prof. Zabol.* **2** (1964) 14.
1341. D. P. Stombaugh, H. S. Teague, W. L. Roller, *I. Anim. Sci.* **6** (1969) 844.
1342. H. F. Smith, *Amer. Industr. Hyg. Quart.* **17** (1956) 145.
1343. C. P. Carpenter, H. F. Smith, Jr., U. C. Pozzani, *J. Industr. Hyg.* **31** (1949) 343.
1344. V. I. Mikhailov, *Probl. Kosmich. Biol. Akad. Nauk SSSR* **4** (1965) 531.
1345. E. M. Boyd, M. L. McLachlin, W. F. Perry, *J. Industr. Hyg.* **26** (1944) 29.
1346. S. D. Silver, F. P. McGrath, *I. Industr. Hyg.* **30** (1948) 7.
1347. K. B. Lehmann, *Arch. Hyg.* **5** (1886) 68.
1348. I. H. Weatherby, *Proe. Soc. Exp. Biol.* **81** (1952) 300.
1349. R. A. Coon, R. A. Jones, L. J. Jenkins, I. Siegel, *Toxicol. Appl. Pharmacol* **16** (1970) 646.
1350. K. S. Warren, S. Schenker, *J. Lab. Clin. Med.* **64** (1964) 442.
1351. W. D. Lotspeich, *Amer. J. Physiol.* **206** (1965) 1135.
1352. P. Vinay, E. Alignet, C. Pichette, M. Watford, G. Lemieux, A. Gougoux, *Kidney Intern.* **17** (1980) 312.
1353. D. Z. Levine, L. A. Nash, *Amer. J. Physiol.* **225** (1973) 380.
1354. K. S. Warren, *J. Clin. Invest.* **37** (1958) 497.
1355. J. Oliver, E. Bourke, *Clin. Sci. Molec. Med.* **48** (1975) 515.
1356. J. Yoshida, K. Nakame, R. Nakamura, *Nippon Chikusangakukaiko* **28** (1957) 185.
1357. D. C. Topping, W. J. Visek, *J. Nat.* **106** (1976) 1583.
1358. B. Toth, *Int. J. Cancer* **9** (1972) 109.
1359. Litten Bionetics, *NTIS, PB-245*, 506 (1975) Washington D.C.
1360. Yara Fertilizer Industry Handbook, 31. May 2005, www.yara.com
1361. T. Wakabayashi, *Ammonia Plant Saf.* **20** (1978) 17.
1362. *Erdöl Kohle Erdgas Petroch.* **37** (1984) no. 4, 143.
Chem. Ing. Tech. **56** (1984) no. 8, A 407.
1363. H. J. Bomelburg, *Plant Oper. Prog.* **1** (1982) no. 3, 175.
1364. *Chem. Ing. Tech.* **56** (1984) no. 4, A 154.
1365. K. Dietrich, *Kältech. Klim.* **22** (1970) no. 6, 184.
1366. H. L. v. Cube, *Kältetechnik* **16** (1964) no. 3, 76.
1367. F. Özvegyi, *Kälte (Hamburg)* **23** (1970) no. 6, 298.
1368. A. Miller, *Kälte (Hamburg)* **20** (1967) no. 6, 275.
1369. H. Borrmann, *Chem. Ing. Tech.* **40** (1968) 1192.
1370. C. Hollmann, *Gas Wärme* **9** (1960) 267.
1371. G. Kurz, *Proc. Eng.* (1976) Oct., 97.
1372. M. N. Park, *Nitrogen 88: Brit. Sulphur International Conference*, Geneva Mar 1988.
1373. "World Ammonia Production, Consumption and Trade", Statistical supplement, *Nitrogen* **217** (1995) 22.
1374. "Five Year Outlook for Ammonia 1995 – 2000", British Sulphur.
1375. F. J. Wollner, *Ammonia Casale's 75th Anniversary Symposium*, Lugano, Switzerland, Nov. 1996.
1376. IFA – Production and International Trade – July 2005.
1377. *Yara Fertilizer Handbook*, 31 May 2005.

1378. D. A. Kramer: "Nitrogen", in Mineral Commodity Profiles, US Department of the Interior, US Geological Survey, Open file report 2004-1290.
1379. S. R. Wilson: Agriculture as a Producer & Consumer of Energy, Farm Foundation & USDA Office of Energy Policy, CF Industries, June 24-25 2004.
1380. BP Global-Reports and publications – Statistical Review of World Energy 2005.
1381. M. Appl, *Ammonia*, Methanol, Hydrogen, Carbon monoxide — Modern Production Technologies CRU Publishing Ltd, London 1997 63 – 65.
1382. "BP Statistical Review of the World's Energy", British Petroleum Co., June 1995.
1383. M. J. McGrath, E. J. Houde, paper presented at AIChE Spring 1999 Meeting, 14-18 March 1999 Houston.
1384. M. Appl, "Is Coal the Future Feedstock for Ammonia", 2nd Casale Symposium for Customers and Licensees, Lugano 30 May – 2 June 2006.
1385. More Ammonia from US Coke and Coal, *FINDS*, A Stokes Engineering Publication Volume XX, Number 3, Third Quarter 2005.
1386. R. Strait, *FINDS*, A Stokes Engineering Publication, Volume XVII, Number 1, First Quarter 2002.
1387. *Nitrogen+Syngas*, **280** (2006) 13–15.
1388. K. Isermann in A. I. More (ed.), *Proc. Br. Sulphur Corp. Int. Conf. Fert. Technol. 4th 1981* 1982, 571.
1389. Nitrex Statistik 1980, Nitrex AG, Zürich.
1390. A. Quispel: *The Biology of Nitrogen Fixation*, Elsevier/North Holland Publ. Comp., Amsterdam – Oxford – New York 1974.
1391. Emerging Technologies no. 1: Nitrogen Fixation. An Analysis of Present Research and Implications for the Future. Technical Insights Inc., Fort Lee, New Jersey.
1392. R. W. F. Hardy, U. D. Havelka, *Science (Washington, D.C.)* 1975, no. 4188, 633.
1393. *Nitrogen* **144** (1983) 32.
1394. *The Energetics of Biological Nitrogen Fixation*, Workshop Summaries I, Plant, Physiology, Am. Soc. of Plant Physiologists, Rockville, Maryland, USA.
1395. D. F. Shanmugam, F. O'Gara, D. Andersen, R. C. Valentine: "Biological Nitrogen Fixation," *Am. Rev. Plant Physiol.* **29** (1978) 263.

1396. H. J. Rehm: "Beiträge der modernen Biotechnologie und Biochemie zur Ernährung," *VCI-Schriftenreihe Chemie + Fortschritt* 1983, no. 3, 18.
1397. *Nitrogen* **146** (1983) 24.
1398. J. R. Postgate: "Fertilizer '83," *Br. Sulfur Corp. 7th Int. Conf.*, London 1983, Proceedings of the Conf., p. 119.
1399. G. J. Leigh, in *Catalytic Ammonia Synthesis*, ed. by J. R. Jennings, 1991, 365 – 387.
1400. J. S. Pathe, C. A. Atkins, R. M. Rumbird, in A. H. Gibson, W. E. Newton (eds.): *Current Perspectives in Nitrogen Fixation*, Australian Academy of Science, Canberra 1981.
1401. V. P. Gutschick, Long term strategies for supplying nitrogen to crops; Informal report LA-6700–19S, Los Alamos Scientific Laboratory, Los Alamos, NM, USA, 1977.
1402. R. F. Michin, S. J. Pate, *J. Exp. Bot.* **24** (1973) 259.
1403. M. J. Merrik, *J. R. Agric. Soc. Engl.* **147** (1986) 202.
1404. M. Tamaguchi: "Biological Nitrogen Fixation in Flooded Rice Fields," in Nitrogen and Rice, International Rice Research Institute, Manila, Philippines, 1979, p. 193 – 204.
1405. K. Gopalakrishna Pillai, D. B. B. Chaudary, K. Krishnamurty: "Bio-Fertilizers in Rice Culture – Problems and Prospects for Large Scale Adaptation," *Fert. News* **25** (1980) 40 – 45.
1406. *Chem. Eng. News* **55** (1977) no. 40, 19.
1407. *Nitrogen* **111** (1978) 41.
1408. G. N. Schrauzer, T. D. Guth, *J. Am. Chem. Soc.* **99** (1977) 7189.
1409. M. W. Kellogg, US 4298589, 1981 (J. R. LeBlanc, R. B. Peterson).
1410. D. L. Johnson, T. M. O'Connor, 31st AIChE Ammonia Safety Symp., Boston 1986.
1411. *Nitrogen* **176** (1988) 25.
1412. L. J. Buividas in A. I. More (ed.): *Proc. Br. Sulphur Corp. Int. Conf. Fert. Technol. 4th 1981* 1982, 601.
1413. *Nitrogen* **147** (1984) 3.
1414. *Nitrogen* **140** (1982) 5.
1415. G. P. Williams: "World-wide Ammonia Benchmarking Study 2000-2001: Multi-client Benchmarking Report" Plant Survey International, Inc 10813 Bland Ridge Court, Petersburg, VA 23805, USA.

Ammonium Compounds

See also: Phosphate Fertilizers

Karl-Heinz Zapp, BASF Aktiengesellschaft, Ludwigshafen, Federal Republic of Germany (Chaps. 1 and 2)

Karl-Heinz Wostbrock, BASF Aktiengesellschaft, Ludwigshafen, Federal Republic of Germany (Chaps. 3 and 4)

Manfred Schäfer, BASF Aktiengesellschaft, Ludwigshafen, Federal Republic of Germany (Chap. 3)

Kimihiko Sato, Asahi Glass Company Ltd., Yokohama, Japan (Sections 3.2.1 and 3.5 in part)

Herbert Seiter, BASF Aktiengesellschaft, Ludwigshafen, Federal Republic of Germany (Chap. 4)

Werner Zwick, BASF Aktiengesellschaft, Ludwigshafen, Federal Republic of Germany (Chap. 4)

Ruthild Creutziger, BASF Aktiengesellschaft, Ludwigshafen, Federal Republic of Germany (Chap. 5)

Herbert Leiter, BASF Aktiengesellschaft, Ludwigshafen, Federal Republic of Germany (Chap. 5)

1.	**Ammonium Nitrate**	299	2.4.	Uses	312
1.1.	Physical and Chemical Properties	300	2.5.	Economic Aspects	312
1.2.	Production	301	3.	**Ammonium Chloride**	312
1.2.1.	From Ammonia and Nitric Acid	301	3.1.	Properties	313
1.2.2.	Conversion of Calcium Nitrate Tetrahydrate	304	3.2.	Production	314
			3.2.1.	Modified Solvay (Ammonia–Soda Ash) Process	314
1.3.	Granulation	305			
1.4.	Grain Stabilization and Surface Treatment (Conditioning)	306	3.2.2.	Direct Reaction Between HCl and NH_3	317
1.5.	Transport Regulations	307	3.2.3.	Reaction of Reciprocal Pairs of Salts	317
1.6.	Safety	307	3.3.	Corrosion	317
1.7.	Uses	308	3.4.	Packaging, Storage, Shipping, and Handling	318
1.8.	Economic Aspects	309	3.5.	Uses	318
2.	**Ammonium Sulfate**	309	4.	**Ammonium Carbonates**	318
2.1.	Properties	309	4.1.	Ammonium Hydrogencarbonate	319
2.2.	Production	310	4.2.	Ammonium Carbamate	319
2.2.1.	From Coke-Oven Gas	310	4.3.	Ammonium Carbonate	320
2.2.2.	From Ammonia and Sulfuric Acid	310	4.4.	Storage, Transport, and Handling	321
2.2.3.	Coproduct in Organic Syntheses	311	4.5.	Uses	321
2.2.4.	From Gypsum	311	5.	**Toxicology and Occupational Health**	321
2.2.5.	Other Processes	311	6.	**References**	322
2.3.	Granulation and Storage	312			

1. Ammonium Nitrate

Ammonium nitrate [*6484-52-2*], NH_4NO_3, is a major chemical product. It is produced most frequently by neutralization of nitric acid with ammonia and is mainly processed into high-quality fertilizers. As a straight fertilizer, it accounts for 24 % of world consumption of nitrogen fertilizers and, in addition, is present in many blended and complex fertilizers, thus making an important contribution to feeding the world's population. Ammonium nitrate is also used as an oxidizing agent and a constituent of many explosives.

1.1. Physical and Chemical Properties

Ammonium nitrate is a colorless salt, M_r 80.05, d_4^{20} 1.725, specific heat capacity 1.70 J g^{-1} K^{-1} between 0 and 31 °C; mp 169.6–170 °C. The reaction forming ammonium nitrate from ammonia and nitric acid is highly exothermic:

$$NH_3\,(g) + HNO_3\,(l) \longrightarrow NH_4NO_3\,(s) \quad \Delta H = -146 \text{ kJ/mol}$$

The salt has five modifications:

I	169.6	to	125.2 °C cubic
II	125.2	to	84.2 °C tetragonal
III	84.2	to	32.3 °C orthorhombic
IV	32.3	to	−16.9 °C orthorhombic
V	below		−16.9 °C tetragonal

The transition at 32.3 °C is of particular significance for storage of fertilizers containing ammonium nitrate: on passing through the transition repeatedly, the fertilizer granules lose strength and finally disintegrate because of the differing densities.

Ammonium nitrate dissolves readily in water; in addition, the salt is hygroscopic. Therefore, care must be taken to avoid moisture during transport and storage. When the salt dissolves in water, heat is absorbed; therefore, ammonium nitrate can be used in freezing mixtures, for example, in mixtures with sodium chloride and ice. The heat of solution in an almost infinite quantity of water is +26.4 kJ/mol at 18 °C, the integral heat of solution to saturation is +16.75 kJ/mol, and the heat of solution in saturated solution is +15 kJ/mol. The physical properties of ammonium nitrate solutions important for concentration by evaporation on an industrial scale are shown in Tables 1, 2, and 3 and Figures 1 and 2 (see below, and next page).

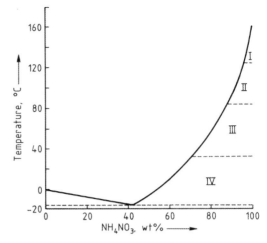

Figure 1. The system $NH_4NO_3 - H_2O$
Roman numerals denote the modifications

The solubility of ammonium nitrate in some nonaqueous solvents is considerable. The salt dissolves in liquid NH_3 and vigorously absorbs NH_3 to form solutions known as Divers liquid. Aqueous ammonium nitrate solutions 50–70 % by weight vigorously absorb NH_3 and may be used for stripping NH_3 from gases. Solutions such as these are also used for ammoniating superphosphate. The salt dissolves in methanol to give solutions of about 20 % at 30 °C and about 40 % at 60 °C. Its solubility in ethanol is about

Table 1. Densities of NH_4NO_3 solutions

t, °C \ c, wt %	20 %	30 %	40 %	50 %	60 %	70 %	80 %	90 %	94 %	97 %	99 %
20	1.0830	1.1275	1.1750	1.2250	1.2785	–	–	–	–	–	–
40	1.0725	1.1160	1.1630	1.2130	1.2660	1.3220	–	–	–	–	–
60	1.0620	1.1045	1.1510	1.2005	1.2525	1.3090	1.3685	–	–	–	–
80	1.0550	1.0935	1.1390	1.1875	1.2395	1.2960	1.3555	–	–	–	–
100	1.0410	1.0820	1.1270	1.1745	1.2265	1.2825	1.3420	1.4075	–	–	–
120							1.3285	1.3930	1.4210	–	–
140								1.3785	1.4065	1.4285	–
160									1.3940	1.4165	1.4325
180										1.4060	1.4225
200											1.4121
220											1.4030

Table 2. Boiling points of aqueous ammonium nitrate solutions at atmospheric pressure

c, wt %	60	80	90	94	96	98	99
bp, °C	113.5	128.5	147	165	182	203	222

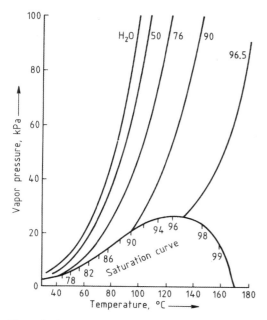

Figure 2. Vapor pressures of ammonium nitrate solutions
The saturation curve shows the vapor pressures of saturated solutions. The other curves show the vapor pressure of unsaturated solutions of definite composition. The salt contents are indicated in weight percent.

Table 3. Molar and specific heat capacities of ammonium nitrate solutions

NH_4NO_3,	Heat capacities,	
	$J\,mol^{-1}\,K^{-1}$	$J\,g^{-1}\,K^{-1}$
2.9	320.8	4.038
9.1	309.6	3.870
15.1	294.4	3.678
28.6	241.6	3.021
47.1	233.6	2.916
64	204.0	2.552

4 % at 20 °C and is minimal in acetone. Ammonium nitrate is insoluble in ether.

Ammonium nitrate is a powerful oxidizing agent. It is stable at normal temperature and pressure. On heating, it decomposes above 170 °C to gases. These reactions are promoted especially by small quantities of chlorine or free acid.

1.2. Production

Ammonium nitrate is produced mainly from ammonia and nitric acid. Ammonium nitrate is also formed in the production of nitrogen–phosphorus (NP) and nitrogen–phosphorus–potassium (NPK) fertilizers on decomposition of crude phosphate with nitric acid, the salt generally remaining in the fertilizers. In Europe the reaction of calcium nitrate with NH_3 and CO_2 is also used to produce NH_4NO_3.

1.2.1. From Ammonia and Nitric Acid

The neutralization of 45–65 % HNO_3 with gaseous NH_3 is accompanied by the release of 100–115 J per mole of NH_4NO_3 [6]. In most processes this considerable heat of reaction is used for partial or complete evaporation of the water. Depending on the pressure and the concentration of the nitric acid, 95–97 % solutions of ammonium nitrate can be obtained. During neutralization, the components must be mixed quickly and thoroughly in the reactor to avoid local overheating, losses of nitrogen, and decomposition of the ammonium nitrate.

One design of a neutralization reactor is shown in Figure 3 [7]. Ammonia and acid enter the inner tube, the actual neutralization zone, through pipes a and b. The neutralizer is filled to

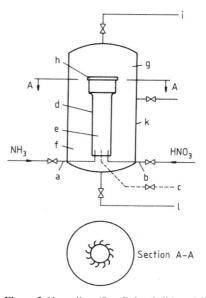

Figure 3. Neutralizer (Soc. Belge de l'Azote) [7]
A) Filling level; a) Ammonia feed pipe; b) Nitric acid feed pipe; c) Optional steam feed pipe; d) Inner tube; e) Inner compartment; f) Outer compartment; g) Steam chamber; h) Neutralizer top; i) Steam outlet; k) Container; l) Solution outlet

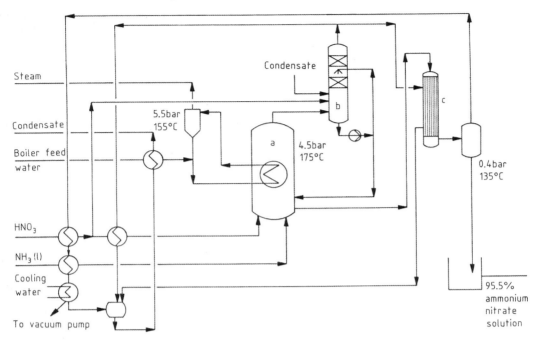

Figure 4. The UCB process
a) Reactor; b) Scrubber; c) Falling film evaporator

mark A with ammonium nitrate solution. While the solution boils vigorously in the outer compartment f, boiling is suppressed in the inner compartment e because of a slight excess pressure. The higher the temperature in the neutralizer (to 180 °C in pressure processes), the more important it is to maintain an exact pH value and to avoid introducing chlorides, heavy metals, and organic compounds. In addition, the quantity of ammonium nitrate at elevated temperature is kept as small as possible. The neutralizers and evaporators are generally constructed from Cr–Ni stainless steels.

A reduced-pressure process is operated by Chemico [6]. Normal-pressure installations use the Uhde process [8], the SBA (Soc. Belge de l'Azote) process [9], etc.; they and are characterized by low reaction temperatures and minimal corrosion. The heat can be best utilized if neutralization is carried out under pressure [10]. Steam can be generated in some processes. Operation under pressure goes back to FAUSER [11]. The pressure processes differ in removal of the heat of reaction.

In the *UCB process* (Fig. 4) [12], a heat exchanger in the pressure reactor uses a part of the heat of reaction to make steam.

Ammonia and 52–63 % HNO_3 are preheated and sprayed into the sump of the reactor. The pressure in the reactor is approximately 0.45 MPa (4.5 bar), the temperature is 170–180 °C, and the pH is 3–5. The pH is kept in this range by controlling the ratio of reactants. The heat exchanger cools the reaction mixture, and neutralization follows a stable course. The 75–80 % NH_4NO_3 solution leaving the reactor is concentrated to 95 % by evaporation in a falling film evaporator. The heat of reaction generates

1) process steam out of the evaporating water from the nitric acid. This is used in the process to preheat boiler feed water and nitric acid and to operate the falling film evaporator.
2) pure steam up to 0.55 MPa (5.5 bar) in the heat exchanger. This can be fed into the steam pool and used for other purposes.

The pH 3–5 reduces nitrogen losses into the process steam. The working conditions are adjusted

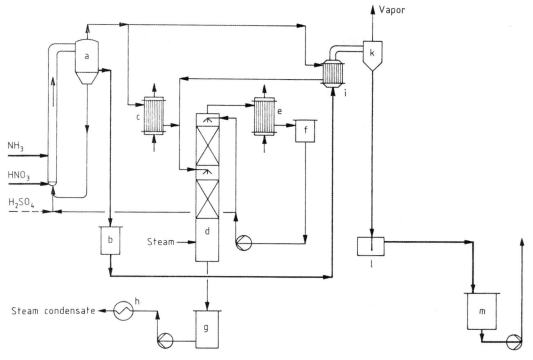

Figure 5. The Stamicarbon process
a) Neutralizer; b) Intermediate tank; c) Surplus steam condenser; d) Ammonia scrubber; e) Condenser; f) Dilute ammonia solution tank; g) Condensate tank; h) Cooler; i) Evaporator; k) Separator; l) Seal pot; m) Storage tank for 95 % ammonium nitrate solution

so that excess process steam does not accumulate.

Another process that works under pressure is the *Stamicarbon process* (Fig. 5) [13]. The neutralizer is a loop reactor that opens into a separator. The reaction solution is circulated without a pump by heat generated.

Nitric acid (60 wt%), preheated ammonia, and a small quantity of sulfuric acid are introduced at the lower end of the loop. The reactor operates at 0.4 MPa (4 bar) and 178 °C. The ammonium nitrate solution formed in the reactor has a concentration of 78 %. The steam removed at the top of the separator is passed through a mist eliminator and is mainly used to concentrate the NH_4NO_3 solution to 95 % in a vacuum evaporator. Excess steam is condensed, and ammonia is recovered from the condensate and returned to the reactor. In a second evaporator, the concentration can be increased to 98 – 99.5 % using fresh steam. The temperature of the ammonium nitrate solution is kept below 180 °C throughout neutralization and evaporation.

The *NSM/Norsk Hydro pressure process* (Fig. 6) [14] uses preheated ammonia and nitric acid.

The pressure in the reactor is between 0.4 and 0.5 MPa (about 4.5 bar), and the temperature ranges from 170 to 180 °C, conditions corresponding to a 70 – 80 % solution. Forced circulation and a thermal siphon effect circulate the solution through the reactor. Some of the heat of reaction is used to generate pure steam in an external boiler; some vaporizes water in the reactor, producing process steam, which is used to concentrate the ammonium nitrate solution to 95 %. The ammonia losses are kept small by washing the process steam with nitric acid, which is added to a circulating ammonium nitrate solution. Further concentration of the NH_4NO_3 solution, to 99.5 %, is carried out with steam in a special vacuum evaporator.

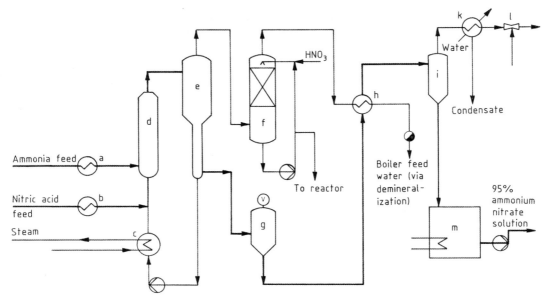

Figure 6. NSM/Norsk Hydro process
a) Ammonia evaporator/superheater; b) Nitric acid preheater; c) Boiler; d) Reactor; e) Reactor separator; f) Scrubber; g) Flashtank; h) Evaporator; i) Separator; k) Condenser; l) Ejector; m) Tank

In the United States the *Stengel process* is significant [15], [16]. It produces anhydrous ammonium nitrate directly.

The ammonia and the ca. 58 % nitric acid are preheated with fresh steam and fed into a packed vertical tube reactor at 0.35 MPa (3.5 bar) and 240 °C. The mixture of NH_4NO_3 and steam is expanded into a vacuum in a centrifugal separator. After stripping with hot air, a 99.8 % NH_4NO_3 melt is discharged, onto a cooled steel belt, solidified, and then broken up or granulated. The steam is removed at the top.

In all these processes the maintenance of the desired pH is most important. Where the reaction temperatures are below 170 °C, the pH is kept between 2.4 and 4 to minimize nitrogen losses. In pressure neutralizers, a higher pH, 4.6 to 5.4, is necessary on account of the higher temperatures and greater danger of decomposition.

1.2.2. Conversion of Calcium Nitrate Tetrahydrate

The production of nitrophosphate fertilizers by digestion of crude phosphate with nitric acid in the Odda process produces as a byproduct calcium nitrate tetrahydrate, $Ca(NO_3)_2 \cdot 4\,H_2O$, in considerable quantities ($\rightarrow$ Fertilizers). The output of nitrophosphates is increasing, while demand for calcium nitrate is declining. In a process introduced some years ago, the calcium nitrate tetrahydrate is treated with ammonia and carbon dioxide to form ammonium nitrate and calcium carbonate:

$$Ca(NO_3)_2 \cdot 4\,H_2O\,(s) + 2\,NH_3\,(g) + CO_2\,(g) \longrightarrow$$
$$2\,NH_4NO_3\,(aq) + CaCO_3\,(s) + 3\,H_2O$$

$\Delta H = -126$ kJ/mol

The accompanying heat is adequate to evaporate all the water. However, a direct procedure is not possible because of the unfavorable equilibrium at elevated temperatures. In the *BASF process* (Fig. 7), heat removal is separated from the reaction of $Ca(NO_3)_2$ with $(NH_4)_2CO_3$: NH_3 and CO_2 are dissolved in a circulated NH_4NO_3 solution, and the heat given off is removed. The calcium nitrate tetrahydrate is also dissolved in an NH_4NO_3 solution. The two solutions are then reacted at ca. 50 °C; the heat produced is minimal.

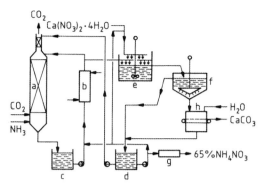

Figure 7. BASF process for conversion of calcium nitrate
a) Absorption tower; b) Condenser; c) NH_4NO_3–$(NH_4)_2CO_3$ solution; d) Cloudy NH_4NO_3 solution; e) Reaction vessel; f) Settling vessel; g) Classification filter; h) Belt filter

The grain size of the calcite precipitated can be influenced by the way the reactants are introduced. After reaction the approx. 65 % NH_4NO_3 solution is separated from $CaCO_3$ on a belt filter and concentrated by evaporation. The $CaCO_3$ still contains small amounts of ammonium compounds and phosphate and is mostly used for the production of calcium ammonium nitrate. If the calcium nitrate is suitably prepared before conversion, quite pure $CaCO_3$ may be produced.

For directly converting calcium nitrate and at the same time removing the heat of reaction, Hoechst developed a special vertical reactor [17]. The gaseous CO_2 is introduced at the bottom, and the ammonia is introduced in three zones, each cooled with water.

1.3. Granulation

Pure ammonium nitrate has a nitrogen content of 35 %. Commercial fertilizer ammonium nitrate is produced in several countries with an N content of 34.5 % or 33.5 %. It generally contains additives to stabilize the grain and improve storage properties.

Various processes have been developed for granulation [18], [19]. *Prilling* is preferred for large plants producing upwards of 1000 t per day, while the other processes are also suitable for lower outputs, around 250 t per day. Prilling starts from melts containing around 0.5 % water. Round or rectangular spray towers up to 70 m tall are used. The towers are concrete, steel, aluminum, etc. Steel frameworks and concrete must be coated with protective paints. Figure 8 [20] shows a diagram of an ammonium nitrate prilling plant.

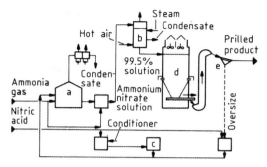

Figure 8. ICI process with prilling tower
a) Ammonium nitrate neutralizers; b) Falling film evaporator; c) Filter; d) Prilling tower; e) Screen

The solution is sprayed through one-component nozzles, perforated plates, or perforated centrifuges at the top of the tower. Cold air is drawn through the tower countercurrently to remove the heat evolved on crystallization. As the droplets fall through the tower, they solidify into round granules, which are discharged at the foot of the prilling tower and subsequently cooled and screened.

There is no need for the granulate to be dried if 99.5 % NH_4NO_3 melt is sprayed. The prilled grain has a 2 – 3 mm diameter, thus smaller than from most other processes. Grain size is affected to a certain extent by the shape of the nozzle, by the perforation cross sections and the rotational speed of the spray centrifuge, and by the viscosity of the melt.

Serious emission problems attributable to ammonium nitrate fumes are often encountered in prilling on account of the large air throughput. Modern plants are generally equipped with special emission control systems. In the CFCA shroud system [21], for example, the cooling air is divided by baffle plates or tubes in the prilling tower into two streams. Only the inner stream, containing most of the pollutants, is cleaned. Usually the off-gases from the neutralization and evaporation stages are also cleaned.

The universally applicable recycle salt process (Fig. 9), developed by BASF in 1928 [22] for *granulating* calcium ammonium nitrate, produces 2 – 4 mm granules. The process can be modified, for example, by arranging a revolving drum after the paddle screw to improve granulation, by carrying out drying or cooling in a fluidized bed, or by subjecting the granulate to hot or cold screening.

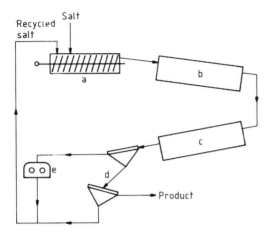

Figure 9. Recycle salt process
a) Granulation screw; b) Drying drum; c) Cooling drum; d) Screens; e) Grinder

The *pan granulation process* has been further developed, e.g., by TVA and Norsk Hydro. Figure 10 [23] shows the plan of the Norsk Hydro process. In the sloping, rotating pan granulator, the 99.5 % ammonium nitrate melt is sprayed onto a moving solid bed and solidifies on the

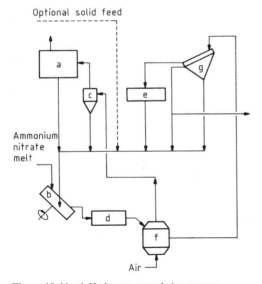

Figure 10. Norsk Hydro pan granulation process
a) Dust filter; b) Pan; c) Cyclone; d) Polishing drum; e) Crusher; f) Cooler; g) Screen

cold particles. The growing granules are graded in the rotating pan and, after reaching the required grain size, roll over the edge. Grain size may be varied by changing the operating parameters: a normal fertilizer grain diameter of 2 – 4 mm and a grain size of 7 – 11 mm for forest fertilization can be made in the same granulator.

Ammonium nitrate can also be granulated in a revolving drum, known as a spherodizer, or in a fluidized bed (e.g., Fisons Nitro-Top process) [24].

In addition to the production of ammonium nitrate containing 33.5 % or 34.5 % of N, each of these processes is also suitable for the production of calcium ammonium nitrate containing 26 % – 28 % N.

The explosives industry requires a porous, low-density (LD) ammonium nitrate that can be impregnated with oil. This LD ammonium nitrate can be produced by spraying with hot air in two-component nozzles into a revolving drum heated with hot air. However, porous LD ammonium nitrate can also be produced by prilling [25]. An ammonium nitrate melt of relatively low concentration (96 %) is sprayed in the prilling tower, and the prilled product is then carefully dried with hot air in two successive dryers. The product contains 0.2 % of water and has a powder density of 800 g/L and a porosity of 33 vol%.

1.4. Grain Stabilization and Surface Treatment (Conditioning)

The crystal transition point at 32 °C and the hygroscopicity adversely affect the storability of the granulated ammonium nitrate. A temperature of 32 °C can often be exceeded, particularly during transport and storage in warm countries: the granules lose strength and finally disintegrate into dust. The modification change and other solid-phase reactions can be inhibited by low moisture content and additives that act as internal stabilizers. The additives are added to the melt before it is sprayed. Examples of additives are anhydrous $CaSO_4$ [26], H_3BO_3 + $(NH_4)_2HPO_4$ + $(NH_4)_2SO_4$ [27], ammonium and potassium polyphosphate [28], and nucleating agents, such as silica gel [29], metal oxides, or kaolin [30]. $Mg(NO_3)_2$ [31] and $Al_2(SO_4)_3$ [32] are important stabilizers. In general, the stabilizers also inhibit the reaction between NH_4NO_3 and $CaCO_3$, which forms the highly hygroscopic calcium nitrate.

Usually, the ammonium nitrate granules are prevented from caking during storage by a final surface treatment (coating). The earlier practice of powdering with inorganic substances to separate the grains mechanically has mostly been replaced by treatment of the grain with effective organic surfactants, such as cationic long-chain fatty amines or anionic alkylarylsulfonates. Anticaking effectiveness is enhanced by addition of nonionic organic substances, such as polyethylene (PE) waxes [33] or paraffins. The fatty amines are applied as melts or oil solutions. One process [34] is multistage: in the first stage the fertilizer granules are coated with a uniform, compact layer of molten $C_{12}-C_{18}$ alkylamine. In the second stage they are cooled to convert the amine into the solid state. In the third and final stage, they are provided with a thin coating of mineral oil.

1.5. Transport Regulations

National and international regulations governing the transport of ammonium nitrate and its mixtures are being based to an ever-increasing extent on the classification of the "United Nations Committee of Exports on the Transport of Dangerous Goods" [35].

A number of ammonium nitrate-based products are to be transported according to regulations for Class 5.1, e.g., technical ammonium nitrate UN No. 1942 or the various ammonium nitrate fertilizers UN No. 2067 – 2070. Main criteria for this class are concentrations of the components and a strict limitation of combustible substances. Mixtures with high concentrations of ammonium nitrate or combustible materials are not allowed to be transported unless they are clearly defined as explosives, Class 1. Mixtures with ammonium nitrate contents below the limits are considered nonhazardous, with the exception of one group that may be subject to self-sustaining decomposition (smoldering) when heated. This group of AN fertilizers is classified in Class 9 (UN No. 2071).

Regulations for the marine transport of the above classes are given in the IMDG Code (Inter-Governmental Maritime Consultant Organisation, London) on pages 5014, 5015, and 9013.

The international regulations for road and rail transport, ADR (Accord européen relatif au transport international des marchandises dangereuses par route) and RID (Règlement international concernant le transport des marchandises dangereuses par chemins de fer) have brought the ammonium nitrate products also into Class 5.1, Subdivision 6. The criteria of composition correspond closely to those of the IMDG Code.

Transport on inland waterways is regulated by ADN (Accord européen relatif au transport international des marchandises dangereuses par voie de navigation interieure) and for the Rhine by ADNR. Although ammonium nitrate products are still placed in a special class (Class III c), these regulations are being redrafted to harmonize the classification.

1.6. Safety

Although stable under normal conditions, ammonium nitrate undergoes a number of decomposition reactions at elevated temperature [36], [37]. The following are significant:

1) Endothermic dissociation and lowering of the pH above 169 °C.

$$NH_4NO_3 \longrightarrow HNO_3 + NH_3 \quad \Delta H = +175 \text{ kJ/mol}$$

2) Exothermic elimination of N_2O on careful heating at 200 °C.

$$NH_4NO_3 \longrightarrow N_2O + 2H_2O \quad \Delta H = -37 \text{ kJ/mol}$$

3) Exothermic elimination of N_2 and NO_2 above 230 °C.

$$4NH_4NO_3 \longrightarrow 3N_2 + 2NO_2 + 8H_2O$$
$$\Delta H = -102 \text{ kJ/mol}$$

4) Exothermic elimination of nitrogen and oxygen, accompanied by detonation.

$$NH_4NO_3 \longrightarrow N_2 + 1/2 O_2 + 2H_2O$$
$$\Delta H = -118.5 \text{ kJ/mol}$$

Pure ammonium nitrate, highly concentrated hot solutions of ammonium nitrate, some mixtures of ammonium nitrate, and fertilizer ammonium nitrate, unless stabilized, are included among explosives detonated by shock waves. Although the heat released is small in comparison with that of an explosive such as hexogen, there are serious risks attending the storage of large quantities. For example, hydrogen ions, chlorides, and heavy metals catalyze the decomposition. The heating of contaminated and tamped ammonium nitrate is particularly dangerous. Following the disasters in Brest and Texas City (1947) [38], [39], where whole ships' cargoes of wax-coated fertilizer ammonium nitrate exploded as a result of fires, the content of inflammable substance has been universally limited to 0.2 % or 0.4 %. In the Federal Republic of Germany, special procedures have been introduced under the *Arbeitsstoffverordnung* (Working Materials Act) [40] for storage, loading, and transport of ammonium nitrate and products containing ammonium nitrate. For example, the storage of ammonium nitrate capable of detonation is permitted only in small batches in specially equipped storage sites. There are similar regulations in many other European countries, whereas in the United States, France, Norway, and England the storage of relatively large amounts is permitted under some conditions.

In Germany inert materials, such as limestone powder, dolomite, or precipitated calcium carbonate, are added to ammonium nitrate for fertilizer use (calcium ammonium nitrate). Such fertilizers, containing no more than 80 % ammonium nitrate, no more than 0.4 % combustible constituents, and no less than 18 % magnesium or calcium carbonate, rate as nondetonatable. The sensitivity of ammonium nitrate and its mixtures to shock is determined by tests in which a sample tamped in a closed steel tube is exposed to a shock [36], [40], [41]. In addition, oil retention [41], a measure of porosity, is a basis for assessing the safety of ammonium nitrate.

To summarize:

1) If products containing ammonium nitrate have solidified during storage, they should not be broken up by blasting
2) Products containing ammonium nitrate must be stored separately from oxidizable, inflammable materials
3) If ammonium nitrate must be heated (during production, evaporation, etc.), the quantity must be kept small and catalysts avoided

1.7. Uses

Ammonium nitrate (AN) is used mainly as fertilizer, either pure, diluted, or in multinutrient mixtures. It is used with urea in liquid fertilizers, which are important in the United States, Eastern Europe, and France.

In the United States, the United Kingdom, and France, ammonium nitrate having a nitrogen content of 33.5 % and higher is used in large quantities in agriculture. A fertilizer containing 32.5 % N is also used in the United States. In the Federal Republic of Germany, ammonium nitrate is used in mixtures with lime, dolomite, ammonium sulfate, or potash. Among these calcium ammonium nitrate is particularly important ($\rightarrow$ Fertilizers).

Calcium Ammonium Nitrate (CAN). Ammonium nitrate solution (ca. 95–97 %) can be granulated with finely divided calcium carbonate from crushed limestone or from the conversion of calcium nitrate. The granules obtained are dried, cooled, screened, and treated to avoid caking. The reaction of ammonium nitrate and limestone to form hygroscopic calcium nitrate is prevented by additives, such as $(NH_4)_2SO_4$, $MgSO_4$, and $FeSO_4$. In Germany, the nitrogen-content of calcium ammonium nitrate was increased in stages from the original level of 20.5 % N to the present level of 27.5 % N. The upper limit under the regulations of the European Economic Community is 28 % N.

Ammonium Sulfate Nitrate (ASN). The mixed sulfate–nitrate fertilizer is produced by adding ammonium sulfate to approximately 95 % ammonium nitrate solution [42] or by neutralizing HNO_3–H_2SO_4 mixtures with ammonia [43]. A slightly hygroscopic product is obtained after granulation. It is a mixture of the double salt $2\,NH_4NO_3 \cdot (NH_4)_2SO_4$ and a little ammonium sulfate [44], containing 26 % N for < 45 % ammonium nitrate. The mixture tends to harden during storage as a result of further reactions. This can be prevented by addition of salts of Mg, Fe, or Al [45].

Potassium Ammonium Nitrate. Potassium ammonium nitrate is produced in the same way

Table 4. Production of fertilizer NH_4NO_3, 10^6 t N, 1981

	Production	Import	Export
Western Europe	4.8	1.0	0.9
Eastern Europe	6.3	–	0.4
North America	1.9	0.1	0.1
South America	0.3	–	–
Africa	0.5	0.2	–
Asia and Oceania	1.0	0.1	–
Total	14.8	1.4	1.4

as ammonium sulfate nitrate, except that a potassium salt (the chloride or sulfate) is added to give, e.g., 20–0–20 (N–P_2O_5–K_2O) fertilizer.

Nitromagnesia. A fertilizer can be produced from ammonium nitrate, ammonium sulfate, and magnesium compounds, such as dolomite, magnesium carbonate, or magnesium sulfate. For example, such a fertilizer may contain 20 % N, 8 % MgO, and generally 0.2 % Cu.

Other Uses. Safety explosives for mining are produced from ammonium nitrate (AN). It is suitable by virtue of its low explosion temperature. In mixtures with NaCl, the explosion temperature is not high enough to set off the much-feared fire damp. For safety explosives, ammonium nitrate must have the following qualities:

NH_4NO_3 content		99.0 %
Ignition residue	max.	0.2 %
Chloride	max.	0.1 %
Nitrite	max.	0.1 %
Sulfate	max.	0.1 %
Water insoluble	max.	0.1 %

In many cases, safety explosives are also based on the reciprocal salt pair $NaNO_3 + NH_4Cl \rightleftharpoons NH_4NO_3 + NaCl$. Where greater explosive power is required, as in mining (rock explosives), porous prilled ammonium nitrate containing approximately 6 % diesel oil is used (ANFO = ammonium nitrate fuel oil).

To a lesser extent, ammonium nitrate is used to produce dinitrogen monoxide, N_2O. The salt must be 99.5 % NH_4NO_3 and be free of organic materials, iron, chlorides, and sulfates.

1.8. Economic Aspects

Table 4 provides a synopsis of the production of ammonium nitrate for fertilizers in various countries.

2. Ammonium Sulfate

Ammonium sulfate [7783-20-2], $(NH_4)_2SO_4$, was produced industrially in the 19th century, mainly from the ammonia in coke-oven gas. At the beginning of this century ammonia was synthesized on an industrial scale, and ammonium sulfate became a widely used fertilizer. Its importance, however, has steadily declined with the development of more concentrated nitrogen fertilizers. Small quantities are used for industrial purposes. Since about 1960, ammonium sulfate has been produced to an increasing extent as a coproduct in organic syntheses. In some countries ammonium sulfate is produced from gypsum.

2.1. Properties

Ammonium sulfate, M_r 132.14, d_4^{20} 1.77, average specific heat capacity 1.423 J g^{-1} K^{-1} between 2 and 55 °C, rhombic bipyramidal. The size and habit of the crystals can be affected significantly by substances in the crystallizing solution, and this can be an important factor in commercial production.

Ammonium sulfate cannot be melted at atmospheric pressure without decomposition, releasing ammonia and leaving bisulfate. However, the ammonia vapor pressure of pure, anhydrous ammonium sulfate is effectively zero up to 80 °C. Above 300 °C, decomposition gives N_2, SO_2, SO_3, and H_2O in addition to ammonia.

The solubility of ammonium sulfate in water is shown in Figure 11. The salt does not form hydrates. The heat of solution on dissolution of 1 mol of salt in 400 mol of water is +9.92 kJ at 18 °C, the integral heat of solution is +6.57 kJ/mol at 30 °C, and the differential heat of solution for saturated solution is +6.07 kJ/mol at 30 °C. Ammonium sulfate deliquesces only above 80 % relative humidity; therefore, the salt can be stored in dry air. The solubility of ammonium sulfate is reduced considerably by addition of ammonia: At 10 °C, from 73 g $(NH_4)_2SO_4$ in 100 g of water, nearly linearly, to 18 g salt in 100 g of 24.5 % aqueous ammonia.

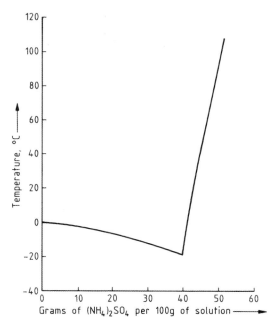

Figure 11. The system $(NH_4)_2SO_4-H_2O$

Calcium sulfate is approximately twice as soluble in ammonium sulfate solution as in water, which favors reaction of gypsum or anhydrite with ammonium carbonate – ammonium sulfate solutions. The salt is insoluble in the usual organic solvents. Dissolution of ammonium sulfate in aqueous solutions of ethanol, propanol, butanol, acetone, pyridine, etc., causes two phases to form, one aqueous and the other the organic solvent.

Special Cr – Ni stainless steel is resistant to ammonium sulfate solutions and to solutions of the salt in sulfuric acid. Ammoniacal solutions do not attack iron or aluminum. If the solutions contain substances that corrode stainless steel (e.g., Cl^-), the vessels can be lined with acid-resistant bricks.

2.2. Production

Ammonium sulfate is produced from

1) Coke-oven gas
2) Ammonia and sulfuric acid
3) Organic syntheses, such as the production of caprolactam
4) Gypsum, ammonia, and carbon dioxide

2.2.1. From Coke-Oven Gas

The production of ammonium sulfate from coke-oven gas has diminished considerably over the past twenty years, because of the partial closure of steel mills and the development of coking processes producing less ammonium sulfate [46].

In the direct process the unpurified coke-oven gas is introduced into sulfuric acid to yield ammonium sulfate contaminated by colored tar products. In the indirect process the ammonia is washed out of the coke-oven gas with water, released with a lime suspension, and finally introduced into sulfuric acid.

2.2.2. From Ammonia and Sulfuric Acid

The heat of reaction between ammonia and sulfuric acid is sufficient to evaporate all the water if an acid concentration of 70 % or higher is used.

$$2\,NH_3\,(g) + H_2SO_4\,(l) \longrightarrow (NH_4)_2SO_4\,(s)$$
$$\Delta H = -274\,\text{kJ/mol}$$

Today, the reaction is usually carried out in saturators, devices that are derived from the evaporation crystallizers commonly used earlier [47–49].

In the saturator process (Fig. 12), neutralization and crystallization are carried out in one and the same apparatus. The sulfuric acid is delivered to the suction side and the ammonia to the pressure side of the forced circulation pump. Crystallization of the metastable solution gives particle sizes generally between 0.5 and

3 mm. The salt is continuously discharged at the lower end of the saturator. The salt is separated in centrifuges, dried, and cooled. The mother liquor is returned to the saturator. Impurities in the sulfuric acid can adversely affect crystallization. Small quantities of phosphoric acid, urea, or inorganic salts are added to promote crystal growth.

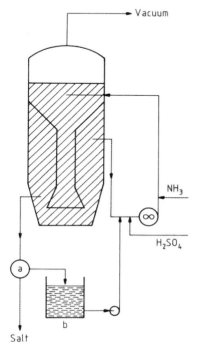

Figure 12. Ammonium sulfate saturator
a) Centrifuge; b) Mother liquor tank

2.2.3. Coproduct in Organic Syntheses

Ammonium sulfate is coproduct in the production of synthetic-fiber intermediates, such as caprolactam, acrylonitrile, and methyl methacrylate, and in the production of formic acid and acrylamide. The most important source is the production of caprolactam, which is required for nylon 6. The conventional caprolactam process produces 2.5 – 4.5 t of ammonium sulfate per tonne of lactam. Of that, 0.3 – 2.8 t is formed in the hydroxylamine–oximation stage and approximately 1.7 t in the Beckmann rearrangement stage. Crystallizers are used in the evaporation of ammonium sulfate solutions that result from caprolactam production. The processes developed by BASF [50], [51], Inventa [50], [52], [53], Toray Industries [54], Snia Viscosa [50], Techni-Chem [52], Kanebo [54], [55], and DSM [56] have reduced the 2.5 – 4.5 t, in some cases, to 1.7 – 1.8 t per tonne of lactam.

2.2.4. From Gypsum

Both anhydrite and gypsum react with NH_3 and CO_2:

$$CaSO_4(\cdot\, 2\,H_2O) + (NH_4)_2CO_3 \longrightarrow$$
$$(NH_4)_2SO_4 + CaCO_3 (+2\,H_2O)$$

The process, developed by BASF during the First World War, has retained its importance where sulfuric acid is in short supply, including India, Pakistan, and Turkey.

Finely ground gypsum is reacted with ammonium carbonate solution in a cascade of stirred vessels. The residence time is several hours. The reaction mixture of calcium carbonate and ammonium sulfate solution is filtered on rotary vacuum filters. The washed calcium carbonate can be used for fertilizing lime, the production of calcium ammonium nitrate, raw material for glass, or filler for rubber or PVC [57]. The slightly cloudy ammonium sulfate solution is filtered in filter presses, acidified with H_2SO_4, and processed in multistage evaporation crystallizers to form coarse-grained ammonium sulfate. (For details and a flowsheet, see [58]) An alternate method for converting gypsum into ammonium sulfate and calcium carbonate is the Continental Engineering Process [59], in which NH_3 and CO_2 are directly introduced into a gypsum slurry in a tall, cylindrical stirred vessel.

In principle, byproduct gypsum from the phosphoric acid process may be used too. However, it may contain phosphoric acid, silicates, fluorosilicates, and organic substances, which make filtering difficult, and may require preliminary washing.

2.2.5. Other Processes

There are several obsolete processes that start from SO_2 and atmospheric oxygen [60], [61]. There has been interest in processes using ammonia to remove SO_2 from power-station exhaust. For example, the Walther process [62] produces a concentrated ammonium sulfite solution by two-stage washing of the dust-free exhaust. The solution is oxidized with atmospheric oxygen and spray-dried to ammonium sulfate powder, which is subsequently granulated [63].

2.3. Granulation and Storage

Typical crystallizers generally produce a 0.5- to 3-mm grain that can be dried to a water content of less than 0.1 %. The free acid content should not exceed 0.03 %, and the N content should not be below 21 %. The grains are protected against caking, generally by small quantities (0.1 %) of surfactants, and are storable.

For use in bulk blending, a procedure for producing the blended fertilizers widely used in the United States, the ammonium sulfate is generally granulated. This is done by compacting, pan granulation, or use of the TVA Ammoniator-Granulator [64]. Granulation can be improved by the use of additives, e.g., calcium salt [65]. The mixed granulation of concentrated urea solutions with ammonium sulfate crystals, e.g., in a pan granulator, gives a hard granulate of high nitrogen content (30 – 43 % N) suitable for bulk blending [66].

2.4. Uses

Almost all of the ammonium sulfate (AS) produced is used as fertilizer; very little is used industrially. In industrial countries ammonium sulfate is almost always a coproduct or byproduct, and it can only be sold as a fertilizer, mostly in the developing countries (Table 5). In industrial countries its low nitrogen content makes transportation cost per unit of nitrogen higher than for other nitrogen fertilizers. In Africa and Asia ammonium sulfate is used especially to fertilize rice, tea, and rubber. In Europe, the United States, and Brazil, it is often a component in blended and complex fertilizers ($\rightarrow$ Fertilizers).

In industry $(NH_4)_2SO_4$ is used for the production of persulfates, flameproofing agents, and fire-extinguishing powders; in tanning; in the photographic, textile, and glass industries; and as a nutrient for yeast and bacterial cultures.

2.5. Economic Aspects

World ammonium sulfate capacity is shown in Table 6 [46]. Table 7 [46] shows the worldwide consumption of ammonium sulfate, which stagnated in 1976. As can be seen from the tables, production capacities are by no means fully utilized.

The transportation costs for shipping ammonium sulfate from the countries in which it is produced, the industrialized countries, into the countries in which it is consumed, the developing countries, reduces any profit. Production in some countries, e.g., China, Mexico, and Spain, has been based on a political decision – to increase domestic employment or preserve limited foreign exchange. In addition, worldwide consumption has been decreasing steadily. These three facts have been the driving force in the industrialized countries behind reducing the coproduction of ammonium sulfate in the production of synthetic fibers. If efforts to reduce sulfur emissions produce additional amounts of ammonium sulfate, which also will have to be sold in the developing world, the pressure on profits will increase further.

3. Ammonium Chloride

Although ammonium chloride [*12125-02-9*], NH_4Cl, M_r 53.49, occurs naturally in volcanic material, production from natural sources is of no significance. The history of industrial NH_4Cl production is intertwined with the development of the soda industry and the large-scale production of synthetic NH_3. The original motivation for producing ammonium chloride was to use it as fertilizer, but it is now used for a wide range of purposes. Ammonium chloride processes are

Table 5. World consumption of ammonium sulfate as fertilizer in 1983, 1000 t N

Western Europe	285
Eastern Europe	818
North America	146
South America	419
Asia	824
Africa	98
Others	30
Total	2620

Table 6. World ammonium sulfate capacity, 1981/82, 1000 t N

	Synthetic	Coproduct	Byproduct and others	Total
World	2900	2289	982	6171
Western Europe	561	913	158	1632
Eastern Europe	210	631	402	1243
North America	219	264	225	708
Central America	395	36	26	457
South America	74	32	1	107
Africa	40	–	39	79
Asia	1387	413	75	1875
Oceania	14	–	56	70

Table 7. Worldwide consumption of ammonium sulfate, 1000 t N

	1976/77	1977/78	1978/79	1979/80	1980/81
World	2814.7	2890.8	2796.0	2916.3	2828.4
Western Europe	461.6	434.0	468.5	471.5	430.8
Eastern Europe	729.9	718.9	684.9	661.7	653.5
North America	208.8	178.7	152.2	172.2	191.2
Central America	325.6	304.8	310.8	341.0	353.4
South America	173.9	226.4	211.0	233.8	238.8
Africa	130.9	138.0	144.3	141.3	128.2
Asia	759.3	858.7	802.4	868.0	769.2
Oceania	24.7	31.3	21.9	26.8	63.3

of historical interest to chemical engineering because they are an early example of chemical processing involving all three phases: solid, liquid, and gas.

3.1. Properties

Solid ammonium chloride has a specific gravity d_4^{20} of 1.530. Its average specific heat $\bar{c}_p$ between 298 and 372 K is 1.63 kJ/kg.

Ammonium chloride has two modifications [67]. The transformation between the two is reversible at 457.6 K (184.5 °C):

α-NH_4Cl (cubic, CsCl type) $\rightleftharpoons$
β-NH_4Cl (cubic, NaCl type) $\rightleftharpoons$

$$\Delta H = +4.3 \text{ kJ/mol}$$

The α modification is the one stable at room temperature. β-NH_4Cl melts at 793.2 K under 3.45 MPa; it sublimes at atmospheric pressure. In fact, NH_4Cl is quite volatile at lower temperatures, dissociating into NH_3 and HCl:

T, K	523.2	543.2	563.2	583.2	603.2	611.2
p, kPa	6.6	13.0	24.7	45.5	81.4	101.3

The solubility of NH_4Cl in water increases with temperature

T, K	273.2	293.2	313.2	333.2	353.2	373.2	389.2
c, wt %	22.9	27.2	31.5	35.6	39.7	43.6	46.6

The integral heat of solution to saturation is +15.7 kJ/mol, and the differential heat of solution at saturation is +15.2 kJ/mol. The solubility in water may be increased by adding ammonia. NaCl tends to salt NH_4Cl out of aqueous ammoniacal solution [68], [69]. The partial pressures of saturated NH_4Cl solutions show that NH_4Cl is weakly hygroscopic

T, K	283.2	293.2	303.2	313.2	323.2	389.2
p, kPa	1.0	1.9	3.3	5.4	8.8	101.3

Moisture causes product caking; however, whether moisture contents less than 0.1 wt % alone are responsible for the caking of ammonium chloride, is questionable. Another possible explanation is sublimation.

Ammonium chloride is very soluble in liquid NH_3 but virtually insoluble in acetone and pyridine. At 292.7 K its solubility in methanol is 3.24 wt %, and its solubility in ethanol is 0.64 wt %.

The crystal form obtained from aqueous solutions can be affected by other substances. This was exploited to produce large crystals, and it is a help in determining impurities arising during NH_4Cl production [70], [71].

3.2. Production

Ammonium chloride is produced commercially by two processes:

1) Modified Solvay process (ammonia – soda ash process or ASAP)
2) Direct reaction between HCl and NH_3

A third process, the reaction of reciprocal pairs of salts [72], has not yet been commercially applied, but is still of interest.

3.2.1. Modified Solvay (Ammonia – Soda Ash) Process

In the Solvay process, ammonia and carbon dioxide are dissolved in aqueous sodium chloride to produce sparingly soluble sodium bicarbonate, which is calcined to sodium carbonate. The ammonia is recovered from the mother liquor by reaction with lime, this reaction also producing the calcium chloride. The source of the lime and the carbon dioxide is limestone. The net reaction is the conversion of the feedstocks rock salt and limestone into sodium carbonate and the byproduct calcium chloride.

Water, carbon dioxide, and ammonia are added only to the extent necessary to compensate for plant losses. The feedstock and product dictate the location of a Solvay plant. For this reason, the plants normally function as largely independent units.

The modified Solvay process (ammonium chloride – soda ash process) is one of the oldest examples of integrated industrial production of two substances. It differs from the Solvay process in that ammonium chloride is also precipitated from the mother liquor. The net reaction is

$$2\,NH_3 + CO_2 + H_2O + 2\,NaCl \longrightarrow 2\,NH_4Cl + Na_2CO_3$$

The amounts of ammonium chloride and soda ash produced are almost equal: two moles (107 g) of NH_4Cl are produced for each mole (106 g) of Na_2CO_3. The modified process, requiring external NH_3 and CO_2, must be incorporated into an integrated system of plants.

Furthermore, the treatment of ammonia-containing waste gases is carried out for environmental reasons, rather than to reduce NH_3 and CO_2 losses. Finally, the energy balance in the modified process is entirely different from that in the traditional Solvay process because of the different feedstocks and products.

Process Description. As a rule, the manufacture of NH_4Cl and Na_2CO_3 is carried out in a continuously operated recycle process in which the concentrations within the cycle may be adjusted according to whether ammonium chloride or sodium carbonate is the primary product [76]. BASF originally developed the process in which NH_4Cl is the primary product (Fig. 13).

The cycle begins with the addition of ammonia and carbon dioxide to the aqueous recycle solution. The ammonium bicarbonate intermediate reacts further with the sodium chloride present in the solution to yield ammonium chloride and sparingly soluble sodium bicarbonate. After thickening, the latter is separated out with a centrifuge; it is washed during centrifugation. It is calcined to produce sodium carbonate.

The sodium bicarbonate remaining in the mother liquor is decomposed by heating to 337.2 K. The very soluble sodium carbonate does not interfere with the crystallization of ammonium chloride. The high temperature of the solution at this point is utilized to dissolve the rock salt feedstock rapidly. This rock salt is introduced via a series of mixing tanks, the quantity dictated by the material balance over the entire cycle. Any insoluble impurities in the rock salt are filtered off. The major impurities are sulfates and other salts of iron, calcium, and magnesium. Disposal of the residues represent an increasing problem. Therefore, pure salt is replacing rock salt as raw material.

Ammonium chloride is recovered from the clear solution in a two-stage crystallization by indirect cooling. The suspension from the crystallizer is thickened in hydrocyclones, and the solid is separated and washed in a centrifuge. The solid from the centrifuge containing 5 – 7 wt % water is dried in a current drier to a moisture content less than 0.1 %. The final product can be coated with an anticaking agent. After the removal of the ammonium chloride, the cycle is complete, and the mother liquor can be reused.

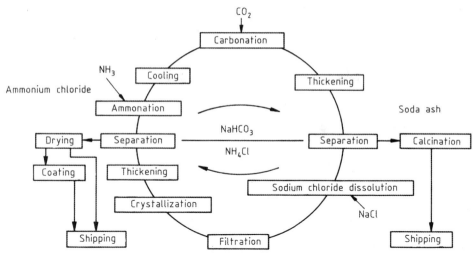

Figure 13. Ammonium chloride – soda ash process

A process that produces soda ash as the primary product has been developed by Asahi Glass [77], [78].

Ammonia is added to the mother liquor prior to the crystallization of ammonium chloride. Crushed raw salt is added with or without washing and dissolved in the solution. The solution is then cooled below 20 °C to crystallize ammonium chloride. The crystals are separated with a centrifugal separator, and the mother liquor is recycled to the carbonation section. The ammonium chloride crystals are dried to a moisture content below 0.3 % in a rotary or fluidized bed drier. Crystal sizes can be kept between 1 and 2 mm for granule or about 200 µm in diameter for high-grade compound fertilizer.

In the carbonation section, the solution is sent into a carbonating tower for the precipitation of sodium bicarbonate through the reaction with blown-in carbon dioxide at 30 – 40 °C. The resulting slurry is separated into sodium bicarbonate crystals and the mother liquor. The mother liquor is sent to the ammonia absorption section. The separated crystals are calcined and converted to dense soda ash.

This process is capable of producing bicarbonate as pure as that from the conventional Solvay process and fertilizer-grade ammonium chloride. The purity of ammonium chloride exceeds 97 wt %.

As shown in Figure 14, the Asahi Glass process includes all the steps shown in Figure 13; however, the sequence differs. The ammonia is added after separation of the sodium bicarbonate. The heat of solution of ammonia serves to provide the heat needed to decompose the residual sodium bicarbonate. However, the energy saving is offset to some extent by the higher energy need in the crystallization because ammonium chloride is more soluble in the ammonia-rich solution [68], [69]. A washing step for the rock salt feed, prior to solution, avoids the filtration before crystallization. With or without washing this process utilizes the raw salt to the fullest extent, an important consideration in Japan where the salt must be imported.

Equipment Variations. Little is known about the process details of the individual process variants. Nevertheless, general characteristics of the process technology, adequate for plant selection, can be provided.

The individual steps of several other important process variants are described below.

If the feed gases are available at sufficiently high pressure, they can be mixed with the recycle solution by gas – liquid contacting devices, such as jet injectors. In the case of ammonia the amount of excess gas is reduced, and the solution is homogenized rapidly. In the case of CO_2 the nucleation rate and crystal growth of $NaHCO_3$ are improved.

316   Ammonium Compounds

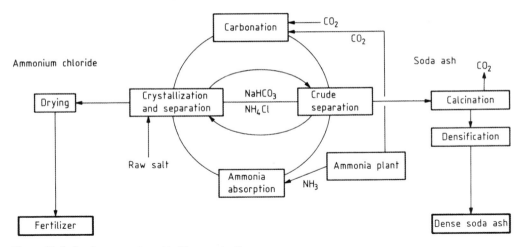

Figure 14. Soda ash – ammonium chloride coproduction process

Further parameters influencing the size of the $NaHCO_3$ crystals are temperature and residence time in the carbonation step. The various carbonation columns used in the production of $NaHCO_3$ differ in the number and type of internal parts such as mixing devices and additional heat exchangers, the positioning and type of the CO_2 nozzles, and the volume needed to produce a given $NaHCO_3$ equivalent. The large $NaHCO_3$ crystals may be separated and washed using a band filter, or after thickening they may be separated by centrifugation.

The important differences in NH_4Cl crystallization lie in the number of stages and type of cooling. Direct cooling during the crystallization, in comparison with indirect cooling, uses more energy and requires the use of additional equipment for refrigerant condensation and recovery. However, it avoids problems associated with indirect cooling, i.e., the operation of heat exchangers, from which a NaCl crust must be periodically removed. Special plate heat exchangers have been developed to solve the fouling problem, and these are to be preferred to shell and tube heat exchangers and coil coolers.

$NaHCO_3$ and NH_4Cl crystallization and removal are the keys to high conversion per pass and product purity.

Ammonium chloride is frequently dried in a drum drier instead of a pneumatic drier.

Pollution Problems. The pollution problems associated with plant operation are not critical. By careful operation the production of wastewater streams can be almost entirely eliminated. However, because of its high ammonia content, any wastewater produced will require treatment.

The scrubbing of waste gases with hydrochloric acid to produce additional ammonium chloride offers a neat solution to the problem of gaseous emissions. Generally ammonia conversions of 98 % and rock salt conversions of 95 % are achieved, the exact values depending on the type of process and the environmental measures.

Grades and Capacity. After closure of several production plants due to reduced market demand, the capacity for NH_4Cl by the modified Solvay process for the western hemisphere and Japan was estimated at ca. 100000 t/a in 1997. Only limited information is available on total capacities and applied technology in China. In Asia, a granular product containing more than 97 % NH_4Cl is usually produced. The ammonia nitrogen content of this fertilizer grade NH_4Cl always exceeds 25 wt %. These plants are primarily for the production of soda ash.

However, if the production of NH_4Cl is of primary interest, the process can be carried out in such a way as to give purities exceeding 99.7 wt % NH_4Cl with < 0.25 wt % NaCl and < 3 ppm Fe, a quality adequate for most industrial purposes. Reagent-grade NH_4Cl can be produced on a large scale with special operating techniques but without additional processing steps. This ammonium chloride contains < 0.01 wt % NaCl and corresponds to the ACS specification [79].

3.2.2. Direct Reaction Between HCl and NH$_3$

The synthesis of NH$_4$Cl from HCl and NH$_3$ is economically attractive if HCl is available as feedstock at low or no cost.

The *Engeclor process* [80], developed by the Brazilian company Engeclor, carries out the reaction in aqueous solution:

Ammonia is fed into the conical section of a saturator while HCl diluted with air is passed into the NH$_4$Cl suspension. The reaction occurs at 353 K, under reduced pressure, and with excess NH$_3$ (pH 8). The suspension is drawn off from the base of the saturator and thickened in hydrocyclones. The NH$_4$Cl is separated in a centrifuge and dried. The mother liquor is recycled to the saturator. The waste gases from the saturator are scrubbed with water.

Schemes have been proposed whereby gaseous feedstocks are fed into nonaqueous solvents. The heat of reaction

$$NH_3(g) + HCl(g) \rightleftharpoons NH_4Cl(s) \quad \Delta H = -176 \text{ kJ/mol}$$

is removed by the evaporation of the solvents, which are subsequently condensed [81]. Fluidized bed processes have also been described [82]. To avoid aerosol formation, carbon dioxide is recommended as carrier gas in the fluidized bed [102].

The annual world production of NH$_4$Cl via the HCl–NH$_3$ process was estimated to be 50000 t in 1997. Ammonium chloride produced by this method generally contains < 0.1 wt % NaCl. The metal content, and especially the heavy metal content, depends on the plant. As a rule the levels are higher than those encountered in the modified Solvay process.

3.2.3. Reaction of Reciprocal Pairs of Salts

The reaction of reciprocal salts [72] is still of interest. Suitable pairs of salts seem to be (NH$_4$)$_2$SO$_4$–KCl, which gives NH$_4$Cl–K$_2$SO$_4$ [73–75][103][104] and (NH$_4$)$_2$SO$_4$–NaCl which gives NH$_4$Cl–Na$_2$SO$_4$ [106]. The basic problem with these salts is the lower purity of NH$_4$Cl due to the high content of sulfate. NH$_4$Cl and KNO$_3$ are obtained by reaction of NH$_4$NO$_3$ and KCl. Different process modifications are described in [105, 107–109]. High-purity KNO$_3$ can be used in manufacturing explosives, and NH$_4$Cl with a purity of 96 % min. can be used as a fertilizer. The preparation of NaNO$_3$ and NH$_4$Cl from NaCl and NH$_4$NO$_3$ has been investigated [110], [111].

3.3. Corrosion

Ammonium chloride is corrosive, as gas, as solid, or in solution, because (1) it is acidic, (2) it complexes metal ions, and (3) it contains the corrosive chloride ion. Therefore, corrosion protection is one of the key factors for production reliability and product quality. Several forms of corrosion are observed.

For solutions, the extent of corrosion largely depends on the NH$_4$Cl concentration, the pH, and the temperature. Equipment made from iron, aluminum, lead, or nonferrous metals is especially prone to stress corrosion cracking.

Older plants were constructed with rubber-lined pipes and containers that had an additional ceramic coating. Today special steels or plastics are usual. Low carbon, pure austenitic, or austenitic–ferritic steel, for example, that designated as SS 360 in the United States, is used up to 313 K. In borderline cases Hastelloy or Ti [83] is used.

The steel must be correctly worked. For example, welding must be carried out under a protective gas, and welded seams always should be posttreated. Up to 333 K polyethylene or fiberglass-reinforced polypropylene can be used where abrasion and erosion are absent. Heat exchangers of the graphite material Diabon have proved satisfactory. Cloths made from polyacrylonitrile are most suitable for filtration. Protection against external corrosion may also be necessary. Protective coatings with an epoxide or phenoxy resin base have proved effective. Such coatings are effective also for the internal protection of vessels or tanks not exposed to much mechanical wear and tear, such as storage silos.

3.4. Packaging, Storage, Shipping, and Handling

On account of its corrosive nature and its tendency to cake, NH_4Cl is best packed in sacks, or in the case of "flowable" goods in big bags, of paper or polyethylene rather than plastic or metal drums.

Despite its low water content (< 0.1 wt %) ammonium chloride tends to cake. There are two methods to enhance the "flowability." First, additives such as soda or B_2O_3 serve to separate the individual particles and preferentially absorb any water present. Product coated with such materials retains the ability to flow freely for a limited storage period and can be stored in silos on a temporary basis. Second, individual particles are encapsulated with hydrophobic chemicals. Fatty acid amines, e.g., stearylamine, and their acetate and hydrochloride derivatives are commonly used.

No special precautions beyond those usual for chemicals are needed for shipping ammonium chloride. Ammonium chloride should not be allowed to come into contact with nitrates or strong acids or alkalies. It is not covered by industrial chemical legislation although European Economic Community guidelines require labeling as a weak poison [84]. Ammonium chloride for the food industry must meet the specifications of the *European Pharmaceutical Handbook* [85] in Europe and those of the Food and Nutrition Board [86] in the United States. The latter restrict the use of anticaking agents.

3.5. Uses

Ammonium chloride is widely used as an effective nitrogen fertilizer for paddy and upland rice, wheat, and other crops in Japan, China, and South East Asia [77]. Most of the annual production of NH_4Cl in Japan is used as high-grade compound fertilizers: chloro-ammonium phosphate, chloro-potash-ammonium phosphate, magnesia-chloro-potash-ammonium phosphate, and nitrogen-potash. Further fertilizer uses are limited by the acidity and high chlorine content.

The industrial uses of technical-grade ammonium chloride are, in order of importance: solid electrolytes in dry cell batteries, one component of quarrying explosives (an especially fine form is used), hardeners for formaldehyde-based adhesives, one component of etching solutions in the manufacture of printed circuit boards, and a component, along with zinc chloride, of fluxes in tin and zinc plating. Minor uses include a rapid fixer additive in photography, cleaner additives, and a nutrient in yeast cultures. Ammonium chloride is also used in tanning, refining of precious metals, textile printing and dying, and in the rubber industry. The addition of ammonium chloride to tiles and bricks prior to firing was proposed. The ammonium chloride serves to control product porosity and accelerates the firing process [87]. The use of NH_4Cl as hardener for formaldehyde-based adhesives has been reduced significantly due to substitution by chloride-free products.

High-purity ammonium chloride is employed in the food and pharmaceutical industries and in a few chemical syntheses.

4. Ammonium Carbonates

Salt of hartshorn was first mentioned at the beginning of the 14th century in English manuscripts. It was produced by dry distillation of antlers, hooves, and leather and consisted of a solid mixture of NH_4HCO_3, $NH_4CO_2NH_2$, and $(NH_4)_2CO_3 \cdot H_2O$. This salt of hartshorn was used as an expanding agent for certain baked goods on account of its decomposition into entirely gaseous products and good handling properties. It was manufactured on a semi-industrial scale at the beginning of the 19th century.

Following the development of the industrial synthesis of ammonia, BASF started large-scale production of ammonium carbonate in 1919. Today, the term salt of hartshorn refers to a mixture of pure NH_4HCO_3 and $NH_4CO_2NH_2$ containing 32.5 wt % NH_3. The following compounds appear in the ternary system $NH_3-CO_2-H_2O$ [88], [89]:

NH_4HCO_3	ammonium hydrogencarbonate
$(NH_4)_2CO_3 \cdot H_2O$	ammonium carbonate
$(NH_4)_2CO_3 \cdot 2 NH_4HCO_3 \cdot H_2O$	ammonium sesquicarbonate
$NH_4CO_2NH_2$	ammonium carbamate

All these substances can be produced from NH_3 and CO_2. The high purity of the feedstocks assures adequate product purity without further processing, provided the materials of construction are suitable.

4.1. Ammonium Hydrogencarbonate

Ammonium hydrogencarbonate [*1066-33-7*], ammonium bicarbonate, NH_4HCO_3, M_r 79.06, ϱ 1.586 g/cm^3. The decomposition pressure of ammonium bicarbonate increases rapidly with temperature [90]:

T, K	298.6	307.4	313.9	318.2	323.2	329.0	332.5
p, kPa	7.85	16.26	26.79	37.06	52.65	82.11	108.54

The solubility of ammonium bicarbonate in water increases with temperature [90]

T, K	273.2	283.2	293.2	303.2	313.2	323.2	333.2
c, wt %	10.6	13.9	17.8	22.1	26.8	31.6	37.2

The reaction forming NH_4HCO_3 is exothermic:

$$NH_3(g) + CO_2(g) + H_2O(l) \rightleftharpoons NH_4HCO_3(s)$$
$$\Delta H = -126.5 \text{ kJ/mol}$$

In the $NH_3 - CO_2 - H_2O$ system, ammonium bicarbonate is the only compound soluble in water without decomposition. Upon dissolution of one mole of salt in 6–8 L H_2O the heat of solution is +26.4 kJ at 288.2 K. The dissociation constant in aqueous solution is $K_d = 1.45$ [91] for the reaction $NH_4HCO_3 \rightleftharpoons NH_4^+ + HCO_3^-$.

Ammonium bicarbonate is best produced in aqueous solution, be it continuously or batch. The process is relatively easy to control. The crystallization is the critical stage. The heat of reaction must be removed, while the residence time, concentration, and temperature profile have to be controlled to yield large, easily separable crystals.

The flow diagram of a continuous process is shown in Figure 15. Ammonia is added to the mother liquor, which is kept cool, to achieve the desired NH_3 concentration of slightly above 10 wt %. Carbon dioxide is added and allowed to react in an absorption column. The resulting solution, which has been warmed by the reaction, is cooled indirectly in a crystallizer. The crystal suspension is withdrawn, thickened in hydrocyclones, and passed to a centrifuge, where the solid is separated. This solid is pneumatically dried, cooled, and conditioned. The mother liquor is recycled. If stainless steel equipment is used, the product meets the specifications of the *European Pharmaceutical Handbook* [92] and the *Food Chemical Codex* [93] without further purification.

The world production in 1997 was estimated at ca. 100 000 t/a, almost half of which was produced in the western hemisphere and the remainder in Asia. The process variants of individual producers usually differ in the absorption/crystallization equipment, and this also results in different particle size distributions. The theory of CO_2 and NH_3 adsorption is discussed in [94], and crystallization with direct cooling is described in [95].

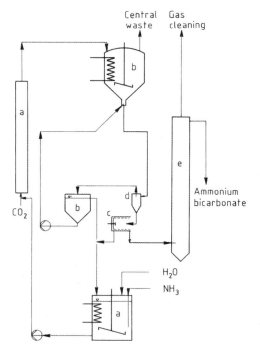

Figure 15. Production of ammonium bicarbonate
a) Gas absorption; b) Crystallization; c) Centrifugation; d) Thickening; e) Drying

4.2. Ammonium Carbamate

Ammonium carbamate [*1111-78-0*], $NH_4CO_2NH_2$, M_r 78.08, melts in a sealed tube at 425.2 K.

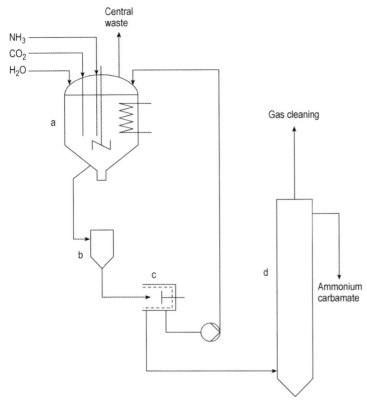

Figure 16. Production of ammonium carbamate
a) Crystallizer; b) Thickening; c) Centrifuge; d) Drying

The dissociation pressures show that the compound is quite volatile and thermally unstable [90].

T, K	283.19	288.10	291.10	294.41	298.10
p, kPa	38.0	55.3	68.50	88.50	114.80
T, K	299.97	304.10	309.07	313.05	318.02
p, kPa	130.40	173.70	242.70	315.30	431.00

The solid consists of dimers [96]. The heat of solution in water is + 15.9 kJ/mol.

Ammonium carbamate is manufactured by two methods. In the first process, which directly gives dry ammonium carbamate, predried NH_3 and CO_2 are fed into a heat exchanger, where they react and are cooled. The carbamate condenses within the pipes and is removed later by partial melting. This carbamate is sold as carbamate rocks.

The second process (Fig. 16), also semibatch, uses a concentrated $(NH_4)_2CO_3$ solution feed. Ammonia and carbon dioxide are fed into the solution, which is kept cool. The carbamate crystals are separated batchwise by centrifugation and dried pneumatically or under a protective CO_2 atmosphere.

A fluidized-bed process has been proposed for carbamate production [97].

Ignoring the large-scale production associated with urea manufacture ($\rightarrow$ Urea), the world carbamate capacity in 1997 was estimated at 4000 t/a.

4.3. Ammonium Carbonate

Ammonium carbonate [506-87-6], $(NH_4)_2CO_3 \cdot H_2O$, M_r 114.11, crystallizes as flat, columnar, prismatic crystals or as elongated flakes. The carbonate melts at 316.2 K, simultaneously decomposing to the carbamate and sesquicarbonate. $(NH_4)_2CO_3 \cdot H_2O$ is only formed under precisely defined conditions, which are difficult to achieve in any production plant. Therefore usually equimolar mixtures of ammonium carbamate and ammonium hydrogencarbonate are used and sold as ammonium carbonate. These mixtures contain the same amount of ammonia as $(NH_4)_2CO_3 \cdot H_2O$. The world carbonate capacity in 1997 was estimated at 7000 t/a.

4.4. Storage, Transport, and Handling

The ammonium carbonates show noticeable dissociation pressures at temperatures as low as 293 K. Both carbamate and carbonate decompose to yield the bicarbonate, resulting in product losses and composition changes. For this reason the substances should be stored in sealed containers and kept as cool as possible. In sealed containers decomposition occurs only to a limited extent because progressive decomposition is prevented by the establishment of equilibrium. The rate of decomposition of NH_4HCO_3 is also a function of particle size [95]. Ammonium hydrogencarbonate has a tendency to cake, and its flowability is maintained by anticaking additives such as corn flour, saccharose, MgO, or $MgCO_3$.

There are, in general, no special requirements with regard to transport, safety precautions, and handling.

4.5. Uses

Ammonium bicarbonate is mainly used as an expanding agent for certain baked goods; it is also used in buffer solutions for neutralization of acids when additional anions are undesirable. Further uses include an ammonization agent for the humic acid in turf, a nitrogen source for yeast cultures, and blowing agents for foam rubber and poly(vinyl chloride). Minor uses include smelling salts and formaldehyde binders in laminates. Ammonium hydrogencarbonate is also used in the production of textiles, ceramics, pigments [98], and leather.

Ammonium carbamate is used as an insecticide for cereals, a neutralization agent in the chemical industry, and a feedstock for ammonium carbonate production.

Ammonium carbonate is used in the manufacture of catalysts [99–101], as blowing agent in foam rubber and foam plastic, and as an additive in photographic developers. In a number of countries it is also used as an expanding agent in certain baked goods.

5. Toxicology and Occupational Health

Ammonium chloride is harmful by ingestion; the oral LD_{50} in rats is 1410 mg/kg [112]. The i.v. LD_{50} in mice is 7–10 mmol/kg [113]. Ammonium chloride causes no skin irritation, but it is irritating to the eyes [114]. Ammonium chloride is not sensitizing in animals [115]. Oral treatment of various animal species with daily doses of 500–1000 mg/kg causes metabolic acidosis [116–118]: plasma and urinary pH and dioxide-combining power decrease, whereas chloride, urea nitrogen, and solids concentrations in the blood and gluconeogenesis increase. The Ames test [119][120] with *S. typhimurium* and *E. coli* and the micronucleus test [121] in mice did not show mutagenic effects. Ammonium chloride was not carcinogenic in rats and mice after chronic exposure in the diet and in drinking water [122–125].

Ammonium bicarbonate is harmful by ingestion. The LD_{50} (rat, oral) is ca. 1570 mg/kg [126] and the LD_{50} (mouse, i.v.) is 3.1 mmol/kg [127]. Rats fed for 8 days at a daily dose of 474 mg/kg showed only a slight increase in the ammonia content of the urine [128]. No mutagenic effects were detected in the Ames test with six different strains of *S. typhimurium* and in the chromosomal aberration test in vitro with a Chinese hamster fibroblast cell line [129].

Ammonium carbonate is virtually nontoxic after a single oral uptake. The LD_{50} (rat, oral) is 2150 mg/kg [126] and the LD_{50} (mouse, i.v.) is 1.02 mmol/kg [127]. Rats fed for 5 weeks with 5 % ammonium carbonate in the diet showed depressed growth and elevated urea concentrations in the blood [130]. Chinchilla rabbits were given 100–200 mg/kg ammonium carbonate by gavage or in their drinking water for 5 to 26 months [131], [132]. The treatment cycle was for three weeks with the test compound followed by one week without. When ammonium carbonate was applied in the drinking water, the only notable effect was a parathyroid hypertrophy. Given by gavage ammonium carbonate leads to enlargement of adrenals, ovaries, mammary glands, and womb, as well as lactation and proliferation of ovarian follicles and corpora lutea. However, because these effects were found only in the gavage study, the relevance to humans seems to be doubtful.

Ammonium carbamate is harmful by ingestion. The LD_{50} (rat, oral) is in the range 690–1470 mg/kg [126] and the LD_{50} (mouse, i.v.) is 0.99 mmol/kg [127]. It is not irritating to rabbit skin, but irritates the rabbit eye [133]. No mutagenic effects were detected in the Ames test using four different strains of *S. typhimurium* [134]. Because ammonium carbamate is quite volatile and thermally unstable, forming ammonia and carbon dioxide, the MAK value of ammonia of 20 ppm (TWA 25 ppm) must be taken into account.

6. References

General References
1. G. D. Honti: *The Nitrogen Industry,* Akadémiai Kiadó, Budapest 1976.
2. J. L. Chadwick: *Fertilizers: Nitrogen, Interim Report No. 127 A 1, Private Report by Process Economics Program,* Menlo Park, Calif., 1980.
3. L. Medard: *Les Explosifs Occasionnels,* **vol. 2,** Technique et Documentation, Paris 1979.
4. *Winnacker-Küchler,* **vol. 2.**
5. G. S. Scott, R. L. Grant: *Ammonium Nitrate, Its Properties and Fire Explosion Hazards,* U.S. Bureau of Mines, J.C. 7463 (1948).
6. Chemico, *Nitrogen* 1967, no. 49, 27–31.
7. Soc. Belge de l'Azote, DE 971418, 1951.
8. J. Morrison, *Oil Gas Int.* **8** (1968) no. 9, 93.
9. Soc. Belge de l'Azote, *Nitrogen* 1960, no. 5, 28–31.
10. D. Wagener, K. H. Laue, *Chem. Ing. Tech.* **50** (1978) no. 6, 421–425.
11. G. Fauser, DE 590469, 1931.
12. A. David, C. Parmentier, J. Passelecq, *Hydrocarbon Process.* **57** (1978) no. 11, 169–175.
13. Stamicarbon, *Fertilizer Know-How from A to Z.*
14. NSM/Norsk Hydro, *A New Low Energy Ammonium Nitrate Process,* Nov. 1980.
15. L. A. Stengel, US 2568901, 1947 (and later patents).
16. A. S. Hester, J. J. Dorsay, J. T. Kaufman, *Ind. Eng. Chem.* **46** (1954) 622–632. J. J. Dorsay, *Ind. Eng. Chem.* **47** (1955) 11–17.
17. G. Langhans, B. Bieniok, *ISMA 1976 Technical Conference The Hague 1976,* Elsevier, Amsterdam 1976, preprints pp. 215–233.
18. I. W. McCamy, M. M. Norton: "Production of Granular Urea, Ammonium Nitrate and Ammonium Polyphosphate – a Process Review," in A. I. More (ed.): *Granular Fertilizers and their Production,* Pap. Int. Conf., The British Sulphur Corporation Ltd., London 1977.
19. *Nitrogen* 1978, no. 116, 34–39.
20. *Nitrogen* 1968, no. 56, 27–29.
21. *Nitrogen* 1977, no. 107, 34–39.
22. BASF, DE 691686, 1928.
23. *Nitrogen* 1975, no. 95, 31–36.
24. *Nitrogen* 1973, no. 86, 33–37.
25. R. Collins, *Nitrogen* 1979, no. 118, 35–37.
26. Oesterreichische Stickstoffwerke AG, AT 306754, 1972.
27. Mississippi Chemical Corp., US 3630712, 1971.
28. Veba-Chemie AG, US 4001377, 1977.
29. Stamicarbon, NL 6709519, 1969.
30. Fisons Fertilizer Ltd., GB 1189448, 1970.
31. R. W. R. Carter, A. G. Roberts, *Proc. Fert. Soc.* 1969, no. 110, 31–32.
32. C. Sjölin, *J. Agric. Food Chem.* **19** (1971) no. 1, 83–95.
33. BASF, DE 1905834, 1969.
34. NSM, US 4150965, 1979.
35. Library of Congress Catalog No. 81-82269, ISBN 0-940394-00-9 (1981).
36. G. Hansen, W. Berthold, *Chem. Ztg.* **96** (1972) 449–455.
37. M. Berthelot, *Rev. Sci.* **6** (1883) 8.
38. V. J. Clancey, *RARDE Memor.* 1962, no. 10, 48–63.
39. *Chem. Age* **61** (1949) 485–490.
40. Verordnung über gefährliche Arbeitsstoffe in der Fassung vom 11. 2. 1982, Anhang II, Nr. 11, Bundesarbeitsblatt 4/1982.
41. EG-Richtlinie (80/876/EWG) vom 15. 7. 1980, Amtsblatt der Europäischen Gemeinschaften Nr. L 250, 7–11.
42. BASF, DE 355037, 1919.
43. Gew. Victor, DE 650381, 1929; DE 614705, 1931.

44. E. Jänecke, *Z. Anorg. Allg. Chem.* **160** (1927) 171–184.
45. I. G. Farbenind., DE 612708, 1930.
46. *Nitrogen* 1983, no. 141, 26–30.
47. H. Svanoe, *Ind. Eng. Chem.* **32** (1940) no. 5, 636–639.
48. Th. Messing, *Chem. Ztg.* **91** (1967) no. 24, 963–967.
49. Standard-Messo, DE-AS 1260440, 1964.
50. *Eur. Chem. News* 1976, no. 734, 24–26.
51. BASF, GB 1372108, 1971; DE-AS 2508247, 1975.
52. *Eur. Chem. News* Caprolactam Supplement, 2 May 1969.
53. *Eur. Chem. News Suppl.* 1972, no. 552, 3–4.
54. *Chem. Eng. News* **51** (1973) no. 15, 14–15.
55. Kanebo Ltd., US 3689477, 1971.
56. S. J. Loyson, G. H. J. Nunnink, *Hydrocarbon Process.* **51** (1972) no. 11, 92–94.
57. H. Hofmann, W. Witte, J. Schlicke, H. J. Wassermann, *Silikattechnik* **32** (1981) no. 9, 269–271.
58. *Ullmann,* 4th ed., **vol. 7,** 523–524.
59. J. F. Witte, J. J. De Wit, R. M. Voncken, *Chem. Weekbl.* **65** (1969) no. 37, 25–30.
60. *Ullmann,* 3rd ed., **vol. 3,** p. 618.
61. *Chem. Eng. News* **38** (1960) no. 35, 44.
62. H. Bechthold, *Tech. Mitt. Krupp Werksber.* **39** (1981) no. 2, 43–48.
63. Buckau-Walther, DE-OS 3108986, 1981 (H. Bechthold).
64. *Nitrogen* 1967, no. 45, 33.
65. Kali und Salz, DE 2603917, 1976 (S. Luther, W. v. Maessenhausen).
66. G. C. Hicks, J. M. Stinson, *Ind. Eng. Chem. Process Des. Dev.* **14** (1975) no. 3, 269–276.
67. *Gmelin,* 8th ed., System 23, 150–191.
68. E. Weitz, *Z. Elektroch.* **31** (1925) 546.
69. E. Weitz, H. Stamm, *Ber. Dtsch. Chem. Ges.* **61** (1928) 1144.
70. A. W. Bamforth, *Chem. Eng. (London)* 1974, 455–457.
71. E. V. Khamskii, N. S. Dregubskii, *J. Appl. Chem. USSR* **46** (1973) 2449–2455.
72. V. E. Pikalow et al., *Khimiko-Farmatsevticheskii Zhurnal* **10** (1976) no. 8, 111–112; *Pharm. Chem. J.* **10** (1976) 1095–1096.
73. Allied Chemical Corp., US 3595609, 1969.
74. Kali und Salz AG, DE 2142114, 1971.
75. Dan Kako Co Ltd., JP 4717693, 1971.
76. Z. Rant, *Die Erzeugung von Soda,* F. Enke Verlag, Stuttgart 1968, pp. 136–137.
77. Asahi Glass, JP 200173, 1953 (S. Uemura et al.).
78. T. Miyata, *Chem. Ind. (London)* 1983, 21 Feb., 142–145; K. Tsunashima, K. Nakaya, The New Asahi (NA) Process for Synthetic Soda Ash Production, Fifth Ind. Miner. Int. Cong. Madrid, Spain, 25–28 April 1982, H 1–H 9.
79. ACS, Reagent Chemicals, 6th ed., ACS, Washington, D.C., 1981.
80. A. W. Bamforth, S. R. S. Sastry, *Chem. Process Engng.* **53** (1972) no. 2, 72–74.
81. F. Shadman, A. D. Randolph, *AIChE J.* **24** (1978) 782–788.
82. V. G. Shlyakhtor, V. V. Streitsor, *Theor. Found. Chem. Engng.* **8** (1974) no. 1, 133–136.
83. A. K. Gorbachev et al., *The Soviet Chem. Ind.* **10** (1978) 1004–1005.
84. Guideline 67/5481 of the Commission of the EEC, 5th Adaption.
85. *European Pharmaceutical Handbook*, vol. 1, Deutscher Apotheken Verlag, Stuttgart 1974.
86. *Food Chemical Codex,* 3rd ed., National Academy Press, Washington, D.C., 1981.
87. Studie van de Relaties tussen chemische, fysische en mineralogische Kenmerken van de Boomse Klin en van der Verhittingsprodukten, Katholische Universität Leuven, Doctor Jos. Decleer 1983, 113–136.
88. E. Terres, H. Weiser, *Z. Elektroch.* **27** (1921) 177–244.
89. E. Jänecke, *Z. Elektrochem.* **35** (1929) 723–727.
90. *Gmelin,* 8th ed., System 23, 42–66.
91. G. M. Marion, G. R. Dutt, *Soil. Sci. Soc. Am. Proc.* **38** (1974) 889–891.
92. *European Pharmaceutical Handbook,* **vol. 1,** Deutscher Apotheken Verlag, Stuttgart 1974.
93. *Food Chemical Codex,* 4th ed., National Academy Press, Washington, D.C., 1996, p. 25–26.
94. V. Rod, M. Rylek, *Collect. Czech. Chem. Commun.* **39** (1974) 1996–2006.
95. J. P. Usyukin, *Khim. Prom.* **8** (1972) 610–611; *Soviet Chem. Ind.* **4** (1973) 267–270.
96. A. Gieren et al., *Angew. Chem.* **85** (1973) 308–309.
97. BASF, DE-OS 3346719, 1985 (K. Kinkel).
98. C. B. Pimentel, *Rev. Quim. Ind.* **39** (1970) 453.
99. Shell Oil Co., US 5234477, 1993.
100. Platinum Plus Inc., WO 9504211, 1995.
101. Meidensha Corp., JP 10 033947, 1998.
102. BASF AG, EP 718238-A1, 1996.
103. S. Deng, CN 1144772 A, 1997.
104. Dalian Chem. Res. Inst. Min. Chem. Ind., CN 1131640-A, 1996.
105. Project and Development India Ltd., IN 170975-A, 1992.

106. Broul, CS 8702-618-A, 1988.
107. Rubezahn, SU 1393-791-A, 1989.
108. Pot-Nitrate Mfr. Pty., ZA 8805-130-A, 1989.
109. Chem Factory Tulufa, CN 8601-635-A, 1987.
110. Dongshen Inst. Fine Chem., *Huaxue Shijie* **38** (1997) no. 5, 237–240.
111. Chem Factory Tulufa, CN 8603-687-A, 1987.
112. BASF AG, unpubl. results, 1983.
113. D. P. Stombaugh et al., *Anim. Sci.* **6** (1969) 844.
114. BASF AG, unpubl. results, 1969.
115. Hoechst AG, unpubl. results, 1987; cited in Hoechst GDS, April 14th 1994.
116. W. D. Lotspeich, *Am. J. Phys.* **206** (1965) 1135.
117. P. Vinay et al., *Kidney Int.* **17** (1980) 312.
118. D. Z. Levine, *Am. J. Physiol.* **6** (1973) 380.
119. Hoechst AG, unpubl. results (87.0392) 1987; cited in Hoechst GDS, April 14th 1994.
120. M. Ishidate et al., *Food Chem. Toxicol.* **22** (1984) 623–636.
121. M. Hyashi et al., *Food Chem. Toxicol.* **26** (1988) 487–500.
122. T. Fujii et al., *Food Chem. Toxicol.* **25** (1987) 359–362.
123. S. Fukushima et al., *Cancer Research* **46** (1986) 1623–1626.
124. M. Arai et al.: Symposia on the 4th Int. Cancer Congress, Budapest, vol. 1 (120), 1986, p. 21–26.
125. A. Flasks, B. D. Clayson, *Br. J. Cancer* **31** (1975) 585–587.
126. BASF AG, unpubl. results, 1989.
127. R. P. Wilson et al., *Am. J. Vet. Res.* **29** (1968) 897.
128. J. Oliver, E. Bourke, *Clin. Sci. Mol. Med.* **48** (1975) 515.
129. M. Ishidate ,Jr., et al., *Food Chem. Toxicol.* **22** (1984) 623.
130. J. S. Finlayson, C. A. Baumann, *J. Nutrition* **59** (1956) 211.
131. I. G. Fazekas, *Orvosi Hetilap* **90** (1949) 777; cited also in WHO Tech. Rep., Ser. No. 683, 1982.
132. I. G. Fazekas, *Endokrinologie* **32** (1954) 45. Cited also in WHO Tech. Rep., Ser. No. 683, 1982.
133. BASF AG, unpubl. results, 1958.
134. BASF AG, unpubl. results, 1991.

Nitrates and Nitrites

WOLFGANG LAUE, Gewerkschaft Victor Chemische Werke (now BASF), Castrop – Rauxel, Federal Republic of Germany

MICHAEL THIEMANN, Uhde GmbH, Dortmund, Federal Republic of Germany

ERICH SCHEIBLER, Uhde GmbH, Dortmund, Federal Republic of Germany

KARL WILHELM WIEGAND, Uhde GmbH, Dortmund, Federal Republic of Germany

See also: For ammonium nitrate, see → Ammonium Compounds, Chap. 1.

1.	**Sodium Nitrate**	325
1.1.	Properties	326
1.2.	Occurrence	326
1.3.	Production	327
1.3.1.	Chile Saltpeter	327
1.3.2.	Synthetic Sodium Nitrate	328
1.4.	**Product Forms, Storage, and Transportation**	330
1.5.	Uses	330
1.6.	Economic Aspects	331
2.	**Potassium Nitrate**	331
2.1.	Properties	331
2.2.	Occurrence	332
2.3.	Production	332
2.3.1.	Bacterial Production of Saltpeter	332
2.3.2.	Converted Saltpeter	332
2.3.3.	Potassium Nitrate from Calcium Nitrate	333
2.3.4.	Potassium Nitrate from Ammonium Nitrate	333
2.3.5.	Potassium Nitrate from Potassium Chloride and Nitric Acid	334
2.4.	**Product Forms, Storage, and Transportation**	336
2.5.	Uses	336
2.6.	Economic Aspects	337
3.	**Calcium Nitrate**	337
3.1.	Properties	337
3.2.	Occurrence	338
3.3.	Production	338
3.3.1.	Calcium Nitrate from Limestone and Nitric Acid	339
3.3.2.	Calcium Nitrate as a Byproduct of the Odda Process	339
3.4.	**Product Forms, Storage, and Transportation**	341
3.5.	Uses	341
3.6.	Economic Aspects	342
4.	**Sodium Nitrite**	342
4.1.	Properties	342
4.2.	Production	343
4.3.	**Product Forms, Storage, and Transportation**	343
4.4.	Uses	343
4.5.	Economic Aspects	344
5.	**Potassium Nitrite**	344
5.1.	Properties	344
5.2.	Production	344
5.3.	Uses	344
6.	**Ammonium Nitrite**	345
6.1.	Properties	345
6.2.	Production	345
6.3.	Uses	346
7.	**Calcium Nitrite**	346
7.1.	Properties	346
7.2.	Production	346
7.3.	Uses	346
8.	**Analysis of Nitrates and Nitrites**	347
9.	**Environmental Aspects**	348
10.	**Toxicology and Occupational Health**	349
11.	**References**	350

1. Sodium Nitrate

Sodium nitrate in the form of Chile saltpeter was the most important inorganic nitrogen fertilizer until about 1920; it is the only abundantly occurring mineral nitrate.

1.1. Properties [1]

Sodium nitrate [7631-99-4], Chile saltpeter, Chilean nitrate, $NaNO_3$, M_r 85.00, has an enthalpy of formation $\Delta H = -466.8$ kJ/mol [2], an enthalpy of fusion $\Delta H = +15.7$ kJ/mol [3] and mp 306.8 °C [3]. For pressure dependence of the melting point see [4]. At 25 °C, anhydrous sodium nitrate forms colorless trigonal crystals of space group D_{3d}^6; $\varrho_{298} = 2.261$ g/cm³ [3]; $C_{p,\,298} = 93.1$ J mol^{-1} K^{-1} [2]; refractive index at 589 nm $n_\omega = 1.587$, $n_\epsilon = 1.336$ [3]. This modification forms trigonal crystals of space group D_{3d}^5 at a transition point of ca. 275 °C.

Sodium nitrate does not form hydrated solid phases (Fig. 1). The salt is hygroscopic and readily dissolves in water; at 25 °C, 92.1 g dissolves in 100 g of water [3]; enthalpy of solution at infinite dilution $\Delta H_{298} = +20.5$ kJ/mol [3]. Sodium nitrate is soluble in liquid ammonia and forms $NaNO_3 \cdot 4NH_3$ below −42 °C. The solubility of sodium nitrate in anhydrous methanol is 2.8 wt % at 25 °C [5].

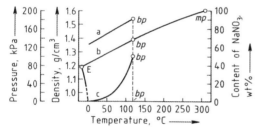

Figure 1. Properties of a saturated solution of sodium nitrate in water
a) Density; b) Content of sodium nitrate; c) Vapor pressure bp = boiling point of saturated solution containing 68.5 wt % sodium nitrate at 101.3 kPa and 119 °C; E = cryohydric point: −17.5 ± 0.5 °C, 38.0 % sodium nitrate.

Sodium nitrate is an oxidant and releases oxygen on heating:

$NaNO_3 \rightleftharpoons NaNO_2 + 1/2\,O_2 \quad \Delta H_{298} = +109\,\text{kJ/mol}$ (1.1)

Molten sodium nitrate remains stable in air up to 500 °C; reaction (1.1) reaches equilibrium with the evolution of oxygen between 600 and 750 °C [6]. The product (sodium nitrite) decomposes (see Section 4.1). Alkali nitrates contain small amounts of nitrites at temperatures as low as 300 °C [7]. In aqueous solution, alkali nitrates have little oxidizing power; they are reduced to ammonia by nascent hydrogen.

1.2. Occurrence

Deposits of sodium nitrate (the mineral nitratine) occur chiefly in arid regions such as Peru, Mexico, Egypt, and South Africa. The only commercial deposits are, however, in Chile. The origin of these deposits has not been determined conclusively. A number of theories have been proposed since the early 1960s [8 – 10].

The Chilean deposits lie between latitudes 19 ° and 26 ° at ca. 800 – 1600 m above sea level. The uppermost bed is 20 – 40 cm thick and consists of unconsolidated weathering products containing only a little salt ("chuca"). The next bed, up to 3 m thick, is called "costra"; it is usually a coarse breccia cemented with clay and salts of the saltpeter deposits. The "caliche" or salt bed proper lies beneath and has a thickness of 0.5 – 5.0 m. It is also brecciated and contains various salts (Table 1) and widely varying amounts of clay, sand, and gravel.

Table 1. Analyses of caliche samples (composition in wt %)

Salt	1	2	3	4	5
$NaNO_3$	34.2	34.4	43.3	28.5	53.5
KNO_3	1.6				17.3
Na_2SO_4	8.4	1.6	25.3	5.4	1.9
$CaSO_4$	6.3	1.6	30.9	2.7	0.5
$MgSO_4$	2.0	5.4		3.4	1.4
NaCl	32.0	4.0		17.2	21.3

In addition to nitratine ($NaNO_3$) and generally a little niter (KNO_3), the rock often contains a wide variety of components: halite (NaCl), gypsum ($CaSO_4 \cdot 2H_2O$), anhydrite ($CaSO_4$), bloedite ($Na_2SO_4 \cdot MgSO_4 \cdot 4H_2O$), glauberite ($Na_2SO_4 \cdot CaSO_4$), polyhalite ($K_2SO_4 \cdot MgSO_4 \cdot 2CaSO_4 \cdot 2H_2O$), and darapskite ($NaNO_3 \cdot Na_2SO_4 \cdot H_2O$). In addition to the salts listed in Table 1, borates, iodates, other iodine compounds, and perchlorates may be present in amounts up to a few tenths of a percent.

Reserves may well support several more centuries of working. In Chile alone, reserves of commercial caliche containing 7 wt % sodium nitrate are estimated at 2.5×10^9 t. Reserves of lower quality are about ten times greater [11].

1.3. Production

1.3.1. Chile Saltpeter

The winning of sodium nitrate from caliche is very complicated. The raw material contains only 7 – 40 wt % sodium nitrate (corresponding to 2 – 7 % N) and requires multistage processing to achieve an acceptable yield. The saltpeter industry in Chile produces ca. 600 000 t of sodium nitrate, 50 000 t of sodium sulfate, and more than 3000 t of iodine yearly [12]; it is one of the world's leading producers of iodine.

As early as 1810, low yields of impure saltpeter were produced by small boileries in Chile. The first improvement was the *Shanks hot-leaching process*, introduced in 1876 [13, 14]. Caliche with a minimum sodium nitrate content of 13 wt % was extracted with mother liquor at ca. 110 °C; after crystallization of sodium chloride at < 5 °C, the product was obtained by crystallization at 22 °C. The yield was 65 – 80 %, and the energy cost was high. The 200 or so plants employing the Shanks process ranged in capacity from 10 000 to 100 000 t/a of sodium nitrate. High energy costs made the Shanks process uncompetitive with synthetic saltpeter after World War I.

The *Guggenheim low-temperature leaching process* was developed in 1923 [8, 15]. By the beginning of the 1930s, two Guggenheim plants with sodium nitrate capacities of 500 000 and 600 000 t/a had been built. The Guggenheim process features relatively low leaching temperatures (40 – 45 °C), 90 % yields, low energy consumption due to good heat utilization, rational methods of caliche mining, and very large plant units. The process can economically handle caliche containing as little as 7 wt % sodium nitrate. Old spoil heaps of spent earth can be processed along with residues from leaching the coarse caliche fraction containing > 7 wt % sodium nitrate.

The isolation of sodium nitrate is based on differences between the solubility curves of the salts contained in the caliche. For example, the solubility of sodium nitrate is four times greater than the solubility of sodium chloride in water at 45 °C (Fig. 2). Operating conditions must be avoided under which insoluble or sparingly soluble double salts containing sodium nitrate are stable. For example, if the Mg^{2+} and Ca^{2+} concentration in the recycle liquor is sufficiently high, darapskite ($NaNO_3 \cdot Na_2SO_4 \cdot H_2O$) forms the insoluble "protective compounds" blödite ($Na_2SO_4 \cdot MgSO_4 \cdot 4 H_2O$) and glauberite ($Na_2SO_4 \cdot CaSO_4$). Darapskite is a sparingly soluble component that accounts for 30 % of the sodium nitrate in caliche and is stable below 55 °C.

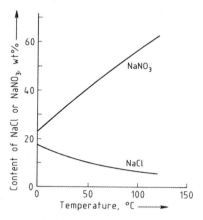

Figure 2. Solubility curves of sodium nitrate and sodium chloride in a saturated aqueous solution containing solid sodium nitrate and sodium chloride [16]

The production of Chile saltpeter from caliche is summarized in Figure 3. The warm (40 – 45 °C) mother liquor recycled to leaching contains ca. 330 g/L of sodium nitrate. The leaching process is performed in a series of vats, each having a capacity of ca. 1000 t of crushed ore. The underflow of each vat is heated before passing to the next one since the dissolution of sodium nitrate is endothermic. Sodium nitrate therefore dissolves preferentially. The product solution obtained after leaching contains 450 g/L of sodium nitrate and is cooled to 5 – 10 °C in heat exchangers and with ammonia refrigerating equipment. The mother liquor obtained after centrifugation of crystallized sodium nitrate is heated to 40 – 45 °C with heat from the condensers in the ammonia chilling loop and then with cooling water from the diesel engines used to generate electric power. Crystalline sodium nitrate obtained after centrifugation contains 3.0 – 3.5 % moisture; it is generally melted at ca. 340 °C and prilled. The granulated product contains only about 1% sodium chloride. Table 2 gives a detailed product analysis [17]. Large modern plants also produce coarse crystals of Chile saltpeter, which do not require further processing by melting.

Table 2. Analysis of saltpeter products

Component	Granulated sodium nitrate,* wt%	Granulated potassium nitrate,* wt%	Synthetic saltpeter, wt%
$NaNO_3$	98.32	67.37	99.70
KNO_3		30.36	0.06
NaCl	0.67	0.97	0.09
$Na_2B_4O_7$	0.24	0.23	
$NaIO_3$	0.03	0.03	
Moisture	0.17	0.21	0.10
Undetermined	0.57	0.83	0.05

* From Chile saltpeter [17].

To recover more than 80% of the sodium nitrate present in caliche, the "spent earth" containing 1.0 – 1.5 wt% sodium nitrate must be leached with water. The resulting dilute liquor is either used for leaching fine caliche (Fig. 3) or placed in large open pans and evaporated by the sun and the hot dry winds of the salt desert; precipitated crystals are removed periodically. Greater utilization of solar energy in the future will allow economical processing of caliche containing as little as 4% sodium nitrate.

Byproducts. Iodine recovery from the mother liquor begins when the sodium iodate content reaches 6 – 9 g/L. The enriched solution is treated with sodium hydrogen sulfite to precipitate free iodine:

$$2\,NaIO_3 + 5\,NaHSO_3 \rightarrow I_2 + 3\,NaHSO_4 + 2\,Na_2SO_4 + H_2O \quad (1.2)$$

Iodine is separated by filtration and purified by sublimation. The side stream after iodine recovery is returned to the circulating mother liquor.

The recovery of Chile niter (KNO_3) as a byproduct is profitable if the caliche contains > 2 – 3 wt% potassium nitrate. This product also accumulates in the mother liquor. To reduce processing costs, the product is refined to a potassium nitrate content of only 30% (Table 2).

1.3.2. Synthetic Sodium Nitrate

The most important method for producing synthetic sodium nitrate is the reaction of tail gases from nitric acid plants with sodium hydroxide or sodium carbonate solution. This method is, however, intended for tail-gas clean-up (removal of nitrogen oxides) rather than for nitrate and nitrite production.

Tail-gas absorption initially yields some sodium nitrate and larger amounts of sodium nitrite. The reactions that produce nitrite alone

$$2\,NaOH + NO_2 + NO \rightarrow 2\,NaNO_2 + H_2O \quad (1.3)$$

$$Na_2CO_3 + NO_2 + NO \rightarrow 2\,NaNO_2 + CO_2 \quad (1.4)$$

proceed faster than the reactions that produce both nitrite and nitrate.

$$2\,NaOH + 2\,NO_2 \rightarrow NaNO_3 + NaNO_2 + H_2O \quad (1.5)$$

$$Na_2CO_3 + 2\,NO_2 \rightarrow NaNO_3 + NaNO_2 + CO_2 \quad (1.6)$$

The countercurrent absorption process takes place in 18–8 stainless steel columns connected in series. In sodium carbonate absorption, the columns are packed with ceramic rings; iron rings are used for sodium hydroxide absorption.

Fresh alkaline solution is circulated into the last column. The liquor is forwarded batchwise to successive columns and finally withdrawn with a content of ca. 500 g/L of sodium nitrate and sodium nitrite. The product liquor contains both nitrate and nitrite, even if nitrite removal is employed (see Chap. 4). Nitrite is decomposed by inversion with nitric acid at 50 °C in a column made of 18–8 steel prior to nitrate recovery:

$$3\,NaNO_2 + 2\,HNO_3 \rightarrow 3\,NaNO_3 + 2\,NO + H_2O \quad (1.7)$$

The highly concentrated nitrogen monoxide can be used to synthesize other compounds (e.g., hydroxylamine) or can be returned to the nitric acid plant.

The residual nitrogen monoxide is stripped from the inverted solution with steam or air at ca. 100 °C. The solution is then neutralized with sodium carbonate or sodium hydroxide, passed through a precoat filter, and concentrated batchwise in multistage iron evaporators. The first stage is operated at 100 – 120 °C ($p = 120 - 150$ kPa in the vapor space). The resulting solution should not crystallize and should still dissolve any salt deposits formed in the second

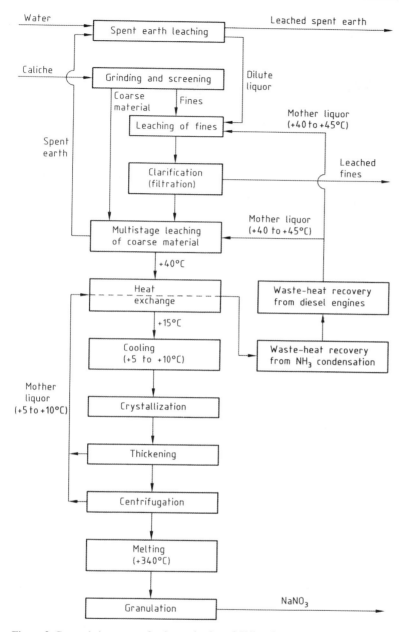

Figure 3. Guggenheim process for the production of Chile saltpeter

stage. The second stage is operated under vacuum ($p = 10 - 20$ kPa) at 65 – 75 °C; the product is concentrated to form a slurry containing ca. 20 % precipitated sodium nitrate. The crystals are separated in a centrifuge without cooling. Depending on the quality required, they are washed more or less intensively with cold water. Sodium nitrate from the centrifuge has a moisture content of 2 – 3 wt % and is dried to ca. 0.1 wt % moisture in drums heated with flue gas. The product is dried, cooled, and stored in bins.

The mother liquor from the centrifuge is recycled to the first evaporator stage along with the wash water. Discharge of the liquor is only necessary if the alkali used contains more chloride than permitted in sodium nitrate by quality standards. Accordingly, the process generally does not produce any wastewater.

1.4. Product Forms, Storage, and Transportation

Chile saltpeter is marketed as both a "granular" product (mainly used as a fertilizer) and a coarse crystalline product. Technical-grade synthetic sodium nitrate is a fine, crystalline white powder with a bulk density of about 1.36 kg/L. It is sold as an untreated product and also as a free-flowing product containing up to 0.1 % alkyl aryl sulfonate as an anticaking agent [18]. Table 2 presents analyses of sodium nitrate.

Sodium nitrate can be transported in bulk form. The technical grade, especially the untreated product, is generally shipped in polyethylene bags. As an oxidizing agent, sodium nitrate must not come in contact with readily oxidizable products during transport and storage.

1.5. Uses

Inorganic nitrates and nitrites overlap in their areas of application. Mixtures of the compounds are sometimes employed. More than half of the sodium nitrate produced worldwide (including Chile saltpeter) is used as a fertilizer for crops such as cotton, tobacco, and vegetables. In contrast to other nitrogen fertilizers, sodium nitrate does not overacidify the soil. In Europe and the United States, sodium nitrate is of minor importance compared to other fertilizers.

The major industrial use of sodium nitrate is in the explosives industry. It is employed along with ammonium nitrate in the production of water-containing slurry and gel explosives, whose consistency is stabilized with rubber additives. Sodium nitrate, like potassium nitrate, is added to ammonium nitrate melts (see Section 2.5). Sodium nitrate is also used in pyrotechnics, for example, in flare mixtures.

Large amounts of sodium nitrate are used in the glass and enamel industry as a refining agent for removing air bubbles from melts.

In metallurgy, molten salt mixtures containing sodium nitrate are employed as heat-transfer baths for quench hardening and tempering of steel, light alloys, and copper alloys. For the range 260 – 550 °C, $NaNO_3$ – KNO_3 melts are used; melts containing ca. 45 % sodium nitrite along with potassium nitrate and sodium nitrate can be used from 150 to 300 °C [19]. The heat of fusion of the salt mixture can be utilized to keep the temperature constant; the solidification point of the ternary system KNO_3 – $NaNO_3$ – $NaNO_2$ can be adjusted precisely by altering the proportion of components. Metal parts with oxide scales can be pretreated by briefly dipping them in molten $NaNO_3$ – $NaOH$ so that they can be electrically descaled in a sodium sulfate solution without the use of acid (more environmentally friendly); similar methods employ potassium nitrate.

Molten mixtures of sodium nitrate, other alkali nitrates, and nitrites function as heat-transfer and heat-storage media in solar technology. The maximum temperature for this application is ca. 600 °C. The thermophysical properties of such melts are described in [20 – 23].

Mixtures of sodium nitrate and borax also serve as an auxiliary in soldering and welding. Sodium nitrate is recommended as a corrosion-inhibiting additive in antifreezes and cooling brines. In the regeneration of spent sulfuric acid by distillation in cast-iron vessels, passivation with sodium nitrate is used to prevent corrosion.

Sodium nitrate is used to promote combustion; it is added to activated carbon when sulfur dioxide is oxidized to sulfur trioxide in exhaust-gas combustion devices and is also used to improve the combustibility of tobacco. Cleaning agents for plugged drain pipes are made from sodium nitrate, aluminum filings, and an alkali; hydrogen produced by the action of water is oxidized by sodium nitrate, and this reaction provides the heat needed for effective cleaning. Molten sodium nitrate serves as a reaction medium, for example, in the electrolytic production of chromium dioxide for magnetic tape. A melt of sodium nitrate and potassium nitrate provides a suitable medium for the thermal decomposition of dissolved ammonium nitrate to dinitrogen monoxide. The same melt can be used for the conversion of urea to cyanuric acid in a method devised by Stamicarbon. Pickling salt contains sodium nitrate.

Other applications of sodium nitrate include the production of dyes, pharmaceuticals, charcoal briquetts, and other nitrates.

1.6. Economic Aspects

The output of Chile saltpeter has decreased continually since the late 1920s (Table 3). Despite competition from synthetic sodium nitrate, output seems to have reached a stable level. In Chile, production capacity increased in 1984 [24], and another plant is planned [25]. The cost of producing Chile saltpeter is far less sensitive to rising energy prices than that of synthetic sodium nitrate, which requires ammonia and caustic soda as feedstocks. Most Chile saltpeter is exported; the most important consumers are Western Europe (120 000 t/a), especially Spain and The Netherlands, and the United States (100 000 t/a) [17].

Table 3. Production of Chile saltpeter (10^3 t)

Year	Production	Year	Production
1830	1	1940	1380
1850	30	1950	1660
1880	220	1960	930
1900	1650	1970	700
1910	2470	1975	800
1920	2530	1979	650
1928	3160	1984	710
1933	530	1987	590

Few production figures are available for synthetic sodium nitrate. Sodium nitrate has not been manufactured in the United States since 1988. Further stagnation of sodium nitrate production is expected in the near future. This forecast could change if solar energy development is stepped up [26].

2. Potassium Nitrate

Potassium nitrate was manufactured in China with the aid of bacteria as early as the Middle Ages. Up to the beginning of the 20th century, its only important use was in the production of gunpowder.

2.1. Properties [27]

Potassium nitrate [7757-79-1], saltpeter, KNO_3, M_r 101.10, has an enthalpy of formation ΔH_{298} = −492.8 kJ/mol [2], an enthalpy of fusion ΔH = +11.9 kJ/mol [3], and mp 334 °C; for pressure dependence of the melting point, see [4]. At 25 °C, anhydrous potassium nitrate forms colorless rhombic crystals of space group D_{2h}^{16}; ϱ_{298} = 2.109 g/cm^3; $C_{p,\,298}$ = 96.36 J mol^{-1} K^{-1} [2]; refractive index at 589 nm n_α = 1.335, n_β = 1.5056, n_γ = 1.5064 [2]. This modification forms rhombic crystals of space group D_{3d}^5 at a transition point of ca. 127.7 °C (enthalpy of formation ΔH = +5.1 kJ/mol) [28, 29]. High-pressure allotropes account for the caking tendency of potassium nitrate [30].

Potassium nitrate does not form solid phases containing water (Fig. 4). Its enthalpy of solution in water at infinite dilution is ΔH_{298} = +34.9 kJ/mol [3]. Potassium nitrate is soluble in liquid ammonia; no ammoniates are known. The solubility of potassium nitrate in anhydrous methanol at 25 °C is 0.34 wt % [5].

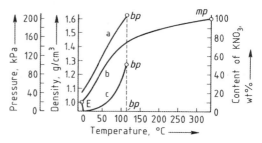

Figure 4. Properties of a saturated solution of potassium nitrate in water
a) Density; b) Content of potassium nitrate; c) Vapor pressure

bp = boiling point of saturated solution containing 77.0 wt % potassium nitrate at 101.3 kPa and 115.6 °C; E = cryohydric point: −2.85 °C, 9.66 % potassium nitrate.

Like sodium nitrate, potassium nitrate liberates oxygen on heating. The endothermic reaction

$$KNO_3 \rightleftharpoons KNO_2 + 1/2\,O_2 \quad \Delta H_{298} = +124\,\text{kJ/mol} \quad (2.1)$$

is reversible. In contact with air, pure molten potassium nitrate is stable up to about 530 °C; reaction (2.1) comes to equilibrium between 650 and 750 °C. Above 750 °C, the nitrite product decomposes with the formation of nitrogen oxides [6]. The $KNO_3 - NaNO_3$ system forms a eutectic with 45% sodium nitrate, *mp* ca. 218 °C. Solid solutions exist above 175 °C with no miscibility gap; at room temperature the system has a miscibility gap from 0.5 to 99.9 wt% sodium nitrate. The composition of the aqueous solution saturated with both salts is described in [31].

2.2. Occurrence

Potassium nitrate (in the form of the mineral niter) occurs as efflorescences on soils. Deposits in China and the East Indies were economically important; in the first half of the 19th century, more than 10 000 t/a of potassium nitrate were produced by leaching such soils (current annual production is still several thousand tonnes). Before the large-scale manufacture of converted saltpeter began, East Indian saltpeter enjoyed a virtual monopoly.

Potassium nitrate is present in tobacco plants, from which it can be extracted with water and recovered to diminish nitrogen oxide evolution during smoking.

2.3. Production

For older processes, see [14, 27].

2.3.1. Bacterial Production of Saltpeter

Bacterial production of potassium nitrate was employed from the 17th to the 19th century in Europe. This was necessary to avoid having to import this raw material, which was indispensable for military purposes during wartime. Nitrogen-rich organic wastes, especially animal feces and urine, were placed in loose, air-permeable piles of earth together with lime and potash; the nitrogen compounds were converted to nitrates by bacterial nitrification. After several years, the mass was leached with water. Potash was added to the crude liquor to convert calcium and magnesium nitrates to potassium nitrate and sparingly soluble calcium and magnesium carbonates. The filtered liquor was evaporated to obtain saltpeter, which was purified by recrystallization.

2.3.2. Converted Saltpeter

From the mid-19th century until the 1950s, the conversion of Chile saltpeter with potassium chloride was the most important potassium nitrate production process (Fig. 5). Potassium chloride was supplied in large quantities after 1860 by the German potash industry.

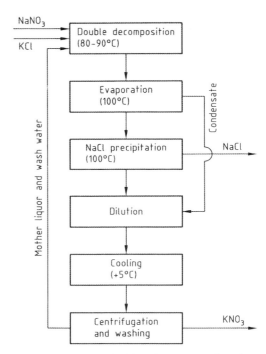

Figure 5. Conversion of sodium nitrate to potassium nitrate

$$NaNO_3 + KCl \rightleftharpoons KNO_3 + NaCl \qquad (2.2)$$

The process is based on the slight increase in solubility of sodium chloride with temperature. The mother liquor from potassium nitrate crystallization in the preceding cycle (point C in Fig. 6) is heated and reacted with stoichiometric quantities of sodium nitrate and potassium chloride. Impure starting materials are generally used in this step. The reaction mixture (point A, Fig. 6) is concentrated at 100 °C with the addition of sodium carbonate; sodium chloride and impurities (e.g., $MgCO_3$) precipitate and are filtered out. The filtrate (point B, Fig. 6) is again diluted with the evaporation condensate (to prevent sodium salts from precipitating on cooling), clarified, cooled to 5 °C to precipitate potassium nitrate, and centrifuged. The mother liquor (point C, Fig. 6) is recycled. The potassium nitrate product is recrystallized for technical purposes. Because of high energy costs and other factors, the conversion process can no longer compete with the processes described in Sections 2.3.3, 2.3.4 and 2.3.5; nevertheless, older plants continue to use it. Physical and chemical fundamentals are described in [32 – 35]; process design may be found in [36, 37].

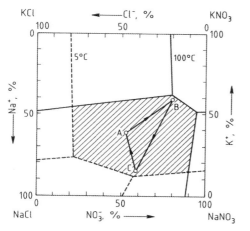

Figure 6. The $NaNO_3$ – KNO_3 – KCl – $NaCl$ – H_2O system

Jänecke's method [34] is used to plot the relative content of cations and anions in percent equivalents in the salt mixture corresponding to the saturated solution. Absolute concentrations or quantities of water are not shown. Values for 5 °C are according to [32], for 100 °C according to [33].
The lines labeled 5 and 100 °C represent the boundaries of fields in each of which one of the four salts occurs as the solid phase in the saturated solution. In the shaded region, sodium chloride precipitates from the saturated solution at 100 °C but potassium nitrate precipitates at 5 °C. (See text for explanation of the working triangle ABC.)

2.3.3. Potassium Nitrate from Calcium Nitrate

A number of processes have been developed for obtaining potassium nitrate from calcium nitrate, which is a byproduct of other processes (e.g., digestion of rock phosphate with nitric acid).

With potassium sulfate as a starting product, the reaction

$$Ca(NO_3)_2 + K_2SO_4 + 2H_2O \rightarrow 2KNO_3 + CaSO_4 \cdot 2H_2O \quad (2.3)$$

yields sparingly soluble gypsum, which is easily separated. This is the basis for the relatively simple Victor process [38], which was employed after 1951 in Germany for several decades to manufacture potassium nitrate [39]. Other methods employ potassium chloride as raw material but have no industrial importance [39].

In another proposed process, limestone is first reacted with nitric acid to form calcium nitrate. Potassium sulfate is then added to the resulting liquor at ca. 100 °C. The gypsum formed is removed and the reaction mixture is cooled. Potassium nitrate crystals are isolated with a purity of 99.5 %, and the mother liquor is recycled to the beginning of the process [40].

A process based on the use of calcium nitrate and ion exchangers has also been proposed [41].

2.3.4. Potassium Nitrate from Ammonium Nitrate

Potassium nitrate is made from ammonium nitrate (available in high purity) and potassium chloride by double decomposition in aqueous solution:

$$NH_4NO_3 + KCl \rightarrow KNO_3 + NH_4Cl \quad (2.4)$$

The ammonium chloride byproduct is easily separated.

The solubility relationships (Fig. 7) in the system of reciprocal salt pairs described in reaction (2.4) [42] lead to a simple reaction design (Fig. 8) [43]. The starting reagents are added in alternate stages. Ammonium chloride precipitates at a higher temperature and from a higher concentration than potassium nitrate. Evaporation is therefore necessary prior to ammonium chloride crystallization, the condensate being recycled before potassium nitrate crystallization. Because the salt solution is aggressive, molybdenum-alloyed carbon steel is used for process equipment. Fertilizer-grade product containing 95 wt % potassium nitrate and 5 wt % ammonium nitrate is manufactured by this process [24, 44].

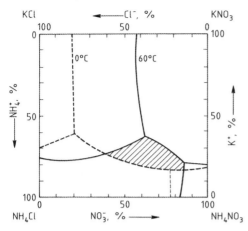

Figure 7. The $NH_4NO_3 - KNO_3 - KCl - NH_4Cl - H_2O$ system (see Fig. 6 for the method of plotting) In the shaded region, ammonium chloride precipitates from the saturated solution at 60 °C but potassium nitrate precipitates at 0 °C. Absolute concentrations are not shown (see [41]).

2.3.5. Potassium Nitrate from Potassium Chloride and Nitric Acid

Production without Chlorine Formation. In an aqueous medium containing 50 % nitric acid, the reaction

$$HNO_3 + KCl \rightleftharpoons KNO_3 + HCl \quad (2.5)$$

proceeds without oxidation of chloride ion to chlorine and/or nitrosyl chloride only if the temperature is below 40 °C (or below 70 °C at 40 % HNO_3). A number of process design options are available: removing hydrogen chloride by absorption, decreasing potassium nitrate solubility by the addition of alcohol, or using cation exchangers. Only the Israel Mining Industry (IMI) process has been introduced on a large scale (Fig. 9) [45]. Solid potassium chloride is reacted with nitric acid at 5 – 10 °C (c) in the presence of an organic solvent (n-butanol or isopentanol). The solvent extracts the hydrogen chloride, and potassium nitrate is precipitated quantitatively from the aqueous phase. The advantage of the low digestion temperature is that hydrogen chloride is extracted from the organic solvent in preference to nitric acid; furthermore, the aqueous solution contains less potassium nitrate. The three phases present (potassium nitrate, aqueous phase, organic phase) are then separated. The

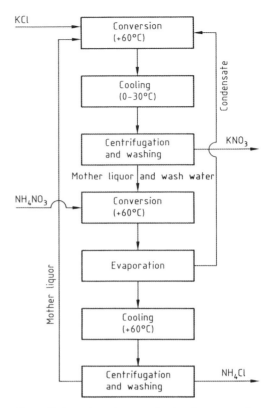

Figure 8. Potassium nitrate production from ammonium nitrate

potassium nitrate is centrifuged and washed with water (e). The aqueous phase is returned to the reactors (c). The purity of the product (> 95 %) [46] is sufficient for use as fertilizer; recrystallization is necessary for the technical grade. Hydrogen chloride is separated as a 22 % aqueous solution from the alcohol (which is not miscible with water in all proportions) by countercurrent extraction with water in three mixer – settler units (f). After extraction, the alcohol, which contains nitric acid, is returned to the conversion reactor. Separation of hydrochloric acid from the solvent has been improved by use of a different solvent that extracts nitric acid preferentially and extracts virtually no hydrochloric acid [47]. Haifa Chemicals uses tributyl phosphate for this purpose. Another proposal for improving the process is to replace the alcohol by an organic acid and an amine [48]. Rigid poly(vinyl chloride) and titanium are used as construction

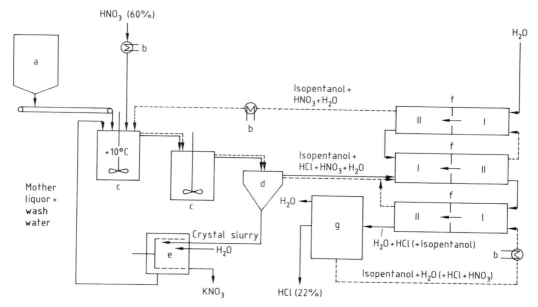

Figure 9. Simplified flow sheet of the IMI process (based on literature data)
a) Potassium chloride supply; b) Cooler; c) Reactor; d) Decanter; e) Centrifuge; f) Extractor (I = mixer, II = separator); g) Multistage evaporator (details not published)
Dashed lines indicate streams containing mainly isopentanol.

materials, with graphite for distillation equipment. A substantial portion of investment is for solvent conditioning.

This process is employed in two potassium nitrate plants in Haifa. Potassium nitrate output in 1987 was 250 000 t, of which ca. 30 000 t was of technical purity and the remainder fertilizer grade [49]. The hydrochloric acid byproduct is utilized for the digestion of rock phosphate.

Production with Chlorine Formation. The South West Potash Corporation (SWP) process is employed on a large scale for the production of potassium nitrate and chlorine [44, 50]. The most important steps follow (Fig. 10) [14]:

1) Potassium chloride is reacted with 65 % nitric acid in an autoclave at 75 °C and ca. 200 kPa to yield potassium nitrate, chlorine, and nitrosyl chloride.

$$4\,HNO_3 + 3\,KCl \rightarrow 3\,KNO_3 + Cl_2 + NOCl + 2\,H_2O \quad (2.6)$$

2) Nitrosyl chloride is oxidized with 80 % nitric acid to yield chlorine and nitrogen dioxide.

$$NOCl + 2\,HNO_3 \rightarrow 1/2\,Cl_2 + 3\,NO_2 + H_2O \quad (2.7)$$

3) Nitrogen dioxide is oxidized with oxygen and absorbed in water to yield 65 % nitric acid, which is recycled to the potassium chloride conversion step.

$$4\,NO_2 + O_2 + 2\,H_2O \rightarrow 4\,HNO_3 \quad (2.8)$$

(Alternatively, nitrogen dioxide can be withdrawn as end product.)

4) The solution of potassium nitrate and nitric acid is concentrated to > 80 % nitric acid by crystallization of potassium nitrate in a vacuum crystallizer. Nitric acid can be obtained in 75 % or 85 % concentration because the azeotropic concentration in the $HNO_3 - H_2O$ system at 80 kPa increases to 82 % nitric acid if the liquid phase contains 40 % potassium nitrate and to 87 % nitric acid at 50 % potassium nitrate [51].

5) The potassium nitrate is centrifuged, dried, melted at 340 – 350 °C, and prilled.

Intense corrosion dictates the use of titanium reactors with acid-resistant linings, whereas tower internals are made of glass or Teflon.

At present, Cedar Chemical Corp. (Trans-Resources) has an annual capacity of 100 000 – 120 000 t of potassium nitrate in an SWP plant at Vicksburg, Mississippi; ca. 20 000 t/a of the output is technical grade. Chlorine, dinitrogen tetroxide, and 65 % nitric acid are obtained as byproducts.

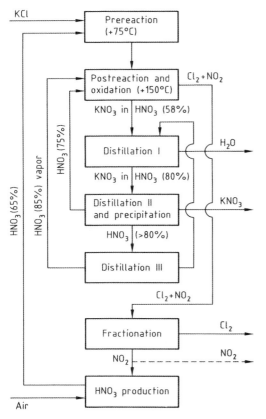

Figure 10. South West Potash Corporation process

2.4. Product Forms, Storage, and Transportation

All technical-grade and most fertilizer-grade potassium nitrate is sold as a fine crystalline product with a bulk density of 1.1 kg/L. Technical grades may be untreated or treated with fatty amines or stearates to prevent caking [18]. Fertilizer-grade potassium nitrate is also sold as prills (bulk density 1.3 kg/L, obtained by atomizing a melt at 340 °C), because prills have better storage qualities and are easier to apply as fertilizer. Table 4 gives the analyses of typical grades. As a rule, the technical product is shipped in plastic or multi-ply paper bags. As an oxidant, potassium nitrate must not be stored or transported with easily oxidized products.

2.5. Uses

About 75 % of potassium nitrate is manufactured with a purity of only 90 % for use as a nonhygroscopic fertilizer. The absence of chlorine is advantageous for growers of tobacco, citrus fruits, and vegetables (especially for hothouse crops). Potassium nitrate can also be used in the production of clear liquid fertilizers and is an important constituent of multinutrient fertilizers.

Almost half of the technical-grade product is employed in metallurgy, mainly in heat-transfer baths in mixtures with other nitrates and nitrites (see Section 1.5). Potassium nitrate is also a component of soldering fluxes and welding electrodes. Large amounts are used in the glass, enamel, and ceramics industries. Glass devices and optical lenses can be improved by techniques based on $Na^+ - K^+$ ion exchange (treatment of sodium-containing glasses in molten KNO_3).

About 10 – 20 % of technical-grade potassium nitrate is consumed in the manufacture of explosives and pyrotechnics. Along with sodium nitrate and ammonium nitrate, potassium nitrate is an oxygen supplier in safety explosives. Because of its low hygroscopicity, potassium nitrate remains an important raw material for the production of matches and primer cords, as well as primer compositions with controlled burning rate for local heating. Another important use in the explosives industry is the addition of 2 – 10 % potassium nitrate to ammonium nitrate to lower the melting point, which facilitates prilling and improves the stability of the resulting granules.

Small amounts of potassium nitrate are employed in food preservation, cheese processing, and for improving the quality of tobacco.

Because of its oxidizing properties, potassium nitrate can be used for desooting in combustion processes. It is also recommended as an agent for controlling noxious odors during the application of sewage sludge or manure (at the same time, it improves the fertilizing action of the waste).

Potassium nitrate has been proposed as an oxidizing component in an acid-based gas gen-

Table 4. Composition (wt %) of potassium nitrate

Component	Technical grade		Fertilizer grade	
	Requirements for black powder *	Sample	Product I	Product II
KNO_3	min. 99.0	99.8	95.0	96.5
Cl^-	max. 0.03	0.015	0.2	1.8
NO_2^-	not specified		0.002	0.2
SO_4^{2-}	not specified	0.02	3.4	0.05
Na^+	max. 0.1	0.02	0.2	0.7
Ca^{2+} and Mg^{2+}	max. 0.07	0.03	1.0	0.1
Heavy metals	max. 0.1		0.02	0.02
Insolubles	max. 0.1	0.05	0.05	0.2
H_2O	max. 0.25	0.06	0.1	0.1

* Not all requirements are shown.

erator system for the rapid inflation of air bags used to prevent injuries in motor vehicle accidents.

Potassium nitrate is also employed as an oxidant in chemical syntheses (e.g., alizarin synthesis).

2.6. Economic Aspects

Except for the small output from deposits still worked in India and the production of Chile saltpeter (which contains only 20 – 30 % KNO_3, see Section 1.3.1), the only source of potassium nitrate for more than 100 years has been synthetic production. Although use of the technical-grade product can be expanded only slightly, a variety of processes have been developed to reduce the cost of manufacturing impure saltpeter and to increase its use as a fertilizer.

Table 5. Potassium nitrate capacities in 1987 (10^3 t)

Country	Fertilizer grade	Technical grade
Israel	250	30
United States	80	20
FRG	0	30
Other West European countries	20	20
Eastern bloc	*	30
India	*	*
Chile (natural saltpeter)	60 – 75	0

* Capacity unknown.

World production capacity in 1987 for synthetic potassium nitrate is about 425 000 t/a, 75 % of which is fertilizer grade and only 25 % technical grade (see Table 5). The largest producer is Trans-Resources with 340 000 t/a (Cedar Chemical and Haifa Chemicals, Israel) [52]. Israel is the largest exporter; its most important customers are the United States, Western Europe, and Latin America. By 1992, Haifa Chemicals will boost its capacity to 300 000 t/a [53]. The annual output of Chile saltpeter is 60 000 – 75 000 t of potassium nitrate. Soquimich, the sole manufacturer in Chile, is the third largest potassium nitrate producer in the world [54].

3. Calcium Nitrate

3.1. Properties [55]

Calcium nitrate [*10124-37-5*], $Ca(NO_3)_2$, M_r 164.09, has an enthalpy of formation of ΔH_{298} = −937.4 kJ/mol [2]. Anhydrous calcium nitrate crystallizes in the cubic system, space group T_h^6; ϱ_{298} = 2.504 g/cm^3 [3]; $C_{p,\,298}$ = 149.5 Jmol^{-1} K^{-1} [2]; refractive index n = 1.595 at 589 nm; mp 561 °C (incipient decomposition).

Anhydrous calcium nitrate is highly hygroscopic and readily soluble in water when heated (enthalpy of solution ΔH_{298} = −18.7 kJ/mol at infinite dilution [56]); see also Figure 11 and Table 6. In addition to the anhydrous salt, three hydrates are known [57, 58]. The tetrahydrate [*13477-34-4*], $Ca(NO_3)_2 \cdot 4\,H_2O$, is the most important industrially: enthalpy of formation ΔH_{298} =+2130 kJ/mol; monoclinic crystals of space group C_{2h}^5; ϱ_{298} = 1.896 g/cm^3; refractive index at 589 nm n_α = 1.465, n_β = 1.498, n_γ = 1.504. The tetrahydrate dissolves in water with strong cooling (enthalpy of solution ΔH_{292} = + 32.7 kJ/mol at infinite dilution).

Calcium nitrate dissolves readily in liquid ammonia, methanol, and ethanol. A number of ammoniates are known.

Table 6. Data for the system Ca(NO$_3$)$_2$ – H$_2$O (see also Fig. 11)

Transition point and temperature, °C		Content of Ca(NO$_3$)$_2$, wt %	Range	Stable hydrate formula	mp, °C
E*	−28.7	42.9			
			E – A	Ca(NO$_3$)$_2$ · 4H$_2$O	42.7
A	+42.6	70.8			
			A – B	Ca(NO$_3$)$_2$ · 3H$_2$O	51.1
B	+50.6	77.2			
			B – C	Ca(NO$_3$)$_2$ · 2H$_2$O	
C	+51.6	78.1			
			above C	Ca(NO$_3$)$_2$	561
bp** +151		79.0			

Water vapor pressure over saturated solution with solid Ca(NO$_3$)$_2$ · 4H$_2$O					
Temperature, °C	0	10	20	30	40
Pressure, kPa	0.36	0.69	1.25	1.99	2.63

* E = cryohydric point.
** Boiling point (101.3 kPa) of the saturated solution.

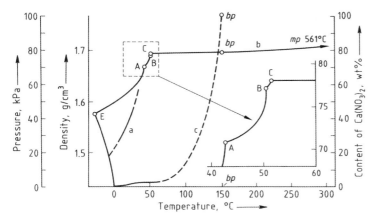

Figure 11. Properties of a saturated solution of calcium nitrate in water (see Table 6 for explanations)
a) Density; b) Content of calcium nitrate; c) Vapor pressure

Thermal decomposition of calcium nitrate

$$Ca(NO_3)_2 \rightarrow CaO + 2NO_2 + 1/2 O_2$$
$$\Delta H = +369 \text{ kJ/mol} \quad (3.1)$$

begins at 500 °C [6]. For the reaction mechanism, see [59].

Double Salts. Several weakly hygroscopic double salts are formed with ammonium nitrate [55]. The double salt NH$_4$NO$_3$ · 5Ca(NO$_3$)$_2$ · 10H$_2$O has a high melting point (102 °C) and is important in the production of granular calcium nitrate. The double salt with urea, Ca(NO$_3$)$_2$ · 4CO(NH$_2$)$_2$, is nonhygroscopic and can be employed as a fertilizer. The system Ca(NO$_3$)$_2$ – KNO$_3$ – H$_2$O includes the phase Ca(NO$_3$)$_2$ · 4KNO$_3$ [60].

3.2. Occurrence

Only small amounts of Ca(NO$_3$)$_2$ · 4H$_2$O occur (nitrocalcite) as efflorescences in limestone caves, for example, in Kentucky.

3.3. Production

Calcium nitrate became the first synthetic nitrogen fertilizer under the name Norge saltpeter. It was made by dissolving limestone in nitric acid and neutralizing the resulting liquor with powdered limestone or lime. This process is still in use, but ammonia is now employed for neutralization.

Calcium nitrate is also an important byproduct of nitric acid digestion of rock phosphates by the Odda process (→ Phosphate Fertilizers). It is either used directly for fertilizer production or converted to ammonium nitrate.

A calcium nitrate solution is also obtained when tail gases from nitric acid plants are absorbed in milk of lime. The initial nitrate – nitrite solution is converted to pure calcium nitrate solution by a reaction analogous to reaction (1.7).

3.3.1. Calcium Nitrate from Limestone and Nitric Acid

In this process, limestone is mixed with 50 – 60 % nitric acid (Fig. 12). Calcium nitrate is formed by the reaction

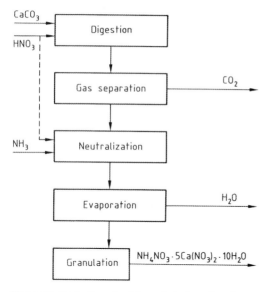

Figure 12. Production of calcium nitrate from limestone

$$CaCO_3 + 2\, HNO_3 \rightleftharpoons Ca(NO_3)_2 + H_2O + CO_2 \qquad (3.2)$$

Water vapor and carbon dioxide are withdrawn from the process. Any remaining free nitric acid is then neutralized with ammonia to give ammonium nitrate; further ammonium nitrate can be added to adjust the content of this compound and to improve the storage qualities of the product. The calcium nitrate solution is then evaporated. Solid calcium nitrate can be subjected to any of the usual fertilizer processes (e.g., granulation, flocculation, or prilling).

The end product contains ca. 15.5 wt % nitrogen (0.9 % ammonium N, 14.6 % nitrate N). The main constituent (63 %) is the easily crystallized phase $NH_4NO_3 \cdot 5\,Ca(NO_3)_2 \cdot 10\,H_2O$; also present are 12 % $Ca(NO_3)_2 \cdot 2\,H_2O$ and 25 % calcium nitrate.

3.3.2. Calcium Nitrate as a Byproduct of the Odda Process

The Odda process is based on the digestion of rock phosphate with nitric acid to give phosphoric acid and calcium nitrate (Fig. 13) [61, 62]:

$$Ca_3(PO_4)_2 + 6\, HNO_3 \rightarrow 2\, H_3PO_4 + 3\, Ca(NO_3)_2 \qquad (3.3)$$

The Ca^{2+} in the digester liquor is treated as a ballast. The $CaO : P_2O_5$ ratio serves as a measure of the ballast content and is decreased from 1.5 to 0.3 by removing the sand, with subsequent crystallization and removal of calcium nitrate. The remaining liquor (nitrophosphoric acid) is neutralized with ammonia, evaporated, granulated, classified, and conditioned (coated).

Digestion of Rock Phosphate. The digestion of rock phosphate with nitric acid is an exothermic reaction and proceeds as follows at ca. 70 °C:

$$Ca_5[(PO_4)_3F] + 10\, HNO_3 \rightarrow 5\, Ca(NO_3)_2 + HF + 3\, H_3PO_4 \qquad (3.4)$$

Usually 50 – 60 % nitric acid is employed in an excess up to 20 % over the stoichiometric amount, depending on the rock phosphate grade and process.

The digestion reaction is governed by three interrelated factors: digestion temperature, particle size, and residence time in the reactor. If the temperature is too high, nitric acid may decompose to form nitrogen oxides. If the temperature is too low and the particles are too large, a longer residence time is necessary.

Crystallization of Calcium Nitrate. The $CaO : P_2O_5$ ratio of the digester liquor is lowered by crystallizing and removing calcium nitrate tetrahydrate $Ca(NO_3)_2 \cdot 4\,H_2O$ to enhance the water solubility of phosphorus pentoxide in

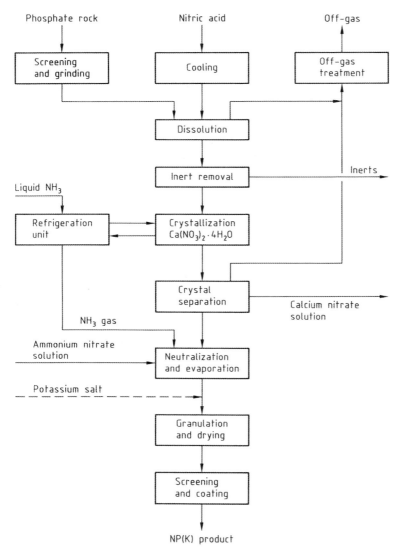

Figure 13. The Odda process

the fertilizer. The quantity of calcium nitrate tetrahydrate obtained depends on the crystallization temperature, which is usually below 0 °C. The tetrahydrate crystals are washed with cold nitric acid and calcium nitrate – ammonium nitrate solution to recover as much of the phosphorus pentoxide content as possible. The crystals should contain less than 0.3 wt % residual phosphorus pentoxide. Experiments have shown that ca. 60 wt % nitric acid is suitable for washing. The acid is precooled to $< 10\,°C$, sometimes as low as $-5\,°C$, to minimize the risk of redissolving the crystals.

Any acid remaining in the crystal filter cake would interfere with the subsequent conversion of calcium nitrate tetrahydrate. It is therefore eliminated by treatment with a solution of calcium nitrate and ammonium nitrate.

The tetrahydrate crystals are redissolved in ammonium nitrate solution. The cooling effect produced in this endothermic process and that resulting from the evaporation of ammonia are used to cool the crystallization liquor.

Either the calcium nitrate – ammonium nitrate solution is processed directly to fertilizers (see Section 3.3.1), or the calcium nitrate is treated with ammonium carbonate to yield ammonium nitrate.

Conversion of Calcium Nitrate. Figure 14 gives the flow sheet for calcium nitrate conversion. Ammonium carbonate is obtained by reacting ammonia and carbon dioxide in a 65 % ammonium nitrate solution in an absorption tower:

$$2\,NH_3 + CO_2 + H_2O \rightarrow (NH_4)_2CO_3 \qquad (3.5)$$

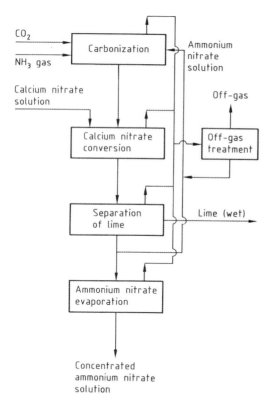

Figure 14. Calcium nitrate conversion

The reaction occurs in the liquid phase and the product contains < 38 wt % ammonium carbonate.

Calcium nitrate and ammonium carbonate are then reacted at ca. 50 °C in the conversion reactor:

$$(NH_4)_2CO_3 + Ca(NO_3)_2 \rightarrow CaCO_3 + 2\,NH_4NO_3 \qquad (3.6)$$

The reactor operates with a small excess of ammonium carbonate. If the excess is too great, the precipitated lime becomes too fine and is difficult to process. With an excess of calcium nitrate, the precipitate becomes slimy and cannot be filtered. Another reason for avoiding an excess of calcium nitrate is to keep the calcium ammonium nitrate obtained by reacting lime with ammonium nitrate from becoming hygroscopic. Calcium ammonium nitrate, with lime as a filler, contains ca. 26 wt % nitrogen.

3.4. Product Forms, Storage, and Transportation

Calcium nitrate for fertilizer use is marketed as a noncaking grade that is usually treated with aryl alkyl sulfonates. Airtight bins are required for the storage of hygroscopic calcium nitrate. Coated or plastic bags are suitable for transportation. Like other nitrates, calcium nitrate must be kept away from oxidizable substances.

3.5. Uses

Calcium nitrate is employed as a nitrogen fertilizer either directly (see Section 3.3.1) or indirectly as calcium ammonium nitrate (see Section 3.3.2).

The combined use of calcium nitrate and urea in fertilizers has also been proposed [63, 64]. Other nitrates are obtained by reacting calcium nitrate with salts of other cations, preferably sulfates (see also Section 2.3.3).

Calcium nitrate is employed as a granulation aid for ground basic slag, ammonium and potassium sulfates, ammonium nitrate, and other fertilizer salts. Because of its high solubility, it is used as an oxidizing additive in water-containing explosives (along with ammonium nitrate).

Addition of calcium nitrate has been suggested as a way of improving the combustion qualities of heating oil and tobacco. It is also used in the preparation of cooling brines and in wastewater treatment (see Chap. 9). Because of its highly endothermic heat of solution, the tetrahydrate $Ca(NO_3)_2 \cdot 4H_2O$ can be used for cooling, for example in medical applications.

3.6. Economic Aspects

The use of calcium nitrate as a fertilizer is declining worldwide. In 1978, only 170 000 t of nitrogen was consumed in this form, most of it in Scandinavia (where the important producers are located), followed by Egypt. No statistics are available on the use of calcium nitrate in the industrial sector.

4. Sodium Nitrite

Sodium nitrite is the most industrially important salt of nitrous acid.

4.1. Properties [65]

Sodium nitrite [7632-00-0], $NaNO_2$, M_r 69.00, has an enthalpy of formation of $\Delta H_{298} = -358 \pm 2$ kJ/mol. The β allotrope is stable at 25 °C and in the pure state forms colorless rhombic crystals of space group C_{2v}^{20}; $\varrho_{298} = 2.168$ g/cm^3 [3]; $C_{p,298} = 63.64$ J mol^{-1} K^{-1}; refractive index at 589 nm $n_\alpha = 1.35$, $n_\beta = 1.46$, $n_\gamma = 1.65$. At 163 °C, this modification adapts a rhombic form of type D_{2h}^{25} (enthalpy of crystallization +1.2 kJ/mol) [66], mp 271 °C.

Sodium nitrite is hygroscopic (Fig. 15) [65]. The anhydrous salt has an enthalpy of solution of $\Delta H_{298} = +13.9$ kJ/mol at infinite dilution [3]. The hydrate $NaNO_2 \cdot 0.5 H_2O$ exists below -5 °C. Sodium nitrite is soluble in liquid ammonia; $2NaNO_2 \cdot NH_3$ forms below -64 °C. The solubility of sodium nitrite in 95 % ethanol at 25 °C is ca. 1.4 %.

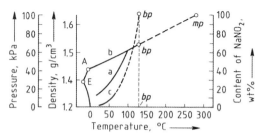

Figure 15. Properties of a saturated solution of sodium nitrite in water
a) Density; b) Content of sodium nitrite; c) Vapor pressure
bp = boiling point of saturated solution containing 68.7 wt % sodium nitrite at 101.3 kPa and 128 °C; A = transition point: -5.1 °C, 41.6 % sodium nitrite; E = cryohydric point: -19.0 °C, 28.6 % sodium nitrite.

Sodium nitrite oxidizes slowly in air. Above 330 °C it decomposes:

$$2\,NaNO_2 \rightarrow Na_2O + NO + NO_2 \qquad (4.1)$$

If the resulting nitrogen oxides are not removed, they react further with sodium nitrite:

$$NaNO_2 + NO_2 \rightleftharpoons NaNO_3 + NO \qquad (4.2)$$

$$2\,NaNO_2 + 2\,NO \rightarrow 2\,NaNO_3 + N_2 \qquad (4.3)$$

so that the end products of decomposition between 330 and 400 °C are sodium nitrate, sodium oxide, and nitrogen. Above 600 °C, the sodium nitrate product decomposes to oxygen and sodium nitrite (reaction 1.1), so that the end products at this temperature are sodium oxide, nitrogen, and oxygen [6].

Aqueous solutions of alkali nitrites reduce strong oxidizing agents (e.g., potassium permanganate solution), but in acidic solution they oxidize potassium iodide to iodine and Fe^{2+} to Fe^{3+}. They oxidize ammonium salts in weakly acidic solution to nitrogen above 50 °C. Urea and amidosulfonic acid are likewise oxidized to nitrogen:

$$CO(NH_2)_2 + 2\,NaNO_2 + 2\,HCl \rightarrow CO_2 + 2\,N_2 + 3\,H_2O$$
$$+ 2\,NaCl \qquad (4.4)$$

$$SO_3H{-}NH_2 + NaNO_2 + HCl \rightarrow N_2 + H_2O + H_2SO_4$$
$$+ NaCl \qquad (4.5)$$

In anhydrous melts at 210–220 °C, excess sodium nitrite oxidizes urea in an exothermic reaction that yields sodium carbonate and nitrogen:

$$CO(NH_2)_2 + 2\,NaNO_2 \rightarrow Na_2CO_3 + 2\,N_2 + 2\,H_2O \qquad (4.6)$$

Equimolar quantities under the same conditions form sodium cyanate:

$$CO(NH_2)_2 + NaNO_2 \rightarrow NaCNO + N_2 + 2\,H_2O \qquad (4.7)$$

The $NaNO_2$ – $NaNO_3$ system forms a eutectic at 37.5 mol % sodium nitrate, mp 227 °C. Mixed crystals precipitate from the melt at all concentrations [67].

Nitrites do not occur in mineral form. Nitrite is formed as a metabolic product by microorganisms that oxidize organic nitrogen-containing substances; small amounts of nitrite are thus found in soil and groundwater.

4.2. Production

Until the beginning of the 20th century, sodium nitrite was produced by the reduction of sodium nitrate with lead, iron, zinc, manganese dioxide [which is oxidized to $BaMnO_4$ in the presence of $Ba(OH)_2$], coke, sulfur dioxide, and calcium carbide. Of these processes, the only one of industrial significance was the reduction of molten sodium nitrate with lead at ca. 400 °C. Other earlier processes are described in [14, 68].

Large-scale production of sodium nitrite is now based on the reaction of nitrogen oxides with sodium carbonate or sodium hydroxide solution. The molar ratio of nitrogen oxides is called the "oxidation degree" (OD):

$$\frac{[NO_2]}{[NO]+[NO_2]} \cdot 100 = OD$$

At OD < 50 %, the reaction is described by

$$2\,NaOH + NO_2 + NO \rightarrow 2\,NaNO_2 + H_2O \quad (4.8)$$

Excess nitrogen monoxide does not react but is oxidized to nitrogen dioxide by ambient oxygen. At OD > 50 % and a high partial pressure of oxygen, the following (slower) reactions also take place:

$$2\,NaOH + 2\,NO_2 \rightarrow NaNO_3 + NaNO_2 + H_2O \quad (4.9)$$

$$2\,NaOH + 3\,NO_2 \rightarrow 2\,NaNO_3 + NO + H_2O \quad (4.10)$$

The reaction with sodium carbonate proceeds analogously. In practice, nitrate formation according to reactions (4.9) and (4.10) cannot be completely prevented even at OD < 50 %.

To obtain nitrite solutions with a minimum nitrate content, the OD must be kept below 50 % [69] and the pH above 8. In an acidic environment, nitrous acid is formed and decomposes to form nitric acid:

$$3\,HNO_2 \rightarrow HNO_3 + 2\,NO + H_2O \quad (4.11)$$

If a pure nitrite solution is desired, the gas leaving the absorption columns of the nitric acid plant must first be freed of entrained nitric acid. Because nitrogen dioxide is more thoroughly and rapidly absorbed in the columns than nitrogen monoxide, the gas mixture subjected to alkaline absorption has OD < 50 % and contains 3 – 4 vol % oxygen. Alkaline absorption is described in Section 1.3.2 [70].

Nitrogen oxides produced by catalytic combustion of ammonia can also be used directly for the production of sodium nitrite. If the oxygen excess is kept small and an intermediate "acid washing" step is used, the OD can be kept lower than 50 %.

4.3. Product Forms, Storage, and Transportation

Sodium nitrite is sold as the salt and in solution. The finely crystalline, slightly yellowish salt is marketed in untreated form and also after treatment with aryl alkyl sulfonates [18]. The salt contains ca. 99.0 % sodium nitrite, 0.6 % sodium nitrate, < 0.1 % sodium chloride and sodium sulfate, and < 0.1 % water.

Because of its hygroscopic nature and toxicity, sodium nitrite can be transported only in packaged form (plastic bags and metal tanks). It must be stored and shipped separately from oxidizable substances, ammonium salts, urea, food, and animal feeds. Sodium nitrite solution is marketed at a density of about 1.3 kg/L; it contains ca. 500 g/L sodium nitrite, 10 – 25 g/L sodium nitrate, and 5 g/L sodium hydrogen carbonate and has a pH of ca. 8. The solution can be stored in iron tanks.

4.4. Uses

Large amounts of sodium nitrite are consumed in the chemical and pharmaceutical industries for the production of nitroso and isonitroso compounds, diazotization reactions (especially for dyes), and the synthesis of pharmaceuticals (e.g., caffeine) and agricultural pesticides (e.g., pyramin).

Other applications of sodium nitrite are in metallurgy and corrosion prevention. It is used as an accelerator in phosphating, added to alkaline pickling solutions for aluminum, applied to steel as a descaling agent, added to drilling oils and abrasives, used for passivating metal surfaces and for creating protective coatings [e.g., of the type $Fe(NH_3)_x(NH_2)_3 \cdot yH_2O$] on carbon steels [71], and added as a corrosion inhibitor to lubricating oils and greases as well as to cooling water loops. Sodium nitrite serves as an oxidizing agent in the alkaline detinning of tinplate scrap [72] and a corrosion inhibitor in heat-transfer baths (see Section 1.5).

Sodium nitrite is used in lubricants for glass-forming equipment [73]. Like sodium nitrate and potassium nitrate, it is used as a heat-transfer medium.

Sodium nitrite is recommended as an additive in concrete and gypsum to improve strength and combat the corrosion of iron reinforcement.

In curing salts used in the food industry, sodium nitrite is still the most reliable agent for protecting against botulism, a dangerous bacterial contaminant of meats. High nitrite concentrations can, however, lead to the formation of carcinogenic nitrosamines; legislative restrictions have, therefore, been imposed on the use of nitrite. Methods have been developed for achieving a satisfactory effect with minimal quantities [74]. The simultaneous addition of other substances (e.g., ascorbic acid) is reported to prevent the formation of nitrosamines [75, 76].

4.5. Economic Aspects

Sodium nitrite is produced as an important intermediate by the chemical industry in many countries, usually in conjunction with nitric acid production. Production statistics are not available.

5. Potassium Nitrite

5.1. Properties [77]

Potassium nitrite [7758-09-0], KNO_2, M_r 85.10, has an enthalpy of formation of $\Delta H_{298} = -375 \pm 5$ kJ/mol. The anhydrous product forms colorless monoclinic crystals of space group C_{s}^{3}, which are stable at 25 °C, $\varrho_{298} = 1.915$ g/cm^3 [3], mp 440 °C [3] (decomposition begins below the melting point).

Potassium nitrite is highly hygroscopic (Fig. 16) [78]; the enthalpy of solution of the anhydrous salt is $\Delta H_{298} = +13.4$ kJ/mol at infinite dilution [2]. The hydrate $KNO_2 \cdot 0.5 H_2O$ exists below −8.9 °C.

Thermal decomposition of potassium nitrite yields nitrogen oxides and is analogous to that of sodium nitrite but begins at ca. 410 °C [6].

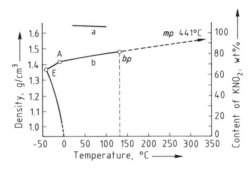

Figure 16. Properties of a saturated solution of potassium nitrite in water
a) Density; b) Content of potassium nitrite
bp = boiling point of saturated solution containing 82.0 wt % potassium nitrite at 100.9 kPa and 132 °C; A = transition point: −8.9 °C, 71.9 % potassium nitrite; E = cryohydric point: −40.2 °C, 64.9 % potassium nitrite.

The system $KNO_2 - KNO_3$ has a eutectic point at 316 °C and 22.5 mol % potassium nitrite [79]. In the quaternary system $KNO_2 - KNO_3 - NaNO_3 - NaNO_2$, the composition with the lowest freezing point (142 °C) is 37.5 mol % potassium nitrite, 37.5 mol % sodium nitrate, and 25 mol % sodium nitrite [23, 80].

5.2. Production

Potassium nitrite can be obtained by reduction of potassium nitrate (similarly to sodium nitrite). A variety of other reactions between sodium nitrate and potassium salts have also been described [77]. Production of potassium nitrite by absorption of nitrogen oxides in potassium hydroxide or potassium carbonate solution is not employed on a large scale because of the high price of these alkalies. Furthermore, the fact that potassium nitrite is highly soluble in water makes the solid salt difficult to recover. The solubility relationships in the system $KNO_2 - KNO_3 - H_2O$ are described in [81].

5.3. Uses

Potassium nitrite is used only in very small quantities, for example, as an oxidizing agent and as a corrosion inhibitor in washing with hot potash for removal of carbon dioxide from gas streams (e.g., synthesis gas). Potassium nitrite is more soluble than sodium nitrite and does not form sparingly soluble bicarbonate in carbon-dioxide-rich solutions.

6. Ammonium Nitrite

Ammonium nitrite is a very unstable compound that is virtually always handled in aqueous solution.

6.1. Properties

Ammonium nitrite [*13446-48-5*], NH_4NO_2, M_r 64.04 [3], has an enthalpy of formation of $\Delta H_{298} = -260 \pm 5$ kJ/mol and forms cubic crystals, $\varrho_{298} = 1.69$ g/cm^3 [3]. The melting point is not well defined because of decomposition.

Ammonium nitrite is highly hygroscopic and readily soluble in water [78] (enthalpy of solution $\Delta H_{285} = +19.3$ kJ/mol at infinite dilution [3]). The saturated solution contains 42.5 wt % ammonium nitrite at the cryohydric point (-27.9 °C), 55.4 % at 0 °C, and 65.0 % at 20 °C. Hydrated solid phases are not known.

Solid ammonium nitrite is stable only in the pure, dry state. Exothermic decomposition begins at 60 °C and may occur explosively:

$$NH_4NO_2 \rightarrow N_2 + 2 H_2O \quad \Delta H = -300 \text{ kJ/mol} \quad (6.1)$$

The saturated solution is stable only up to 30 °C; dilute solutions release nitrogen according to reaction (6.1) above 60 °C. Decomposition involves nitrous acid as an intermediate [82]; the stability of the aqueous solution is enhanced by the addition of ammonia, calcium hydroxide, or potassium iron(II) cyanide.

Dilute ammonium nitrite solutions can be inverted with nitric acid below 30 °C [83]:

$$3 NH_4NO_2 + 2 HNO_3 \rightarrow 3 NH_4NO_3 + 2 NO + H_2O \quad (6.2)$$

Mixed crystals of ammonium nitrite and ammonium nitrate precipitate from an aqueous solution of $NH_4NO_2 - NH_4NO_3$ containing > 47 % ammonium nitrite at 20 °C [84]. The most highly concentrated solution in this system at 20 °C contains only 6 % water, 48 % ammonium nitrite, and 46 % ammonium nitrate.

6.2. Production

Ammonium nitrite is produced and utilized industrially solely in the form of a weakly ammoniacal solution with pH > 7.5. When ammonium nitrite solution is prepared, the simultaneous formation of ammonium nitrate must be avoided because the two compounds cannot be separated from one another.

Ammonium nitrite is produced on an industrial scale from nitrogen oxides and ammonia or ammonium carbonate. The reaction is similar to that used for the formation of sodium nitrite. Because of the instability of ammonium nitrite (reaction 6.1), however, care must be taken to keep the pH between 8 and 9 and the temperature between 0 and +5 °C.

The heat balance of ammonium nitrite formation depends strongly on the feed ammonium compound:

$$2 NH_3 + NO_2 + NO + H_2O \rightarrow 2 NH_4NO_2$$
$$\Delta H_{298} = -77.5 \text{ kJ} \quad (6.3)$$

$$(NH_4)_2CO_3 + NO_2 + NO \rightarrow 2 NH_4NO_2 + CO_2$$
$$\Delta H_{298} = -47 \text{ kJ} \quad (6.4)$$

$$2 NH_4HCO_3 + NO_2 + NO \rightarrow 2 NH_4NO_2 + CO_2 + H_2O$$
$$\Delta H_{298} = -13.5 \text{ kJ} \quad (6.5)$$

Use of ammonium bicarbonate appears advantageous because no water is formed in the reaction to dilute the solution. However, since it has a low solubility the use of ammonium carbonate is generally preferred.

The process is usually carried out under an overpressure of 0.1 – 0.6 kPa [85]. The $NO_2 - NO$ gas mixture is first cooled to 0 °C to decrease the degree of oxidation (see Section 4.2) to 35 – 40 % by condensation of dilute nitric acid, and then fed into absorption columns connected in series and packed with ceramic Raschig rings or fitted with sieve trays. The circulating solution is held at ca. 0 °C by coolers located at the bottoms of the columns and outside them. To prevent crystallization of ammonium bicarbonate, a metered amount of aqueous ammonia solution is added along with the ammonium carbonate. The final product solution has a pH of 8 and contains ca. 150 – 200 g/L ammonium nitrite, 20 g/L ammonium nitrate, 20 g/L ammonium carbonate, and 2 – 5 g/L ammonia. Ammonium nitrite mist is formed in the absorption step, a final gas scrub is therefore used to recover ammonium carbonate solution for reuse [86].

The ammonium nitrite yield relative to the nitrogen oxides inlet to ammoniacal absorption is less than 90 %. Losses include nitrate formation, nitrite decomposition (reaction 6.1), and tail-gas and mist losses.

6.3. Uses

Ammonium nitrite can be used to passivate steam boiler equipment and chemical plants because it does not leave any nonvolatile residues behind. It is also used as a blowing agent in the manufacture of hollow rubber articles. In the laboratory, very pure nitrites can be prepared from pure ammonium nitrite and the hydroxides of other cations.

7. Calcium Nitrite

7.1. Properties [87]

Calcium nitrite, [13780-06-8], Ca(NO$_2$)$_2$, M_r 132.09, has an enthalpy of formation of ΔH_{298} = -750 ± 10 kJ/mol, ϱ_{273} = 2.26 g/cm^3 [78], mp 398 °C (decomposition begins below melting point).

Anhydrous calcium nitrite is hygroscopic and dissolves readily in water (Fig. 17) [78]; the enthalpy of solution for 1 mol in 800 mol of water is ΔH_{293} = -8.5 kJ/mol. The monohydrate crystallizes from solution below 129 °C to form silky hexagonal needles (ϱ_{293} = 2.23 g/cm^3).

Below 34.6 °C, the tetrahydrate Ca(NO$_2$)$_2 \cdot$ 4H$_2$O crystallizes in tetragonal form: ϱ_{293} = 1.72 g/cm^3, enthalpy of solution for 1 mol in 800 mol of water ΔH_{293} = $+33.5$ kJ/mol.

Calcium nitrite decomposes to calcium oxide and a mixture of nitrogen monoxide and nitrogen dioxide at temperatures as low as 250 °C [6]. The hydrates are decomposed by atmospheric carbon dioxide, with the formation of nitrogen oxides. A saturated solution of calcium nitrite boils at 133 °C (101.3 kPa) with decomposition.

7.2. Production

Calcium nitrite is formed when a mixture of gaseous nitrogen dioxide and nitrogen monoxide (OD < 50 %) is absorbed in a suspension of calcium hydroxide. This process is used to clean up the final tail gases from low-pressure nitric acid plants when no demand exists for sodium nitrate or sodium nitrite. The spent absorption liquor contains calcium nitrite and generally some calcium nitrate. The nitrite can be recovered as the tetrahydrate Ca(NO$_2$)$_2 \cdot$ 4H$_2$O by concentration and cooling. This product has a lower solubility than calcium nitrate tetrahydrate Ca(NO$_3$)$_2 \cdot$ 4H$_2$O. The nitrite is not, however, recovered on a large scale; the calcium nitrite solution is usually inverted with nitric acid to obtain calcium nitrate.

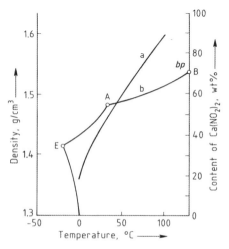

Figure 17. Properties of a saturated solution of calcium nitrite in water
a) Density; b) Content of calcium nitrite
bp = boiling point of saturated solution at 101.3 kPa and 133 °C with decomposition; A = first transition point: Ca(NO$_2$)$_2 \cdot$ 4H$_2$O + Ca(NO$_2$)$_2 \cdot$ H$_2$O at + 34.6 °C and 55.1 % Ca(NO$_2$)$_2$; B = second transition point: Ca(NO$_2$)$_2 \cdot$ H$_2$O + Ca(NO$_2$)$_2$ at + 129 °C and 71 % Ca(NO$_2$)$_2$; E = cryohydric point: ice + Ca(NO$_2$)$_2 \cdot$ 4H$_2$O at -20.0 °C and 34.2 % Ca(NO$_2$)$_2$.

7.3. Uses

No large-scale application exists for calcium nitrite. Calcium nitrite is reacted with the sulfates of other cations. The corresponding nitrite solutions are obtained after removal of the precipitated gypsum by filtration. Calcium nitrite mixed with calcium nitrate and calcium chloride is recommended as a noncorrosive salt for thawing ice on roads (along with NaNO$_2$). Calcium nitrite is proposed as a corrosion-inhibiting additive in concrete that also improves workability at low temperature.

8. Analysis of Nitrates and Nitrites

Nitrate Determination [88, 89]. For *volumetric determination*, nitrate is reacted in strongly alkaline solution with Devarda's alloy (50 % Cu + 45 % Al + 5 % Zn) or in neutral solution with Arnd's alloy (65 % Cu + 35 % Mg) and thus converted to ammonia. The ammonia is distilled into a receiver containing sulfuric acid and determined by acidimetric methods. Ammonium and nitrite are also detected and must, therefore, be determined separately. The method is not applicable in the presence of urea, urea – aldehyde condensates, cyanamide, or protein.

Gravimetrically, nitrate is determined as nitron nitrate after precipitation with nitron in acetic acid solution. Perchlorate, chlorate, nitrite, and other anions precipitate along with nitrate. The method cannot be used in the presence of urea – aldehyde condensates and protein.

At low concentrations (ppm range), nitrate is determined *photometrically* with 2,4-dimethylphenol, brucin, or sodium salicylate. Nitrite interferes and must be removed.

The *potentiometric method* with a nitrate ion-sensitive electrode is suitable for the determination of nitrate in single samples and for automated continuous monitoring (e.g., of wastewater) [90]. Perchlorate, iodide, chlorate, and bromide ions as well as high nitrite concentrations interfere with the method.

Nitrate can be determined in the presence of nitrite by *titrimetry* after reduction with iron(II) sulfate; the end point is indicated potentiometrically [91].

Prior to nitrate determination, nitrite can be destroyed in a weakly acidic medium with ammonium salts, urea, sodium azide, amidosulfonic acid, or hydroxylamine (cf. reactions 4.4 and 4.5).

Ion chromatography is an effective method for determining and separating nitrate [92]. The nitrate ion is detected either with a conductivity detector or a UV detector (excitation wavelength 254 nm). This technique has the advantages of high precision, rapidity, and independence of flow rate [93].

Nitrite Determination. High nitrite concentrations are determined by titration at ca. 35 °C with potassium permanganate in sulfuric acid solution. To avoid loss of nitrite, the solution is allowed to flow with stirring into acidified potassium permanganate solution until it becomes colorless [88]. Other ions that can be oxidized by potassium permanganate in sulfuric acid solution must be considered in evaluating the results.

Low contents of nitrite are determined photometrically with sulfanilamide and *N*-(1-naphthyl)ethylenediammonium dichloride as a red azo dye [94]. The reaction is specific for nitrite and can be automated [95]. This method is recommended for series analyses (e.g., of water samples).

Determination of Cations.

Sodium as the primary constituent can be determined gravimetrically by precipitation with uranyl magnesium acetate. This method can be employed even with a large excess of potassium [89]. Flame photometry and atomic absorption spectroscopy can be used for lower concentrations and in the trace range.

Potassium is precipitated as sparingly soluble potassium tetraphenyl borate and determined as such by weighing after treatment with sodium tetraphenyl borate in weakly alkaline or acidic solution [96 – 98]. Volumetric determination of potassium with sodium tetraphenyl borate and an electrometric end point indication is also important [98 – 100]. The chloroplatinate method is employed in the United States [89, 100].

Thermometry measures the heat (ca. −50 kJ/mol) released when potassium is precipitated as the perchlorate [101]. Potassium is also determined by conductometric titration with sodium tetraphenyl borate [98].

The potassium content can be determined quickly in process analysis by flame photometry [89] and atomic absorption spectroscopy. In the plant, β or γ radiation of ^{40}K solid salts or a solution can be utilized for analysis, product inspection, and process control [102].

Ammonium is determined acidimetrically by distillation of ammonia from alkaline solution. Trace amounts can be determined by photometry after formation of indophenol or reaction with Nessler's reagent [94]. Potentiometric determination with an ammonia electrode is suitable for all concentration ranges [90].

Calcium and magnesium in nitrates and nitrites are determined volumetrically by titration with the demineralized disodium salt of ethylenediaminetetraacetic acid. Other divalent cations must first be masked or precipitated [103].

Determination of Impurities. *Water content* can be measured either by the Karl Fischer method or by determination of the weight loss at 105 °C [89].

High concentrations of *chloride* are determined volumetrically by titration with silver nitrate [89]; trace amounts are measured by titration with mercury(II) nitrate and diphenylcarbazone as indicator [100]. A photometric technique is based on the determination of iron(III) rhodanide formed after liberation of rhodanide ion from mercury(II) rhodanide by chloride ion [95]. Very low concentrations can be determined turbidimetrically as silver chloride at 420 nm. This has the advantage that toxic mercury is not involved in the analysis.

Chlorate can be determined iodometrically in strong hydrochloric acid solution. *Perchlorate* is determined gravimetrically as potassium perchlorate [89]. *Iodate* can be determined iodometrically in weakly acidic solution. (Nitrite ion must be destroyed before iodometric analyses are performed.)

Boric acid can be determined by acidimetric titration, either directly or after isolation as methyl borate in the presence of mannitol or glycerol [89].

9. Environmental Aspects

Nitrate Removal. Nitrates are undesirable in wastewater for two reasons: uncontrolled denitrification can interfere with the final clarification step in biological wastewater treatment plants, and high nitrate concentrations can cause contamination of drinking water.

In 1970 the World Health Organization (WHO) issued guidelines recommending that the maximum limit for nitrate concentrations should be 50 mg/mL, this was later reduced to 45 mg/mL. The WHO also recommended a maximum acceptable concentration of 100 mg/L [105]. In the United States the EPA proposed regulations for drinking water on May 22, 1989 [106]. Public comment closed August 21, 1989, and the final rule was scheduled for December 1990. A maximum containment level for total nitrate and nitrite of 10 mg/L (as nitrogen) is proposed. As early as July 1980, the Council of the European Community issued a directive concerning the quality of water for human consumption [107]. The maximum permissible nitrate concentration was set at 50 mg/L, with a guide level of only 25 mg/L. Member countries agreed that drinking-water quality would meet this standard within five years (for the members Portugal, Spain, and Greece, who joined the EC later, the appointed five-year period started in 1988). To achieve this, some waterworks dilute their water with low-nitrate water; the trend to equip waterworks with nitrate removal facilities is also increasing. Because this equipment is very expensive, nitrate removal from wastewater before it reaches the treatment plant and reduction in nitrate loading of surface waters (agricultural) are desirable goals; see also → Fertilizers, Chap. 9.1. Three methods have been described for the removal of nitrates from wastewater: reverse osmosis, anion exchange, and denitrification.

Reverse Osmosis. The principles of reverse osmosis are described elsewhere. Pressure is used to force water through a membrane that is impermeable to any dissolved solutes or suspended solids. Despite its high cost, reverse osmosis is becoming more and more important in wastewater treatment; the cost depends on the nature of the inlet wastewater stream (pretreatment). If the wastewater has a high suspended solid content or the dissolved salts might crystallize, reverse osmosis can only be used to a limited extent because the membranes may become plugged. This technique results in the production of a concentrated nitrate solution, which is then dealt with by other methods.

Anion Exchange. Units based on anion-exchange resins [108] can only be used in special cases, because of their inadequate selectivity for nitrate versus chloride ions. The same is true for a process developed especially for the removal of nitrate from wastewater in cellulose nitrate production, which utilizes water-insoluble secondary amines in the chloride form as liquid ion exchangers [109]. In contact with wastewater, these substances exchange chloride for nitrate. In the next step, the amines exchange their nitrate ions for chloride ions from a potassium chloride solution to yield pure potassium nitrate as a by-product. This technique, however, requires virtually chloride-free wastewater. A similar procedure has been described in which the amine is used in the sulfate form and ammonium nitrate is obtained as byproduct after stripping with ammonia gas [110].

Denitrification. Denitrification is currently the most important process for nitrate removal [111 – 114]. Nitrate is reduced to nitrogen by facultatively anaerobic bacteria (e.g., *Achromobacter, Denitrobacillus, Nitrococcus,* or *Spirillum*) under anaerobic conditions. These bacteria can use nitrate and nitrite as a source of oxygen.

$$2\,NO_3^- + 10\,H^+ + 10\,e^- \rightarrow 2\,OH^- + 4\,H_2O + N_2 \quad (9.1)$$

$$2\,NO_2^- + 6\,H^+ + 6\,e^- \rightarrow 2\,OH^- + 2\,H_2O + N_2$$

To maintain metabolism, an adequate supply of carbohydrates must be present or added. The carbohydrates are oxidized to carbon dioxide and water.

Denitrification is the sole metabolic route for converting fixed nitrogen to the molecular form. The required anaerobic conditions are considered satisfied if the BOD_5 value (five-day biological oxygen demand) is more than 1.5 times the oxygen fixed in the nitrate.

Nitrite Removal. Nitrites must be removed from wastewater because they are toxic (see Chap. 10) and also because they interfere with the oxygen economy of natural waters and water treatment plants. Nitrite ions can be eliminated by oxidation to nitrate, by reduction to elemental nitrogen, or by denitrification (see Nitrate Removal).

Oxidation to Nitrate. Nitrite is reacted with, for example, sodium hypochlorite in a weakly acidic medium (pH 3 – 4):

$$NaNO_2 + NaOCl \rightarrow NaNO_3 + NaCl \quad (9.2)$$

This process has the advantage of a high reaction rate with virtually 100 % conversion. It can be used as a continuous method through relatively simple redox potential measurements and automation. It has the drawback, however, of increasing the sodium chloride content of water [115].

Another option is oxidation with hydrogen peroxide:

$$NaNO_2 + H_2O_2 \rightarrow NaNO_3 + H_2O \quad (9.3)$$

This reaction occurs at pH 3 – 3.5 and can be effected only in batch operation because a stepwise measurement and hydrogen peroxide metering cycle must be used. The process has the advantages of low cost and no salting (Table 7) [115].

Table 7. Comparison of nitrite removal processes using hydrogen peroxide and sodium hypochlorite [110]

	Hydrogen peroxide	Sodium hypochlorite
Batch operation	yes	yes
Continuous operation	no	yes
pH	2.0 – 4.0	3.0 – 3.5
Concentration of oxidizing agent, wt %	35	12.5
Oxidizing agent required, kg/kg nitrite	2.25	13.5
Cost of oxidizing agent, €/kg nitrite	1.93	3.24
Salt produced, kg/kg nitrite	0	2.5 NaCl
Loss of active agent on storage	no	yes

Nitrite is also oxidized to nitrate by *Nitrobacter* species under aerobic conditions:

$$NO_2^- + 1/2\,O_2 \rightarrow NO_3^-$$

Reduction to Nitrogen. Nitrites can be reduced to nitrogen with amidosulfuric acid:

$$NaNO_2 + NH_2SO_2OH \rightarrow N_2 + NaHSO_4 + H_2O \quad (9.4)$$

This reaction is of value for high nitrite levels and is carried out at ca. pH 4. Urea is also employed as reducing agent (reactions 4.4. and 4.5).

10. Toxicology and Occupational Health [116]

Nitrates and nitrites cause relaxation of smooth muscle cells [117, pp. 224 – 244], which can be so drastic that the contractile system is totally inhibited (protein phosphorylation). These compounds have a vasodilatory action, allowing blood to pool in the veins so that the heart is not completely filled. The systolic blood pressure therefore drops, and the pulse rate increases.

Dilation of blood vessels in the brain leads to headaches. Acute circulatory collapse and fainting are possible. Dilation of vessels in the retina leads to vision disorders.

Acute, severe poisoning is associated with a critical drop in blood pressure, vomiting, and cyanosis, and ends in death due to respiratory and circulatory failure. Methemoglobin produced by reaction of inorganic nitrite or amyl nitrite with hemoglobin is responsible for this effect. Nitrates are converted to nitrites by microbes in the intestines. Methemoglobin prevents oxygen transport because its central heme group contains a Fe^{3+} ion which is incapable of binding oxygen. Only hemoglobin (Hb) with its central Fe^{2+} ion can bind oxygen reversibly:

$$Hb \cdot Fe^{2+} \cdot O_2 \rightleftharpoons Hb \cdot Fe^{3+} \cdot OH^-$$

In the presence of nitrite, coupled oxidation occurs [117, p. 650]. Oxygen is transferred from the hemoglobin to nitrite with the formation of nitrate and methemoglobin (Fe^{3+}). In intact erythrocytes methemoglobin is reduced by enzymes (diaphorase, methemoglobin reductase).

Methemoglobin formation occurs more readily in premature and newborn babies than in adults. The intestinal flora of babies differs from that of adults and has a higher reducing capacity; nitrate is therefore more easily reduced to nitrite. In the fetus hemoglobin is more readily oxidized than in adults and the erythrocyte reductase system is not fully developed [118]. Vegetables fertilized with nitrate or tap water containing nitrate can cause methemoglobinemia in babies.

Severe intoxication is expected at an oral dosage level > 2 g. A single oral dose of 4 g of nitrite is lethal in humans [119]. In the event of poisoning, the stomach should be pumped and a physician notified [117, p. 250], [120, 121].

Nitrosamines cause tumor formation in animals that depends on dosage and route of administration. Nitrosamines are classified as risk factors for tumor formation in humans.

In the stomach secondary and tertiary amines can react with nitrite to form nitrosamines. Secondary amines are formed during cooking (roasting, frying) of proteins and alcoholic fermentation, and are constituents of flavoring substances. Low nitrosamine concentrations are found in pickled meat, alcoholic beverages, and tobacco smoke. Tetracycline contains groups that can be nitrosylated. The drug aminophenazone has been withdrawn from the market on account of its nitrosamine content (N-nitrodimethylamine).

11. References

1. *Gmelin*, **21**, 982 – 1048.
2. J. Barin, O. Knacke: *Thermochemical Properties of Inorganic Substances*, Springer Verlag, Berlin-Göttingen-Heidelberg-New York 1973.
3. R. C. Weast, M. J. Astle: *Handbook of Chemistry and Physics*, 63rd ed., CRC Press, Boca Raton, Florida 1982/83.
4. R. Schamm, K. Tödheide, *High Temp. High Pressures* **8** (1976) no. 1, 65 – 71.
5. F. Winkler, H.-H. Emons, W. Klauer, H. Kühn, *Z. Chem.* **12** (1972) no. 5, 190 – 191.
6. K. H. Stern, *J. Phys. Chem. Ref. Data* **1** (1972) no. 3, 747 – 772.
7. R. N. Kust, J. D. Burke, *Inorg. Nucl. Chem. Lett.* **6** (1970) 333 – 335.
8. J. J. Lehr: *Handbuch der Pflanzenernährung und Düngung*, **vol. 2**, Springer Verlag, Wien 1968, pp. 1003–1013.
9. G. E. Ericksen: "Geology and Origin of the Chilean Nitrate Deposits," *U.S. Geol. Surv. Prof. Paper* **1188** (1981) 41.
10. D. E. Garret: "The Chemistry and Origin of the Chilean Sodium Nitrate Deposits," *Proceedings of the Sixth International Symposium on Salt*, Toronto, May 24 – 28, 1983.
11. US-Bureau of Mines Minerals: Commodity Summaries 1979.
12. *Mining Annual Review* (1988) 296.
13. *Gmelin*, **21**, 95 – 118.
14. *Ullmann*, 3rd ed., **15**, 52 – 67.
15. F. A. Henglein, *Chem. Ztg.* **82** (1958) 287 – 294.
16. *Gmelin*, **21**, 372.
17. *Nitrogen*, **148** (1984) March-April, 19 – 21.
18. F. Wolf, J. Holzweissig, *Chem. Techn. (Leipzig)* **20** (1968) 477 – 480.
19. P. Lesage, M. Goetz, *Trait. Therm.* **114** (1977) 29–36.
20. D. A. Nissen, *J. Chem. Eng. Data* **27** (1982) 269 – 273.
21. D. J. Rogers, G. J. Janz, *J. Chem. Eng. Data* **27** (1982) 424 – 428.
22. T. Asahina, M. Kosaka, K. Tajiri, *Sol. World Congr., Proc. Bienn. Congr. Int. Sol. Energy Soc., 8th*, **vol. 3**, Pergamon Press, Oxford 1983, pp. 1716 – 1720.

23. Y. Takahashi, R. Sakamoto, M. Kaminoto, *Int. J. Thermophys.* **9** (1988) no. 6, 1081 – 1090.
24. C. Wahba (ed.): *World Directory of Fertilizer Manufacturers*, 6th ed., The British Sulphur Corporation, London 1986.
25. *Chem. Mark. Rep.* **236** (1989) no. 7, 7.
26. M. Smart, G. Russell, M. Tashiro: "Nitrogen Products", *Chem. Economics Handbook*, SRI-International, August 1987, 757.8002T.
27. *Gmelin*, **22**, 269 – 323.
28. A. Mustajoki, *Ann. Acad. Sci. Fenn. Ser. A 6* **99** (1962) 1 – 11.
29. P. Weidenthaler, *J. Phys. Chem. Solids* **25** (1964) 1491 – 1493.
30. E. Rapport, G. C. Kennedy, *J. Phys. Chem. Solids* **26** (1965) 1995 – 1997.
31. *Gmelin*, **22**, 1111.
32. W. Reinders, *Z. Anorg. Allgem. Chem.* **93** (1915) 202 – 212.
33. E. Cornec, H. Krombach, *Ann. Chim. (Paris)* **12** (1929) 235 – 253.
34. E. Jänecke, *Z. Anorg. Chem.* **71** (1911) 1 – 18.
35. A. M. Babenko, *Zh. Prikl. Khim. (Leningrad)* **48** (1975) no. 8, 1752 – 1756.
36. F. Chemnitus, *Chem. Ztg.* **53** (1929) no. 9, 85 – 86.
37. Yu. N. Dulepov, V. P. Napol'skikh, V. A. Postnikov, *Khim. Promst. (Moscow)* 1978 no. 9, 699 – 701.
38. Gewerkschaft Victor, DE 974 061, 1952 (H. Schmalfeldt, W. Vollmer).
39. *Ullmann*, 4th ed., **20**, 342 – 344.
40. S. J. Gohil, V. P. Mohandas, J. M. Joshi, J. R. Sanghavi, *Res. Ind.* **31** (1986) no. 3, 238 – 240.
41. U. K. Tipnis, B. T. Mandalia, *Res. Ind.* **32** (1987) no. 2, 82 – 84.
42. E. Jänecke, *Angew. Chem.* **41** (1928) 916 – 924.
43. Allied Chemical Corp., US 3 595 609, 1969 (J. L. Beckham).
44. *Phosphorus Potassium* (1971) no. 51, 50 – 52.
45. *Phosphorus Potassium* (1971) no. 51, 52 – 55.
46. N. P. Finkelstein, S. H. Garnett, L. Kogan: "Process for the Manufacture of Chloride Free Potassium Fertilizers," *Int. Potash Technol. Conf., 1st*, (1983) 571 – 576.
47. Haifa Chemicals, US 4 364 914, 1982 (S. Manor, M. Bar-Gori, A. Alexandrow, M. Kreisel).
48. A. Eyal, J. Mizrahi, A. Baniel, *Ind. Eng. Chem. Process Des. Dev.* **24** (1985) no. 2, 387 – 390.
49. *Phosphorus Potassium* (1988) no. 153, 14 – 16.
50. *Chem. Eng.* **74** (1967) no. 24, 118 – 119.
51. American Metal Climax, Inc., US 3 211 525, 1959 (J. D. Buehler et al.).
52. Securities and Exchange Commission S1-SEC-Registration, Feb. 6, 1989.
53. *Chem. Week* **142** (1988) no. 4, 34.
54. *Phosphorus Potassium* (1986) no. 141, 16.
55. *Gmelin*, **28**, 341 – 384. *Gmelin*, **28**, 1318 – 1321.
56. J. Boerio, P. A. G. O'Hare, *J. Chem. Thermodyn.* **8** (1976) 725 – 729.
57. W. W. Ewing, N. L. Krey, H. Law, E. Lange, *J. Am. Chem. Soc.* **49** (1927) 1958 – 1973.
58. W. W. Ewing, A. N. Rogers, J. Z. Miller, E. Mc Govern, *J. Am. Chem. Soc.* **54** (1932) 1335 – 1343.
59. M. Doumeng, *Rev. Chim. Miner.* **7** (1970) 897 – 925.
60. L. V. Opredelenkova et al., *Ukr. Khim. Zh. (Russ. Ed.)* **37** (1971) no. 2, 131 – 136.
61. *Phosphorus Potassium* (1988) no. 155, 28 – 33.
62. *Phosphorus Potassium* (1990) no. 166, 21 – 24.
63. S. Lanyi et al., *Bul. Inst. Politech. Bucureşti Ser. Chim.* **50** (1988) 55 – 60.
64. A. Crispoldi, A. Moriconi, M. Chiappafreddo, EP 278 562 A2, 1988.
65. *Gmelin*, **21**, 960 – 982.
66. M. I. Kay, *Ferroelectrics* **4** (1972) 235 – 243.
67. V. I. Kosyakov, Z. G. Bazarova, A. N. Kirgintsev, *Izv. Akad. Nauk SSSR Ser. Khim.* **197** (1971) no. 3, 643 – 645.
68. *Gmelin*, **21**, 263 – 266.
69. S. D. Deshpande, S. N. Vyas, *Ind. Eng. Chem. Prod. Res. Dev.* **18** (1979) no. 1, 69 – 71.
70. G. C. Inskeep, T. H. Henry, *Ind. Eng. Chem.* **45** (1953) 1386 – 1395.
71. A. P. Aholzin, Yu. Ya. Kharitonov, S. G. Kovalenko, *Werkst. Korros.* **38** (1987) 417 – 421.
72. Goldschmidt, DE 1 621 581, 1966.
73. Glass. Eng. Res. Inst., SU 729 150, 1978.
74. Monsanto, DE 2 713 259, 1977.
75. J. Sander, *Umsch. Wiss. Tech.* **74** (1974) 780.
76. Unilever, GB 1 440 183, 1972.
77. *Gmelin*, **22**, 263 – 268.
78. J. Bureau, *Ann. Chim. (Paris)* **8** (1937) 89 – 139.
79. A. N. Kruglov, O. M. Lutova, *Zh. Neorg. Khim.* **21** (1976) 2767 – 2769.
80. S. Berul et al., *Izv. Sekt. Fiz. Khim. Anal. Inst. Obsch. Neorg. Khim. Akad. Nauk SSSR* **25** (1954) 218 – 235.
81. Y. S. Shenkin, S. A. Ruchnova, E. V. Gorozhankin, *Zh. Neorg. Khim.* **20** (1975) 1717 – 1719.

82. E. Abel, *Monatsh. Chem.* **81** (1950) no. 4, 539 – 542.
83. B. V. Stamicarbon, NL 7 608 957, 1976.
84. P. I. Protsenko, A. A. Ugryumova, T. E. Kutsenko, *Zh. Prikl. Khim (Leningrad)* **44** (1971) no. 1, 30 – 34.
85. M. A. Miniovich, N. I. Smalii, *Khim. Promst. (Moscow)* **50** (1974) no. 5, 349 – 353.
86. Bayer, DE 2 015 156, 1970.
87. *Gmelin*, **28**, 338 – 341.
88. W. Fresenius, G. Jander: *Handbuch der analytischen Chemie*, Part III: Quantitative Analyse, vol. 5a: Stickstoff, Springer Verlag, Berlin-Göttingen-Heidelberg 1957.
89. Verband Deutscher Landwirtschaftlicher Untersuchungs- und Forschungsanstalten (VDLUFA): "Die Untersuchungen von Düngemitteln," *Methodenbuch (Loseblattsammlung)*, **vol. 2**, J. Neumann-Neudamm, Melsungen 1976.
90. Gesellschaft deutscher Chemiker: *Vom Wasser*, **vol. 49**, Verlag Chemie, Weinheim, Germany 1978, pp. 139–171.
91. W. D. Treadwell, H. Vontobel, *Helv. Chim. Acta* **20** (1937) 573 – 589.
92. J. L. Veuthey, J. P. Senn, W. Haerdi, *J. Chromatogr.* **445** (1988) no. 1, 183 – 188. G. Schwedt, C. Byung Seo, S. Dreyer, E. U. Ruhdel: "Nitrite, Nitrate in Waters. Photometric Rapid Testing and Ion Chromatography in Comparison," *Chromatographie, Spektroskopie*, Vogel-Verlag, Würzburg 1987, pp. 46 – 47.
93. *Nitrogen* (1984) no. 151, pp. 39 – 40.
94. Gesellschaft deutscher Chemiker: *Deutsche Einheitsverfahren zur Wasser-, Abwasser- und Schlamm-Untersuchung (Loseblattsammlung)*, Verlag Chemie, Weinheim, Germany 1979.
95. Technicon Industrial Systems: *Industrial Method*, No. 99–70 W/B$^+$ Tarrytown 1974; No. 102–70 W/C$^+$, Tarrytown 1978.
96. P. Raff, W. Brotz, *Fresenius Z. Anal. Chem.* **133** (1951) 241 – 248.
97. ISO Standards 2051, 2485.
98. DIN-Norm 10 400.
99. H. J. Schmidt, *Fresenius Z. Anal. Chem.* **157** (1957) 321 – 337.
100. H. Tollert: "Analytik des Kaliums," *Die chemische Analyse*, **vol. 51**, Enke Verlag, Stuttgart 1962.
101. G. Peuschel, F. Hagedorn, *Kali Steinsalz* **6** (1972) 4–11.
102. K. C. Scheel, *Angew. Chem.* **66** (1954) 102 – 106.
103. G. Schwarzenbach, H. Flaschka: *Die Komplexometrische Titration*, Enke Verlag, Stuttgart 1965.
104. D. J. de Renzo: *Nitrogen Control and Phosphorus Removal in Sewage Treatment*, Noyes Data Corp., Park Ridge, N.Y. 1978.
105. *Nitrogen* (1987) no. 167, 14 – 17.
106. National Primary and Secondary Drinking Water Regulations, Fed. Reg. 54 : 97 : 22062, May 22nd, 1989.
107. EEC Directive July 15th, 1980, 80/778/EEC, Aug. 30th, 1980.
108. A. B. Mindler: "Removal of Ammonia and Nitrates from Wastewaters," in C. Calmon, H. Gold (eds.): *Ion Exchange and Pollution Control*, vol. 1, CRC, Boca Raton, Fla, 1979, pp. 217 – 223.
109. T. K. Mattila, T. K. Letho, *Ind. Eng. Chem. Process Res. Dev.* **16** (1977) no. 4, 469 – 472.
110. Kemira Oy, DE 2 350 962, 1973.
111. N. Matsche, *Muench Beitr. Abwasser- Fisch-Flussbiol.* **43** (1989) 308 – 330.
112. J. P. van der Hoek, A. Klapwijk, *Water Supply* **6** (1988) 57 – 62.
113. O. Novak, K. Svardal, *Wien Mitt.: Wasser-Abwasser Gewaesser* **81** (1989) F1 – F54.
114. B. P. Gayle, G. D. Boardman, J. H. Sherrard, R. E. Benoit, *J. Environ. Eng. (N.Y.)* **115** (1989) no. 5, 930 – 943.
115. M. Rodenkirchen, *Galvanotechnik* **74** (1983) no. 8, 924 – 928.
116. N. J. Sax: *Dangerous Properties of Industrial Materials*, 6th ed., Van Nostrand Reinhold Comp., New York 1984, pp. 2002 – 2005.
117. W. Forth, D. Henschler, W. Rummel (eds.): *Allgemeine und spezielle Pharmakologie und Toxikologie*, 4th ed., B.J.-Wissenschaftsverlag, Mannheim 1983.
118. W. Wirth, G. Hecht, C. Gloxhuber: *Toxikologie-Fibel*, Thieme, Stuttgart 1967.
119. G. Fülgraff, D. Palm (eds.): *Pharmakotherapie/Klinische Pharmakologie*, 4th ed., Fischer-Verlag, Stuttgart 1982, p. 368.
120. H. Thaler, *Dtsch. med. Wochenschr.* **101** (1976) 1740 – 1742.
121. H.-H. Roth, *Dtsch. Apoth. Ztg.* **119** (1979) 1097.

Phosphate Fertilizers

GUNNAR KONGSHAUG, Norsk Hydro Research Centre, Porsgrunn, Norway (Chaps. 1 – 4 and 13 – 15)
BERNARD A. BRENTNALL, British Sulphur, London, United Kingdom (Chap. 5)
KEITH CHANEY, Levington Agriculture, Levington, United Kingdom (Chap. 6)
JAN-HELGE GREGERSEN, Norsk Hydro Research Centre, Porsgrunn, Norway (Chap. 7)
PER STOKKA, Norsk Hydro Research Centre, Porsgrunn, Norway (Chap. 7)
BJØRN PERSSON, Hydro Supra, Landskrona, Sweden (Chap. 8)
NICK W. KOLMEIJER, Hydro Agri Rotterdam, Vlaardingen, The Netherlands (Chaps. 9 and 10)
ARNE CONRADSEN, Hydro Landbruk, Porsgrunn, Norway (Chap. 11)
TORBJØRN LEGARD, Norsk Hydro Research Centre, Porsgrunn, Norway (Sections 12.1 and 12.2)
HARALD MUNK, Landwirtschaftliche Versuchsanstalt, Kamperhof Mülheim Ruhr, Federal Republic of Germany (Sections 12.3 and 12.4)
ØYVIND SKAULI, Norsk Hydro Research Centre, Porsgrunn, Norway (Chap. 16)

See also: Fertilizers

1.	History	354
2.	Terminology	355
3.	Composition of Phosphate Fertilizers	355
4.	Phosphate Rock	357
5.	Economic Aspects	358
5.1.	Phosphate Rock	358
5.2.	Phosphate Fertilizer Consumption	359
5.3.	Phosphate Fertilizer Production	359
6.	Phosphorus Uptake by Plants	361
7.	Chemical Equilibria in Fertilizer Production	363
8.	Superphosphates	367
8.1.	Single Superphosphate	367
8.1.1.	Chemistry	367
8.1.2.	Production	368
8.1.3.	Fluorine Recovery	370
8.1.4.	Granulation	371
8.2.	Triple Superphosphate	371
8.3.	Double Superphosphate	371
8.4.	PK Fertilizers	371
9.	Ammonium Phosphates	372
9.1.	Fertilizer Grades and Applications	372
9.2.	Production	372
9.2.1.	Ammonium Phosphate Powder	372
9.2.2.	Granular Ammonium Phosphates	374
9.2.3.	Off-Gas Treatment	376
10.	Compound Fertilizers by the Sulfur Route	376
10.1.	Granulation of Mixtures of Dry Materials	377
10.2.	Granulation of Dry Materials with Additives Producing Chemical Reactions	378
10.3.	Slurry Granulation	378
10.4.	Melt Granulation	380
11.	Compound Fertilizers by the Nitro Route	381
11.1.	Chemistry	382
11.2.	Product Specification	382
11.3.	Nitrophosphate Process with Calcium Nitrate Crystallization (Hydro)	382
11.4.	Nitro Process with Calcium Nitrate Crystallization (BASF)	385
11.5.	Nitro Process with Ion Exchange (Kemira Superfos)	387
11.6.	Nitro Process with Sulfate Recycle (DSM)	387
11.7.	Effluent Control of Nitrophosphate Process	388
12.	Other Straight Phosphate Fertilizers	388
12.1.	Phosphate Rock for Direct Application	388
12.2.	Partially Acidulated Phosphate Rock	389
12.3.	Basic and BOF Slag Fertilizers	390
12.4.	PK Mixed Fertilizers with Basic Slag	392
13.	Energy Consumption	392
14.	Effluents from Phosphate Fertilizer Production	392
15.	Heavy Metals in Phosphate Fertilizers	392
16.	Safety in Storage, Handling, and Transport	393
17.	References	394

The article was coordinated by GUNNAR KONGSHAUG.

Abbreviations used in this article:

AN   ammonium nitrate, NH_4NO_3
BOF   basic oxygen furnace (slag)
BPL   bone phosphate of lime
DAP   diammonium phosphate, $(NH_4)_2HPO_4$
DCP   dicalcium phosphate, $CaHPO_4$
MAP   monoammonium phosphate, $NH_4H_2PO_4$
MCP   monocalcium phosphate, $Ca(H_2PO_4)_2$
PAPR   partially acidulated phosphate rock
SSD   self-sustaining decomposition
SSP   single superphosphate
TPL   total phosphate of lime
TSP   triple superphosphate (known as concentrated superphosphate in North America)

1. History [1]

Farmers have always been anxious to improve crop yields. Some thousand years ago, Chinese farmers used calcined bones and the Incas in Peru used phosphoguano to increase crop output. In Europe, bones have been applied for centuries to French vineyards.

Several seventeenth century publications in Europe mention the beneficial effect of bones as a fertilizer for plant growth.

The German alchemist HENNING BRANDT discovered phosphorus in 1669 by isolating it from urine. In 1769 the Swedish scientist J. G. GAHN discovered that calcium phosphate is the main component of bones. About 30 years later, the conclusion was reached that the fertilizing effect of bones is due mainly to calcium phosphate and not to organic material. In 1797 the British physician GEORGE PEARSON gave the name superphosphate to the phosphate compound (calcium dihydrogenphosphate) found in bone; this name was later applied to fertilizers. Field trials demonstrated that bones should be crushed and applied in very small pieces.

Merchants then moved into the fertilizer business and established local powder mills for bone grinding. Attempts were made to improve fertilizer efficiency by composting bones in earth, animal waste, or plant waste; by boiling bones in water; or by treating them with steam under pressure.

Increased understanding of the fertilizer effect of phosphorus and a rapid increase in the use of bones in the early nineteenth century led to the idea of using chemical treatment of bones to improve fertilizer efficiency. Developments occurred in many countries. HEINRICH W. KÖHLER of Bohemia was probably the first to suggest and file a patent for using acids (especially sulfuric acid) in the processing and commercial production of phosphate fertilizers (1831).

In 1840 JUSTUS VON LIEBIG's theory on phosphorus uptake in plants contributed greatly to acceptance of the product and to rapid worldwide growth of the phosphate fertilizer industry. In the early 1840s the lack of bones as a raw material led to the export of phosphoguano from Peru. The discovery of low-grade mineral phosphates in France and England eased the raw material situation, but the development of the phosphate industry was secured by the discovery of large sedimentary phosphate deposits in South Carolina. The deposits were rediscovered as phosphate rock in 1859; mining began in 1867, and in 1889 the mine supplied 90% of the worldwide phosphate fertilizer production.

The production of ammonium phosphate fertilizers by ammoniation of phosphoric acid began around 1917 in the United States. The Haber–Bosch process boosted this product line, and in 1926 the IG Farbenindustrie in Germany announced the development of a series of multinutrient (compound) fertilizers based on crystalline ammonium phosphate.

Separation of calcium sulfate from the superphosphate slurry by increasing the sulfuric acid/rock ratio and use of phosphoric acid for acidulation led to the development of concentrated (triple) superphosphates and commercialization in ca. 1890.

The treatment of phosphate rock with nitric acid (nitrophosphate process) was developed in the late 1920s by the Norwegian ERLING B. JOHNSON. The IG Farbenindustrie, DSM in The Netherlands, and Norsk Hydro in Norway commercialized this route for complex fertilizer production in the 1930s.

In Europe, calcium silicophosphate fertilizer is produced as a byproduct of the steel industry. Iron ore may contain phosphorus, which can be removed by slagging out with lime. The product is sold under the name Thomas phosphate or basic slag. Many attempts have been made to produce similar fertilizer products by thermal treatment of phosphate rock with additives but most were unsuccessful due to high energy costs. Small amounts of fused magnesium phosphate and calcined defluorinated phosphate are produced in Brazil, China, Korea, and Japan [2].

2. Terminology

Phosphorus Content. The phosphorus-containing component of phosphate rock is apatite. *Fluorapatite* [*1306-05-4*] is the most common phosphate rock mineral. The correct formula of fluorapatite is $Ca_{10}F_2(PO_4)_6$, but it can be simplified to $Ca_5F(PO_4)_3$.

In commercial trading of phosphate rock, the phosphorus content is calculated as the weight percentage of tricalcium phosphate, $Ca_3(PO_4)_2$, and expressed as the bone phosphate of lime (BPL) or the total phosphate of lime (TPL). Tricalcium phosphate is not present as such in phosphate rock, but to simplify the relationship between BPL and fluorapatite, the formula of fluorapatite is sometimes expressed as $3\,Ca_3(PO_4)_2 \cdot CaF_2$.

The phosphorus in fertilizers is supplied as orthophosphate (referred to here as phosphate), PO_4^{3-}, but the content is generally expressed as the weight percentage of phosphorus pentoxide (P_2O_5) or, incorrectly, as phosphoric acid. Similarly, the contents of potassium, magnesium, and calcium are given as the weight percentage of their oxides: potassium oxide (K_2O), magnesium oxide (MgO), and calcium oxide (CaO). In the Scandinavian countries (except Iceland) and Ireland, however, the nutrient contents are expressed as elements (P, K, Mg, Ca).

$$BPL = 2.185 \times P_2O_5 = 5.008 \times P$$
$$P_2O_5 = 0.458 \times BPL = 2.291 \times P = 0.724 \times H_3PO_4$$
$$P = 0.200 \times BPL = 0.436 \times P_2O_5$$
$$= 0.316 \times H_3PO_4$$
$$PO_4^{3-} = \text{(ortho)phosphate ion}$$
$$HPO_4^{2-} = \text{hydrogen(ortho)phosphate ion}$$
$$H_2PO_4^- = \text{dihydrogen(ortho)phosphate ion}$$
$$H_3PO_4 = \text{phosphoric acid}$$

Types of Fertilizer. The production routes for phosphate fertilizers are summarized in Figure 1.

Superphosphate fertilizers (Chap. 8) are produced by treating phosphate rock with acid to give calcium dihydrogenphosphate.

The product obtained with sulfuric acid is called *single (or normal) superphosphate* (SSP). It is produced as superphosphate powder (run-of-pile), which later can be granulated to form 2 – 5 mm particles. Single superphosphate contains mainly calcium dihydrogenphosphate and calcium sulfate.

The most common product is, obtained by treatment of phosphate rock with phosphoric acid, is called *triple superphosphate* (TSP) (or concentrated superphosphate in North America). Triple superphosphate is produced either by use of run-of-pile powder as an intermediate or by a direct slurry granulation process.

Of minor importance is *double superphosphate*, which is a mixture of single and triple superphosphates. The term double superphosphate may cause some confusion because it was used earlier as a name for triple superphosphate.

Partly acidulated phosphate rock (PAPR, see Section 12.2) is produced in the same way as superphosphates, but with less sulfuric acid, to obtain ca. 50 % water-soluble P_2O_5. Annual production is negligible compared to other superphosphates.

Ammonium phosphate fertilizers (Chap. 9) contain ammonium dihydrogenphosphate [*7722-76-1*], $NH_4H_2PO_4$ (referred to in the fertilizer trade as monoammonium phosphate, MAP), and diammonium hydrogenphosphate [*7783-28-0*], $(NH_4)_2HPO_4$ (referred to as diammonium phosphate, DAP).

Compound (multinutrient) NP or NPK fertilizers are made by acidulating rock with sulfuric acid (Chap. 10) or nitric acid (Chap. 11).

Ground phosphate rock (Section 12.1) may also be used directly as a fertilizer. Basic slag (Section 12.2) is a fertilizer byproduct of the steel industry.

3. Composition of Phosphate Fertilizers

The major chemical components of the most common phosphate fertilizers are given in Table 1.

In phosphate rock, the F^- in fluorapatite may be replaced by OH^- and Cl^-; PO_4^{3-} by CO_3^{2-},

Phosphate Fertilizers

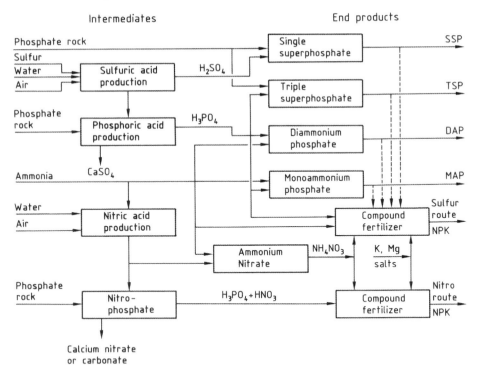

Figure 1. Primary production routes for phosphate fertilizers

SO_4^{2-}, CrO_4^{2-}, and SiO_4^{4-}; and Ca^{2+} by Na^+, K^+, Mg^{2+} and heavy metals. Possible metal compounds in phosphate fertilizers have been reported [3–7]. Their formation depends on process conditions and concentrations.

Dissolved iron and aluminum in superphosphate precipitate slowly as complex salts. High concentrations of aluminum and iron form amorphous aluminum and iron phosphate in superphosphates, which revert to crystalline calcium metal phosphate during production and storage. The most common components are $CaFe_2(HPO_4)_4 \cdot nH_2O$ and $CaAl_2(HPO_4) \cdot nH_2O$; $(Fe,Al)CaH(PO_4)_2 \cdot nH_2O$ may also be present. The same calcium metal phosphates are formed when phosphoric acid containing free calcium ions is ammoniated.

When phosphoric acid is ammoniated, the metals always seem to be present as metal ammonium phosphates in crystalline form, amorphous form, or as pyrophosphate gel.

Crystalline form:
$FeNH_4(HPO_4)_2$
$AlNH_4(HPO_4)_2$
Amorphous form:
$FeNH_4(HPO_4)_2 \cdot nH_2O$
$AlNH_4(HPO_4)_2 \cdot nH_2O$
$Mg(NH_4)_2(HPO_4)_2 \cdot nH_2O$
$AlNH_4F_2(HPO_4) \cdot nH_2O$
$FeNH_4F_2(HPO_4) \cdot nH_2O$
Pyrophosphate gel:
$(Mg, Al, Fe)NH_4FHP_2O_7 \cdot nH_2O$

Phosphate solubilities are measured in different ways to indicate plant availability ($\rightarrow$ Fertilizers, Chap. 6.3.):

1) Extraction with water
2) Extraction with citrate solution (neutral ammonium citrate solution)
3) Extraction with 2 % citric acid
4) Extraction with 2 % formic acid
5) Determination of total phosphate content

Calcium dihydrogenphosphate $Ca(H_2PO_4)_2$ (monocalcium phosphate, MCP), MAP, and

Table 1. Main chemical components of commercial fertilizers

Chemical composition	Type of fertilizer								
	Single super-phosphate (SSP)[a]	Triple super-phosphate (TSP)[a]	MAP	DAP	S-route NPK[b]	N-route NPK[c]	PAPR[d]	Basic slag	Phosphate rock
P_2O_5[e], wt%	21	48	53	46	15	12	21	14	29
P_2O_5, % (min.-max.)	18–22	46–50	48–61	46–53	5–30	5–23	18–24	13–15	27–33
$NH_4H_2PO_4$, MAP, P:P[f] (total)	0	0	0.80–0.89	0.01	0–0.89	0.74	0	0	0
$(NH_4)_2HPO_4$, DAP, P:P (total)	0	0	0.05	0.84–0.93	0–0.93	0.01	0	0	0
$Ca(H_2PO_4)$, MCP, P:P (total)	0.81–0.91	0.81–0.91	0	0	0–0.71	0	0.45	0	0
$CaHPO_4$, DCP, P:P (total)	0	0	0.01	0.01	0–0.2	0.2	0	0	0
Fluorapatite, P:P (total)	0.04	0.04	0	0	0–0.04	0	0.5	0	1.0
Calciumsilicophosphate, P:P (total)	0	0	0	0	0	0	0	1.0	0
Metal phosphates (mainly Al, Fe, Mg), P:P (total)	0.05–0.15	0.05–0.15	0.05–0.2	0.05–0.2	0.05–0.2	0.05	0.05	0	0
CaO, wt%	32	22	0.7	0.7	0.7–15	1.5–7	34	>35	47
$CaO:P_2O_5$, weight ratio	1.52	0.46	0.01	0.01	0.01–0.5	0.3	1.6	>2.5	1.6
N, wt%	0	0	10–12	16–21	0–30	12–30	0	0	0
$NO_3:NH_4$, molar ratio			0	0	0–0.9	0.7–0.9			
K_2O, wt%	0–30	0–30	0	0	0–21	0–21	0	0–30	0

[a] Range for P_2O_5 given without K addition. [b] Compound NP and NPK fertilizers based on the sulfuric acid route; the P component may be based on MAP, DAP, SSP, TSP, or a mixture; [c] Compound NP and NPK fertilizers based on the nitrophosphate route; specific values given for 75% water solubility P_2O_5. [d] Partially acidulated phosphate rock. [e] Estimated average P_2O_5 content in given type of fertilizer. [f] P:P (total) denotes the ratio of the phosphorus content in the specified chemical component to the total phosphorus content of the fertilizer.

DAP are water soluble; calcium hydrogenphosphate, $CaHPO_4$ (dicalcium phosphate, DCP), is citrate soluble (not water soluble).

Calcium metal phosphates as crystalline precipitates are citrate soluble.

Metal ammonium phosphates as amorphous precipitates or pyrophosphate gel are citrate soluble. Metal ammonium phosphates as crystalline precipitates are not usually citrate soluble.

Superphosphates normally contain >90% water-soluble and >98% citrate-soluble P_2O_5. Ammonium phosphates contain >85% watersoluble and >99% citrate-soluble P_2O_5. Compound NPK fertilizers contain >70–75% water-soluble and >99% citrate-soluble P_2O_5.

In partly acidulated rock fertilizers, the P_2O_5 solubility in water and citrate is ca. 45–50% and the solubility in citrate solution somewhat higher.

Ground phosphate rock has no citrate-soluble P_2O_5 but solubility in 2% citric acid may be 5–53% and solubility in 2% formic acid may be as high as 86%. Basic slag (calcium silicophosphate) has no water-soluble P_2O_5, but citrate solubility is reported to be up to 90% and the citric acid solubility up to 97% [8].

4. Phosphate Rock

Phosphate rock is virtually the sole raw material for phosphate fertilizers. The primary source is sedimentary phosphate rock (phosphate precipitated from seawater and bones) but magmatic (igneous) phosphate rocks are also important.

Phosphate rock contains calcium phosphates as apatites, mainly fluorapatites. It is mined in almost 30 countries, and phosphate fertilizers are produced in 26 countries in Europe, 13 in America, 9 in Africa, 13 in Asia, and 2 on the Australian continent [9].

About 90% of mined phosphorus is used in fertilizers, being divided almost equally among superphosphates, ammonium phosphates, and compound fertilizers (including nitrophosphates).

5. Economic Aspects

5.1. Phosphate Rock

Phosphate rock reserves of varying composition and quality are widely distributed. Exploitation of deposits occurs in many countries, but large-volume production for captive use and export is limited to a few countries. Total world rock production in 1989 was 161.4×10^6 t and was divided as follows (10^6 t):

United States	49.9
Former Soviet Union	35.0
People's Republic of China	18.5 *
Morocco	18.0
Jordan	6.7
Tunisia	6.6
Israel	3.9
Brazil	3.7
Togo	3.4
Republic of South Africa	3.0
Senegal	2.3
Syria	2.3

* Down from 24.8×10^6 t in 1988 due to market disruption

The four major producers of rock account for 75 % of world output. The output of former Soviet and Chinese phosphate rock mines was essentially used for domestic consumption.

Since the early 1970s, phosphate rock has become a less important export product in comparison with processed phosphate products (mainly phosphoric acid, ammonium phosphates, and triple superphosphate). World export of phosphate rock was 43.5×10^6 t in 1989 (down from a peak of 53.4×10^6 t in 1979) and was divided as follows (10^6 t):

Morocco	12.4
United States	8.3
Jordan	5.9
Togo	3.3
Former Soviet Union	3.2
Israel	2.4
Syria	1.7
Senegal	1.4
Nauru	1.2
Republic of South Africa	1.1
Tunisia	1.1

The four leading export countries accounted for 69 % of world trade.

The shift away from phosphate rock as the major vehicle for trade in phosphates results from the development of vertically integrated industries at or near mine sites. The key element in these operations is phosphoric acid capacity, which allows a concentrated, gypsum-free product to be transported to market.

The shift toward captive processing of phosphate rock is reflected in a significant change in the pattern of phosphate rock consumption and world trade. The total trade in rock continues to decrease and appears to have fallen below 40×10^6 t/a in 1990 for the first time since 1970. On the other hand, the tonnages delivered to downstream processing industries are now substantial. Total worldwide consumption of phosphate rock in 1989 was 158.9×10^6 t and was divided as follows (10^6 t):

Western Europe	17.0
Eastern Europe	42.2
Former Soviet Union	31.8
North America	42.4
Africa	14.4
Morocco	5.2 *
Republic of South Africa	2.1
Tunisia	5.1
Central America	2.4
Mexico	2.4
Australasia	2.5
Australia	1.7
Middle East	3.9
Iraq	1.1
Israel	1.4
Jordan	1.0
Asia	28.7
People's Republic of China	18.8
India	3.0
Indonesia	1.1
Japan	1.6
South Korea	1.5
South America	4.6
Brazil	4.1

* Down from 10.6×10^6 t in 1988 as a result of market disruption.

A comparison of phosphate rock consumption statistics with production and export statistics reflects the development of downstream processing facilities in many rock-exporting countries. Demand for phosphate rock has thus receded in a number of regions, notably Western Europe and Japan.

Phosphate rock statistics are generally expressed as tonnes of product. The phosphorus content of the rock varies. Rock of sufficiently high phosphorus content (grade) to be acceptable commercially contains 30–40 % P_2O_5 (66–87 BPL).

The majority (ca. 90 %) of phosphate rock produced is converted into fertilizer. Smaller, nonfertilizer end uses account for ca. 10 % of phosphate rock consumption; the most important are the manufacture of detergents and animal feed additives.

5.2. Phosphate Fertilizer Consumption

Worldwide consumption of phosphate fertilizers continues to grow, but during the 1980s growth was erratic and less dynamic than in the 1960s and 1970s. Consumption (in 10^6 t P_2O_5) was as follows (in 10^6 t):

1960	9.7
1965	14.0
1970	19.7
1980	31.6
1985	33.0
1986	34.8
1987	36.4
1988	37.7
1989	36.6

A number of factors will contribute to the future consumption of phosphate fertilizers:

1) The agricultural sector in the export-oriented countries of the developed world has suffered consistently from overproduction and mounting subsidy bills. In addition to poor farm economics, government programs have been aimed at reducing crop production.
2) In many developed countries, years of regular phosphate application have led to a buildup of phosphate soil reserves. As farm economics have deteriorated, farmers have been content to exploit the soil reserves by reducing phosphate application.
3) Environmental concern about the consequences of overfertilization and runoff of nutrients in groundwater has led to a more accurate application of nutrients in many countries.

In Western Europe, phosphate fertilizer consumption peaked in 1979 at 6.8×10^6 t of P_2O_5; pressures on agriculture from the Common Agricultural Policy (CAP), the General Agreement on Tariffs and Trade (GATT) negotiations, and environmental groups indicate that consumption levels are unlikely to recover and may stabilize close to 5×10^6 t of P_2O_5.

In Central Europe and the former Soviet Union, phosphate fertilizer consumption increased steadily during most of the 1980s in line with economic policies. A peak of 11.6×10^6 t of P_2O_5 was recorded in 1988. Changed economic circumstances may lead to reduction.

Consumption of phosphate fertilizer in North America peaked in 1980 at 5.6×10^6 t of P_2O_5. In the latter half of the 1980s, annual use fluctuated between 4.3 and 4.6×10^6 t. Decline in consumption is not anticipated. At 2.4×10^6 t P_2O_5 in 1989, phosphate fertilizer usage in South America was down some 0.4×10^6 t P_2O_5 over 1980. Demand in Brazil has proved the single most important variable, fluctuating by as much as 1×10^6 t/a in the 1980s. In Africa, annual demand fluctuated between 1.3×10^6 t (1981) and 0.9×10^6 t (1989) due to civil conflict, economic crises, and drought.

Asia had consistent growth in phosphate fertilizer consumption during the 1980s. This reflects the constantly increasing demand for food by the rapidly growing population. Chinese consumption almost doubled during the 1980s, reaching 5.3×10^6 t of P_2O_5 in 1989. In the Middle East, P_2O_5 consumption (including Australasia) in 1989 stood at 12.2×10^6 t, 33 % of the world total.

5.3. Phosphate Fertilizer Production

Changes in the phosphate industry toward vertical integration and the trading of processed P_2O_5 products rather than phosphate rock have resulted in significant shifts in phosphate fertilizer product patterns. The importance of low-grade phosphate fertilizers (i.e., products with a low P_2O_5 content) has declined significantly. These include single superphosphate (typically 18–22 % P_2O_5) and basic slag (typically 13–15 % P_2O_5). Not only does the low nutrient content make such products more expensive to transport, handle, and apply, but production economics have made some of them unattractive.

High-nutrient fertilizers now dominate the world phosphate fertilizer market and fall into three categories:

1) Ammonium phosphates
2) Triple (or concentrated) superphosphate
3) Binary and ternary compound (or complex) fertilizers

The ammonium phosphate category comprises two main fertilizer products: diammonium phosphate, DAP ($N-P_2O_5$ content: 18–46; i.e., 18 wt % N and 46 wt % P_2O_5) and monoammonium phosphate, MAP ($N-P_2O_5$ content: 11–53).

Triple (or concentrated) superphosphate is a straight, single-nutrient phosphate product (typical P_2O_5 content 48 wt %).

Compound (multinutrient) fertilizers are produced in various combinations (→ Fertilizers, Chap. 3.1.). The NPK fertilizers contain all three main fertilizer nutrients, i.e., nitrogen, phosphorus, and potassium (typical $N-P_2O_5-K_2O$ content in weight percent: 16-16-16, NP fertilizers contain no potash (nutrient content typically 22-20-0), PK fertilizers are produced mainly in Western Europe (nutrient content typically 0-25-25).

In 1989, world phosphate fertilizer production amounted to 38.6×10^6 t of P_2O_5. Production breakdown follows (in 10^6 t):

Single superphosphate	7.0
Triple superphosphate	4.6
Ammonium phosphates	13.7
Compound fertilizers	10.3
Other low-analysis phosphates	3.0

Single superphosphate retains its importance in some developing countries and in Eastern Europe. In 1989 the People's Republic of China accounted for 2.5×10^6 t of P_2O_5 production, Eastern Europe for almost 2×10^6 t, and India for 0.5×10^6 t. As economic conditions in Central Europe change, single superphosphate will probably be displaced by more concentrated products.

Production of *triple superphosphate* (known as concentrated superphosphate in North America) is widely dispersed. More than 1×10^6 t of P_2O_5 was produced in Eastern Europe in 1989, but this is likely to decrease. Producers in the United States accounted for ca. 0.8×10^6 t of P_2O_5, and North African and Middle Eastern industries for an additional 1.1×10^6 t. Not all triple superphosphate is applied directly to soil. It is often blended at the distributor or even the farm level with other nutrients and applied as compound fertilizer. The same is true of single superphosphate, although to a lesser degree.

Triple superphosphate is the least important export product. In 1989, world trade amounted to 1.5×10^6 t of P_2O_5 (3.3×10^6 t of product). Approximately 0.9×10^6 t of these exports originated in North Africa and the Middle East, and 0.25×10^6 t in the United States.

Ammonium phosphates were developed in the United States primarily as components of blended fertilizers (bulk blends): DAP is the most important product, world production amounting to 9.3×10^6 t of P_2O_5 in 1989, 5.8×10^6 t of this being produced in North America. In 1989, U.S. exporters shipped ca. 3.6×10^6 t. Morocco is now the second most important exporter (0.5×10^6 t P_2O_5), and output is growing. The production of MAP is concentrated primarily in the former Soviet Union and North America. In 1989, world output was 4.4×10^6 t of P_2O_5, of which 2.5×10^6 t was manufactured in the former Soviet Union and 1.3×10^6 t in North America. These products have become so attractive in terms of cost effectiveness on the world market that an increasing proportion of ammonium phosphate is being used as a direct application fertilizer, particularly in less-developed countries.

Compound fertilizers are a development of the European fertilizer industry. They may be produced by treating phosphate rock with sulfuric acid (the sulfur, wet process phosphoric acid, or ammonium phosphate route) or with nitric acid (the nitrophosphate route). In addition, fertilizer components such as ammonium phosphate may be mixed physically to produce blended compound fertilizers. Blending often occurs at the wholesale or retail level; the total volume of blended compounds is therefore not reflected in consumption (delivery) statistics.

In 1989, the chemical production of compound fertilizers exceeded 2×10^6 t of P_2O_5 in both Eastern and Western Europe and 1×10^6 t in North America. However, considerable pressure has been exerted on European industries from imported DAP and triple superphosphate (which will lead to continuing rationalization of this sector).

Although a significant amount of regional trade in P_2O_5 fertilizers such as NPKs occurs within Western Europe, between the former Soviet Union and Central Europe, interregional trade in phosphate fertilizers is now dominated by vertically integrated producers. The United States dominates the international market for DAP (see above).

6. Phosphorus Uptake by Plants
[10–12]

Role in Plant Nutrition. Phosphorus is essential for vital growth processes in plants because it is a constituent of nucleic acids. It is also a constituent of phospholipids. Phosphorus compounds (coenzymes) are involved in respiration, energy transfer, and the efficient utilization of nitrogen. See also → Fertilizers, Chap. 2.1.2.

Phosphorus is of special importance in root development and in the ripening of seeds and fruit. The application of phosphate fertilizer to soils that are low or deficient in available phosphate improves root development and seedling growth, giving the crops a better start.

Phosphorus is taken up by plants as the dihydrogenphosphate ($H_2PO_4^-$) or the hydrogenphosphate (HPO_4^{2-}) ion.

Phosphorus Deficiency Symptoms. In phosphorus deficiency the growth of plant tops and roots is greatly restricted: shoots are short and thin, growth is upright and spindly; leaves are small, and defoliation is premature; lateral shoots are few in number, and lateral buds may die or remain dormant; blossoming is greatly reduced, resulting in poor yields of grain and fruit. These symptoms also apply to nitrogen deficiency, but with phosphate deficiency the leaf color is generally a dull, bluish green, usually with a purple tint (rather than yellow or red). Leaf margins may also show brown scorching.

Forms of Phosphate in the Soil. Most phosphate in the soil is present in the solid phase; only a small amount is dissolved in the soil solution. Solid-phase phosphate occurs partly in organic form (30–50 %), but in mineral soil it exists mainly in inorganic form (50–70 %). The three major components of *organic soil phosphate* are nucleic acids, phospholipids, and phytin with its derivatives. Organic phosphate is continuously released as HPO_4^{2-} ions by microbial degradation of organic soil matter. However, this phosphate release does not always coincide with maximum plant uptake.

Most *inorganic soil phosphates* fall into two groups: those containing calcium, and those containing iron and aluminum. The calcium compounds of greatest importance are apatites with the general formula $Ca_5(F, Cl, OH, 1/2 CO_3)(PO_4)_3$; calcium hydrogenphosphate, $CaHPO_4$ (dicalciumphosphate, DCP); and calcium dihydrogenphosphate, $Ca(H_2PO_4)_2$ (monocalciumphosphate, MCP).

Fluorapatite is the most insoluble of these compounds and is therefore unavailable to plants. In contrast, MCP and DCP are readily available for plant growth. Except in recently fertilized soil, these two compounds are present in very small quantities because they readily react to form more insoluble compounds.

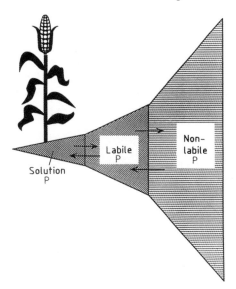

Figure 2. Main soil phosphate fractions

The iron and aluminum phosphates in soil are probably hydroxyphosphates. They are very stable in acid soil and are extremely insoluble.

Three inorganic soil phosphate fractions are important in terms of plant nutrition (Fig. 2):

1) Phosphate in soil solution: dissolved phosphates available for plant uptake

2) Phosphate in the labile pool: often absorbed on mineral soil particles and can rapidly go into soil solution
3) Phosphate in the nonlabile fraction: only very slowly released into the labile pool

Phosphate dissolved in soil solution can be taken up by plants. Phosphate removed from this pool is replenished by phosphate from the labile pool. Nonlabile (insoluble) phosphate is released very slowly into the labile pool–too slowly to make a significant contribution to the crop in one growing season.

Effect of Soil on Phosphate Availability. The pH determines the phosphate form in soil solution. In very acid solution, only $H_2PO_4^-$ ions are present. As pH increases the HPO_4^{2-} ions and then the PO_4^{3-} ions dominate. At pH 7.0 the proportions of $H_2PO_4^-$ and HPO_4^{2-} ions are roughly equal:

$$\underset{\text{Acid solution}}{H_2PO_4^-} \underset{H^+}{\rightleftharpoons} HPO_4^{2-} \underset{H^+}{\rightleftharpoons} \underset{\text{Very alkaline solution}}{PO_4^{3-}}$$

Although the $H_2PO_4^-$ ion is generally considered to be more available than the HPO_4^{2-} ion, both can be taken up by plants. However, in soil this relationship is complicated by other compounds and ions.

In acid soil, soluble iron, aluminum, and manganese react with $H_2PO_4^-$ ions to form insoluble phosphates that are unavailable for plant growth:

$$\underset{\text{Soluble}}{Al^{3+} + H_2PO_4^- + 2 H_2O} \longrightarrow \underset{\text{Insoluble}}{2 H^+ + Al(OH)_2H_2PO_4}$$

In most acid soils the concentration of iron and aluminum ions exceeds that of $H_2PO_4^-$ ions, and only minute quantities of phosphate remain immediately available to plants. Phosphate can also be fixed by hydrous oxides, kaolinite, montmorillonite, and illite. Alkaline soil containing an excess of exchangeable calcium (e.g., calcium carbonate) also precipitates phosphates:

$$\underset{\text{Soluble}}{Ca(H_2PO_4)_2 + 2 CaCO_3} \longrightarrow \underset{\text{Insoluble}}{Ca_3(PO_4)_2 + 2 CO_2 + 2 H_2O}$$

Because insoluble phosphates are formed in both acid and alkaline soils, maximum phosphate availability is obtained at a soil pH of 6.0–7.0. Even in this range, phosphate availability may be low, and added soluble phosphates can be readily converted into unavailable forms.

Use of Phosphate as a Fertilizer. Phosphate ions are relatively immobile in soil compared to nitrate, and are taken up by plants from the soil solution at the root surface. The critical time for phosphate supply to the plant is in the seedling stage, when seed reserves of phosphate have been exhausted and the root system has not developed sufficiently to supply phosphate needs.

Two strategies are employed with phosphate fertilizers:

1) Sufficiently high phosphate levels are built up in soil to meet the crop requirement for phosphate ions. Crops with shallow root systems (e.g., potatoes) require a higher concentration of phosphate in soil solution to sustain growth than crops with extensive root systems (e.g., cereal).
2) A small reservoir of phosphate ions is provided near the site at which seedling roots develop (a hot spot of phosphate-enriched soil). Enriched hot spots can be created only if the phosphate fertilizer dissolves rapidly in the soil. For most soil the fertilizer must be water- or citrate-soluble.

In some situations the association of roots with microorganisms (endotropic mycorrhizae) increases phosphate uptake by a number of crops. However, at present this seems to be of only agronomic importance in soil with a very low phosphate content. Similarly, root exudates can influence chemical and microbial activity in the root zone and can increase phosphate uptake where soil phosphate supply is marginal.

High-yield crops used in modern agriculture demand high levels of soluble phosphate in the soil. This can be achieved only by adding materials (fertilizers) containing soluble phosphates. The natural soil phosphate level is usually not sufficient.

The quantity of soluble phosphate fertilizer added for a particular crop depends on the phosphate removed in the harvested yield, the response of the crop to the fertilizer, and the available phosphate level of the soil in which the crop is grown. The amounts of phosphate removed by crops are as follows (kilogram of P_2O_5 per tonne of fresh material) [13]:

Sugar beets, swede roots, cabbage, carrots, onions	0.7–0.9
Potato, kale, maize, French beans, beetroot, cauliflower, silage grass, cereal straw, vining peas, broad beans	1.0–1.6
Brussel sprouts	2.1–2.6
Grass, hay, cereal grain, dried peas	5.9–8.8
Field beans	11
Oilseed rapeseed	16

7. Chemical Equilibria in Fertilizer Production

In acidic soil, fluorapatite in porous phosphate rocks may slowly be broken down to form water-soluble phosphate ions. In fertilization, a strong acid is used to break down the apatite to hydrogenphosphates.

The system of phosphate fertilizers can be understood by means of acid–base and solid–liquid chemical equilibria. The primary acid–base equilibria involve phosphoric acid, sulfuric acid, nitric acid, and ammonia. The main solid–liquid equilibria involve ammonium phosphates, calcium phosphates, and water.

Fertilizer Acids. Most important is the phosphoric acid equilibrium [14]:

$H_3PO_4 \rightleftharpoons H^+ + H_2PO_4^-$ $K = 10^{-2.13}$

$H_2PO_4^- \rightleftharpoons H^+ + HPO_4^{2-}$ $K = 10^{-7.2}$

$HPO_4^{2-} \rightleftharpoons H^+ + PO_4^{3-}$ $K = 10^{-12.4}$

These equilibria can be represented in different ways: as a function of mole percent and pH (→Phosphoric Acid and Phosphates, [15] or as a function of logarithmic concentration and pH (also called a Bjerrum diagram, Fig. 3) [16–21].

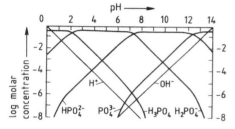

Figure 3. Phosphoric acid equilibria in solution [17]

Sulfuric acid and nitric acid are also important acids in phosphate fertilizer production. The equilibrium reactions are [14]

$H_2SO_4 \rightleftharpoons H^+ + HSO_4^-$ $K => 1$

$HSO_4^- \rightleftharpoons H^+ + SO_4^{2-}$ $K = 10^{-1.99}$

$HNO_3 \rightleftharpoons H^+ + NO_3^-$ $K = 10^{1.37}$

These equilibria can also be illustrated in a logarithmic diagram.

Ammonium Phosphates. Ammoniation of phosphoric acid involves the following reactions [14]:

$NH_4^+ \rightleftharpoons H^+ + NH_3$ $K = 10^{-9.24}$

$NH_4^+ + H_2PO_4^- \rightleftharpoons NH_4H_2PO_4$ (MAP)

$2\,NH_4^+ + HPO_4^{2-} \rightleftharpoons (NH_4)_2HPO_4$ (DAP)

These equilibria can be represented in a logarithmic diagram by combination of the ammonia and phosphoric acid constants (Fig. 4).

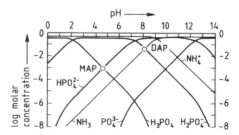

Figure 4. Ammonium phosphate equilibria in solution

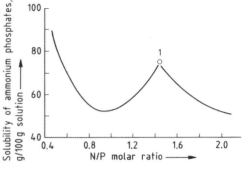

Figure 5. Effect of NH_3/H_3PO_4 molar ratio on the solubility of ammonium phosphates at 75 °C [23]

The maximum concentrations of NH_4^+ and $H_2PO_4^-$ ions necessary for MAP production are present in the pH range 2.5–7.0 (parallel lines at the top of Fig. 4). The optimum point for MAP production lies where the concentrations of H_3PO_4, (phosphoric acid) and HPO_4^{2-} (the DAP component) are minimized. This is the point where the concentrations of H_3PO_4 and HPO_4^{2-} equal each other, their concentrations lines cross at pH 4.65. A 1 % aqueous solution of MAP has a pH of 4.5 [22].

The maximum concentrations of NH_4^+ and HPO_4^{2-} ions necessary for DAP production are present at about pH 8.0 (parallel lines at the top of Fig. 4). The optimum point for DAP production lies where the concentrations of $H_2PO_4^-$ (the MAP component) and NH_3 (ammonia) are minimized. This is the point where the concentrations of $H_2PO_4^-$ and NH_3 equal each other, their concentrations lines cross at pH 8.1. A 1 % aqueous solution of DAP has a pH of 8.0 [22].

The concentration curves of HPO_4^{2-} and $H_2PO_4^-$ in Figure 4 give the fractions of DAP and MAP present as a function of pH in the neutralization process.

In fertilizer production, high pH must be avoided to minimize formation of free ammonia in the liquid, which is a potential source of loss.

Crystallization. Neutralization and evaporation are used in most fertilizer-producing processes. Knowledge of solubility and crystallization in the system is important. In neutralizing phosphoric acid, the water solubility of a mixture of H_3PO_4, MAP, and DAP is usually presented as shown in Figure 5.

In Figure 6 the N/P molar ratio is plotted as a function of MAP–DAP content.

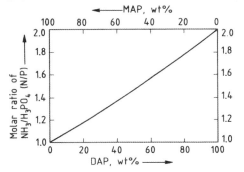

Figure 6. N/P molar ratio as a function of MAP–DAP content

If only water, MAP, and DAP are considered, point 1 in Figure 5 shows that at 75 °C, a saturated solution consists of 75 wt % MAP–DAP, and 25 wt % water. The N/P molar ratio is 1.45, corresponding to 52 wt % MAP and 48 wt % DAP (Fig. 6). The composition at point 1 is therefore 25 wt % water, 39 wt % MAP, and 36 wt % DAP. For this composition a reduction in temperature below 75 °C results in crystallization.

A more informative way of presenting the water–MAP–DAP system is to use a phase diagram (Fig. 7). Points 1 in Figures 5 and 7 represent the same composition. The temperature curves in Figure 7 give the crystallization temperature for a given concentration of MAP, DAP, and water. Point 1 is on the borderline between the crystallization areas for MAP and DAP; MAP and DAP will both crystallize as the temperature is reduced below 75 °C.

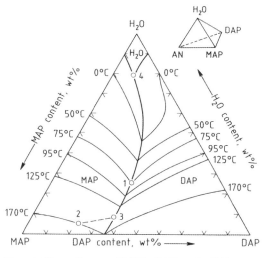

Figure 7. Phase diagram of MAP–DAP–water [24]

A solution given by point 2 in Figure 7 (73 % MAP, 22 % DAP, 5 % water) starts to crystallize at 170 °C. If the temperature is reduced, MAP begins to crystallize. As MAP crystallizes, the liquid composition changes along a straight line through point 2, with its origin at the MAP corner toward point 3. When the liquid phase composition reaches point 3, both MAP and DAP crystallize. Further cooling results in the crystallization of MAP and DAP. The liquid phase composition moves along the borderline between the MAP and DAP crystallization areas. At point 4 the last liquid phase crystallizes at the eutectic temperature well below −20 °C (to give ice, MAP, and DAP).

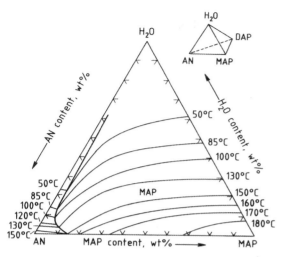

Figure 8. Phase diagram of AN – MAP – water [25]

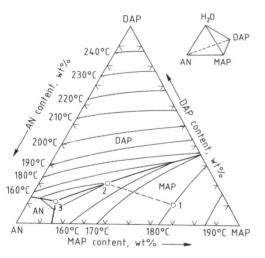

Figure 9. Phase diagram of AN – MAP – DAP [26]

Compound Fertilizers. In compound fertilizers the nitrogen content is usually increased by the addition of ammonium nitrate (AN). A system without DAP can be represented by the AN – MAP – water phase diagram shown in Figure 8. The MAP area is dominant. Compositions with a high MAP content and a low water content tend to supercool (undercool).

If the liquid phase of the compound fertilizer melt also contains DAP, a quaternary tetrahedral phase diagram must be used. Figures 7 and 8 represent two sides of the tetrahedron. Figure 9 gives the AN – MAP – DAP diagram representing the "base" of the tetrahedron. An anhydrous melt is difficult to obtain in compositions with a high DAP content.

For a melt with the composition given by point 1, MAP starts to crystallize if the temperature is reduced below 180 °C (Fig. 9). Only MAP crystallizes until the temperature reaches 140 °C (point 2); then both MAP and DAP begin to crystallize. On further cooling the composition of the liquid phase follows the borderline between the MAP and DAP crystallization areas. The last liquid crystallizes at the eutectic temperature of ca. 130 °C (point 3).

With increasing water content, the system becomes a quaternary system. To simplify the tetrahedral phase diagram, the water level is kept constant while the three other components are varied. This corresponds to cutting the tetrahedron at a given water content. Figure 10 shows a section through such a tetrahedron, where the water content for all compositions is 10 wt %.

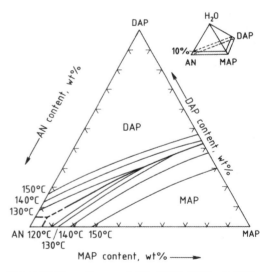

Figure 10. Phase diagram of AN – MAP – DAP with 10 % water [27]

Three primary crystallization areas exist. The lines represent compositions at which the liquid phase is in equilibrium with two solid phases. For one composition the liquid phase is in equilibrium with the three solid phases, AN(s), MAP(s), and DAP(s).

Calcium Phosphates. The phosphate component in superphosphates, MCP, is produced as follows:

$$Ca_5F(PO_4)_3 + 7 H_3PO_4 \rightleftharpoons 5 Ca(H_2PO_4)_2 + HF$$

Phosphoric acid is produced by the sulfuric acid treatment of phosphate rock, gypsum ($CaSO_4$) is formed as a byproduct. In triple superphosphate production, gypsum is separated from phosphoric acid prior to MCP formation. If calcium sulfate ends up in the product, it can be considered to be an inert solid.

In nitrophosphate processing, ca. 50 % of the remaining calcium in the phosphoric acid mother liquor reacts to give DCP, $CaHPO_4$, during neutralization.

Both superphosphate and nitrophosphate processing involve calcium phosphate solid – liquid equilibria:

$$Ca^{2+} + 2 H_2PO_4^- \rightleftharpoons Ca(H_2PO_4)_2 \text{ (s) MCP}$$

$$Ca^{2+} + HPO_4^{2-} \rightleftharpoons CaHPO_4(s) \text{ DCP}$$

Because these equilibria depend on ion concentration and temperature, the phase diagram described by GMELIN may be used to determine which calcium phosphate precipitates (Fig. 11) [28].

MCP is produced under the conditions used in superphosphate production (high liquid P_2O_5 concentration); DCP is produced under the conditions used in nitrophosphate processing (high temperature, lower liquid P_2O_5 concentration).

When the compound-fertilizer route is based on superphosphates, MCP is partially converted to DCP and MAP during neutralization with ammonia:

$$Ca(H_2PO_4)_2 + NH_3 \longrightarrow CaHPO_4 + NH_4H_2PO_4$$

In compound-fertilizer granulation, both MCP and DCP are in the solid state and can be considered as inert.

Additives in Compound-Fertilizer Production. Potassium salts may be added in compound-fertilizer production:

$$KCl + NH_4NO_3 \rightleftharpoons KNO_3 + NH_4Cl$$

$$K_2SO_4 + 2 NH_4NO_3 \rightleftharpoons 2 KNO_3 + (NH_4)_2SO_4$$

To achieve steady state, the potassium salt has to be dissolved in a multicomponent melt. This may affect salt crystallization and melt viscosity (Fig. 12). If this reaction is not controlled, the liquid phase may increase during granulation or storage.

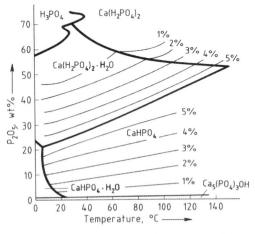

Figure 11. Phase diagram for calcium phosphates [28] The lines indicate the minium CaO content in the liquid necessary for calcium phosphate precipitation

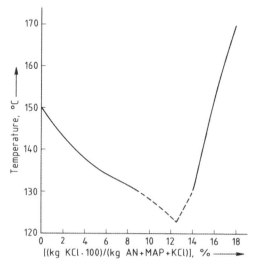

Figure 12. Crystallization points for potassium chloride dissolved in an 86 % AN – 14 % MAP melt with 0.5 % water [29]

Other salts added to MAP- and DAP-containing melts (e.g., ammonium sulfate, magnesium sulfate, dolomite, borax, micronutrients) can be considered as inert salts in phosphate fertilizer production.

8. Superphosphates

For definitions, see Chapter 2.

8.1. Single Superphosphate

8.1.1. Chemistry

Single superphosphate is produced by acidulation of finely ground phosphate rock with sulfuric acid. Many chemical reactions occur when the phosphate mineral is mixed with sulfuric acid (70 %). The mixture is liquid for 2 – 10 min but solidifies in the next 5 – 10 min.

After 40 – 60 min the superphosphate product is transported to storage where reaction is completed after 3 – 10 days. The reaction rates depend on the type and particle size of the ground phosphate rock, the type and concentration of trace elements in the rock, and the concentration and amount of sulfuric acid.

The primary overall reaction for fluorapatite is

$$2\,Ca_5F(PO_4)_3 + 7\,H_2SO_4 \longrightarrow 3\,Ca(H_2PO_4)_2 + 7\,CaSO_4 + 2\,HF$$

It has two consecutive stages:

$$Ca_5F(PO_4)_3 + 5\,H_2SO_4 \longrightarrow 3\,H_3PO_4 + 5\,CaSO_4 + HF$$

$$Ca_5F(PO_4)_3 + 7\,H_3PO_4 \longrightarrow 5\,Ca(H_2PO_4)_2 + HF$$

The first reaction is complete in 5 – 20 min. Components of the rock, such as calcium fluoride, calcium carbonate, silica, iron, and aluminum, all affect reaction rate.

Silica reacts with hydrogen fluoride from the above reaction to give fluosilicic acid, which dissociates to silicon tetrafluoride:

$$6\,HF + SiO_2 \longrightarrow H_2SiF_6 + 2\,H_2O$$

$$H_2SiF_6 \longrightarrow SiF_4(g) + 2\,HF$$

Phosphate rock contains 2 – 5 % fluorine, of which 10 – 30 % is rapidly evolved as silicon tetrafluoride during acidulation. The SiF_4-enriched off-gas is used for the production of fluosilicic acid. The rest of the fluorine remains in the superphosphate product as unreacted calcium fluoride, water-soluble calcium silicofluoride, or free fluosilicic acid.

Practical experience shows that phosphates rich in carbonate are acidulated more quickly and evolve more silicon tetrafluoride. The silicon tetrafluoride yield increases with higher sulfuric acid concentration, higher temperature, more finely ground rock, and longer mixing time. However, some of these conditions are harmful for the production of a good superphosphate.

The effects of particle size and acid concentration on the reaction rate for Morocco rock have been studied in the laboratory [30]. In the example shown in Figure 13 the free sulfuric acid is consumed after 60 min. The concentration of free phosphoric acid reaches a maximum of ca. 58 %. After 1 h, approximately 80 % of the phosphate is water-soluble, about 70 % of this being free phosphoric acid. The amount of water-soluble phosphate increases to 92 % after one week and to 93 % after three weeks. At this time, 14 % is free phosphoric acid (corresponding to 3 % P_2O_5 in superphosphate).

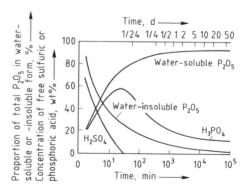

Figure 13. Reaction of Morocco phosphate rock with 76 % sulfuric acid

On a technical scale the mixture goes to storage for curing (i.e., complete reaction) after 0.5 – 2 h. The mixture solidifies to form a gel consisting of colloidal MCP and containing anhydrous calcium sulfate crystals. The gel structure gives the fresh superphosphate a high plasticity.

Superphosphate becomes fluid under pressure. These properties make handling difficult but facilitate granulation.

With increasing storage time and decreasing content of free acid, the gel structure becomes crystalline and plasticity decreases.

Byproducts. Superphosphate contains anhydrous calcium sulfate (gypsum). Small amounts of the dihydrate occur after prolonged storage when the content of free acid is very low. Hydration of calcium sulfate is inhibited strongly in the highly viscous aqueous phase. The stability limit for $CaSO_4 \cdot 2H_2O$ is reached rather early because hydration results in concentration of the liquid phase (Fig. 14).

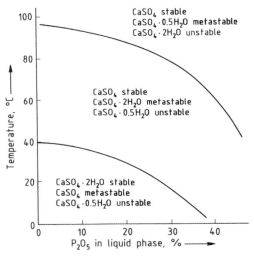

Figure 14. Different forms of gypsum in superphosphate [31]

When the concentration of free phosphoric acid decreases below a certain limit, citrate-soluble DCP is formed:

$Ca(H_2PO_4)_2 \rightleftharpoons CaHPO_4 + H_3PO_4$

Formation of DCP can be reduced or avoided by the use of more concentrated acid or by drying the superphosphate product to increase the acid concentration in the liquid phase.

The iron and aluminum compounds in phosphate rock can form water-soluble compounds during acidulation, but they react to produce water-insoluble compounds when the amount of free acid is reduced.

8.1.2. Production [31]

The production of superphosphate consists of grinding the phosphate rock, mixing the reaction components, acidulation, curing in a den, transportation to storage for final curing, and granulation or further processing to compound fertilizers.

Grinding Phosphate Rock. Phosphate rock is ground before being mixed with sulfuric acid. The particle size depends on different factors: reactive phosphate rock can be ground more coarsely; the use of concentrated sulfuric acid demands a more finely ground rock. Modern continuous superphosphate plants with a short curing time also need a more finely ground rock, in general >90 % through 100 mesh (<150 μm).

Batch Processes. Up to about 1970 the batch process dominated because it allowed a more accurate feed of phosphate rock and sulfuric acid. Mixing time could also be adjusted to give a good quality product. The primary disadvantage was the need for manual cleaning.

Examples of batch processes are the Beskow den process (developed in Europe, Fig. 15) and the Sturtevant den process (developed in the United States).

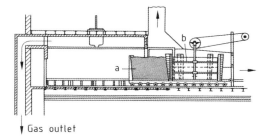

Figure 15. Beskow den batchwise superphosphate process
a) Reaction vat; b) Rotating cutter for removing superphosphate product

Beskow Den Process. The chamber floor is mounted on rollers. The curved back wall of the chamber is fixed to the floor and the side walls can be folded away. The front wall consists of two folding doors that are opened prior to emptying the chamber. The chamber wagon is then pulled toward a fixed revolving cutter equipped with vertical knives that scrape off the superphosphate. Beskow wagons can hold 50–100 t of product. Mixing of the components must be rapid and thorough so that the mixture remains thin enough to run over the surface of the chamber contents.

Figure 16. The TVA mixer for superphosphate production

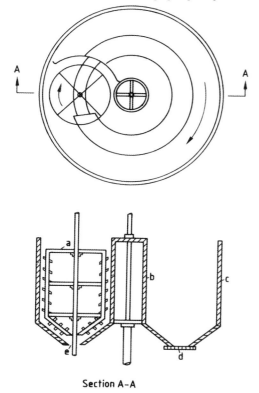

Figure 17. Moritz–Standaert superphosphate den process [32]
a) Cutter; b) Inner cylinder; c) Outer cylinder; d) Stationary bottom; e) Superphosphate outlet

Continuous Processes. Continuous processes are now dominant. Sulfuric acid is measured and controlled by magnetic flow meters. Different types of paddle mixers are used.

The *Tennessee Valley Authority (TVA)* has developed a simple cone mixer (Fig. 16) without moving parts. The acid is fed in by two to eight tangential nozzles, and phosphate rock is fed into the center of the swirling acid. This mixer is usually combined with a den belt and used mainly to produce triple superphosphate.

The *Moritz–Standaert den process* was often used in Western Europe and involves a movable, circular den (Fig. 17).

Another common European process is the *Broadfield den process* (Fig. 18). The Broadfield den consists of a slat conveyor mounted on rollers, with a long stationary box over it and a revolving cutter at the end. A variable-speed drive is used for the conveyor, giving a retention time of ca. 30 min [32]. A similar process was developed by *Nordengren*. Here, a reaction chamber is built into the top of the den to allow sufficient time for solidification of the superphosphate mixture [32].

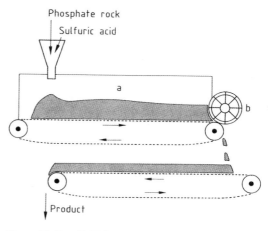

Figure 18. Broadfield den process
a) Den; b) Cutter

In the *Kuhlmann den process* (Fig. 19), denning is accomplished by transporting a thin layer of the mixture on a long, flexible belt conveyor. The den is troughed for more than half its length to form its own side walls. A hood is provided to collect the off-gas [32].

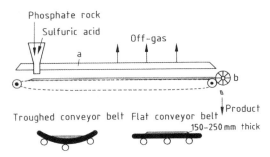

Figure 19. Kuhlmann den superphosphate process
a) Hood; b) Cutter

A high-speed mixer used in combination with a den has been developed by *Agrimont* in Italy (Fig. 20). The very brief and intense mixing gives a short curing time and a high-quality superphosphate.

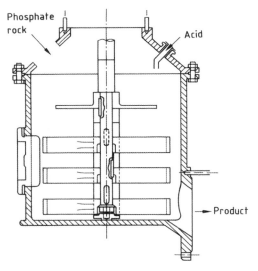

Figure 20. Agrimont mixer for superphosphate production (courtesy of Agrimont Italy)

8.1.3. Fluorine Recovery [33]

Off-gases from the mixer and the den contain air, steam, carbon dioxide, and silicon tetrafluoride. The fluorine content is normally 0.5–2 vol % but depends on process conditions and air flow.

When the off-gases are washed in scrubbers, silicon tetrafluoride reacts with water to form fluosilicic acid [*16961-83-4*], which is used as a commercial product (→ Fluorine Compounds, Inorganic, Chap. 5.2.):

$$3\,SiF_4 + 2\,H_2O \longrightarrow 2\,H_2SiF_6 + SiO_2$$

The silica is at first colloidal and therefore impossible to filter. It solidifies in an uncontrollable fashion.

The first cleaning units consisted of Venturi scrubbers and spinning disk scrubbers (Fig. 21). Hydro Supra has now developed a process giving clean fluosilicic acid (20–25 % H_2SiF_6 containing <100 mg of P_2O_5 per liter) suitable as a raw material for the production of aluminum fluoride from alumina (Fig. 22). The first stage is a Venturi scrubber (a), where solid P_2O_5 is absorbed. The fluosilicic acid produced in absorption towers (b) is fed to an aging tank (c) where silica particles grow. The silica particles are then filtered from the product acid (d) and mixed with the Venturi liquid (e) before being recycled to the process.

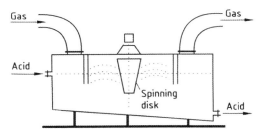

Figure 21. Spinning disk scrubber

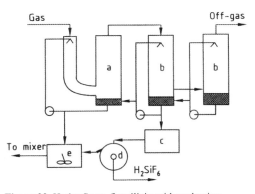

Figure 22. Hydro Supra fluosilicic acid production
a) Venturi scrubber; b) Absorption tower; c) Aging tank; d) Filter; e) Recycle tank

8.1.4. Granulation [34]

After 3–6 weeks storage, superphosphate may be crushed, bagged, and sold as powder superphosphate. The product is dusty and has a high caking tendency. This disadvantage can be avoided by granulation to 2–5-mm particles. Granulated superphosphates have been on the market since 1950 and now account for all commercial superphosphates in the Western world.

Run-of-pile powder superphosphate is granulated with steam or water in a drum (length 6–12 m, diameter 3–5 m). Another possibility is use of a pan granulator with a diameter of 3–5 m.

Freshly produced superphosphate is more easily granulated but may cause handling problems due to its plasticity (see Section 8.1.1).

After granulation, the superphosphate product is dried in rotating drums, cooled in a similar drum, and screened to a product size of 2–5 mm. The fine undersize particles are directly recirculated to the granulator, whereas the coarse oversize particles are crushed before recycling. The final product can be cooled further before storage and dispatch.

Dust formed in the dryer, the cooler, the screens, and the crushers is separated from the off-gases in dust-collecting equipment.

8.2. Triple Superphosphate [35]

Single superphosphate contains up to 50 % calcium sulfate. If sulfuric acid is replaced by phosphoric acid for acidulation of phosphate rock, triple superphosphate is produced with a higher content of MCP and without calcium sulfate. The principal chemical reaction is

$$Ca_5F(PO_4)_3 + 7 H_3PO_4 \longrightarrow 5 Ca(H_2PO_4)_2 + HF$$

Triple superphosphate has similar properties to single superphosphate but lower plasticity. Optimal process conditions and product quality are reached with a phosphoric acid concentration of 50–54 % P_2O_5. The same equipment is used as for single superphosphate production (see Section 8.1.2). However, the mixing time is shorter, due to faster chemical reaction (10–20 s). The reaction heat is one-third that for single superphosphate. The same temperature (80–100 °C) is reached, but less water vapor and silicon tetrafluoride are evolved.

Granulation based on run-of-pile powder triple superphosphate is similar to granulation of single superphosphate (Section 8.1.4).

A slurry granulation process can also be used. The Dorr–Oliver process route is usually employed. Phosphoric acid (40 % P_2O_5) is mixed with ground phosphate rock in a large tank reactor with a retention time of 3–5 h. The slurry is then pumped directly to a granulator. As in the run-of-pile route, the granulation process consists of drying, crushing, recycling, and cooling.

8.3. Double Superphosphate

In the production of double superphosphate, phosphate rock is treated with a mixture of sulfuric and phosphoric acids. The phosphate content of double (enriched) superphosphate thus corresponds to mixtures of single superphosphate and triple superphosphate containing 18–45 % P_2O_5. The product is used mainly as an intermediate for PK products [36].

By mixing sulfuric and phosphoric acids, more dilute phosphoric acid can be used. This reduces the energy requirement for concentration of phosphoric acid. The balance between sulfuric acid and phosphoric acid depends on the acid concentration [32].

8.4. PK Fertilizers [36]

Triple superphosphate or double superphosphate can be granulated together with potassium chloride, potassium sulfate, and magnesium sulfate. The granulation process is the same as for single superphosphate (see Section 8.1.4), but chloride salts release hydrogen chloride gas during drying. The hydrogen chloride gas must be recovered in a scrubber.

The PK fertilizers are normally applied in the fall.

9. Ammonium Phosphates

9.1. Fertilizer Grades and Applications

Fertilizer-grade ammonium phosphates are normally produced from nonpurified wet process phosphoric acid (WPPA, or the so-called sulfur route) and consequently have a purity of only 70–85 %. In spite of relatively large deviations from the stoichiometric N/P ratio of the pure products, in the fertilizer grade they are called monoammonium phosphate (MAP), diammonium phosphate (DAP), and ammonium polyphosphate (APP). Triammonium phosphate, $(NH_4)_3PO_4$, has no significance as a fertilizer because it is unstable at 20 °C.

The nutrient content of solid ammonium phosphate fertilizers depends greatly on the quality of the phosphoric acid (Table 2).

Table 2. Composition of ammonium phosphate fertilizers

Type	N, wt%	P_2O_5, wt%
MAP from fertilizer acid	10–12	57–48
MAP from purified acid	12	61
DAP from fertilizer acid	16–18	48–46
DAP from purified acid	21	53

Granular MAP and DAP are used as straight fertilizers or applied after mechanical blending with other granular fertilizers (bulk blending). Ammonium phosphate salts are also formed in situ or mixed into processes for the production of compound NP or NPK fertilizers from phosphoric acid or nitrophosphates.

Solid MAP is used extensively in the United States for the manufacture of suspension fertilizers (see → Fertilizers, Chap. 3.2.3.).

Ammonium polyphosphates are normally a constituent of liquid fertilizers (see → Fertilizers, Chap. 3.2.2.1.). The solid products were also expected to have a very important potential in connection with their high plant nutrient value and other special features, but in practice this did not turn out to be true. Only small quantities of solid products have been produced with a polyphosphate content up to 25 % of the ammonium phosphate [37], [38].

Pure MAP and DAP are used as fertilizers in irrigated greenhouses (completely water-soluble compounds).

9.2. Production

For a long time solid ammonium phosphate fertilizers were produced only in granular form in an integrated neutralization–granulation process [39]. This is still the case for DAP, but since the mid-1960s, considerable capacity has existed for producing MAP in powder form. This product is used as a relatively concentrated, low-cost intermediate for compound-fertilizer manufacture.

Crystallization processes for technically pure ammonium phosphates are not described here.

The high ammonia vapor pressure of DAP has important consequences for its industrial production. The same applies to the pronounced maximum solubility of a mixture of MAP and DAP at an N/P molar ratio of ca. 1.4 [40] (eutectic mixture containing 17 % water, fp 110 °C). The point at which the borderline in Figure 7 crosses the 110 °C isotherm corresponds to the composition 46 % MAP, 37 % DAP, and 17 % H_2O. This is equal to an N/P molar ratio of 1.4 (Fig. 6).

9.2.1. Ammonium Phosphate Powder

MAP Powder. The processes used for the production of MAP powder (particle size 0.1–1.5 mm) are relatively simple. All processes are claimed to give good storage, handling, and granulation properties when used for the manufacture of MAP, DAP, or other NP/NPK compound fertilizers.

In the *PhoSAI process* of Scottish Agricultural Industries [41] (Fig. 23), ammonia reacts with phosphoric acid (minimum concentration 42 % P_2O_5) in a stirred-tank reactor (d) to produce a slurry with an N/P ratio of 1.4 (point of maximum solubility) under atmospheric pressure. The slurry flows to a pin mixer (e) in which the N/P ratio is brought back to ca. 1 by the addition of more concentrated phosphoric acid. During this step the solubility decreases and more water vaporizes due to the heats of reaction and crystallization. A solid product is formed that typically contains 6–8 % moisture. This is screened (f) and the oversize particles are ground, after which the product can be stored without further treatment.

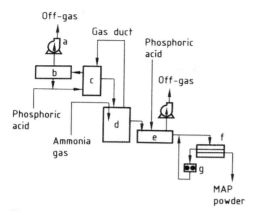

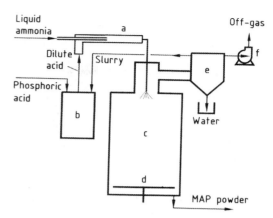

Figure 23. PhoSAI process for nongranular MAP
a) Fan; b) Separator; c) Gas scrubber; d) Reactor; e) Pin mixer; f) Screen; g) Oversize crusher

Figure 25. Swift process for MAP powder
a) Pipe reactor; b) Surge tank; c) MAP tower; d) Discharge scraper; e) Scrubber system; f) Fan

In the *Minifos process* [42] of Fisons (now Hydro Fertilizers), ammonia and phosphoric acid (45–54 % P_2O_5) react (Fig. 24, b) at 0.21 MPa to give an N/P ratio of 1. The liquid product is flash-sprayed in a natural-draught tower (c), where the droplets solidify to form a powder with 6–8 % residual moisture.

More recent MAP powder plants are based on the use of pipe reactors (including those designed by Hydro Fertilizers).

Considerable quantities of MAP powder are produced from "sludge acids" from phosphoric acid clarification with solids contents up to 20 %.

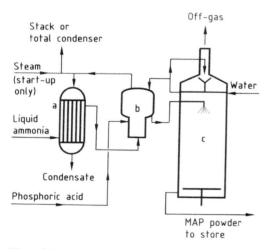

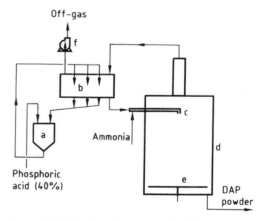

Figure 24. Hydro Minifos process for MAP powder
a) Ammonia evaporator; b) Pressure reactor; c) Spray tower

Figure 26. ERT – Espindesa process for DAP powder
a) Buffer vessel; b) Scrubber; c) Pipe reactor; d) DAP tower; e) Discharge scraper; f) Fan

Alternative processes based on the use of a pipe reactor instead of a tank reactor were developed by Gardinier S.A. [39], Swift Agricultural & Chemical Corporation (Fig. 25) [43], and ERT – Espindesa [44], [45]. Product humidity is reported to be as low as 2 %, which results in less caking.

DAP Powder. Based on the favorable results obtained with a pipe reactor for MAP powder manufacture, ERT – Espindesa experimented with the production of DAP powder [44], [45] (Fig. 26). The DAP product has a remarkably high water- and citrate-soluble P_2O_5 content, probably due to the short residence time in the reactor, which does not yield the insoluble

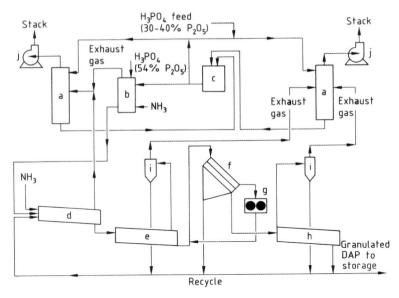

Figure 27. The TVA process for granulated DAP
a) Scrubber; b) Preneutralizer; c) Surge; d) Granulator; e) Dryer; f) Screens; g) Crusher; h) Cooler; i) Cyclone; j) Fan

salts that are normally formed in long-retention-time DAP plants. The same effect has been observed with MAP formed in a pipe reactor.

9.2.2. Granular Ammonium Phosphates

Since the 1960s, granular ammonium phosphates have belonged to some of the most important fertilizer products [46]. They are applied directly or used as intermediates in bulk blends and for the manufacture of compound fertilizers. More than five times as much DAP is produced as MAP. Since the beginning of large-scale production in the 1930s, considerable developments have been made to optimize raw material and energy efficiencies, to reduce the capital cost of the plants, to improve their reliability, and make them more environmentally friendly.

Granular ammonium phosphate plants consist of a wet and a dry section, the dry section being a granulation loop ($\rightarrow$ Fertilizers, Chap. 5.).

Granular DAP. In the 1950s the *Dorr–Oliver process* was most commonly used. Phosphoric acid and ammonia react in two or three vessels in series at atmospheric pressure under conditions that take advantage of solubility and vapor pressure properties.

Initially, three reactor vessels were used. Neutralization to MAP was carried out in the first. In the second, more ammonia was injected until ca. 80 % DAP was formed. Final ammoniation occurred in the third reactor but was later carried out in the blunger (pugmill) of the granulation loop. Because of the low solubility of DAP the process must be run with a high slurry water content to maintain fluidity. High recycle ratios are also required to avoid overgranulation in the blunger where the slurry and recycle are mixed to give a moist granulate containing 4–6 % water. Typically, these recycle ratios are between 11 : 1 and 8 : 1, depending on where final ammoniation occurs. The granules are dried in a concurrent rotary dryer and screened. The fines, smaller granules, and broken oversize particles are recycled to the pugmill. Cooling of the on-size product (1.5–3 mm) is not essential but is recommended to minimize loss of ammonia and ensure good storage stability.

Off-gases from reactors, granulator, and dryer are scrubbed with filter-strength (28 % P_2O_5) acid. Stronger acid (40–52 % P_2O_5) is used in the reactor section.

In the early 1960s the TVA [47], [48] (Fig. 27) introduced a modified process that became very

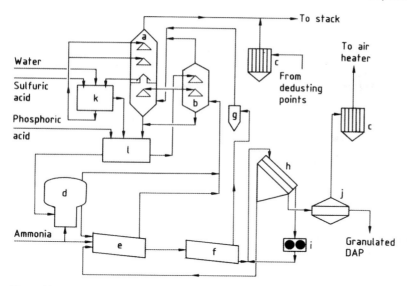

Figure 28. Hydro Fertilizers DAP granulation process with pressure reactor
a) Tail-gas scrubber; b) Ammonia scrubber; c) Bag filter; d) Pressure reactor; e) Granulator; f) Dryer; g) Cyclone; h) Screens; i) Crusher; j) Fluidized-bed cooler; k), l) Surge tanks

popular because the recycle ratio dropped to ca. 5:1.

Acid reacts with ammonia in a preneutralizer (b) where the N/P ratio is controlled at about 1.4. The heat of reaction raises the slurry temperature to the boiling point (ca. 115 °C) and evaporates water. The hot slurry contains 15–20 wt % water and is pumped to a rotary drum granulator (d) where it is sprayed on a relatively thick bed of recycled material. Additional ammonia is sparged underneath this bed to increase the molar ratio to 2.0. Additional heat is generated, resulting in the evaporation of more moisture. The decrease in solubility obtained in going from a molar ratio of 1.4 to 2.0 assists granulation.

Further reduction of the recycle ratio (e.g., to 4:1) is attained by carrying out preneutralization to an N/P ratio of 1.4 in a pressure reactor (Hydro Fertilizers) [47], [49] operating at 0.1 MPa overpressure; so that the boiling point of the reaction mixture is elevated by ca. 20 °C compared to operation at atmospheric pressure (Fig. 28). Thus, the steep temperature dependence of the solubility of ammonium phosphate in water allows operation at a lower water content while maintaining the ammonium phosphate in solution. The amount of water fed to the granulator is therefore minimized.

In an attempt to further simplify DAP plants and reduce the recycle ratio, TVA, CROS S.A., and ERT–Espindesa have tried to use pipe reactors [50] (also called T reactors or pipe-cross reactors) directly releasing slurry in the rotary granulator, instead of applying a preneutralization tank.

CROS and ERT–Espindesa [51] report successful operation, but experiences in several U.S. plants were disappointing because of high heat input in the granulator.

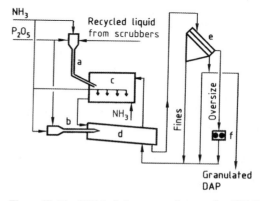

Figure 29. The AZF dual pipe process for granulated DAP
a, b) Pipe reactors; c) Granulator; d) Dryer–cooler; e) Screen; f) Mill

CdF – Chimie AZF [51–53] originally had the same experience, but succeeded by introducing a second pipe reactor spraying into the dryer (Fig. 29).

Approximately half of the phosphoric acid is fed to a pipe reactor (a) that operates at N/P = 1.4 and releases its product into the granulator (c). The remaining acid is fed into a second pipe reactor (b) at N/P = 1.1 that sprays into the dryer (d). Extra ammonia is introduced underneath the rolling bed in the rotary drum granulator.

The pipe reactor (b) in the dryer produces MAP powder. Part of it is carried away into the cyclones of the dedusting loop; the remainder crystallizes on the DAP product.

The recycle ratio of this process is lower than with a conventional preneutralizer; fuel consumption of the dryer is also reduced.

Granular MAP. Granular MAP is produced in the same types of plants as DAP. The most common process is the TVA, with a preneutralizer and rotary drum ammoniation (Fig. 27). The preneutralizer is operated at an N/P ratio of 0.6, again a point of high solubility, and more ammonia is added in the granulator to increase the molar ratio to 1.0.

In contrast to DAP production, ammonia recovery by acid scrubbing is not necessary, but all off-gases are scrubbed to recover dust and fumes. Granular plants can thus be simpler, but most units are designed so that both products can be made in the same plant.

The pipe reactor processes [43] have been very successful in MAP production. They operate with a very slight ammonia loss, and little or no additional heat is required for drying. In some plants, preneutralizers have been replaced by pipe reactors.

9.2.3. Off-Gas Treatment

An essential process step in all ammonium phosphate plants is scrubbing of the off-gas. The preneutralizer and pipe reactors give ammonia slip due to vapor pressure and "mechanical losses" caused by imperfect mixing. Absorption of ammonia in the rotary drum or blunger is not 100 %; ammonia is released from DAP in the dryer and cooler due to its high vapor pressure.

Off-gases may also contain dust and ammonium fluoride aerosols.

In an environmentally friendly plant, all off-gases are treated [54]. In DAP plants, relatively large quantities of ammonia must be recovered. This is normally done by using filter-strength (28 % P_2O_5) phosphoric acid, which is subsequently sent to the preneutralizer or pipe reactor. Dilute sulfuric acid scrubbing may also be employed. Sulfuric acid can also increase the N/P ratio of the product.

Passage of large quantities of air through scrubbers that operate on wet process phosphoric acid results in the stripping of considerable quantities of fluorine-containing gases, which must be absorbed in tail-gas scrubbers. In the production of granular MAP, off-gas treatment is simpler due to the lower ammonia slip.

In modern powder MAP plants, entrained dust is recovered in a scrubber with circulating process water that is sent to the reactor section.

10. Compound Fertilizers by the Sulfur Route

Compound fertilizers (also called complex or mixed fertilizers) contain more than one plant nutrient element. They may be ternary (N + P + K) or binary (N + P, P + K, or N + K). They may also contain considerable quantities of magnesium or trace elements.

The reason for applying compound fertilizer is convenience: farmers no longer have to apply several straight (single-element) fertilizers separately. However, even if compound fertilizers are used, nitrogen is often still applied separately because supplementary dressings of nitrogen may be needed during the growing season. Also, some straight nitrogen materials (e.g., urea – ammonium nitrate solution) are sometimes considerably cheaper than the same amount of nitrogen in compound fertilizers.

The first compound fertilizers were powder mixtures (e.g., of ammonium sulfate, superphosphate, and potash). In the 1930s, producers began to ammoniate superphosphate and thus introduced a cheap form of nitrogen, while improving the physical properties of the superphosphate.

To lower transportation costs, higher-grade products were used, which, however, caused serious caking problems. Increased mechanical application called for free-flowing material, and in the 1950s granulation therefore started to become popular.

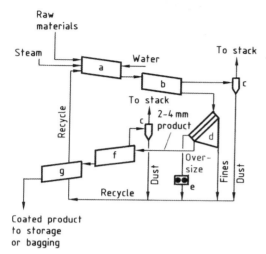

Figure 30. Typical plant for granulation of dry materials
a) Rotary drum granulator; b) Dryer; c) Cyclone; d) Screen; e) Crusher; f) Cooler; g) Coating drum

The concentration of compound fertilizers increased as single superphosphate was replaced by triple superphosphate, and ammonium nitrate or urea replaced ammonium sulfate. Later, even higher grades were attained by using phosphoric acid and ammonia. Production was no longer a simple, mainly mechanical operation, but rather a complex chemical processing operation.

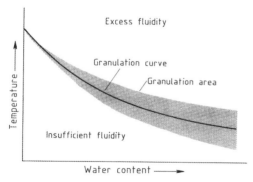

Figure 31. Relationship of temperature and water content for the granulation of fertilizers

Bulk blends (see → Fertilizers, Chap. 5.5.) became popular in some countries as a method by which a local dealer with a simple mixing plant could supply whatever mix the farmer needed. Granular materials with an approximately equal size distribution (e.g., DAP, compacted potassium chloride, ammonium nitrate, or urea) are used. These compounds are shipped in bulk to the blenders who offer prescription mixing, custom application, and other services to the farmers.

Dry mixing of nongranular materials is no longer used and therefore not described here.

10.1. Granulation of Mixtures of Dry Materials

Agglomeration with Steam or Water.
Mixtures of dry powders or ground components can be granulated by agglomeration with the aid of steam or water.

Several devices are used, but rotary drum granulation is the most popular method (Fig. 30).

Incoming materials are screened to remove coarse lumps, which are crushed. The materials are fed in by batchwise weighing or with continuous weighing belts. In the rotary drum granulator (a) the mixture of recycle and raw materials is adjusted for agglomeration by injecting steam underneath the bed and, if necessary, by spraying water on top. Granulation is performed so that optimum quantities of on-size granules (in Europe mostly 2 – 4 mm, in the United States normally 1 – 3.3 mm) are formed, with a minimum of oversize particles [55]. Obtaining more than 60 – 70 % of the product in the desired size range is difficult, but the recycle ratio for such operations is normally <1 : 1. Optimum granulation is obtained at a defined "liquid-phase" content (i.e., the water content plus the salts that dissolve in this amount of water). Because the solubility of salts increases with temperature, a higher temperature means that less water is required (Fig. 31). Although such a curve can be used to predict the optimum point at which granulation occurs, it does not describe granulation efficiency, which depends on the degree of plasticity of the mixture. Pure inorganic salts (e.g., KCl, NH_4NO_3) have little plasticity and are relatively difficult to granulate. In contrast, fresh single superphosphate shows excellent agglomeration properties (see section 8.1.2). The binding tendency of ammonium phosphates is highly dependent on their impurity content [56].

Steam granulation is normally better than the use of water alone because the damp granules are hotter, have a lower moisture content, and, therefore dry more easily. The product is usually more dense and stronger. The curve shown in Figure 31 depends not only on the composition of the salt mix, but also on the physical properties of the raw materials. The particle size of the feed and the potash coating may affect binding ability [57].

After the granulator, the remainder of the plant is similar for all granulation processes ($\rightarrow$ Fertilizers).

The grades produced in such a plant depend on the raw materials. The product can be very low in concentration (N-P_2O_5-K_2O = 9-9-9) if it is based on single superphosphate and ammonium sulfate, and very concentrated (19-19-19) if it is based on urea and nongranular MAP or DAP.

Compaction and Extrusion [58–62].
Roll compaction is described elsewhere ($\rightarrow$ Fertilizers, Chap. 5.3.5.) and is used relatively little for the manufacture of compound fertilizers. Compaction is used for materials that are difficult to granulate (e.g., straight KCl). It operates at a low moisture level (0.5 – 1.5 %) and at ambient temperature, so that heat-sensitive materials can be admixed.

Particles made by compaction tend to be angular compared to the spherical granules produced by other shaping techniques. A further disadvantage is that the compacted particles may react with each other, resulting in disintegration.

10.2. Granulation of Dry Materials with Additives Producing Chemical Reactions

Chemically reacting materials can be introduced to increase the temperature. This allows granulation to proceed with a minimum water content and with reduced amounts of steam or water.

Ammoniation of superphosphates by injection of liquid or gaseous ammonia underneath the bed or by use of aqueous solutions of ammonium nitrate or urea containing free ammonia is such a technique. The ammonia initially neutralizes free acid in the superphosphate and then reacts with the water-soluble MCP to form non-water-soluble, but citrate-soluble P_2O_5 (mainly DCP) [63–65].

Whether the decrease in water-soluble P_2O_5 content is acceptable depends on farmers' requirements. The United Kingdom and The Netherlands traditionally prefer water-soluble P_2O_5, whereas in the United States, France, and Germany "available" P_2O_5 is fully accepted as an equally valuable plant nutrient.

When nongranular MAP is present in the formulation it is often beneficial in terms of improving the granulability of the material that is to be ammoniated in the granulator. This increases the solubility of the ammonium phosphate, and heat is generated due to the chemical reaction of ammonia with MAP.

If heat released during ammoniation of the superphosphate or MAP is insufficient, sulfuric or phosphoric acid and more ammonia can be added directly to the granulator [66]. The development of the continuous ammoniator granulator by TVA has been very important in this context [65], [67]; see $\rightarrow$ Fertilizers, Chap. 5.3.2. This technique has been used very successfully in the granulation of solid urea-based NPK grades [66].

The TVA process was subsequently extended with a preneutralizer [68], [69], a reactor in which a slurry is produced by reacting ammonia or ammoniating solution with sulfuric or phosphoric acid [70]. It is used for formulations in which the heat of reaction becomes too great for release in the granulator.

The preneutralizer operates at the optimum N/P ratio of 1.4 to minimize introduction of water. Preneutralizers have recently been replaced by pipe reactors, which allow direct release into the granulator of slurry or melt with an even lower water content. An alternative is to introduce all or part of the P_2O_5 in the form of MAP powder; the crystallizing ammonium nitrate solution then provides the necessary plasticity. For solid urea-based NPK grades the pipe reactor process is a viable method.

10.3. Slurry Granulation

In slurry granulation, all or most of the ingredients enter the granulation system in slurry form (Fig. 32). See also $\rightarrow$ Fertilizers, Chap. 5.2.

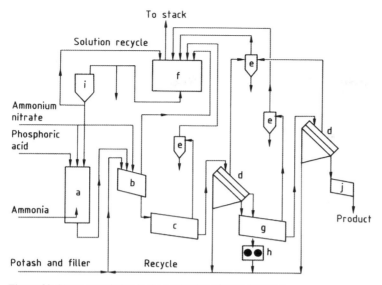

Figure 32. Slurry granulation in the production of NPK fertilizer
a) Preneutralizer; b) Blunger; c) Dryer; d) Screen; e) Cyclone; f) Scrubber; g) Cooler; h) Crusher; i) Cyclone; j) Coating drum

The slurry is usually manufactured by neutralization of nitric, phosphoric, or sulfuric acid–or mixtures of these acids–with ammonia. Potash may be added to the neutralized slurry before granulation or in the granulator. If potassium chloride is required, premixing with slurries containing ammonium nitrate is beneficial for product quality because the reaction

$$KCl + NH_4NO_3 \longrightarrow KNO_3 + NH_4Cl$$

is virtually complete before the product goes to storage.

The granulator may be of the rotary drum or pugmill (blunger) type (→ Fertilizers, Chap. 5.3.), and recycled material (fines, broken oversize, and part of the on-size) is added in a proportion that ensures good granulation efficiency.

If the preneutralizer is used only to produce ammonium phosphate, the slurry must contain 15–20 % water so that it remains sufficiently fluid even when operating at the optimum N/P molar ratio. This creates conditions of excessive fluidity in the granulator, particularly when high-concentration grades such as 17-17-17 or 23-23-0 are produced. Large amounts of recycle are therefore required to control granulation, particularly when ammonium nitrate is added in the form of a concentrated solution rather than a solid.

As in the manufacture of granular DAP or MAP (see Section 9.2.2), processes have been modified to reduce the recycle ratio and thus save energy and boost the capacity of existing units or save capital when constructing new units. Employment of a cooled recycle by positioning the screening units downstream from the cooler allows granulation to occur at lower temperature and thus higher water content (Fig. 31) [71].

In a plant that produces grades containing both ammonium nitrate and ammonium phosphate, the phosphoric and nitric acids can be neutralized separately or together (coneutralization). Although separately neutralized ammonium nitrate can be concentrated easily (e.g., up to 95–98 %), the ammonium phosphate slurry still contains 15–20 wt % water. However, addition of ammonium nitrate to the ammonium phosphate increases its fluidity which allows the water content of the slurry to be decreased [72], [73]. This is the reason why nitric and phosphoric acids are often coneutralized. The coneutralized liquor is pumped to the granulator, if necessary, after further evaporation.

Other techniques to reduce the recycle ratio include the use of a pressure reactor to increase the boiling point and thus increase the solubility of ammonium phosphate. Fluid slurries are therefore obtained at a low water content [74], [75].

A pipe reactor can also be used that operates at a higher temperature and a lower water content than the preneutralizer. The pipe discharges directly into the granulator. If the ventilation capacity is sufficient to remove all the process steam, the recyle requirement decreases [76]. Pipe reactors have been fitted in blungers [77], rotary drum granulators [76], [78], [79], dryers, or both in rotary drum granulators and dryers [80], [81].

The use of a relatively high recycle in a slurry granulation plant is not always a disadvantage [82], [83]. Granule formation takes place largely by layering. The water is on the outside of the granule which facilitates drying. Because only a small portion of the granules can be exported, the screening operation is not very critical and the size-distribution quality is good. Finally, plants operating with a high recycle ratio and where granulation is "water-balance controlled" are normally more stable than low-recycle granulation plants whose capacity is determined by granulation efficiency, which depends on fine tuning by the operators.

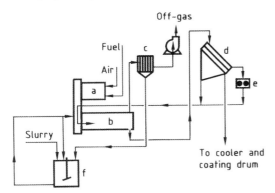

Figure 33. C & I – Girdler spherodizer granulation process for production of compound fertilizers
a) Air heater; b) Spherodizer granulator; c) Dust removal; d) Screen; e) Crusher; f) Surge tank

Slurry granulation is also carried out in spherodizers (Fig. 33) – spray-drum granulators in which granulation and drying are combined (→ Fertilizers, Chap. 5.2., Chap. 5.3.) [84].

Neutralized slurry containing all the potash is sprayed against a falling curtain of recycle solids produced by flights in a rotating drum. Hot air is blown in cocurrently with the slurry spray. Granulation takes place by layering. Granulation efficiency is high and the need for recycle is relatively low. The granules are well rounded, hard, and generally considered to be of excellent quality.

Drying during granulation is also a feature of the Scottish Agricultural Industries double-drum granulation process with a very large internal recycle (→ Fertilizers) [85]. Granulation and drying of compound fertilizers in fluidized-bed and spouted-bed equipment are also reported [86].

10.4. Melt Granulation

See also → Fertilizers.

Mixtures of ammonium nitrate and ammonium phosphate maintain their fluidity at very low moisture content provided the temperature is sufficiently high (Figs. 9 and 10). The concentrated melt can therefore be allowed to solidify and dry by the heat of crystallization.

If the acid is concentrated and/or preheated, the heat of neutralization may be sufficient to drive out all the water. If not, the N – P solution can be concentrated by using external heat in an evaporator. Removal of the water before granulation is thermally more efficient than using a conventional dryer. Hydro Fertilizers operated such a melt granulation plant for several years (Fig. 34) [87].

Another way to produce melts is to use a pipe reactor operating at high temperature [88]. The pipe produces virtually anhydrous melts, which are blown directly onto a cooled recycle bed in the granulator (Fig. 35).

Normally phosphoric-acid-containing solutions mixed with nitrates are not ammoniated in pipe reactors because chlorine and organic impurities may reduce the safety limits for ammonium nitrate explosion. The practical use of pipe reactors is thus limited to processing of either pure ammonium nitrate or ammonium phosphates and sulfates.

The TVA ran a demonstration plant in which 15 – 25 % of the P_2O_5 was condensed to nonorthophosphates, mainly pyrophosphates [89].

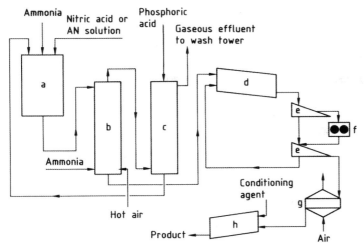

Figure 34. Hydro Fertilizers melt granulation process for the production of NPK fertilizer
a) Reactor; b) Dehydrator; c) Scrubber; d) Granulator; e) Screen; f) Mill; g) Cooler; h) Coating

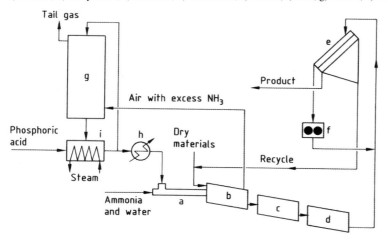

Figure 35. Pipe reactor – drum granulator process
a) Pipe reactor; b) Drum granulator; c) Dryer; d) Cooler; e) Screens; f) Crusher; g) Scrubber; h) Heater; i) Heat exchanger

Potassium salts and other solid raw materials can be added to the melts to produce NPK products.

Methods to form particles from melts include pan, pugmill, and rotary drum granulation, as well as prilling [90] and flaking.

In a conventional melt granulation loop, the recycle requirement is determined not by the water balance, but by the need to remove sufficient heat. The recycle ratio is so low that the plant will be "granulation-efficiency" controlled.

11. Compound Fertilizers by the Nitro Route

Nitrophosphates are nitrogen- and phosphate-containing fertilizers that are produced by digestion of phosphate rock with nitric acid. Examples of known production processes are the mixed acid and Odda processes.

The nitrophosphate process was invented by ERLING B. JOHNSON of Odda, Norway [91] to avoid the diluting effect of the sulfate ion in superphosphate. The process had been developed to the commercial stage by the 1930s, and

Norsk Hydro introduced its own technology in 1938. However, not until the sulfur shortage in the early 1950s did nitrophosphates become of interest, and a number of variants of the process were developed [92][93][94].

Approximately 10 % of the world's phosphate fertilizers are produced by the nitrophosphate route.

11.1. Chemistry

The overall acidulation reaction is given by

$$Ca_5F(PO_4)_3 + 10\,HNO_3 \longrightarrow 5\,Ca(NO_3)_2 + 3\,H_3PO_4 + HF$$

The reaction is normally carried out with an excess of 10–20 % nitric acid [95], [96]; acid-insoluble materials in the phosphate rock may be partly removed.

Most nitrophosphate processes include removal of calcium nitrate from the solution to increase the nutrient content of the product, to avoid precipitation of water-insoluble DCP during neutralization and to obtain a product that meets thermal stability requirements for transportation.

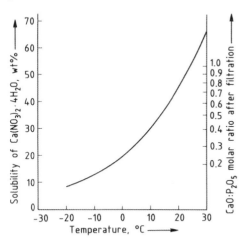

Figure 36. Crystallization points for calcium nitrate in the digestion liquor (digestion of phosphate rock with nitric acid)

About 60 % of the calcium nitrate crystallizes at 20 °C and 80–85 % at −5 °C (Fig. 36) [97], [98]. Alternatively, calcium can be precipitated as calcium sulfate by addition of ammonium sulfate to the solution. If the reason for using the nitrophosphate process is to remain independent of a sulfur source, sulfur can be recycled in the Merseburg process:

$$CaSO_4 + 2\,NH_3 + CO_2 + H_2O \longrightarrow CaCO_3 + (NH_4)_2SO_4$$

Calcium can be precipitated as calcium carbonate with carbon dioxide:

$$Ca(NO_3)_2 + 2\,NH_3 + CO_2 + H_2O \longrightarrow CaCO_3 + 2\,NH_4NO_3$$

Care must be taken to suppress formation of tricalcium phosphate, which is citrate insoluble:

$$2\,CaHPO_4 + CaCO_3 \longrightarrow Ca_3(PO_4)_2 + H_2O + CO_2$$

Another nitrophosphate process is the ion-exchange method in which dissolved calcium ions are replaced by potassium ions from a potassium-loaded ion-exchange resin [99]. An advantage of this process is that the resulting phosphoric acid–potassium nitrate solution can be used for making chloride-free NPK fertilizers.

11.2. Product Specification

High-grade fertilizers containing 50–90 % water-soluble P_2O_5 are produced, depending on the amount of calcium removed. Citrate-soluble P_2O_5 accounts for 99–100 % of the total phosphorus. The N/P_2O_5 ratio obtained by the normal nitrophosphate route ranges from about 0.7 to 5.0 in actual plant operation. Above a ratio of 2, ammonium nitrate from external sources must be added. If no ammonium sulfate is added, ca. 55–60 % of the nitrogen is in the form of ammonium and 40–45 % in the form of nitrate.

Today's nitrophosphate products are of a quality comparable to those based on the sulfur route [100].

11.3. Nitrophosphate Process with Calcium Nitrate Crystallization (Hydro)

Norsk Hydro has carried out extensive research and development on nitrophosphate process technology. The principles of the Odda process have been retained, i.e., phosphate rock digestion and calcium nitrate crystallization.

The process consists of three main parts:

1) Mother liquor production (phosphoric acid with nitric acid, Fig. 37)
2) Production of prilled or granulated products
3) Conversion of calcium nitrate to ammonium nitrate and calcium carbonate

The calcium nitrate can also be processed to granulated calcium nitrate fertilizers.

Digestion. Phosphate rock is dissolved in an excess of 60 % nitric acid in two stirred-tank reactors in series (Fig. 37). Because of the heat of reaction, the digestion temperature increases to 50–70 °C.

The main reactions for phosphate rock containing silica, carbonates, and metal oxides are:

$$Ca_5F(PO_4)_3 + 10\, HNO_3 \longrightarrow 3\, H_3PO_4 + 5\, Ca(NO_3)_2 + HF$$

$$6\, HF + SiO_2 \xrightarrow{-2H_2O} H_2SiF_6 \rightleftharpoons SiF_4 + 2\, HF$$

$$YCO_3 + 2\, HNO_3 \longrightarrow Y(NO_3)_2 + CO_2 + H_2O$$
$$Y = Ca\ or\ Mg$$

$$X_2O_3 + 6\, HNO_3 \longrightarrow 2\, X(NO_3)_3 + 3\, H_2O$$
$$X = Al\ or\ Fe$$

Acidulation of phosphate rock liberates small amounts of nitrogen oxides (NO_x), water vapor, hydrogen fluoride, and silicon tetrafluoride. These gases are vented to a scrubbing system.

The amount of NO_x formed is highly dependent on the amount of iron, sufides, organic material, and other reducing agents in the rock. Urea is added to the first digester to reduce the amount of NO_x formed. No urea is found in the final product [101]. An antifoaming agent is also added to the first digester.

Some phosphate rocks contain large amounts of acid-insoluble material, which can be removed with sand traps or hydrocyclone (Fig. 37, d) systems. The insolubles are normally washed in a drum (e) and filtered on a belt before disposal.

Crystallization is accomplished by cooling. The digestion liquor is cooled and forms $Ca(NO_3)_2 \cdot 4H_2O$ crystals (Fig. 37, f). Crystallization begins at ca. 23 °C, and solubility decreases rapidly with decreasing temperature. The end temperature of the crystal slurry is adjusted (normally to 0 to −5 °C) to obtain the desired composition (CaO/P_2O_5 ratio) in the final product.

Crystallization is carried out batchwise to avoid buildup of deposits on the cooling coils. Cooling is performed by circulating cold ammonia–water solution through coils. Some of the heat removed is used to vaporize the ammonia required in the process. The rest of the heat load is removed by mechanical refrigeration or by cold water.

Calcium Nitrate Filtration. The crystal slurry enters a feed tank and is fed continuously to filters (Fig. 37, g), where $Ca(NO_3)_2 \cdot 4H_2O$ crystals are separated from the mother liquor. In a second filter stage, the crystals are washed with nitric acid and water. The wash acid is returned to the first digester.

The mother liquor is pumped to the neutralizers. The calcium nitrate crystals are flushed to a melting tank (i) before being processed in the conversion plant or to calcium nitrate fertilizers.

Neutralization. The mother liquor contains phosphoric acid, nitric acid, some hydrofluoric acid, dissolved calcium and magnesium nitrates, small amounts of dissolved impurities (e.g., iron, aluminum, silicon), and suspended insoluble particles (e.g., quartz). Neutralization takes place in one or two reactors by addition of gaseous ammonia under strict pH control (Fig. 38, a). Nitric acid or ammonium nitrate solution is added to adjust the N/P_2O_5 ratio in the final product.

When the acids are neutralized with ammonia, calcium and most of the dissolved impurities precipitate as fluorides and phosphates:

$$Ca(NO_3)_2 + 2\, HF + 2\, NH_3 \longrightarrow 2\, NH_4NO_3 + CaF_2$$

$$Ca(NO_3)_2 + H_3PO_4 + 2\, NH_3 \longrightarrow 2\, NH_4NO_3 + CaHPO_4$$

Ammonium nitrate and ammonium phosphates remain in solution. Ammonium nitrate solution is added to give the desired N/P_2O_5 ratio.

Due to the heat of reaction, water and ammonia are evolved during operation. Ammonia is normally recovered as ammonium nitrate in the off-gas treatment system.

Nitrophosphate Evaporation. The nitrophosphate liquor from the neutralizers is concentrated in two stages in evaporators under reduced pressure (Fig. 38, b). The water content in the concentrated slurry is ca. 2.5 % for granulation and 0.5 % for prilling.

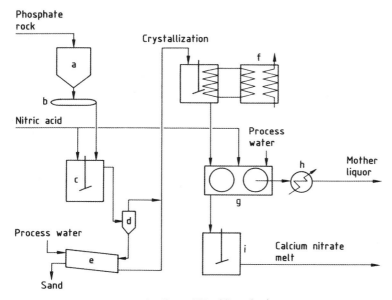

Figure 37. Nitrophosphate mother liquor (NP acid) production
a) Rock bin; b) Belt scale; c) Digestion; d) Hydrocyclone; e) Washing; f) Cooling system; g) Calcium nitrate filtration; h) Heater; i) Melting tank

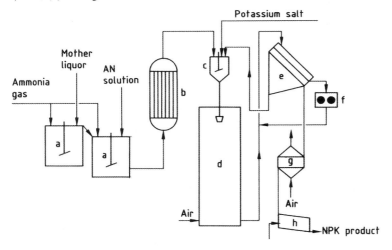

Figure 38. Nitrophosphate neutralization, evaporation, and prilling
a) Neutralization; b) Nitrophosphate evaporator; c) Mixing vessel; d) Prilling tower; e) Screen; f) Mill; g) Cooler; h) Coating drum

During evaporation, some ammonia escapes with the water vapor. Ammonia is recovered by scrubbing with acidic ammonium nitrate solution or by indirect condensation and subsequent stripping from the condensate. Prior to ammonia recovery, fluorine is removed from the vapor by scrubbing to avoid plugging and corrosion.

Mixing and Prilling/Granulation. The concentrated nitrophosphate liquor flows by gravity to the mixing–prilling section (Fig. 38, c, d). For the production of NPK, concentrated nitrophosphate liquor is mixed with potassium chloride (or sulfate) in a high-speed mixer. Recycled material from the dry handling system containing dust, crushed oversize particles, and fines is also added to the mixer.

In the prilling process [102–104] the mixer overflows to a rotating prill bucket from which the slurry is sprayed into the prill tower. Fans at the top of the tower cause cold air to flow countercurrent to the droplets and result in solidification. The solid prills fall onto a rotating tower bottom and are scraped toward the edge by a scraper. The percentage of off-size normally varies between 10 and 20 %.

The granulation process is similar to that presented for the sulfur route (see Figs. 31, 32, 33, 34).

After screening (e), coarse particles are crushed (f) and recycled, together with fine particles, to the mixer. The product flows to a rotating drum or a fluidized-bed cooler (g) to adjust product temperature before being sent to storage. Solid and liquid anticaking agents are added to the product in the coating drum (h).

Salt Preparation. The primary steps of salt preparation are weighing, drying, and screening. The main salts added to the process are kieserite, dolomite, potassium chloride, and potassium sulfate. These fairly pure salts contain only small quantities of other metals or insolubles.

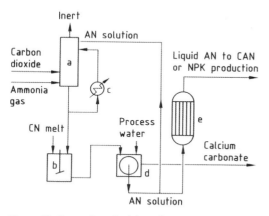

Figure 39. Conversion of calcium nitrate
a) Absorption; b) Conversion; c) Cooler; d) Filtration; e) Evaporation
AN = ammonium nitrate; CAN = calcium ammonium nitrate; CN = calcium nitrate

Calcium Nitrate Conversion. The calcium nitrate byproduct can be either used for the production of calcium nitrate fertilizers or converted into ammonium nitrate solution and calcium carbonate [105–108].

Calcium nitrate crystals from the filters are melted before being reacted with ammonium carbonate solution. The slurry is filtered on a vacuum filter, and the filter cake is washed with water.

In the Norsk Hydro process (Fig. 39), absorption of ammonia and carbon dioxide, and subsequent precipitation of calcium carbonate, take place in separate units. Ammonium carbonate is formed by reaction of gaseous ammonia and carbon dioxide in a recirculated stream of 60 % ammonium nitrate solution.

Because of the heat of reaction, external coolers (c) are installed to control the absorber temperature and to reach the desired temperature for the ammonium carbonate solution entering the conversion reactor.

Precipitation of calcium carbonate takes place in the conversion reactor (b):

$$\text{Ca(NO}_3)_2 + (\text{NH}_4)_2\text{CO}_3 \longrightarrow 2\,\text{NH}_4\text{NO}_3 + \text{CaCO}_3$$

To ensure complete precipitation the reactor is operated with a slight excess of carbon dioxide. Because the heat of reaction is relatively low, no cooling is required.

The reactor slurry can be filtered on a rotary vacuum filter or a belt filter (d); the filter cake should be washed with hot water. The calcium carbonate is used both for technical and agricultural purposes. Another alternative is to make calcium ammonium nitrate fertilizer (28 % N).

Because the ammonium nitrate solution contains a slight excess of ammonia, it is neutralized with nitric acid prior to evaporation (e).

11.4. Nitro Process with Calcium Nitrate Crystallization (BASF)

The Odda process has been used by BASF since 1953 at Ludwigshafen and until recently at its subsidiary in Köln.

The company designed and built a new 700 000-t/a plant at Antwerp, which went into operation in 1985; BASF now offers the process for license. Contracts have been awarded to Agrolinz (Austria) and Bharuch (India).

The principles of the BASF process (Fig. 40) are similar to those of the Norsk Hydro process; the following description focuses on the

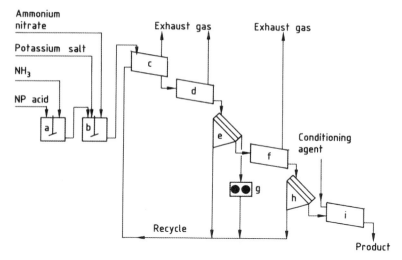

Figure 40. The BASF granulation process for production of NPK fertilizers
a) Neutralization; b) NPK slurry; c) Granulation; d) Dryer; e) Hot screen; f) Cooling; g) Crusher; h) Cold screen; i) Finishing

main features specific to the BASF process [98], [109].

Digestion is carried out in a cascade of reactors. The temperature must be kept above 68 °C to ensure complete acidulation and below 72 °C to prevent corrosion. Residence time is 2 h. Foaming is usually controlled by specially designed stirrers, but antifoaming agents may be required. Only rock particles larger than 4 mm — usually <10 % of the total — require crushing.

The process employs sedimentary rocks containing 68–75 % tricalcium phosphate in the form of fluorapatite.

Crystallization and Separation of Calcium Nitrate. Crystallization is carried out in tanks with integral cooling coils. The cooling medium is chilled brine. In more recent systems the crystallizers are operated batchwise, but the cooling coils are connected in series.

The following conditions should be satisfied to achieve optimum crystal separation:

1) High separation efficiency with regard to $Ca(NO_3)_2 \cdot 4 H_2O$ and nitrophosphate solution
2) Low sensitivity to changing process conditions
3) Minimal costs for operation, maintenance, and repair

The shear centrifuges used in older plants operate in conjunction with downstream hydrocyclones and only partially meet these requirements. The centrifuges are subjected to a great deal of wear and are extremely sensitive to the penetration of sand from the digestion stages. The screen slots must be wide; they allow large amounts of salt to slip through, which cannot be recovered adequately with downstream hydrocyclones. A sharp boundary between the separation zone and the washing zone cannot be achieved. Belt filters are therefore used in new plants and meet the above-mentioned requirements more satisfactorily.

Neutralization and Evaporation. The N/P_2O_5 ratio of the nitrophosphate solution from the mother liquor is adjusted with concentrated ammonium nitrate obtained by the conversion of separated calcium nitrate. The slurry is then ammoniated in two stages at 120–130 °C. Overammoniation, which would result in ammonia loss and cause reversion of phosphate to a citrate-insoluble form, is prevented by an automatic pH monitor.

If necessary, the slurry is evaporated at atmospheric pressure in heated, single-stage evaporators with forced circulation.

Granulation. Final correction of the N/P ratio and incorporation of the potassium salt are carried out in mixers. The slurry is then sprayed into a drum granulator (spherodizer) or a pugmill granulator. The spherodizer type of drum operates with slurries containing 15–20 % water, and only a small amount of material is recycled. In standard drum or pugmill units a water content <8 % is required, and the granulation and drying units are separate. Production also involves screening, crushing, cooling, and coating units.

Calcium Nitrate Conversion. As in the Norsk Hydro process (see page 385), calcium nitrate conversion is carried out by treatment with ammonium carbonate. An ammonium carbonate solution is made by absorbing carbon dioxide and ammonia into ammonium nitrate solution. A belt filter is again used for separation. The calcium carbonate either can be used for making lime ammonium nitrate fertilizer or can be used in the cement industry. The ammonium nitrate solution passes through pressure filters to remove any calcium carbonate present due to malfunction upstream; it is then concentrated in steam-heated evaporators.

11.5. Nitro Process with Ion Exchange (Kemira Superfos)

The ion-exchange process (Fig. 41) permits the production of high-grade, chloride-free NPK fertilizers based on the cheapest possible raw materials [99].

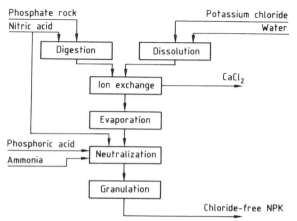

Figure 41. Kemira Superfos ion-exchange NPK process

Acidulation of calcium phosphate with nitric acid gives a solution containing calcium ions, phosphoric acid, and nitrate ions:

$$Ca_5F(PO_4)_3 + 10\,HNO_3 \longrightarrow 5\,Ca^{2+} + 3\,H_3PO_4 + 10\,NO_3^- + HF$$

If the acidulated solution is exposed to a potassium-loaded cation exchanger, calcium ions are exchanged for potassium ions. The cation exchanger is regenerated with potassium chloride solution. The product solution contains only the valuable constituents potassium, phosphate, and nitrate; the byproduct solution from regeneration contains the unwanted calcium and chloride, which are removed.

Production. The product solution is treated in the usual way to give granular NPK fertilizers.

The process was developed in the late 1960s, and an industrial plant began operating at the end of 1971 in Fredericia, Denmark.

Conventional chloride-free NPK fertilizers based on potassium sulfate contain ammonium nitrate, ammonium phosphate, and potassium sulfate. In NPK fertilizers based on ion exchange, the potassium sulfate is replaced by potassium nitrate.

Products based on ion exchange can be made with a higher total content of nitrogen, phosphorus, and potassium nutrients because they do not contain "ballast" sulfate. A typical N-P_2O_5-K_2O grade is 17-17-17; the corresponding grade in potassium-sulfate-based processes is 15-15-15.

The presence of potassium nitrate instead of chloride or sulfate has a favorable effect on physical properties. Ammonium chloride formation, which often causes caking problems, does not occur. The presence of nitrate instead of chloride or sulfate facilitates production of hard, uniformly sized granules with good storage properties in a spherodizer.

11.6. Nitro Process with Sulfate Recycle (DSM)

Phosphate rock is acidulated with nitric acid at 65 °C (Fig. 42, c). Phosphoric acid may also be added. Calcium nitrate in the slurry is then reacted with ammonium sulfate at 55 °C (d) to form calcium sulfate dihydrate [110].

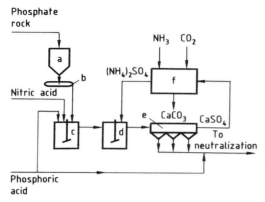

Figure 42. The DSM nitro process for production of NPK fertilizer with sulfate recycle
a) Rock bin; b) Belt weigher; c) Digestion; d) Reactor; e) Filter; f) Conversion plant (see Fig. 39)

The slurry is filtered (e) and the gypsum filter cake is fed to the ammonium sulfate recovery section where 52–53 % ammonium carbonate solution is used to produce a slurry of ammonium sulfate solution and solid calcium carbonate. A 40 % ammonium sulfate solution is obtained for recycle, after filtration of the calcium carbonate (Fig. 39). The mother liquor from the filter is mixed with phosphoric acid to increase the P_2O_5 water solubility of the final product. After neutralization the remaining moisture is removed by steam evaporators before the product is mixed with other nutrients and prilled or granulated.

A feature of this process is the absence of straight nitrogen fertilizer coproduct, although low N/P_2O_5 ratios are obtained only by the addition of other P_2O_5 materials (phosphoric acid, triple superphosphate, or MAP).

The use of a mixture of acids to produce nitrophosphates was developed to eliminate calcium nitrate as a final coproduct. Some of the calcium nitrate formed during acidulation of the phosphate rock with nitric acid remains in the slurry after filtration and ends up as DCP in the final product. To increase the P_2O_5 water solubility above 30 %, large amounts of phosphoric acid must be added.

11.7. Effluent Control of Nitrophosphate Process

Gaseous emissions from the nitrophosphate process are especially low in fluorine; <1 % of the fluorine in the rock evolves during digestion. The concentration of nitrous gases can be reduced by addition of urea [101]. The NO_x concentration of the stack gas is <150 ppm. Furthermore, ammonia-containing effluents can be collected and treated before recycling to the process as ammonium nitrate or gaseous ammonia.

Emissions from the dry section of the plant handling solid materials are treated in cyclone or filter systems. For granulation units, treatment similar to that described for the sulfur route and MAP production is required (Chaps. 9 and 10). The prilled product consists of even particles with a smooth surface; dust generation from a prilling process is therefore considerably lower than in a granulation plant. The dust collected in these systems is recycled to the process.

Dust emission from the prilling tower is very low in nitrophosphate production, compared to the production of straight fertilizers (e.g., urea or ammonium nitrate). Thus treatment systems are not needed for the cooling air.

12. Other Straight Phosphate Fertilizers

12.1. Phosphate Rock for Direct Application

The agronomic effect of directly applied phosphate rock is assumed to be related to the degree of carbonate substitution for phosphate within the apatite structure. Carbonate fluorapatite is less resistant against weathering and may thus release P_2O_5 into the soil [111]. These rocks are often called soft phosphate rocks.

Only sedimentary rocks are suitable for direct application. Rock from North Carolina (30.1 wt % P_2O_5) and Tunisia (28.7 wt % P_2O_5) is most reactive (86.3 and 74.6 % P, respectively, extracted by 2 % formic acid) [112], [113].

World use of rock for direct application (excluding China) in 1989 was ca. 1.2×10^6 t P_2O_5 (i.e., about 3.5 % of total phosphate use). The main users worldwide are the former Soviet Union, France, Malaysia, and Sri Lanka.

Application. Direct application of reactive rock is most useful for gradually building up the phosphate status of an acid soil or for maintaining phosphate levels in soil that already has a high status.

Reactive rocks can be used to provide an "insurance" for crops that are not particularly responsive to phosphate (e.g., perennial grazed grassland, established tree crops, and paddy rice). Crops that have a high calcium uptake (e.g., legumes) can often extract more phosphate from rock sources. Rock is least satisfactory for small-seeded annual crops or crops grown at low soil temperature.

Phosphate rock is not readily soluble in soil and can therefore only be used to build up soil reserves. However, the slow dissolution means that phosphate ions are still being released several years after application [114]. The best chemical test of the potential dissolution rate of phosphate rock is provided by extraction with 2 % formic acid.

Dissolution is aided if the phosphate rock is "soft" or if it is finely ground and applied as a powder.

Dissolution also improves if the soil is moist and acidic (pH<5.5), but the proportion of phosphate taken up by a crop decreases with increasing soil acidity [115]. Models of dissolution rate in soil are described in [116], [117].

Production. Run-of-mine rock is ground, screened, and applied directly to soil without further treatment. The rock must be ground to a particle size of 0.15 – 0.075 mm [118]. Handling problems associated with this finely ground, dusty material can be alleviated by granulation.

The minigranulation process developed by IFDC compacts finely ground phosphate rock into spherical granules by use of a binder [119]. Binders include soluble salts, mineral acids, and organic materials. Granules made with a soluble salt binder disintegrate in soil and are almost as effective as finely ground rock.

12.2. Partially Acidulated Phosphate Rock [118]

Partially acidulated phosphate rock (PAPR) consists of a mixture of immediately available MCP and gypsum (calcium sulfate). The amount of MCP present is determined by the degree of acidulation. Typically, PAPR is produced by using 50 % of the acid required for full conversion to superphosphate.

Very little PAPR is in commercial production. The main use is in New Zealand where ca. $(75-100) \times 10^3$ t of P_2O_5 are produced in PAPR form.

Application. A PAPR mixture has potential agronomic advantages because it combines the immediate effect of water-soluble MCP with the prolonged release of the rock. In a responsive situation, the water-soluble fraction is most beneficial if the fertilizer is granulated and placed near the crop roots, but such placement reduces the value of the rock component.

The agronomic value of PAPR is affected by the "softness" of the original rock as well as by soil and crop factors that influence the agronomic value of directly applied rock. This value is often surprisingly high under laboratory conditions [120].

The PAPR is most valuable in moist, acid soil with moderate phosphate status where the crop is not particularly responsive. The crop should then obtain its immediate phosphate needs from MCP, whereas the rock fraction maintains or builds up the general phosphate status over time.

Partial acidulation improves the agronomic performance of almost all rocks but probably has economic advantages over full acidulation (i.e., production of superphosphates) only in areas where crops show little or no immediate response and in soils where the rock component dissolves rapidly enough to maintain the soil at a satisfactory phosphate status.

Production. Sulfuric and phosphoric acids are most commonly used for partial acidulation. The former is generally preferred for economic reasons. Two processes have been evaluated for the production of sulfuric-acid-based PAPR [121].

The *run-of-pile (ROP) process* is simpler and may be operated in one or two stages. Ground phosphate rock, sulfuric acid, and water (if required) are fed continuously to a mixer (usually of the pugmill type). Retention time in the mixer is 30 – 60 s. Continuous operation is preferable to batch operation to avoid poor product handling as well as excessive agglomeration and caking. Other materials can be added to the mixer to improve acidulation or provide additional nutrients.

Depending on the characteristics of the rock and the degree of acidulation, the product can be transferred directly from the mixer to storage without curing or denning. It is typically stored for two weeks before being reclaimed, crushed to pass through a 4-mm screen, and prepared for dispatch.

In the *single-step acidulation – granulation process* the feed includes recycled material (product screen undersize and some product). Depending on the rock and the operating conditions, the recycle to product ratio in the feed is ca. 2. Retention time in the granulator is usually 5 – 8 min.

The moist, plastic product leaving the granulator is fed to a rotary drum dryer. After being dried, the granules are screened in a closed circuit with a crusher to obtain a product size of 1 – 4 mm.

Table 3. Composition of basic and BOF slag fertilizers

Composition or property	Basic slag (Thomas phosphate), wt %	BOF slag (Thomas kalk), wt %
Typical values		
P_2O_5 (citric acid soluble)	13 – 15	4 – 8
$CaO + MgO$	45 – 50	45 – 55
SiO_2	8 – 12	ca. 12
$Fe(FeO + Fe_2O_3)$	8 – 12	12 – 18
Mn, Al, and trace elements	2 – 4	3 – 6
German regulation requirements		
Minimum citric acid soluble P_2O_5	10	3
Minimum CaO		35
< 0.16 mm	> 75	
< 0.63 mm	> 96	
< 0.315 mm		> 80
< 1.0 mm		> 97
Color	brownish gray	
Density	ca. 3.2 g/cm³	
Bulk density	ca. 1.6 g/cm³	
Tapped density	ca. 2.1 g/cm³	

12.3. Basic and BOF Slag Fertilizers

The origin of basic fertilizer slag and basic oxygen furnace (BOF) slag is apatite from magmatic iron ores. The slag is separated during steel production by addition of lime. Excess lime and silica (sand) can be added to increase the fertilizer value and give porous crystals of calcium silicophosphate, $Ca_5P_2SiO_{12}$ (silicocarnotite). Slag composition varies with origin and process; BOF slag has a lower P_2O_5 content. Typical values and German fertilizer regulation requirements are given in Table 3.

Since the mid-1960s the production of phosphorus-containing fertilizer slag has declined because the steel industry requires high-quality ores with a low phosphorus content. Today, only iron ores from Kiruna (Sweden) and Lorraine (France) yield phosphorus-containing fertilizer slag. Annual EC production of basic slag is ca. 220 000 t of P_2O_5 (1.6×10^6 t of slag) and of BOF slag ca. 12 000 t of P_2O_5 (220 000 t of slag).

Application. The fertilizing effect of slag increases with its silica content. Many years ago, extra sand was sometimes added to liquid slag either during the slagging process or while the slag stream was being poured off. At present, nearly all basic slag fertilizers have a high citric acid solubility (>80 %).

Production. Basic slag is produced as a by-product in the basic Bessemer process for melting phosphorus-containing iron ores.

In all steel-making processes based on blowing with air or oxygen, oxidation of phosphorus-containing iron in the converter produces both steel and phosphorus-containing slag. All elements that are initially reduced in the blast furnace along with the iron oxide are subsequently oxidized in the converter at high temperature. They accumulate in the slag to form calcium silicophosphate. At the end of the blowing process, liquid slag floats on the molten steel surface. This slag consists mainly of a lime – phosphate – silicate melt, as well as certain amounts of iron and manganese oxides.

Addition of phosphate rock to the converter as a replacement for lime has frequently been recommended to increase the P_2O_5 content. Such addition cannot be used if the phosphate contains fluorine because this severely reduces the solubility of the basic slag in 2 % citric acid due to apatite formation. Use of fluorine-free phosphates gives the desired increase in the citric-acid-soluble phosphate fraction of the slag [122].

In the basic Bessemer process for oxidizing carbon, phosphorus, silicon, etc., oxygen is introduced in the form of air blown through the bottom of the converter. The use of pure oxygen instead of air has led to a series of other steel production processes. A basic slag containing bound P_2O_5 is formed by blowing the oxygen through a lance from above (Linz – Donawitz Arbed Centre process, LDAC) or through nozzles on the floor of the converter (Oxygen Bottom Maximilianshütte process, OBM). The oxygen blast can be interrupted before complete decarburization of the melt (ca. 0.7 %) occurs; most of the "first slag" is then poured off. The "second slag", formed after final blowing of the melt, remains in the converter and is reprocessed with the next melt (two-slag process). The first slag resembles basic Bessemer slag, but has a lower iron and a higher P_2O_5 content (5 – 7 % Fe, 18 – 22 % P_2O_5). It is marketed as basic slag [123]. In the one-slag process the charge is blown until dephosphorization is complete, and mixed slag with a lower phosphate content is produced.

Separation of the Slag. The converter containing liquid slag floating on the molten steel is tilted. Slag flows into a trolley where it solidifies to a solid block, which is then taken to the slag heap where it cools. When the slag solidifies, considerable segregation occurs in the interior of the block, resulting in virtually complete separation of lime-rich phosphates, and iron and manganese oxides [124].

The LDAC and BOF slags are poured into flat beds for technical reasons. This allows rapid cooling and thus speeds up further processing.

The slag begins to solidify even at high temperature and thickens rapidly. Complete dissolution of the added sand is therefore difficult. Moreover, the melting point of silicophosphate in the slag increases with increasing silica content. Sand is therefore sometimes dried and mixed with ground coke before being added. Thorough mixing results in reaction of the coke with iron oxides in the slag.

Constitution. The shaded area in the phase diagram in Figure 43 shows the composition of commercial basic Bessemer slags, when the iron oxides and manganese oxide present are ignored. The line connecting $3\,CaO \cdot P_2O_5$ and $2\,CaO \cdot SiO_2$ clearly separates the lime-rich part of the ternary system from the rest because the melting points that occur on this line are all much higher than those of the neighboring phosphorus-rich compositions. The orthophosphate $3\,CaO \cdot P_2O_5$ and the orthosilicate $2\,CaO \cdot SiO_2$ are extremely stable and therefore dominate the crystal phases in the basic slag. The two phases have similar crystal structures, leading to formation of a solid solution series during crystallization. Calcium silicophosphates are almost 100 % soluble in 2 % citric acid and are present in concentrations of 60 – 80 wt % in normal basic slag or in slag produced by the LDAC or the OBM process [125].

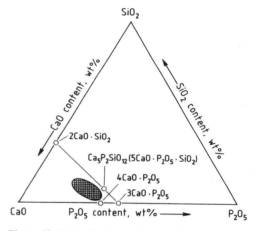

Figure 43. Phase diagram for calcium silicophosphates

The crystal phase changes that occur during cooling contribute largely to the physical properties of the slag.

Size Reduction. The blocks of slag disintegrate during the course of weeks or months. The resulting particle sizes vary greatly between extremely fine powder and larger pieces. Rapid cooling in a slag bed prevents spontaneous breakdown of the slag so that further grinding is necessary.

The slag is ground in ball mills or tube mills to a fineness of at least 75 % <0.16 mm. In the mill, efficient dust removal is necessary because prolonged inhalation of basic slag phosphate dust can damage the respiratory tract owing to the presence of silicates and silicophosphates (basic slag is covered by the "2. Verordnung zur änderung der Gefahrstoffverordnung," Bundesgesetzblatt dated April 23, 1990, Part 1).

Attempts to manufacture granulated basic slag phosphate directly on a large scale have so far not been successful for economic reasons [125].

12.4. PK Mixed Fertilizers with Basic Slag

Increasing quantities of granulated mixtures of basic slag and potash salts are used as PK mixed fertilizers with $P_2O_5 : K_2O$ ratios between 1 : 1 and 1 : 3. They are usually produced by agglomerating the powdered basic slag and potassium salts in double-shaft granulators, followed by granulation by rolling in cylindrical drums and drying with hot gases. Inorganic substances are used as granulation aids [126].

The addition of fused phosphates or DCP is legally permitted to allow the P_2O_5 content of the slag to be adjusted to maintain a constant P/K ratio [127].

13. Energy Consumption [128]

Phosphate rock is the raw material for phosphate fertilizers. Approximately 5 t of ore must be mined and beneficiated to produce 1 t of commercial phosphate rock with an average phosphorus content of 14 % (32 % P_2O_5). This process requires 0.9 GJ/t or 2.9 GJ/t of P_2O_5. If the rock is calcined, the total energy requirement is doubled. Sulfuric acid is required for the production of phosphoric acid, which is an intermediate for 80 % of the phosphate fertilizers worldwide. The energy contribution for sulfuric acid varies from an export of 8 GJ/t of P_2O_5 to a requirement of 7 GJ/t of P_2O_5 depending on the raw materials.

Production of phosphoric acid (54 % P_2O_5) from sulfuric acid and uncalcined rock requires between 10.8 GJ/t of P_2O_5 (hemihydrate) and 12.9 GJ/t of P_2O_5 (dihydrate).

Processing the acid to ammonium phosphate fertilizers requires a further 1.5 – 1.75 GJ/t of P_2O_5. Processing to triple superphosphate requires a further 2.6 – 3.5 GJ/t of P_2O_5.

Complex fertilizers based on the nitrophosphate route require a minimum of 5.5 GJ/t of P_2O_5 to process the rock to complex phosphate fertilizers.

14. Effluents from Phosphate Fertilizer Production

Disposal of gypsum-containing effluents in the sea is the main environmental concern in phosphate fertilizer production. Some companies have established land disposal and eliminated all gypsum effluents.

Emissions from phosphate fertilizer production vary considerably, normally depending on the age of the plant (Table 4).

Table 4. Emission levels from phosphate fertilizer plants

Emitted substance	Modern plants	Old plants
To air		
NH_3, kg N/t N	0.01	1–10
NO_x, kg N/t N	0.3	0.5–4
F, kg F/t P	0.01	<0.12
To water		
$NH_4^+ + NO_3^-$, kg N/t N	0.01	0.6–5
PO_4^{3-}, kg P/t P	0.15	1.0–4
F, kg F/t P	0.06	0.9

Nutrient losses during fertilizer production are <1 % of the nutrients handled in old plants and <0.04 % in modern plants. Thus, the main environmental concerns associated with fertilizers are not at the production site but on the farm (→ Fertilizers, Chap. 9.2.1.).

15. Heavy Metals in Phosphate Fertilizers

Many elements that do not appear to be necessary for plant nutrition occur in the raw materials used for fertilizer production and thus in small amounts in fertilizers.

Table 5 lists the heavy metals found in phosphate rocks and compares typical contents with those found in soil. Contents vary widely with origin and type of rock [129].

In the sulfur route, 80 – 90 % of the mercury and lead and 30 – 50 % of the cadmium end up in the gypsum waste. In the other processes, almost all the trace elements end up in the phosphate fertilizer.

Table 5 also gives the input of heavy metals to a topsoil after a century's use of phosphate

Table 5. Heavy-metal content of phosphate rock and soil [129]

Element	Average content*, mg/kg P	Range, mg/kg rock	Average content, mg/kg		Input to soil after 100 years of fertilizer use, mg/kg soil**
			Rock	Soil	
Arsenic	45	1–300	7	6	0.04
Cadmium	170	0.01–120	25	0.35	0.14
Chromium	1000	0.3–460	150	70	0.83
Cobalt	13	0.5–6	2	8	0.01
Copper	200	6–80	30	30	0.17
Lead	40	3–40	6	35	0.03
Manganese	200	6–300	30	1000	0.17
Mercury	0.2	0.01–0.10	0.03	0.06	0.0017
Molybdenum	33	1–10	5	1.2	0.03
Nickel	230	1–85	35	50	0.19
Zinc	660	3–800	100	90	0.55

* 15% P in rock. ** Topsoil (20 cm soil), soil volume weight 1.2 kg/L, annual application of 20 kg P/ha for 100 years.

fertilizer. This input is small compared with the levels naturally present in the average soil, with the notable exception of cadmium.

Cadmium is highly toxic to humans but less so to plants. Concern exists that the use of phosphate fertilizers will slowly increase the cadmium content of arable land and that within about a century this might eventually result in an unacceptably high cadmium level in agricultural produce.

The average daily cadmium intake in the European diet is 20 µg; WHO recommends a maximum daily limit of 70 µg.

Table 6. Phosphorus and cadmium content of phosphate rocks [129]

Type of rock	Phosphorus content, wt%	Cadmium content	
		mg/kg rock	mg/kg P
Magmatic origin			
Kola, Soviet Union	17.2	0.15	0.9
Palfos, Republic of South Africa	17.2	0.15	0.9
Sedimentary origin			
Bou Craa, Morocco	15.9	35	220
Togo	15.7	55	350
Youssofia, Morocco	14.6	40	274
Jordan	14.6	5	34
Texas Gulf, USA	14.4	40	278
Florida, USA	14.4	8	56
Negev, Israel	14.2	20	140
Khouribga, Morocco	14.2	16	113
Khneifiss, Syria	13.9	6	43
Gafsa, Tunisia	13.2	50	380

Fertilizers can contain cadmium if sedimentary rock phosphate is used as a raw material. Table 6 shows typical values for cadmium content in some important rock phosphates.

No commercial process is currently available for removing cadmium from phosphate fertilizers. Several companies are, however, carrying out research into the development of cadmium removal processes. The most promising processes involve separating cadmium from phosphoric acid for the sulfur route and from the mother liquor for the nitro route.

16. Safety in Storage, Handling, and Transport

Phosphate fertilizers without free phosphoric acid are regarded as nontoxic, with no dangerous properties.

Phosphate substances that are handled as fertilizers do not present a specific health risk nor do they have undesirable chemical effects. Their transport is therefore not covered by dangerous goods regulations.

Phosphatic raw materials and many fertilizer products are transported in bulk. The physical stability of the contents of the ship hold in marine transport is a risk factor during loading and voyages in rough seas. The International Maritime Organization (IMO) takes these risks into consideration and specifies values for cohesiveness and angle of repose [130].

Phosphate compounds are present in compound NP or NPK fertilizers, often with nitrates as the nitrogen source. Nitrates have some sensitivity to heat. Different grades of these fertilizers are listed in the UN's "Recommendations on the Transport of Dangerous Goods" and in the specific codes for all transport modes.

Self-sustaining decomposition (SSD) is linked predominantly to NPK grades but may also be observed in NK grades. The thermal stability of fertilizer products is tested in a standardized procedure (the trough test) [131]. If the product exhibits SSD after the heat source is removed, the product is assigned to Class 9. If the velocity of SSD exceeds 25 cm/h, bulk transport is prohibited. For NPK grades SSD may be favored by a low water solubility of phosphorus in KCl-based, nitrate-containing commodities.

17. References

1. K. D. Jacob: *Superphosphate: Its History, Chemistry, and Manufacture,* U.S. Department of Agriculture and Tennessee Valley Authority (TVA), Washington, D.C. 1964, pp. 8–94.
2. J. Ands: "Thermal Phosphate", in F. T. Nielsson (ed.): *Manual of Fertilizer Processing,* Marcel Dekker, New York 1987, pp. 93–124.
3. K. D. Jacob: *Superphosphate: Its History, Chemistry and Manufacture,* U.S. Department of Agriculture and Tennessee Valley Authority (TVA), Washington, D.C., 1964, pp. 8–94.
4. E. F. Dillard, A. W. Frazies, T. C. Woodis, F. P. Achorn: "Precipitated Impurities in 18-46-0 Fertilizers Prepared from Wet-Process Phosphoric Acid," Bulletin Y-162 TVA/OACD-83/12, National Fertilizer Development Center and TVA, Muscle Shoales, Ala, 1981.
5. E. F. Dillard, A. W. Frazier: "Precipitated Impurities in MAP and Their Effect on Chemical and Physical Properties of Suspension Fertilizers," Bulletin Y-183 TVA/OACD-83/17, National Fertilizer Development Center and TVA, Muscle Shoales, Ala, 1983.
6. G. M. Lloyd, F. P. Achorn, R. M. Scheib: *Diammonium Phosphate Quality-1988,* American Institute of Chemical Engineers, Clearwater, Fla., 1988, pp. 1–43.
7. E. F. Dillard, J. R. Burnell, J. Gautney: "Stable Suspension Fertilizers from MAP," paper presented at the *American Chemical Society Meeting,* Washington, D.C., Aug. 26–31, 1990.
8. J. Ands: "Thermal Phosphate" in F. T. Nielsson (ed.): *Manual of Fertilizer Processing,* Marcel Dekker, New York 1987, pp. 93–124.
9. *World Fertilizer Atlas,* 7th ed., British Sulphur. Corp., London 1983.
10. J. Archer: "Phosphorus," in *Crop Nutrition and Fertilizer Use,* Farming Press, Ipswich 1985, pp. 57–64.
11. N. C. Brady: "Supply and Availability of Phosphorus and Potassium," in *The Nature and Properties of Soils,* 8th ed., Macmillan Publ. Co., New York 1974, pp. 456–475.
12. K. Mengel, E. A. Kirby: "Phosphorus," in *Principles of Plant Nutrition,* 3rd ed., International Potash Institute, Bern 1982, pp. 387–409.
13. Ministry of Agriculture, Fisheries and Food (MAFF): *Phosphate and Potash for Rotations,* Booklet 2496, Her Majesty's Stationery Office, London 1985.
14. G. H. Aylward, T. J. V. Findlay: *SI Chemical Data,* John Wiley & Sons, Sydney 1971, pp. 122–123.
15. A. F. Hollemann, E. Wiberg: *Lehrbuch der anorganischen Chemie,* De Gruyter, Berlin 1976, p. 404.
16. K. S. Førland: *Kjemisk Likevekt,* Tapir, Trondheim 1974, pp. 1–60.
17. G. Hägg: *Kemisk Reaktionslära,* Almqvist & Wiksell, Stockholm 1969, pp. 67–105.
18. J. N. Butler: *Ionic Equilibrium,* Addison–Wesley, Reading 1964, pp. 206–260.
19. L. G. Sillen, "Graphic Presentation of Equilibrium Data," in I. M. Kolthoff, P. I. Elving, E. B. Sandell (eds.): *Treatise on Analytical Chemistry,* vol. 1, Interscience, New York 1959, pp. 277–291.
20. L. G. Sillen, P. W. Lange, C. O. Gabrielson: *Fysikalsk-Kemiska Räkneuppgifter,* Almqvist & Wiksell, Stockholm 1951, pp. 193–213.
21. L. G. Sillen, P. W. Lange, C. O. Gabrielson: *Problems in Physical Chemistry,* Prentice-Hall, New York 1952, pp. 193–213.
22. *Ullmann,* 4th ed., **18,** p. 332.
23. C. H. Davis, R. G. Lee: TVA-Publication, presented at the *American Chemical Society Meeting,* Atlanta, Ga, Nov. 1967.
24. P. Stokka: *Three Component Phase Diagram for MAP, DAP and Water,* Norsk Hydro Internal Report, Porsgrunn 1985.
25. P. Stokka: "Recent Developments in the Pan Granulation Process," *Proc. Int. Fert. Ind. Assoc. Tech. Conf.,* Edmonton, Canada, 1988, pp. 5.1–5.12.

26. P. Stokka: *Three Component Phase Diagram for AN, MAP, DAP at Water Content* $\leq 1\%$, Norsk Hydro Internal Report, Porsgrunn 1985.
27. P. Stokka: *Three Component Phase Diagram for AN, MAP, DAP at Water Content 10%*, Norsk Hydro Internal Report, Porsgrunn 1985.
28. *Gmelin*, 8th ed., **Calcium (System no. 28)** 1127.
29. P. Stokka: *Viscosity of AN/MAP Melt with KCl*, Norsk Hydro Internal Report, Porsgrunn 1986.
30. R. J. Nunn, T. P. Dee, *Proc. ISMA Tech. Conf.*, paper LE 388, Cambridge 1953.
31. "Super Flo Process for Superphosphate," *Phosphorus Potassium* **1** (1962) 34.
32. K. D. Jacob: *Superphosphate: Its History, Chemistry, and Manufacture,* U.S. Department of Agriculture and Tennessee Valley Authority (TVA), Washington, D.C., 1964, pp. 116 – 250.
33. P. Monaldi, G. Venturino: "Process for Recycling H_2SiF_6 Solutions Recovered by Gas Washing to the Superphosphate Den," *Proc. ISMA Tech. Conf.*, 1976, Paper 1, 1 – 16.
34. "The Market Role of GTSP – Key Features of Plant Design," *Proc. Br. Sulphur Corp. Int. Conf. Fert.*, 1977, Paper 21.
35. "Triple Superphosphate Manufacture," *Phosphorus Potassium* **80** (1975) 33.
36. A. Sinte Maartensdijk: "Direct Production of Granulated Superphosphates and PK-Compounds from Sulphuric Acid, Phosphoric Acid, Rock Phosphate and Potash," *Proc. ISMA Tech. Conf.*, 1976, Paper 12, 200 – 214.
37. *Proc. Annu. Meet. Fert. Ind. Round Table*, 1968, 123 – 127;1976, 93 – 98, 120 – 130. I. W. McCamy, M. M. Norton, *Proc. Br. Sulphur Corp. Int. Conf. Fert. Technol.*, 1977, 68 – 94.
38. B. R. Parker et al., *Proc. ISMA Tech. Conf.*, Orlando 1978, Paper 19, 255 – 282.
39. *Phosphorus Potassium* **81** (1976) 37 – 38.
40. *Proc. Annu. Meet. Fert. Ind. Round Table* **16** (1966) 69 – 73. I. C. Brosheer, J. F. Anderson, *J. Am. Chem. Soc.* **68** (1946) 902 – 904.
41. I. A. Brownlie, R. Graham, *Proc. ISMA Tech. Conf.*, Helsinki, 1963, Paper 1, 2 – 26.
42. J. D. C. Hemsley, *Proc. Annu. Meet. Fert. Ind. Round Table*, 1968, 115 – 123.
43. *Phosphorus Potassium* **49** (1970) 18 – 19.
44. *Phosphorus Potassium* **157** (1988) 36.
45. L. M. Marzo, J. L. Lopez-Nino, *Proc. Fert. Soc.* **245** (1986) 1 – 13.
46. E. L. Newman, L. H. Hull, *Farm Chem.* **I** (1965) no. 6, 48, 49;**II** (1965) no. 8, 25 – 30; **III** (1965) no. 9, 48, 50, 70.
47. G. H. Wesenberg in F. T. Nielsson (ed.): *Manual of Fertilizer Processing,* Marcel Dekker, New York 1987, pp. 258 – 265.
48. R. R. Heck, *Proc. ISMA Tech. Conf.*, Brussels, 1968, V.
49. D. M. Ivell, N. D. Ward, *Proc. Int. Fert. Ind. Assoc. Tech. Conf.*, Paris 1984, TA/84/3, Paper 3, 3.1 – 3.10.
50. G. H. Wesenberg in F. T. Nielsson (ed.): *Manual of Fertilizer Processing,* Marcel Dekker, New York 1987, pp. 251 – 258.
51. P. Chinal, Y. Cotonea, *Proc. Int. Fert. Ind. Assoc. Tech. Conf.*, Paris 1984, TA/84/1, Paper 1, 1.1 – 1.19.
52. A. Constantinides, J. Lainé, D. Bellis, *Proc. Tech. Conf.*, Port el Kantaoui 1986, TA/86/12.
53. P. Chinal, Y. Cotonea, C. Debayeux, J. F. Priat, *Proc. Annu. Meet. Fert. Ind. Round Table*, London, April 17, 1986.
54. G. M. Hebbard, *Proc. Annu. Meet. Fert. Ind. Round Table* vol. 129, Washington, D.C. 1979.
55. A. T. Brook, *Proc. Fert. Soc.*, 1957, no. 47.
56. O. Bognati, L. Buriani, I. Innamorati, *Proc. ISMA Tech./Agr. Conf.* Stresa, 1967, Paper IX.
57. H. Rug, K. Kahle, *Proc. Fert. Soc.* 1990, no. 297.
58. A. Seixas, J. D. Ribeiro Marcal, J. Correia, *Proc. Int. Fert. Ind. Assoc. Tech. Conf.*, Paris 1984, TA/84/11.
59. S. Maier, *Proc. Br. Sulphur Corp. Int. Conf. Fert. Technol.*, 1977, 283 – 295.
60. W. B. Pietsch, *Proc. Br. Sulphur Corp. Int. Conf. Fert. Technol.*, 1983, 467 – 479.
61. R. Zisselmar, *Proc. Fert. Soc.* 1985, no. 238.
62. A. Stephenson, *Proc. Fert. Soc.* 1985, no. 238.
63. F. G. Keenan, *Ind. Eng. Chem.* **22** (1930) no. 12, 1378, 1382.
64. L. M. White, J. O. Hardesty, W. H. Ross, *Ind. Eng. Chem.* **27** (1935) no. 5, 562 – 567.
65. L. D. Yates, F. T. Nielsson, G. C. Hicks, *Farm. Chem.* **117** (1954) no. 7, 38 – 48;no. 8, 34 – 41.
66. D. M. Ivell, *IFDC-FADINAP-FAI Workshop*, Madras, India 1988.
67. Tennesse Valley Authority, US 2 741 545, 1953 (F. T. Nielsson).
68. G. H. Wesenberg in F. T. Nielsson (ed.): *Manual of Fert. Processing,* Marcel Dekker, New York 1987, 251 – 258.
69. R. R. Heck, *Proc. ISMA Tech. Conf.*, Brussels, 1965, Paper V.
70. *Phosphorus Potassium* **49** (1970) 18 – 19.
71. M. D. Pask, I. Podilchuk, *ISMA/ANDA Tech. Conf.*, Sao Paolo, Brazil 1975.

72. G. Hunter, J. L. Hawksley, *Proc. FAI–ISMA Semin. Technol. Comp. Fert. Based Urea, Use Benefic, Low Grade Phosphate Rock,* New Delhi 1975, III/3, 1–34.
73. J. L. Hawksley, *Proc. Br. Sulphur Corp. Int. Conf. Fert. Technol.* 1977, 110–124.
74. K. J. Barnett, D. M. Ivell, *Proc. Fert. Soc.* 1983, no. 216.
75. D. M. Ivell, N. D. Ward, *Proc. Int. Fert. Ind. Assoc. Tech. Conf.,* Paris 1984, TA/84/3.
76. B. R. Parker, M. M. Norton, I. W. McCamy, D. G. Salladay, *Proc. Int. Fert. Ind. Tech./Econ. Conf.,* Orlando 1978, 255–282.
77. R. J. Milborne, D. W. Philip, *Proc. Fert. Soc.* 1986, no. 244.
78. F. G. Membrillera, J. L. Toral, F. Codina: *Proc. Br. Sulphur Corp. Int. Conf. Fert. Technol.* 1977, Paper 12.
79. I. S. Mangat, J. M. Toral, *Proc. Int. Fert. Ind. Assoc. Tech./Econ. Conf.,* Orlando 1978, 239.
80. P. Chinal, Y. Cotonea, C. Debayeux, J. F. Priat, *Proc. Fert. Soc.* 1986, no. 245.
81. P. Chinal, Y. Cotonea, *Proc. Int. Fert. Ind. Assoc. Tech. Conf.,* Paris 1984, TA/82/2.
82. D. W. Leyshon, I. S. Mangat, *Proc. ISMA Tech. Conf.,* The Hague 1976, TA/76/21, 368.
83. I. S. Mangat, *Proc. Br. Sulphur Corp. Int. Conf. Fert. Technol.* 1977, Paper 11.
84. E. Peletti, J. C. Reynolds, *Proc. Br. Sulphur Corp. Int. Conf. Fert. Technol.,* 1977, 95–109.
85. J. W. Baynham, *Proc. ISMA Tech. Conf.,* Edinburgh 1965, Paper 12.
86. Y. F. Berquin, *Proc. Br. Sulphur Corp. Int. Conf. Fert. Technol.,* 1977, 296–303.
87. S. J. Porter, W. F. Sheldrick, *Proc. Annu. Meet. Fert. Ind. Round Table* **16** 1966, 104–109.
88. *TVA-Bulletin Y-107,* 1976, 11th Demonstration, 44–48.
89. R. G. Lee, R. S. Meline, R. D. Young, *Ind. Eng. Chem. Process Des. Dev.* **11** (1972) no. 1, 90–94.
90. T. A. Mitchel, B. H. Fotherfill, W. J. Kelly, *Proc. ISMA Tech. Conf.,* Sandelfjord 1970, Paper 6.
91. Odda Smelteverk, US 1 816 285, 1929 (E. B. Johnson).
92. G. Langhans, B. Bieniok, *Proc. Tech. Conf. ISMA 1976* 1977, paper 13, 215–233.
93. K. Tesche, "New Experiences with the Kampka Nitro Process," *Proc. FAI–ISMA Semin. Technol. Comp. Fert. Based Urea Use Benefic Low Grade Phosphate Rock* ITE/71/16 (1971).
94. J. Kofiseh, L. Hellmer, H. P. Bethke: "Nitrophosphate Process Using Direct Cooled Continuous Crystallization," *UNIDO Second Interregional Fert. Symposium,* Kiev 1971.
95. S. I. Vol'Gkovich, A. A. Sokolovskii, *Russ. Chem. Rev.* **43** (1974) no. 3, 224–234.
96. A. V. Slack: "Phosphoric Acid," *Fert. Sci. Technol. Ser.* **1** (1968) 654 ff.
97. J. F. Steen, S. G. Terjesen: "Norsk Hydro Nitrophosphate Process," *UNIDO Second Interregional Fert. Symposium,* Kiev 1971.
98. L. Diehl, K. F. Kummer, H. Oertel: *Nitrophosphates with Variable Water Solubility: Preparation and Properties,* The Fert. Society of London, April 1986.
99. K. C. Knudsen: *Production of Chloride-Free NPK Fertilizer and Feed Grade Dicalcium Phosphate,* The Fertilizer Society of London, Oct. 1985.
100. O. Kjøhl: "Product Quality Requirements in Bulk Shipments of Fertilizers," *New Dev. Phosphate Fert. Technol. Proc. Tech. Conf. ISMA, 1976.*
101. R. Ringbakken, O. Lie, G. T. Mejdell: "Urea as an Agent in the Destruction/Recovery of NO_x in the Nitric Acid and Nitrophosphate Fertilizers Production," *Proc. ISMA Tech. Conf.,* Prague 1974, Paper 22, 22.1–22.20.
102. Norsk Hydro, US 3 617 235, 1967 (I. A. Friestad, O. Skanli).
103. Norsk Hydro, US 3 561 678, 1967 (I. A. Friestad).
104. Norsk Hydro, US 3 900 164, 1973 (I. A. Friestad).
105. Hoechst, DE 1 016 723, 1955 (F. Moosbrugger et al.).
106. BASF, DE 1 064 536, 1957 (K. Huberich).
107. Chemische Fabrik Kalk, DE 1 065 436, 1957 (H. Nees, K. Geiersberger).
108. J. F. Steen, T. Heggebø, E. Aasum: "The Norsk Hydro Nitrophosphate Process," *Manual of Fertilizers Processing,* chap. 15, Marcel Dekker, New York 1987, pp. 393–419.
109. *Phosphorus Potassium* **155** (1988) 25–33.
110. Stamicarbon BV, Technical information, Jan. 1991.
111. G. H. McClellan: "Mineralogy and Reactivity of Phosphate Rock," *Seminar on Phosphate Rock for Direct Application,* IFDC publication, Muscle Shoales, Ala., 1978, pp. 57–81.
112. "Potential for Use of Unacidulated and Partially Acidulated Phosphate Rock," *Phosphorus Potassium* **168** (1990) 15–19.
113. "Ground Phosphate Rock as a Direct Application Fertilizer in New Zealand," *Phosphorus Potassium* **128** (1983) 17–21.

114. F. E. Khawawneh, E. C. Doll: "The Use of Phosphate Rock for Direct Application," *Adv. Agron.* **30** (1978) 195–206.
115. N. S. Bolan, M. J. Hedley: "Dissolution of Phosphate Rocks in Soils," *Fert. Res.* **24** (1990) 125–134.
116. G. J. D. Kirk, P. H. Nye: "A Simple Model for Predicting the Rate of Dissolution of Sparingly Soluble Calcium Phosphates in Soil," *J. Soil. Sci.* **37** (1986) 541–554.
117. A. D. Mackay et al.: "A Simple Model to Describe the Dissolution of Phosphate Rock in Soils," *Soil Sci. Soc. Am. J.* **50** (1986) 291–296.
118. "Partial Acidulation of Phosphate Rock," *Phosphorus Potassium* **150** (1987) 48–53.
119. O. W. Livingston: "Minigranulation – A Method for Improving the Properties of Phosphate Rock for Direct Application," *Seminar on Phosphate Rock for Direct Application,* IFDC publication, Muscle Shoales, Ala., 1978, pp. 367–377.
120. L. L. Hammond et al.: "Agronomic Value of Unacidulated and Partially Acidulated Phosphate Rocks Indigenous to the Tropics," *Adv. Agron.* **40** (1986) 89–140.
121. J. J. Schultz: "Sulphuric Acid-Based Partially Phosphate Rock – its Production, Cost and Use," *IFDC Tech. Bull.* IFDC-T-31, Ala., April 1986.
122. Fisons, DE-AS 1 592 644, 1966 (G. G. Brown, D. C. Harper).
123. G. Trömel, E. Görl, *Arch. Eisenhüttenwes.* **35** (1964) 287–298.
124. G. Trömel, *Stahl Eisen* **63** (1943) 21–30.
125. IRSID, FR 1 309 365, 1961. ATH, DE 1 237 592, 1961 (M. T. Blauel, T. Kootz, A. Michel).
126. Kali & Salz, DE 973 396, 1953 (H. Keitel, W. Jahn-Held, W. Appel).
127. Kali & Salz, DE 2 263 334, 1972 (A. Singewald, W. Jahn-Held).
128. M. S. Mudahar, T. P. Hignett: "Energy and Fertilizer. Policy Implications and Options for Developing Countries," *IFDC Tech. Bull.,* IFDC T-20, Ala., 1982.
129. O. C. Bøckman, O. Kaarstad, O. H. Lie, I. Richards: *Agriculture and Fertilizers. Fertilizers in Perspective,* Agricultural Group, Norsk Hydro, Oslo 1990.
130. International Maritime Organization: Code of Safe Practice for Solid Bulk Cargoes, Appendix C, 1987, p. 94.
131. International Maritime Organization: Code of Safe Practice for Solid Bulk Cargoes, Appendix D 4, 1987, p. 121.

Potassium Compounds

See also: Fertilizers

HEINZ SCHULTZ, Kali und Salz AG, Kassel, Federal Republic of Germany (Chaps. 1, 4, 6–10; Sections 3.3, 5.1, 5.2, 5.4–5.6, 5.8–5.12)

GÜNTER BAUER, Kali und Salz AG, Kassel, Federal Republic of Germany (Chap. 2; Sections 3.1 and 3.2)

ERICH SCHACHL, Kali und Salz AG, Kassel, Federal Republic of Germany (Chap. 2; Sections Sections 3.1 and 3.2)

FRITZ HAGEDORN, Kali und Salz AG, Kassel, Federal Republic of Germany (Sections 5.3 and 5.7)

PETER SCHMITTINGER, Hüls Aktiengesellschaft, Werk Lülsdorf, Niederkassel, Federal Republic of Germany (Chaps. 11–13)

1.	Introduction	400
1.1.	Occurrence	400
1.2.	History	400
2.	Potash Salt Deposits	401
2.1.	Minerals	401
2.2.	Geology of Potash Deposits	403
3.	Mining of Potash Salts	407
3.1.	Shaft Mining	407
3.2.	Extraction, Conveying, and Haulage	407
3.3.	Solution Mining	408
4.	Treatment of Potash Ores	409
4.1.	Intergrowth and Degree of Liberation	409
4.2.	Grinding	410
5.	Potassium Chloride	412
5.1.	Properties	412
5.2.	Production by Crystallization from Solution	413
5.2.1.	Phase Theory	413
5.2.2.	Hot Leaching Process	415
5.2.3.	Processing of Carnallite	419
5.2.4.	Equipment	421
5.3.	Flotation	425
5.3.1.	Potash Ores Suitable for Flotation	426
5.3.2.	Carrier Solutions	426
5.3.3.	Flotation Agents	426
5.3.4.	Theory	427
5.3.5.	Flotation Equipment	428
5.3.6.	Processes	430
5.4.	Electrostatic Separation	432
5.4.1.	Theoretical Basis	433
5.4.2.	Equipment and Processes	433
5.5.	Heavy-Media Separation	435
5.6.	Debrining and Drying	436
5.7.	Process Measurement and Control	436
5.8.	Waste Disposal and Environmental Aspects	439
5.9.	Granulation	440
5.10.	Quality Specifications	441
5.11.	Toxicology and Occupational Health	443
5.12.	Economic Aspects and Uses	443
6.	Potassium Sulfate	445
6.1.	Properties	445
6.2.	Raw Materials	446
6.3.	Production	446
6.3.1.	From Potassium Chloride and Sulfuric Acid (Mannheim Process)	446
6.3.2.	From Potassium Chloride and Magnesium Sulfate	447
6.3.3.	From Potassium Chloride and Langbeinite	449
6.3.4.	From Potassium Chloride and Kainite	449
6.3.5.	From Potassium Chloride and Sodium Sulfate	450
6.3.6.	From Potassium Chloride and Calcium Sulfate	450
6.3.7.	From Alunite	450
6.3.8.	From Natural Brines and Bitterns	450
6.4.	Granulation	451
6.5.	Quality Specifications	451
6.6.	Toxicology and Occupational Health	452
6.7.	Economic Aspects and Uses	452
7.	Potash–Magnesia	452
8.	Production of Potassium Salts from Other Raw Materials	453
8.1.	The Dead Sea	453
8.2.	The Great Salt Lake	453
8.3.	Searles Lake	454
8.4.	Other Sources	454
9.	Storage and Transportation	454
10.	Analysis of Potassium Compounds	455
11.	Potassium Hydroxide	455
11.1.	Properties	455
11.2.	Production	455
11.3.	Quality Specifications	456
11.4.	Economic Aspects and Uses	456

Ullmann's Agrochemicals, Vol. 1
© 2007 Wiley-VCH Verlag GmbH & Co. KGaA, Weinheim
ISBN: 978-3-527-31604-5

12.	Potassium Carbonate	457	12.5. Economic Aspects and Uses	460
12.1.	Properties	457	13. Potassium Hydrogencarbonate	460
12.2.	Production	457	13.1. Properties and Production	460
12.3.	Quality Specifications and Analysis	460	13.2. Uses	461
12.4.	Storage and Transportation	460	14. References	461

1. Introduction

1.1. Occurrence

Potassium occurs in nature only in the form of its compounds. It is one of the ten most common elements in the earth's crust. Several widely distributed silicate minerals contain potassium, in particular, the feldspars and micas. The weathering of these minerals produces soluble potassium compounds, which are present in seawater and occur in extensive salt deposits. Potassium is important in the metabolism of plants and animals, and is therefore found in ash from plant materials and in the bodies of animals. Potassium compounds are obtained almost entirely by the mining of salt deposits.

1.2. History

Before the discovery and exploitation of potassium salt deposits, the production of potassium compounds consisted almost entirely of potash (K_2CO_3) obtained from natural sources such as wood ash, residues from distilleries, Bengal saltpeter, wool grease, and mother liquors from sea salt production. Quantities were small and were used only for the production of soap, glass, and explosives.

In 1840, J. VON LIEBIG laid the foundations of the theory of mineral fertilizers in his paper "The Application of Organic Chemistry to Agriculture and Physiology" (→ Fertilizers, Chap. 2.). This spread the knowledge that potassium was one of the most important plant nutrition elements. In 1851, some deep mine workings in Stassfurt struck minerals containing potassium and magnesium, although these could not be used as fertilizers. In 1861, ADOLPH FRANK started the first plant using the process that he developed for producing from carnallite a potassium salt that could be employed as a fertilizer. This soon led to the development of other processes and to the establishment of many potash mines and works. Many attempts were made up to the end of World War I to hinder the building of new factories in Germany, which had a world monopoly in the production of potash fertilizers, and attempts were made to ration production and supply, and to regulate prices. In spite of this, 69 factories were in existence in 1910, and 198 in 1918. When World War I ended, Alsace was returned to France, and the potash works that were built there shortly after 1900 became French property, so that Germany lost her monopoly in potash.

The German national assembly of 1919 enacted the potash regulations and the so-called closure order, which reduced the number of operating factories to 29 by the year 1938. This also caused production to be concentrated in a few large potash companies [17, 18].

After the end of World War II, ca. 60 % of German production capacity went to the area that was later to become the German Democratic Republic, and became the VEB Kombinat Kali. After German unification in 1990, the newly formed Mitteldeutsche Kali AG took over these operations. In the Federal Republic of Germany, several groups of works were formed, leading in 1970 to the formation of the company Kali und Salz AG following a series of amalgamations. This company owns seven potash works, all located in the former Federal Republic of Germany [19, 20].

Most of the French potash works in Alsace were nationalized after World War I. During World War II, production was greatly increased. Several of the deposits have now become exhausted, and many of the works have been closed. Production has been concentrated in two large works, although these are likely to be closed down soon after the end of this century [21].

Potash production in Spain began in 1926 with the start up of a factory in Catalonia. Other factories were established later both there and in Navarra.

The kainite deposits in Sicily have been worked since 1959–1960; the salts are used for potassium sulfate production [22].

The Soviet Union began potash production in 1931 at a plant in the Northern Urals. In 1939, plants that had been operating in Eastern Poland since 1920 were taken over by the Soviet Union. In 1963, the first plant was started to exploit a very extensive deposit in White Russia. The CIS today has several large operations in the Urals and White Russia, giving it the largest capacity of all potash-producing countries.

In the United States, potash production began during World War I because the United States economy could no longer buy German potash fertilizers. Potassium salts were obtained from Searles Lake in California and in northern Nebraska. Most of these operations ceased production in the 1920s. After the discovery of potash deposits in the area of Carlsbad, New Mexico, a large number of potash works were founded since 1931 [23]. Since the foundation of the Canadian potash industry, which has the United States as its main outlet, the number of works in the Carlsbad region and their production rate have continually declined. New potash works were started in southern Utah at Moab in 1964 and at the Great Salt Lake in 1970 [24].

The most important potash deposit in North America was discovered during World War II in Saskatchewan, Canada. After initial problems due to its great depth and the presence of water-bearing overlying rock, which was difficult to deal with, several potash works were founded in the early 1960s, and today Canada is second only to the CIS as a potash producer. Two more potash plants were established in the 1980s in New Brunswick on the east coast of Canada.

In 1986, a potash plant began production at Sergipe in Brazil.

In 1974, a potash mine was opened in Yorkshire, United Kingdom.

Potash production in Palestine began on the north bank of the Dead Sea in 1931. Potash plants have now been producing since 1952 at the southern end on the Israeli side and since 1982 on the Jordanian side.

Potash ores are treated today by three basic processes: leaching–crystallization, flotation, and electrostatic treatment. Gravity separation is of minor importance because of the small density differences between salt minerals.

The oldest process is leaching–crystallization. In this process, salt solutions were originally cooled in open vessels. Vacuum cooling of the solutions in crystallizers was introduced in 1918 in the United States, so that much less energy was required and cooling times were reduced from days or weeks to minutes. Flotation was introduced in 1935 in the United States. This proved so efficient, especially for the treatment of sylvinite ores, that it is now the main potash treatment process worldwide. The electrostatic process was first used on a large scale in the German potash industry in 1974. It is now widely used in Germany for treating complex hard salts.

2. Potash Salt Deposits

2.1. Minerals

Potash salt deposits were formed by the evaporation of seawater [25]. Their composition is often affected by secondary changes in the primary mineral deposits. More than 40 salt minerals are now known, which contain some or all of the small number of cations Na^+, K^+, Mg^{2+}, and Ca^{2+}; the anions Cl^- and SO_4^{2-}; and occasionally Fe^{2+} and BO_3^{3-}, as well. Most of these are listed in Table 1 [26].

The more important salt minerals are halite, anhydrite, sylvite, carnallite, kieserite, polyhalite, langbeinite, and kainite. Gypsum occurs at the edges of salt deposits and in the overlying strata. Bischofite, tachhydrite, glauberite, thenardite, glaserite, and leonite occur additionally in some deposits [4, 27, 28].

Other minerals, not described in detail here, are useful in elucidating difficult geological questions with regard to the origin of salt deposits. In special geochemical investigations, small amounts (ppm) of Rb^+ and Cs^+ in place of K^+; Sr^{2+} replacing Ca^{2+}; Mn^{2+} replacing Fe^{2+}; Br^- replacing Cl^-; etc., are important [29, 30]. The individual minerals can be identified microscopically (grains or thin sections) and by X-ray analysis.

Potassium salt deposits always consist of a combination of several minerals (Table 2). The German term *Hartsalz* (hard salt) refers to the greater hardness of sulfate-containing potash minerals in potash deposits.

Table 1. Principal salt minerals

Mineral	CAS registry no.	Formula	Crystal system/ crystal class	Refractive indices $n_\alpha : n_\beta : n_\gamma$ or $n_{opt.} : n_C$; optical activity	Density, g/cm^3	Hardness (Mohs)
Anhydrite	[14798-04-0]	$CaSO_4$	rhombic D_{2h}-mmm	1.570 : 1.575 : 1.614 opt. +	2.96	3.8
Ascharite	[13768-64-4]	$Mg_2[B_2O_5] \cdot H_2O$	monoclinic C_{2h}-2/m	1.575 : 1.646 : 1.650 opt. −	2.70	3
Astrakhanite	[15083-77-9]	$Na_2Mg[SO_4]_2 \cdot 4H_2O$	monoclinic C_{2h}-2/m	1.483 : 1.486 : 1.487 opt. −	2.23	3
Bischofite	[13778-96-6]	$MgCl_2 \cdot 6H_2O$	monoclinic C_{2h}-2/m	1.495 : 1.507 : 1.528	1.60	1.5
Boracite (stassfurtite: lumpy form of boracite)	[1303-91-9]	$Mg_3[Cl/B_7O_{13}]$	rhombic C_{2v}-mm 2	1.662 : 1.667 : 1.673 opt. +	2.95	7
Carnallite	[1318-27-0]	$KMgCl_3 \cdot 6H_2O$	rhombic D_{2h}-mmm	1.467 : 1.475 : 1.495 opt. +	1.60	2.7
D'Ansite	[12381-13-4]	$Na_{21}Mg[Cl_3(SO_4)_{10}]$	cubic Td-43 m	1.489	2.65	
Epsomite	[14457-55-7]	$MgSO_4 \cdot 7H_2O$	rhombic D_2-222	1.432 : 1.455 : 1.461 opt. −	1.68	2.5
Glaserite	[13932-19-9]	$K_3Na[SO_4]_2$	trigonal D_{3d}-3 m	1.491 : 1.498 opt. +	2.70	2.7
Glauberite	[13767-89-0]	$CaNa_2[SO_4]_2$	monoclinic C_{2h}-2/m	1.515 : 1.532 : 1.536 opt. −	2.85	3
Gypsum	[13397-24-5]	$CaSO_4 \cdot 2H_2O$	monoclinic C_{2h}-2/m	1.521 : 1.523 : 1.530 opt. +	2.32	2
Halite (rock salt)	[14762-51-7]	$NaCl$	cubic O_h-m3m	1.5443	2.168	2.5
Kainite	[1318-75-2]	$(KMg[ClSO_4])_4 \cdot 11H_2O$	monoclinic C_{2h}-2/m	1.494 : 1.506 : 1.516 opt. −	2.13	3
Kieserite	[14567-64-7]	$MgSO_4 \cdot H_2O$	monoclinic C_{2h}-2/m	1.518 : 1.531 : 1.583 opt. +	2.57	3.7
Koenenite	[12252-18-5]	$[Mg_7Al_4(OH)_{22}] [Na_4(CaMg)_2Cl_{12}]$	trigonal D_{3d}-3 m	1.55 : 1.58 opt. +	2.15	1
Langbeinite	[14977-37-8]	$K_2Mg_2[SO_4]_3$	cubic T-23	1.534	2.83	4.2
Leonite	[15226-80-9]	$K_2Mg[SO_4]_2 \cdot 4H_2O$	monoclinic C_{2h}-2/m	1.479 : 1.483 : 1.488 opt. +	2.20	2.7
Löweite	[16633-52-6]	$Na_{12}Mg_7[SO_4]_{13} \cdot 15H_2O$	trigonal C_{31}-3	1.495 : 1.478 opt. −	2.34	2.5 – 3
Mirabilite (Glauber's salt)	[14567-58-9]	$Na_2SO_4 \cdot 10H_2O$	monoclinic C_{2h}-2/m	1.394 : 1.396 : 1.398 opt. −	1.49	1.7
Polyhalite	[15278-29-2]	$K_2MgCa_2[SO_4]_4 \cdot 2H_2O$	triclinic C_1-1	1.548 : 1.562 : 1.567 opt. −	2.78	3 – 3.6
Rinneite	[15976-45-1]	$K_3Na[FeCl_6]$	trigonal D_{3d}-3m	1.588 : 1.589 opt. +	2.35	3
Schoenite	[15491-86-8]	$K_2Mg[SO_4]_2 \cdot 6H_2O$	monoclinic C_{2h}-2/m	1.461 : 1.463 : 1.476 opt. +	2.03	2.6
Sylvite	[14336-88-0]	KCl	cubic O_h-m3m	1.4903	1.99	2
Syngenite	[13780-13-7]	$K_2Ca[SO_4]_2 \cdot H_2O$	monclinic C_{2h}-2/m	1.501 : 1.517 : 1.518	2.58	2.5
Tachhydrite	[12194-70-6]	$CaMg_2Cl_6 \cdot 12H_2O$	trigonal D_{3d}-3m	1.520 : 1.512	1.67	2
Thenardite	[13759-07-4]	Na_2SO_4	rhombic D_{2h}-mmm	1.471 : 1.477 : 1.484 opt. +	2.67	2.7
Vanthoffite	[15557-33-2]	$Na_6Mg[SO_4]_4$	monoclinic C_{2h}-2/m	1.485 : 1.4876 : 1.489 opt. −	2.69	3.6

Table 2. Marine salt rocks: mineral constituents *

Salt rock	Main components	Secondary components
Rock salt	Na	A, Po, Ki, La, clay minerals, etc.
Anhydrite	A	Na, Dol, Mag, gypsum, Sy, C, clay minerals, borates
Carnallitite	C, Na	Ki, Sy, A
Sylvinite	Sy, Na	C, Ki, A
Hard salt		
kieseritic	Na, Ki, Sy	C, A, La, Po, borates
langbeinitic	Na, La	Ki, A, Sy, C
anhydritic	Na, Sy, A	Ki, C
Kainitite	Na, Kai	Ki, Sy, C
Bischofitite	Bi, C and/or Ta	Na, Ki, Po, A, borates
Tachhydritite	Ta, Bi	C, Na
Claystone	Clay minerals, quartz, mica, A, Dol, Mag	Na, Sy, C, Bi, Ta, La, coenenite, rinneite

* Abbreviations: A = anhydrite; Bi = bischofite; C = carnallite; Dol = dolomite; Kai = kainite; Ki = kieserite; La = langbeinite; Mag = magnesite; Na = halite; Po = polyhalite; Sy = sylvite; Ta = tachhydrite.

2.2. Geology of Potash Deposits

Potash deposits occur worldwide in almost all geological systems. The most important deposits were formed in the Devonian, Carboniferous, Permian, Cretaceous, and Tertiary periods [4, 25, 27, 31–37]. All major potash deposits are of marine origin. Bodies of seawater became isolated from the open ocean when bars formed under the water surface, and under arid climatic conditions, the seawater became concentrated, finally depositing the dissolved salts. The important feature is that exchange between normal seawater and concentrated salt solution generally does not occur. During sedimentation, the less soluble salts were deposited first, and the most soluble salts last. In most salt-forming sea basins, this process was repeated many times, resulting in cyclical salt formation [36, 38].

A complete salt deposition cycle begins with basic carbonates (limestone, dolomite, and sometimes magnesite), followed by sulfates (gypsum and anhydrite), rock salt, and finally potassium and magnesium salts (Table 1). Intermediate layers of clay sediments are the result of repeated influxes of fresh water, which can lead to partial redissolution of the salt deposits. Also, eolian (airborne) transportation into the basin can occur. A recent example of a salt deposition basin is provided by the Kara-Bugas, a lagoon on the eastern side of the Caspian Sea.

In the course of the earth's history, the deposited salts underwent many changes, often leading to the formation of a modified mineral constitution. The original formations were affected by relatively low-temperature thermal and hydraulic metamorphoses, and sometimes by volcanic action, producing the mineral compositions that exist today (Table 2) [39, 40].

Rock movements have changed the original horizontal stratification in many salt deposits. Salt migration, folding, and upward movement to form diapirs have led to tilting of the potash layer, sometimes to an acute angle; to thinning; or to local accumulation. As a result of movements, salt deposits often came into contact with groundwater and were partially or completely redissolved.

Table 3 lists the most important potash salt deposits, showing the main potash minerals present, the geological systems, and the geographical locations. Table 4 shows the distribution of potash deposits, with estimates of the reserves [4, 37, 41].

Table 3. Important potash deposits, mineral composition [a], geologic age [b]

Country	Geological age								
	Rect	Plei	Tert	Cret	Jura	Perm	Carb	Dev	Sil	Camb
Australia	B									
Brazil				SC						
Canada						S		SC		
CIS		KS		LKSC	SC	S			S	
Congo			SC							
Germany						HSC				
Ethiopia		SCK								
France			S							
United Kingdom						S				
Israel	B									
Italy			K							
Jordan	B									
Netherlands						C				
Spain			S							
Thailand				CS						
United States	B					SL	S	SC	S	

[a] S = sylvinite; C = carnallite; H = hard salt; L = langbeinite; K = kainite; B = brine.
[b] Rect = recent; Plei = Pleistocene; Tert = Tertiary; Cret = Cretaceous; Jura = Jurassic; Perm = Permian; Carb = Carboniferous; Dev = Devonian; Sil = Silurian; Camb = Cambrian.

Table 4. Minable potash reserves in units of 10^6 t K_2O

Canada, Saskatchewan (conventional mining only)	4500–6000
New Brunswick	60–80
United States	100–150
Brazil	10–40
Chile and Peru	30–50
Congo	ca. 20
Germany	400–800
United Kingdom	30–50
Italy	10–20
France	ca. 20
Spain	20–30
CIS	2000–3000
Dead Sea (Israel and Jordan)	100–200
China	10–100
Thailand	up to 160
Laos	up to 20
World *	7500–10 000

* If reserves only extractable by solution mining are included (particularly those in Saskatchewan, but also including those in the CIS), the figure for the total minable reserves increases by a factor of 4 or 5. Other deposits of potash salts are either unimportant or of minor local importance compared with the above figures. These exist in Australia, Ethiopia, Iran, Libya, Morocco, Poland, and Tunisia.

The concept of a reserve presupposes the possibility of economical extraction, which depends on the presence of a useful potash content together with usable quantities of other materials. Other important factors are the type of deposit, a uniform and usable seam thickness at a depth that is neither too great nor too low (water problems), economic workability, the possibility of solution extraction, and above all, proximity to consumers and profitability. Losses incurred during the extraction process (10–20 %) must be deducted from the total size of the reserve.

The two largest known potash deposits in the world, which are in Saskatchewan (Canada) and White Russia, are of Devonian origin. The Permian deposits (Germany, United States, and CIS) were for a long period the classical salt deposits and were the most important potash reserves, but these lost their economic importance after World War II.

However, the known potential of extractable potash deposits is so large that the world supply is guaranteed for many hundreds of years.

Cambrian. In the central part of the Angara-Lena Trough, ca. 800 km north-northeast of Irkutsk (CIS) in the vast eastern Siberian basin, an evaporite series more than 2 km thick contains a potash deposit. A sylvinite series up to 28 m thick is situated at a depth of 600–900 m in the interior of this Nepa potash basin. At the edges it merges into carnallitite, with over- and underlying carnallite–halite. The potash content of the best sylvinite seam, up to 13 m thick, is 19–30 % K_2O, with an extremely low $MgCl_2$ content [4, 27, 37].

Silurian. In the center of the Michigan basin is a sylvinite zone having a total thickness of 28 m within a 900-m-thick evaporite series of the Salina group covering an area of 34 000 km^2. A solution-mining pilot plant (see Section 3.3) is in operation [4, 37].

Devonian. *Canada.* The western Canadian basin between the Rocky Mountains and Hudson Bay accomodates eight Devonian salt-bearing units, of which the Prairie Evaporite is the most important both in extent and in economic importance, having four potash seams, partly sylvinitic and partly carnallitic. The upper limit of the Prairie deposit is inclined to the south, beginning at a depth of 600 m in the northeast and ending at >3000 m in the southwest. Of the total area (ca. 200 000 km^2) of the potash seams, ca. 50 000 km^2 is extractable by mining. Other areas can only be exploited by solution mining. The clay-bearing sylvinites, each 3–5 m thick, contain 25–30 % K_2O, but no potassium magnesium sulfate. At present, ten plants are in operation [4, 27, 37].

United States. The potash deposit described above, which is of great importance for Canada, stretches across the border to Montana and North Dakota. The potash-bearing area covers more than 30 000 km^2, but because it is at a depth of >1000 m and up to 3500 m, extraction by conventional mining is impossible [4, 37].

CIS. The potash deposits of Soligorsk and Starobinsk lie within the Pripyat marshes, 25 000 km^2 in area, ca. 120 km south of Minsk and 800 km from the Baltic harbor of Ventspils. Potash extraction began in 1963, and output reached 4.2×10^6 t of K_2O by 1980. The Upper Salt is 3200 m thick and contains ca. 60 potash zones in the upper half. These can be grouped into four workable potash-bearing formations. Each of these consists of interbedded layers of salt clays, rock salt, sylvinite seams, and some carnallitite. Only stage II (thickness: 1.8–4.4 m, K_2O: 17.7 %, insolubles: 5 %) and the lower part of stage III (thickness: up to 2.8 m, K_2O: 13.4 – 16.4 %, insolubles: 9 %) are extracted. Potash is mined at depths of 350 – 950 m, where level seams with only slight deformation exist [4, 28, 37].

Carboniferous. *United States.* The potash deposit of the Paradox formation at Moab, Utah, is extracted by Cane Creek Mine. The valuable sylvinite deposit is at present extracted by solution mining. The extent of the salt deposit is ca. 25 000 km^2, of which 15 000 km^2 is potash bearing. It contains about ten potash seams, of which four are workable, containing sylvinite (25 % K_2O) and some carnallitite. The salt has been formed by tectonic action into anticlines, sometimes diapiric, stretching from northwest to southeast [4, 27, 37]. Because the deposit is undulating, conventional mining is complicated and made hazardous by the presence of oil and gas.

Canada. In the eastern provinces of Canada on the Atlantic coast (i.e., New Brunswick, Nova Scotia, and Newfoundland) are a number of small sedimentary basins with evaporites of the Lower Carboniferous Windsor group. Gypsum, anhydrite, and slightly to strongly folded rock salt are present, along with several mainly tectonically bounded small structures containing potash salts. In New Brunswick a high-quality sylvinite seam occurs that contains 25 – 30 % K_2O up to 40 m thick. At present, potash is extracted by PCA (Potash Corporation of America) 7 km north of Sussex at a depth of 460 – 760 m, and by PMC (Potacan Mining Corporation) 25 km south-southwest of Sussex at a depth of 600 – 1000 m [4, 27, 37].

Permian. The Permian period was one of immense salt accumulation, which took place in three vast evaporite basins: (1) the Central European Basin, (2) the East European Basin, and (3) the American Midcontinental Basin, as well as numerous smaller sedimentation areas in Europe, Asia, North and South America, and the Arctic [27].

1) *The Central European Basin* extends from Yorkshire to Central Poland and Lithuania, and from the River Main in northern Germany to the northern part of the North Sea [42]. In the Zechstein, seven sedimentation periods are distinguished, of which the three lowest–the Werra, Stassfurt, and Leine series–are of economic importance for potash extraction.

Germany. In the Werra and Fulda areas, the Hessen and Thuringia potash seams of the Werra series are mined (hard salt and carnallitite in level deposits at a depth of 400 – 1000 m with a thickness of 2 – 5 m, K_2O: 9 – 12 %). The Stassfurt potash seam of the Stassfurt series is mined in the Harz – Unstrut – Saale area (hard salt and carnallitite at a depth of 500 – 1000 m and a thickness of 5 m, K_2O: 20 %). The potash seams Ronnenberg and Riedel of the Leine series are mined in the Hanover area in salt diapirs (sylvinite in inclined deposits, depth: 350 – 1500 m, thickness: 2 – 40 m, K_2O: 12 – 30 %). Hard salts of the Stassfurt series are at present extracted to only a minor extent. Finally, potash is mined on the Massif of Calvörde near Zielitz (depth: 600 – 900 m, Ronnenberg sylvinite inclined at < 18 – 25°, thickness: up to 10 m, K_2O: 14 – 20 %) [4, 37].

United Kingdom. In Yorkshire, a level deposit of carnallitic sylvinite is extracted, which correlates with the German Riedel seam both petrographically and stratigraphically (average thickness: 5 m, K_2O: ca. 27 %, depth: 1100 m) [4] *The Netherlands.* In northeast Holland, the carnallitic potash salts of the Leine series have been drilled at various places. The plans are to extract carnallitite by solution mining [4].

Poland. The Zechstein series in Poland correlates with that in Germany. The carnallitic formations of the Stassfurt seam have been drilled, and the Leine deposits also contain some carnallitite and sylvinite.

2) *The Eastern European Basin* extends from the Barents Sea to the Caspian Sea and covers an area of 1.5×10^6 km^2.

CIS. Potash deposits are known in several parts of the western foothills of the Urals from the Pechora Depression to the Orenburg District, and in the region of the Caspian Depression. The most important reserve is in the area of Solikamsk and Berezniki on the upper Kama, 200 km north of the town of Perm. It has been mined since the late 1920s. Potash seams occur in the upper part of a salt series >400 m thick that is part of the Kungurian. These seams are sylvinitic below and sylvinitic – carnallitic above, interlayered with rock salt. The lower section, consisting of four sylvinite seams of total average thickness 20 m, with a K_2O content of 15 – 20 %, is mined at a depth of 300 – 500 m in predominantly level deposits [4, 27, 37].

3) *The Midcontinental Basin of North America* stretches from North Dakota to Texas but is broken up into numerous smaller basins. Horizons of rock salt are present in almost the whole of the Permian system. The deposition of potash in large quantities apparently took place only during a small part of the Upper Permian period in a subregion of the Delaware basin of New Mexico and Texas [27].

United States. In Carlsbad, New Mexico, sylvinite and langbeinite are mined in the 50–140-m-thick McNutt potash zone of the Salado formation in five out of a total of eleven potash seams up to 3 m thick in a level deposit at a depth of 250 – 575 m. The sylvinite contains 12 – 30 % K_2O, and the langbeinite 7 – 9 % K_2O. Massive and disseminated polyhalite, carnallite, kieserite, kainite, leonite, glauberite, and thenardite are also present [4, 27, 37].

4) *Brazil.* In the Amazon basin at Nova Olinda, 150 km south of Manaus, drilling of a level deposit of Lower Permian sylvinite at a depth of 1000 – 1150 m has been carried out. The thickness is up to 4.5 m, and the K_2O content 17 %. Extractable reserves amount to ca. 35×10^6 t of K_2O [4, 37].

Jurassic. *Turkmenistan/Uzbekistan.* Up to six seams of sylvinite can be found at a depth of 200 – 1200 m in the Gaurdak formation in the areas of Gaurdak – Tyubegatan, Kugitang, and Okuzbulak; and carnallite may be found at Karlyuk. The seams of sylvinite, up to 6 m thick, contain 12.5 – 25 % K_2O, sometimes with as much as 12 % insoluble material. A deposit at Karshi in Uzbekistan is to be exploited, and solution mining on a pilot scale has been started at Karlyuk [4, 37].

Cretaceous. Along the east coast of Brazil and the west coast of Africa from Gabon to Angola, Lower Cretaceous potash-bearing evaporites occur. Their similar formations indicate the existence of a common sedimentation area before the two southern continents began to drift apart during the tectonic movements of the Cretaceous period [4, 37].

Brazil, Sergipe. The Taquari-Vassouras deposit of the Muribeca formation has been mined since 1985. It has a basal layer of tachhydrite, above which is a thin layer of carnallitite and two seams of sylvinite, 10 m and 8 m thick, containing 15 % K_2O, with a rock salt bed 4 – 10 m thick between the two seams. The reserves of sylvinite amount to 525×10^6 t, with $16\,000 \times 10^6$ t of carnallitite [4, 37].

Congo. Salt deposits have been formed in ten sedimentation cycles, of which the second highest contains four sylvinite seams. The other cycles contain large masses of carnallitite, and, as in Sergipe, tachhydrite. Mining of the sylvinite in the third layer (depth of 370 m, 3 – 16 % K_2O) began in 1969 but ceased in 1976 when the mine became flooded in the space of two days [4, 37].

Thailand. The lower part of the Maha-Sarakham formation below the Khorat Plateau northeast of Bangkok consists of evaporites with a potash sequence up to 90 m overall thickness at depths of 90 – 530 m. The K_2O content of the carnallitite, which can be 15 m thick, is usually ca. 10 %, but can occasionally reach 14 %. Sylvinite, with a maximum thickness of 3.7 m and a K_2O content of 18 %, has been found in only a few of the 48 exploration boreholes, with tachhydrite in some locations. Test mining has begun in the concession areas, which cover 300 km^2 [4, 37].

Tertiary. *France.* The Lower Oligocene deposit in Alsace contains two sylvinite seams in a marl – rock salt series. The upper layer has a thickness of up to 2 m and contains 19 – 25 % K_2O; the lower, up to 5.5-m-thick layer, with 15 – 23 % K_2O, also contains 15 % insolubles (clay, anhydrite, and dolomite). Mining is carried out at comparatively high rock temperatures at a depth of 500 – 1000 m in flat or slightly inclined seams that have been disturbed by faults [4, 37].

Spain. Deposits are located in two areas of the Ebro basin. In Catalonia and Navarra, potash salts lie above rock salt. These deposits are up to 10 m thick in Catalonia and up to 15 m in Navarra. Above this occurs an interbedded deposit of rock salt, carnallitite, marl, and anhydrite. Only the sylvinite seams A and B are mined. These are up to 4 m in total thickness at a depth of 1020 m, some deposits being level and some inclined. The crude salt contains 12.5 – 14 % K_2O [4, 37].

Italy. The Miocene evaporites starts with an anhydrite – polyhalite – halite series followed by rock salt with up to six kainite interbeds and a roof of kieserite – sylvinite. Since 1976, only the kainite seams, which are up to 32 m thick, and contain 12 – 14 % K_2O, have been mined at ca. 800-m depth in seams inclined at 25 – 65° [4, 37].

CIS. In the eastern Galician foredeep of Carpathia, potash salts have been known since 1804 (mainly langbeinite and kainite, with a little sylvinite) and have been mined since 1864 at Stebnik and Kalusch. The individual lenticular deposits occur in five horizons of 1 – 5 m thickness, the kainite containing 10 – 12 % K_2O and the sylvinite 8 – 19 %. The deposit is inclined at 45° [4].

Pleistocene. In the Danakil depression, Ethiopia, is a salt basin of ca. 40×100 km, 100 m below the surface. Boreholes have revealed a potash zone containing sylvinite, carnallitite, and kainite at a depth of ca. 740 m [4].

Salt Lakes and Subterranean Brines. *Israel and Jordan.* Potassium chloride, magnesium chloride, rock salt, soda, chlorine, and bromine are obtained from the Dead Sea [4, 37].

United States. Potassium and sodium sulfate are extracted from the Great Salt Lake [4, 37]. Potassium and magnesium chloride are obtained from subterranean brines from the Great Salt Lake Desert [4]. Potassium chloride and sulfate, sodium borate, and boric acid are obtained from subterranean brines in Searles Lake [4, 37].

Australia. Langbeinite is obtained from subterranean brines at Lake McLeod.

China. Tsaerhan Lake, a dry lake, ca. 1100 km northeast of Lhasa, the largest salt lake in the Chaidamu basin, appears to be the most important potential reserve of potassium chloride in China [4, 37].

3. Mining of Potash Salts

3.1. Shaft Mining

Mineral salts are very soluble, and therefore any flow of water into the mine from the overlying strata, which are normally water bearing, must be prevented. This causes difficulties when sinking a shaft, and severe accidents have been caused by influxes of water. The freezing technique, especially deep freezing, is comparatively safe and is used to sink most mineshafts. The shaft itself is protected from water-bearing rocks by the use of tubbings, which are segments bolted together to form rings and are usually made from cast iron or steel-reinforced concrete. The shafts generally have a diameter of 5 – 7 m, and the depth can exceed 1000 m.

3.2. Extraction, Conveying, and Haulage

In most potash mines, the salt is mined from subhorizontal deposits. Generally, rooms are created by removing the salt, and pillars are left between these to prevent the cover rocks from collapsing. This enables an extraction rate of 25 – 60 % to be achieved. For cost reasons the mined-out rooms are not backfilled. In some mines, the total ore is extracted, which causes substantial subsidence of the overlying strata (Alsace).

In steeply dipping deposits (e.g., in the salt domes of northern Germany), roof mining was originally carried out. This was later replaced by floor mining and then by funnel mining, which is now being used increasingly in numerous variations [43]. Entry drifts are driven one above the other at intervals of 15 – 20 m, and the remaining potash salt is mined by sublevel stoping. Material loosened by explosives falls via the lowest funnel-shaped region into the main haulage level underneath. The mined-out room, 100 – 250 m in height, is usually backfilled with salt waste after mining. Funnel mining is much safer and cheaper than roof mining, because both ore stoping and backfilling take place under gravity, and the mined room need not be reentered. During drilling of the blast holes, the miners are protected by horizontal pillars between the sublevels.

Drilling and blasting operations are carried out with the help of trackless vehicles. Large holes with a diameter of ca. 40 cm and a length of 7 m are drilled by mobile drilling equipment, and another mobile drilling rig is used to drill blast holes around the larger hole in a predetermined pattern. The explosive (generally ammonium nitrate with addition of oil) is brought to the workplace in tanks carried by diesel vehicles and is blown into the shot holes by compressed air.

In predominantly horizontal sylvinite deposits, the most frequent method of extraction is by cutting with heavy machinery (180 t per unit) at high cutting rates. These machines produce material suitable for transport by conveyor belt, enabling continuous extraction in 30–60-m sections.

Extraction by borers with two to four rotors is mainly used to produce long pillars in a room and pillar system (Canada), the length of the chambers being as much as 1000 m. In longwall mining, however, which is usually carried out as a caving operation, two or three drum shearers are used (e.g., in Alsace and the CIS). These machines enable daily outputs between 1500 and 4000 t to be achieved.

The recovered material is transported by trackless diesel or cable-fed electric front-end loaders with a capacity of 15 t. Transport along underground roads is increasingly by conveyor belts but also by electric or diesel trains with 30-t-capacity wagons or by dumper trucks with a hopper capacity of 40 t. For transportation by conveyor belt, the material must first be broken into suitably sized pieces.

The network of roadways for conveying, traveling, and ventilation usually extends >100 km in large potash works. A radio communication system is generally used.

Since the introduction of very heavy machinery and diesel-powered vehicles, extra attention must be given to ventilation. Powerful fans supply fresh air at up to 30 000 m³/min.

In hoisting shafts, the skips have a capacity of up to 25 t. These operate automatically and supply large intermediate storage bins in the filling station, so that the continuously operating manufacturing plants can be supplied with ore at a steady rate. Average daily throughput can be ca. 30 000 t of ore.

Improvements in mining methods and the introduction of new techniques have enabled the output per worker underground to be increased considerably. In Germany during 1965–1974 the output of potash ore increased by 20 %, despite a >50 % reduction in the number of employees.

3.3. Solution Mining

Solution mining is an alternative to conventional mining for the extraction of potash ore. The advantages of this method are that the high expense of sinking a shaft is eliminated and reserves can be exploited where conventional mining is impossible (e.g., at great depth). Also, this method can be used where existing mine workings are available but conventional mining methods are no longer feasible, even though extensive reserves may still exist.

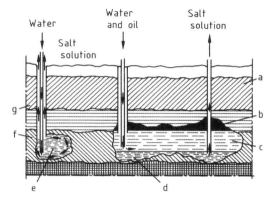

Figure 1. Solution mining
a) Overlying rock; b) Oil cushion; c) Partly unsaturated salt solution; d) Saturated salt solution; e) Cavity produced by dissolution; f) Potash deposit; g) Sodium chloride layer

Since 1964, Kalium Chemicals in Saskatchewan has operated a plant in which brine extracted from a potash deposit at a depth of ca. 1500 m is used to produce very pure potassium chloride. The process is based on a series of patents [44], but details have not yet been published. Water or an unsaturated solution of potassium chloride is passed through a system of boreholes into the potash seam, which is 20–25 m thick; potassium chloride and sodium chloride are dissolved. The almost saturated solution is pumped to the surface and fed to the production plant. Rock salt above the potash seam is protected from dissolution by an oil or air cushion (see Fig. 1). The brine produced passes through multiple-effect evaporators in which sodium chloride crystallizes. Potassium chloride is then crystallized in a series of vacuum coolers [45].

In Utah (United States), where it was necessary to terminate a conventional mining operation due to severe geological and technological problems, operation was resumed in 1972 by using solution mining. Shafts and underground cavities were flooded, providing a route for the brine formed when water was fed in to dissolve the salt. The brine was passed into surface ponds where solar evaporation caused a mixture of potassium chloride and sodium chloride to crystallize, which was treated in a flotation plant to produce 60 % K_2O potassium chloride [46].

In Canada a conventional mining operation was also converted to the solution mining method after penetration of water led to complete flooding of the mine and forced operations to cease. Water is passed via boreholes into the flooded mine and is converted into a concentrated brine, which is withdrawn and cooled in a pond during the very cold Canadian winter. The potassium chloride that crystallizes is recovered and processed to give a salable product [47].

In the former German Democratic Republic, extensive research into the solution mining of carnallite or potassium chloride from carnallitic deposits has been carried out. An experimental plant with a KCl capacity of ca. 50 000 t/a was operated for a long period (see Section 5.2.3) [48].

4. Treatment of Potash Ores

4.1. Intergrowth and Degree of Liberation [49, 50]

The salt minerals in potash ores are intergrown to varying extents. Before the minerals can be separated and the useful components recovered, the ore must be sufficiently reduced in size so that individual components are accessible to the processing method to be used. In the hot leaching process, sylvite is extracted, and therefore it must first be liberated (i.e., it must not be occluded inside other minerals). To achieve this, it is sufficient to break down the ore to a particle size of 4–5 mm or less.

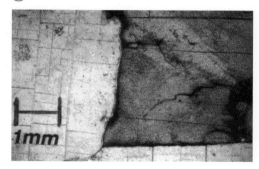
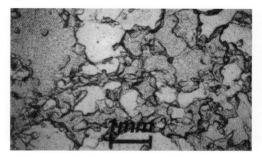

Figure 2. Thin sections of sylvinite ores
A) Coarsely intergrown (potash works in Lanigan, Canada);
B) Finely intergrown (Kaliwerk Sigmundshall, Germany)

For the mechanical treatment processes (i.e., flotation, electrostatic treatment, and gravity separation), liberation of the minerals must be complete (i.e., individual grains must consist as much as possible of pure minerals). The extent to which the minerals in the potash ore are intergrown can vary greatly from deposit to deposit (see Fig. 2), which means that the crude salt must be size-reduced to varying degrees before further processing.

The degree of intergrowth of individual minerals can be determined by examination of a thin section. A photomicrograph shows the sizes and shapes of individual minerals in relation to each other [51]. The disadvantage of this method is that a thin section gives only a two-dimensional view of a relatively small region of the salt mineral, and a very small sample of the substance is examined. The three-dimensional arrangement of minerals present and their distribution are not observed. For these reasons, and also because of the high cost of preparing thin sections, this method is now of only minor importance for the industrial processing of potash ores.

More usually, the degree of liberation in size-reduced samples is determined. The degree of liberation of a mineral means the percentage ratio of fully liberated mineral particles to the total content of the mineral in the sample.

Degree of liberation:

$$\frac{\% \text{ Free mineral}}{\% \text{ Free} + \% \text{ Intergrown mineral}} \cdot 100$$

The degree of liberation L depends not only on the grain size achieved by the grinding operation but also on the type of grinding. It can be given for particular ranges of grain size so that the way in which it depends on grain size can be determined, or it can be expressed as an average for the total sample (the integral degree of liberation $\bar{L}$). The arithmetic mean of the degree of liberation of each range of grain sizes is obtained from the following formula:

$$\bar{L} = \frac{\sum\limits_{i=1}^{n} L_i p_i a_i}{100 a}$$

where p_i is the mass fraction of the ith size range in percent; a_i the percentage of useful mineral in the ith size range; and a the percentage of useful mineral in the total sample. Apart from the fraction of useful mineral a, which is determined by chemical analysis, the liberated, nonintergrown fraction of useful mineral in each grain size range must be determined. Two methods for doing this are possible:

1) Visual estimation (by counting under the microscope) of the proportion of intergrown particles
2) Heavy-medium separation of the free or almost free grains

The first method is easy to apply to salt minerals because the intergrowth effects are readily recognized owing to the transparency of the grains. For coarse-grained materials such as those usually found in ground products from coarsely intergrown potash ores, this process cannot be used.

The separation of free or nearly free mineral grains from a size fraction is carried out by using heavy liquids of appropriate density, such as tetrabromoethane–toluene mixtures. A float–sink separation is carried out to determine the fraction of free mineral grains. This method cannot be used for salt particles with a grain size <0.5 mm because of the agglomeration of fine grains. In this case, the method of counting under the microscope must be used, or the degrees of liberation of the coarser size ranges must be extrapolated.

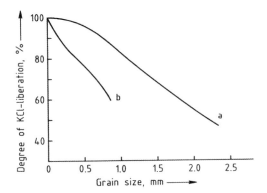

Figure 3. Liberation curves of sylvinite ores
a) Coarsely intergrown sylvinite ore (from New Brunswick, Canada); b) Finely intergrown sylvinite ore (from a north German salt deposit)

The results are expressed in liberation curves, which give the degree of liberation as a function of grain size (Fig. 3).

4.2. Grinding [52, 53]

Potash salts are easily size-reduced. Therefore, fines may be formed, which can cause problems in later stages of processing. Great care must be taken in selection of the equipment for various stages of grinding.

The maximum grain size to which the potash ore is ground depends on the processing method used and the degree of intergrowth of the ore. For the hot leaching process, an upper grain size limit of 4–5 mm is adequate. For mechanical processing, the ore must be ground to a degree of liberation >75 %. For German sylvinite ores and hard salts, this is achieved by grinding to a maximum grain size of 0.8–1.0 mm. For the much coarser sylvinite ores of New Mexico, a maximum grain size limit of 2.4 mm is sufficient. The sylvinite ores of Saskatchewan are even more coarsely intergrown, so that size reduction to < 9 mm would give adequate liberation. However, such large crystals cannot be treated by conventional flotation, and the material is therefore normally ground to < 2.3 mm. One large Canadian manufacturer produces a coarse crystalline product by grinding the potash ore to < 9 mm, removing grains < 1.7 mm, and treating this fine material by conventional flotation. The remaining fraction (1.7 – 9 mm) is treated in a heavy-medium separation plant (see Section 5.5), giving a product with 60 % K_2O (95 % KCl) and a size distribution of the granular commercial grade.

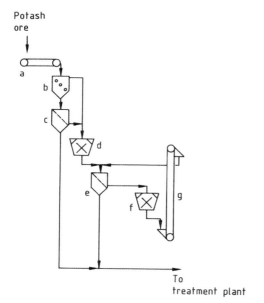

Figure 4. Two-stage grinding system
a) Conveyor belt; b) Grid; c) First screening stage; d) First grinding stage; e) Second screening stage; f) Second grinding stage; g) Bucket elevator

The preliminary size reduction of potash ore is carried out by underground mobile crushers, usually in the vicinity of a mining operation. When the ore is mined by continuous mining this gives sufficient size reduction for it to go directly to the haulage line. Further size reduction to the grain size required for processing is carried out in two stages on the surface. An initial size reduction with impact or hammer mills is always carried out to produce 4 – 12-mm particles, depending on the raw material and processing method to be used. A coarse grinding plant usually includes two grinding and screening stages (see Fig. 4).

The final fine grinding stage is carried out either by wet grinding in rod mills or by dry grinding with rollers or impact crushers (Figs. 5 and 6). Wet grinding with rod mills in a recirculating system with classification is standard for most flotation plants. Classification is by spiral classifiers, vibrating screens, curved screens, or cyclones. Wet screening produces only a small amount of fines and has the further advantage of providing a scrubbing effect that facilitates the removal of clay from clay-bearing ores; a necessary step before the flotation process.

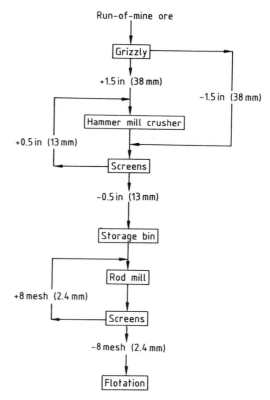

Figure 5. Wet grinding and screening of coarsely intergrown potash ore for flotation [5]
Reprinted with permission of the Society for Mining, Metallurgy, and Exploration, Inc.

The production of a fine product by dry grinding is rarely used in flotation plants. It is indispensable as a preparation for electrostatic processing, which is not compatible either with the changes to the mineral surfaces caused by aqueous solutions or with excess moisture. The grinding operation must be carried out carefully to give a product low in fines. Roller mills and impact mills can be used. Roller mills have the disadvantage that throughputs are relatively low and maintenance costs are high. Although this was the preferred method in the early days of electrostatic processing owing to its gentle grinding action, it has now been largely replaced by impact grinding, which is also used for size reduction of the middle product from electrostatic separation [54].

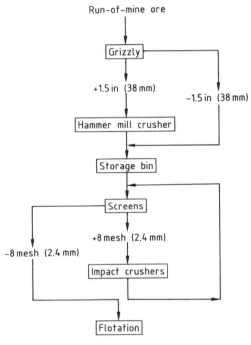

Figure 6. Dry grinding and screening of coarsely intergrown potash ore for flotation [5]
Reprinted with permission of the Society for Mining, Metallurgy, and Exploration, Inc.

5. Potassium Chloride

Potassium chloride [7447-40-7], KCl, mineral name sylvite [14336-88-0], forms colorless nonhygroscopic crystals. It occurs in many salt deposits (see Section 2.2) mixed with halite and other salt minerals. Natural sylvite is usually opalescent or milky white, as are crystals obtained from an aqueous solution. Sylvite is often colored red by hematite. With magnesium chloride it forms the double salt carnallite [1318-27-0], $KCl \cdot MgCl_2 \cdot 6 H_2O$, which is also commonly found in salt deposits. Potassium chloride is produced in large quantities from mined potash ores and from salt-containing surface waters. More than 90 % of the potassium chloride produced is used in single- or multi-nutrient fertilizers, either directly or after conversion to potassium sulfate ($\rightarrow$ Fertilizers, Chap. 3.). The remainder has various industrial uses and is the raw material for the manufacture of potassium and its compounds.

5.1. Properties

Potassium chloride crystallizes in the cubic system, usually as actual cubes. Some physical properties described are as follows:

Relative molecular mass	74.55
Melting point	771 °C
Crystal system and type	cubic 0_h^5
Refractive index n_D^{20}	1.4903
Density	1.987 g/cm^3
Specific heat c_p	693.7 J kg^{-1} K^{-1}
Heat of fusion	337.7 kJ/kg
Enthalpy of formation ΔH^0	-436.7 kJ/mol
Entropy S^0	82.55 J mol^{-1} K^{-1}
Dielectric constant (at 10^6 Hz)	4.68
Thermal coefficient of expansion (15–25 °C)	33.7×10^{-6} K^{-1}

Solubilities in water at various temperatures appear in Table 5, and the phase diagram of the system KCl–H$_2$O is shown in Figure 7.

Table 5. Solubility of potassium chloride in water (g/100 g) [16]

Temperature, °C	0	10	20	30	40	50
Solubility	28.1	31.2	34.2	37.2	40.2	43.1
Temperature, °C	60	70	80	90	100	
Solubility	45.9	48.6	51.3	53.8	56.2	

In the system KCl–H$_2$O, the only solid phases formed are KCl and ice. The cryohydric point (ice + KCl) is -10.7 °C (29.7 mol K$_2$Cl$_2$/1000 mol H$_2$O). The boiling point of the saturated solution at 1.013 bar is 108.6 °C (71.6 mol K$_2$Cl$_2$/1000 mol H$_2$O).

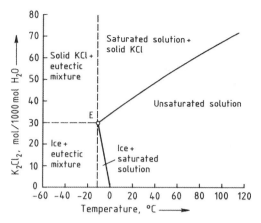

Figure 7. Solubility curves for potassium chloride in water
E = cryohydric point : Ice – potassium chloride solution

5.2. Production by Crystallization from Solution

5.2.1. Phase Theory

The salt deposits were formed by the evaporation of seawater, which contains the principal ions Na^+, K^+, Mg^{2+}, Ca^{2+}, Cl^-, and SO_4^{2-}. With water, these ions constitute a six-component system. The concentration of Ca^{2+} in the most interesting region of the system is negligibly small, so that the system reduces to quinary. It is nevertheless very complicated, with 23 different salts being formed between 0 and 100 °C, depending on the temperature and the ratio of concentrations of the components.

Understanding how the salt deposits were formed and how they behave in dissolution processes requires knowledge of the solution equilibria.

The theoretical foundations for this were laid by VAN'T HOFF et al., who between 1896 and 1906 investigated the formation of oceanic salt deposits [55]. Many investigators have continued this work up to the present.

J. D'ANS critically evaluated all published data up to 1933, expressing his results in graphical form [56]. In the same book, he described experimental methods for determining solution equilibria and gave recommendations for the graphical representation of experimental results [57].

Much of the equilibrium data published up to 1967 are given in [58,59].

In the years following World War II, in the Kaliforschungs-Institut (Potash Research Institute) of Hanover, AUTENRIETH carried out comprehensive and detailed research into the stable and metastable equilibria of most relevance to potash production (especially from hard salt), giving the results in a form suitable for practical application [60–66].

The intensive investigations carried out into the quinary system make it the most thoroughly investigated system with more than four components. However, only parts of the system that are of most relevance to potash production have been thoroughly investigated. An obstacle in the application of equilibrium data to practical problems is that such a complex system is very difficult to represent in a two-dimensional diagram. However, by fixing parameters, working with projections on a plane, and using diagrams showing lines of constant parameters, even nonexperts can work with them.

The best-known region is that in which the solutions are saturated with sodium chloride. This is also the most important region for potassium chloride manufacture, both by the hot leaching process and by flotation, because in both cases a solution saturated with sodium chloride is used. The most important part of the so-called NaCl saturation space at 25 °C is shown in Figure 8 as a three-dimensional view.

Each point in the interior of this space corresponds to a solution saturated with NaCl, in which the concentrations of $MgCl_2$, K_2Cl_2, and $MgSO_4$ are given by the distance of the point from the axes. For practical reasons, the concentration figures are given in moles per 1000 mol of H_2O, and the concentrations of KCl and NaCl are given in double moles. If one or more of the salt concentrations increase, and if the saturation concentration of another salt is exceeded, this salt separates out. Its identity depends on the concentration ratio in the solution. Such two-salt solutions (saturated with two salts) lie on two-salt surfaces, which form the boundaries of the saturation space. Of the twelve other salts whose saturation spaces form the boundary of the NaCl saturation space, seven can be seen on the figure because their two-salt surfaces lie in the concentration range of Figure 8.

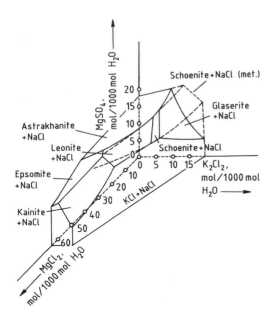

Figure 8. Three-dimensional view of the quinary system (saturated with NaCl) at 25 °C with 0–65 mol MgCl$_2$/1000 mol H$_2$O, showing stable and metastable regions

The KCl–NaCl two-salt surface on the front side of Figure 8 is of special significance for the potash industry, because it allows solutions to be made up that are saturated with both salts at 25 °C. Such solutions are important in the treatment of sylvinitic potash ores.

In many instances, a salt does not crystallize spontaneously when its concentration exceeds that indicated on a two-salt surface. Instead, highly supersaturated solutions are formed, which can remain stable for hours or days, depending on temperature and composition. These supersaturated solutions are termed metastable. They become stable saturated solutions by crystallization of a salt. In Figure 8, continuations of the stable schoenite–NaCl two-salt surface to the right and left into the unstable region are shown as broken lines. In potash manufacture, stable solution equilibria are seldom attained. This is particularly true for hard salt processing in which NaCl-saturated solutions with high MgSO$_4$ content often lead to the undesired crystallization of double sulfates such as schoenite, leonite, langbeinite, and glaserite. Here, the rates of dissolution, nucleation, and crystallization of these salts as a function of temperature and composition of the solutions are especially important [64–66].

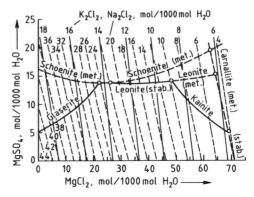

Figure 9. 25 °C isotherms of the quinary system saturated with NaCl
Stable (—) and metastable (---) surfaces saturated with KCl and NaCl. Concentrations of K$_2$Cl$_2$ (—) and Na$_2$Cl$_2$ (---) are indicated by lines of equal concentration

For practical application of equilibrium data, the boundary surface of the NaCl saturation space is projected, for example, in the direction of the K$_2$Cl$_2$ axis in the MgSO$_4$–MgCl$_2$ plane. In Figure 9, the two-salt surface NaCl–KCl for the stable and metastable 25 °C isotherm is shown. The indicated K$_2$Cl$_2$ and Na$_2$Cl$_2$ lines of constant concentration enable the complete composition to be read off for each of the solutions shown here. By using this diagram, the dissolution and crystallization processes possible in this part of the quinary system can be described quantitatively. To control the crystallization of potassium chloride from 90 °C solutions in an industrial plant, for example, the most important boundary surfaces of the 90 °C isotherm of the system (Fig. 10) are additionally required.

If any of the components of the quinary system are present in such small quantities that they have a negligible effect on the process, the mathematical treatment can be simplified by dealing only with the remaining subsystem. The following subsystems are of importance:

1) Na$^+$, K$^+$, Mg^{2+}, Cl$^-$, and H$_2$O with NaCl saturation (see Fig. 17) for the conversion of carnallite into potassium chloride and bischofite
2) K$^+$, Mg^{2+}, Cl$^-$, and H$_2$O (see Fig. 16) for the decomposition of carnallite

3) Na^+, K^+, Cl^-, and H_2O [57] for the selective dissolution of NaCl (e.g., from crystalline product obtained from hard salt in the hot leaching process)
4) K^+, Mg^{2+}, Cl^-, SO_4^{2-}, and H_2O [61] for the production of potassium sulfate and potash – magnesia
5) Na^+, Mg^+, Cl^-, SO_4^{2-}, and H_2O [57] for the production of thenardite or Glauber's salt

Two different processes are used, depending on the composition of the ore. In the *sylvinite* hot leaching process, the other salts present in addition to KCl and NaCl play only a minor role in the process solutions. In *hard salt* leaching, process solutions contain appreciable amounts of $MgCl_2$ and $MgSO_4$. In the case of carnallite-containing hard salts, preliminary carnallite decomposition must be carried out (see Section 5.2.3.) if the amount of carnallite present exceeds a critical level.

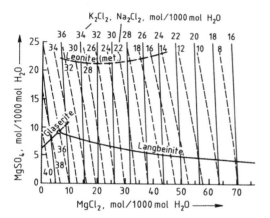

Figure 10. 90 °C isotherms of the quinary system saturated with NaCl
Stable (—) and metastable (---) surfaces saturated with KCl and NaCl. Concentrations of K_2Cl_2 (—) and Na_2Cl_2 (---) are indicated by lines of equal concentration

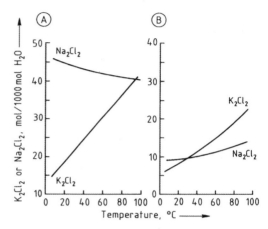

Figure 11. Solubility curves for KCl and NaCl (schematic) A) Sylvinite leaching (with nonevaporative cooling); B) Hard salt leaching

5.2.2. Hot Leaching Process

The hot leaching process is the oldest industrial process used to produce potassium chloride from potash ore. It was first used in 1860 in Stassfurt and since then has been developed further in Germany, where it is still the dominant process. It is especially suitable for treating very finely intergrown ores or ores that contain other salt minerals or insoluble minerals in addition to the sylvite and halite. It enables a high-purity product with a uniform grain size to be produced. In many plants, especially in Canada, where flotation is the main production process, small hot leaching plants are also operated, in which the product fines (<0.2 mm) are recrystallized, or potassium chloride is extracted and crystallized from the flotation residues or thickened clay slurries. These procedures give a considerable improvement in total yield and result in a very pure, completely water-soluble product.

The different solubility properties of sodium chloride and potassium chloride are shown in Figure 11. The solubility of potassium chloride is lower in hard salt solutions than in sylvinite solutions. The difference between the potassium chloride contents of saturated solutions at low and high temperatures is less for solutions of hard salt than for solutions of sylvinite, so that the amount of potassium chloride that can be crystallized from a given amount of solution is smaller, which has a marked effect on the energy requirement. Furthermore, an important difference between the two solution types is that the solubility of sodium chloride in sylvinite solutions decreases with increasing temperature, whereas it increases in hard salt solutions. This is apparent in Figure 11B, which shows the behavior of process solutions in a carnallitic hard salt plant with a magnesium chloride content of ca. 240 g/L. This dependence of the solubility of sodium chloride on temperature and magnesium salt content explains why the sodium chloride contents of crystallized products differ, depending on whether they came from the treatment of sylvinite or hard salt.

The Process (Fig. 12). The potash ore, ground to a fineness of <4–5 mm, is stirred in a continuous dissolver with leaching brine heated to just below its boiling point. The leaching brine is the mother liquor from the crystallization stage of a previous cycle of the process. The quantity of leaching brine required is determined by the amount of potassium chloride in the ore. The potassium chloride should be extracted from the ore as completely as possible, and the resulting product solution should be as nearly saturated as possible. The residue consists of two fractions of different particle size. The coarse fraction is removed from the dissolver and debrined. The fine fraction (fine residue or slime) is removed from the dissolver along with the crude solution, which is clarified with the aid of clarifying agents. The slime that separates is filtered off, and the filtrate from the coarse and fine residues is recycled to the recirculating brine. The residues are washed with water or plant brines low in potassium chloride to remove the adhering crude solution, which has a high potassium chloride content. The residues are then disposed of by dumping (see Section 5.8).

The hot, clarified, crude solution is cooled by evaporation in vacuum equipment. Potassium chloride and sodium chloride crystallize as the water is removed. The sodium chloride content of the crystals formed can be controlled by complete or partial replacement of the evaporated water during crystallization. The crystals formed are separated from the mother liquor and processed further. The mother liquor is heated and recycled to the dissolver as leaching brine.

The leaching process is usually carried out in two stages in a main dissolver and a secondary dissolver. The ground ore is first added to the main dissolver where it is mixed with the already partially saturated solution from the secondary dissolver. This causes the solution to be almost completely saturated, and it is then removed from the leaching equipment. The partly extracted ore is next fed to the secondary dissolver where it comes in contact with fresh leaching brine, and the potassium chloride that was not extracted in the main dissolver is taken up by the solution, which is then fed to the main dissolver. The leaching process in the main dissolver can be cocurrent or countercurrent (Fig. 13).

The residence time in the leaching equipment is insufficient to give complete KCl–NaCl equilibrium. The crude solution always contains less potassium chloride and more sodium chloride than corresponds to equilibrium. Another reason for the potassium chloride concentration to be maintained below saturation is that complete recovery of potassium chloride from the ore is possible only if the hot solution at the end of the leaching process still has a small capacity for dissolving potassium chloride. Also, the crude solution cools by 1–2 °C as it travels from the main dissolver to the overflow from the following hot clarification stage. If the solution leaving the main dissolver were completely saturated in potassium chloride, the potassium chloride would begin to crystallize at this point, leading to losses in the slime.

For the crystallization process the potassium chloride content should be maintained as near to saturation as possible in the leaching equipment, so that the concentration difference between the mother liquor and the hot crude solution is as great as possible. The greater this difference, the smaller is the amount of solution used, and the lower is the energy consumption.

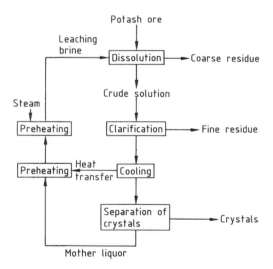

Figure 12. Overall schematic of a hot leaching process

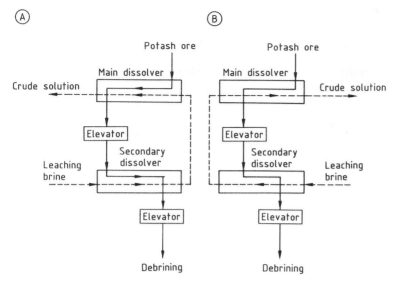

Figure 13. Schematic arrangements of a hot leaching apparatus
A) Cocurrent flow; B) Countercurrent flow

Also, a high concentration of potassium chloride in the solution leads to a high potassium chloride content in the crystalline product obtained on cooling. If the solution is unsaturated with respect to potassium chloride, a corresponding amount of sodium chloride in excess of its saturation concentration is taken up by the solution to compensate for the missing potassium chloride. This means that, on cooling, a quantity of sodium chloride crystallizes, causing a decrease in the potassium chloride content of the product even in sylvinitic solutions in which the solubility of sodium chloride increases with decreasing temperature.

Cooling the crude solution by evaporation increases the concentration of salts present below saturation, and the salts with which the solution is saturated crystallize. Unwanted crystallization of sodium chloride can be prevented by adding water to the vacuum cooling equipment, especially in the case of solutions from sylvinite ore leaching. If the required potassium chloride content in the crystalline product cannot be obtained by adding water to the solution, which is often the case when hard salt is being processed, the same result can be obtained by treating the product with cold water. A product containing 60 % K_2O (95 % KCl) is usually required. By this cold-water treatment technique, it is even possible to obtain a product with 62 % K_2O (98 % KCl). The spent solution is recycled to the process.

For KCl of analytical or pharmaceutical quality, potassium chloride produced by the hot leaching process must be purified by single or multiple recrystallization.

Processing of Hard Salt. Unlike the sylvinitic potash ores, whose principal constituents are sodium and potassium chloride, hard salts contain not only the alkali chlorides but also large amounts of kieserite, usually with varying amounts of carnallite, langbeinite, and anhydrite. Therefore, process brines produced by the leaching of hard salt are characterized by high contents of magnesium chloride and magnesium sulfate, which make potassium chloride production more difficult.

Magnesium sulfate comes mainly from kieserite, which is very soluble (Fig. 14) but has a slow rate of dissolution that becomes even slower if large amounts of dissolved $MgCl_2$ are present. The amount of magnesium sulfate that dissolves depends on the grain size of the ore fed to the dissolving equipment, the kieserite content of the ore, the time for which the crude salt – solution mixture is stirred, the temperature, and the magnesium chloride content of the brine.

Magnesium chloride results from reaction of dissolved magnesium sulfate with potassium chloride to give sulfate-containing double salts, and also from any carnallite (KCl · MgCl$_2$ · 6H$_2$O) present in the hard salt. In general, the MgCl$_2$ and MgSO$_4$ contents are kept constant in the circulating brines, so that the rate at which magnesium is taken up from the ore is balanced by the rate at which it leaves the system in the residues and products. If the ore contains large amounts of carnallite, part of the circulating brine must be removed continuously from the system to prevent the MgCl$_2$ level becoming too high.

operating conditions (Figs. 9 and 10). The extent of supersaturation with double sulfates must therefore be controlled to prevent uncontrolled crystallization of double salts and consequent introduction of impurities into the product or disturbance of the process. To achieve this, the circulating brine, or part of it, is fed to a reactor in which it is agitated intensively in the presence of nuclei (20–40 wt %) of the double salt to be removed until the brine is no longer supersaturated with respect to it [67].

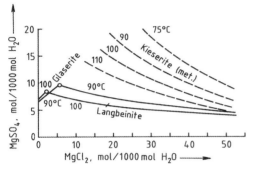

Figure 14. Metastable solubility of kieserite in the quinary system saturated with KCl and NaCl between 75 and 110 °C, and solubilities of langbeinite and glaserite at 90 and 100 °C

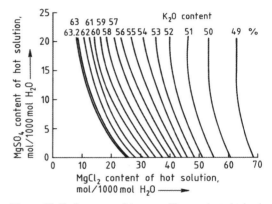

Figure 15. K$_2$O content of the crystalline product obtained by cooling an equilibrium solution saturated with NaCl and KCl from 90 to 25 °C without evaporation while maintaining saturation

Because of the large amounts of magnesium salts in process brines, the hard salt leaching process has three important disadvantages compared with the sylvinite process:

1) In solutions of hard salt, the range of solubilities is considerably less than in solutions of sylvinite (Fig. 11). To extract a given amount of potassium chloride, much more liquor, and hence more energy, is required compared with sylvinite.
2) With solutions of hard salt, crystallization of double sulfates such as schoenite, leonite, and langbeinite often occurs. These salts can appear in the residue or the product, which leads either to potassium losses in the residue or to a lower K$_2$O content of the product. Also, double salts can crystallize in pipelines, vessels, and pumps, interfering with the process and in extreme cases bringing it to a standstill. The particular double salt that crystallizes depends on the MgCl$_2$ content of the solutions, the temperature, and other

3) Another disadvantage of the hard salt leaching process is that, with higher magnesium salt contents, the temperature dependence of the solubility of NaCl becomes unfavorable (Fig. 11), that is, when the potassium chloride is crystallized by cooling, considerable amounts of sodium chloride can also crystallize (Fig. 15). In practice, equilibrium is not completely reached when the KCl–NaCl is dissolved, and cooling of the solution often occurs by water removal, so that the K$_2$O content of the crystals formed is usually only a little more than 40 %. Since the usual potassium chloride for fertilizers (excluding special products) has a minimum potash content of 60 % K$_2$O, this primary product was formerly crystallized in a second leaching plant. Alternatively, a product can be made with 60 % K$_2$O directly, if the crystallization of sodium chloride during vacuum cooling is

prevented by addition of sufficient water before or during crystallization of the solution to ensure that only potassium chloride crystallizes [67]. The K_2O content in the product of a primary crystallization can be increased to the required level by treating the product with water or a plant solution unsaturated with respect to sodium chloride. Both the excess water added during vacuum cooling and the water used for treating the product must be removed from the recirculating system of the plant. Both cause a loss of yield whose extent depends on operating conditions. To avoid this, excess water must be evaporated from process liquors.

5.2.3. Processing of Carnallite

Carnallite, $KCl \cdot MgCl_2 \cdot 6H_2O$, is the most abundant potassium mineral in salt deposits and occurs widely in mixtures with halite or with halite and kieserite in the form of carnallitite ore (see Section 2.2). In the early days of the potash industry in Germany, it was the preferred starting material for the production of potassium chloride. Today, sylvinitic ores and hard salts are used almost exclusively, because the extraction and processing of carnallite ore are considerably more difficult and expensive for the following reasons:

1) Carnallitite ore has unfavorable mechanical properties that make mining more difficult.
2) The K_2O content of pure carnallite is ca. 17%, compared with ca. 63% for pure sylvite.
3) Whereas the separation and purification of sylvite from sylvinite ore can usually be carried out by flotation, which does not involve a phase change, the extraction of potassium chloride from carnallitite ore necessitates dissolution or decomposition of the carnallite, and a high energy consumption for decomposition or purification of the decomposition product, depending on the process.
4) The treatment of carnallite generates large quantities of concentrated magnesium chloride solution, which must be disposed of.

For these reasons, mined carnallitite ore is today seldom used as a raw material. However, carnallite is often a major component in mixed salts of the hard salt type, and hence influences the choice of processing method.

Large quantities of a carnallite–halite mixture obtained by solar evaporation of water from the Dead Sea are used for the production of potassium chloride in Israel and Jordan.

Theoretical Basis. The theory of the production of potassium chloride from carnallite and carnallite-containing mixed salts is based on the $K_2Cl_2 - MgCl_2 - H_2O$ system shown in Figure 16, which is valid between -3 and $117\,°C$. In VAN'T HOFF coordinates (moles of salt per 1000 mol H_2O) the points representing the composition of water ($P^O_{H_2}$), bischofite ($P_{bischofite}$), and carnallite ($P_{carnallite}$) are indicated. All possible mixtures of water and carnallite are shown on the straight line between $P^O_{H_2}$ and $P_{carnallite}$, the molar ratio $K_2Cl_2 : MgCl_2$ here being always 1 : 2, as in carnallite. The curve from L_4 via R and E to point L_0 represents an arbitrary isotherm, and indicates compositions in which solutions are in equilibrium with the corresponding solid phase. Point L_0 gives the solubility of KCl in water, and the solutions L_1 are in equilibrium with KCl as the solid phase. Solution E is in equilibrium with the solid phases KCl and carnallite, and the solutions L_2 are in equilibrium with carnallite only. Solution R is in equilibrium with bischofite and carnallite. The solutions L_3 are in equilibrium with bischofite, and point L_4 indicates the solubility of bischofite in water.

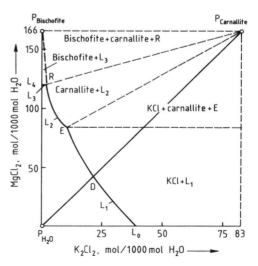

Figure 16. The system $K_2Cl_2 - MgCl_2 - H_2O$ (not to scale) represented by using van't Hoff coordinates

If carnallite is dissolved in water, the composition of the solution follows the straight line $P_{H_2O} - P_{carnallite}$, which intersects curve L_1 at point D. Here, the solution is saturated with KCl, and further addition of carnallite results in dissolution of $MgCl_2$ and crystallization of KCl until point E is reached. This incongruent solubility is the basis for the simple method of processing carnallite (i.e., cold decomposition by mother liquor).

Table 6. Composition of stable saturated solutions in Figure *

Point	Temperature, °C	Density, g/cm³	Concentration, mol/1000 mol H_2O			
			K_2Cl_2	$MgCl_2$	$MgSO_4$	Na_2Cl_2
E_{25}	25	1.275	5.8	70.8		4.4
Q_{25}	25	1.291	5.8	68.0	5.2	4.2
E_{105}	105	1.328	13.2	93.7		4.4
Q_{105}	105	1.325	13.7	92.9	1.0	4.3

* Solutions Q_2^5 and Q_1^{05} are saturated with $MgSO_4$ and NaCl, and correspond to solutions E_2^5 and E_1^{05}, which are saturated with NaCl.

Carnallitic potash ores or crystallized products from solar evaporation always contain so much halite that solutions produced during processing are saturated with sodium chloride. Figure 17 is a section from the quaternary system $K_2Cl_2 - NaCl - MgCl_2 - H_2O$ saturated with NaCl, showing the 25 °C and 105 °C isotherms. The kieserite content in the ore in the region of the solutions E has in practice a negligible effect on the composition of the solutions. Some analyses of equilibrium solutions of the quaternary system are given in Table 6.

Processing Methods. Many processes for the treatment of carnallite have been described in the literature and used [10], but only a few are important today. By far the most important is cold decomposition by mother liquor. The complete dissolution process is still used occasionally.

Cold decomposition is carried out at ambient temperature (e.g., 25 °C). Carnallitite ore is mixed and agitated with water or with a solution of low $MgCl_2$ content such that the composition of the mixture corresponds to point B in Figure 17. This causes the crystallization of an amount of potassium chloride corresponding to the line $B - E_{25}$, with formation of a solution of composition E_{25}. The very fine potassium chloride produced still contains fine, salted-out sodium chloride, undissolved halite, and sometimes kieserite and clay, depending on the composition of the carnallitite ore used. Approximately 85 % of the potassium chloride contained in carnallite can be obtained as a crystalline product by this process. The yield can be increased by evaporative concentration of the decomposition mother liquor E_{25} or Q_{25} (analyses are given in Table 7). During evaporation, sodium chloride and sometimes magnesium sulfate crystallize and must be removed. The synthetic carnallite that crystallizes when the concentrated solution is cooled is fed to the cold decomposition process.

Potassium chloride from the decomposition, which consists of very finely divided particles and is rather impure, must be purified. Purification by flotation is difficult owing to the fineness of the decomposition product (see, however, Section 5.3.6). For this reason, purification is nearly always carried out by the hot leaching process, which yields a very pure, completely water-soluble, coarse-grained product.

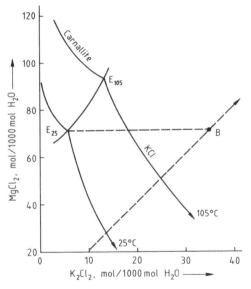

Figure 17. Quaternary system ($K_2Cl_2 - MgCl_2 - Na_2Cl_2 - H_2O$) saturated with NaCl
See text for explanation

The mother liquor E_{25} or Q_{25} from the decomposition process can take up a certain amount of sodium chloride (see Table 7). If the carnallitite used in the cold decomposition contains only small amounts of halite and if water is used for the decomposition, potassium chloride can be produced that requires no further purification, because sodium chloride has dissolved in the decomposition liquor. In Israel, a large proportion of the carnallite – halite mixture recovered from the Dead Sea by solar evaporation is so completely freed from halite by fractional crystallization and hydraulic classification that a grade of potassium chloride with > 60 % K_2O can be produced directly by cold decomposition. This so-called cold crystallization process is carried out so that the crystals produced have the same grain size as those from the crystallization plant of the hot leaching process and no further increase in grain size is required. A schematic diagram showing various processes used in the treatment of carnallite from the Dead Sea is given in Figure 18.

In the electrostatic treatment of carnallitic hard salts (see Section 5.4.2), concentrates can be produced that contain various amounts of halite together with sylvite and carnallite. These concentrates are reacted with suitable process brines or with water to bring about cold decomposition of the carnallite. The K_2O content of the product, and hence its potential use, depend on the halite content of the concentrate.

The complete dissolution process is used in an experimental plant in central Germany in which carnallite is extracted from a salt deposit by solution mining (see Section 3.3) [48].

The crude salt is dissolved by a solution in which the magnesium chloride content is adjusted so that at the chosen temperature of ca. 80 °C, no decomposition of carnallite takes place, and hence no crystallization of potassium chloride, and the carnallite is completely dissolved. The solution, which is almost saturated with carnallite, is pumped out, evaporated, and cooled, causing crystallization of the carnallite, which is then treated by the cold decomposition process to produce potassium chloride.

For mixed salts containing less than ca. 15 % carnallite, carnallite decomposition is generally dispersed with, and the ore is treated directly by the hot leaching process. The magnesium chloride that enters the process in the carnallite must be removed from the brine circuit in appropriate process brines. If the carnallite content of the ore is high, a carnallite decomposition stage is carried out before the hot leaching process.

5.2.4. Equipment

Leaching. The choice of leaching equipment depends on the properties of the material to be leached and the throughput required. The techniques used in plants in which the potash is completely leached (e.g., in many German and French plants) differ fundamentally from those in plants (e.g., in Canada) where only the fine materials from a screening operation, cyclone fines, and slimes are recrystallized or extracted.

Potash ores are leached at rates of up to 1000 t/h salt, producing up to 2000 m^3/h solution. Screw dissolvers, up to 14 m long and 3 m in diameter, are widely used (Fig. 19). High material transfer is achieved by fitting partitions in the upper part of the dissolver at intervals of 2 – 3 m. These are immersed in the suspension of crystals and force the solution to flow perpendicular to the main flow direction. At the outlet of the dissolver, the solids are scooped from the solid – liquid mixture by an elevator system with perforated buckets from which liquid drains during conveying; the residual water content is ca. 15 – 20 %. Alternatively, bucket wheels can be used.

Elevators or bucket wheels remove only the coarse residue (ca. 75 % of the total residue) from the screw dissolver. The amount of potassium chloride-containing solution adhering to the coarse residue is usually lowered to 2 – 4 % by centrifugation.

The fine residue (ca. 25 % of the total residue), which consists of very fine salt and insoluble components of the ore (mostly anhydrite and clay), flows with the raw solution from the screw dissolver into the hot clarifier, where flocculation agents are added and sedimentation occurs. Sealed, insulated circular clarifiers are used. The fine residue is removed from the hot clarifier as a suspension with a solids content of ca. 50 % and debrined on rotary filters to a residual water content of 10 – 16 %. To avoid losses of potassium chloride, the solution adhering to the residue must be removed. For small quantities of residue, this is done by washing on the filter with water or a process brine having a low potassium chloride content. For large amounts of residue, two- or three-stage water washing is used; the filter cake is slurried with the water, and the re-

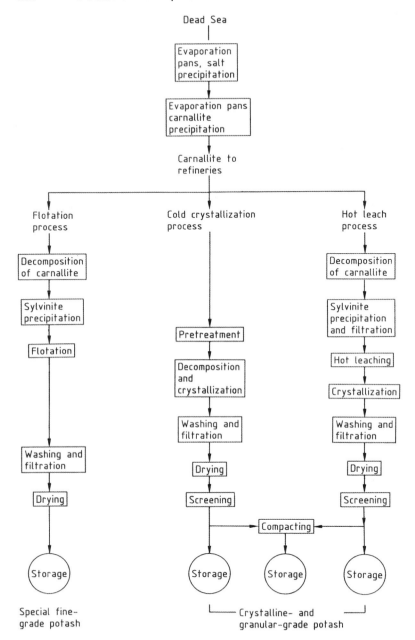

Figure 18. Flow diagram for potassium chloride production by the Dead Sea Works Ltd. (Israel) [4] Reproduced from [4] with permission

sulting suspension is filtered. This treatment is repeated once or twice.

In North America, where the flotation process predominates and only small quantities of ore are treated by the hot leaching process, a series of agitated vessels is used. The leaching process takes place in either cocurrent or countercurrent flow. Between the stages of countercurrent leaching, solid–liquid separation is carried out with hydrocyclones or hydroclassifiers [68]. The layout of a leaching plant including crystallization is shown in Figure 20.

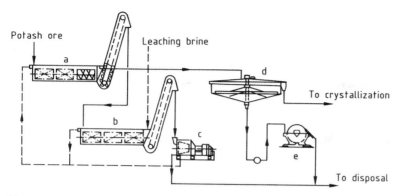

Figure 19. Leaching system with residue debrining
a), b) Screw leachers with draining elevators; c) Vibratory screen centrifuge; d) Hot clarifier; e) Rotary filter

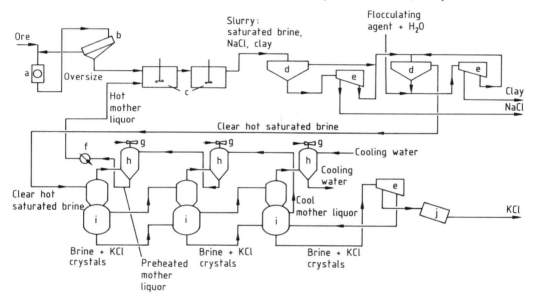

Figure 20. Leaching–crystallization process for production of potassium chloride from potash ore [8]
Reproduced from [8] with permission
a) Crusher; b) Screen; c) Leach tanks; d) Thickener; e) Centrifuge; f) Heater; g) Steam ejector; h) Barometric condenser; i) Vacuum cooler–crystallizer; j) Dryer

Crystallization [69, 70]. The hot solution from the clarifier of the leaching plant is almost saturated with potassium chloride and is cooled by expansion evaporation in vacuum equipment to cause crystallization. The vapors are condensed in surface or barometric condensers, with process brines used for higher temperatures and cooling water for lower temperatures. The amount of heat that can be recovered depends on the number of stages in the cooling system. In the past, vacuum cooling plants were constructed with up to 24 stages, to give maximum possible heat recovery. However, the number of stages in a modern plant is usually between four and eight.

The high cooling rate of a vacuum plant results in a high degree of supersaturation that produces very fine crystals, unless the design of the plant and the crystallizing conditions are optimized to give a coarse product. Several suitable crystallizers are now available.

If it is not important to have coarse crystals (e.g., if they are to be reprocessed), stirred crystallizers or conical-based evaporators are used.

424   Potassium Compounds

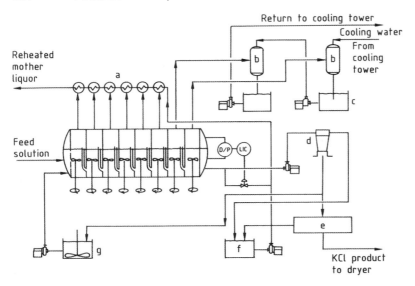

Figure 21. Eight-stage horizontal crystallizer
a) Surface condensers; b) Barometric condenser; c) Hot well; d) Cyclone; e) Centrifuge; f) Filtrate tank; g) Slurry holding tank.
Reprinted with permission from R. M. McKercher et al., *Proc. 1st Int. Potash Technol. Conf.,* Oct. 3–5, 1983, Pergamon Press.

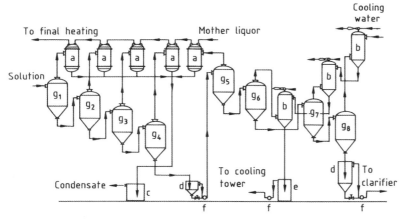

Figure 22. Conical-based evaporator–crystallizer plant with eight evaporation stages
a) Surface condensers; b) Barometric condensers; c) Barometric collection vessel for condensate; d) Barometric collection vessel for salts and brines; e) Barometric collection vessel for cooling water; f) Pumps; $g_1 - g_8$) Evaporators

The former consist of horizontal cylindrical vessels divided into chambers by vertical walls. The solution or suspension flows from chamber to chamber through openings in the walls as it is cooled. These crystallizers are provided with mechanical stirrers or air agitation (Fig. 21).

The conical-based evaporator consists of a vertical cylinder with a lower conical section. The solution is sprayed into the top, and much of the water evaporates. A certain liquid level is maintained in the evaporator, into which liquid from the nozzle falls as a supersaturated solution. Several evaporators are arranged in series one above the other so that the suspension flows through each stage in the direction of decreasing pressure without the need for intermediate pumps (Fig. 22). The advantage of this arrangement lies in its very simple and economical construction, with no moving parts. However, the product is very finely divided.

The aim of the crystallization process is usually a dust-free product with optimum grain size. Two main requirements must be met to achieve this: the solution must have only a slight degree of supersaturation, and this should be removed by crystallization on seed crystals that are already present. The slight degree of supersaturation is achieved by providing internal circulation within each crystallization stage, in which the cooled solution is mixed with the added hot solution in an exactly controlled ratio. The continuous presence of seed crystals is ensured by keeping the solids content in each crystallizer at ca. 30 wt %.

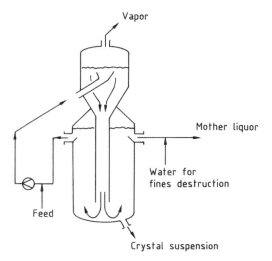

Figure 23. Oslo crystallizer

These conditions are fulfilled by two types of crystallizer: the fluidized-bed (Oslo) type and the draft-tube agitated type.

In the Oslo crystallizer (Fig. 23), the solution is supersaturated in a separate evaporator and flows into a crystal slurry vessel below, where the supersaturation is removed by crystallization on KCl crystals already present. To reduce the degree of supersaturation in the evaporator, a large amount of clarified solution is pumped from the crystal slurry vessel into the evaporator, together with the incoming hot feed solution. Clarified solution for the next stage is removed from the vessel at another takeoff point. A crystal suspension is also removed from the vessel at a rate corresponding to the rate of crystallization, and is fed either to the next stage as seed crystals or to the debrining stage. Crystallizers of this type are constructed by companies such as Lurgi, Swenson, and Struther-Wells.

In draft-tube agitated crystallizers, a state of supersaturation is created and removed in the same crystallizer. This type is manufactured by Swenson (DTB=draft tube baffle), Standard-Messo, and Kali und Salz. In the Swenson and Standard-Messo crystallizers, the entire suspension is brought into continuous contact with the evaporator surface by a propeller stirrer that provides internal circulation; immediately after supersaturation is achieved, a large quantity of salt is made available to remove the state of supersaturation. The coarser crystals collect preferentially in the lower part of the crystallizer, where they are removed as a suspension and fed to the next stage or to the debrining stage. The hot solution is passed through a clarifying zone and is fed to the next stage.

All the processes described so far are operated cocurrently (i.e., the crystallizing salt is transported with the cooling solution from one stage to the next). In contrast, the Kali und Salz process (Fig. 24) operates countercurrently [71]. The salt produced in each crystallizer is passed through a classifier at the bottom of the crystallizer as soon as a predetermined minimum crystal size has been reached; then, unlike the other systems, it is passed to the next hottest stage. The classifier is fitted with equipment that enables the upward flow rate, and hence crystal concentration and product crystal size, to be controlled independently of the throughput rate of the solution. Mixing of the cooler solution with the hotter solution in a crystallizer is prevented by transferring the salt with the help of the solution from the stage into which it is pumped.

5.3. Flotation

Since the early 1900s, ores of many different kinds have been processed by froth flotation.

Investigations into the flotation of potash ores began in the early 1930s in the United States and the Soviet Union [72,73]. The first full-scale plant began production in 1935 in Carlsbad (New Mexico, United States) [74].

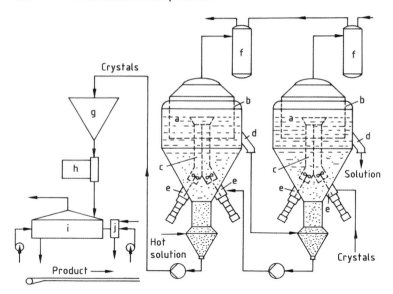

Figure 24. Kali und Salz countercurrent crystallization process
a) Crystallizers; b) Bells; c) Draft tubes; d) Liquor overflow; e) Stirrers; f) Condenser; g) Thickener; h) Centrifuge; i) Dryer; j) Combustion chamber

The first reports of investigations into the flotation of potassium salts in Germany appeared in 1939 [75]; the first full-scale plant began operating in 1953 [76, 77].

At present, most of the worldwide production of potash fertilizers is by flotation. This is the most widely used process in Canada, the United States, the CIS, and France. Small hot leaching plants are often attached to the flotation plants for treating slimes and intermediate products. The flotation process is not widely used in Germany, where a combination of electrostatic treatment with the hot leaching process predominates.

5.3.1. Potash Ores Suitable for Flotation

Various types of crude potash salts can be treated by flotation:

Sylvinite ores are mixtures of sylvite (KCl) and halite (NaCl) in varying ratios. They represent the majority of potash ores treated. The sylvinites of Canada, the United States, and the CIS also contain up to 8 % clay components [78].

Hard salts contain kieserite ($MgSO_4 \cdot H_2O$), as well as sylvite and halite, and sometimes also anhydrite ($CaSO_4$).

Mixed salts consist of a mixture of sylvinite ore or hard salt with carnallite ($KCl \cdot MgCl_2 \cdot 6 H_2O$).

Polymineral salts contain not only sylvite, halite, and kieserite, but also langbeinite ($K_2SO_4 \cdot 2 MgSO_4$), kainite ($4 KCl \cdot 4 MgSO_4 \cdot 11 H_2O$), polyhalite ($K_2SO_4 \cdot MgSO_4 \cdot 2 CaSO_4 \cdot 2 H_2O$), and clay.

5.3.2. Carrier Solutions

In the flotation of water-soluble salts the carrier liquids are salt solutions that are saturated with the components of the raw material.

Thus, sylvinite ore flotation is carried out in a KCl–NaCl solution. For the flotation of hard salt, the brine also contains various amounts of magnesium sulfate and magnesium chloride.

5.3.3. Flotation Agents

The *collectors* are the true flotation agents, which selectively coat the surface of the component to be floated. For the flotation of sylvite, straight-chain primary aliphatic amines are used in the form of their hydrochlorides or acetates [79]. Mixtures of amines of various chain lengths, which largely eliminate the effects of pulp temperature variation, are extremely useful. A typical flotation amine has, for example, the following composition: 5 % $C_{14}-NH_3Cl$, 30 % $C_{16}-NH_3Cl$, 65 % $C_{18}-NH_3Cl$, iodine number: 4.

Foamers contribute to the dispersion of long-chain amines and to the stabilization and homogeneous distribution of amine micelles [80]. The following substances are preferred: aliphatic alcohols with chain lengths $>C_4$, terpene alcohols, alkylpolyglycol ethers, and methyl isobutyl carbinol, which is used mainly in Canada and the United States.

Extenders for sylvite flotation are nonpolar materials, especially oils of various types. They are probably incorporated in the micelles and increase their hydrophobic properties. Extenders are especially effective in the flotation of coarse particles [81].

Clay depressants are used in salt flotation to block clay components, which would otherwise bind large amounts of flotation agent. Clay contents of 1.5 – 2 % can be controlled by treatment with these depressants. If larger amounts are present, additional steps must be taken (clay flotation or classification). Clay depressants include guar and starch products, carboxymethyl cellulose, and polyacrylamide [78].

5.3.4. Theory

The combinations of reagents used for sylvite flotation have been found by empirical investigation.

Theoretical studies of the separation of potassium and sodium chloride by flotation have led to various interpretations.

The theories developed up to 1961 are thoroughly discussed in [82], which also describes experimental results that have contributed greatly to the understanding of salt flotation.

The most important theories are reviewed below:

According to the *exchange theory* [83–85], hydrophobic properties can be imparted to a mineral if the polar group of a collector can be incorporated into the crystal lattice in place of an ion. In the flotation of sylvinite ore, for example, exchange between the K^+ ion (radius: 0.135 nm) and the NH_3^+ group of the amine (radius: 0.143 nm) is assumed to occur, whereas the Na^+ ion (radius: 0.095 nm) is too small (Fig. 25). This does not account for the fact that kieserite ($MgSO_4 \cdot H_2O$), for example, is readily floated by amines in water although the Mg^{2+} ion has a radius of only 0.065 nm.

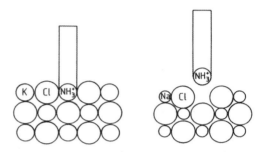

Figure 25. Incorporation of a polar group of the collector into the crystal lattice [83]
Reprinted with permission of the Society for Mining, Metallurgy, and Exploration, Inc.

The *structure theory* [86] postulates that the interanion distance in the crystal lattice of the salt matches the intermolecular distance in the lattice of the amine hydrochloride within ± 20 %. In sylvite, this condition for oriented growth of the collector on the crystal is fulfilled (Fig. 26), whereas the interionic distance of 0.398 nm in the halite lattice is too small. However, langbeinite has good flotation properties, although these steric requirements are not met [86].

According to the *hydration theory* [87], flotation of a mineral is impossible if it has a positive heat of solution, whereas salts with negative heats of solution can be floated. If heat is evolved when a salt dissolves, more energy is produced by hydration of the ions than is required for breakdown of the lattice. This suggests that such salts are strongly bonded to water molecules, preventing adhesion of the collector. A comment on this theory is given in [88].

Another hydration theory is based on new information on the hydration of ions in dilute solutions [89]. This differentiates between positive local hydration (in which the ion reduces the mobility of the neighboring water molecules) and negative local hydration (water molecules close to the ions are more mobile than in pure water).

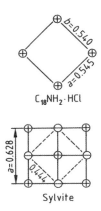

Figure 26. Lattice analogy between the amine and the salt [86] (distances in nm)

According to [90], in the case of positive local hydration (e.g., with Na^+ ions), the Cl^- ions are shielded by water molecules, so that the bond between a cationic amine and sodium chloride can only be weak. In the case of negative local hydration (e.g., with K^+ ions), formation of a bond with the collector is unhindered. Application of these theories, developed for dilute electrolyte solutions, to the hydration state of salt surfaces is difficult, but should be qualitatively valid [90].

The G zone theory (G = Grenzflächen = interfaces) [91] applies both to the flotation of sylvinite with amines and to the flotation of kieserite with an anionic surfactant (the sodium salt of a highly sulfated fatty acid). In both cases, adsorption of the collector proceeds by the same mechanism.

Amines and the anionic surfactant, in a saturated solution of salts that can be floated with these reagents, are present either as micelles or in true solution. In the usual carrier brines, flocculation (amines) or formation of very small droplets occurs (anionic surfactant).

Based on experimental results, zones are assumed to be formed on the surfaces of crystals suspended in a mixed-salt solution (carrier liquid) in which ions from the lattice are present in the form of a saturated solution.

When collectors are introduced into the suspension, they migrate to the zones where their solubility is greatest due to their thermodynamic tendency to dissolve (i.e., to the boundary zones of those particles that they cause to float). The bonding of the collectors in these zones is reinforced if their solubility in the surrounding carrier brine is low. The success of flotation depends on the strength of this bond (Fig. 27).

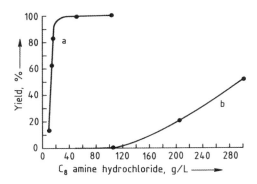

Figure 27. Flotation of potassium chloride with octylamine in saturated solutions of KCl and KCl + NaCl (Feed material: 20 % KCl, 80 % NaCl; 0.1–0.315 mm)
a) Carrier liquid saturated in KCl and NaCl (low solubility of octylamine); b) Carrier liquid saturated only in KCl (high solubility of octylamine)

Froth cannot be produced in the usual carrier liquids either by collectors or by foaming agents. The collision of an air bubble with a crystal coated with the collector material causes the latter to spread over the surface of the bubble, stabilizing it and causing the crystal to float.

Large-scale operations of sylvite and kieserite flotation has thus been placed on a good unified theoretical foundation.

5.3.5. Flotation Equipment

The flotation equipment used in the potash industry resembles that used for flotation of other ores.

Mechanical flotation cells operate with a rotor–stator system that causes both thorough mixing of the pulp and thorough distribution of the air, which can be drawn in by suction or injected from a compressed air network.

In recent decades, high-capacity flotation has been introduced almost everywhere. In this process, several stirrers operate in a single trough [92].

Agitair high-capacity flotation equipment is used in Germany, and the preconcentrate is purified in individual cells of the Denver type. Wemco high-capacity machines are used by the French potash industry, and Mechanobr high-capacity cells by the former Eastern-bloc countries.

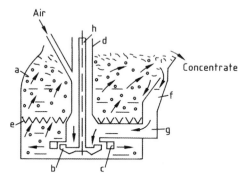

Figure 28. Mechanical flotation cell with fluidized bed
a) Flotation cell; b) Stirrer; c) Stator; d) Tube of the stirrer system; e) Grid; f) Circulation box; g) Circulation pipe; h) Shaft

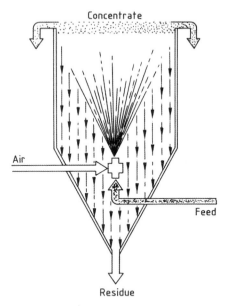

Figure 29. Principle of pneumatic free-jet flotation Reproduced from [95] with permission

Modified Mechanobr M 7 fluidized-bed units (Fig. 28), manufactured by Mescerjakov, are standard equipment in the CIS [93,94]. Typical features include the presence of a grid and circulation of the suspension via a box on the front wall of the cell, connected to the agitator by a circulation pipe.

Pneumatic flotation equipment operates without stirrers. In the free-jet flotation process developed in Germany [95], the conditioned pulp is pumped through an aerator fitted with a porous gas distributor or annular nozzle, either outside or inside the separation vessel, and flows as a free jet into the separation vessel. Air bubbles with the adhering minerals rise to the top and form a froth that flows over the edge of the vessel into a channel (Fig. 29). The pneumatic flotation process has been tested successfully for a special case of sylvite flotation. It is used industrially for the recovery of kieserite from leaching residues.

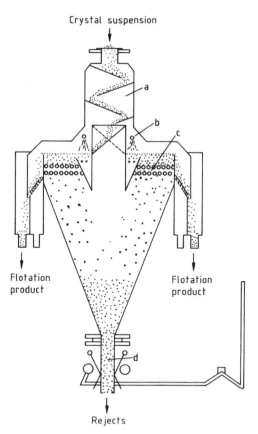

Figure 30. Pneumatic coarse-grain flotation cell (Gogorchimproect type) [96]
a) Feeding device; b) Spray nozzle; c) Aeration system (perforated rubber tubes); d) Device for removing reject material

In the CIS, a pneumatic flotation cell has been developed for the treatment of coarse sylvinite ores with a maximum grain size of 3–5 mm [96,97] (Fig. 30). The conditioned pulp is fed from above onto a layer of froth immediately above the perforated-tube aeration system. The hydrophobic sylvine crystals are retained in the froth, so that mechanical damage of the bubble – crystal combination is kept to a minimum. The flotation rate is very high.

5.3.6. Processes

Sylvinite Ore Flotation. Pure sylvinite of German origin can be floated without serious problems. The components of the ore are relatively strongly intergrown [50] and must be broken down by grinding until each grain consists as far as possible of only one component.

The ore is first ground to a particle size of < 4 mm, screened to remove fines, and then slurried with carrier brine and ground to a grain size of < 0.8–1.0 mm in a circulating system that includes rod mills, spiral classifiers, or wet screens (Fig. 31).

The solids content of the flotation pulp is adjusted to 30–40 wt %, and it is then mixed with flotation agents (ca. 40 g of oil, 20 g of foaming agent, and 40–80 g of collector per tonne of crude salt). As much of the desired product as possible is then extracted by the three-stage high-capacity rougher flotation. The rougher tails are classified in hydrocyclones, debrined (the coarse fraction in centrifuges, and the fine fraction in rotary filters), and dumped. Since the filtrate still contains finely divided salt, it is clarified in a thickener and recycled for slurrying the ore.

The material floated in the rougher flotation is then concentrated in a cleaner flotation process, usually in three stages. This does not produce a marketable product (with > 60 % K_2O). The concentrate is therefore separated from the brine and washed with water in salt washing equipment.

This washing process dissolves more halite than sylvite, producing a potash salt that meets quality requirements. The washing liquor is removed by centrifuges, and the product is dried and conveyed to silos.

The cleaner flotation process also produces some intermediate-quality product with a fairly high K_2O content, whose coarse particles contain most of the residual intergrown material from the ore. They are therefore recovered by hydrocyclones and fed to the screening – grinding stage for size reduction. The energy consumption for the entire process is ca. 10 kW · h/t ore.

The treatment of sylvinite ores from Canada, the United States, and the CIS is more difficult owing to the clay content of up to 8 %. The clay components disintegrate in the brine, forming very fine slimes that absorb large quantities of flotation agent. The flotation process is there-

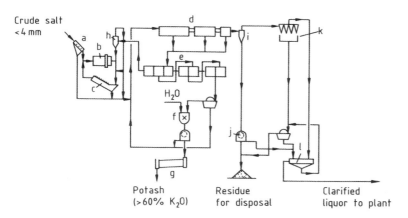

Figure 31. Flow diagram of the flotation of potash ores
a) Fine screen; b) Wet grinding; c) Classification; d) Rougher flotation (high volume); e) Cleaner flotation (three stage); f) Water washing for concentrate and debrining; g) Drying; h) Classification of intermediate-quality product; i) Residue (tails) collection; j) Residue debrining; k) Cyclone; l) Clarification of brine for recirculation

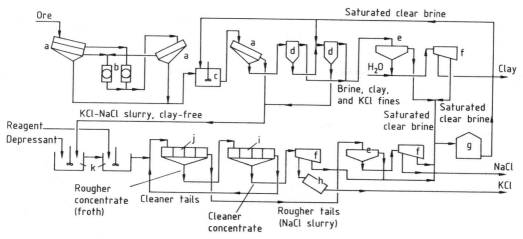

Figure 32. Schematic of flotation of potash ores
a) Screen; b) Crushers; c) Scrubber; d) Classifier; e) Thickener; f) Centrifuge; g) Brine tank; h) Dryer; i) Cleaner flotation cells; j) Rougher flotation cells; k) Conditioners
Reproduced from [98] with permission

fore sometimes preceded by a desliming stage (Fig. 32). The clay is first detached from the surface of the salt by vigorous agitation in scrubbers, so that it can be floated in a separate operation. The clay is usually removed by multistage classification [98]. The fine material is then thickened and washed with water to reduce losses of K_2O. Up to 1.5 – 2 % of residual clay in the crude salt can be handled by adding clay depressants (see Section 5.3.3).

An advantage of clay-bearing sylvinite ores is that they are usually sufficiently liberated at grain sizes of < 3 – 5 mm. Sometimes, coarse crystals are floated separately from fine crystals.

Hard-salt flotation is similar to the sylvinite flotation process (Fig. 31). However, the flotation product also contains kieserite, which cannot be removed by washing because it dissolves only slowly in water.

Sylvite, which is intergrown with kieserite, must therefore be liberated as fully as possible, which leads to increased grinding costs. Because of the increased production of fines, more flota-tion agent is needed. Also, in warm summer months, potassium-containing double salts (leonite, schoenite) can separate from the circulating brine. Since these salts end up in the residue, the yield is reduced.

Mixed-Salt Flotation. Carnallite ($KCl \cdot MgCl_2 \cdot 6H_2O$) in the mixed salt is usually decomposed by a brine with low $MgCl_2$ content. The potassium chloride produced has a grain size of < 0.04 mm and is accompanied by sylvinite ore or hard salt components.

In the 1980s, this complex salt mixture was successfully treated by flotation alone for the first time [99, 100].

Potassium chloride produced by the decomposition is first floated with a fairly small amount of amine in pneumatic or stirred flotation cells; then the sylvite component of the sylvinite ore or hard salt is floated with a further measured amount of amine, and a discardable residue is obtained (Fig. 33). The flotation froth is then treated by flotation in cleaner cells to give a concentrate with > 55 % K_2O. A final washing process gives a potash fertilizer salt containing > 60 % K_2O.

Polymineral salts are rarely treated by flotation. Potash ore from Stebnik (CIS) contains not only sylvite, halite, and clay materials, but also large amounts of kainite, langbeinite, and polyhalite [101]. Clay depressants are added first, followed by flotation of halite with a mixture of $C_7 - C_9$ fatty acids. Sodium hydroxide solution is used to adjust pH [101].

Schoenite can be floated from a salt mixture by using coconut oil acids as the collector [102].

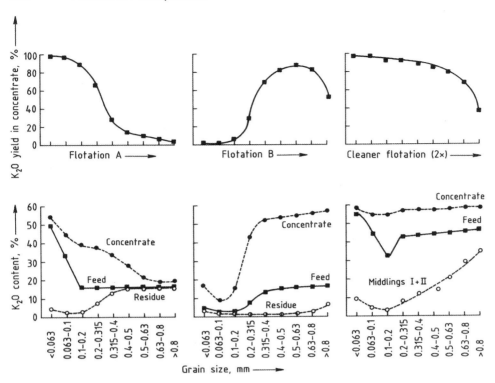

Figure 33. Flotation of fine salt (< 0.1 mm) from carnallite decomposition, showing K_2O content of each size fraction [100] Yields are based on amounts of K_2O in each fraction

The flotation of anhydrite to produce a salt mixture of kieserite, langbeinite, and polyhalite is recommended in [103].

A combination of cold leaching and flotation for the treatment of salt mixtures containing langbeinite and polyhalite is described in [104].

5.4. Electrostatic Separation

Electrostatic separation depends on the directional movement of electrically charged bodies or particles in an electric field. The processes used differ according to the various methods of producing or separating the charges and according to the separation equipment used. The separation of mixtures of salt minerals (e.g., in the treatment of crude potash ores) invariably involves the selective exchange of electric charges that takes place on contact between the various mineral particles, followed by separation in free-fall separators. Although roll separators are widely used for treating other minerals, they have not been used for salts, mainly because of their low throughput.

The first investigations into the industrial-scale separation of potassium chloride and sodium chloride were carried out after World War II by the International Minerals & Chemical Corp. in Carlsbad, New Mexico [105]. The ore, which contains alkali-metal chlorides and various amounts of clay and sulfate minerals, was ground, heated to ca. 500 °C, cooled to ca. 110 °C, and separated in a free-fall separator with a field strength of 2 – 6 kV/cm. Results were not encouraging and the method was abandoned.

Research in the Kaliforschungs-Institut in Hannover in 1956 led to an industrial breakthrough. The addition of organic and inorganic reagents much improved the electrical charge exchange between the mineral components, and the separating temperature could be reduced to < 100 °C [106, 107]. The potash works of Neuhof-Ellers, a subsidiary of Kali und Salz, was the first industrial plant to use this process to produce kieserite in 1974. In the following years, plants for the electrostatic production of kieserite and potash concentrates, and for the dry removal of residues, were installed in three other factories, with capacities up to ca. 1000 t/h [108]. Investigations aimed at the introduction of this process for processing sylvinite ores from the potash deposits in Saskatchewan, Canada, have been carried out by PCS Mining, Saskatoon [109].

5.4.1. Theoretical Basis

The basis of the process is the mutual selective exchange of electrical charge between the salt components, which occurs on contact. The direction, selectivity, and intensity of the charge exchange can be influenced by a large number of reagents (conditioning agents) [110]. In addition to this chemical conditioning, treatment with air of specific relative humidity is necessary. This is usually controlled by means of the air temperature and should be between 5 and 25 % [110, 111]. By the appropriate choice of conditioning agent and relative humidity, the charging properties of the individual mineral components in a potash ore can be controlled so that the desired components are recovered. In this way, an ore of complex composition can be completely separated into its components [112].

The mechanism of charging by contact depends on the transfer of electrons between the touching mineral surfaces, which must have suitable surface properties (i.e., surface energies appropriate for the exchange of charge). These energies are influenced (or created) by chemical conditioning agents and partial water vapor pressure [113–115].

5.4.2. Equipment and Processes

Before the separation process, the potash ore must be size-reduced to give complete liberation of its components. Fine particles (< 0.1 mm) behave nonselectively in an electrostatic field, so the grinding process must be carried out as carefully as possible (e.g., by impact grinding). The conditioning agents (20 – 100 g/t ore) are added to the ore in a mixer or introduced in the vapor state into the fluidized-bed dryer that heats the salt to the separation temperature (25 – 80 °C),

whereby the relative humidity is adjusted to a suitable value for selective charging of the salt particles [116–118]. Alternatively, a rotary dryer can be used. Depending on the conditioning agent used and the relative humidity and temperature, sylvite generally becomes positively charged, whereas halite and kieserite can be positive or negative. Accompanying minerals such as langbeinite, carnallite, or kainite can be separated individually or together with other mineral components [119–122].

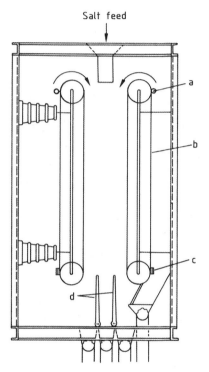

Figure 34. Vertical belt separator
a) Rotary brush; b) Rubber belt; c) Stationary brush; d) Flaps

Separation of the mixture of charged minerals is carried out in free-fall separators with an electrical field strength of 4 – 5 kV/cm. For a distance between the electrodes of 25 cm, the applied voltage is 100 – 125 kV. Formerly, vertical belt separators (Fig. 34) were common, but they are now used only for special applications. The electrodes consist of rotating rubber belts with conducting coatings. Brushes on the side opposite the electric field remove salt fines, which otherwise settle on the electrode surface, forming a coating that weakens the electric field and hinders separation. The separator used exclusively today is the tube free-fall separator (Fig. 35), which was developed by the potash industry [123]. It consists of two opposed rows of steel tubes, ca. 2 m long. The tubes rotate on their axes, and salt fines are removed by brushes on the side remote from the falling salt. The maximum working length of separators of this type is ca. 10 m. The charged salt mixture leaving the fluidized bed is fed to the top of the separator and falls through the electric field between the electrodes. This causes the particles to move sideways, the direction depending on the sign of the charge. At the bottom of the separator, adjustable flaps enable a cathodic fraction, an anodic fraction, and a middle fraction to be recovered. These fractions are then treated further or removed from the process as product and waste (or middlings).

The particle sizes for complete liberation should not exceed 1.5–2 mm for sylvite and halite, or 1.2 mm for kieserite, because the movement of particles in the electric field of the separator is determined by both the horizontal electrical force and the vertical effect of gravity. Since the electrical force depends on the charge on the particle, which in turn is a function of the surface area, the surface–volume ratio is a very important factor in determining the extent of sideways movement in the electric field. For fine materials, the effect of the electrical force is the greater; the converse is true for coarse particles.

Modern tube free-fall separators have a throughput of $20-30\,t\,h^{-1}\,m^{-1}$. Their energy consumption is very low because for this throughput the current is only ca. 2 mA.

Generally, even for a single-stage separation step two free-fall separators are arranged one above the other so that the concentrate and waste material from the upper separator bypass the lower one, while the middlings flow directly to the lower separator where further separation occurs. This is shown schematically in Figure 36. In most cases, multistage separation or treatment is necessary, in which the concentrate produced in the first stage is purified or concentrated in another single stage or in multistage separation (Fig. 37).

Electrostatic separation has thus far been used industrially on a large scale only for the treatment of hard salts with the principal components sylvite, halite, kieserite, and various percentages of carnallite. Three basic methods of separation into the individual components are

1) Separation of sylvite in the first stage [105, 111, 124], and separation of kieserite and halite in the second stage
2) Separation of kieserite in the first stage [125, 126], and separation of sylvite and halite in the second stage
3) Separation of halite in the first stage [127–129], and separation of kieserite and sylvite in the second stage

In all cases, carnallite appears in the sylvite fraction.

The first stage of method 2 is used to produce kieserite on a large scale, and method 3 is used in several variations for the production of a dry halite residue and of kieserite and

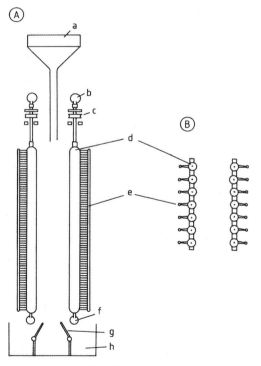

Figure 35. Tubular free-fall separator
A) Cross section; B) Plan view a) Salt feed; b) Upper bearing; c) Motor; d) Tubes; e) Brushes; f) Lower bearing; g) Flaps; h) Receiver for products

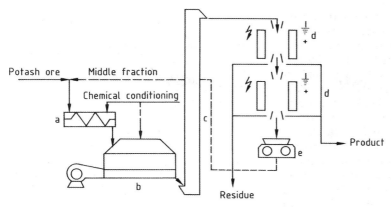

Figure 36. Schematic of single-stage electrostatic separation
a) Mixer; b) Fluidized-bed preheater; c) Elevator; d) Free-fall separator; e) Middle-fraction grinding

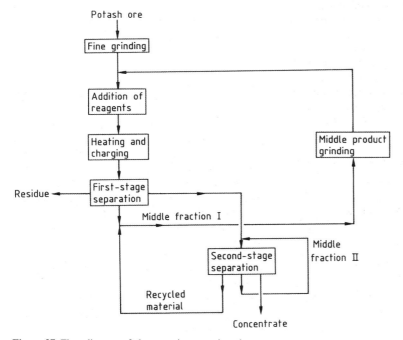

Figure 37. Flow diagram of electrostatic separation plant

potash concentrates [108]. So far, these plants have been operated almost entirely in association with leaching and flotation plants because electrostatic treatment alone gives unsatisfactory yields.

5.5. Heavy-Media Separation

Heavy-media separation is not used widely in the potash industry because of the generally unfavorable extent of intergrowth and the small density difference between the main components sylvite ($\varrho = 1.99$) and halite ($\varrho = 2.17$). However, the process can be used for very coarsely intergrown high-quality sylvinite ores as mined in Saskatchewan. International Minerals & Chemical Corporation (IMCC) operates a plant of this type in Esterhazy, Saskatchewan [130]. The heavy medium is a suspension of magnetite in a saturated salt solution whose density is adjusted to 2.10. Small salt crystals (< 1.7 mm) are very difficult to separate from magnetite; therefore, only materials with a particle size of 1.7 – 8 mm are treated by this process. Separation is carried out in hydrocyclones in two stages. A process scheme is given in Figure 38.

IMCC also operates a heavy-media separation plant in Carlsbad, New Mexico, where langbeinite ($\varrho = 2.83$) is recovered from a potash ore in which sylvite and halite are the other main components.

5.6. Debrining and Drying

The products and waste from all the treatment processes except for the dry electrostatic process are obtained as suspensions with various solids contents and must be debrined. The two main aims of debrining potassium chloride product are (1) to achieve as low a moisture content as possible to minimize drying costs and (2) to remove as much of the adhering brine as possible to maximize product purity. As the brine adhering to the residues contains potassium chloride, its recovery minimizes yield losses.

Suspensions must often be thickened in circular thickeners or hydrocyclones before debrining.

The choice of equipment is determined mainly by the particle size of the material to be treated. The extent to which adhering brine can be removed by washing with water is also important.

Pan filters are used where debrining of a fine-grained product is combined with water washing to increase the K_2O content and when the product is to be stored intermediately. Residual moisture content in this case is 12 – 14 %.

Drum filters are generally used for debrining fine residues or when washing of the filter cake is necessary; they give a residual moisture content of 9 – 11 %. Alternatively, belt filters are used because they have a high capacity and allow the filter cake to be washed with recovery of the washing liquids.

The most commonly used debrining apparatus consists of centrifuges of various designs [131]. In potash works in Canada, the United States, Jordan, and the CIS, screen-bowl and solid-bowl centrifuges of 1400-mm diameter are in general use. Screen-bowl centrifuges have throughputs of 60 – 110 t/h and achieve a residual moisture content of 3 – 8 %. They are used to treat both products and residues. Solid-bowl centrifuges mostly are used for fine residues, and achieve throughputs of 70 – 120 t/h and a residual moisture content of 6 – 8 %. In European and some Canadian factories, large-diameter (> 900 mm) pusher centrifuges are used, with throughputs of 40 – 50 t/h and a residual moisture content of 3 – 6 %. For coarse products and residues, vibratory screen and screw screen centrifuges are generally used. These have a diameter of 900 – 1200 mm and a throughput of 35 – 70 t/h, and give a residual moisture content of 2 – 4 %.

The products are usually dried in drum dryers with diameters up to 3 m and lengths of 20 m [132]. They are heated with oil or gas in cocurrent flow at throughputs up to 120 t/h. The dryers are fitted with internals to promote heat exchange and prevent caking of the salt. Exhaust gases are dedusted, first by cyclones and then by electrostatic filters, wet scrubbers, or fabric filters. The main reasons for the widespread use of drum dryers are their ruggedness, safe operation, lack of sensitivity to throughput variations, and adaptability to differing grain sizes and moisture content. Figure 39 gives a schematic diagram of a drying plant of this type.

Since the early 1960s, fluidized-bed dryers have been used to an increasing extent. At first, they were used mainly for coarse products (> 0.5 mm) with low initial moisture content. Later, improvements in fluidized-bed technology enabled products with grain size down to 0.1 mm and initial moisture content up to 8 % to be dried. Their most important advantages compared with drum dryers are improved heat and mass transfer, more efficient use of energy, and much smaller floor space requirement. Cyclones and bag or electrostatic filters are used for dedusting the exhaust gases from a fluidized-bed dryer. A typical plant is shown in Figure 40.

5.7. Process Measurement and Control

The methods normally used in the chemical industry for measuring and controlling process parameters are also used in potash plants. Special methods include the analytical determination of potassium by means of its natural radioactivity

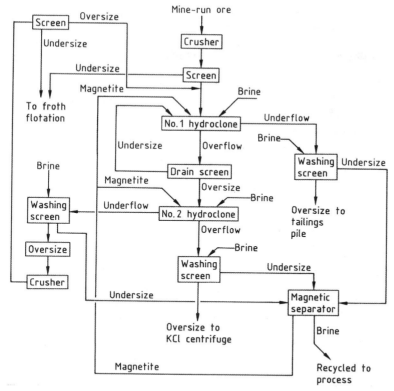

Figure 38. Heavy-media separation of potash ore [130]
Reprinted from [130] with permission of John Wiley & Sons, Inc.

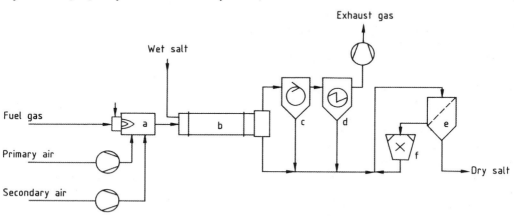

Figure 39. Drying plant for potassium chloride
a) Combustion chamber; b) Rotary dryer; c) Deduster cyclone; d) Electrostatic filter; e) Screen; f) Grinding equipment

[133], the use of flame photometry for so-called ratio analysis, and mineral analysis by infrared spectrometry.

The natural mixture of potassium isotopes includes ca. 0.012 % of the radioactive isotope ^{40}K, which is both a β and a γ emitter. Both types of radiation are measured to determine potassium content. The β emissions are usually measured in the laboratory, whereas on-line measurement of γ emissions is widely used for

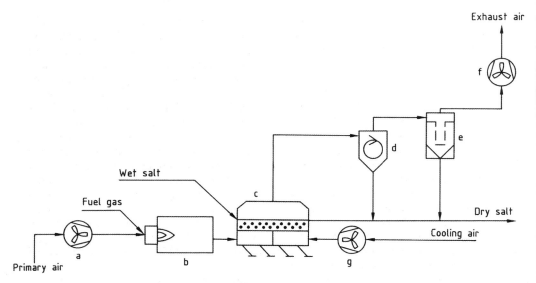

Figure 40. Fluidized-bed dryer with dedusting equipment
a) Fan; b) Combustion chamber; c) Fluidized-bed dryer; d) Deduster cyclone; e) Suction bag filter; f) Exhaust fan; g) Fan for cooling air

process control. Owing to the great penetrating power of γ rays, the reliability of measurement is strongly dependent on the geometry of the measuring equipment. Equipment for the radiometric determination of potassium in bulk products is shown in Figure 41 [134].

The on-line method of ratio analysis [135, 136] of KCl – NaCl mixtures is based on measuring the potassium and sodium contents of a sample with a double-beam flame photometer, and calculating the ratio from the measured data. Since this ratio is independent of the weight and dilution of the sample wet materials with varying water content can be analyzed without weighing the sample. Ratio analysis is used for determining residual NaCl content in flotation concentrates that have been water washed. The NaCl content is used to determine the amount of washing water required, and enables losses of K_2O due to excessive use of washing water to be avoided. The method has also been used for controlling the K_2O content in products from leaching – crystallization plants.

Infrared spectroscopy is suitable for continuous determination of carnallite or kieserite in crude potash salts. Reflection photometry can be used to measure water of crystallization, and hence the content of these minerals. In some circumstances, the carnallite content can be determined even when kieserite is present [136].

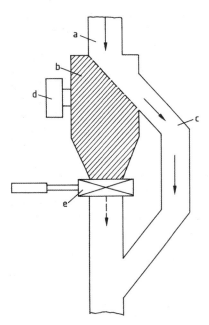

Figure 41. Equipment for radiometric determination of potassium in solid materials
a) Inlet for material to be analyzed; b) Container for material to be analyzed; c) Bypass; d) Detector; e) Time-controlled valve

5.8. Waste Disposal and Environmental Aspects [137, 138]

The main environmental problem of the potash industry is disposal of process waste. The total world production of potash ore is ca. 250×10^6 t/a, whose processing necessitates the disposal of ca. 200×10^6 t of waste without damage to the environment [137]. In the Canadian province of Saskatchewan alone, 300×10^6 t of solid waste has been generated during the last 30 years, covering an area of ca. 35 km^2 with solid and liquid materials [139].

The composition of the waste depends on the type of ore treated. Waste from the treatment of sylvinite consists mainly of halite. Waste materials from hard salt treatment are halite and kieserite, and from carnallitite ore processing, halite and magnesium chloride, which is always produced in the form of an aqueous solution. Salt solutions that must be disposed of are also generated during the production of potassium sulfate from potassium chloride and magnesium sulfate, and during the recovery of kieserite from residues from the treatment of hard salt by dissolving the halite.

Four methods for disposing of waste are dumping, backfilling, pumping into the ground, and discharge into natural water systems.

The disposal of waste by dumping is by far the most important method. Salt solutions that run off the dumped materials must be demonstrated not to harm the environment when they are absorbed into the ground. Salt solutions can originate from the brine adhering to wet residues or from the carrier liquid for transporting solid waste to the dump, or they can form when atmospheric precipitation dissolves the salt from waste material. If the ground underneath the dump is not impermeable, it must be sealed by layers of clay or plastic sheeting, or a combination of the two. The salt-containing runoff water is collected in ditches at the edges of the dump, and as much as possible is returned to the recirculating brine system in the plant. Excess brine is disposed of along with other liquid waste. In Germany, solid waste is formed into steep conical heaps after drying or debrining as fully as possible. This reduces the amount of salt-containing runoff water formed by atmospheric precipitation and also minimizes the ground area required. In most Canadian installations, filtration residues (tails) are slurried with brines and pumped as a suspension into a large lagoon surrounded by dykes. Flat deposits are formed over a very large area. The brine that runs off is pumped back to the plant and reused for slurrying solid waste (Fig. 42) [140, 141]. If the salt solution enters groundwater-bearing layers despite sealing, boreholes are sunk, and salt-bearing groundwater is pumped back into the lagoon. Attempts to cover dumped waste material to prevent the formation of salt-containing water by the leaching effect of rainfall have in the long run proved unsuccessful [139].

Under certain geological conditions, and if mining methods are suitable, solid residues can be transported underground for backfilling. Since the bulk density of the residue is much lower than that of the potash ore, often only a part of the residue can be accommodated by the space left after extraction of the ore. Backfilling is the main method of waste disposal in North German salt works where the salt beds are steeply inclined, as well as the potash works of New Brunswick in Canada. In most potash works, where the potash is mined from level deposits, backfilling is not possible for technical and economic reasons.

Pumping salt solutions back into the ground is possible if certain geological requirements are met. The formation used for this purpose must possess sufficient porosity and permeability, and must have no contact with formations that could provide a water supply or contain salt deposits. Salt solutions are generally pumped under pressure through lined boreholes into the porous formation.

Since 1926, in the Werra potash region of Germany, large quantities of brine have been pumped into a porous dolomite layer 20 – 25 m thick. Injection wells have an absorption capacity up to 1000 m^3/h at head pressures up to 11 bar [142]. Since most of this waste brine comes from kieserite production, the amount produced has decreased drastically with the introduction of electrostatic ore treatment [143]. In Saskatchewan, excess brine is pumped into deep formations of dolomite or sandstone. The capacity of these wells is up to ca. 200 m^3/h at head pressures up to 60 bar. Kalium Chemicals, Saskatchewan, has been disposing of waste brine since 1979 in caverns produced by solution mining.

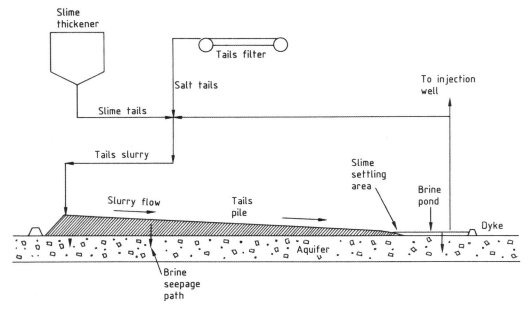

Figure 42. Conventional potash waste handling in Saskatchewan [139]
Reproduced with permission

The possibility of disposing waste brine in rivers and lakes depends very much on the location. Sea disposal, as practiced in the United Kingdom and by one of the potash works in New Brunswick, presents few major problems if the outfall is sufficiently remote from the coast. The potash works on the Dead Sea can dispose of waste brine in the Dead Sea itself without harmful consequences. The Potash Company of America in Saskatchewan discharges its solid and dissolved wastes into the southern end of the salt-containing Lake Patience, which is isolated from the rest of the lake by a dam.

For most potash works, waste brine cannot be disposed in natural salt water. If underground disposal is impossible, disposal into flowing natural water is the only alternative. In all countries, this is subject to ever-stricter regulation. One particular problem, for which a solution in the foreseeable future is being sought, is the high salt content in the Werra River due to brine from the potash works in the eastern part of the Werra region [143].

The only significant atmospheric pollution caused by potash works is salt dust emitted by the drying plant and from the handling of the ore and products during production and supply. Dust removal from waste gases from dryers is discussed in Section 5.6. The dusts produced during production and conveying are usually removed by air extraction, and trapped in fabric filters or wet scrubbers. During the drying of products containing magnesium chloride brines, hydrolysis of the magnesium chloride can lead to the emission of hydrogen chloride, which can be removed from the gas by wet scrubbing or absorption in calcium hydroxide in combination with a woven filter [144]. This procedure can also greatly reduce the level of sulfur dioxide if it is present in the exhaust gas.

5.9. Granulation [145–149]

Potassium chloride can be produced in a wide range of crystal sizes, depending on the composition of the potash ore, its degree of intergrowth, and the process used. Different particle-size distributions are needed for various applications. To meet these requirements, the potash industry offers products with standardized size distributions (see Section 5.10) obtained by screening to give the various fractions. The resultant distribution of the product among the various standardized grades does not always correspond to market requirements. Demand for products with a particle size of ca. 1 – 5 mm (coarse and granular grades) exceeds their normal production rate in a potash works. Most potash works must therefore increase the proportion of coarse product by granulating part of the primary product.

The main reason for the high demand for granulated potassium chloride is the technique of applying fertilizers that was developed mainly in North America and Western Europe in the postwar years. This technique, which is now the most widely used, requires coarse particles with a rather narrow size range. This is needed both for single nutrient fertilizers and for bulk-blended materials, which are widely used in North America. Also, granulated potassium chloride has a lower tendency to cake or form dust than the fine product.

Two methods of granulation are commonly used in the fertilizer industry: agglomeration of molten or wet material in rotating drums or dishes, or compaction of dry material in roll presses. The latter is the method most often used for potash fertilizers.

The starting material for compaction usually consists of a mixture of fine material from the production process with recycled material from the grinding and screening system of the compaction plant. Although dusts from the plant can also be compacted, a high proportion of fines (< 0.1 mm) must be avoided because they cause problems in the compaction equipment. Amines from the flotation process on the surfaces of the particles can also interfere with compaction and must be destroyed or rendered inactive by heat or chemical treatment.

Roll presses used in the potash industry usually have 60 – 125-cm-long rolls with diameters of 60 – 100 cm. The feed material, which is usually at 100 – 120 °C, is generally predensified by force feeders that feed it into the nip between rolls, where it is deaerated and compressed, with plastic deformation of each particle, to produce a dense sheet of material. The compression force is 40 – 50 kN/cm of roller length. The sheet of material is fed to a prebreaker that size-reduces it for ease of transportation. It then goes to a grinding – sieving plant, which produces either a granular product or granular and coarse products. Undersize material is recycled to the compaction press ($\rightarrow$ Fertilizers, Chap. 5.3.5.). A typical flow diagram for a compaction – granulation plant is given in Figure 43.

The compaction product consists of irregularly shaped angular particles. Handling causes abrasion of the edges and corners to form unwanted dust. The granular material is therefore usually treated with water after screening and then dried in a fluidized-bed dryer at 180 – 200 °C. This smooths the corners and edges, and gives a dense surface to the granules. Another method of dust reduction is to treat the granules with liquids that bind the dust formed by abrasion (see Section 5.10).

5.10. Quality Specifications

The main application of potassium chloride is in potash fertilizers, either as a single-nutrient fertilizer or as the potash component in mixed or complex fertilizers. In English-speaking areas and on the international market it is often called muriate of potash (MOP).

Single-nutrient fertilizers, which were formerly the most common type, have been replaced largely by complex or mixed fertilizers. Also, the finely divided fertilizers formerly used have increasingly been replaced by granulated products. Single potash fertilizers often contain the additional nutrient magnesium sulfate.

Potassium chloride must have differing grain structure and nutrient content depending on its intended use, e.g., for the production of granulated NPK or PK fertilizers, for application as suspension or liquid fertilizers, for bulk blending with other components ($\rightarrow$ Fertilizers, Chap. 3.), or as a single-nutrient fertilizer. The potash industry has developed internationally accepted quality standards.

The three main grades differ in grain size:

Standard	0.2 – 0.8 mm
Coarse	0.8 – 2.0 mm
Granular	1.2 – 3.5 mm

Other products, such as those designated soluble, special standard, or special fine, can contain a high proportion of grains <0.2 mm. Grain-size distributions of the principal grades are listed in Table 7. Products from various manufacturers can diverge considerably from these specifications.

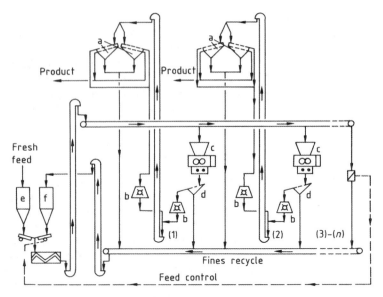

Figure 43. Flow sheet of a modern potash compaction–granulation plant [149]
a) Double screens; b) Hammer mills; c) Compactors; d) Screens; e) Fresh feed vessel; f) Recycled materials vessel
Reproduced from [149] with permission

Table 7. Typical grain-size distribution in potash fertilizers (cumulative percentage retained)

Mesh width, mm	Tyler mesh	Cumulative subsieve fraction		
		Granular	Coarse	Standard
+3.36	+ 6	2–12		
+2.38	+ 8	30–45	2–20	
+1.68	+ 10	75–90	25–50	
+1.19	+ 14	95–98	70–90	5–15
+0.84	+ 20	99	90–98	20–45
+0.60	+ 28		99	50–75
+0.42	+ 35			70–90
+0.30	+ 48			85–95
+0.21	+ 65			96–98
+0.15	+100			97–99

The granular and coarse products are used for bulk blending with other granulated fertilizers or as single-nutrient fertilizers. The standard and other fine materials are used in the manufacture of granulated multinutrient fertilizers, and for suspension and liquid fertilizers.

The potassium chloride content, which for fertilizers is usually given as % K_2O (100 % KCl = 63 % K_2O), is generally at least 60 % K_2O (95 % KCl) for this application. The main impurities include sodium chloride, or magnesium sulfate, and sometimes anhydrite or clay minerals, depending on the raw materials and production method. Fertilizer-quality potassium chloride produced by flotation is often colored red to red-brown by hematite inclusions or salt clay.

For single-nutrient fertilization, especially in European agriculture, potash fertilizers with a low K_2O content are often used. These generally contain a guaranteed level of water-soluble magnesium salts. In Germany, for example, two granulated potash fertilizers with a guaranteed MgO content are widely used: 40er Kornkali with MgO (40 % K_2O, 6 % MgO) and Magnesia-Kainit coarse (11 % K_2O, 5 % MgO, 24 % Na_2O).

About 5 % of the potassium chloride produced worldwide is used as an industrial chemical, mainly for the production of potassium hydroxide (see Chap. 11) and chlorine by chloralkali electrolysis. Material of this quality is produced by recrystallizing low-purity potassium chloride. A distinction is made between industrial-grade and chemical-grade material, depending on the impurity content and area of use. Typical analyses of these products are

Industrial grade:	KCl:	99.0–99.5 %
	NaCl:	0.8– 0.3 %
Chemical grade:	KCl:	99.8–99.9 %
	NaCl:	0.05–0.03 %

The remaining impurities consist mainly of bromide and alkaline-earth sulfates, depending on the raw material and production process.

Dust content and free-flowing properties are important quality criteria, and depend on chemical composition and grain size. With granulated materials, dust formation by abrasion occurs due to handling during manufacture and transport even if the materials are specially treated during production (see Section 5.9). To bind the dust, the granulated material is treated with conditioning agents such as polyglycols, mineral oil products, vegetable oils and waxes, or mixtures of these. The addition rates are usually 0.3 – 5.0 kg/t [150, 151]. Various methods have been developed to assess or measure the abrasion resistance and dust content of the granulated product. Abrasion resistance is tested by subjecting samples of the granules to a process of screening or rolling in a rotating drum with steel spheres or rods, and weighing the fine material formed after a given time [152]. Dust content is determined by air blowing a sample under precisely defined conditions and weighing the dust collected [150].

Fine potassium chloride (standard quality or finer) tends to cake on storage or transport over long distances. Anticaking agents are therefore generally added, usually mixtures of aliphatic amines, or sometimes higher fatty acids. Salt products that do not contain alkaline-earth sulfates are treated with potassium hexacyanoferrate(II). Addition rates for organic anticaking agents are 50 – 300 g/t, but potassium hexacyanoferrate(II) is effective even at 15 – 25 g/t [150, 151]. Anticaking agents can affect the wettability of potassium chloride, which in turn can affect its granulation properties when complex fertilizers are being produced. The anticaking treatment must therefore suit the user's requirements exactly [153].

5.11. Toxicology and Occupational Health

No toxic hazards are associated with the normal handling of potassium chloride. According to USP XVII of 1970, the usual therapeutic dose (e.g., for treating potassium deficiency) taken orally would be 1 – 10 g/d. The LD_{50} (oral, guinea pig) is 2500 mg/kg.

Protective measures for storage and handling and personal protection such as breathing apparatus or gloves are unnecessary.

5.12. Economic Aspects and Uses

A review of world potash production since 1900 is given in Table 8. The figures show the total output from the potash industry including potassium sulfate and potassium products for industrial use [4, 154, 155].

Germany was the sole producer up to World War I, and was later joined by France and the United States, and then Spain, the Soviet Union, and Poland. After World War II, the leading producers were the central and western European countries and the United States. In the 1960s, the Soviet potash industry grew strongly, and the Soviet Union became the leading producer. Also in the 1960s, the first potash works in Saskatchewan was started up. In a few years, several large potash works were in operation there, and Canada became the second largest producer after the Soviet Union. The capital investment in the Soviet Union and Canada and the rapidly increasing use of fertilizers in agriculture in the 1960s and 1970s led to a steep increase in world potash production. Since 1980, the average annual increase in world potash production has been only 0.7 %.

Almost two-thirds of world potash production is exported. All the potash-producing countries are exporters except for Brazil and China. Canada is by far the largest exporter. World exports of potash in 1989 (in 10^3 t K_2O) were as follows [154]:

Canada	6 808
France	517
former East Germany	2 251
West Germany	1 339
Israel	1 007
Italy	64
Jordan	755
Spain	354
United Kingdom	251
United States	482
former USSR	3 407
Total	17 235

The economic situation, particularly in developed countries, greatly influences the extent and regional distribution of exports. Both the quantity exported and its distribution among consumers are greatly affected by the state of their

Table 8. World potash production (in 1000 t K$_2$O) by country

Country	Year												
	1900	1910	1920	1930	1940	1950	1960	1970	1980	1985	1986	1987	1988	1989
Germany	322	905	924	1381	1746									
West Germany						906	1978	2306	2737	2583	2161	2201	2290	2186
former East Germany						1200	1598	2419	3422	3465	3485	3510	3510	3200
France			194	506	517	896	1580	1765	1894	1745	1629	1539	1502	1195
Spain				25	88	159	265	521	658	659	702	740	766	742
Italy							25	152	102	143	109	117	126	154
United Kingdom									306	337	391	429	460	463
Middle and Western Europe	322	905	1118	1912	2351	3161	5446	7163	9119	8932	8477	8536	8654	7940
United States				44	56	344	1168	2394	2467	2240	1245	1171	1218	1461	1580
Canada									3173	7300	6637	6698	7267	8327	7360
North America				44	56	344	1168	2394	5640	9540	7882	7869	8485	9788	8940
Poland				1	70									
former Soviet Union						221	312	1084	4087	8064	10367	10228	10888	11300	10231
Eastern Europe				1	70	221	312	1084	4087	8064	10367	10228	10888	11300	10231
Israel					45		83	546	790	1163	1240	1253	1244	1273
Jordan										545	662	722	805	792
Dead Sea					45		83	546	790	1708	1902	1975	2049	2065
Brazil											11	37	48	109
Congo							123							
China								26	20	40	24	40	54	56
Others								149	20	40	35	77	102	165
Total	322	905	1163	2038	2961	4641	9007	17585	27533	28929	28511	29961	31893	29341

agriculture, especially in developed regions, and by the demand for or availability of convertible currency in the exporting or importing country. Transport costs for potash fertilizers have a considerable bearing on total cost to the consumer, and logistical considerations also influence the direction and size of exports or imports. Finally, fluctuations in the rate of exchange of currencies between the countries concerned are very important.

Because of the conditions described above, certain special regional relationships developed. The agricultural requirements of the former COMECON countries were satisfied by the Soviet and East German potash industries only. In western Europe, the market was supplied almost entirely by western European producers. In North America, Canadian and United States producers were in a dominating position. Canada had a good export market in Asia, as did Europe and Jordan. The Latin American market was supplied mainly by Canadian producers, together with former East Germany and the former Soviet Union. The political and economic changes in the former Eastern Bloc, the unification of Germany, and the collapse of the dollar have all led to changes in the supply–demand relationships described above, although the distance between the producer and the consumer is still of overriding importance. The future development of potash exports will be influenced greatly by imports into China and Brazil. Both countries have only minimal production and are compelled to import fertilizers on a large scale. Problems associated with their internal economies and with foreign exchange have thus far limited imports [156, 157].

The capacities of the potash producers in various countries for 1990–1991 (in 10^3 t K$_2$O) were [155]:

Middle and Western Europe	8 700
Germany	5 700
France	1 500
Spain	750
Italy	250
United Kingdom	500
North America	13 028
United States	1 838
Canada	11 190
former Soviet Union	12 880
Western Asia	2 220
Israel	1 380
Jordan	840
Brazil	150
China	50
Others	200
Total	37 218

Estimated world demand for potash fertilizers in the business year 1990 – 1991 by regions (in 1000 t K_2O) was as follows:

Europe	7 520
Eastern Europe	2 120
Western Europe	5 400
former Soviet Union	5 600
America	7 320
North America	5 100
Central America	350
South America	1 870
Asia	4 705
Western Asia	155
Southern Asia	1 360
Eastern Asia	3 190
Africa	522
Oceania	260
World	25 927

In using these figures, it should be borne in mind that the capacities quoted by individual producers are generally too high and that only ca. 95 % of the total potash production is used in the form of fertilizers. Hence, the total consumption of products of the potash industry is ca. 1.5×10^6 t of K_2O higher than the figure given above. Nevertheless, considerable overcapacity exists worldwide, as can be seen by comparing the two tables.

For the use of potassium compounds in fertilizers, see → Fertilizers.

Industrial-grade and chemical-grade potassium chloride are used mainly for the electrolytic production of potassium hydroxide. Other important uses include the production of drilling fluids for the oil industry, aluminum smelting, metal plating, production of various potassium compounds, and applications in the food and pharmaceutical industries [158].

6. Potassium Sulfate [159]

Potassium sulfate [*7778-80-5*], K_2SO_4, mineral name arcanite, forms colorless, nonhygroscopic crystals. It occasionally occurs in nature in the pure state in salt deposits (e.g., in Germany, the United States, and the CIS) but is more widely found in the form of mineral double salts in combination with sulfates of calcium, magnesium, and sodium (see Table 1). Potassium sulfate is, after potassium chloride, the most important potassium-containing fertilizer, being used mainly for special crops. Potassium sulfate constitutes ca. 5 % of the world demand for potash fertilizers.

6.1. Properties

Potassium sulfate forms orthorhombic crystals, which transform to the trigonal modification at 583 °C. Some properties of potassium sulfate are listed below:

M_r	174.25
fp	1069 °C
Crystal system and type	orthorhombic D_{2h}^{16}
Phase change at	583 °C
Crystal system and type at > 583 °C	trigonal D_{3d}^3
Refractive indices n_D^{20}	1.4933; 1.4946; 1.4973
Density	2.662 g/cm^3
Specific heat capacity c_p	752.9 J kg^{-1} K^{-1}
Heat of fusion	197.4 kJ/kg^1
Heat of transformation (orthorhombic/trigonal)	48.5 kJ/kg^1
Enthalpy of formation ΔH^0	− 1438 kJ/mol
Entropy S^0	175.6 J mol^{-1} K^{-1}
Dielectric constant (at 4×10^8 Hz)	6.3
Thermal coefficient of expansion (cubic)	130×10^{-6} K^{-1}

Apart from the naturally occurring double-salt minerals mentioned above, potassium sulfate also forms double salts and mixed crystals with ammonium sulfate and the sulfates of beryllium, magnesium, calcium, strontium, barium, and lead. It is reduced to potassium sulfide or potassium polysulfides by reducing agents such as hydrogen and carbon monoxide at high temperature.

The solubility of potassium sulfate in water is listed in Table 9. The cryohydric point is − 1.51 °C (7.1 g K_2SO_4/100 g H_2O), and the boiling point of the saturated solution is

101.4 °C (24.3 g K_2SO_4/100 g H_2O). The solid phases formed in the system K_2SO_4–H_2O are K_2SO_4, $K_2SO_4 \cdot H_2O$, and ice. In aqueous ammoniacal solution, the solubility decreases rapidly with increasing ammonia concentration [160]. Potassium sulfate is virtually insoluble in industrial organic solvents.

Table 9. Solubility of potassium sulfate in water (g/100 g) [15]

Temperature, °C	0	10	20	30	40	50
Solubility	7.35	9.24	11.1	12.9	14.8	16.6
Temperature, °C	60	70	80	90	100	
Solubility	18.4	20.0	21.5	22.8	24.0	

6.2. Raw Materials

Potassium sulfate is produced from single or mixed minerals or brines, or by the reaction of potassium chloride with sulfuric acid or sulfates [161]. The economically important minerals include the deposits of hard salt in Germany, langbeinite in New Mexico (United States), and kainite in Sicily. In the United States, sulfate-containing crystalline products from the evaporation of water from the Great Salt Lake (Utah) and Searles Lake (California) are also used for the production of potassium sulfate.

Potassium chloride is usually converted to potassium sulfate by reaction with sulfuric acid, but SO_2–air mixtures can also be used. Reactions with sodium sulfate and gypsum are also of interest. Small amounts of potassium sulfate are obtained during the production of alumina from alunite.

6.3. Production

The choice of production method and the location of a potassium sulfate plant depend on having a plentiful economic supply of the starting materials and being able to utilize or dispose of the byproducts or waste. Most production plants are located on salt deposits from which at least some of the raw materials can be obtained. Plants in which potassium chloride is reacted with sulfuric acid with liberation of hydrogen chloride are usually located in regions with a demand for hydrochloric acid, (e.g., for a acidification of petroleum boreholes), or they operate in conjunction with a chemical works having a process that uses hydrogen chloride. In such cases, the exploitability of hydrogen chloride determines the capacity of the potassium sulfate plant.

6.3.1. From Potassium Chloride and Sulfuric Acid (Mannheim Process)

The reaction of sulfuric acid with potassium chloride takes place in two stages:

$$KCl + H_2SO_4 \longrightarrow KHSO_4 + HCl$$

$$KCl + KHSO_4 \longrightarrow K_2SO_4 + HCl$$

The first reaction step is exothermic and proceeds at relatively low temperature. The second is endothermic and must be carried out at higher temperature. The relationship between total reaction time and temperature is shown in Figure 44. In practice, the process is operated at 600–700 °C. To minimize the chloride content of the product, a small excess of sulfuric acid is used, which is later neutralized with calcium carbonate or potassium carbonate, depending on the purity requirements for the product.

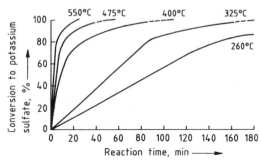

Figure 44. Temperature dependence of the reaction between potassium chloride and sulfuric acid
Reproduced from [162] with permission

The reaction is usually carried out in so-called Mannheim furnaces (Fig. 45) [162].

The furnace has a closed dish-shaped chamber, with diameter up to 6 m, heated externally by an oil or gas burner. Potassium chloride and sulfuric acid are fed into the chamber in the required ratio at an overhead central point. The mixture reacts with evolution of heat and is mixed by a slowly moving stirrer fitted with stirring arms

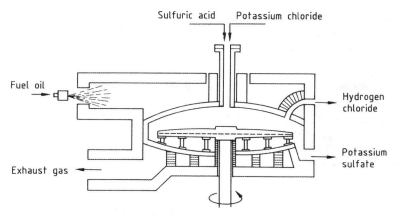

Figure 45. Schematic diagram of a Mannheim furnace
Reproduced from [162] with permission

with scrapers (rabbles), which propels the mixture from the center of the chamber to the periphery. Potassium sulfate leaves the reaction chamber at this point and is neutralized and cooled. It normally contains 50–52 % K_2O and 1.5–2 % chloride. Hydrogen chloride gas formed is absorbed in water to form hydrochloric acid or used in gaseous form.

The Mannheim process is the most widely used method of producing potassium sulfate due to its simplicity, high yield, and the many ways in which the byproduct can be utilized. Hydrogen chloride is used to produce dicalcium phosphate, vinyl chloride, or calcium chloride if it cannot be sold as hydrochloric acid.

Disadvantages of the process include high energy consumption, severe corrosion, and high capital cost. In the United States, reductions in corrosion and energy consumption are achieved by using the Cannon process, in which the reaction is carried out in a directly fired fluidized bed.

Another variation is the Hargreaves process, which is also used in the United States. Briquetted potassium chloride is heated in reaction chambers in a stream of sulfur dioxide from the combustion of sulfur, excess air, and water vapor. The yield and the degree of conversion are both ca. 95 %.

The Mannheim and Hargreaves processes are also used to produce sodium sulfate from sodium chloride and sulfuric acid. Mannheim furnaces can be used to produce potassium and sodium sulfates alternately. Research into the reaction of potassium chloride with sulfuric acid in a liquid–liquid extraction process did not result in the construction of a production plant [163, 164].

6.3.2. From Potassium Chloride and Magnesium Sulfate [165, 166]

In a process used mainly in Germany, the sulfate required is provided by kieserite, $MgSO_4 \cdot H_2O$, a component of German hard salt deposits. The reaction can be represented by the following overall equation:

$$2\,KCl + MgSO_4 \longrightarrow K_2SO_4 + MgCl_2$$

Kieserite reacts very slowly and must be ground finely before reaction. Alternatively, it can first be recrystallized to give epsomite, $MgSO_4 \cdot 7\,H_2O$.

The basis of the process is explained in Figure 46. The fundamental relationships for the single-stage process of KUBIERSCHKY and the two-stage process of KOELICHEN and PRZIBYLLA are shown as broken lines on the isotherm diagram.

For the single-stage process, the most favorable mixing ratio of the starting materials is given by point C. In the presence of sufficient water, this mixture reacts to form potassium sulfate and a sulfate mother liquor (point m). This solution has the highest magnesium chloride content attainable by direct reaction, which determines the yield. The magnesium content of solution m reaches a maximum at 25 °C, and the

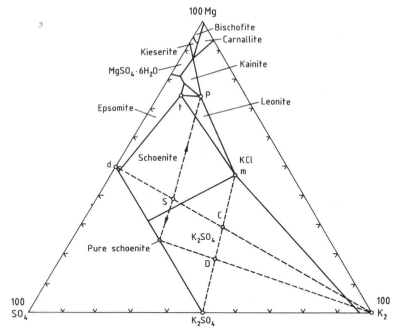

Figure 46. Isothermals of the system $K_2 - Mg - Cl_2 - SO_4^{2-} - H_2O$ at 25 °C according to JÄNECKE

process is therefore carried out at this temperature. The single-stage process achieves a theoretical potassium yield of only 46.1 % and sulfate yield of 67.5 %.

For this reason, the two-stage process is now used exclusively. In this process, the starting materials are first mixed in the presence of a definite quantity of water corresponding to point S to form schoenite, $K_2SO_4 \cdot MgSO_4 \cdot 6 H_2O$. So-called potash – magnesia liquor, which has a high magnesium chloride content (point P), is also formed. The schoenite is reacted with additional potassium chloride (point D) to form potassium sulfate and sulfate mother liquor:

$2 KCl + 2 MgSO_4 + x H_2O \longrightarrow$
$K_2SO_4 \cdot MgSO_4 \cdot 6 H_2O + MgCl_2$ (aq.)
$2 KCl + K_2SO_4 \cdot MgSO_4 \cdot 6 H_2O + x H_2O \longrightarrow$
$2 K_2SO_4 + MgCl_2$ (aq.)

This process gives a theoretical potassium yield of 68 % and sulfate yield of 83.7 %.

The flow diagram of the process is shown schematically in Figure 47.

In the first stage, called the potash – magnesia stage, schoenite or leonite, $K_2SO_4 \cdot MgSO_4 \cdot 4 H_2O$, is produced by stirring solid epsomite or finely ground kieserite with potassium chloride in sulfate mother liquor recycled from the second stage. The suspension produced is filtered on rotary filters; the potash – magnesia brine, which contains 180 – 200 g/L magnesium chloride, is removed; and the solid crystalline product, also known as potash – magnesia, is fed to the next stage, sometimes after being washed with sulfate mother liquor, where it is stirred with potassium chloride solution at ca. 70 °C. The temperature of the mixture is 35 – 40 °C, and solid potassium sulfate is formed. This is thickened, debrined by centrifuges, and dried in drum or fluidized-bed dryers.

If the sulfate reaction is carried out in a classifying crystallizer at a high solids content [167], a very pure, coarsely crystalline product is obtained with K_2O content of 53 % and chloride content of < 0.5 %.

In the industrial process, sodium chloride is always present. If the molar ratio of $Na_2 : K_2$ in the sulfate mother liquor exceeds 2 : 5, glaserite, $3 K_2SO_4 \cdot Na_2SO_4$, is formed instead of potassium sulfate. To prevent this, the potassium chloride used must be of adequate purity.

The yield is determined by losses to the waste brine. The higher the magnesium chloride con-

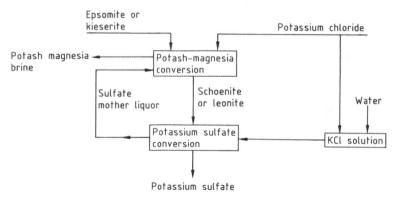

Figure 47. Flow diagram of the two-stage production of potassium sulfate from potassium chloride and magnesium sulfate

tent of the potash–magnesia brine is, the lower is the potassium content, and therefore the potassium loss. Excess potassium-rich sulfate mother liquor must be recovered.

6.3.3. From Potassium Chloride and Langbeinite [168]

Large deposits of langbeinite, $K_2SO_4 \cdot 2\,MgSO_4$, can be found in New Mexico (United States). Langbeinite can be converted to potassium sulfate according to the following overall equation:

$$K_2SO_4 \cdot 2\,MgSO_4 + 4\,KCl \longrightarrow 3\,K_2SO_4 + 2\,MgCl_2$$

The potash ore also contains halite and varying amounts of sylvite, from which langbeinite is separated by gravity separation, flotation, and dissolution of halite, giving various crystal sizes. The coarser langbeinite fraction is sold as potash–magnesium fertilizer, and the finer fraction is reacted with potassium chloride to produce potassium sulfate. A flow diagram for the process is given in Figure 48.

The potassium sulfate formed is granulated and marketed in three different grain sizes: granular (0.8–3.4 mm), standard (0.2–1.6 mm), and special standard. The latter has a high content of grains < 0.2 mm.

6.3.4. From Potassium Chloride and Kainite [161]

In Sicily, kainite, $KCl \cdot MgSO_4 \cdot 2.75\,H_2O$, is obtained from a potash ore by flotation. It is then converted into schoenite at ca. 25 °C by stirring with mother liquor containing the sulfates of potassium and magnesium from the later stages of the process. Schoenite is filtered off and decomposed with water at ca. 48 °C. This causes magnesium sulfate and part of the potassium sulfate to dissolve and most of the potassium sulfate to crystallize. The crystals are filtered and dried. The sulfate mother liquor is recycled to the kainite–schoenite conversion stage. The mother liquor produced there, which still contains ca. 30 % of the potassium used, is treated with gypsum, $CaSO_4 \cdot 2\,H_2O$, causing sparingly soluble syngenite [*13780-13-7*], $K_2SO_4 \cdot CaSO_4 \cdot H_2O$, to precipitate. Syngenite is decomposed with

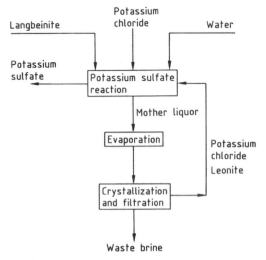

Figure 48. Flow diagram of the production of potassium sulfate from langbeinite

water at ca. 50 °C, which dissolves potassium sulfate and reprecipitates gypsum. The potassium sulfate solution is recycled to the schoenite decomposition stage, and gypsum is reused to precipitate syngenite. A simplified flow diagram of the process is given in Figure 49.

6.3.5. From Potassium Chloride and Sodium Sulfate [164]

The production of potassium sulfate from potassium chloride and sodium sulfate takes place in two stages, with glaserite, $Na_2SO_4 \cdot 3\,K_2SO_4$, as an intermediate, according to the following equations:

$$4\,Na_2SO_4 + 6\,KCl \longrightarrow Na_2SO_4 \cdot 3\,K_2SO_4 + 6\,NaCl$$
$$Na_2SO_4 \cdot 3\,K_2SO_4 + 2\,KCl \longrightarrow 4\,K_2SO_4 + 2\,NaCl$$

Potassium chloride and sodium sulfate are reacted at 20–50 °C in water and recycled process brines to form glaserite, which is filtered and then reacted with more potassium chloride and water to form potassium sulfate. Because the mother liquor from the glaserite stage has a high potassium and sulfate content, the maximum potassium yield is 73 %, and the maximum sulfate yield is 78 %. The yield can be increased considerably by cooling the mother liquor to produce more crystals and by including a final evaporation stage.

A production plant in the CIS uses the glaserite process [166], and an experimental plant is operating in Canada [169].

Alternatively, sodium sulfate solution can be used to charge an anion exchanger with sulfate, which reacts with potassium chloride solution to give a high yield of potassium sulfate [170, 171]. This process is used at Quill Lake, Saskatchewan, Canada, which has a high sodium sulfate content.

6.3.6. From Potassium Chloride and Calcium Sulfate

Processes based on gypsum, $CaSO_4 \cdot 2\,H_2O$, have often been proposed, because it is so readily available. Two processes have been tested in experimental plants:

1) Reaction with potassium chloride in strongly ammoniacal solution

2) Reaction with anion exchangers

Potassium chloride reacts with gypsum in water to give syngenite. If the reaction is carried out in a concentrated solution of ammonia at low temperature, potassium sulfate with a very low syngenite content is obtained [163, 172, 173].

$$CaSO_4 \cdot 2\,H_2O + 2\,KCl \longrightarrow K_2SO_4 + CaCl_2\,(aq.)$$

By carrying out the reaction in two or more stages, high concentrations of calcium chloride in the waste brine can be produced, and hence very high yields. The ammonia required for the reaction medium must be recovered by distillation. This process has attracted some interest [174].

The anion-exchange process is carried out in two stages [175]. First, the ion-exchange resin (R) is treated with a suspension of gypsum, charging it with sulfate. The charged resin is then treated with a concentrated solution of potassium chloride, to replace the dissolved chloride by sulfate. Potassium sulfate crystallizes from the solution, sometimes after the addition of solid potassium chloride. The process takes place according to the following equations:

$$R_2Cl_2 + CaSO_4 \longrightarrow R_2SO_4 + CaCl_2$$
$$R_2SO_4 + 2\,KCl \longrightarrow K_2SO_4 + R_2Cl_2$$

6.3.7. From Alunite [176]

Alunite [12588-67-9], $K_2SO_4 \cdot Al_2(SO_4)_3 \cdot 4\,Al(OH)_3$, occurs in several extensive deposits. On being heated to 800–1000 °C, it decomposes with liberation of sulfur trioxide to form a mixture of alumina and potassium sulfate. The latter can be extracted from the mixture. A plant in the CIS uses this process.

6.3.8. From Natural Brines and Bitterns

In the United States, large quantities of potassium sulfate are produced from the brines of the Great Salt Lake and smaller amounts from the brines of Searles Lake (see Section 8.3). Extensive investigations and project studies have been carried out in various countries into the extraction of potassium sulfate from the mother liquors (bitterns) produced when salt is extracted from seawater, and from concentrated brines in Tunisia and the Atacama desert of Chile [164].

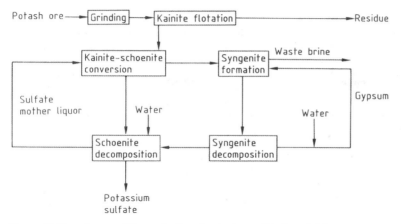

Figure 49. Flow diagram of the production of potassium sulfate from kainite

6.4. Granulation

The demand for granulated potassium sulfate has increased greatly for the same reasons that the demand for potassium chloride has increased. The method of production with compaction rolls is widely used (see Section 5.9), although potassium sulfate does not behave under pressure like potassium chloride, which undergoes plastic deformation with merging of the grain boundaries. However, the granulated product is much more dense and solid if the potassium sulfate is wetted before compaction by adding up to 2 % water, or if steam is introduced through force feeders located above the compaction rolls. Also, the pressure used must be considerably higher than that for potassium chloride [177, 178]. A compaction plant for potassium sulfate is otherwise similar to that for potassium chloride, although the throughput is much lower.

6.5. Quality Specifications

The important quality specifications for agricultural-grade potassium sulfate, known in the English-speaking world and on the international market as sulfate of potash (abbreviated to SOP), are the K_2O and chloride content. All producers guarantee a minimum K_2O content of 50 % (92.5 % K_2SO_4), typical values being 50.5–51.0 %. A completely water-soluble potassium sulfate with K_2O content of > 52 % is supplied for the production of liquid fertilizers.

The maximum permissible chloride content is 3 %, according to EC Guidelines, but the usual commercial upper limit is 2.5 %. The chloride content depends very much on the production method, and some producers offer potassium sulfate with chloride content < 0.5 %.

The most common impurities, apart from alkali chlorides, are the sulfates of calcium and magnesium.

The demand for granulated potassium sulfate as a single-nutrient fertilizer and for bulk blending has greatly increased. The crystal size and size distribution of the products granular, coarse, and standard are the same as those for potassium chloride (see Section 5.10).

Potassium sulfate does not cake as readily as potassium chloride and is therefore not treated with anticaking agents. Dust-reducing agents are added at the rate of 5 kg/t to the fine and granulated products; the same additives are used as for potassium chloride (see Section 5.10).

Industrial-grade potassium sulfate with a K_2SO_4 content of 99.6–99.9 %, purified by recrystallization, is supplied as a raw material for the production of other potassium compounds and as an auxiliary material or reagent for various branches of industry. Specifications with respect to chemical purity and crystal size distribution are suited to the individual user's requirements and can sometimes be very detailed.

6.6. Toxicology and Occupational Health

No health hazard is associated with potassium sulfate if it is handled in accordance with regulations. According to Swiss law relating to toxic substances, it is a Class 4 material (LD_{50}: 500–5000 mg/kg). Otherwise, the information given in Section 5.11 for potassium chloride applies.

6.7. Economic Aspects and Uses

Potassium sulfate accounts for ca. 5 % of world production by the potash industry, expressed in K_2O units. It is produced in 11 countries. Of the total, ca. two-thirds comes from Belgium and Germany. Other European producers are Italy, Spain, Finland, and Sweden. Potassium sulfate is also produced by the CIS. Other production plants are in the United States, Japan, South Korea, Taiwan, the Philippines, and China. Present world production and consumption both exceed 1.5×10^6 t K_2O [179, 180].

A list of the most important exporting countries (exports for 1985–1986 in 10^3 t K_2O) is given below:

Western Europe	586
Belgium	256
Finland	18
West Germany	227
Italy	54
Spain	31
former East Germany	61
United States	74
Asia/Middle East	16
Israel	3
South Korea	8
Taiwan	5
Total	737

The consumption (1985–1986 in 10^3 t K_2O) amounted to:

Western Europe	558
Eastern Europe	106
Africa	121
North America	115
Latin America	51
Asia/Middle East	271
Oceania	6
Total	1228

Potassium sulfate is up to twice as expensive as potassium chloride based on K_2O content due to the costs of the raw material (potassium chloride) and processing. It is therefore only used as a potash fertilizer for applications where it performs much better than potassium chloride.

The sulfur content of potassium sulfate is an advantage where there is a deficiency of sulfur in the soil. Also, it has only a slight oversalting effect on soil in arid or semiarid areas. It is useful for fertilizing crops that are sensitive to chloride or whose quality is improved by chloride-free fertilization. This applies particularly to tobacco, vegetables, potatoes, vines, and citrus or various other fruit [179, 181].

Potassium sulfate is used in industry for the production of other potassium compounds, accelerators for rapid-setting cements, synthetic rubbers, desensitizers for explosives, lubricants, powdered fire extinguishers, dyes, explosives, and pharmaceuticals.

7. Potash – Magnesia

Crops sensitive to chloride can be fertilized with potassium sulfate alone or by fertilizers containing the sulfates of potassium and magnesium. These consist either of a mixture of the two sulfates or of the double sulfates schoenite, leonite, or langbeinite in dehydrated form.

Formerly, large quantities of a mixture of fine-grained potassium sulfate and kieserite were marketed under the name potash–magnesia (K_2O: 27–30 %, MgO: 9–12 %). This has since been almost completely replaced by a granulated potash–magnesia fertilizer, which is produced by first mixing potassium sulfate and kieserite at 95 °C with hot synthetic langbeinite. The hot mixture is granulated in a drum granulator, and the product is quickly cooled to $< 60 °C$. It contains 29–30 % K_2O and 10 % MgO.

Langbeinite from the potash deposits in Carlsbad, New Mexico (United States), can be used in the pure state directly as a potash–magnesia fertilizer. Crude langbeinite also contains halite and varying amounts of sylvite, and the method of purification utilizes the low rate of dissolution of langbeinite in water. If the sylvite content is low, the potash ore is first ground to < 6 mm and then simply washed with water to dissolve the halite. The langbeinite that remains is dried and screened to obtain the commercial size gradings. For higher sylvite content, an additional stage is required to separate the components by a combination of gravity separation and flotation. The langbeinite produced has a K_2O content of 20 – 22 % and an MgO content of 18 – 19 %, and is marketed as Sulpomag or K-Mag.

8. Production of Potassium Salts from Other Raw Materials

Potassium salts occur not only in salt deposits, but also in solution in many types of inland lakes and in seawater. Where the concentration of the salt solution is high enough and the climatic and topographical conditions are suitable, potassium chloride or potassium sulfate can be produced by solar evaporation. Typical analyses of some of the more important sources are listed in Table 10.

Table 10. Composition of some natural brines compared with seawater (in wt %)

	Dead Sea	Wendover brine	Great Salt Lake	Seawater
K^+	0.6	0.6	0.7	0.04
Na^+	2.9	9.4	7.6	1.08
Mg^{2+}	3.4	0.4	1.1	0.13
Ca^{2+}	1.3		0.016	0.04
Cl^-	17.0	16.0	14.1	1.94
SO_4^{2-}	0.04	0.2	2.0	0.27
Br^-	0.5		0.01	0.006
H_2O	74.3	73.3	74.5	96.5

8.1. The Dead Sea

Potash production by solar evaporation began at the northern end of the Dead Sea in 1931. Later, production switched to the southern end but was interrupted by the war of 1947 – 1948. In 1952, potash production was resumed in Sodom by the Israeli State *Dead Sea Works*. The capacity was built up in several stages and now amounts to $> 2 \times 10^6$ t of potassium chloride in all the usual commercial grades [182, 183].

Brine from the lake is concentrated in evaporation ponds with a total area of ca. 90 km², from which the crystallized salts are recovered. Most of the dissolved sodium chloride separates out first in the primary evaporation ponds. A mixture of carnallite and sodium chloride then crystallizes in the main production ponds. This is removed as a suspension by suction dredgers and pumped to the works where the crystals are filtered off and treated by the carnallite cold decomposition process (see Section 5.2.3). Most of the resulting NaCl – KCl mixture is then treated in a hot leaching plant (see Section 5.2.2), where potassium chloride with > 60 % K_2O is produced in crystallizers. Part of the crop of crystals from the main production ponds is treated by a cold crystallization process developed by the Dead Sea Works. This decomposition process directly yields a product containing > 60 % K_2O.

A potash works was started by the *Arab Potash Co.* in 1982 on the southeast bank of the Dead Sea near Safi, Jordan. Evaporation and carnallite production are carried out in a pond system with a total area of 100 km². The process resembles that of the Dead Sea Works. Magnesium chloride is removed from carnallite in a cold decomposition process, and the salt mixture formed is then treated in a hot leaching plant to give potassium chloride [182, 184].

8.2. The Great Salt Lake [185]

The Great Salt Lake is the result of the shrinkage by evaporation of the former Lake Bonneville and lies in the eastern part of the basin. It has a high salt content and is the reason for the existence of several plants that produce sodium chloride and, since 1968, potassium salts.

At the western end of the Great Salt Lake, in the Great Salt Lake Desert near Wendover, are the Bonneville Salt Flats. Here, under a salt crust, are porous sediments containing brines, which are regenerated by water from atmospheric precipitation. Potassium chloride has been produced from these brines since 1937. They are collected by a system of ditches and evaporated in ponds. A mixture of potassium and sodium chloride crystallizes, from which potassium chloride is obtained by flotation [186].

Unlike the Wendover brines, the Great Salt Lake brines contain considerable amounts of sulfate (see Table 10). The process used in Wendover cannot therefore be used here. The opportunity exists of obtaining substantial quantities of valuable potassium sulfate rather than potassium chloride, which has been carried out since 1970 in the Great Salt Lake Minerals & Chemicals Corporation (GSLM & CC) plant at Ogden on the east bank. Sodium chloride is first crystallized in the 56-km^2 pond system until the solution is saturated in potassium salts. Further solar evaporation then takes place in the main production ponds, producing a mixture containing varying proportions of kainite, carnallite, and schoenite, with small amounts of sodium chloride [187]. This mixture is converted into schoenite in the plant by treatment with recycled process brine. The sodium chloride that is not dissolved by this reaction must be removed before further treatment. This is carried out by flotation of a side stream [102]. Schoenite is then decomposed by water, which produces very pure potassium sulfate. The brine from this decomposition stage has a high potassium content and is recycled to the first stage of the process. The brine produced by the reaction at this stage is recycled to the evaporation pond. The sulfate content of the main crop of crystals is higher than the potassium content, so that further potassium sulfate can be produced by addition of potassium chloride from an external source [188].

The level of the Great Salt Lake increased so much during the 1970s and 1980s that the pond system of the Great Salt Lake Minerals & Chemicals Corporation (GSLM & CC) overflowed during 1984, and production had to stop [189]. The lake level then fell, the pond system was reinstated, and production started again in 1989.

8.3. Searles Lake [190]

In Trona, on the northwest bank of the almost dry Searles Lake, brines are obtained that contain not only sodium and potassium chlo-rides, but also considerable quantities of sulfate, carbonate, and borate ions. Recycled process brines are first added, and evaporation produces sodium chloride and the double salt burkeite [*12179-88-3*], $Na_2CO_3 \cdot 3\,Na_2SO_4$. Potassium chloride is obtained by vacuum cooling of the potassium- and borate-containing mother liquor. Part of the chloride is reacted with part of the burkeite to form glaserite, $Na_2SO_4 \cdot 3\,K_2SO_4$, an intermediate stage in potassium sulfate production. Another reaction used for potassium sulfate production is that of potassium borate in the end brines with sulfuric acid, to form potassium sulfate and boric acid:

$$K_2B_{10}O_{16} \cdot 8\,H_2O + 6\,H_2O + H_2SO_4 \longrightarrow K_2SO_4 + 10\,H_3BO_3$$

8.4. Other Sources

In addition to the sources mentioned under potassium sulfate, other salt lakes exist whose potassium content would appear to offer the possibility of extracting potassium salts. In 1969 at Lake McLeod in Western Australia, a works produced langbeinite, $K_2SO_4 \cdot 2\,MgSO_4$, for a short period of time but ceased operations for unknown reasons. In China, in the Qinghai Province, potassium chloride production has been carried out for several years at Tsarhan Lake by using solar evaporation. Large increases in production are planned [191].

Seawater has a low potassium content (Table 10) so that economical extraction of potassium salts is not possible. The production of sea salt by solar evaporation in salt gardens yields residual brines with increased potassium content. In some places (e.g., India), small quantities of low-percentage potassium salts are produced from these bitterns. However, economical production of potash fertilizers of marketable quality is not possible from the amounts of mother liquor available from even the largest sea salt producers.

9. Storage and Transportation

($\rightarrow$ Fertilizers, Chap. 8.1.)

The demand for potash fertilizers fluctuates greatly throughout the year, but because potash plants need to produce at as steady a rate as possible, large storage capacities are needed to accommodate periods of low demand. Therefore, potash plants usually have high-capacity product storage facilities. Also, in seaports, where fertilizers are loaded onto ships, the largest potash companies or their subsidiaries have large storage capacities. In both cases, long storage sheds are used, usually with walls sloping to match the angle of repose of the potash salt. The sheds are usually filled by means of conveyor belts located under the shed roof. They are emptied either by bucket loaders or scrapers that move the salt into a channel under the floor of the silo or onto a conveyor belt at the side, which carries it via sloping bands or elevators to the loading plant. More recently, especially where there is a shortage of land, round silos have been used, often arranged in rows. These too are filled from above by conveyor belts and are emptied through openings at ground level. The majority of potash fertilizer is transported in bulk in self-discharging wagons with a capacity up to 100 t, by rail, truck, or inland waterway. Transport from potash plants remote from a seaport or the main consuming area is usually by special trains that run on a fixed timetable between the potash plant and the seaport or intermediate storage facility. For example, the transport of potash fertilizers from Saskatchewan to a cargo-handling plant in Vancouver uses continuous-loop train tracks that enable 10 000 t to be delivered in 102 wagons with a capacity of 98 t each [192].

A small proportion of potash salts is supplied in sacks, usually containing 50 kg. The sacks are filled either by the supplier or at the loading plant at the seaport by automatic sack-filling machines. The sacks are usually paletized.

10. Analysis of Potassium Compounds

Potassium is usually determined gravimetrically. In the United States, it is precipitated as the hexachloroplatinate [193]. Precipitation as the tetraphenylborate is another widely used method, being the standard ISO method for fertilizers, and can be either a gravimetric or a volumetric procedure (ISO 5318 and 5310) [194]. Precipitation as the perchlorate or tartrate is seldom used. Flame photometry is used for both laboratory and process control analysis. X-ray fluorescence can be used for the analysis of solids or brines.

A review of methods recommended for potassium determination in fertilizers is given in → Fertilizers.

11. Potassium Hydroxide [195–198]

11.1. Properties

Pure, solid potassium hydroxide [1310-58-3], KOH, caustic potash, M_r 56.11, ϱ 2.044 g/cm^3, mp 410 °C, bp 1327 °C, heat of fusion 7.5 kJ/mol, is a hard, white substance. It is deliquescent and absorbs water vapor and carbon dioxide from the air. Potassium hydroxide dissolves readily in alcohols and water (heat of solution 53.51 kJ/mol). The solubility of KOH (g KOH/100 g H$_2$O) in water is shown below:

Temperature, °C	0	10	20	30	50	100
Solubility	97	103	112	126	140	178

The mono-, di-, and tetrahydrates are formed with water. Aqueous potassium hydroxide is a colorless, strongly basic, soapy, caustic liquid, whose density depends on the concentration:

Concentration, wt %	10	20	30	40	50
Density, g/cm^3	1.092	1.188	1.291	1.395	1.514

Technical caustic potash (90–92 % KOH) melts at ca. 250 °C; the heat of fusion is ca. 6.7 kJ/mol.

11.2. Production

Today, potassium hydroxide is manufactured almost exclusively by potassium chloride electrolysis. The diaphragm, mercury, and membrane processes are all suitable for the production of potassium hydroxide, but the mercury process is preferred because it yields a chemically pure 50 % potassium hydroxide solution.

In the *diaphragm process*, a KCl-containing, 8–10 % potassium hydroxide solution is initially formed, whose salt content can be reduced to ca. 1.0–1.5 % KCl by evaporation to a 50 % liquor. Further purification is complicated, and the quality of liquor from mercury cells cannot be achieved.

In the *mercury process* a very pure KCl brine must be utilized, because even traces (ppb range) of heavy metals such as chromium, tungsten, molybdenum, and vanadium, as well as small amounts (ppm range) of calcium or magnesium, lead to strong evolution of hydrogen at the amalgam cathode. The very pure potassium hydroxide solution running off the decomposers is cooled, freed from small amounts of mercury in precoated filters, and in some cases sent immediately to the consumer as a 45 – 50 % liquor in drums, tank cars, or barges.

Since about 1985, new cell rooms for the manufacture of potassium hydroxide solution have used the *membrane process*. At present, the cell liquor has a low chloride content (10 – 50 ppm); the KOH concentration is 32 %. Before dispatch, it is concentrated to 45 – 50 % by evaporation.

Nonelectrochemical processes have been proposed for the manufacture of chlorine and potassium hydroxide from KCl by thermal decomposition of potassium nitrite in the presence of Fe_2O_3 [199].

This method involves reacting KCl with NO_2 to obtain Cl_2 and potassium nitrite, reacting the KNO_2 with iron(III) oxide and oxygen to give potassium ferrate ($K_2Fe_2O_4$), and reacting the ferrate with water to produce KOH. Another method consists of reacting an aqueous solution of KCl with NO_2 and O_2 to give Cl_2 and KNO_3, which is reacted with water in the presence of Fe_2O_3 to produce KOH.

Largely water-free, ca. 90 – 95 % potassium hydroxide (caustic potash) is obtained by evaporating potassium hydroxide solution. The residual content of 5 – 10 % H_2O is bound as a monohydrate.

Suitable evaporation processes are single- or multistage falling-film evaporators [200], Badger single-tube evaporators, or boilers connected in cascade. Heating is carried out with steam or by means of heat-transfer agents (salt melts, Dow-therm). Flash evaporators are used as the final stage in large-capacity plants [201].

To counter the strong corrosiveness of the potassium hydroxide solution and retain the purity of the caustic potash, the equipment is made largely from high-purity nickel (LC 99.2) or is silverplated. The equipment is often protected by polarization.

For dispatch, caustic potash comes on the market poured directly into drums or packed in polyethylene bags after cooling; in blocks, molded pieces, flakes, prills, and as a powder. Potassium hydroxide is classified as a corrosive material:

UN no.	1814	(for aqueous solution)
UN no.	1813	(for dry material)
GGVS/GGVE	Class 8	
RID/ADR	Class 8	

Handling is described in [202].

11.3. Quality Specifications

Potassium hydroxide solution is supplied in pure quality [total alkalinity 49.7 – 50.3 %, KOH 48.8 % (min.), NaOH 0.5 % (max.), CO_3^{2-} 0.1 % (max.)] or in technical quality [total alkalinity 49.7 – 50.3 %, NaOH 1.0 % (max.), CO_3^{2-} 0.3 % (max.)]. The contents of Cl^-, SO_4^{2-}, Fe^{2+}, and Ca^{2+} are < 30 ppm. Solid caustic potash produced from amalgam liquor has a total alkalinity (calculated as KOH) of 89 – 92 %, NaOH 1.5 % (max.), CO_3^{2-} 0.5 % (max.), Cl^- 0.01 % (max.). The values for SO_4^{2-}, Fe^{2+}, and Ni^{2+} are < 50 ppm. Caustic potash from diaphragm electrolysis has a Cl^- content of 2.5 – 3.0 % and higher content of heavy metals. For analysis, see 12.3.

11.4. Economic Aspects and Uses

Pure-quality potassium hydroxide is used as a raw material for the chemical and pharmaceutical industry, in dye synthesis, for photography as a developer alkali, and as an electrolyte in batteries and in the electrolysis of water. Technical-quality KOH is used as a raw material in the detergent and soap industry; as a starting material for inorganic and organic potassium compounds and salts (e.g., potassium carbonate, phosphates, silicate, permanganate, cyanide); for the manufacture of cosmetics, glass, and textiles; for desulfurizing crude oil; as a drying agent; and as an absorbent for carbon dioxide and nitrogen oxides from gases.

World production is estimated at ca. 700 – 800 × 10^3 t/a. Main producers are the United States [203], Germany, Japan, and France. Other important producer countries are Belgium, the United Kingdom, Italy, Spain, South Korea, India, Israel, former Yugoslavia, former Czechoslovakia, Sweden, and Romania.

12. Potassium Carbonate [195–198]

Potassium carbonate was produced in antiquity and used for many purposes. In the Old Testament, potash is mentioned in Jeremiah (written in the 7th century B.C.). ARISTOTLE describes the extraction of wood ash with water; the Romans manufactured soap from fat and potash. LAVOISIER identified potash as potassium carbonate.

The production of potash from wood ash for the manufacture of glass and soap was a flourishing industry in the Middle Ages in areas having a plentiful supply of wood such as Russia, and also in Scotland. Since 1860, potash salts have replaced wood as a raw material for the manufacture of potassium carbonate.

In Anglo-American usage, the term potash today includes potassium carbonate as well as all potassium salts, such as KCl, K_2SO_4, and $K_2SO_4 \cdot MgSO_4 \cdot xH_2O$, that are used as fertilizers; the potassium content is given as K_2O.

Potassium carbonate occurs in small amounts in a few African lakes (e.g., Lake Chad and in the vicinity of Lake Victoria), as well as in the Dead Sea.

12.1. Properties

Anhydrous potassium carbonate [584-08-7], K_2CO_3, M_r 138.21, ϱ 2.428 g/cm^3, mp 891 °C, is a white, hygroscopic, powdery material that deliquesces in moist air. It is readily soluble in water with the formation of an alkaline solution. The solubility of K_2CO_3 (g K_2CO_3/100 g H_2O) in water is given below:

Temperature, °C	0	10	20	30	40	50
Solubility	105.5	108.0	110.5	113.7	116.9	121.2
Temperature, °C	60	70	80	90	100	
Solubility	126.8	133.1	139.8	147.5	155.7	

On addition of acid, potassium carbonate reacts with the evolution of carbon dioxide:

$$K_2CO_3 + H_2SO_4 \longrightarrow K_2SO_4 + CO_2 + H_2O$$

K_2CO_3 forms several hydrates, of which $K_2CO_3 \cdot 1.5 H_2O$ is the stable phase in contact with the saturated solution from 0 °C to ca. 110 °C. This hydrate (M_r 165.24, ϱ 2.155 g/cm^3) crystallizes in glassy, virtually dust-free crystals. It is also hygroscopic and deliquesces in moist air. It is completely dehydrated at 130 – 160 °C.

12.2. Production

From Caustic Potash and Carbon Dioxide. The most important process for the manufacture of potassium carbonate begins with electrolytically produced potassium hydroxide solution. The almost chemically pure solution obtained by the mercury process (see 11.2) is reacted with carbon dioxide or CO_2-containing off-gases (flue gas, lime kiln gas).

$$2 KOH + CO_2 \longrightarrow K_2CO_3 + H_2O$$

Solid potassium carbonate is then obtained by crystallization (under vacuum and with cooling) from liquors or in the fluidized-bed process.

In the continuous crystallization process (Fig. 50), the filtered, fresh carbonate solution is mixed with mother liquor and concentrated in several preliminary evaporators connected in series until the hydrate $K_2CO_3 \cdot 1.5 H_2O$ finally precipitates in the crystallizer after cooling under vacuum [204]. The mother liquor is separated from the crystal suspension in hydrocyclones and centrifuges, filtered, and fed back to the process. The crystals are dried at ca. 110 – 120 °C in rotary kilns or fluidized-bed dryers and packed for sale as potash hydrate, or they are calcined at 200 – 350 °C to give 98 – 100 % K_2CO_3. Impurities such as soda, sulfate, silicic acid, and iron that concentrate in the mother liquors can be partially removed [205] by removing a partial stream of the mother liquor, which is either used for brine purification in the electrolysis process or sold as a low-grade potassium carbonate solution, or by crystallizing the double-salt $NaKCO_3$ at elevated temperature in a separate crystallization and drying process.

The resulting potassium carbonate is very pure and meets the requirements of USP, BP, DAB, and JP if the process is operated in appropriate manner.

Starting from potassium carbonate solution, prills can be produced in a combined reactor, in which spray drying and fluidized-bed granulation take place simultaneously [206].

In the fluidized-bed process, aqueous potassium hydroxide solution is sprayed into a

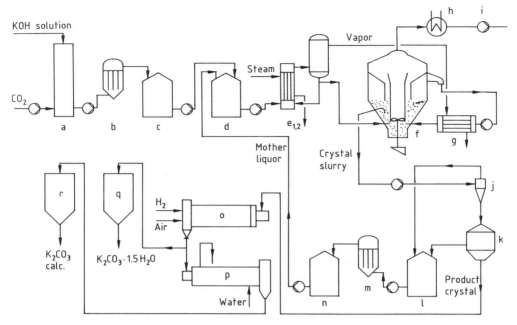

Figure 50. Preparation of potassium carbonate with continuous crystallization
a) Carbonization; b) Crude liquor filter; c) Fresh liquor tank; d) Mixed liquor tank; e_1), e_2) Preliminary evaporation; f) Vacuum/cooling crystallization (Chemietechnik Messo system); g) Preheater; h) Vapor condenser; i) Vacuum pump; j) Hydrocyclone; k) Centrifuge; l) Centrifuge liquor tank; m) Filter for mother liquor; n) Mother liquor tank; o) Drying or calcining rotary kiln; p) Cooling device for calcined K_2CO_3; q) Storage for hydrated potash; r) Storage for calcined potash

fluidized-bed reactor from above and exposed to a countercurrent of CO_2-containing hot gas (Fig. 51) [207, 208]. Carbonization and calcination take place in the same reactor. Hard, spherical potassium carbonate prills are formed having a high packing density. The prills are discharged and sieved. The coarse grains are ground and returned to the reactor together with the very fine grains, where they act as crystallization seeds. The salable, dust-free, medium grains are cooled and packed. Because no mother liquor is formed, the quality of the potassium carbonate depends on that of the raw materials. Compared to the crystallization process the chlo-ride, soda, and sulfate contents are usually higher, but the investment and production costs are lower.

Amine Process. In the Mines de Potasse d'Alsace process, potassium chloride is reacted under pressure in autoclaves with carbon dioxide in precarbonated isopropylamine solution. Potassium hydrogencarbonate precipitates and is filtered off, carefully purified of amine by intensive washing, and dried. It can be converted to potassium carbonate by calcination. Free amine, containing carbon dioxide, is recovered from the mother liquor by distillation and recycled. The chloride, predominantly present in the mother liquor as amine chlorohydrate, is reacted with hydrated lime to give free amine and an aqueous solution of calcium chloride [209].

The use of triethylamine [210], hexamethylenimine [211], or piperidine [212] is also patented. All the processes have the disadvantage that calcium chloride liquor is obtained, which can be utilized today only to a small extent and therefore represents an environmental pollutant.

Nepheline Decomposition Process [213]. In the CIS, considerable amounts of potassium carbonate are formed as a byproduct in the nepheline decomposition process for aluminum hydroxide production. The mineral nepheline is decomposed with limestone by sintering at 1300 °C:

$$(Na, K)_2 \cdot Al_2O_3 \cdot 2\,SiO_2 + 4\,CaCO_3 \xrightarrow{1300\,°C} (Na, K)_2O \cdot Al_2O_3 + 2(2\,CaO \cdot SiO_2) + 4\,CO_2$$

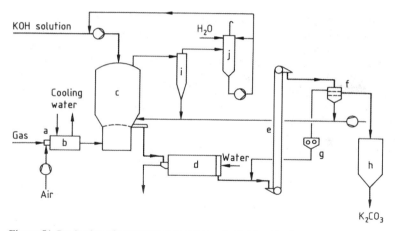

Figure 51. Production of potassium carbonate by the fluidized-bed process
a) Burner; b) Gas cooler; c) Fluidized-bed reactor; d) Cooler; e) Elevator; f) Screen; g) Mill; h) Silo; i) Cyclone; j) Exhaust gas scrubber

Alumina, portland cement, soda, and potash are obtained from the product in a complex process. The sinter product is leached with an Na_2CO_3–NaOH solution. After filtration, a filter cake is obtained that is processed to give portland cement and an aluminate solution containing silicic acid. After precipitation of the silicic acid as alkaline aluminum silicate the purified aluminate solution is reacted with carbon dioxide:

$$2\,(Na, K)AlO_2 + CO_2 + 3\,H_2O \longrightarrow$$
$$2\,Al(OH)_3 + (Na, K)_2CO_3$$

The aluminum oxide hydrate is filtered off, and the carbonate solution is concentrated by fractional crystallization in a three-stage process and separated into sodium carbonate and $K_2CO_3 \cdot 1.5\,H_2O$.

Feldspar ($KAlSi_3O_8$) and leucite ($KAlSi_2O_6$) can also be decomposed analogously and used for alumina, cement, and potassium carbonate manufacture [214].

The magnesia process (Engel–Precht process) is of limited interest:

$$3\,(MgCO_3 \cdot 3\,H_2O) + 2\,KCl + CO_2 \longrightarrow$$
$$2\,(MgCO_3 \cdot KHCO_3 \cdot 4\,H_2O) + MgCl_2$$

In hot water the double salt ($MgCO_3 \cdot KHCO_3 \cdot 4\,H_2O$) decomposes under pressure into magnesium carbonate and dissolved potassium carbonate.

Other processes. Le Blanc process:

$$K_2SO_4 + CaCO_3 + 2\,C \longrightarrow CaS + K_2CO_3 + 2\,CO_2$$

Formate process :

$$K_2SO_4 + Ca(OH)_2 + 2\,CO \longrightarrow 2\,HCOOK + CaSO_4$$
$$K_2SO_4 + Ca(OH)_2 + 2\,CO \longrightarrow 2\,HCOOK + CaSO_4$$
$$HCOOK + KOH + 1/2\,O_2 \longrightarrow K_2CO_3 + H_2O$$

"Piesteritz" process :

$$K_2SO_4 + 2\,CaCN_2 + 2\,H_2O \longrightarrow$$
$$2\,KHCN_2 + CaSO_4 + Ca(OH)_2$$
$$2\,KHCN_2 + 5\,H_2O \longrightarrow K_2CO_3 + 4\,NH_3 + CO_2$$

These processes are uneconomical today because of high energy consumption and poor product quality, and are no longer used.

In the decomposition of chromium ores with potassium hydroxide solution, a chromate-containing potassium carbonate is obtained as by-product. The production of potassium permanganate yields considerable amounts of potassium carbonate solution [215].

Organic raw materials, such as sunflower stalks, molasses, and suint, are used to a small extent for potash manufacture. They are ashed, leached with water, and processed to potash by fractional crystallization and calcination [216].

Ion-Exchange Process [217, 218]. An acidic ion exchanger loaded with ammonium ions is charged with KCl solution, K^+ being absorbed and an ammonium chloride solution running off.

The ion exchanger is then eluted with an excess of ammonium carbonate solution (regeneration of the exchanger). The eluate, a K_2CO_3 – $(NH_4)_2CO_3$ solution, is separated by thermal cleavage to give ammonia and carbon dioxide. The ammonium chloride is reacted with magnesium hydroxide to give magnesium chloride and ammonia, which is recycled.

12.3. Quality Specifications and Analysis

Depending on the intended use, potassium carbonate is offered in varying commercial forms and degrees of purity: as granules, as powder, and as potassium carbonate hydrate ($K_2CO_3 \cdot 1.5 H_2O$). The material manufactured from mercury-process potassium hydroxide solution is of high purity, particularly with respect to chloride content. In the amine process, the chloride content is higher and the sodium carbonate content lower, while nepheline decomposition gives high sodium carbonate contents and relatively high sulfate contents (Table 11).

Analysis. The *total alkalinity* includes $K_2CO_3 + KOH + Na_2CO_3$; it is determined with 0.5 N H_2SO_4 by potentiometric titration or with a methyl orange indicator (change to brown-red). Sodium is determined by flame photometry. The *chloride content* is determined by turbidity measurement after addition of $AgNO_3$. The *sulfate content* is determined by ion chromatography or gravimetrically after precipitation as barium sulfate. The *metal content* is determined by atomic absorption spectroscopy or photometrically by complex formation (Fe^{2+} as sulfosalicylate, Si^{4+} as the molybdato complex, Cu^{2+} as pyrrolidinodithiocarbamate, and Ni^{2+} as the diacetylglyoxime complex). Test methods for *photographic-grade* potassium carbonate, anhydrous are described in ISO 3623-1976 (E).

12.4. Storage and Transportation

Potassium carbonate is stored in bunkers; the ventilation air must be dry because of the hygroscopicity of the product.

Silo vehicles and bulk containers are used for dispatch to bulk customers. Smaller amounts are packed in polyethylene valve sacks of 25 – 50 kg. The material is not hazardous; for pharmaceutical use it is classified as GRAS (generally recognized as safe) by FDA [219].

12.5. Economic Aspects and Uses

The glass industry is the most important consumer of K_2CO_3. Large amounts are also required for potassium silicate manufacture.

Potassium carbonate is used for many organic syntheses. Numerous inorganic and organic potassium salts are manufactured from potassium carbonate (potassium phosphate, bromide, iodide, dichromate, cyanide, and ferrocyanide); in addition it is a starting material for drying, neutralization, and condensation agents. As a regenerable absorbent for carbon dioxide, hydrogen sulfide, and sulfur oxides, it is attaining importance in environmental protection. Potassium carbonate is used as a fertilizer for acidic soil.

Other users are the electrical industry, the dye industry, the printing trade, the textile industry, the leather goods industry, and the ceramic industry. Soft soap manufacture has lost its earlier importance as a customer. Potash solutions are used as fire retardants and as cooling brines (freezing point $-36\,°C$ at 576 g/L=40.5 wt % K_2CO_3).

The food industry uses potassium carbonate as a leavening agent in baked goods, as a debiterizing agent for cocoa beans, and as an additive for drying raisins. Potash in DAB quality is frequently used in the pharmaceutical industry as a raw material and auxiliary.

The most important producer countries for potassium carbonate are the CIS, France, Germany, the United States, and Japan. Other producers are Israel, Spain, India, South Korea, Belgium, Italy, former Yugoslavia, and China.

13. Potassium Hydrogencarbonate
[195–197]

13.1. Properties and Production

Potassium hydrogencarbonate [298-14-6], $KHCO_3$, M_r 100.12, ϱ 2.17 g/cm^3, is a white, crystalline powder that is sparingly soluble in

Table 11. Analyses of calcined potassium carbonate of varying origin (data in %, remainder H_2O)

	From potassium hydroxide solution		Diaphragm process	Amine process	Nepheline decomposition
	Mercury process				
	Crystallization	Fluidized bed			
K_2CO_3	98.0 – 99.8	98.5 – 99.5	97 – 99	99	97.5 – 98.5
Na_2CO_3	0.1 – 0.5	0.1 – 0.5	0.5 – 1.0	0.01	0.1 – 1.0
Cl	0.001 – 0.002	0.004 – 0.013	0.2 – 1.0	0.19	0.01 – 0.03
SO_4	0.003 – 0.005	0.005 – 0.013	0.005 – 0.010		0.25 – 0.60
Si + Ca	0.005	0.006	0.010	0.009	
Fe	0.0003 – 0.0005	0.0002 – 0.0006	< 0.0005	< 0.0024	0.0007 – 0.0021

water and insoluble in alcohol. When heated above 120 °C, it decomposes into potassium carbonate, water, and carbon dioxide.

It is manufactured industrially by passing carbon dioxide into concentrated potassium carbonate solutions or exposing these to a countercurrent of purified, cold flue gas in trickle towers (overcarbonization). Because of its low water solubility (22.4 g of $KHCO_3$ in 100 mL of H_2O at 20 °C) it precipitates in crystalline form, is separated by centrifugation, and dried at ca. 110 °C. In some potassium carbonate production processes, potassium hydrogencarbonate is obtained as a precursor (see 12.2).

The total alkalinity of the industrial material, calculated as $KHCO_3$, is at least 98 – 100 % [$KHCO_3$ 98 % (min.), Na^+ 0.1 % (max.), Cl^- 0.01 % (max.), SO_4^{2-} 0.02 % (max.); for analysis see potassium carbonate, p. 98].

13.2. Uses

Potassium hydrogencarbonate is used in the manufacture of fire extinguisher powders, in the food industry as a leavening agent, and in the chemical and pharmaceutical industry for the manufacture of high-purity potassium carbonate and other pure potassium salts. Producing countries are the United States, Germany, and France.

14. References

General References
1. *Fertilizer Manual,* International Fertilizer Development Center, Muscle Shoals, Ala., 1979, pp. 225 – 247.
2. *Kirk-Othmer,* 3rd ed., **18,** 920 – 950.
3. *Winnacker-Küchler,* 4th ed., **2,** 268 – 333.
4. The British Sulphur Corporation, *World Survey of Potash Resources,* London 1985.
5. V. A. Zandon in N. L. Weiss (ed.): *SME Mineral Processing Handbook,* Society of Mining Engineers, New York 1985, section 22.
6. R. M. McKercher (ed.): *"Potash Technology," 1st International Potash Technology Conference,* Saskatchewan, Canada, Oct. 3 – 5, 1983, Pergamon Press, Toronto 1983.
7. W. H. Eatock: "Potash Refining in Saskatchewan," *Min. Eng. (Littleton, Colo.)* **34,** (1982) no. 9, 1350 – 1353.
8. M. P. Kurtanjek: "Mining and Beneficiating Potash, Recent Developments and Trends outside North America," *Phosphorus Potassium* **128** (1983) Nov./Dec., 26 – 31.
9. *Gmelin,* **System no. 22.**
10. F. Serowy: *Verarbeitungsmethoden der Kalirohsalze,* W. Knapp, Halle/Saale 1952.
11. H. Schubert: *Aufbereitung fester mineralischer Rohstoffe,* VEB Deutscher Verlag für Grundstoffindustrie, Leipzig, vol. I, 1968; vol. II, 1986; vol. III, 1984.
12. D. Fulda et al.: *Kali, das bunte, bittere Salz,* VEB Deutscher Verlag für Grundstoffindustrie, Leipzig 1990.
13. A. Heinz, R. v. d. Osten (eds.): *ABC Kali und Steinsalz,* VEB Deutscher Verlag für Grundstoffindustrie, Leipzig 1982.
14. I. Barin: *Thermochemical Data of Pure Substances,* VCH, Weinheim 1989.
15. M. Broul, J. Nývlt, O. Söhnel: *Solubility in Inorganic Two-component Systems,* Elsevier, Amsterdam 1981.
16. *Ullmann,* 4th ed., **13,** 447 – 496.

Specific References

17. R. Slotta: "Die Kali- und Steinsalzindustrie," *Technische Denkmäler in der Bundesrepublik Deutschland,* vol. 3, Deutsches Bergbaumuseum, Bochum 1980.
18. D. Hoffmann: *Elf Jahrzehnte Deutscher Kalisalzbergbau,* Verlag Glückauf, Essen 1972.
19. Kaliverein e.V.: *Die Kaliindustrie in der Bundesrepublik Deutschland,* 6th ed., 1988.
20. *Phosphorus Potassium* **166** (1990) March/April, 17–19.
21. R. Weissenberger: *Chronique des mines de potasse d'Alsace,* Ziegler, Bergholtz 1985.
22. *Phosphorus Potassium* **168** (1990) July/Aug., 12–13.
23. J. W. Turrentine: *Potash in North America,* Reinhold Publ., New York 1943.
24. *Phosphorus Potassium* **68** (1973) Nov./Dec., 39–43.
25. O. Braitsch: *Entstehung und Stoffbestand der Salzlagerstätten,* Springer Verlag, Berlin 1962.
26. *Ullmann,* 4th ed., **13,** 450.
27. M. A. Zharkov: *Paleozoic Salt Bearing Formations of the World,* Springer Verlag, Berlin 1984.
28. E. A. Vysotsky, V. Z. Kislik: "Epochs of Bischofite Deposition in Geological History," *Int. Geol. Rev.* **29** (1987) no. 2, 134–139.
29. R. Kühn, *Kali Steinsalz* **1** (1955) no. 9, 3–16.
30. R. Kühn, *Geol. Jahrb.* **90** (1972) 127–220.
31. H. Borchert: *Ozeane Salzlagerstätten,* Borntråger, Berlin 1959.
32. F. Lotze: "Steinsalz und Kalisalze," *Die wichtigsten Lagerstätten der "Nicht-Erze",* **vol. III, part 1,** Borntråger, Berlin 1938.
33. F. Lotze: *Steinsalz und Kalisalze,* 2nd ed., **part 1,** Borntråger, Berlin 1957.
34. F. Lotze: *Die Salzlagerstätten in Zeit und Raum,* Arbeitsgem. Forsch. Landes Nordrhein-Westfalen, no. 195, Westdeutscher Verlag, Köln-Opladen 1969.
35. G. Richter-Bernburg: "Salzlagerstätten," in Bentz-Martini (ed.): *Lehrbuch der Angewandten Geologie,* vol. 2, part 1, Enke Verlag, Stuttgart 1968, pp. 916–1061.
36. G. Richter-Bernburg, *Z. Dtsch. Geol. Ges.* **105** (1953) 593–645.
37. M. Brongersma, *Mar. Geol.* **11** (1972) 123–144.
38. C. Kippenberger et al.: *Untersuchungen über Angebot und Nachfrage mineralischer Rohstoffe, XX Kali,* BGR Hannover-DIW Berlin, Schweizerbart, Stuttgart 1986.
39. J. D'Ans, *Naturwissenschaften* **34** (1947) 295–301.
40. J. D'Ans, R. Kühn, *Kali Steinsalz* **3** (1960) 69–84.
41. H. Mayrhofer: "World Reserves of Mineable Potash Salts Based on Structural Analysis," *Proceedings of the 6th International Symposium on Salt,* **vol. 1,** The Salt Institute, Alexandria, USA, 1983.
42. P. A. Ziegler, *Geological Atlas of Western and Central Europe,* Elsevier, Amsterdam 1982.
43. E. Messer, *Kali Steinsalz* **5** (1970) 244–251.
44. Pittsburgh Plate Glass Co., US 3 058 729, 1962 (J. B. Dahms, B. P. Edmonds); CA 627 308, 1963 (J. B. Dahms); US 4 007 964, 1977 (E. L. Goldsmith); US 3 262 741, 1966 (B. P. Edmonds, J. B. Dahms); US 3 433 530, 1969 (J. B. Dahms, B. P. Edmonds); US 4 329 287, 1980 (E. L. Goldsmith).
45. *Phosphorus Potassium* **138** (1985) July/Aug., 32–33.
46. D. Jackson, *Eng. Min. J.* **174** (1973) no. 7, 59–68.
47. *Phosphorus Potassium* **168** (1990) July/Aug., 23–28.
48. G. Duchrow, I. Fitz, N. Grüschow, *Phosphorus Potassium* **167** (1990) May/June, 26–32.
49. H. Schubert: *Kali, das bunte, bittere Salz,* 1VEB Deutscher Verlag für Grundstoffindustrie, Leipzig 1990 72–84.
50. R. Kühn, *Kali Steinsalz* **5** (1970) 307–317.
51. R. Kühn, *Erzmetall* **8** (1955) Suppl. B 93–B 107, B 115.
52. H. Autenrieth, O. Braun, W. Otto: *Winnacker-Küchler,* 4th ed., **2,** 281–283.
53. V. A. Zandon in N. L. Weiss (ed.): *SME Mineral Processing Handbook,* Society of Mining Engineers, New York 1985, 22-4–22-5.
54. J. Götte, *Kali Steinsalz* **10** (1990) 261–264.
55. J. H. van't Hoff: *Untersuchungen über die Bildungsverhältnisse der ozeanischen Salzablagerungen,* Akad. Verlagsges., Leipzig 1912.
56. H. D'Ans: *Die Lösungsgleichgewichte der Systeme der Salze ozeanischer Salzablagerungen,* Verlagsges. für Ackerbau, Berlin 1933.
57. J. D'Ans, *Z. Elektrochem.* **56** (1952) 497–505.
58. *Gmelin,* System no. 22.
59. A. B. Sdanowsky, E. I. Lyakhowskaya, R. E. Schleymowitch: *Handbuch der Löslichkeit der Salzsysteme,* Gaskhimisdat, Leningrad, vol. I, 1953; vol. II, 1954; vol. III, 1961; vol. IV, 1963 (Russ.).

60. H. Autenrieth, *Kali Steinsalz* **1** (1953) no. 2, 3–17.
61. H. Autenrieth, *Kali Steinsalz* **1** (1954) no. 7, 3–22.
62. H. Autenrieth, *Kali Steinsalz* **1** (1955) no. 11, 18–32.
63. H. Autenrieth, *Kali Steinsalz* **2** (1958) no. 6, 181–200.
64. H. Autenrieth, *Kali Steinsalz* **5** (1969) no. 5, 158–165.
65. H. Autenrieth, *Rev. chim. minér.* **7** (1970) 217–229.
66. H. Autenrieth, *Kali Steinsalz* **5** (1970) no. 9, 289–306.
67. G. Peuschel, *Kali Steinsalz* **9** (1986) no. 9, 296–303.
68. W. P. Wilson, A. G. McKee: *Proceedings of the 4th Symposium on Salt,* **vol. I,** The Northern Ohio Geological Society, Cleveland, Ohio, 1974, 517–525.
69. W. P. Wilson: *Proceedings of the 3rd Symposium on Salt,* **vol. II,** The Northern Ohio Geological Society, Cleveland, Ohio, 1969, 20–29.
70. J. H. Wolf: R. M. McKercher (ed.): *"Potash Technology,"* 1st International Potash Technology Conference, Saskatchewan, Canada, Oct. 3–5, 1983, Pergamon Press, Toronto1983, 711–716.
71. H. Domning, *Kali Steinsalz* **7** (1977) no. 4, 155–160.
72. W. H. Coghill, J. R. De Vaney, J. B. Clemmer, S. R. B. Cooke, Report of Investigations of the US-Bureau of Mines, Report no. 3271, 1935.
73. A. S. Kusin, *Kalii* **6** (1937) 17–27.
74. R. A. Pierce, L. D. Anderson, *Eng. Min. J.* **143** (1942) 38–41.
75. Kreller, *Kali, Verw. Salze Erdöl* **33** (1939) 35–37, 46–47, 53–57.
76. E. Rüsberg, *Chem.-Ing.-Tech.* **27** (1955) 1–4.
77. O. Karsten, in W. Gründer: *Erzaufbereitungsanlagen in Westdeutschland,* Springer Verlag, Berlin 1955, 343–345.
78. V. A. Arsentiev, J. Leja, *CIM Bull.* **3** (1977) 154–158.
79. Du Pont, US 2 088 325, 1937 (J. E. Kirby).
80. C. M. Aleksandrovic, *Freiberg. Forschungsh.* **A 544** (1975) 73–81.
81. H. Schubert, *Aufbereit. Tech.* **7** (1967) 365–368.
82. A. Singewald, *Chem.-Ing.-Tech.* **33** (1961) 376–393, 558–572, 676–688.
83. D. W. u. M. C. Fuerstenau, *Min. Eng. (Littleton, Colo.)* **8** (1956) 302–307.
84. A. F. Taggart, *Elements of Ore Dressing,* John Wiley & Sons, New York 1951.
85. A. M. Goudin, Testimony in Transcript of Evidence, Civil Action no. 1829, District of New Mexico, p. 1255.
86. R. Bachmann, *Erzmetall* **8** (1955) Suppl. B 109–B 116.
87. J. Rogers, J. H. Schulman: *Second International Congress of Surface Activity,* **vol. II,** Reprints, Butterworths, London 1957, 330–338.
88. A. Singewald, *Erzmetall* **12** (1959) 121–135.
89. O. J. Somojlov: *Struktur von wäßrigen Elektrolytlösungen,* B. G. Teubner, Leipzig 1961.
90. H. Schubert, *Aufbereit. Tech.* **6** (1966) 305–313.
91. F. Hagedorn, *Kali Steinsalz* **10** (1991) 315–328.
92. D. Uhlig, *Neue Bergbautech.* **5** (1975) 145–155.
93. H. Köhler et al., *Neue Bergbautech.* **16** (1986) 45–50.
94. N. F. Mescerjakov, Y. W. Rjabov, V. N. Kuznetzov, *Freiberg. Forschungsh.* **A 594** (1978) 33–54.
95. A. Bahr, K. Legner, H. Lüdke, F. W. Mehrhoff, *Aufbereit. Tech.* **1** (1957) 1–9.
96. N. F. Mescerjakov: *Flotacionnye maschiny,* Isdatel'stwo Nedra, Moskau 1972.
97. H. Schubert, *Neue Bergbautech.* **4** (1974) 223–228.
98. *Phosphorus Potassium* **145** (1986) 29–33.
99. Kali & Salz, DE 3 435 124, 1987 (F. Hagedorn, G. Peuschel, A. Singewald).
100. F. Hagedorn, *Kali Steinsalz* **9** (1986) 232–238.
101. H. Köhler, W. Kramer, *Neue Bergbautech.* **11** (1981) 362–366.
102. R. B. Tippin, *Chem. Eng. (N.Y.)* **184** (1977) no. 15, part 1, 73–75.
103. VEB Kali, DD 220 237, (L. Herrmann et al.).
104. S. Mildner, R. Ecke, DD 35 637, 1965.
105. I. M. Le Baron, W. C. Knopf, *Min. Eng. (Littleton, Colo.)* **10** (1958) 1081–1083.
106. H. Autenrieth, *Kali Steinsalz* **5** (1969) 171–177.
107. Kali-Forschungsanstalt, DE 1 056 551, 1957 (H. Autenrieth).
108. G. Fricke, *Kali Steinsalz* **9** (1986) 287–295.
109. D. Larmour in: R. M. McKercher (ed.): *"Potash Technology,"* 1st International Potash Technology Conference, Saskatchewan, Canada, Oct. 3–5, 1983, Pergamon Press, Toronto 1983597–602.

110. R. Bock, *Chem.-Ing.-Tech.* **53** (1981) 916–924.
111. Kali & Salz, DE 1 249 783, 1966 (A. Singewald, G. Fricke).
112. A. Singewald, U. Neitzel: R. M. McKercher (ed.): *"Potash Technology,"* 1st International Potash Technology Conference, Saskatchewan, Canada, Oct. 3–5, 1983 Pergamon Press, Toronto 1983, 589–595.
113. L. Ernst, *Kali Steinsalz* **9** (1986) 275–286.
114. L. Ernst, *Ber. Bunsenges. Phys. Chem.* **93** (1989) 857–863.
115. L. Ernst, *Ber. Bunsenges. Phys. Chem.* **94** (1990) 1435–1439.
116. Kali & Salz, DE 3 603 166, 1986 (G. Fricke, I. Giesler, R. Diekmann).
117. Kali & Salz, DE 3 603 165, 1986 (H. Balzer, H. Burghardt, F. Maikranz).
118. Kali & Salz, DE 3 603 167, 1986 (U. Neitzel, G. Fricke).
119. Kali & Salz, DE 1 077 611, 1959 (H. Autenrieth, G. Peuschel).
120. Kali & Salz, DE 1 142 802, 1961 (H. Autenrieth, G. Peuschel, G. Weichart).
121. Kali & Salz, DE 2 007 677, 1970 (A. Singewald, G. Fricke, D. Jung).
122. Kali & Salz, DE 2 052 993, 1970 (A. Singewald, G. Fricke, D. Jung).
123. Kali & Salz, DE 1 154 052, 1960 (H. Autenrieth, H. Dust).
124. Kali & Salz, DE 1 076 593, 1957 (H. Autenrieth).
125. Kali & Salz, DE 1 261 453, 1967 (A. Singewald, G. Fricke).
126. Kali & Salz, DE 1 667 814, 1968 (G. Fricke, A. Singewald).
127. Kali & Salz, DE 1 283 772, 1967 (H. Autenrieth, H. Wirries).
128. Kali & Salz, DE 1 792 120, 1968 (A. Singewald, G. Fricke).
129. Kali & Salz, DE 1 953 534, 1969 (A. Singewald, G. Fricke).
130. W. B. Dancy: *Kirk-Othmer,* 3rd ed., **18**, 931–933.
131. T. E. Burus, B. J. Clarke, W. B. Coome, A. H. Newcombe: R. M. McKercher (ed.): *"Potash Technology,"* 1st International Potash Technology Conference, Saskatchewan, Canada, Oct. 3–5, 1983, Pergamon Press, Toronto 1983, 541–546.
132. R. Diekmann: Lecture held at *Int. Potash Technol. Conf., 2nd,* Hamburg, May 26–29, 1991.
133. E. Weps, *Kali Steinsalz* **8** (1981) 177–183.
134. Kali & Salz, DE 3 434 190, 1984 (O. Pfoh, C. Radick, H. Thenert).
135. F. Hagedorn, *Kali Steinsalz* **7** (1977) 161–164.
136. T. Fleischer, *Kali Steinsalz* **9** (1986) 304–313.
137. H. J. Scharf: "Environmental Aspects of K-Fertilizers in Production, Handling and Application," *Development of K-Fertilizer Recommendations,* International Potash Institute, Worblaufen-Bern 1990, 395–402.
138. *Phosphorus Potassium* **148** (1987) March/April, 30–35.
139. M. D. Haug, K. W. Reid: Lecture held at *Int. Potash Technol. Conf., 2nd,* Hamburg, May 26–29, 1991.
140. J. E. Tallin, D. E. Pufahl: R. M. McKercher (ed.): *"Potash Technology,"* 1st International Potash Technology Conference, Saskatchewan, Canada, Oct. 3–5, 1983, Pergamon Press, Toronto 1983 755–760.
141. K. W. Reid, G. A. Maki: Lecture held at *Int. Potash Technol. Conf., 2nd.,* Hamburg, May 26–29, 1991.
142. H. E. Schroth, *Phosphorus Potassium* **67** (1973) Sept./Oct., 38.
143. A. Singewald, *Die Weser* **57** (1983) no. 516, 1–8.
144. N. Knöpfel: Lecture held at *Int. Potash Technol. Conf., 2nd.,* Hamburg, May 26–29, 1991.
145. H. Stahl, *Aufbereit. Tech.* **20** (1980) 525–533.
146. W. B. Pietsch: R. M. McKercher (ed.): *"Potash Technology,"* 1st International Potash Technology Conference, Saskatchewan, Canada, Oct. 3–5, 1983, Pergamon Press, Toronto 1983, 661–669.
147. L. Medemblik: R. M. McKercher (ed.): *"Potash Technology,"* 1st International Potash Technology Conference, Saskatchewan, Canada, Oct. 3–5, 1983, Pergamon Press, Toronto1983, 653–659.
148. A. S. Middleton, D. A. Cormode, J. E. Scotten: R. M. McKercher (ed.): *"Potash Technology,"* 1st International Potash Technology Conference, Saskatchewan, Canada, Oct. 3–5, 1983, Pergamon Press, Toronto 1983, 647–651.
149. *Phosphorus Potassium* **173** (1991) May/June 28–36.
150. K. Kahle, G. Leib: Lecture held at *Int. Potash Technol. Conf., 2nd.,* Hamburg, May 26–29, 1991.
151. L. I. Skvirski, A. A. Chityakov, Z. L. Kozel: Lecture held at *Int. Potash Technol. Conf., 2nd.,* Hamburg, May 26–29, 1991.

152. H. Rieschel, K. Zech, *Phosphorus Potassium* **115** (1981) Sept./Oct., 33–39.
153. H. Rug, K. Kahle, *Phosphorus Potassium* **170** (1990) Nov./Dec., 23–27.
154. International Fertilizer Industry Association (IFA): *Potash Statistics 1989,* Paris.
155. Prognose-Arbeitsgruppe Weltbank/FAO/Unido, 1991, unpublished.
156. C. Childers, *Phosphorus Potassium* **169** (1990) Sept./Oct., 16–20.
157. O. Walterspiel, *Kali Steinsalz* **10** (1989) 168–174.
158. *Phosphorus Potassium* **165** (1990) Jan./Febr., 18–19.
159. *Gmelin,* System no. 22, Suppl. vol., pp. 1280–1338.
160. J. Näther, H. H. Emons, *Bergakademie* **21** (1969) 310–313.
161. *Phosphorus Potassium* **156** (1988) July/Aug., 27–34.
162. *Phosphorus Potassium* **122** (1982) Nov./Dec., 36–39.
163. N. P. Finkelstein, S. H. Garnett, L. Kogan: R. M. McKercher (ed.): *"Potash Technology," 1st International Potash Technology Conference, Saskatchewan, Canada, Oct. 3–5, 1983,* Pergamon Press, Toronto 1983,, 571–576.
164. U. Neitzel, *Kali Steinsalz* **9** (1986) 257–261.
165. H. Autenrieth, O. Braun, W. Otto: *Winnacker-Küchler,* 4th ed., **2,** 320–322.
166. H. Scherzberg, G. Döring: Lecture held at *Int. Potash Technol. Conf., 2nd.,* Hamburg, May 26–29, 1991.
167. Kali & Salz, DE 3 418 147, 1984 (E. Menche, H. G. Diehl, H. Eberle).
168. W. B. Dancy: *Kirk-Othmer,* 3rd ed., **18,** 945.
169. D. K. Storer: R. M. McKercher (ed.): *"Potash Technology," 1st International Potash Technology Conference, Saskatchewan, Canada, Oct. 3–5, 1983,* Pergamon Press, Toronto, 1983,577–582.
170. R. Phinney, EP 0 199 104, 1986.
171. Kali & Salz, DE 3 607 641, 1986 (S. Vajna, G. Peuschel).
172. Société d'Études Chimiques pour L'Industrie et l'Agriculture (SECPIA), DE 956 304, 1954 (J. Lafont).
173. J. A. Fernandez Lozano, A. Wint, *Chem. Eng. J. (Lausanne)* **23** (1982) 53–61.
174. *Phosphorus Potassium* **167** (1990) May/June, 11.
175. Superfos A/S, Vedbaek, DK, DE 3 331 416, 1983 (K. C. B. Knudsen); US 4 504 458, 1983 (K. C. B. Knudsen).
176. *Chem. Eng. (N.Y.)* **81** (1974) 98–99.
177. Kali & Salz, DE 2 810 640, 1978 (N. Knöpfel, F. Wartenpfuhl, A. Hollstein).
178. A. Hollstein, *Kali Steinsalz* **7** (1979) 498–500.
179. *Phosphorus Potassium* **141** (1986) Jan./Feb., 17–21.
180. *Phosphorus Potassium* **151** (1987) Sept./Oct., 16–21.
181. G. Kemmler, *Kali Steinsalz* **9** (1985) 167–169.
182. The British Sulphur Corporation, *World Survey of Potash Resources,* London 1985 pp. 62–64.
183. *Phosphorus Potassium* **24** (1966) 40–44.
184. A. M. Amarin, K. Manasrah, *Proc. IFA-NFC Joint Middle East-South Asia Fertilizer Conference,* Lahore, Pakistan, Dec. 3–6, 1988.
185. *Kirk-Othmer,* 2nd ed., Suppl. vol., 438–467.
186. The British Sulphur Corporation, *World Survey of Potash Resources,* London 1985., pp. 38–39.
187. U. Neitzel, *Kali Steinsalz* **5** (1971) 327–334.
188. P. Behrens, Industrial Processing of Great Salt Lake Brines, Utah Geological and Mineral Survey Bulletin 116, 1980, 223–228.
189. *Phosphorus Potassium* **132** (1984) July/Aug., 6.
190. The British Sulphur Corporation, *World Survey of Potash Resources,* London 1985, pp. 42–43.
191. The British Sulphur Corporation, *World Survey of Potash Resources,* London 1985, p. 70.
192. *Phosphorus Potassium* **173** (1991) May/June, 26–27.
193. W. Horwitz (ed.): *Official Methods of Analysis of the AOAC,* 11th ed., Association of Official Analytical Chemists (AOAC), Washington D.C. 1970.
194. Verband Deutscher Landwirtschaftlicher Untersuchungs- und Forschungsanstalten (LUFA): "Die Untersuchung von Düngemitteln," *Methodenbuch,* **vol. II,** Verlag J. Neumann-Neudamm, Melsungen 1972, method 4.1.

General References
195. *Kirk-Othmer,* 3rd ed., **18,** 936–939.
196. J. Ford: "Caustic potash," *Encycl. Chem. Process. Des.* **7** (1978) 22–34.
197. *Ullmann,* 4th ed., **13,** 489–496.
198. Hüls, Handbook KOH-, K2CO3-, KHCO3-Products, Marl 1992.

Specific References
199. N. Takeuchi, *Soda to Enso* **39** (1988) no. 461, 277–290.
200. Bertrams, Concentration Plants for NaOH-, KOH-, Na2S- and CaCl2-liquors, Muttenz, Switzerland, 1979.
201. Sulzer-Escher-Wyss, US 4 927 494, 1990 (R. Winkler et al.).
202. Oxy-Occidental Chem. Corp., Caustic Potash Handbook, Irving 1987.
203. Oil Paint Drugs, Chemical Marketing Reporter, 28th May, 1990.
204. Messo Chemietechnik, Brochure, Mass Crystallization, Duisburg, Germany, 1990.
205. Mannesmann, DE 3 816 061, 1989 (R. Schmitz).
206. VEB Kombinat Kali, DD 255 328 A, 1986 (K. Will, G. Elberling).
207. Bertrams, Fluid Bed Processes, Muttenz, Switzerland, 1979.
208. Diamond Shamrock Corp., Company brochure, Cleveland, Ohio, 1969.
209. *Inf. Chim.* **99** (1971) Aug./Sept., 125.
210. Kali-Chemie, DT 1 220 401, 1962 (P. Schmid).
211. J. N. Shokin et al., *Khim. Promst. (Moscow)* **9** (1978) 685.
212. FMC Corp., BE 616 193, 1962 (A. B. Gency, M. J. McCarthy).
213. D. M. Ginzburg, A. A. Tripolskii, *Tr. Gos. Nauchno-Issled. Proektn. Inst. Osnon. Khim.* **30** (1973) 26.
214. IMC Corp., US 3 073 443, 1960 (R. E. Snow).
215. A. Schmidt, *Angewandte Elektrochemie,* Verlag Chemie, Weinheim 1976, p. 183.
216. Lemar Developments, AU 563 487, 1987 (B. W. Levy).
217. *Chem. Age Int.,* 29th Sept., 1972.
218. Dynamit Nobel, DT 1 812 769, 1968 (D. Labriola et al.).
219. FDA, *Fed. Regist.* **48** (1983) no. 224, 52 440-3.

Potassium Nitrate → Nitrates and Nitrites

Urea

JOZEF H. MEESSEN, DSM Stamicarbon, Geleen, The Netherlands (Chaps. 1–7)

HARRO PETERSEN, BASF Aktiengesellschaft, Ludwigshafen, Federal Republic of Germany (Chap. 8)

1.	**Physical Properties**	468
2.	**Chemical Properties**	468
3.	**Production**	469
3.1.	**Principles**	469
3.1.1.	Chemical Equilibrium	469
3.1.2.	Physical Phase Equilibria	472
3.2.	**Challenges in Urea Production Process Design**	473
3.2.1.	Recycle of Nonconverted Ammonia and Carbon Dioxide	474
3.2.2.	Corrosion	476
3.2.3.	Side Reactions	477
3.3.	**Description of Processes**	478
3.3.1.	Conventional Processes	478
3.3.2.	Stripping Processes	478
3.3.2.1.	Stamicarbon CO_2-Stripping Process	478
3.3.2.2.	Snamprogetti Ammonia- and Self-Stripping Processes	481
3.3.2.3.	ACES Process	483
3.3.2.4.	Isobaric Double-Recycle Process	484
3.3.3.	Other Processes	484
3.4.	**Effluents and Effluent Reduction**	485
3.5.	**Product-Shaping Technology**	486
4.	**Forms Supplied, Storage, and Transportation**	487
5.	**Quality Specifications and Analysis**	488
6.	**Uses**	489
7.	**Economic Aspects**	489
8.	**Urea Derivatives**	490
8.1.	**Thermal Condensation Products of Urea**	490
8.2.	**Alkyl- and Arylureas**	490
8.2.1.	Transamidation of Urea with Amines	490
8.2.2.	Alkylation of Urea with Tertiary Alcohols	491
8.2.3.	Phosgenation of Amines	491
8.2.4.	Reaction of Amines with Cyanates (Salts)	492
8.2.5.	Reaction with Isocyanates	492
8.2.6.	Acylation of Ammonia or Amines with Carbamoyl Chlorides	492
8.2.7.	Aminolysis of Esters of Carbonic and Carbamic Acids	492
8.3.	**Reaction of Urea and Its Derivatives with Aldehydes**	493
8.3.1.	α-Hydroxyalkylureas	493
8.3.2.	α-Alkoxyalkylureas	494
8.3.3.	α,α'-Alkyleneureas	495
8.3.4.	Cyclic Urea – Aldehyde Condensation Products	496
9.	**References**	498

Abbreviations:

CRH, % critical relative humidity
ΔH_S, kJ/mol integral heat of solution
m, mol/kg urea molality, moles of urea per kilogram of water
$P^O_{(s)H^2}$, Pa water vapor pressure of a saturated urea solution
P_v, Pa vapor pressure

Urea [*57-13-6*], $CO(NH_2)_2$, M_r 60.056, plays an important role in many biological processes, among others in decomposition of proteins. The human body produces 20–30 g of urea per day.

In 1828, WÖHLER discovered [1] that urea can be produced from ammonia and cyanic acid in aqueous solution. Since then, research on the preparation of urea has continuously progressed. The starting point for the present industrial production of urea is the synthesis of BASAROFF [2], in which urea is obtained by dehydration of ammonium carbamate at increased temperature and pressure:

$$NH_2COONH_4 \rightleftarrows CO(NH_2)_2 + H_2O$$

In the beginning of this century, urea was produced on an industrial scale by hydration of cyanamide, which was obtained from calcium cyanamide:

$$CaCN_2 + H_2O + CO_2 \rightarrow CaCO_3 + CNNH_2$$

$$CNNH_2 + H_2O \rightarrow CO(NH_2)_2$$

After development of the NH₃ process (HABER and BOSCH, 1913, →Ammonia, Chap. 2. →Ammonia, Chap. 3. →Ammonia, Chap. 4.), the production of urea from NH_3 and CO_2, which are both formed in the NH_3 synthesis, developed rapidly:

$$2\,NH_3 + CO_2 \rightleftharpoons NH_2COONH_4$$

$$NH_2COONH_4 \rightleftharpoons CO(NH_2)_2 + H_2O$$

At present, urea is prepared on an industrial scale exclusively by reactions based on this reaction mechanism.

1. Physical Properties [3], [4]

Pure urea forms white, odorless, long, thin needles, but it can also appear in the form of rhomboid prisms. The crystal lattice is tetragonal–scalenohedral; the axis ratio $a:c = 1:0.833$. The urea crystal is anisotropic (noncubic) and thus shows birefringence. At 20 °C the refractive indices are 1.484 and 1.602. Urea has an *mp* of 132.6 °C; its heat of fusion is 13.61 kJ/mol.

Physical properties of the melt at 135 °C follow:

ϱ	1247 kg/m³
Molecular volume	48.16 m³/kmol
η	3.018 mPa·s
Kinematic viscosity	2.42×10^{-6} m²/s
Molar heat capacity, C_p	135.2 J mol⁻¹ K⁻¹
Specific heat capacity, c_p	2.25 kJ kg⁻¹ K⁻¹
Surface tension	66.3×10^{-3} N/m

In the temperature range 133–150 °C, density and dynamic viscosity of a urea melt can be calculated as follows:

$$r = 1638.5 - 0.96\,T$$

$$\ln \eta = 6700/T - 15.311$$

The density of the solid phase at 20 °C is 1335 kg/m³; the temperature dependence of the density is given by 0.208 kg m⁻³ K⁻¹.

At 240–400 K, the molar heat capacity of the solid phase is [5]

$$C_p = 38.43 + 4.98 \times 10^{-2}T + 7.05 \times 10^{-4}T^2$$
$$- 8.61 \times 10^{-7}T^3$$

The vapor pressure of the solid phase between 56 and 130 °C [6] can be calculated from

$$\ln P_v = 32.472 - 11755/T$$

Hygroscopicity. The water vapor pressure of a saturated solution of urea in water $P^O_{(s)H^2}$ in the temperature range 10–80 °C is given by the relation [7]

$$\ln P_{(s)H_2O} = 175.766 - 11552/T - 22.679 \ln T$$

By starting from the vapor pressure of pure water $P^O_{H^2}$, the critical relative humidity (CRH) then can be calculated as

$$\text{CRH} = (P_{(s)H_2O}/P_{H_2O})\,100$$

The CRH is a threshold value, above which urea starts absorbing moisture from ambient air. It shows the following dependence on temperature:

25 °C	76.5 %
30 °C	74.3 %
40 °C	69.2 %

At 25 °C, in the range of 0–20 mol of urea per kilogram of water, the integral heat of solution of urea crystals in water ΔH_s as a function of molality m is given by [8]:

$$\Delta H_s = 15.351 - 0.3523m + 2.327 \cdot 10^{-2}m^2$$
$$- 1.0106 \cdot 10^{-3}m^3 + 1.8853 \cdot 10^{-5}m^4$$

Urea forms a eutectic mixture with 67.5 wt % of water with a eutectic point at −11.5 °C.

The solubility of urea in a number of solvents, as a function of temperature is summarized in Table 1 [9], [10].

2. Chemical Properties

Upon *heating*, urea decomposes primarily to ammonia and isocyanic acid. As a result, the gas phase above a urea solution contains a considerable amount of HNCO, if the isomerization reaction in the liquid phase

$$CO(NH_2)_2 \rightleftharpoons NH_4NCO \rightleftharpoons NH_3 + HNCO$$

Table 1. Solubility of urea in various solvents (solubility in wt % of urea)

Solvent	Temperature, °C					
	0	20	40	60	80	100
Water	39.5	51.8	62.3	71.7	80.2	88.1
Ammonia	34.9	48.6	67.2	78.7	84.5	90.4
Methanol	13.0	18.0	26.1	38.6		
Ethanol	2.5	5.1	8.5	13.1		

has come to equilibrium [11]. In dilute aqueous solution, the HNCO formed hydrolyzes mainly to NH_3 and CO_2. In a more concentrated solution or in a urea melt, the isocyanic acid reacts further with urea, at relatively low temperature, to form biuret ($NH_2-CO-NH-CO-NH_2$), triuret ($NH_2-CO-NH-CO-NH-CO-NH_2$), and cyanuric acid $(HNCO)_3$ [12]. At higher temperature, guanidine $[CNH(NH_2)_2]$, ammelide $[C_3N_3(OH)_2NH_2]$, ammeline $[C_3N_3OH(NH_2)_2]$, and melamine $[C_3N_3(NH_2)_3]$ are also formed [13], [14].

Melamine can also be produced from urea by a catalytic reaction in the gas phase. To this end, urea is decomposed into NH_3 and HNCO at low pressure, and subsequently transformed catalytically to melamine.

Urea reacts with NO_x, both in the gas phase at 800–1150 °C and in the liquid phase at lower temperature, to form N_2, CO_2, and H_2O. This reaction is used industrially for the removal of NO_x from combustion gases [15], [16].

Reactions with Formaldehyde. Under *acid conditions*, urea reacts with formaldehyde to form among others, methyleneurea, as well as dimethylene-, trimethylene-, tetramethylene-, and polymethyleneureas. These products are used as slow-release fertilizer under the generic name ureaform [17] (→Fertilizers, Chap. 4.4.2.1.). The reaction scheme for the formation of methyleneurea is given below:

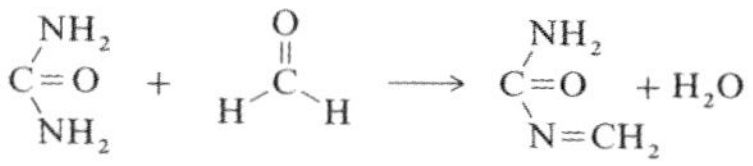

Urea + Formaldehyde → Methyleneurea + H_2O

Methyleneurea reacts with additional molecules of formaldehyde to yield dimethyleneurea and other homologous products.

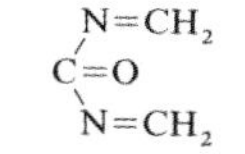

Dimethyleneurea

The reactions of urea with formaldehyde under *basic conditions* are used widely for the production of synthetic resins (→Amino Resins, Chap. 7.1.). As a first step, methylolurea instead of methyleneurea is formed:

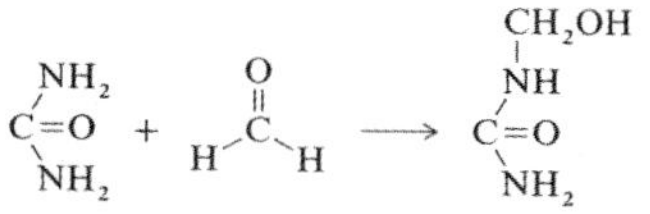

Monomethylolurea

This product subsequently reacts with formaldehyde to dimethylol urea, $CO(NHCH_2OH)_2$, and further polymerization products. Since urea is also the raw material for the production of melamine, from which melamine–formaldehyde resins are produced, it is the most important building block in the production of amino resins.

When urea is applied as fertilizer to soil, it *hydrolyzes* in the presence of the enzyme urease to NH_3 and CO_2, after which NH_3 is bacteriologically converted into nitrate and, as such, absorbed by crops [17].

3. Production

3.1. Principles

3.1.1. Chemical Equilibrium

In all commercial processes, urea is produced by reacting ammonia and carbon dioxide at elevated temperature and pressure according to the Basaroff reactions:

$2 NH_3 (l) + CO_2 (l) \rightleftharpoons NH_2COONH_4$
$\Delta H = -117 \text{ kJ/mol}$ (1)

$NH_2COONH_4 \rightleftharpoons NH_2CONH_2 + H_2O$
$\Delta H = +15.5 \text{ kJ/mol}$ (2)

A schematic of the overall process and the physical and chemical equilibria involved is shown in Figure 1. In the first reaction, carbon dioxide and ammonia are converted to ammonium carbamate; the reaction is fast and exothermic. In the second rection, which is slow and endothermic, ammonium carbamate dehydrates to produce urea and water. Since more heat is produced in the first reaction than consumed in the second, the overall reaction is exothermic.

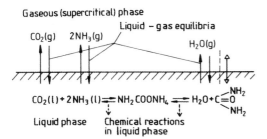

Figure 1. Physical and chemical equilibria in urea production

Processes differ mainly in the conditions (composition, temperature, and pressure) at which these reactions are carried out. Traditionally, the composition of the liquid phase in the reaction zone is expressed by two molar ratios: usually, the molar $NH_3 : CO_2$ and the molar $H_2O : CO_2$ ratios. Both reflect the composition of the so-called initial mixture [i.e., the hypothetical mixture consisting only of NH_3, CO_2, and H_2O if both Reactions (1) and (2) are shifted completely to the left].

First attempts to describe the chemical equilibrium of Reactions (1) and (2) were made by FREJACQUES [19]. Later descriptions of the chemical equilibria can be divided into regression analyses of measurements [20], [21] and thermodynamically consistent analyses of the equilibria [20], [22]. As far as the most important consequences of these equilibria on urea process design are concerned, the methods correspond closely to each other: The achievable conversion per pass, dictated by the chemical equilibrium as a function of temperature, goes through a maximum (Figs. 2 and 3). This effect is usually attributed to the fact that the ammonium carbamate concentration as a function of temperature goes through a maximum. This maximum in the ammonium carbamate concentration can be explained, at least qualitatively, by the respective heat effects of Reactions (1) and (2). However, this mechanism cannot explain the observed conversion maximum fully and quantitatively; other contributing mechanisms have been suggested [23].

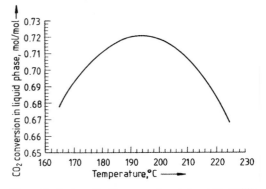

Figure 2. Carbon dioxide conversion at chemical equilibrium as a function of temperature
$NH_3 : CO_2$ ratio = 3.5 mol/mol (initial mixture); $H_2O : CO_2$ ratio = 0.25 mol/mol (initial mixture)

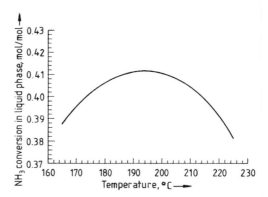

Figure 3. Ammonia conversion at chemical equilibrium as a function of temperature
$NH_3 : CO_2$ ratio = 3.5 mol/mol (initial mixture); $H_2O : CO_2$ ratio = 0.25 mol/mol (initial mixture)

The influence of the composition of the initial mixture on the chemical equilibrium can be explained qualitatively by Reactions (1) and (2) and the law of mass action:

1) Increasing the $NH_3 : CO_2$ ratio (increasing the NH_3 concentration) increases CO_2 conversion, but reduces NH_3 conversion (Figs. 4 and 5).
2) Increasing the amount of water in the initial mixture (increasing the $H_2O : CO_2$ ratio) results in a decrease in both CO_2 and NH_3 conversion (Figs. 6 and 7).

In these cases, too, a full quantitative description cannot be derived simply from the law of mass action and Reactions (1) and (2). Other, not yet fully understood reaction mechanisms probably contribute to the chemical equilibria to a minor extent.

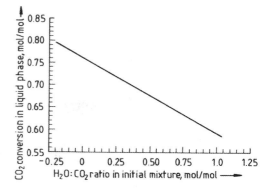

Figure 6. Carbon dioxide conversion at chemical equilibrium as a function of $H_2O : CO_2$ ratio
$T = 190\,°C$; $NH_3 : CO_2$ ratio = 3.5 mol/mol (initial mixture)

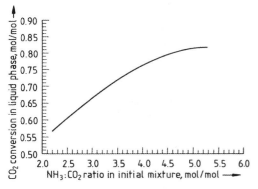

Figure 4. Carbon dioxide conversion at chemical equilibrium as a function of $NH_3 : CO_2$ ratio
$T = 190\,°C$; $H_2O : CO_2$ ratio = 0.25 mol/mol (initial mixture)

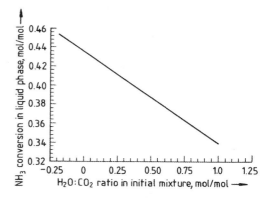

Figure 7. Ammonia conversion at chemical equilibrium as a function of $H_2O : CO_2$ ratio
$T = 190\,°C$; $NH_3 : CO_2$ ratio = 3.5 mol/mol (initial mixture)

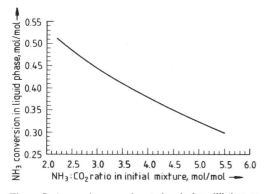

Figure 5. Ammonia conversion at chemical equilibrium as a function of $NH_3 : CO_2$ ratio
$T = 190\,°C$; $H_2O : CO_2$ ratio = 0.25 mol/mol (initial mixture)

In Figures 2, 4, and 6, the conversion at chemical equilibrium is expressed as CO_2 conversion, that is, the amount of CO_2 in the initial mixture converted into urea (plus biuret), if no changes occur in overall NH_3, CO_2, and H_2O concentrations in the liquid phase. This way of representing the chemical equilibrium is consistent with the presentation usually found in the traditional urea literature. However, it is based on the arbitrary choice of CO_2 as the key component. Historically, this may be justified by the fact that in early urea processes, CO_2 conversion was more important than NH_3 conversion. For the present generation of stripping processes, however, giving a higher weight to CO_2 conversion is not justified. Comparing, e.g., Figs. 4 and 5, shows that an arbitrary choice of one of the two feedstock components as yardstick to evaluate optimum reaction conversion can easily lead to faulty conclusions.

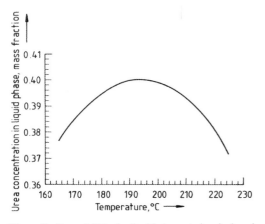

Figure 8. Urea yield in the liquid phase at chemical equilibrium as a function of temperature
$NH_3 : CO_2$ ratio = 3.5 mol/mol (initial mixture); $H_2O : CO_2$ ratio = 0.25 mol/mol (initial mixture)

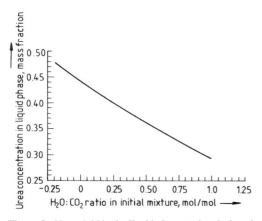

Figure 9. Urea yield in the liquid phase at chemical equilibrium as a function of $H_2O : CO_2$ ratio
$T = 190\,°C$; $NH_3 : CO_2$ ratio = 3.5 mol/mol (initial mixture)

Ultimately, project economics (investment and consumptions) will dictate the choice of process parameters in the reaction section. Without going into such time- and place-dependent economic considerations, one can argue that the urea yield (i.e., the concentration of urea in the liquid phase) is a better tool for judging optimum process parameters than CO_2 or NH_3 conversion. Figure 8 illustrates that urea yield as a function of temperature also goes through a maximum; the location of this maximum is of course composition dependent. Figure 9 again shows the detrimental effect of excess water on urea yield; thus, one of the targets in designing a recycle system must be to minimize water recycle.

Figure 10 shows that the urea yield as a function of $NH_3 : CO_2$ ratio reaches a maximum somewhat above the stoichiometric ratio (2 : 1). This is one of the reasons that all commercial processes operate at $NH_3 : CO_2$ ratios above the stoichiometric ratio. Another important reason for this can be found from the physical phase equilibria in the $NH_3 - CO_2 - H_2O$ – urea system.

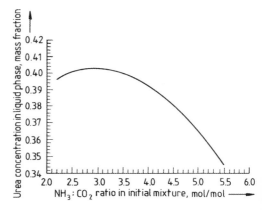

Figure 10. Urea yield in the liquid phase at chemical equilibrium as a function of $NH_3 : CO_2$ ratio
$T = 190\,°C$; $H_2O : CO_2$ ratio = 0.25 mol/mol (initial mixture)

3.1.2. Physical Phase Equilibria

In urea production, the phase behavior of the components under synthesis conditions is important. In all commercial processes, conditions are such that pressure and temperature are well above the critical conditions of the feedstocks ammonia and carbon dioxide; i.e., both components are in the supercritical state. The chemical interaction between NH_3 and CO_2 (mainly the formation of ammonium carbamate) results in a strongly azeotropic behavior of the "binary" system $NH_3 - CO_2$. An approach to the description of the phase equilibria if urea and water are added to the $NH_3 - CO_2$ system was given by KAASENBROOD and CHERMIN [24]. If a less volatile solvent C (water) is added to an azeotropic system A–B (NH_3–CO_2) at a pressure where both components A and B are supercritical, then the $T-X$ liquid and gas planes for the ternary system thus formed assume a special shape owing to the peculiar path described by the boiling points of the changing solutions (Fig. 11). Sections through the liquid plane for constant solvent content are analogous to the liquid line for the binary system. The liquid plane for the ternary systems appears as a ridge in the $T-X$ space. If the peak points of this ridge are linked up, the top ridge line is obtained. The points on this line do not have the same A : B ratio as the maximum for the binary azeotrope, because A and B are not soluble in solvent C to the same extent. The A : B ratio changes and the boiling point increases as the percentage of C increases.

Analogous to the description of Figure 11, the equilibria in the NH_3–CO_2–H_2O–urea system under urea synthesis conditions show a maximum in temperature at a given pressure as a function of NH_3 : CO_2 ratio. A full description of the phase equilibria in this system is even more complex than the aforementioned hypothetical A–B–C system, since the solid–liquid (S–L) and solid–gas (S–G) equilibria interfere with the liquid–gas (L–G) equilibria.

The strongly azeotropic behavior of the NH_3–CO_2 system, and the associated temperature maximum (or pressure minimum) in the ternary and quaternary systems with water and urea, are of practical importance in the realization of commercial urea processes. Carbon dioxide is less soluble than ammonia in water and urea melts. As a result, the pressure gradient at constant temperature is much steeper on the CO_2-rich side of the top ridge line. Moreover, this difference in solubility also causes the pressure minimum (or temperature maximum) to shift toward higher NH_3 : CO_2 ratios as the amount of solvent (water and urea) increases. Practically, this means that in order to achieve relatively low pressures at a given temperature, the NH_3 : CO_2 ratio in all commercial processes is chosen well above the stoichiometric ratio (2 : 1). In some processes, this ratio is chosen on the pressure minimum (on the top ridge line, i.e., at a ratio of ca. 3 : 1), whereas in other processes an even greater excess of ammonia is used.

3.2. Challenges in Urea Production Process Design

Like any process design, a urea plant design has to fulfil a number of criteria. Most important items are product quality, feedstocks and utilities consumptions, environmental aspects, safety, reliability of operation and a low initial investment. Since the urea process already has half a century of commercial scale history, it will be clear that compromises between the aforementioned, partly conflicting, criteria are well established. Also resulting from the age of urea process design is the observation that a process can only be successful if acceptable and competing solutions to all of these criteria can be combined into one process design. Apart from applying straightforward normal engineering approaches,

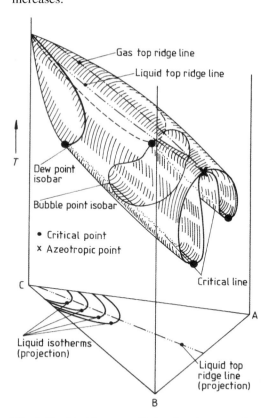

Figure 11. Liquid–gas equilibrium in a ternary system with binary azeotrope at constant pressure
The system A–B forms a binary azeotrope; C is a solvent for both A and B. The pressure is such that both A and B are supercritical, whereas the pressure is below the critical pressure of C.

the challenge of finding an optimum synergy between partly conflicting criteria, focuses in urea plant design essentially on a few peculiarities:

1) The thermodynamic limit on the conversion per pass through the urea reactor, combined with the azeotropic behavior of the NH_3-CO_2 system, necessitates a cunning recycle system design.
2) The intermediate product ammonium carbamate is extremely corrosive. A proper combination of process conditions, construction materials, and equipment design is therefore essential.
3) The occurrence of two side reactions – hydrolysis of urea and biuret formation – must be considered.

3.2.1. Recycle of Nonconverted Ammonia and Carbon Dioxide

The description of the chemical equilibria in Section 3.1.1 indicates that the conversion of the feedstocks NH_3 and CO_2 to urea is limited. An important differentiator between processes is the way these nonconverted materials are handled.

Once-Through Processes. In the very first processes, nonconverted NH_3 was neutralized with acids (e.g., nitric acid) to produce ammonium salts (such as ammonium nitrate) as coproducts of urea production. In this way, a relatively simple urea process scheme was realized. The main disadvantages of the once-through processes are the large quantity of ammonium salt formed as coproduct and the limited amount of overall carbon dioxide conversion that can be achieved. A peculiar aspect of this historic development is a partial "revival" of these combined urea–ammonium nitrate production facilities (UAN plants, see Section 3.3.3).

Conventional Recycle Processes. Once-through processes were soon replaced by total-recycle processes, where essentially all of the nonconverted ammonia and carbon dioxide were recycled to the urea reactor. In the first generation of total-recycle processes, several licensors developed schemes in which the recirculation of nonconverted NH_3 and CO_2 was performed in two stages. Figure 12 is a typical flow sheet of these, now called conventional, processes. The first recirculation stage was operated at medium pressure (18–25 bar); the second, at low pressure (2–5 bar). The first recirculation stage comprises at least a decomposition heater (d), in which carbamate decomposes into gaseous NH_3 and CO_2, while excess NH_3 evaporates simultaneously. The off-gas from this first decomposition step was subjected to rectification (e), from which relatively pure NH_3 (at the top) and a bottom product consisting of an aqueous ammonium carbamate solution were obtained. Both products are recycled separately to the urea reactor (c). In these processes, all nonconverted CO_2 was recycled as an aqueous solution, whereas the main portion of nonconverted NH_3 was recycled without an associated water recycle. Because of the detrimental effect of water on reaction conversion (see Figs. 6, 7, and 9), achieving a minimum CO_2 recycle (and thus maximum CO_2 conversion per reaction pass) was much more important than achieving a low NH_3 recycle. All conventional processes therefore typically operate at high $NH_3 : CO_2$ ratios (4–5 mol/mol) to maximize CO_2 conversion per pass. Although some of these conventional processes, partly equipped with ingenious heat-exchanging networks, have survived until now (see Section 3.3.1), their importance decreased rapidly as the so-called stripping processes were developed.

Stripping Processes. In the 1960s, the Stamicarbon CO_2-stripping process was developed, followed later by other processes (see Section 3.3.2). Characteristic of these processes is that the major part of the recycle of both nonconverted NH_3 and nonconverted CO_2 occurs via the gas phase, such that none of these recycles is associated with a large water recycle to the synthesis zone. Another characteristic difference between conventional and stripping processes in terms of the recycle scheme, can be found in the way heat is supplied to the recirculation zones. The energy balance of the conventional processes is shown in Figure 13. In this first-generation urea process, the heat supplied to the urea synthesis solution was used only once; therefore, this type of process can be referred to as an $N=1$ process. Such a process required about 1.8 t of steam per tonne of urea.

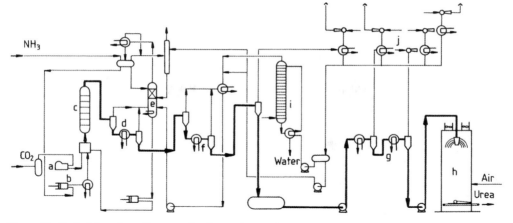

Figure 12. Typical flow sheet of a conventional urea plant
a) CO_2 compressor; b) High-pressure ammonia pump; c) Urea reactor; d) Medium-pressure decomposer; e) Ammonia–carbamate separation column; f) Low-pressure decomposer; g) Evaporator; h) Prilling; i) Desorber (wastewater stripper); j) Vacuum condensation section

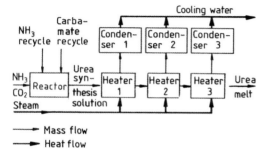

Figure 13. Conceptual diagram of the heat balance of a conventional urea process
Heat to each subsequent heater is supplied in the form of steam; the heat is used only once ($N = 1$).

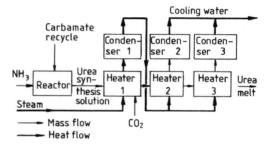

Figure 14. Conceptual diagram of the heat balance of a stripping plant
Heat supplied to the first heater (the stripper) is recovered in the first condenser (high-pressure carbamate condenser) and subsequently used again in the low-pressure heaters (decomposers and water evaporators); the heat is effectively used twice ($N = 2$).

The energy balance of a stripping plant is shown in Figure 14. As in conventional plants, heat must be supplied to the urea synthesis solution to decompose unconverted carbamate and to evaporate excess ammonia and water. However, a distinct difference in the heat balance with respect to the conventional process is that only the heat in the first heater (the high-pressure stripper) is imported. This heat is recovered in a high-pressure carbamate condenser (unconverted ammonia and carbon dioxide are condensed to form ammonium carbamate) and reused in the low-pressure heaters. The heat supplied is effectively used twice; thus, the term $N = 2$ process is used. The average energy consumption of the stripping process is 0.8 – 1.0 t of steam per tonne of urea.

In the 1980s, some processes were described that aim at a greater reduction of energy consumption by a further application of this multiple effect to $N = 3$ (Fig. 15) [25–29]. As can be seen from Figures 14 and 15, the steam requirement for process heating is reduced in these types of processes. However, whether the total energy consumption for the process is also reduced is doubtful, if the full capabilities of a $N = 2$ type of process are exploited and if the total energy supply scheme, including the energy supply to the carbon dioxide compressor drive, are taken into consideration [30]. Moreover, it seems that the emphasis in urea technology now is shifting from low energy consumption toward other factors, such as more durable construction materials, more modern process control systems, and simple process design [31].

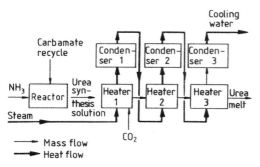

Figure 15. Further heat integration of a stripping plant in conceptual form
Heat supplied to the first heater (the stripper) is effectively used three times ($N = 3$).

3.2.2. Corrosion [32]

Urea synthesis solutions are very corrosive. Generally, ammonium carbamate is considered the aggressive component. This follows from the observation that carbamate-containing product streams are corrosive whereas pure urea solutions are not. The corrosiveness of the synthesis solution has forced urea manufacturers to set very strict demands on the quality and composition of construction materials. Awareness of the important factors in material selection, equipment manufacture and inspection, technological design and proper operations of the plant, together with periodic inspections and nondestructive testing are the key factors for safe operation for many years.

Role of Oxygen Content. Since the liquid phase in urea synthesis behaves as an electrolyte, the corrosion it causes is of an electrochemical nature. Stainless steel in a corrosive medium owes its corrosion resistance to the presence of a protective oxide layer on the metal. As long as this layer is intact, the metal corrodes at a very low rate. Passive corrosion rates of austenitic urea-grade stainless steels are generally between <0.01 and (max.) 0.10 mm/a. Upon removal of the oxide layer, activation and, consequently, corrosion set in unless the medium contains sufficient oxygen or oxidation agent to build a new layer. Active corrosion rates can reach values of 50 mm/a. Stainless steel exposed to carbamate-containing solutions involved in urea synthesis can be kept in a passivated (noncorroding) state by a given quantity of oxygen. If the oxygen content drops below this limit, corrosion starts after some time – its onset depending on process conditions and the quality of the passive layer. Hence, introduction of oxygen and maintenance of a sufficiently high oxygen content in the various process streams are prerequisites to preventing corrosion of the equipment. Although some alternatives have been mentioned [33], [34], the use of oxygen to influence the redox potential has become common practice in urea manufacture ever since it was initially suggested [35], [36].

From the point of view of corrosion prevention, the condensation of $NH_3 - CO_2 - H_2O$ gas mixtures to carbamate solutions deserves great attention. This is necessary because – notwithstanding the presence of oxygen in the gas phase – an oxygen-deficient corrosive condensate is initially formed on condensation. In this condensate the oxygen is absorbed only slowly. This accounts for the severe corrosion sometimes observed in cold spots inside gas lines. The trouble can be remedied by adequate isolation and tracing of the lines.

When condensation constitutes an essential process step – for example, in high-pressure and low-pressure carbamate condensers – special technological measures must be taken. These measures can involve ensuring that an oxygen-rich liquid phase is introduced into the condenser, while appropriate liquid–gas distribution devices ensure that no dry spots exist on condensing surfaces.

Not only condensing but also stagnant conditions are dangerous, especially where narrow crevices are present, into which hardly any oxygen can penetrate and oxygen depletion may occur.

Role of Temperature and Other Process Parameters in Corrosion. *Temperature* is the most important technological factor in the behavior of the steels employed in urea synthesis. An increase in temperature increases active corrosion, but more important, above a critical temperature it causes spontaneous activation of passive steel. The higher-alloyed austenitic stainless steels (e.g., containing 25 wt % chromium, 22 wt % nickel, and 2 wt % molybdenum) appear to be much less sensitive to this critical temperature than 316 L types of steel.

Sometimes, the $NH_3 : CO_2$ ratio in synthesis solutions is also claimed to have an influence on the corrosion rate of steels under urea synthesis conditions. Experiments have showed that under practical conditions this influence is not measurable because the steel retains passivity. Spontaneous activation did not occur. Only with electrochemical activation could 316 L types of steel be activated at intermediate $NH_3 : CO_2$ ratios. At low and high ratios, 316 L stainless steel could not be activated. The higher-alloyed steel type 25 Cr 22 Ni 2 Mo showed stable passivity, irrespective of the $NH_3 : CO_2$ ratio, even when activated electrochemically. Of course, these results depend on the specific temperature and oxygen content during the experiments.

Material Selection. Corrosion resistance is not the only factor determining the choice of construction materials. Other factors such as mechanical properties, workability, and weldability, as well as economic considerations such as price, availability, and delivery time, also deserve attention.

Stainless steels that have found wide use are the austenitic grades AISI 316 L and 317 L. Like all Cr-containing stainless steels, AISI 316 L and 317 L are not resistant to the action of sulfides. Hence it is imperative in plants using the 316 L and 317 L grades in combination with CO_2 derived from sulfur-containing gas, to purify this gas or the CO_2 thoroughly.

In stripping processes, the process conditions in the high-pressure stripper are most severe with respect to corrosion.

In the *Stamicarbon CO_2-stripping process*, a higher-alloyed, but still fully austenitic stainless steel (25 Cr 22 Ni 2 Mo) was chosen as construction material for the stripper tubes. This choice ensures better corrosion resistance than 316 L or 317 L types of material but still maintains the advantages of workability, weldability, reparability, and the cheaper price of stainless steel-type materials.

In the *Snamprogetti stripping processes*, titanium usually is chosen for this critical application, although mechanically bonded bimetallic 25 Cr 22 Ni 2Mo – zirconium tubes have also been suggested to improve corrosion resistance [37], [38].

In the *ACES process*, duplex alloys (ferritic–austenitic) are used as construction material for the stripper tubes.

3.2.3. Side Reactions [39]

Three side reactions are of special importance in the design of urea production processes:
Hydrolysis of urea

$$CO(NH_2)_2 + H_2O \rightarrow NH_2COONH_4 \rightarrow 2\,NH_3 + CO_2 \quad (3)$$

Biuret formation from urea:

$$2\,CO(NH_2)_2 \rightarrow NH_2CONHCONH_2 + NH_3 \quad (4)$$

Formation of isocyanic acid from urea:

$$CO(NH_2)_2 \rightarrow NH_4NCO \rightarrow NH_3 + HNCO \quad (5)$$

All three side reactions have in common the decomposition of urea; thus, the extent to which they occur must be minimized.

The *hydrolysis reaction* (3) is nothing but the reverse of urea formation. Whereas this reaction approaches equilibrium in the reactor, in all downstream sections of the plant the NH_3 and CO_2 concentrations in urea-containing solutions are such that Reaction (3) is shifted to the right. The extent to which the reaction occurs is determined by temperature (high temperatures favor hydrolysis) and reaction kinetics; in practice, this means that retention times of urea-containing solutions at high temperatures must be minimized.

The *biuret reaction* (4) also approaches equilibrium in the urea reactor [20], [22]. The high NH_3 concentration in the reactor shifts Reaction (4) to the left, such that only a small amount of biuret is formed in the reactor. In downstream sections of the plant, NH_3 is removed from the urea solutions, thereby creating a driving force for biuret formation. The extent to which biuret is formed is determined by reaction kinetics; therefore, the practical measures to minimize biuret formation are the same as described above for the hydrolysis reaction.

Reaction (5) shows that formation of isocyanic acid from urea is also favored by low NH_3 concentrations. This reaction is especially relevant in the evaporation section of the plant. Here, low pressures are applied, resulting in a transfer of NH_3 and HNCO into the gas phase and, consequently, low concentrations of these constituents in the liquid phase. Together with the relatively high temperatures in the evaporators, this shifts Reaction (5) to the right. The extent to which this reaction occurs is again determined by kinetics. The HNCO removed via the gas in the evaporators collects in the process condensate from the vacuum condensers, where low temperatures shift Reaction (5) to the left, again forming urea. As a result of this mechanism of chemical entrainment, attempts to minimize entrainment from evaporators with physical (liquid – gas) separation devices are destined to be unsuccessful.

3.3. Description of Processes

3.3.1. Conventional Processes

As explained in Section 3.2, conventional processes have generally been replaced by stripping processes. Only two conventional processes may still have some importance in the near future.

Urea Technologies Inc. (UTI) Heat Recycle Process (see Fig. 16). Ammonia (containing passivating air), recycled carbamate, and about 60 % of the feed CO_2 are charged to the top of an open-ended coil reactor (c) operating at 210 bar. Ammonium carbamate is formed within the coil, exits the coil at the bottom, and then flows up and around it–the exothermic heat of carbamate formation in the coil driving the endothermic dehydration of carbamate to urea outside the coil. The reactor is claimed to achieve a uniform temperature profile in this way. In the reactor, a relative high $NH_3 : CO_2$ ratio (4.2 : 1) is applied. The reactor effluent is depressurized and subcooled, and the flashed gases are released before the first decomposer (f). Gases leaving the first decomposer separator (g) are mixed with about 40 % of the feed carbon dioxide and partially condensed in the heat recovery section (i – k). The two combined gas streams are then further condensed and form the carbamate recycle flow. Urea as an 86 – 88 % solution is concentrated by evaporation (i) before granulation or prilling. The process is applied in eight small-scale and two medium-scale plants.

MTC (Mitsui Toatsu Corporation) Conventional Processes of Toyo Engineering Corporation. The conventional processes developed by Toyo Engineering Corporation (TEC) were successfully commercialized until the mid-1980s. The continuous evolution of these processes is reflected in their sequential nomenclature:

TR – A	Total-Recycle A Process
TR – B	Total-Recycle B Process
TR – C	Total-Recycle C Process
TR – CI	Total-Recycle C Improved Process
TR – D	Total-Recycle D Process

Partial-recycle versions of these processes were also realized. These TEC MTC conventional processes were applied in more then 70 plants. However, the licensor of these processes has announced a stripping process (the ACES process; see Section 3.3.2.3); this probably means the end of the TEC MTC conventional process line.

3.3.2. Stripping Processes

3.3.2.1. Stamicarbon CO_2-Stripping Process
(Figs. 17 and 18)

The synthesis stage of the Stamicarbon process consists of a urea reactor (c), a stripper for unconverted reactants (d), a high-pressure carbamate condenser (e), and a high-pressure reactor off-gas scrubber (f). To realize maximum urea yield per pass through the reactor at the stipulated optimum pressure of 140 bar, an $NH_3 : CO_2$ molar ratio of 3 : 1 is applied. The greater part of the unconverted carbamate is decomposed in the stripper, where ammonia and carbon dioxide are stripped off. This stripping action is effected by countercurrent contact between the urea solution and fresh carbon dioxide at synthesis pressure. Low ammonia and carbon dioxide concentrations in the stripped urea solution are obtained, such that the recycle from the low-pressure recirculation stage (h, j) is minimized. These low concentrations of both ammonia and carbon dioxide in the stripper effluent can be obtained at relatively low temperatures of the urea solution because carbon dioxide is only sparingly soluble under such conditions.

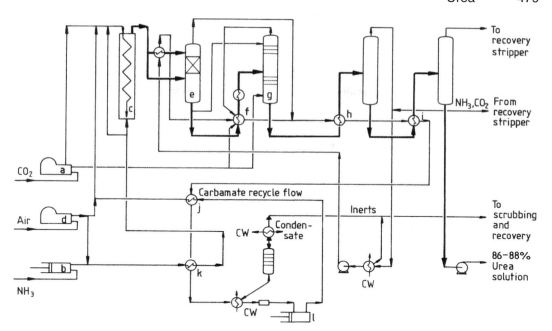

Figure 16. UTI heat recycle urea process
a) CO_2 compressor; b) High-pressure ammonia feed pump; c) Urea reactor with internal coil; d) Air compressor; e) Liquid distributor (used as first flash vessel); f) First decomposer; g) First separator; h) Second decomposer; i) Urea concentrator; j) Carbamate heater; k) Ammonia heater; l) High-pressure carbamate recycle pump
CW = Cooling water

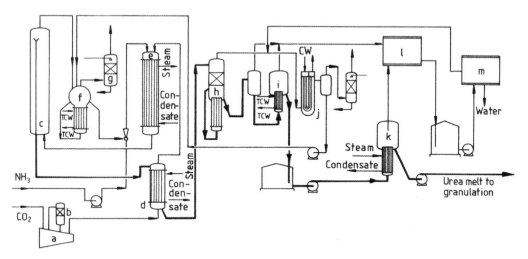

Figure 17. Stamicarbon CO_2-stripping urea process (The process suitable for combination with a granulation plant is shown here; combination with prilling is also possible.)
a) CO_2 compressor; b) Hydrogen removal reactor; c) Urea reactor; d) High-pressure stripper; e) High-pressure carbamate condenser; f) High-pressure scrubber; g) Low-pressure absorber; h) Low-pressure decomposer and rectifier; i) Pre-evaporator; j) Low-pressure carbamate condenser; k) Evaporator; l) Vacuum condensation section; m) Process condensate treatment
CW = Cooling water; TCW = Tempered cooling water

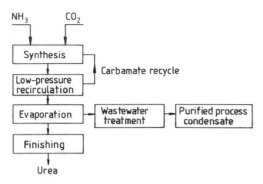

Figure 18. Functional block diagram of the Stamicarbon CO_2-stripping urea process

Condensation of ammonia and carbon dioxide gases, leaving the stripper, occurs in the high-pressure carbamate condenser (e) at synthesis pressure. As a result, the heat liberated from ammonium carbamate formation is at a high temperature. This heat is used for the production of 4.5-bar steam for use in the urea plant itself. The condensation in the high-pressure carbamate condenser is not effected completely. Remaining gases are condensed in the reactor and provide the heat required for the dehydration of carbamate, as well as for heating the mixture to its equilibrium temperature. In an improvement to this process, the condensation of off-gas from the stripper is carried out in a prereactor, where sufficient residence time for the liquid phase is provided. As a result of urea and water formation in the condensing zone, the condensation temperature is increased, thus enabling the production of steam at a higher pressure level [40].

The feed carbon dioxide, invariably originating from an associated ammonia plant, always contains hydrogen. To avoid the formation of explosive hydrogen – oxygen mixtures in the tail gas of the plant, hydrogen is catalytically removed from the carbon dioxide feed (b). Apart from the air required for this purpose, additional air is supplied to the fresh carbon dioxide input stream. This extra portion of oxygen is needed to maintain a corrosion-resistant layer on the stainless steel in the synthesis section. Before the inert gases, mainly oxygen and nitrogen, are purged from the synthesis section, they are washed with carbamate solution from the low-pressure recirculation stage in the high-pressure scrubber (f) to obtain a low ammonia concentration in the subsequently purged gas. Further washing of the off-gas is performed in a low-pressure absorber (g) to obtain a purge gas that is practically ammonia free. Only one low-pressure recirculation stage is required due to the low ammonia and carbon dioxide concentrations in the stripped urea solution. Because of the ideal ratio between ammonia and carbon dioxide in the recovered gases in this section, water dilution of the resultant ammonium carbamate is at a minimum despite the low pressure (about 4 bar). As a result of the efficiency of the stripper, the quantities of ammonium carbamate for recycle to the synthesis section are also minimized, and no separate ammonia recycle is required.

The urea solution coming from the recirculation stage contains about 75 wt % urea. This solution is concentrated in the evaporation section (k). If the process is combined with a prilling tower for final product shaping, the final moisture content of urea from the evaporation section is ca. 0.25 wt %. If the process is combined with a granulation unit, the final moisture content may vary from 1 to 5 wt %, depending on granulation requirements. Higher moisture contents can be realized in a single-stage evaporator, whereas low moisture contents are economically achieved in a two-stage evaporation section.

When urea with an extremely low biuret content is required (at a maximum of 0.3 wt %), pure urea crystals are produced in a crystallization section. These crystals are separated from the mother liquor by a combination of sieve bends and centrifuges and are melted prior to final shaping in a prilling tower or granulation unit.

The process condensate emanating from water evaporation from the evaporation or crystallization sections contains ammonia and urea. Before this process condensate is purged, urea is hydrolyzed into ammonia and carbon dioxide (l), which are stripped off with steam and returned to urea synthesis via the recirculation section. This process condensate treatment section can produce water with high purity, thus transforming this "wastewater" treatment into the production unit of a valuable process condensate, suitable for, e.g., cooling tower or boiler feedwater makeup. Since the introduction of the Stamicarbon CO_2-stripping process, some 125 units have been built according to this process all over the world.

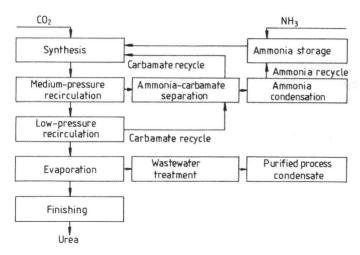

Figure 19. Functional block diagram of the Snamprogetti self-stripping process

3.3.2.2. Snamprogetti Ammonia- and Self-Stripping Processes [41–47]

In the first generation of NH_3- and self-stripping processes, ammonia was used as stripping agent. Because of the extreme solubility of ammonia in the urea-containing synthesis fluid, the stripper effluent contained rather large amounts of dissolved ammonia, causing an ammonia overload in downstream sections of the plant. Later versions of the process abandoned the idea of using ammonia as stripping agent; stripping was achieved only by supply of heat ("thermal" or "self"-stripping). Even without using ammonia as a stripping agent, the $NH_3 : CO_2$ ratio in the stripper effluent is relatively high, so the recirculation section of the plant requires an ammonia–carbamate separation section, as in conventional processes (see Fig. 19).

The process uses a vertical layout in the synthesis section. Recycle within the synthesis section, from the stripper (h) via the high-pressure carbamate condenser (f), through the carbamate separator (e) back to the reactor (b), is maintained by using an ammonia-driven liquid–liquid ejector (c) [43], [45] (see Fig. 20). In the reactor, which is operated at 150 bar, an $NH_3 : CO_2$ molar feed ratio of ca. 3.5 is applied. The stripper is of the falling film type [46]. Since stripping is achieved thermally, relatively high temperatures (200–210 °C) are required to obtain a reasonable stripping efficiency. Because of this high temperature, stainless steel is not suitable as construction material for the stripper from a corrosion point of view; titanium and bimetallic zirconium – stainless steel tubes have been used [37], [38].

Off-gas from the stripper is condensed in a kettle-type boiler (f) [44]. At the tube side of this condenser the off-gas is absorbed in recycled liquid carbamate from the medium-pressure recovery section. The heat of absorption is removed through the tubes, which are cooled by the production of low-pressure steam at the shell side. The steam produced is used effectively in the back end of the process.

In the medium-pressure decomposition and recirculation section, typically operated at 18 bar, the urea solution from the high-pressure stripper is subjected to the decomposition of carbamate and evaporation of ammonia (i). The off-gas from this medium-pressure decomposer is rectified. Liquid ammonia reflux is applied to the top of this rectifier (j); in this way a top product consisting of pure gaseous ammonia, and a bottom product of liquid ammonium carbamate are obtained. The pure ammonia off-gas is condensed (k) and recycled to the urea synthesis section. To prevent solidification of ammonium carbamate in the rectifier, some water is added to the bottom section of the column to dilute the ammonium carbamate below its crystallization point. The liquid ammonium carbamate – water mixture obtained in this way is also recycled to

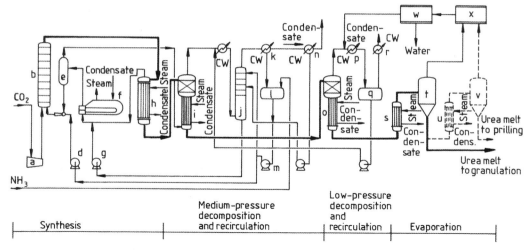

Figure 20. Schematic of the Snamprogetti self-stripping process
a) CO_2 compressor; b) Urea reactor; c) Ejector; d) High-pressure ammonia pump; e) Carbamate separator; f) High-pressure carbamate condenser; g) High-pressure carbamate pump; h) High-pressure stripper; i) Medium-pressure decomposer and rectifier; j) Ammonia–carbamate separation column; k) Ammonia condenser; l) Ammonia receiver; m) Low-pressure ammonia pump; n) Ammonia scrubber; o) Low-pressure decomposer and rectifier; p) Low-pressure carbamate condenser; q) Low-pressure carbamate receiver; r) Low-pressure off-gas scrubber; s) First evaporation heater; t) First evaporation separator; u) Second evaporation heater; v) Second evaporation separator; w) Wastewater treatment; x) Vacuum condensation section
CW = Cooling water

the synthesis section. The purge gas of the ammonia condensers is treated in a scrubber (n) prior to being purged to the atmosphere.

The urea solution from the medium-pressure decomposer is subjected to a second lowpressure decomposition step (o). Here, further decomposition of ammonium carbamate is achieved, so that a substantially carbamate-free aqueous urea solution is obtained. Off-gas from this low-pressure decomposer is condensed (p) and recycled as an aqueous ammonium carbamate solution to the synthesis section via the medium-pressure recovery section.

Concentrating the urea–water mixture obtained from the low-pressure decomposer is performed in a single or double evaporator (s–v), depending on the requirements of the finishing section. Typically, if prilling is chosen as the final shaping procedure, a two-stage evaporator is required, whereas in the case of a fluidized-bed granulator a single evaporation step is sufficient to achieve the required final moisture content of the urea melt. In some versions of the process, heat exchange is applied between the off-gas from the medium-pressure decomposer and the aqueous urea solution to the evaporation section. In this way, the consumption of low-pressure steam by the process is reduced.

The process condensate obtained from the evaporation section is subjected to a desorption–hydrolysis operation to recover the urea and ammonia contained in the process condensate.

Up to now, about 70 plants have been designed according to the Snamprogetti ammonia- and self-stripping processes.

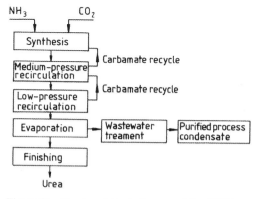

Figure 21. Functional block diagram of the ACES urea process

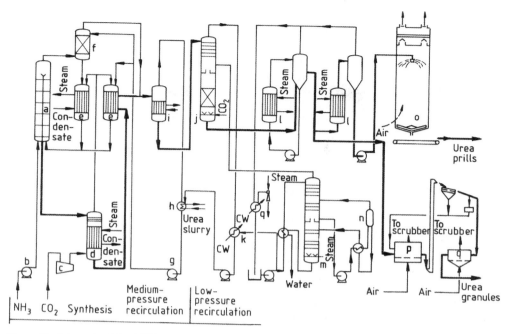

Figure 22. Schematic of the ACES process
a) Urea reactor; b) High-pressure ammonia pump; c) CO_2 compressor; d) Stripper; e) High-pressure carbamate condensers; f) High-pressure scrubber; g) High-pressure carbamate pump; h) Medium-pressure absorber; i) Medium-pressure decomposer; j) Low-pressure decomposer; k) Low-pressure absorber; l) Evaporators; m) Process condensate stripper; n) Hydrolyzer; o) Prilling tower; p) Granulation section; q) Surface condensers
CW = Cooling water

3.3.2.3. ACES Process [29], [48], [49]

The ACES (i.e., Advanced Process for Cost and Energy Saving) process has been developed by Toyo Engineering Corporation. Its synthesis section consists of the reactor (a), stripper (d), two parallel carbamate condensers (e), and a scrubber (f) – all operated at 175 bar (see Figs. 21 and 22).

The reactor is operated at 190 °C and an $NH_3 : CO_2$ molar feed ratio of 4 : 1. Liquid ammonia is fed directly to the reactor, whereas gaseous carbon dioxide after compression is introduced into the bottom of the stripper as a stripping aid. The synthesis mixture from the reactor, consisting of urea, unconverted ammonium carbamate, excess ammonia, and water, is fed to the top of the stripper. The stripper has two functions. Its upper part is equipped with trays where excess ammonia is partly separated from the stripper feed by direct countercurrent contact of the feed solution with the gas coming from the lower part of the stripper. This prestripping in the top is said to be required to achieve effective CO_2 stripping in the lower part. In the lower part of the stripper (a falling film heater), ammonium carbamate is decomposed and the resulting CO_2 and NH_3 as well as the excess NH_3 are evaporated by CO_2 stripping and steam heating. The overhead gaseous mixture from the top of the stripper is introduced into the carbamate condensers (e). Here, two units in parallel are installed, where the gaseous mixture is condensed and absorbed by the carbamate solution coming from the medium-pressure recovery stage. Heat liberated in the high-pressure carbamate condensers is used to generate low-pressure steam in one of the condensers and to heat the urea solution from the stripper after the pressure is reduced to about 19 bar in the shellside of the second carbamate condenser. The gas and liquid from the carbamate condensers are recycled to the reactor by gravity flow. The urea solution from the stripper, with a typical NH_3 content of 12 wt %, is purified further in the subsequent medium- and low-pressure decomposers (i, j), operating at 19 and 3 bar, respectively. Ammonia and carbon dioxide separated from the urea solution here are recovered through stepwise absorption in the low- and medium-pressure absorbers (h, k). Condensation heat in the medium-pressure absorber is transferred directly to the aqueous urea solution feed in the final concentration section. In this final concentration section (l), the purified urea solution is concentrated further either by a two-stage evaporation up to 99.7 % for urea prill production or by a single evaporation up to 98.5 % for urea granule production. Water vapor formed in the final concentrating section is condensed in surface condensers (q) to form process condensate. Part of this condensate is used as an absorbent in the recovery sections, whereas the remainder is purified in the process condensate treatment section by hydrolysis and steam stripping, before being discharged from the urea plant.

The highly concentrated urea solution is finally processed either through the prilling tower (o) or via the urea granulator (p). Instead of concentration via evaporation, the ACES process can also be combined with a crystallization section to produce urea with low biuret content.

Until now, the ACES process has been used in seven urea plants.

3.3.2.4. Isobaric Double-Recycle Process
[26], [28], [50]

The isobaric double-recycle (IDR) stripping process, developed by Montedison, is characterized by recycle of most of the unreacted ammonia and ammonium carbamate in two decomposers in series, both operating at the synthesis pressure. A high molar $NH_3 : CO_2$ ratio (4 : 1 to 5 : 1) in the reactor is applied. As a result of this choice of ratio, the reactor effluent contains a relatively high amount of nonconverted ammonia. In the first, steam-heated, high-pressure decomposer, this large quantity of free ammonia is mainly removed from the urea solution. Most of the residual ammonia, as well as some ammonium carbamate, is removed in the second high-pressure decomposer where steam heating and CO_2 stripping are applied. The high-pressure synthesis section is followed by two lower-pressure decomposition stages of traditional design, where heat exchange between the condensing off-gas of the medium-pressure decomposition stage and the aqueous urea solution to the final concentration section improves the overall energy consumption of the process. Probably because of the complexity of this process, it has not achieved great popularity so far. The IDR process or parts of the process are used in four revamps of older conventional plants.

3.3.3. Other Processes

Urea – Ammonium Nitrate (UAN) Production. Mixtures of urea (mp 133 °C) and ammonium nitrate (mp 169 °C) with water have a eutectic point at −26.5 °C [51]. As a result solutions with high nitrogen content can be made with solidification temperatures below ambient temperature. These mixtures, called UAN solutions, are used as liquid nitrogen fertilizers.

UAN solutions can be made by mixing the appropriate amounts of solid urea and solid ammonium nitrate with water or, alternatively, in a production facility specially designed to produce UAN solutions. In this latter category the Stamicarbon CO_2-stripping technology is especially suitable. In a partial-recycle version of this process, unconverted ammonia emanating from the stripped urea solution and from the reactor off-gas is neutralized with nitric acid. The ammonium nitrate solution thus formed and the urea solution from the synthesis section are mixed to yield a product solution with the desired nitrogen content (32 – 35 wt %) directly. Such a plant designated for the production of UAN solutions is cheaper than the separate production of urea and ammonium nitrate in investment and in operating costs, because evaporation, final product shaping for both urea and ammonium nitrate, and wastewater treatment sections are not required.

Integrated Ammonia – Urea Production. Both feedstocks required for urea production, ammonia and carbon dioxide, are usually obtained from an ammonia plant. Since an ammonia plant is a net heat (steam) producer and a urea plant is a net heat (steam) consumer, it is normal practice to integrate the steam systems of both plants. Since both processes usually contain a process condensate treatment section where volatile components are removed by steam stripping, the advantages of combining these sections have been explored [52–54]. Several attempts for further integration of mass streams of both processes have been published [55–59]. Despite the claimed reduction in both capital and raw material cost, these highly integrated process schemes have not gained acceptance mainly because of their increased complexity.

3.4. Effluents and Effluent Reduction

Gaseous Effluents. There are potentially two sources for air pollution from a urea plant: (1) gaseous ammonia emission from continuous process vents, and (2) urea dust and ammonia emissions from the finishing section (prilling or granulation).

Gaseous Emissions from Process Vents. Noncondensable gases enter the urea process as contaminants in the raw materials, as process air introduced for corrosion protection, and as air leaking into the vacuum sections of the process. At places where these noncondensable gases are vented, proper measures should be taken to minimize ammonia losses. The present state of the art allows reduction of these losses to <0.2 kg of ammonia per tonne of urea produced. This is realized by using conventional absorption techniques. Special attention is required to avoid explosive gas mixtures originating from combustibles (hydrogen, methane) present in the carbon dioxide and ammonia feedstocks, and the air introduced for anticorrosion purposes. Catalytic combustion of the hydrogen present in carbon dioxide, and dosing of nitrogen or excessive amounts of combustibles have been suggested to avoid the risks of formation of explosive gas mixtures [60].

Gaseous Emissions from the Finishing Section (Prilling/Granulation). In the prilling processes, urea dust is produced, mainly from evaporation and subsequent sublimation of urea, but also partly via a chemical mechanism: formation of ammonia and isocyanic acid in the urea melt, evaporation of these components, followed by sublimation of this ammonia–isocyanic acid mixture to form urea dust in the colder air. Urea dust formed in this way is typically very fine (0.5–2 µm). Removal of this fine urea dust from the prilling tower exhaust gas is a technical challenge. Because of the small particle size, dry cyclones cannot be used. Instead, wet impingement type devices have proved successful in removing a major part of the dust from the air.

Many new urea plants use granulation instead of prilling as the finishing technique. The main purpose is to improve the size and strength of the product, but less difficulty in controlling dust emissions is also an advantage. The dust produced in these granulation devices (fluidized-bed granulation or drum granulation) is typically much coarser than the fume-like dust produced in a prilling tower.

This results basically from the much shorter contact time of liquid urea with air. Because of the coarser dust and the smaller air quantities to be handled, wet scrubbers provide adequate and much simpler dust emission control for granulation units compared to the apparatus required for emission control in prilling.

Liquid Effluents. The process condensate produced from the evaporation or crystallization sections of the plant contains 3–8 wt % ammonia and 0.2–2 wt % urea. Two techniques are known to remove these pollutants:

1) Biological treatment [61–63]
2) Chemical hydrolysis and steam stripping to remove ammonia from the condensate

Biological treatment seems to have gained only slight acceptance. The method involving chemical hydrolysis and steam stripping recycles urea (in the form of ammonia) and ammonia to the synthesis section for the production of urea, whereas with biological treatment the urea and ammonia present in the feed to wastewater treatment are lost to urea production. Several chemical hydrolysis and steam-stripping systems are described below.

Stamicarbon System. In the Stamicarbon system [64] first the bulk of ammonia is removed by pre-desorption of the process condensate, followed by hydrolysis of the urea with steam at 170–230 °C to form ammonia and carbon dioxide via ammonium carbamate. The hydrolyzer is a vertical bubble-washer column, operated in countercurrent with respect to gas and liquid flow. Ammonia remaining in the process condensate is then removed by further steam stripping (desorption). Since both the pre-desorber and the desorber operate at low pressure (1–5 bar), low-pressure steam as produced in the urea synthesis section can be used as stripping agent. The combination of pre-desorption and countercurrent operation of the hydrolyzer ensures that the chemical equilibrium of the hydrolysis reaction does not limit the minimum achievable urea content in wastewater to concentrations <1 ppm. Also, the remaining NH_3 concentration is <1 ppm.

Snamprogetti System. The Snamprogetti system [65] also includes a system of pre-desorption, hydrolysis, and final desorption with steam stripping. In this process, however, the hydrolyzer is built as a horizontal column with cross-flow operation with respect to gas and liquid flow. Hydrolysis is carried out at somewhat higher temperature (230–236 °C) and pressure (33–37 bar) than in the Stamicarbon process. Also, this process claims minimum achievable NH_3 and urea concentrations <1 ppm in the liquid effluent.

Toyo Engineering Corporation also offers a system of pre-desorption and hydrolysis, followed by final desorption, as part of its ACES and conventional urea processes. Urea and ammonia concentrations <5 ppm in the effluent are claimed.

UTI System. In the UTI system [66], desorption and hydrolysis take place in a single column, in which the hydrolysis of urea back to ammonium carbamate is carried out concurrently with the stripping of ammonia from the liquid phase. Because of this concept, both the steam required for steam stripping and the steam required to bring the process condensate to a sufficiently high temperature for hydrolysis must be at medium pressure. A small amount of carbon dioxide is fed to the bottom of the column. The system is claimed to achieve residual NH_3 concentrations <10 ppm.

Other Systems. Some systems have been proposed [52–54] in which the wastewater treatment sections of the urea and ammonia plants are combined. Like the systems described above, they also remove urea by hydrolysis to ammonia and carbon dioxide, which are subsequently removed by transferring them into a steam-containing gas phase. The principal difference between these systems and the aforementioned methods is that the ammonia and carbon dioxide produced are recycled to the ammonia plant reforming system, rather than to the urea synthesis section.

3.5. Product-Shaping Technology

Prilling. For a long time, prilling has been used widely as the final shaping technology for urea. In prilling processes, the urea melt is distributed in the form of droplets in a prilling tower. This distribution is performed either by showerheads or by using a rotating prilling bucket equipped with holes. Urea droplets solidify as they fall down the tower, being cooled countercurrently with up-flowing air. The prilling process has several drawbacks:

1) The size of the product is limited to a maximum average diameter of about 2.1 mm. Larger-size product would require uneconomically high prilling towers; moreover, larger droplets tend to be unstable.
2) Very fine dust is formed in the prilling process (see Section 3.4). Removal of this dust is a technically difficult and expensive operation.
3) The crushing strength and shock resistance of prills are limited, making prilled product less suitable for bulk transport over long distances. This problem can be largely overcome with appropriate techniques to improve the physical properties, such as seeding [67] to improve shock resistance or addition of formaldehyde to improve crushing strength, and to suppress the caking tendency.

Granulation [29], [68–70]. The drawbacks of prilling have initiated the development of several granulation techniques. These techniques deviate from the prilling technique in that the urea melt is sprayed on granules, which gradually increase in size as the process continues. The heat of solidification is removed by cooling air or, for some granulation techniques, evaporation of water. Since the contact time between liquid urea and air in these processes is much lower than in prilling, the dust formed in granulation processes is much coarser; therefore, it can be removed much more easily from the cooling air.

Drum granulation systems have been developed, for example, by C & I Girdler,

Kaltenbach – Thuring, and Montedison. *Pan granulation processes* have also been developed, for example, by Norsk Hydro and the Tennessee Valley Authority (TVA). More recently *spouted-bed* and *fluidized-bed granulation techniques* were introduced. A spouted-bed technique was developed by Toyo Engineering Corporation; however, the fluidized-bed technology from Hydro Agri seems to be most successful at this time.

All granulation processes require the addition of formaldehyde or formaldehyde-containing components. Also common to all granulation techniques is that they yield products with larger diameters compared to prilling. However, their capabilities in this respect differ from each other to quite an extent.

Although the improvements brought about by granulation are beyond doubt, prilling techniques still have a place because of the lower investment and lower variable costs associated with prilling compared to granulation.

4. Forms Supplied, Storage, and Transportation

Forms Supplied. Urea may be supplied either in solid form or as a liquid. For *liquid compound fertilizers*, urea is a favorite ingredient. It is generally used in combination with ammonium nitrate as an aqueous solution to obtain liquids containing 32 – 35 wt % nitrogen. These solutions are designated as UAN-32 to UAN-35.

The *solid forms* are generally classified as granular or prilled products, because of the differences in handling properties. Prilled product is considered less suitable for bulk transportation because prills have lower crushing strength, a lower shock resistance, and a higher caking tendency than granules. Because of this, prilled products are usually marginally cheaper than granulated product. Granulated product usually also has a larger diameter (2.0 – 2.5 mm) than prills (1.5 – 2.0 mm), making granules more suitable for bulk blending to produce compound fertilizers.

Special Grades. The majority of urea is designated as "fertilizer grade"; however, some special forms have found limited application:

Technical Grade. Technical-grade urea should be without additions; color, ash-, and metal content are sometimes also specified. For urea used to produce urea – formaldehyde resins, its content of pH-controlling trace components is important. Because of this, technical-grade urea at present is mostly traded as a performance product, rather than being bound to narrow specification limits. The fitness of the product for use is judged by application-specific tests.

Low-Biuret Grade. A maximum biuret content up to 1.2 wt % is considered acceptable for nearly all fertilizer applications of urea. Only for the relative small market segment of foliar spray to citrus crops is a lower biuret content (max. 0.3 %) desirable.

Feed Grade. Some urea is also used directly as a feed component for cattle. Urea used for this purpose should be free of additions. Feed-grade urea is supplied in the form of microprills with a mean diameter of about 0.5 mm.

Slow-Release Grades. Studies show that only 50 – 60 % of fertilizer nitrogen applied to soil is usually recovered by crop plants. Several attempts have been made to increase this percentage by slowing the release of fertilizer to the ground via coating or additions [71].

Urea Supergranules. Granulated product with a very large diameter (up to 15 mm) has found limited application for deep placement in wetland rice [72] and forest fertilization.

Storage. The shift from bagged to bulk transport and storage of prilled and granulated urea has called for warehouse designs in which large quantities of urea can be stored in bulk. These warehouses should be designed in such a way that the product suffers little degradation. Degradation may result from: (1) segregation of fines; (2) disintegration; and (3) absorption, loss, or migration of water.

Segregation of fines can be avoided through uniform product spreading during pouring. *Disintegration* can be minimized by:

1) Providing the product pouring system with a pouring height adjuster
2) Design of "product-friendly" reclaiming systems, because reclaiming the product by means of payloaders and tractor shovels invariably leads to product disintegration

Caking and subsequent product disintegration at unloading are known to result from water absorption. What is not commonly known, however, is that excessive drying of the product during storage also leads to a higher caking tendency and that migration of water from warm product in the bulk of a pile to the cold surface leads to crust formation. Thus, attempts to decrease water absorption through refrigeration or air conditioning, dehumidification, or space heating may cause the air in the warehouse to become too dry or may result in too great a temperature difference between the product and the surrounding air. Instead, the warehouse (especially the roof) should be airtight and thoroughly insulated. The caking tendency of urea can be reduced by addition of small amounts of formaldehyde (up to 0.6 wt %) to the urea melt or by addition of surfactants to the solid product.

Transportation. Urea prills and granules are transported by bulk transport in trucks, ships, rail cars, etc. To withstand numerous and rapid loading and unloading operations, product for bulk transport should have a high initial physical stability. Great demands are made, especially on the shock resistance of the product, e.g., at seaport loading and unloading facilities. In addition, a number of "good housekeeping" rules should be adhered to:

1) Do not load or unload during rain
2) Make sure that the means of transport is clean and dry
3) Close the ship's hold when rain is imminent
4) Do not replace the air above the product or ventilate the holds
5) Cover the product (e.g., by polyethylene sheeting) during prolonged transport
6) Product should be spread rather then poured solely from one point to prevent dust coning due to segregation
7) Restrict the pouring height to avoid unnecessary disintegration

Liquid Fertilizer Transport. Liquid fertilizers are transported by tank cars, railway tanks, ships, and pipelines. Although liquid fertilizer is generally accepted as the most economic form to distribute, the solid form is still the most popular by far. Distribution of large quantities of liquid fertilizer requires a complex infrastructure and is limited at present to large farm units in developed countries. Transport and storage of UAN solutions in carbon steel lines and tanks require the addition of a corrosion inhibitor to the solution.

5. Quality Specifications and Analysis

Typical quality specifications for fertilizer-grade urea are summarized in Table 2. The capabilities of a modern urea plant are better than the typical trade data given in this table.

Table 2. Typical product specifications for fertilizer-grade urea

Specification	Prilled product	Granulated product
Nitrogen content, wt %	min. 46	min. 46
Biuret content, wt %	max. 1	max. 1
Water content, wt %	max. 0.3	max. 0.25
Crushing strength, bar	20 – 25	30 – 60
Shock resistance, wt %	min. 85	100
Product size		
1.0 – 2.4 mm, wt %	90 – 95	
1.6 – 4.0 mm, wt %		95
Bulk density (loose), kg/m^3	730	750 – 790

The total *nitrogen content* is usually determined by digesting urea with sulfuric acid to yield ammonium sulfate. The ammonia content is then determined by distillation and titration. Alternatively, the total N content may be determined by the Kjeldahl method or by using a method based on hydrolyzing urea with urease followed by titration of the ammonia formed.

The *water* content is usually determined with Karl Fischer reagent.

Biuret is determined by the formation of a violet-colored complex of biuret with copper(II) sulfate in an alkaline medium and subsequent measurement of the absorbance of the colored solution at 546 nm.

Crushing strength is defined as the force required per unit cross-sectional area of a granule to crush the granule or, if it is not crushed, the force at which it is deformed by 0.1 mm. A single granule is subjected to a force that is increased at a constant rate, the force at breakage (or at 0.1-mm deformation) being recorded.

The *shock resistance* of granules is defined as the weight percentage of a sample that is not crushed when subjected to a specified shock load. To determine shock resistance, a sample of prills or granules is shot against a metal plate by means of compressed air under normalized conditions. The amount of nondamaged product that remains after the test is determined.

The *granulometry* of the product is measured by conventional sieve techniques.

6. Uses

Urea is used for soil and leaf fertilization (more than 90 % of the total use); in the manufacture of urea – formaldehyde resins; in melamine production; as a nutrient for ruminants (cattle feed); and in miscellaneous applications.

Soil and Leaf Fertilization. Worldwide, urea has become the most important nitrogenous fertilizer. Urea has the highest nitrogen content of all solid nitrogenous fertilizers; therefore, its transportation costs per tonne of nitrogen nutrient are lowest. Urea is highly soluble in water and thus very suitable for use in fertilizer solutions (e.g., "foliar feed" fertilizers). Urea is also used as a raw material for the production of compound fertilizers. Compound fertilizers may be produced by mixing in urea melts or urea solutions before shaping the compound fertilizers or by mixing solid urea prills or granules with other fertilizers (bulk blending). In the latter case, the product sizes must match to prevent segregation of the products during further handling.

Urea – Formaldehyde Resins A significant proportion of urea production is used in the preparation of urea – formaldehyde resins. These synthetic resins are employed in the manufacture of adhesives, molding powders, varnishes, and foams. They are also used to impregnate paper, textiles, and leather.

Melamine Production. At present, nearly all melamine production is based on urea as a feedstock. Since ammonia is formed as a coproduct in melamine production from urea (see Chap. 1), integration of the urea and melamine production processes is beneficial.

Feed for Cattle and other Ruminants. Because of the activity of microorganisms in their cud, ruminants can metabolize certain nitrogen-containing compounds, such as urea, as protein substitutes. In the United States this capability is exploited on a large scale. In Western Europe, by contrast, not much urea is used in cattle feed.

Other Uses. On a smaller scale, urea is employed as a raw material or auxiliary in the pharmaceutical industry, the fermenting and brewing industries, and the petroleum industry. Urea can be used for the removal of NO_x from flue gases. Urea is also used as a solubilizing agent for proteins and starches, and as a deicing agent for airport runways.

7. Economic Aspects

The predicted growth in *demand* for urea until 1997 is slightly more than 3 % per year, bringing the total urea demand in 1997 to some 89×10^6 t/a. Most of the growth will occur in Asia, with China and India in the lead. A little more than 7 % of the worldwide demand for urea is from industry, in which Europe takes a leading role, ahead of North America and the industrialized countries of Asia.

The predicted growth in *capacity* up to 1997 will be around 1.5 % per year, resulting in a total installed capacity of some 100×10^6 t by 1997. Between 1994 and 1997, new plants will account for an increase in capacity of ca. 6.5×10^6 t, whereas closure of old plants will result in a reduction of the installed capacity by some 2×10^6 t. The world's capacity utilization can thus be calculated to be ca. 89 %.

Because of geopolitical changes during the early 1990s, different economic laws have become valid in the former Soviet Union and Eastern Europe, causing those countries to supply to the world market large amounts of prilled urea at prices far below the cost to many producers [e.g., sales prices, free on board (FOB) Black Sea $ 75 per tonne in 1993]. This situation is not expected to last long, although FOB prices at Black Sea ports will generally remain lower compared to other places. In the 1990s urea prices have shown large fluctuations. For instance, in 1994 – 1995 lows of $ 100 per tonne and highs of $ 250 per tonne have been reported. Predictions of future developments of urea prices are therefore highly speculative.

To cope with tough competition on the world market, producers have the tendency to build plants with very high single line capacities (2000 t/d or more), in which operating reliability is of extreme importance. Uninterrupted operating periods of more than one year are often achieved.

Furthermore, producers are increasingly shifting production facilities (both new plants and relocations) to places where natural gas is plentiful and cheap. Europe seems to have lost its competitive edge in export markets due to its expensive feedstocks.

The investment costs (in 10^6 \$) for a present state-of-the-art total-recycle urea plant are estimated to be:

1000-t/d plant (single line) 43
1500-t/d plant (single line) 52
2000-t/d plant (single line) 62

8. Urea Derivatives

8.1. Thermal Condensation Products of Urea

Thermolysis of urea gives biuret, triuret, and cyanuric acid, and in a special process, melamine is produced.

Biuret [*108-19-0*], imidodicarbonic diamide, $H_2NCONHCONH_2$, mp 193 °C, is produced by heating urea in inert hydrocarbons at 110–125 °C or in the melt at 127 °C [73], [74].

Triuret [*556-99-0*], diimidotricarbonic diamide, $H_2NCONHCONHCONH_2$, mp 231 °C, is produced by decomposing urea in a thin film at 120–125 °C [75]. It is also obtained by treating 2 mol of urea with 1 mol of phosgene in toluene at 70–80 °C [76], [77].

Cyanuric acid [*108-80-5*], 1,3,5-triazine-2,4,6($1H,3H,5H$)-trione (**1**), is formed on heating urea in the presence of zinc chloride, sulfuryl chloride, or chlorosulfuric acid [78].

$$6\,H_2NCONH_2 \xrightarrow{-3\,NH_3} 3\,H_2NCONHCONH_2 \xrightarrow{-3\,NH_3}$$

1

Melamine [*108-78-1*], 1,3,5-triazine-2,4,6-triamine (**2**) is produced industrially from urea [79].

$$6\,H_2NCONH_2 \xrightarrow{NH_3} \mathbf{2} + 3\,CO_2 + 6\,NH_3$$

2

8.2. Alkyl- and Arylureas

Most simple substituted alkylureas are crystalline products. Tetramethyl- and tetraethylurea and some cyclic ureas are liquids. Alkylated and arylated ureas are used in the production of plant protection agents, in pharmaceutical and dye chemistry, as plasticizers, and as stabilizers. Alkylureas and polyalkyleneureas are used as additives in the production of aminoplastics.

Various processes can be used for the production of substituted ureas, the most important of which are listed in Table 3.

Table 3. Production processes for substituted ureas

Starting materials	Product
Urea and amines	ureas substituted at one or both nitrogen atoms and cyclic ureas
Urea and tertiary alcohols	ureas substituted at one nitrogen atom
Phosgene and amines	ureas substituted at one nitrogen atom
Isocyanates and NH_3 or amines	ureas substituted at one or both nitrogen atoms
Carbamoyl chloride and NH_3 or amines	ureas substituted at one or both nitrogen atoms
Esters of carbonic or carbamic acids	ureas substituted at one or both nitrogen atoms
Urea and aldehydes or ketones	ureas substituted at one or both nitrogen atoms and cyclic ureas

8.2.1. Transamidation of Urea with Amines

The transamidation of urea with amines is one of the most important industrial production processes for substituted ureas. Monosubstituted ureas are produced by condensation of urea with a sufficiently basic amine in a 1 : 1 molar ratio in the melt at 130–150 °C [80]. Symmetrically disubstituted ureas can be produced from 2 mol of amine and 1 mol of urea at 140–170 °C [81]. Instead of the amines, amine salts can also be reacted with urea in the melt or by prolonged boiling in aqueous solution [82].

Monomethylurea [598-50-5], $CH_3NHCONH_2$, mp 102 °C, is produced industrially by passing monomethylamine into a urea melt [83]. Monomethylurea is used for the synthesis of theobromine.

Phenylurea [64-10-8], $C_6H_5HNCONH_2$, mp 147 °C, can be produced by heating an aqueous solution of aniline hydrochloride and urea to its boiling point [84].

Symmetric N,N'**-dimethylurea** [96-31-1], $CH_3HNCONHCH_3$, mp 105 °C, is used for the synthesis of caffeine by the Traube method and for the production of formaldehyde-free easy-care finishing agents for textiles [84].

Hexamethylenediurea [2188-09-2], $H_2NCONH(CH_2)_6NHCONH_2$, is obtained by heating a mixture of hexamethylenediamine with an excess of urea at 130–140 °C, with elimination of ammonia [85], [86].

N,N'**-Diphenylurea** [102-07-8], carbanilide, $C_6H_5HNCONHC_6H_5$, mp 238 °C, can be produced in high yields by heating 2 mol of aniline with 1 mol of urea in glacial acetic acid [86], [87].

Polyalkyleneureas can be obtained by treating di-, tri-, and tetraalkylenamines in concentrated aqueous solution with urea at ca. 110 °C. They are used, for example, to modify melamine–formaldehyde impregnating resins and binders for derived timber products. Examples of polyalkyleneureas include the following:

1,2-Ethylenediurea,
$H_2NCONH-CH_2CH_2-HNCONH_2$, mp 198 °C
1,3-Propylenediurea,
$H_2NCONH-CH_2CH_2CH_2-HNCONH_2$, mp 186 °C
Diethylenetriurea, $H_2NCONH-CH_2CH_2-N(CONH_2)-CH_2CH_2-HNCONH_2$, mp 215 °C
Dipropylenetriurea, $H_2NCONH-(CH_2)_3-N(CONH_2)-(CH_2)_3-HNCONH_2$

2-Hydroxypropylene-1,3-diurea,
$H_2NCONH-CH_2CH(OH)-CH_2-HNCONH_2$, mp 147 °C

Cycloalkyleneureas. *2-Imidazolidinone* [120-93-4], ethyleneurea (**3**), mp 131 °C [88], [89]; *2-oxohexahydropyrimidine* [65405-39-2], propyleneurea (**4**), mp 260–265 °C [90]; and *2-oxo-5-hydroxyhexahydropyrimidine*, 5-hydroxypropyleneurea (**5**), are produced industrially in the melt above 180 °C, preferably at 200–230 °C, by condensation of 1,2-ethylenediamine or 1,3-propylenediamine with urea and elimination of ammonia. These cycloalkyleneureas are used in the form of their N,N'-dihydroxymethyl compounds for easy-care finishes for cellulose-containing textiles [91].

8.2.2. Alkylation of Urea with Tertiary Alcohols

Urea can be alkylated with tertiary alcohols in the presence of sulfuric acid [92], [93].

tert-**Butylurea**, $(CH_3)_3CHNCONH_2$, mp 182 °C, is produced by treating urea with 2 mol of *tert*-butanol in the presence of ca. 2 mol of concentrated sulfuric acid at 20–25 °C with ice cooling [92–94].

8.2.3. Phosgenation of Amines

Symmetrically disubstituted ureas are produced in good yields by passing phosgene into solutions of amines in aromatic hydrocarbons [95]. In some cases, aqueous solutions or suspensions of amines can also be reacted with phosgenes [96]. The phosgenation of mixed aliphatic–aromatic amines is carried out industrially in the presence of sodium hydroxide at 40–60 °C [97]. For the production of substantive dyes, aminosulfone or aminocarboxylic acid groups are bonded by means of phosgenation [98]

$$2\,RNH_2 + COCl_2 \rightarrow RNH-CO-NHR + 2\,HCl$$

Tetramethylurea [632-22-4], $(CH_3)_2NCON(CH_3)_2$, bp 156.5 °C, is produced from dimethylamine and phosgene and is used as an aprotic solvent [99].

Asymmetric Diphenylurea, $(C_6H_5)_2NCONH_2$, mp 189 °C, is produced from diphenylamine, phosgene, and ammonia:

$$COCl_2 + (C_6H_5)_2NH + 3\,NH_3$$
$$\rightarrow (C_6H_5)_2NCONH_2 + 2\,NH_4Cl$$

Symmetric dimethyldiphenylurea, mp 121–127 °C, is obtained by treating phosgene with monomethylaniline and sodium hydroxide.

$$COCl_2 + 2\,HN{\overset{CH_3}{\underset{C_6H_5}{\diagup}}} + 2\,NaOH \rightarrow {\overset{CH_3}{\underset{C_6H_5}{\diagup}}}NCON{\overset{CH_3}{\underset{C_6H_5}{\diagdown}}} + 2\,NaCl + 2\,H_2O$$

Symmetric dialkyldiarylureas are used under the name Centralite as plasticizers and stabilizers for nitrocellulose and propellants [97].

8.2.4. Reaction of Amines with Cyanates (Salts)

The salts of aliphatic or aromatic amines react with potassium cyanate at 20–60 °C to give substituted ureas in high yields [100], [101].

$$R-NH_2 \cdot HX + KNCO + H_2O$$
$$\rightarrow R-HNCONH_2 + KX + OH^-$$

4-Ethoxyphenylurea [150-69-6] (dulcin) is produced from potassium cyanate and *p*-phenetidine hydrochloride in aqueous solution at room temperature [100]. This production method can be applied to aminosulfonic and aminocarboxylic acids, whereby the betaine-like salts formed by these acids react with potassium cyanate to give ureas that are substituted at only one NH_2 group [101]. Sulfonamides react with potassium cyanate to give potassium salts of the corresponding sulfonylureas, from which the sulfonylureas are obtained by acidification [102]:

$$RSO_2NH_2 + KNCO \rightarrow RSO_2N^-CONH_2 \rightarrow$$
$$K^+$$
$$RSO_2HNCONH_2$$

8.2.5. Reaction with Isocyanates

Symmetrically disubstituted ureas can also be produced from isocyanates by prolonged heating in aqueous solution [103], [104]:

$$RNCO \xrightarrow{H_2O} RHNCOOH \xrightarrow{RNCO} RHNCONHR + CO_2$$

When ammonia or primary or secondary amines are reacted with isocyanates, the corresponding substituted ureas are obtained in almost quantitative yield [104–106]. This process is particularly suitable for the production of unsymmetrically substituted ureas [105], [106].

$$RNCO + HN{\overset{R^1}{\underset{R^2}{\diagup}}} \longrightarrow RHNCON{\overset{R^1}{\underset{R^2}{\diagdown}}} \quad R^1, R^2 = H,\,alkyl,\,aryl$$

8.2.6. Acylation of Ammonia or Amines with Carbamoyl Chlorides

Ammonia and primary or secondary amines react with carbamoyl chlorides to give the corresponding urea derivatives in good yields [107].

Tetraphenylurea, $(C_6H_5)_2NCON(C_6H_5)_2$, [108] is obtained in quantitative yield by heating diphenylcarbamoyl chloride with diphenylamine.

8.2.7. Aminolysis of Esters of Carbonic and Carbamic Acids

Esters of carbonic and carbamic acids (carbonates and carbamates) react with amines at elevated temperatures to give symmetrically disubstituted ureas [109–112].

8.3. Reaction of Urea and Its Derivatives with Aldehydes

8.3.1. α-Hydroxyalkylureas

The industrial production of α-hydroxyalkylureas is limited to the addition of formaldehyde or glyoxal to urea, monoalkylureas, symmetrical dialkylureas, and cyclic ureas. It involves acid- or base-catalyzed additions that are generally equilibrium reactions [113–130].

Urea can bond with up to 4 mol of formaldehyde. However, only monohydroxymethyl- and N,N'-dihydroxymethylurea can be isolated in pure form [131], [132]. Tri- and tetrahydroxymethylureas are formed only as nonisolable intermediates, for example, in the synthesis of trimethoxymethylurea (**6**), [133], [134]; N,N'-dialkoxymethyl-4-oxomethyltetrahydro-1,3,5-oxadiazine (**7**) [135], [136]; and N,N'-dihydroxymethyl-2-oxo-5-alkyltetrahydro-1,3,5-triazines (**8**) [137–140].

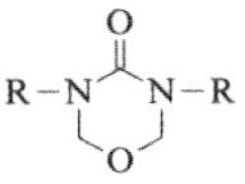

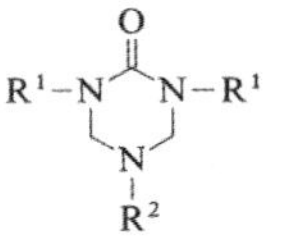

R = Alkoxymethyl
R¹ = Hydroxymethyl
R² = Alkyl

N,N'-Dihydroxymethylurea [140-95-4], HOCH$_2$HNCONHCH$_2$OH [131], [132] is produced industrially by charging 2 mol of formaldehyde per mole of urea to a stirred vessel. The solution is neutralized with triethanolamine. Urea is added with cooling, and the temperature must not exceed 40 °C in this slightly exothermic reaction. After a few hours the reaction mixture is cooled to room temperature and dihydroxymethylurea crystallizes out. The product is dried in a spray-drying tower.

Hydroxymethyl derivatives of cyclic ureas can be produced by reaction of these substances with formaldehyde in an alkaline medium [141].

Hydroxymethyl derivatives of urea and cyclic ureas (**9**)–(**15**) are used in easy-care finishes for textiles [91].

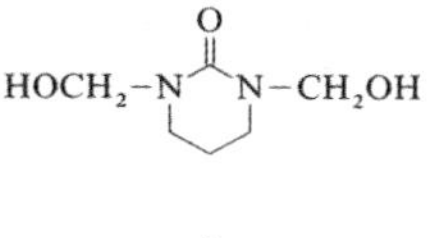

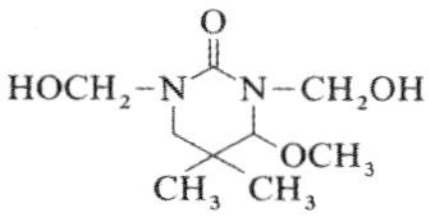

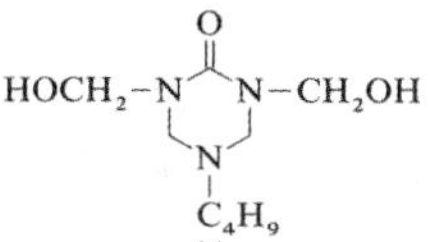

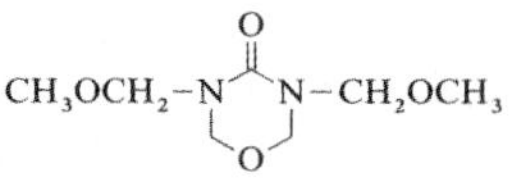

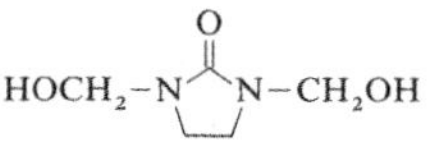

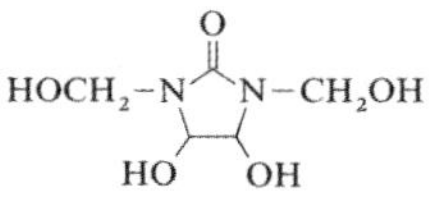

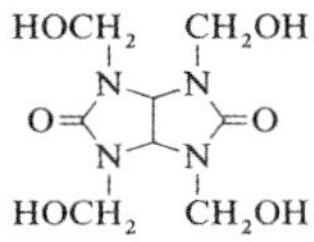

The addition of higher aldehydes to urea, mono- and symmetrically disubstituted ureas, and cyclic ureas generally gives unstable α-hydroxyalkyl compounds. Electron-withdrawing and electron-donating substituents next to the α-hydroxyalkyl group affect the stability of these compounds and also their ability to undergo condensations.

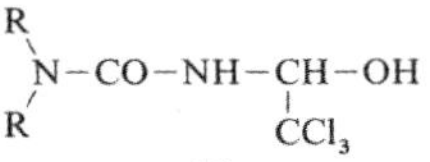

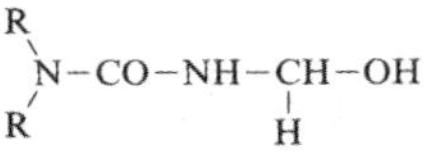

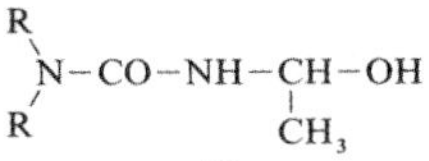

For example, the chloral–urea derivatives (**16**) exhibit considerable differences in reactivity compared with the N-hydroxymethyl (**17**) and N-α-hydroxyethyl compounds (**18**). These differences are exemplified by a decrease in the H-acidity of the OH groups [142]. Chloral compounds (**16**) can form alkali-metal salts, whereas the corresponding salts of N-hydroxymethyl (**17**) and N-α-hydroxyethyl compounds (**18**) are unknown. Condensation of chloral compounds with nucleophiles is possible only under extreme reaction conditions. However, N-α-hydroxyethyl compounds can be converted smoothly with alcohol into N-α-alkoxyethyl compounds in basic and sometimes even in neutral media. The reactivity of N-hydroxymethyl compounds lies between that of the chloral and N-α-hydroxyethyl compounds. The 4-hydroxycycloalkyleneureas (cyclic N-hemiacetals), which are obtained by treating suitable aldehydes with ureas, are stable. For example, 2-oxo-4,5-dihydroxyimidazolidines are formed by cyclization of urea, or its mono- or symmetrically disubstituted derivatives, with glyoxal [143–145]:

$$\begin{array}{c} H_2N-CO-NH_2 \\ + \\ OHC-CHO \end{array} \longrightarrow \begin{array}{c} HN\underset{HO}{\overset{O}{\diagup}}NH \\ OH \\ \mathbf{19} \end{array} \underset{\longleftarrow}{\overset{2CH_2O}{\longrightarrow}} \begin{array}{c} HOCH_2-N\underset{HO}{\overset{O}{\diagup}}N-CH_2OH \\ OH \\ \mathbf{20} \end{array}$$

$$\begin{array}{c} H_2N-CO-NH_2 \\ + \\ OHC-CHO \end{array} + 2CH_2O \longrightarrow \uparrow$$

N,N'-Dihydroxymethyl-2-oxo-4,5-dihydroxyimidazolidine (20) is produced by hydroxymethylation of 4,5-dihydroxy-2-oxoimidazolidine (**19**) in weakly acidic to weakly alkaline aqueous solution or, more elegantly, by direct reaction of urea with glyoxal and formaldehyde in the appropriate molar ratio in weakly acidic to neutral solution at 40 – 80 °C, sometimes in the presence of catalytically active buffers [130].

Compound **20** and its derivative in which the OH groups are partly acetalized with methanol are used as formaldehyde-free cross-linking agents for easy-care finishes for cellulose-containing textiles.

2-Oxo-4-hydroxyhexahydropyrimidines also belong to the group of α-hydroxyalkylureas. These compounds can be produced industrially by cyclocondensation of urea with active enolizable aldehydes (see Section 8.3.4).

8.3.2. α-Alkoxyalkylureas

Condensation of N-hydroxymethylureas with alcohols to give N-alkoxymethylureas (ureidoalkylation of alcohols) is of great industrial importance. Pure hydroxymethylureas and an excess of alcohol are reacted in the presence of catalytic amounts of acid. The nature and quantity of the acid catalyst depend on the reactivity of the N-hydroxymethyl compound, its stability to hydrolysis, and the formation of byproducts and polycondensation products. At elevated temperature the reaction can be carried out under weakly acidic conditions, whereby the equilibrium position must be adjusted by variation of the concentration and the molar ratios [91], [146], [147]. Sometimes, ureidomethylation of alcohols is better at room temperature in the presence of strong acids.

The alcohol-modified urea – formaldehyde condensation products are used as resins for heat- or acid-curing coatings. Besides the water-soluble or almost solvent-free resins, which are becoming increasingly important for environmental reasons, a wide range of aminoplastic resins for coatings exist that are readily soluble in common paint solvents. To convert urea – formaldehyde resins to resins that are soluble in organic solvents, the highly polar N-hydroxymethyl groups obtained in the initial reaction between urea and formaldehyde are acetalized with alcohols, mainly butanol and isobutanol, as well as ethanol and methanol or their mixtures.

The alcohol-modified urea – formaldehyde resins are produced industrially by passing aliphatic alcohols into aqueous solutions of urea and formaldehyde or solutions of hydroxymethylated ureas in the presence of small quantities of acid at 90 – 100 °C so that the water formed and excess alcohol distill off. The molar ratios vary between 2 and 4 mol of formaldehyde and 2 and 5 mol of alcohol per mole of urea. The process is carried out in a reactor equipped with an adequately dimensioned heat exchanger, a vacuum pump, a distillation column, and for alcohols that are sparingly soluble in water, a water separator.

Some *4-hydroxycycloalkyleneureas* are so reactive that they can be converted into the N-α-alkoxy compounds (**21**), (**22**) even in a neutral medium by heating with alcohol [148]:

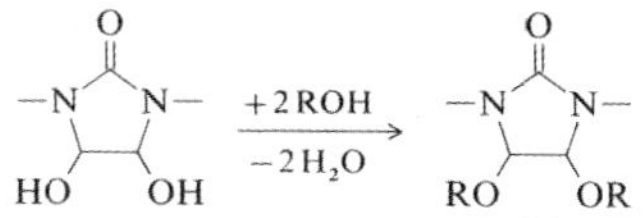

To shift the equilibrium in the N-α-ureidoalkylation of alcohols in the direction of the N-α-alkoxyalkyl compounds, the water formed during the reaction must be removed. In industrial production processes an aprotic entrainer (e.g., an aromatic) is used. In the ureidomethylation of alcohols that are immiscible or sparingly miscible with water, an excess of alcohol is used and water is removed by azeotropic distillation.

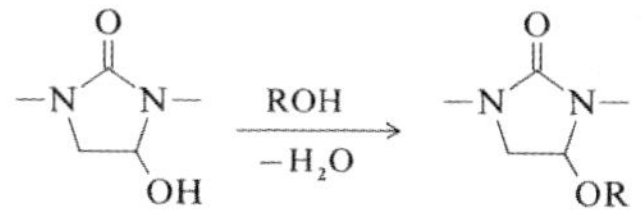

Higher-boiling alcohols can also be ureidomethylated by transacetalization of the N-methoxymethyl compounds.

Polymerizable compounds are obtained by ureidoalkylation of unsaturated alcohols (e.g., allyl alcohol [149], [150]). Tetraallyloxymethyltetrahydroimidazo[4,5-d]imidazole-2,5(1H,3H)-dione (**23**) has achieved importance as a polymerizable coating component [151], [152]:

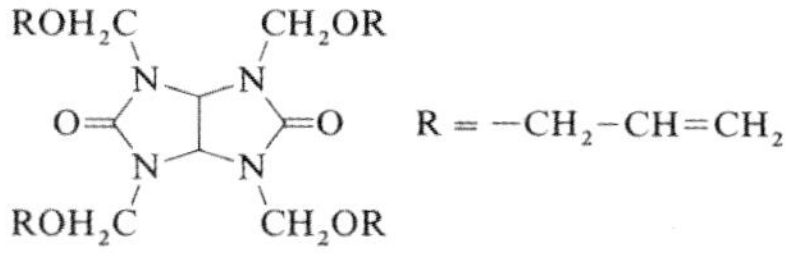

8.3.3. α,α'-Alkyleneureas

α,α'-Alkyleneureas are obtained by condensation of urea or its derivatives with aldehydes in a weakly acidic medium. The aldehyde group first adds to the urea to form an α-hydroxyalkylurea, which then reacts with a second molecule of urea or with another α-hydroxyalkylurea to give linear or branched α,α'-alkyleneureas:

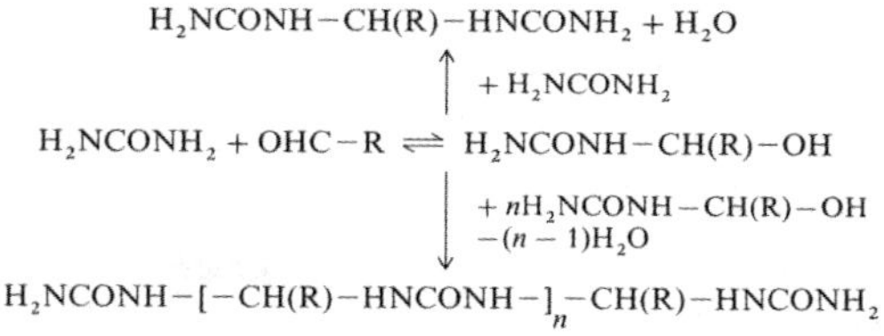

$$H_2NCONH-CH(R)-HNCONH_2 + H_2O$$
$$\uparrow + H_2NCONH_2$$
$$H_2NCONH_2 + OHC-R \rightleftharpoons H_2NCONH-CH(R)-OH$$
$$\downarrow + nH_2NCONH-CH(R)-OH$$
$$-(n-1)H_2O$$
$$H_2NCONH-[-CH(R)-HNCONH-]_n-CH(R)-HNCONH_2$$

Reaction of equimolar quantities of urea and formaldehyde in acidic solution gives polymethyleneureas as a result of stepwise ureidomethylations:

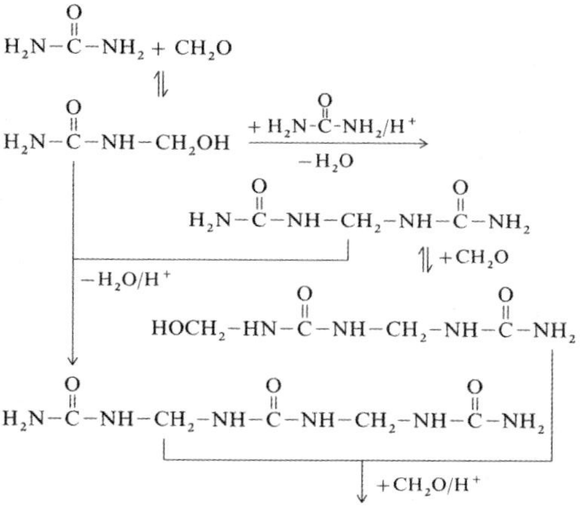

KADOWAKI obtained polymethyleneureas containing up to five urea groups joined by methylene bridges by means of a stepwise synthesis [135]. Because of their extreme insolubility, no higher polymethyleneureas have yet been isolated. Polymethyleneureas are used as slow-release nitrogen fertilizers, for example.

Isobutylidenediurea (**24**), which is sparingly soluble in water, is obtained by condensation of urea with isobutyraldehyde in a molar ratio of 2 : 1 in a weakly acidic medium:

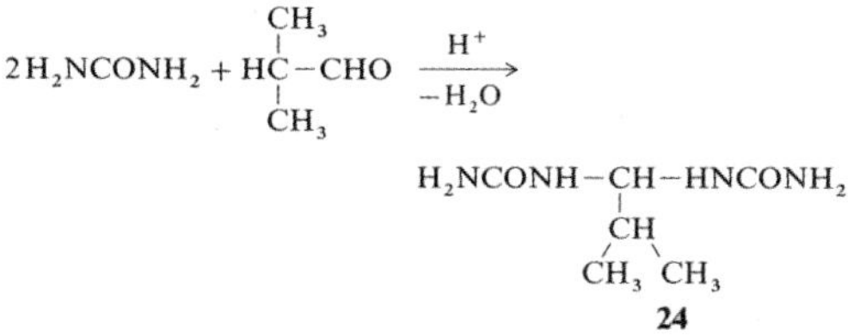

Isobutylidenediurea is also a slow-release nitrogen fertilizer used in various special fertilizer formulations. Isobutylidenediurea is produced by a continuous process. According to a patent published by Mitsubishi Chemical Industries [153], urea is charged continuously with a screw feed via a belt weigher and is reacted with a stoichiometric quantity of isobutyraldehyde in the presence of semiconcentrated sulfuric acid in a mixer. In the last section of the mixer the reaction product is neutralized by injecting dilute aqueous potassium hydroxide solution. The product is dried by using plate driers and processed by sieving, filtering, and grinding.

8.3.4. Cyclic Urea – Aldehyde Condensation Products

Almost all cyclizations of urea and its derivatives with aldehydes involve an α- or a vinylogous ureidoalkylation [113–115], [146]. If the urea bears a nucleophilic substituent on the second nitrogen atom, cyclocondensation occurs [154]. Saturated and unsaturated cyclic ureas with five, six, seven, or eight ring atoms, bicyclic and polycyclic heterocycles (with both uncondensed rings and rings anellated in the 1,2- or 1,3-position), and spiro compounds can be produced this way [154], [155].

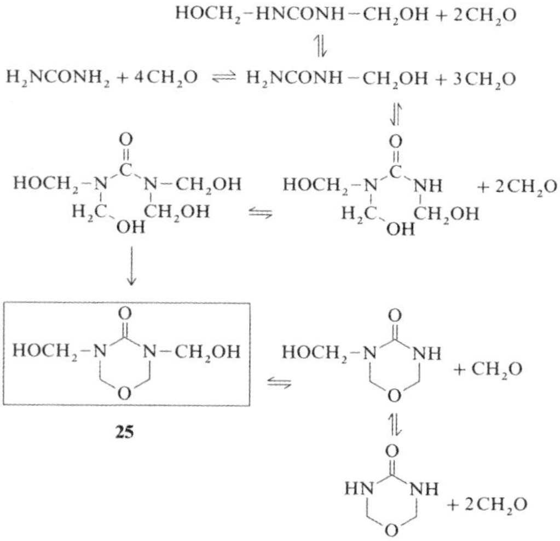

Industrially important reactions are those of urea with formaldehyde, acetaldehyde, isobutyraldehyde, and their mixtures. Treatment of urea with formaldehyde in a molar ratio of 1 : ≥4 gives an equilibrium mixture of hydroxymethyl derivatives. On ureidomethylation of a hydroxymethyl group bonded to the second nitrogen atom of the urea, cyclocondensation to hydroxymethylated 4-oxotetrahydro-1,3,5-oxadiazines (**25**) occurs [135], [136].

4-Oxo-3,5-dialkoxymethyltetrahydro-1,3,5-oxadiazines are used as cross-linking agents for easy-care finishing of textiles. These compounds can be obtained directly by condensation of urea with formaldehyde and alcohols [135], [136]:

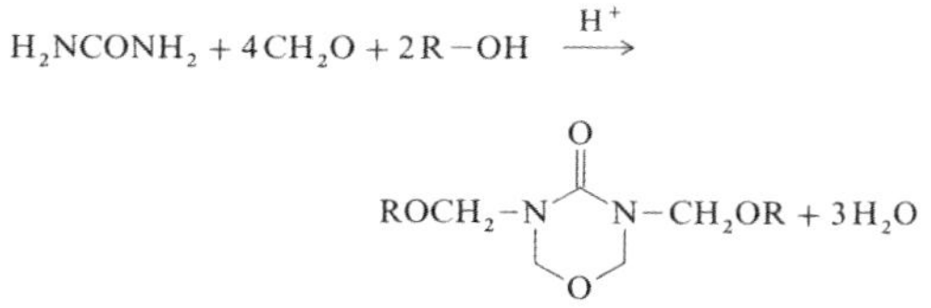

The hydroxymethyl and methoxymethyl derivatives of *2-oxo-5-alkylhexahydro-1,3,5-triazines* have achieved importance as cross-linking agents for easy-care finishing of cellulose-containing fabrics. These compounds are obtained by cyclizing ureidomethylation of urea with formaldehyde and a primary amine [137–140], [154]:

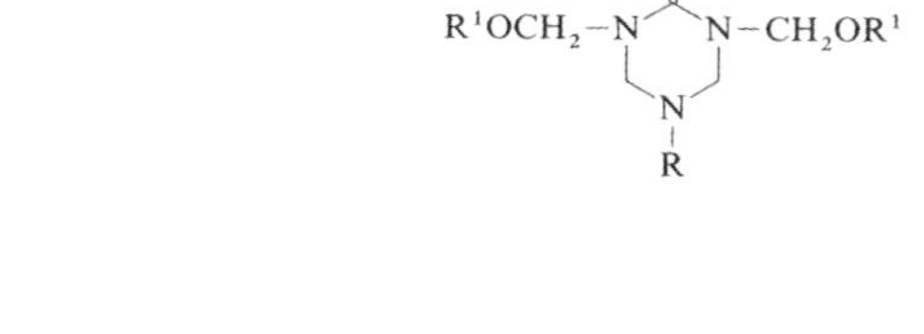

Bicyclic Ureas. 4,5-Dihydroxy- or 4,5-dialkoxyimidazolidin-2-ones can be converted into bicyclic ureas, such as *tetrahydroimidazo[4,5-d]-imidazole-2,5(1H,3H)-dione* (**26**), by means of a double α-ureidoalkylation with urea:

Tetrahydroimidazo[4,5-*d*]imidazole-2,5(1*H*, 3*H*)-dione [496-46-8], acetylenediurea (**26**), can be obtained directly by condensation of glyoxal with excess urea in an acidic medium [156], [157]. A solution of urea is acidified to pH <3 with sulfuric acid, and a 40 % glyoxal solution is added slowly. At the end of the reaction the molar ratio of glyoxal to urea is adjusted to at least 1 : 2.5. Reaction temperature should not exceed 70 °C. The precipitated product is neutralized by decantation and stirring with water.

In the form of its tetrahydroxymethyl derivative (**15**) the product has achieved importance as a cross-linking agent in easy-care textile finishes [158]. The saturated and unsaturated *N*-alkoxymethyl compounds (**23**) are used as cross-linking agents in the paint and coating industry.

The tetrachloro compound (**27**) is used as a chlorine-transfer agent for mild chlorination reactions and as a bleaching agent for textiles [159–161].

The synthetic possibilities for cyclizing α-ureidoalkylations with compounds containing enolizable carbonyl groups are manifold. Aldehydes and ketones are particularly important because they react not only as nucleophiles but also as carbonyl components in these cyclocondensations. Under mild conditions (i.e., in weakly acidic media), aliphatic and aromatic aldehydes generally react only as carbonyl components in reactions with urea and its derivatives, giving linear or branched condensation products. In a more strongly acidic medium and at elevated temperature, however, an α-ureidoalkylation at nucleophilic α-carbon atom in the aldehyde occurs. (see right column)

In the cyclocondensation of urea or its mono- or symmetrically disubstituted derivatives with aldehydes, which have at least one activated hydrogen in α-position to the carbonyl group, 2-oxo-4-hydroxyhexahydropyrimidines (**29**) are formed in the presence of an acid [113], [154], [162], [163]. Symmetrical di-α-hydroxyalkyl compounds (**28**) and poly-α-alkyleneureas are formed as intermediates. They all contain the groups necessary for cyclocondensation involving an α-ureidoalkylation, i.e., the NH component, 1 mol of aldehyde as the carbonyl component, and the second mole of aldehyde as the nucleophilic component with activated α-hydrogen. In cyclocondensation, a bond is formed between the nucleophilic α-carbon atom of one α-hydroxyalkyl group and the carbon α to the urea nitrogen in the second α-hydroxyalkyl group to give 2-oxo-4-hydroxyhexahydropyrimidines (**29**).

Preparation of 2-Oxo-4-hydroxy-5,5-dimethyl-6-isopropylhexahydropyrimidine (*mp* 220–222 °C). One liter of a 30 % aqueous urea solution (5 mol) is treated with 100 mL of 50 % sulfuric acid in a stirred tank reactor with reflux cooling. Then, 10 mol isobutyraldehyde is added with stirring. After being heated at 85 °C for 2 h, the product is filtered and washed with water.

If these cyclocondensations are carried out in the presence of alcohols, 2-oxo-4-alkoxyhexahydropyrimidines (**30**) are formed [163], [164].

These cyclocondensations can also be carried out between urea and two different aldehydes, at least one of which must have an enolizable carbonyl group. For example, the cyclocondensation of 1 mol of urea with 1 mol of formaldehyde and 1 mol of isobutyraldehyde in the presence of an acid gives 2-oxo-4-hydroxy-5,5-dimethylhexahydropyrimidine (**31**) [164]

$$R^1-HN-\overset{\overset{O}{\|}}{C}-NH-R^2 + CH_2O + H\overset{\overset{CH_3}{|}}{\underset{CH_3}{|}}C-CHO \xrightarrow{H^+}$$

$$R^1-N\underset{\underset{CH_3}{|}\ \underset{CH_3}{|}}{\overset{\overset{O}{\|}}{\bigcirc}}N-R^2 + H_2O$$
$$\text{OH}$$
31

$$2H_2NCONH_2 + 2CH_3CHO \longrightarrow \underset{\textbf{34}}{\overset{O}{\underset{H_3C}{\bigcirc}}}\overset{HN\ \ NH}{\underset{NHCONH_2}{}}$$

$$\downarrow$$

$$2H_2NCONH_2 + CH_3CH=CH-CHO$$

The 2-oxo-4-hydroxy-5,5-dimethylhexahydropyrimidines are cyclic N-hemiacetals, whose OH groups can undergo nucleophilic substitutions similar to those undergone by N-hydroxymethylureas. 2-Oxo-4-hydroxy- and 2-oxo-4-alkoxy-5,5-dialkylhexahydropyrimidines are used in the form of their N,N'-dihydroxymethyl compounds in noncrease finishes for cellulose-containing textiles [91], [165].

Preparation of N,N'-Dihydroxymethyl-2-oxo-4-hydroxy(methoxy)-5,5-dimethylhexahydropyrimidine. Urea is treated with formaldehyde in the molar ratio 1 : 1 at pH >9 and 50 – 60 °C in the presence of an excess of methanol. Isobutyraldehyde is added in the presence of a strong mineral acid to bring about cyclocondensation. The reaction mixture is rendered alkaline, and hydroxymethylation is carried out with formaldehyde.

2-Oxo-4-hydroxyhexahydropyrimidines (**32**), which are not or only mono-substituented in the 5-position, react with ureas to give 2-oxo-4-ureidohexahydropyrimidines (**32**).

$$\underset{\textbf{32}}{\overset{O}{\underset{H_3C}{\bigcirc}}\overset{HN\ \ NH}{\underset{OH}{}}} + H_2NCONH_2 \xrightarrow{-H_2O} \underset{\textbf{33}}{\overset{O}{\underset{H_3C}{\bigcirc}}\overset{HN\ \ NH}{\underset{NHCONH_2}{}}}$$

2-Oxo-4-ureido-6-methylhexahydropyrimidine [*1129-42-6*], crotonylenediurea (**33**), is obtained either by condensation of urea with crotonaldehyde in the presence of acid [166] or by the industrially more straightforward route involving condensation of urea with acetaldehyde in the molar ratio 1 : 1 in the presence of acid [167], [168]. 2,7-Dioxo-4,5-dimethyldecahydropyrimido[4,5-*d*]pyrimidine is formed as a byproduct in the second route [168–170].

2-Oxo-4-ureido-6-methylhexahydropyrimidine is produced industrially in a continuous process employing a stirred tank cascade. A 70 % urea solution is treated with acetaldehyde at a molar ratio of 1 : 1 in the presence of a catalytic quantity of 75 % sulfuric acid. The exothermic reaction is kept at 38 – 60 °C by controlled cooling. The pH is initially kept above 3 and in the final reactors below 2 by addition of sulfuric acid. Average residence time in the cascade is 40 min. In the last stirred tank, the reaction mixture is neutralized with aqueous potassium hydroxide solution. Drying is carried out in a spray tower.

2-Oxo-4-ureido-6-methylhexahydropyrimidine is used as a slow-release nitrogen fertilizer [171]. This fertilizer is characterized by its extreme insolubility in water and is therefore not washed out of the soil by rain or irrigation. It decomposes as a result of acid hydrolysis induced by humic acids during the growth period of plants, bringing about mineralization of the nitrogen it contains.

2-Oxo-4-hydroxy-5,5-dimethylhexahydropyrimidyl-N,N'-bisneopentals can be produced from urea, formaldehyde, and isobutyraldehyde in acid-catalyzed cyclo- and linear condensations. These bisneopentals react further according to a Claisen – Tishchenko reaction to give soft and hard resins with good light stability for the paint and coatings industry, depending on the molar ratios of the starting materials [172], [173].

9. References

1. F. Wöhler, *Ann. Phys. Chem.* **2** (1828) no. 12, 253 – 256.
2. A. I. Basaroff, *J. Prakt. Chem.* **2** (1870) no. 1, 283.

3. J. Berliner, *Ind. Eng. Chem.* **28** (1936) no. 5, 517–522.
4. L. Vogel, H. Schubert, *Chem. Tech. (Leipzig)* **32** (1980) no. 3, 143–144.
5. A. A. Kozyro, S. V. Dalidovich, A. P. Krasulin, *J. Appl. Chem. (Leningrad)* **59** (1986) no. 7, 1353–1355. (*Zh. Prikl. Khim. (Leningrad)* **59** (1986) no. 7, 1456–1459.).
6. A. P. Krasulin, A. A. Kozyro, G. Ya. Kabo, *J. Appl. Chem. (Leningrad)* **60** (1987) no. 1, 96–99. (*Zh. Prikl. Khim. (Leningrad)* (1987) no. 1, 104–108).
7. G. Midgley, *The Chem. Eng.* (1977) Dec., 856–866.
8. E. P. Egan, B. B. Luff, *J. Chem. Eng. Data* **11** (1966) no. 2, 192–194.
9. A. Seidell: *Solubilities of Organic Compounds,* 3rd ed., Van Nostrand Company, New York 1941, **vol. 2.**
10. A. Seidell, W. F. Linke: *Solubilities of Inorganic and Organic Compounds,* Suppl. 3rd. ed., Van Nostrand, New York 1952.
11. V. I. Kucheryavyi, G. N. Zinov'ev, L. K. Skotnikova, *J. Appl. Chem. (Leningrad)* **42** (1969) no. 2, 409–410 (*Zh. Prikl. Khim.* **42** (1969) no. 2, 446–447).
12. R. I. Spasskaya, *J. Appl. Chem. (Leningrad)* **46** (1973) no. 2, 407–409. (*Zh. Prikl. Khim.* **46** (1973) no. 2, 393–396).
13. G. Ostrogovich, R. Bacaloglu, *Rev. Roum. Chim.* **10** (1965) 1111–1123.
14. G. Ostrogovich, R. Bacaloglu, *Rev. Roum. Chim.* **10** (1965) 1125–1135.
15. W. R. Epperly, *Chem. Tech. (Heidelberg)* **21** (1991) no. 7, 429–431.
16. A. Lasalle, et al., *Ind. Eng. Chem. Res.* **31** (1992) no. 3, 777–780.
17. B. K. Banerjee, P. C. Srivastava, *Fert. Technol.* **16** (1979) nos. 3–4, 264–288.
18. *Kirk-Othmer,* 3rd. ed., **vol. 2,** pp. 440–469.
19. M. Frejacques, *Chim. Ind. (Paris)* **60** (1948) no. 1, 22–35.
20. S. Inoue, K. Kanai, E. Otsuka, *Bull. Chem. Soc. Jpn.* **45** (1972) no. 5, 1339–1345 (Part I); *Bull. Chem. Soc. Jpn.* **45** (1972) no. 6, 1616–1619 (Part II).
21. D. M. Gorlovskii, V. I. Kucheryavyi, *J. Appl. Chem. (Leningrad)* **54** (1981) no. 10, 1898–1901. (*Zh. Prikl. Khim.* **53** (1980) no. 11, 2548–2551).
22. W. Durisch, S. M. Lemkowitz, P. J. van den Berg, *Chimia* **34** (1980) no. 7, 314–322.
23. S. M. Lemkowitz, J. Zuidam, P. J. van den Berg, *J. Appl. Chem. Biotechnol.* **22** (1972) 727–737.
24. P. J. C. Kaasenbrood, H. A. G. Chermin, paper presented to The Fertilizer Society of London, 1st Dec., 1977.
25. Stamicarbon, EP 0 212 744, 1987 (J. H. Meessen, R. Sipkema).
26. Montedison, EP 0 132 194, 1985 (G. Pagani).
27. Toyo Engineering Corporation, GB 2 109 372, 1983 (S. Inoue et al.).
28. *Eur. Chem. News,* Aug. 9 (1982) 15–16.
29. T. Jojima, B. Kinno, H. Uchino, A. Fukui, *Chem. Eng. Prog.* **80** (1984) no. 4, 31–35.
30. E. Dooyeweerd, J. Meessen, *Nitrogen* **143** (1983) May–June, 32–38.
31. *Nitrogen* **194** (1991) Nov.–Dec., 22–28.
32. R. De Jonge, F. X. C. M. Barake, J. D. Logemann, *Chem. Age India* **26** (1975) no. 4, 249–260.
33. Montedison, EP 0 096 151, 1983 (G. Pagani, G. Faita, U. Grassini).
34. Urea Casale, EP 0 504 621, 1992 (V. Lagana).
35. Stamicarbon, US 2 727 069, 1955 (J. P. M. van Waes).
36. Stamicarbon, US 3 720 548, 1973 (F. X. C. M. Barake, C. G. M. Dijkhuis, J. D. Logemann).
37. C. Miola, H. Richter, *Werkst. Korros.* **43** (1992) 396–401.
38. Snamprogetti, US 4 899 813, 1990 (S. Menicatti, C. Miola, F. Granelli).
39. E. M. Elkanzi, *Res. Ind.* **36** (1991) no. 4, 254–259 (Eng.).
40. Unie van Kunstmestfabrieken, EP 0 155 735, 1985 (K. Jonckers).
41. Snamprogetti, GB 1 542 371, 1979 (U. Zardi, V. Lagana).
42. V. Lagana, G. Schmid, *Hydrocarbon Process.* **54** (1975) no. 7, 102–104 (Eng.).
43. U. Zardi, F. Ortu, *Hydrocarbon Process.* **49** (1970) no. 4, 115–116 (Eng.).
44. Snamprogetti, GB 1 506 129, 1978.
45. Snamprogetti, US 3 954 861, 1976 (M. Guadalupi, U. Zardi).
46. Snamprogetti, US 3 876 696, 1975 (M. Guadalupi, U. Zardi).
47. *Nitrogen* **185** (1990) May–June, 22–29.
48. Toyo Engineering Corp., US 4 301 299, 1981 (S. Inoue, H. Ono).
49. Toyo Engineering Corp., GB 2 109 372, 1983 (S. Inoue et al.).
50. Montedison, GB 1 581 505, 1980 (G. Pagani).
51. A. Seidell, W. F. Linke: *Solubilities of Inorganic and Organic Compounds,* Suppl. 3rd ed., D. Van Nostrand Co., New York 1952, p. 393.
52. Snamprogetti, US 4 327 068, 1982 (V. Lagana, U. Zardi).

53. Unie van Kunstmestfabrieken, US 4 410 503, 1983 (P. J. M. van Nassau, A. M. Douwes).
54. The M. W. Kellogg Co., US 5 223 238, 1993 (T. A. Czup-pon).
55. Snamprogetti, US 4 235 816, 1980 (V. Lagana, F. Saviano).
56. Snamprogetti, GB 1 520 561, 1978 (G. Pagani).
57. Snamprogetti, GB 1 470 489, 1977 (A. Bonetti).
58. Snamprogetti, GB 1 560 174, 1980 (V. Lagana, F. Saviano).
59. V. Lagana, *Chem. Eng. (N.Y.)* **85** (1978) no. 1, 37–39.
60. Snamprogetti, GB 2 060 614, 1981 (V. Lagana, F. Saviano, V. Cavallanti).
61. S. K. Saxena, A. Ali, *Indian J. Environ. Prot.* **9** (1989) no. 11, 831–833 (Eng.).
62. R. M. Krishnan, *Fert. News* (1992) May, 53–57.
63. Ting-Chia Huang, Dong-Hwang Chen, *J. Chem. Technol. Biotechnol.* **55** (1992) no. 2, 191–199.
64. Unie van Kunstmestfabrieken, EP 0 053 410, 1982 (J. Zuidam, P. Bruls, K. Jonckers).
65. Snamprogetti, EP 0 417 829, 1991 (F. Granelli).
66. I. Mavrovic: "Pollution Control in Urea Plants," *Br. Sulphur's 12th Int. Conf. Nitrogen 88,* Geneva, March 27–29, 1988.
67. Unie van Kunstmestfabrieken, EP 0 037 148, 1981 (M. H. Willems, J. W. Klok).
68. *Fert. Int.* **296** (1991) April, 32–37.
69. R. M. Reed, J. C. Reynolds, *Chem. Eng. Prog.* **69** (1973) no. 2, 62–66.
70. *Nitrogen* **95** (1975) May/June, 31–36.
71. R. D. Hauck, M. Koshino: *Fertilizer Technology & Use,* 2nd ed. Soil Science Society of America, Madison 1971, pp. 455–494.
72. N. K. Savant, E. T. Craswell, R. B. Diamond, *Fert. News* **28** (1983) no. 8, 27–35.
73. G. Wiedemann, *Justus Liebigs Ann. Chem.* **68** (1848) 325.
74. Kurzer, *Chem. Rev.* **56** (1956) 95–197.
75. *Zh. Prikl. Khim. (Leningrad)* **42** (1969) no. 13, 713.
76. IG Farbenind., DE 689 421, 1940.
77. G. Kränzlein, H. Keller, H. Schiff, *Justus Liebigs Ann. Chem.* **291** (1896) 374.
78. R. C. Haworth, F. G. Mann, *J. Chem. Soc.* 1943, 603.
79. American Cyanamid, US 2 760 961, 1956 (Mackey).
80. A. Fleischer, *Ber. Dtsch. Chem. Ges.* **9** (1876) 995.
81. A. v. Baeyer, *Justus Liebigs Ann. Chem.* **131** (1864) 252.
82. T. L. Davis, K. C. Blanchard, *J. Am. Chem. Soc.* **45** (1923) 1816. US 1 785 730, 1927 (T. L. Davis).
83. Knoll, DE 896 640, 1942.
84. T. L. Davies, K. C. Blanchard, *Org. Synth.* **3** (1923) 95.
85. Du Pont, US 2 145 242, 1937 (H. W. Arnold).
86. *Houben-Weyl,* **VIII, III,** 151.
87. A. Sonn, *Ber. Dtsch. Chem. Ges.* **47** (1914) 2440.
88. Schweizer, *J. Org. Chem.* **15** (1950) 471.
89. *Houben-Weyl,* **VIII, III,** 164.
90. McKay, Coleman, *J. Am. Chem. Soc.* **72** (1950) 3205.
91. H. Petersen: "Chemical Processing of Fibers and Fabrics, Functional Finishes," Part A, in M. Lewin, S. B. Sello (eds.): *Handbook of Fiber science and Technology,* vol. 2, Marcel Dekker, New York 1983, pp. 48–327.
92. *Org. Synth.* **29** (1949) 18.
93. *Houben-Weyl,* **VIII, III,** 153.
94. L. J. Smith, O. M. Emeron, *Org. Synth. Coll.* **III** (1955) 151.
95. G. M. Dyron, *Chem. Rev.* **4** (1927) 138.
96. W. Hentschel, *J. Prakt. Chem.* **27** (1883) no. 2, 499.
97. CIOS, XXVII/80, 15.
98. Bayer, DE 116 200, 1899. *Friedländer* **6,** 200. BASF, DE 46 737, 1888. *Friedländer* **2,** 450. Bayer, DE 131 513, 1901. *Friedländer* **6,** 968. Bayer, DE 216 666, 1908. *Friedländer* **9,** 372. Bayer, DE 493 811, 1926. Ges. Chem. Ind., *Friedländer* **16,** 1059.
99. A. Lüttringhaus, H. W. Dirksen, *Angew. Chem. Int. Ed. Engl.* **3** (1964) 260.
100. J. Berlinerblau, *J. Prakt. Chem.* **30** (1884) no. 3, 103.
101. Höchst, DE 205 662, 1906. *Friedländer* **9,** 395.
102. Du Pont, US 2 390 253, 1943 (C. O. Henke); *Chem. Abstr.* **40** (1946) 1876.
103. A. Wurtz, *Hebd. Seances Acad. Sci.* **27** (1848) 242.
104. A. Wurtz, *Justus Liebigs Ann. Chem.* **71** (1849) 329.
105. A. W. Hofmann, *Justus Liebigs Ann. Chem.* **74** (1850) 14.
106. A. Wurtz, *Justus Liebigs Ann. Chem.* **80** (1851) 347.
107. W. Michler, *Ber. Dtsch. Chem. Ges.* **9** (1876) 396, 711.
108. W. Michler, C. Escherich, *Ber. Dtsch. Chem. Ges.* **12** (1879) 1162.

109. *Houben-Weyl, VIII,* III, 160.
110. H. Eckenroth, *Ber. Dtsch. Chem. Ges.* **18** (1885) 516.
111. C. A. Bischoff, A. von Hedenström, *Ber. Dtsch. Chem. Ges.* **35** (1902) 3437.
112. T. Wilm, G. Wischin, *Justus Liebigs Ann. Chem.* **147** (1868) 162.
113. H. Petersen, *Angew. Chem.* **76** (1964) 909.
114. H. Petersen, *Test. Rundsch.* **16** (1961) 646.
115. H. Petersen, *Melliand Textilber.* **43** (1962) 380.
116. L. E. Smythe, *J. Phys. Chem.* **51** (1947) 369.
117. L. E. Smythe, *J. Am. Chem. Soc.* **73** (1951) 2735; **74** (1952) 2713; **75** (1953) 574.
118. L. Bettelheim, J. Cedwall, *Sven. Kem. Tidskr.* **60** (1948) 208.
119. G. A. Growe, C. C. Lynch, *J. Am. Chem. Soc.* **70** (1948) 3795; **72** (1950) 3622.
120. J. I. de Jong, J. de Jonge, *Reol. Trav. Chim. Pays-Bas* **71** (1952) 643, 661, 890; **72** (1953) 88.
121. J. Lemaitre, G. Smets, R. Hart, *Bull. Soc. Chim. Belg.* **63** (1954) 182.
122. J. Ugelstadt, J. de Jonge, *Reol. Trav. Chim. Pays-Bas* **76** (1957) 919.
123. M. Okano, Y. Ogata, *Bull. Soc. Chim. Belg.* **63** (1954)182.
124. R. Kvéton, F. Hanousek, *Chem. Listy* **48** (1954) 1205; **49** (1955) 63.
125. N. Landquist, *Acta Chem. Scand.* **9** (1955) 1127, 1459; **10** (1956) 244; **11** (1957) 776.
126. B. R. Glutz, H. Zollinger, *Helv. Chim. Acta* **52** (1969) 1976.
127. P. Eugster, H. Zollinger, *Helv. Chim. Acta* **52** (1969) 1985.
128. H. Petersen, *Textilveredlung* **2** (1967) 744.
129. H. Petersen, *Textilveredlung* **3** (1968) 160.
130. H. Petersen, *Chem. Ztg.* **95** (1971) 625.
131. A. Einhorn, A. Hamburger, *Ber. Dtsch. Chem. Ges.* **41** (1908) 24.
132. A. Einhorn, A. Hamburger, *Justus Liebigs Ann. Chem.* **361** (1908) 122.
133. H. Petersen, *Text. Res. J.* **40** (1970) 335.
134. BASF, GB 1 241 580, 1968 (H. Petersen).
135. H. Kadowaki, *Bull. Chem. Soc. Jpn.* **11** (1936) 248; *Chem. Zentralbl.* (1936 II)3535.
136. Sumitomo Chem. Co., DE 1 123 334, 1962 (T. Oshima).
137. A. M. Paquin, *Angew. Chem.* **60** (1948) 267.
138. W. J. Burke, *J. Am. Chem. Soc.* **69** (1967) 2136.
139. DuPont, US 2 304 624, 1942 (W. J. Burke).
140. DuPont, US 2 321 989, 1942 (W. J. Burke).
141. BASF, GB 1 077 344, 1965. BASF, CH 478 287, 1966. DuPont, US 2 436 355, 1946.
142. H. Petersen in: *Kunststoff-Jahrbuch,* 10th. ed., Wilhelm Pansegrau Verlag, Berlin 1968, p. 46.
143. H. Pauli, H. Sauter, *Ber. Dtsch. Chem. Ges.* **63** (1930) 2063.
144. BASF, DE 910 475, 1951 (H. Scheuermann, B. von Reibnitz, A. Wörner); *Chem. Zentralbl.* (1955) 702.
145. H. Petersen, *Textilveredlung* **3** (1968) 51.
146. H. Petersen, *Chem. Ztg.* **95** (1971) 692.
147. H. Petersen, *Text. Res. J.* **41** (1971) 239.
148. BASF, US 4 219 494, 1980 (H. Petersen).
149. E. R. Atkinson, A. H. Bump, *Ind. Eng. Chem.* **44** (1952) 333.
150. Societe Nobel Française, DE 962 119, 1952 (P. A. Talet); *Chem. Zentralbl.* (1955) 1625.
151. BASF, DE 1 049 572, 1959 (H. Willersinn, H. Scheuermann, A. Wörner).
152. BASF, DE 1 067 210, 1960 (H. Willersinn, H. Scheuermann, A. Wörner).
153. Mitsubishi Chem. Ind., US 3 322 528, 1961; DE 1 543 201, 1965 (M. Hamamoto, Y. Sakaki).
154. H. Petersen, *Synthesis* 1973, 243–292.
155. H. Petersen, *Angew. Chem.* **79** (1967) 1009.
156. BASF, US 2 731 472, 1956 (B. von Reibnitz).
157. H. Petersen, *Synthesis* 1973, 254.
158. BASF, DE 859 019, 1941 (J. Lintner, H. Scheuermann); *Chem. Zentralbl.* (1953) 4779.
159. Biltz, Behrens, *Ber. Dtsch. Chem. Ges.* **43** (1910) 1984.
160. US 2 638 434, 1953; US 2 649 381, 1953.
161. Chemische Werke Hüls, DE 1 020 024, 1957 (H. Steinbrink, I. Amende).
162. BASF, DE 1 230 805, 1962 (H. Petersen, H. Brandeis, R. Fikentscher).
163. G. Zigeuner, W. Rauter, *Monatsh. Chem.* **96** (1965) 1950.
164. BASF, DE 1 231 247, 1962 (H. Petersen, H. Brandeis, R. Fikentscher).
165. H. Bille, H. Petersen, *Text. Res. J.* **37** (1967) 264; *Textilveredlung* **2** (1967) 243.
166. A. M. Paquin, *Kunststoffe* **37** (1947) 165.
167. BASF, DE 1 223 843, 1962 (H. Petersen, H. Brandeis, R. Fikentscher).
168. G. Zigeuner, E. A. Gardziella, G. Bach, *Monatsh. Chem.* **92** (1961) 31.
169. G. Zigeuner, M. Wilhelmi, B. Bonath, *Monatsh. Chem.* **92** (1961) 42.
170. G. Zigeuner, W. Nischk, *Monatsh. Chem.* **92** (1961) 79.
171. BASF, DE 1 081 482, 1959; US 3 190 741, 1963 (J. Jung, H. Müller von Blumencron, C. Pfaff, H. Scheuermann).
172. BASF, EP 0 002 793, 1977; US 4 220 751, 1980 (H. Petersen, K. Fischer, H. Klug, W. Trimborn).
173. BASF, DE 2 757 220, 1977; EP 0 002 794, 1977; US 4 243 797, 1981 (H. Peterson et al.).

Plant Growth Regulators

Hubert Sauter, BASF Aktiengesellschaft, Ludwigshafen, Federal Republic of Germany

Bernd Zeeh, BASF Aktiengesellschaft, Ludwigshafen, Federal Republic of Germany

1.	Introduction	503
2.	Classes of Growth Regulators	505
2.1.	Auxins and Auxin-Type Compounds	505
2.2.	Antiauxins	505
2.3.	Gibberellins	506
2.4.	Gibberellin Antagonists	506
2.5.	Cytokinins	509
2.6.	Abscisic Acid	509
2.7.	Ethylene-Releasing Compounds	510
2.8.	Other Growth Regulators	510
3.	Defoliants and Desiccants	512
4.	References	513

1. Introduction

Plant growth regulators, also known as bioregulators, are natural or synthetic substances that act in small amounts to influence developmental, metabolic, and growth processes in plants without having lethal effects. In agriculture and horticulture, they are employed to obtain specific advantages, which include increased yields, modification of growth forms, improved quality, modification of plant constituents, and facilitation of harvesting. In the classification of these substances, it is impossible to avoid a certain degree of overlap with the herbicides (→ Weed Control) because higher doses of some agents can also cause damage to the plant. In a broader sense, substances such as defoliants and desiccants are also classified as growth regulators or bioregulators.

Phytohormones. Plants synthesize a system of plant hormones (phytohormones) that regulate metabolic processes. The signal effects of this system on metabolic processes are comparable to those of animal hormone systems, but differ as regards organ specificity. Animal hormones are synthesized at specific sites and produce an effect in a particular target organ, whereas phytohormones are formed at many sites and act on many different plant organs (multilocal); they therefore have a wide spectrum of activity. Phytohormones interact with one another, producing antagonistic, additive, or synergistic effects.

Five groups of substances are known to be involved in the transmission of external and internal signals from one part of the plant to another:

1) Auxins
2) Gibberellins
3) Cytokinins
4) Abscisic acid
5) Ethylene

Depending on the stage of plant development, varying amounts of hormones are formed in the plant organs in response to environmental stimuli. The stimulatory or inhibitory effect of phytohormones enables the plant to continually regulate vital processes, e.g., growth, differentiation, flowering, maturation, and the shedding (abscission) of leaves. It is often not the type but the concentration of the phytohormone that determines whether a given developmental process is stimulated or inhibited. Obviously, there is a chance that active substances can be used to specifically influence these complicated events, especially in cultivated plants.

History. Scientific findings in the field of plant physiology not only give insight into the molecular basis of plant development, but also provide technical applications that are becoming increasingly important for the cultivation of plants. The use of plant growth regulators in agriculture and horticulture began in the 1930s. They include both natural phytohormones, such as auxin (3-indolylacetic acid) and gibberellic acid (GA_3), as well as synthetic substances, such as daminozide, chlormequat, and ethephon. Examples of these substances and their applications are listed in chronological order in Table 1. Some are of considerable economic importance (e.g., auxin-type substances, chlormequat), whereas others are no longer in use (e.g., ethylene, acetylene).

Table 1. Plant growth regulators

Year of introduction	Agent (trade name)	Application
1932	ethylene, acetylene	induction of flowering in pineapples
1936	auxin	production of seedless fruit
1943	1-naphthylacetic acid	thinning out of apples
1944	auxin-type substances e.g., 2,4-D *	selective herbicides
1948	maleic hydrazide	inhibition of the growth of grass
1959	gibberellin	improvement in the quality of grapes
1959	chlormequat (Cycocel)	strengthening the stalks of cereals
1962	daminozide	improvement in the quality of apples and cherries
1966	cytokinins	improvement in the fructification of grapes
1971	ethephon (Ethrel)	increase in the latex flow in rubber trees
1973	glyphosine (Polaris)	increase in the sugar yield from cane
1979	mepiquat (Pix)	prevention of excessive growth of cotton
1987	inabenfide (Seritard)	strengthening the stalks of rice
1989	triapenthenol (Baronet)	improvement in the stability of rape

* 2,4-Dichlorophenoxyacetic acid.

Uses. Irrespective of whether a plant is wild or cultivated, its development normally falls short of its capacity, i.e., its genetically defined potential is not fully exploited. Limiting factors are temperature; shortage of light, water, and nutrients; competing plants; insect damage; and microorganisms.

Increased yields can be achieved in plant cultivation by measures such as adequate fertilization and irrigation, use of high-yield varieties, optimal soil treatment, and reduction of diseases, pests, and weeds with agricultural pesticides. Additional increases in plant production can also be obtained by using growth regulators. Some regulators only change the outer form of plants (morphoregulators). Substances that influence other properties of cultivated plants are also being increasingly investigated. These properties include plant constituents (sugar, protein), resistance to stress (dryness, heat, frost, salty soil), and rhythm of plant development (fruit set, induction of flowering, acceleration of maturity, and shedding of fruit and leaves, i.e., abscission).

The development of a marketable growth regulator generally takes longer than the development of other plant treatment agents; determination of the dose rate and the best time for application is often critical and requires extensive outdoor experimentation under varying conditions.

Growth regulators have to meet the same requirements for approval and control as other plant treatment agents (see → Fungicides, → Insect Control, → Weed Control). Detailed information about the product properties of the active agents (e.g., toxicity, formulation, solubility, application, and trade names) are given in [17], [18].

Growth regulators are usually applied in agriculture by spraying them on foliage. Although seed treatment has environmental advantages, this method is rarely implemented because application has to take place at a defined time before sowing. In horticulture, the agent is often added during watering and is taken up through the roots.

Economic Aspects [19]. Compared with the three main groups of plant protection agents (herbicides, insecticides, and fungicides), growth regulators have a relatively small share of the world market (ca. 4 %). The growth rate in sales is, however, comparatively high. At the end of the 1980s the entire market volume for growth regulators was ca. US 750×10^6 (end-user level). The defoliants and desiccants (agents for facilitating harvesting, especially of cotton, tobacco, and potatoes) alone accounted for about 40 % of sales and were predominantly agents with commodity status. The application of growth-retarding agents in European cereal and rape cultivation and in American cotton is another economically important sector.

The most important agents (see Table 1) are maleic hydrazide, chlormequat, ethephon, daminozide, gibberellic acid, glyphosine, mepiquat, and thidiazuron.

2. Classes of Growth Regulators

Growth regulators belong to so many different chemical families that classification according to their chemistry does not seem very useful. The classification system used here is based on mechanisms of action; many growth regulators exert effects that are similar or antagonistic to those of native phytohormones, allowing a rough qualitative classification [1, p. 10].

The names used for growth regulators, like those for other agents in pharmaceutics and plant protection, are numerous and confusing. Manufacturers not only use trade names, but also apply to national and international authorities (e.g., ISO) for a common name. The common name is normally an abbreviation of the chemical name and is used to denote the pure active agent.

2.1. Auxins and Auxin-Type Compounds

Auxins and auxin-type growth substances exert the typical effects of the natural phytohormone auxin (3-indolylacetic acid): the promotion of longitudinal growth by cell elongation, stimulation of cell division in the cambium and roots, and control of enzyme activity.

Auxins have not gained much practical importance as growth-stimulating agents because of their narrow therapeutic dose range. A small overdose can cause the growth-regulating effect to become a herbicidal effect. Auxin-type substances are much more important as herbicides than as growth regulators ($\rightarrow$ Weed Control). The following auxins are of significance as growth regulators.

3-Indolylacetic acid [*87-51-4*], $C_{10}H_9NO_2$, M_r 175.2, was isolated from human urine by KÖGL in 1934 [20].

Tryptophan is the starting material for the biosynthesis of this substance [15, p. 39]. It is used to stimulate root growth in cuttings, in the production of seedless fruit, and in the induction of flowering in pineapples. However, indolylacetic acid has been largely replaced by other carboxylic acids.

3-Indolylbutyric acid [*133-32-4*], IBA, $C_{12}H_{13}NO_2$, M_r 203.2, mp 123–125 °C, colorless crystals.

In the plant this acid is degraded to indolylacetic acid, presumably by β-oxidation. It is made from indole, γ-butyrolactone, and sodium hydroxide and is used in the same way as indolylacetic acid.

Trade names: Hormodin (MSD Agvet), Seradix (Rhône-Poulenc).

1-Naphthaleneacetic acid [*86-87-3*], $C_{12}H_{10}O_2$, M_r 186.2, mp 134–135 °C, colorless crystals, LD_{50} rat oral $\approx$ 2520 mg/kg.

This substance is produced by the hydrolysis of 1-naphthaleneacetonitrile, which is obtained by the chloromethylation of naphthalene and subsequent reaction with sodium cyanide. It is used for the same purposes as indolylacetic acid and to thin out apples, olives, oranges etc. and to inhibit germination of potatoes during storage.

Trade names: Celmone (Excel Ind.); Thin'n Stop-Drop (Platte Chem), Fruitone-N, NAA 800, Tree-Hold (Rhône-Poulenc); Olive Stop (UAP Special Products).

1-Naphthaleneacetamide [*86-86-2*], $C_{12}H_{10}NO$, M_r 185.2, mp 182–184 °C, LD_{50} rat oral $\approx$ 1000 mg/kg. It is used in the same way as naphthaleneacetic acid.

Trade names: Amid-Thin W, Rootone (Rhône-Poulenc, Greenwood Chemical).

2.2. Antiauxins

In a narrower sense, antiauxins are defined as substances that compete with auxins for specific receptors [21]. Only agents that are antiauxins in a broader sense have achieved practical importance; they inhibit the transport of auxins in the plant and, thus, strongly affect growth and morphology [22]. Frequently, these effects are exploited in herbicides, as in the case of the morphactins derived from fluorene-9-carboxylic acid (→ Weed Control). The following antiauxins are used as growth regulators.

N-m-Tolylphthalamic acid [85-72-3], N-m-t, $C_{15}H_{13}NO_3$, M_r 255.3, $LD_{50} \approx 5230$ mg/kg (oral, male albino rats).

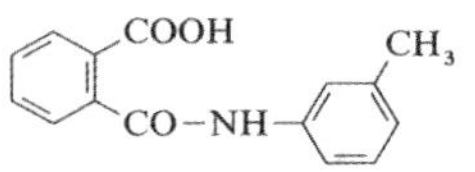

This substance is used to improve the yield of tomatoes, beans, and cherries by increasing fructification.
Trade names: Tomaset, Duraset (Uniroyal, Makhteshim-Agan).

2.3. Gibberellins

Natural gibberellins (GA) have a wide spectrum of phytohormonal effects, especially in the regulation of growth by cell elongation and cell division and in the induction of hydrolytic enzymes. More than 70 natural gibberellins, each containing 19 or 20 carbon atoms, have been found [23]. The nomenclature of these substances (GA_1, GA_2 ...) follows the order of their discovery. Not all the gibberellins are biologically active, some are precursors in the biogenesis of active (usually C_{19}) gibberellins and others are inactivated metabolites.

The biosynthesis of these diterpene compounds involves mevalonate as the starting product which is converted to geranylgeranyl pyrophosphate [24]. The latter is cyclized to give the key substance *ent*-kaurene. Stepwise oxidation then yields the acid, which is converted to the GA_{12} aldehyde by subsequent oxidation and ring contraction. Further biosynthetic steps give the gibberellins, which all have a carboxyl group at position 10 of the skeleton.

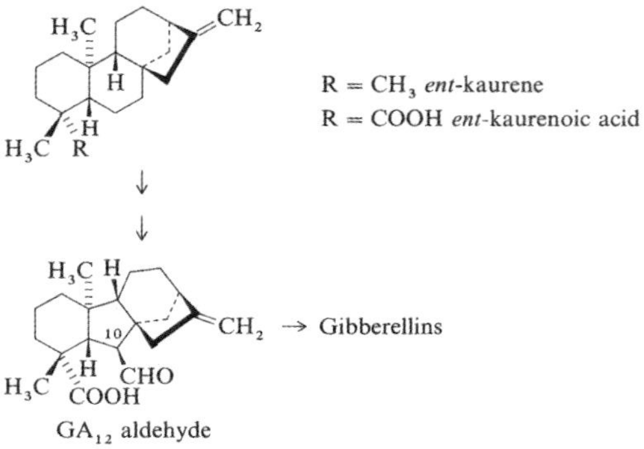

R = CH_3 *ent*-kaurene
R = COOH *ent*-kaurenoic acid

→ Gibberellins

GA_{12} aldehyde

Gibberellic acid [77-06-5], GA_3, $C_{19}H_{22}O_6$, M_r 346.4, mp 233–235 °C. This substance is produced biotechnologically from the fungus *Gibberella fujikuroi* (*Fusarium moniliforme*).

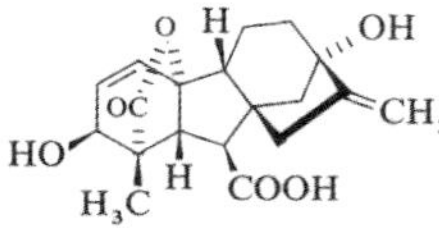

It is used in the United States at concentrations of 1–1000 ppm to produce seedless dessert grapes, to increase fruit and vegetable yields (celery and artichokes), and in the production of vegetable seeds.

Oral doses up to 1500 mg/kg produce no toxic symptoms in mice.
Trade names: Pro-Gibb, Pro Vide (Abbott Labs); Activol, Berelex, Grocel (ICI); Gibrel (MSD Agvet).

Promalin (Abbott), a mixture of GA_4/GA_7 with 6-benzyladenine, is also available for improving the quality and yield of apples.

2.4. Gibberellin Antagonists

Substances that inhibit gibberellin biosynthesis are of very great practical significance. Onium compounds of nitrogen, phosphorus, and sulfur form an important group. They inhibit gibberellin biosynthesis by blocking the cyclization steps leading to formation of *ent*-kaurene [25]. Macroscopic effects include growth retardation, stunted and compact growth, intense green coloring and, frequently, improved lodging resistance of the plants [26]. The spectrum of activity in different plant varieties depends mainly on the individual structure of the onium compound [27].

Chlormequat [*999-81-5*], chlorocholine chloride (CCC), $C_5H_{13}Cl_2N$, M_r 158.1, *mp* ca. 245 °C (decomp.), LD_{50} rat oral ≈ 670 mg/kg.

The biological effects of chlormequat were first described by TOLBERT [28]. Chlormequat is the most important inhibitor of gibberellin biosynthesis. It is synthesized from trimethylamine and 1,2-dichloroethane.

It is used primarily in the cultivation of wheat in Europe. Application leads to strong, compact growth, increased chlorophyll content, and increased stalk diameter which prevents the cereal from falling over (especially wheat), before it is harvested [29], [30]. A mixture of chlormequat and ethephon has proved useful for barley. In the growing of ornamental plants, chlormequat is used to shorten the stems of poinsettias, azaleas, pelargoniums, and lilies.
Trade names: Cycocel, CCC (American Cyanamid, BASF) and as a mixture with ethephon: Terpal C (BASF).

Mepiquat [*24307-26-4*], N,N-dimethylpiperidinium chloride [31], [32], $C_7H_{16}NCl$, M_r 149.7, *mp* 350 °C, LD_{50} rat oral ≈ 1600 mg/kg.

Mepiquat is produced from N-methylpiperidine and chloromethane. It prevents excessively lush vegetative growth of cotton [33–35] and therefore facilitates mechanical harvesting and results in earlier maturation, often in combination with a higher yield.

Mepiquat is used in combination with ethephon (see page 510) to stabilize barley stalks.
Trade names: Pix (BASF); mepiquat mixed with ethephon: Terpal (BASF).

Chlorphonium chloride [*115-78-6*], tri-(n-butyl)-2,4-dichlorobenzylphosphonium chloride [36], $C_{19}H_{32}Cl_3P$, M_r 397.8, *mp* 114–120 °C, LD_{50} rat oral ca. 180 mg/kg

This substance is made from 2,4-dichlorobenzyl chloride and tributyl phosphine. It is used to obtain compact growth in ornamental flowers (lilies and chrysanthemums).
Trade name: Phosfon (Mobil Chemical) (discontinued).

Heterocyclic nitrogen compounds (pyrimidines, pyridines, diazetines, and triazoles) also inhibit gibberellin biosynthesis and are attaining increasing importance. These agents all have a sp^2-hybridized nitrogen atom which can complex the central iron atom in some cytochrome P450 enzymes. They block the oxidative steps of GA synthesis that lead to *ent*-kaurenoic acid [27].

Ancymidol [*12771-68-5*][37], $C_{15}H_{16}N_2O_2$, M_r 256.3, *mp* 110–111 °C, LD_{50} rat oral ≈ 4500 mg/kg.

This substance is structurally related to pyrimidine fungicides, such as fenarimol (→ Fungicides, Chap. 8.6.4.). The fungicidal activity of these substances is based on the inhibition of cytochrome P450 enzymes in fungal ergosterol biosynthesis.

Ancymidol is produced by the metallation (butyllithium) of 5-bromopyrimidine and subsequent reaction with cyclopropyl 4-methoxyphenyl ketone [38].

Ancymidol reduces the length between internodes in many different plant varieties when applied to the leaves or the soil. It produces compact growth and shortened branches in ornamentals (e.g., chrysanthemums, poinsettias, tulips).
Trade names: Reducymol, A-Rest (Elanco).

Inabenfide [*82211-24-3*] [39], $C_{19}H_{15}ClN_2O_2$, M_r 338.8, mp 210–212 °C, LD_{50} rat and mouse > 15 000 mg/kg.

This substance is used to shorten and increase the lodging resistance of paddy rice. It is preferentially absorbed via the roots. For biology, see [40]; for mode of action, see [41].
Trade name: Seritard (Chugai).

Tetcyclacis [*65245-23-0*] [42], $C_{13}H_{12}ClN_5$, M_r 273.7, mp 190 °C, white crystals, LD_{50} rat oral ≈ 258 mg/kg.

This substance is made from quadricyclane by the cycloaddition of azodicarboxylate and 4-chlorophenyl azide, followed by saponification, decarboxylation, and oxidation [43].

Tetcyclacis is highly active and has a wide spectrum of activity. However, it cannot be used under field conditions in the treatment of foliage because of its photolability. It is applied to rice by seed soaking or soil drenching. Tetcyclacis produces shorter, more compact seedlings which have better root development that are more suitable for mechanical planting.
Trade name: Kenbyo (BASF).

Pachlobutrazol [*76738-62-0*] [44], $C_{15}H_{20}ClN_3O$, M_r 293.8, mp 160 °C, LD_{50} rat oral 5346 mg/kg (Bonzi).

Pachlobutrazol is the diastereomer with a 2 *RS*, 3 *RS* configuration. It is used as a growth-retarding agent for ornamentals, amenity trees, apple and pear orchards, and for the reduction of lodging in grass seed. For a description of biology, see [45].
Trade names: Bonzi, Clipper, Cultar, Parley (ICI).

Uniconazole [*83657-22-1*] [46], [47], $C_{15}H_{18}ClN_3O$, M_r 291.8, mp 147–164 °C, LD_{50} rat oral 2020 mg/kg (male).

Uniconazole, like pachlobutrazol, is used in very low amounts (12 g/ha) in ornamentals, trees, shrubs, and rice.
Trade names: Romika, Sumiseven, Sumishort (Sumitomo); Ortho-Prunit, Ortho-Sumagic, Ortho-Prism (Valent U.S.A.).

Triapenthenol [*76608-88-3*] [48], [49], $C_{15}H_{25}H_3O$, M_r 263.4, mp 135.5 °C, LD_{50} rat oral > 5000 mg/kg.

Triapenthenol is used to increase the stability of rape and in grass seed production. It also has a wide spectrum of activity as a growth-retarding agent, especially in dicotyledons (soybean, peanuts, potatoes, and orchard trees).

Trade name: Baronet (Bayer).

Dioxocyclohexane carboxylic acid derivatives with the following structure constitute another group of inhibitors of gibberellin biosynthesis [50], [51].

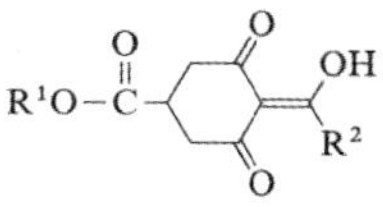

These compounds inhibit the oxidation steps after GA_{12} aldehyde [52].

Cimectacarb [*95266-40-3*] (proposed common name for CGA 163 935) [51], $C_{13}H_{16}O_5$, M_r 252.3, *mp* 36 °C, LD_{50} rat oral 4460 mg/kg.

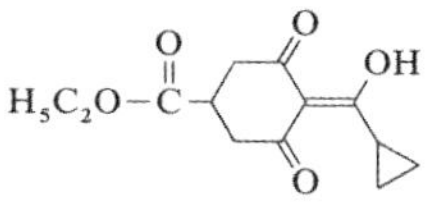

Cimectacarb is a growth-retarding agent for grain, rape, and grass [53].
Trade names: Omega, Vision (Ciba-Geigy).

Prohexadione- calcium [*127277-53-6*],[54], $C_{10}H_{10}O_5Ca$, M_r 250.3, *mp* 360 °C, LD_{50} rat oral > 5000 mg/kg is a new growth retardant from Kumiai Chemical Industry. It is active after foliar application in cereals, turf grasses, rape and ornamental plants.

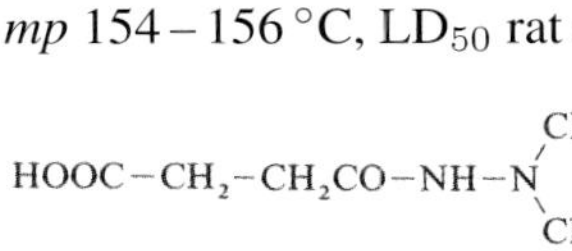

Daminozide [*1596-84-5*], *N,N*-dimethylaminosuccinamide [55], $C_6H_{12}N_2O_3$, M_r 160, *mp* 154 – 156 °C, LD_{50} rat oral ca. 8400 mg/kg.

This substance is produced from succinic anhydride and 1,1-dimethylhydrazine. It is used in fruit farming to accelerate ripening (cherries) and improve both fructification (grapes) and fruit quality (about 30 % of the apples grown in the USA are treated with daminozide). It is also used to regulate linear growth of ornamentals.

Daminozide does not inhibit GA synthesis, but GA transport [56].
Trade names: Alar, B-Nine, Kylar (Uniroyal); Dazide (Fine Agrochemicals).

2.5. Cytokinins [57]

Cytokinins are a series of natural or synthetic adenine derivatives which have a side chain on the exocyclic nitrogen atom and exert certain phytohormonal effects. The typical activity profile of cytokinins includes the stimulation of cell division, slowing down of plant senescence, stimulation of bud formation and germination of seeds, and the promotion of assimilate transport.

The natural hormone zeatin [*1637-39-4*] and synthetic derivatives, e.g., 6-benzylaminopurine (benzyladenine) [*1214-39-7*] and 6-furfurylaminopurine (kinetin) [*525-79-1*], have been used in experimental work.

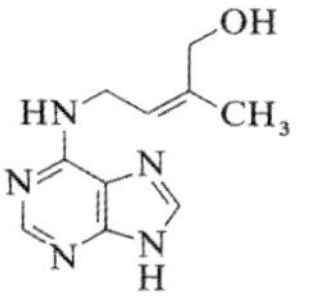

Zeatin

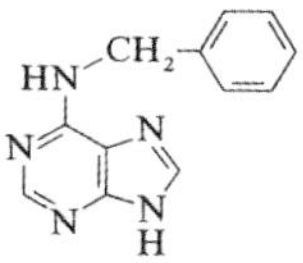

Benzyladenine

In spite of numerous patents for the application of cytokinins, these substances have little practical significance. A mixture of benzyladenine and gibberellic acids [see Section 2.3, Promalin (Abbott)] is commercially available. Some aryl and heteroaryl ureas also exhibit cytokinin-like activity [58]. Thidiazuron (see Chap. 3) which is used as a defoliant for cotton plants, is also a member of this group of compounds.

2.6. Abscisic Acid [58], [59]

Abscisic acid is a phytohormone which acts very much like a physiological antagonist of the cytokinins. It regulates the dormancy of seeds and buds, the fall of fruit and foliage, and the winter rest of plants. This acid also plays a role in stress situations that are caused by environmental factors. In the case of water shortage, for example, the abscisic acid content of the plant increases, resulting in decreased transpiration due to the closing of the stomata.

d-(+)-Abscisic acid [*21293-29-8*], $C_{15}H_{20}O_4$, M_r 264.3, mp 160–162 °C, $[\alpha]_D$ +430°, is colorless and sensitive to light. The (±) acid melts at 191 °C.

Experimental work with abscisic acid and its synthetic analogues is designed to exploit the use of these substances against stress caused by water shortage or cold [60]. These compounds are not yet, however, used in practice.

2.7. Ethylene-Releasing Compounds

Although the effects of ethylene on plant physiology have long been known, its classification as a phytohormone has been the subject of much controversy [61]. Endogenous ethylene is widely distributed in plant tissues (foliage, roots, flowers, seeds) and is synthesized during the ripening of fruit. However, unlike other phytohormones, it is not transported within the plant. Ethylene is probably only formed in the vicinity of its site of action. Its precursor, 1-aminocyclopropane-1-carboxylic acid, is the form that is transported. Ethylene is now defined as a plant hormone because it exerts many effects that regulate plant growth. Contrary effects are often observed, e.g., the inhibition and stimulation of metabolic processes or growth, depending on the type of plant. Thus, ethylene inhibits internodal growth in cereals, but stimulates the growth of sugar cane.

Ethylene is synthesized in the plant from carbon atoms 3 and 4 of methionine, $CH_3SCH_2CH_2CH(NH_2)COOH$, via the intermediates *S*-adenosylmethionine and 1-aminocyclopropane-1-carboxylic acid.

Ethylene gas cannot easily be used outdoors. Consequently, agricultural preparations have been developed that only release ethylene after they have been absorbed by the plant.

Ethephon [*16672-87-0*], 2-chloroethylphosphonic acid [62], $C_2H_6ClO_3P$, M_r 144.5, mp 74–75 °C (decomp.), LD_{50} rat oral ca. 4000 mg/kg, water soluble.

This acid is synthesized from phosphorus trichloride and ethylene oxide [63]. It decomposes to form ethylene in aqueous solutions at pH 4 [64].

$$PCl_3 + 3\,CH_2\!-\!CH_2 \xrightarrow{0-10\,°C} P(OCH_2CH_2Cl)_3 \xrightarrow{160\,°C}$$

$$ClCH_2CH_2PO(OCH_2CH_2Cl)_2 \xrightarrow{HCl/H_2O} ClCH_2CH_2PO_3H_2$$

$$ClCH_2CH_2PO_3H_2 \xrightarrow{H_2O} HCl + CH_2\!=\!CH_2 + H_3PO_4$$

Ethephon is one of the most economically important growth regulators. It accelerates ripening (tomatoes, cherries, and citrus fruit), stabilizes stalks of barley and rice, induces flowering of pineapple plants, and increases yields in rubber plantations (stimulation of latex flow) [65].

Trade names: Ethrel, Florel, Cerone, Chipco, Prep (Rhône-Poulenc); Flordimex (Chemie AG Bitterfeld-Wolfen); Arvest Etheverse (C.F.P.I.); combination with mepiquat, Terpal (BASF); combination with chlormequat, Terpal C (BASF).

Etacelasil [*37894-46-5*] [66], $ClCH_2CH_2Si(OCH_2CH_2OCH_3)_3$, $C_{11}H_{25}ClO_6Si$, M_r 316.9, colorless liquid, bp 85 °C at 1.33×10^{-3} mbar, LD_{50} rat oral 2000 mg/kg.

It is used specially to facilitate the harvesting of olives (abscission agent). Depending on the type of olive, harvesting should be carried out 6–10 days after application (1–2 g of agent per liter).

Trade name: Alsol (Ciba-Geigy).

2.8. Other Growth Regulators

Maleic hydrazide [*123-33-1*], LD_{50} rat oral (as the sodium salt) ca. 6950 mg/kg. This substance was the first synthetic growth inhibitor and was introduced in 1948 under the code name MH-30 (US Rubber) [67], [68].

Maleic hydrazide inhibits cell division but does not influence cell elongation [69]. It is rapidly transported in the plant.

Maleic hydrazide is an economically important growth regulator. It is used to inhibit the growth of lateral shoots in tobacco (USA), to inhibit the growth of grass, and to inhibit the germination of potatoes and onions. When used on ornamental lawns, the grass frequently turns reddish brown because anthocyanin formation is apparently stimulated.

Trade names: Retard, Sucker Stuff, Super Sprout Stop, Super Sucker Stuff (Drexel Chemical); De-Cut, Fair-2, Fair-Plus, Super De-Sprout (Fair Products); Royal MH-30, Royal MH-30 SG, Royal Slo Gro (Uniroyal).

Mefluidide [*53780-34-0*] [70], $C_{11}H_{13}F_3N_2O_3S$, M_r 310.3, *mp* 183–185 °C, LD_{50} rat oral ca. 28 000 mg/kg, LD_{50} mouse oral ca. 1920 mg/kg.

Mefluidide is used to inhibit the growth of grass (roadsides, playing fields, embankments, and green areas). This water-soluble agent is absorbed by grass mainly through its leaves. It inhibits growth due to cell elongation and largely suppresses seed head formation. Mefluidide also accelerates the ripening of sugar cane and is used in mixtures with herbicides.

Trade names: Embark (3 M), Compo grass regulator (BASF).

Amidochlor [*40164-67-8*], $C_{15}H_{21}ClN_2O_2$, M_r 296.8, *mp* 148–149 °C, LD_{50} rat oral 3100 mg/kg [71].

This substance is used as a growth inhibitor for turf grass.

Trade name: Limit (Monsanto)

Dikegulac sodium [*52508-35-7*], $C_{12}H_{17}NaO_7$, M_r 292.6, *mp* > 300 °C, LD_{50} rat oral ca. 31 000 mg/kg (male), ca. 18 000 mg/kg (female). This substance is an intermediate in the production of ascorbic acid and was introduced as a herbicide and growth regulator in 1973 by Hoffmann–La Roche [71], [72].

It is used to inhibit growth and promote the formation of lateral shoots in ornamentals (e.g., azaleas), and to inhibit the growth of shrubs and hedges (reduction of mechanical cutting).

Trade names: Atrinal (Dr. Maag), Atrimmec (PBI/Gordon).

Glyphosine [*2439-99-8*], *N,N*-bis(phosphonomethyl)glycine [73], $C_4H_{11}NO_8P_2$, M_r 263.1, colorless, *mp* 203 °C, LD_{50} rat oral ca. 3900 mg/kg.

Glyphosine accelerates maturation and increases the sucrose content of sugar cane (2–5 kg/ha are applied about 4–8 weeks before harvesting) by reducing the inversion of sucrose [74]. It is produced from glycine and chloromethylphosphonic acid. Glyphosine is being gradually replaced by sodium glyphosate (Polado).

Trade name: Polaris (Monsanto)

Glyphosate [*1071-83-6*], *N*-(phosphonomethyl)glycine, $C_3H_8NO_5P$, M_r 169.1, *mp* 230 °C (decomp.), solubility in water 1.2 % at 25 °C. The sodium salt of glyphosate is available as " Polado" and is recommended for the same applications as glyphosine.

Trade name: Polado (Monsanto) [75].

Fosamine ammonium [*25954-13-6*], ammonium *O*-ethyl carbamoylphosphonate [76],

$C_3H_{12}N_2O_4P$, M_r 170, mp 175 °C, LD_{50} rat oral ca. 24 000 mg/kg (41.5 % product).

$$H_2N-CO-\underset{\underset{NH_4^+}{O^-}}{\overset{\overset{O}{\|}}{P}}-OC_2H_5$$

It is used to inhibit the growth of shrubs (e.g., in woods). It inhibits the formation of shoots if applied in the fall and causes growth inhibition or defoliation if sprayed on the leaves in spring.
Trade name: Krenite (Du Pont).

Folicysteine [*8064-47-9*] is a mixture of 5 % L-cysteine derivatives and 0.1 % folic acid in stabilized, buffered solution. It stimulates the growth and development of plants when used in low concentrations.
Trade name: Ergostim (Montedison).

Mixtures of long-chain fatty alcohols (e.g., octanol and decanol) are used in the United States to control the lateral shoots of tobacco plants by inhibiting the development of axillary buds. In ornamentals, this "chemical pruning" leads to more abundant flowering and improved branching (azaleas and rhododendrons).
Trade names: Antak, Sucker Plucker (Drexel); Off-Shoot (Cochran); Royaltac (Uniroyal).

3. Defoliants and Desiccants

The premature shedding of leaves is induced by defoliants to allow or facilitate harvesting. Defoliants are used primarily in American cotton cultivation and to a lesser extent in soybean, vines, and tomatoes.

Desiccants are substances that artificially accelerate the drying of parts of plants. This effect is exploited in the harvesting of cotton plants and in the destruction of potato foliage.

Some of these harvesting aids are also used as herbicides. Examples are paraquat and diquat, sodium chlorate, dimethylarsinic acid, arsenic acid (H_3AsO_4), and endothal ($\rightarrow$ Weed Control). Thus, the classification of these substances as growth regulators is arbitrary. Other harvesting aids follow.

Butifos [*78-48-8*] [77], tributyl trithiophosphate, $(C_4H_9S)_3PO$, $C_{12}H_{27}OPS_3$, M_r 314.5, colorless or slightly yellow liquid, bp 150 °C at 0.4 mbar, LD_{50} rat oral 200 mg/kg. It is made from butanethiol and phosphorus oxychloride in the presence of a base. Butifos is a very effective defoliant for cotton. About half of the American cotton plantations are treated with organophosphorus defoliants about 1 – 2 weeks before harvesting.
Trade name: DEF (Mobay).

Merphos, tribufos [*150-50-5*], tributyl trithiophosphite [78], $(C_4H_9S)_3P$, $C_{12}H_{27}PS_3$, M_r 298.5, LD_{50} rat oral ca. 348 – 712 mg/kg, insoluble in water. It is made from phosphorus trichloride and tributanethiol and is used as a defoliant for cotton. The leaves fall in a relatively green state, leaving open cotton plants that can easily be mechanically harvested. In addition, fungal diseases (e.g., cotton boll rot) are suppressed by the better airing of the lower parts of the plant.
Trade name: Folex (Rhône-Poulenc).

Buminafos (proposed common name) [*51249-05-9*], dibutyl 1-butylaminocyclohexane phosphonate [79], $C_{18}H_{38}NO_3P$, M_r 343.5, liquid.

$$\underset{}{\text{cyclohexyl}}\!\!\begin{array}{c}NH-C_4H_9\\P(OC_4H_9)_2\\\|\\O\end{array}$$

Cyclohexanone and butylamine are reacted to give the Schiff base, which is then reacted with tributyl phosphite. Buminafos is used to destroy potato foliage, to defoliate cotton plants, and as a contact herbicide (nonselective).
Trade name: Trakephon (Chemie AG Bitterfeld-Wolfen).

Dimethipin [*55290-64-7*] [80], $C_6H_{10}O_4S_2$, M_r 210, mp 166 – 168 °C, LD_{50} rat oral ca. 1150 mg/kg.

$$\underset{O_2}{\overset{}{S}}\!\!\begin{array}{c}CH_3\\CH_3\end{array}$$

Dimethipin is produced from 3-chloro-2-butanone and ethane dithiol with subsequent oxidation. It is used as a defoliant for cotton plants

and as a potato vine desiccant. It enhances the maturation process and reduces seed moisture in the harvesting of rice, rape seed, flax, and sunflower.
Trade name: Harvade (Uniroyal).

Thidiazuron [*41118-83-6*][81], $C_9H_8N_4OS$, M_r 220.3, *mp* 217 °C (decomp.), LD_{50} rat oral ≈ 4000 mg/kg.

Thidiazuron is produced from 5-amino-1,2,3-thiadiazol and phenyl isocyanate. It is used to defoliate cotton plants and has cytokinin activity (see Section 2.5) [82].
Trade name: Dropp (Schering).

4. References

General References
1. W. Draber: "Natürliche und synthetische Pflanzenwachstumsregulatoren," in R. Wegler (ed.): *Chemie der Pflanzenschutz- und Schädlingsbekämpfungsmittel*, vol. 4, Springer Verlag, Berlin 1977, p. 1.
2. D. S. Letham, P. B. Goodwin, T. J. V. Higgins (eds.): *Phytohormones and Related Compounds: A Comprehensive Treatise*, vol. I und II, Elsevier, Amsterdam 1978.
3. W. Draber, J. Stetter: "Plant Growth Regulators," in *Chemistry and Agriculture*. Special Publication no. 36, The Chemical Society, Burlington House, London 1979.
4. N. B. Mandava (ed.): "Plant Growth Substances," *ACS Symp. Ser.* **III** (1979).
5. L. G. Nickell: *Plant Growth Regulators (Agricultural Uses)*, Springer Verlag, Berlin 1982.
6. J. S. McLaren (ed.): *Chemical Manipulation of Crop Growth and Development*, Butterworths, London 1982.
7. T. H. Thomas (ed.): *Plant Growth Regulator Potential and Practice*, The British Plant Growth Regulator Group and The British Crop Protection Council, BCPC Publ., Croydon, U.K., 1982.
8. L. G. Nickell (ed.): *Plant Growth Regulating Chemicals*, vol. I and II, CRC Press, Boca Raton, Fla. 1983.
9. R. L. Ory, F. R. Rittig (eds.): "Bioregulators, Chemistry and Uses," *ACS Symp. Ser.* **257** (1984).
10. M. Bopp (ed.): *Plant Growth Substances 1985*, Springer Verlag, Berlin 1985.
11. N. Takahashi (ed.): *Chemistry of Plant Hormones*, CRC Press, Boca Raton, Fla., 1986.
12. A. F. Hawkins, A. D. Stead, N. J. Pinfield (eds.): *Plant Growth Regulators for Agricultural and Amenity Use*, BCPC Publ., Thornton Heath, U.K., 1987.
13. P. J. Davies (ed.): *Plant Hormones and their Role in Plant Growth and Development*, M. Nijhoff Publ., Dordrecht 1987.
14. J. A. Roberts, R. Hooley: *Plant Growth Regulators*, Blackie, Glasgow 1988.
15. T. C. Moore: *Biochemistry and Physiology of Plant Hormones*, 2nd ed. Springer Verlag, Berlin 1989.
16. M. Luib, P. E. Schott: "Einsatz von Bioregulatoren" in G. Haug, G. Schuhmann, G. Fischbeck (eds.): *Pflanzenproduktion im Wandel*, VCH Verlagsgesellschaft, Weinheim 1990, chap. 10.

Specific References
17. C. R. Worthing, S. B. Walker (eds.): *The Pesticide Manual – A World Compendium*, 7th ed., British Crop Protection Council, The Larenham Press, Larenham, U.K., 1983.
18. C. Sine (ed.): *Farm Chemicals Handbook 1990*, Meister Publ., Willoughby, Ohio.
19. BASF Aktiengesellschaft, Marketing Report: Crop Protection, Ludwigshafen.
20. F. Kögl, D. F. Kostermans, *Hoppe Seyler's Z. Physiol. Chem.* **228** (1934) 90.
21. L. J. Andus: *Plant Growth Substances*, **vol. I,** L. Hill, London 1972, 179 ff. .
22. C. C. McBready, *Annu. Rev. Plant Physiol.* **17** (1966) 283.
23. N. Takahashi, I. Yamaguchi, H. Yamane: "Gibberellins," in [11] p. 57.
24. J. E. Graebe, *Ann. Rev. Plant Physiol.* **38** (1987) 419.
25. W. Rademacher: "Gibberellins, Metabolic Pathways and Inhibitors of Biosynthesis," in P. Boeger, G. Sandmann (eds.): *Target Sites for Herbicide Action.* CRC Press, Boca Raton, Fla., 1989, p. 127.
26. W. Rademacher: "Inhibitors of Gibberellin Biosynthesis, Applications in Agriculture and Horticulture," in N. Takahashi et al. (eds.): *Gibberellins, Proceedings of the Gibberellin Symposium Tokyo 1989*, Springer Verlag, Berlin 1990.

27. H. Sauter, *ACS Symp. Ser.* **257** (1984) 9.
28. N. E. Tolbert, *Plant Physiol.* **35** (1960) 380; *J. Biol. Chem.* **235** (1960) 475.
29. J. Jung, *Arzneim. Forsch.* **30** (1980) 1974.
30. A. I. Zadocev, G. R. Pikus, A. L. Grincenko: *Chlorcholinchlorid in der Pflanzenproduktion,* VEB Dtsch. Landwirtschaftsverlag, Berlin 1977.
31. BASF, DE 2 207 575, 1972. (B. Zeeh et al.).
32. B. Zeeh, K. H. König, J. Jung, *Kem.-Kemi* **1** (1974) no. 9, 621.
33. J. Jung, B. Würzer, H. von Amsberg, 170. Meeting Am. Chem. Soc. Abstr. Paper 69, Pest. Div. 1975.
34. S. Behrendt et al., *Landwirtsch. Forsch. Sonderh.* **35** (1979) 277.
35. P. E. Schott, F. R. Rittig in J. S. McLaren (ed.): *Chemical Manipulation of Crop Growth and Development,* Butterworths, London 1982, p. 415.
36. Mobil Oil, US 3 268 323, 1961 (L. E. Goyette).
37. Eli Lilly, US 3 868 244, 1974 (H. M. Taylor, I. D. Davenport, R. E. Hackler).
38. Eli Lilly, DE 1 770 288, 1968 (J. D. Davenport, R. E. Hackler, H. M. Taylor).
39. Chugai, EP 38 998, 1980 (N. Shirakawa et al.).
40. K. Nakamma, *Jpn. Pestic. Inf.* **51** (1987) 23.
41. T. Miki et al., *Plant Cell. Physiol.* **31** (1990) 201.
42. J. Jung, H. Koch, N. Rieber, B. Würzer, *Z. Acker- Pflanzenbau* **149** (1980) 129–136.
43. BASF, DE 2 615 878, 1976 (R. Platz et al.). BASF, DE 3 001 580, 1980 (N. Rieber et al.).
44. ICI, EP 15 639, 1979 (B. Sugavanam).
45. B. G. Lever, S. J. Shearing, J. J. Batch: "PP 333 – A New Broad Spectrum Growth Retardant," in *British Crop Protection Conference – Weeds 1982,* vol. 1, BCPC Publ., Croydon, U.K., 1982, p. 3.
46. Sumitomo, EP 54 431, 1980 (Y. Funaki et al.).
47. H. Oshio, K. Izumi: "S-3307, A New Plant Growth Retardant – Its Biological Activities, Mechanism and Mode of Action," in P. Macgregor (ed.).: *Plant Growth Regulators in Agriculture,* FFTC Book Series no. 34, Taipei, Taiwan 1986, p. 202.
48. Bayer, EP 15 387, 1979 (W. Reiser et al.).
49. K. Luerssen, W. Reiser, *Pestic. Sci.* **19** (1987) 153.
50. Kumiai, JP 71 26483 (K. Motojima et al.).
51. Ciba-Geigy, EP 0 126 713, 1983 (H. G. Brunner).
52. I. Nakayama: "Mode of Action of Novel Plant Growth Regulators, Dioxocyclohexanecarboxylic Acid Derivatives," in N. Takahashi et al. (eds.): *Gibberellins, Proceedings of the Gibberellin Symposium,* Tokyo 1989, Springer Verlag, Berlin 1990.
53. E. Kerber, G. Leypoldt, A. Seiler: "CGA 163 935, A New Plant Growth Regulator for Small Grain Cereals, Rape and Turf," in *Brighton Crop Protection Conference – Weeds – 1989,* vol. 2, BCPC Publ., Croydon, U.K., 1989 p. 83.
54. T. Miyazawa et al.: "Prohexadione-calcium, a New Plant Growth Regulator for Cereals and Ornamental Plants," in *British Crop Protection Conference –Weeds 1991,* vol. 3, BCPC Publ., Farnham, United Kingdom, 1991, p. 967.
55. J. A. Ridell, H. A. Hageman, C. M. J'Anthony, W. L. Hubbard, *Science (Washington D.C.)* **136** (1962) 391. H. J. Brooks, *Nature (London)* **203** (1964) 1303.
56. K. Takeno, R. L. Legge, R. P. Pharis, *Plant Physiol.* **67 (suppl.)** (1981) 581.
57. R. Horgan, B. Jeffcoat (eds.): *Cytokinins: Plant Hormones in Search of a Role,* British Plant Growth Regulator Group, Bristol 1987.
58. T. Okamoto, K. Shudo, Y. Isogai: "Structural and Biological Links Between Urea and Purine Cytokinins," in J. Miyamoto, P. C. Kearney (eds.): *Pesticide Chemistry: Human Welfare and the Environment,* vol. I, Pergamon Press, Oxford 1983, p. 333.
59. In [4] p. 196.
60. W. Rademacher, R. Maisch, L. Liessegang, J. Jung in [12]p. 53. A. Flores, A. Gran, F. Laurich, K. Dörffling, *J. Plant Physiol.* **132** (1988) 362.
61. K. Lürssen, *Chem. Unserer Zeit* **15** (1981) 122. J. A. Roberts, G. A. Tucker (eds.): *Ethylene and Plant Development,* Butterworths, London 1985.
62. Amchem. Prods., DE 1 667 968, 1968 (C. D. Fritz, W. Evans, A. R. Cooke). GAF, DE 2 050 245, 1969 (D. I. Randall, C. Vogel, R. W. Wynn). DE 2 053 967, 1969 (D. I. Randall, C. Vogel).
63. Hoechst, DE 2 061 610, 1970 (G. Stähler).
64. A. R. Cooke, D. I. Randall, *Nature (London)* **218** (1968) 974.
65. J. R. Sterry, *Meded. Rijksfac. Landbouwwet. Gent* **34** (1969) 462. L. Ferte, J. Pecheur, *Phytiatr. Phytopharm.* **19** (1970) 113.
66. Ciba-Geigy, DE 2 149 680, 1971 (W. Föry, H.-P. Fischer).
67. D. L. Schoene, O. L. Hoffmann, *Science (Washington D.C.)* **109** (1949) 588.

68. US Rubber, US 2 614 916, 1949 (O. L. Hoffmann, D. L. Schoene).
69. T. Hoffmann, E. V. Parups, *Residue Rev.* **25** (1969) 59. L. D. Nooden, *Physiol. Plant.* **22** (1969) 260. J. Bonaly, *C. R. Hebd. Seances Acad. Sci. Ser. D.* **273** (1971) 150.
70. 3 M., US 3 639 474, 1969 (J. K. Harrington et al.).
71. Monsanto, US 3 901 685, 1971 (K. W. Ratts).
72. P. Bocion, W. H. De Silva, G. A. Hüppi, W. Szkrybalo, *Nature (London)* **258** (1975) 142.
73. Monsanto, US 3 556 762, 1968 (P. C. Hamm).
74. C. A. Porter, L. E. Ahlrichs, *Rep. Hawaii. Sugar Technol.* **30** (1971) 71. L. E. Ahlrichs, C. A. Porter, Proc. 11th Brit. Weed Contr. Conf., Brighton 1972, p. 1215. E. G. Jaworski, *J. Agric. Food Chem.* **20** (1972) 1195.
75. Monsanto, US 3 799 758, 1971 (J. E. Franz).
76. DuPont, US 3 619 166, 1969 (B. Quebedaux, jr.). DuPont, DE 1 923 273, 1969 (W. P. Langsdorf, jr., B. Quebedaux, jr.).
77. Chemargro, US 2 943 107, 1959 (K. H. Rattenbury, J. R. Costello). Pittsburgh Coke & Chem., 2 965 467, 1956 (F. X. Markley).
78. Virginia Carolina Chem., US 2 955 803, 1958 (L. E. Goyette). Chemagro, US 3 089 807, 1961 (L. Trademan, F. R. Yagelowich).
79. Chemiekombinat Bitterfeld, DD 94 280, 1971 (E. Günther et al.).
80. Uniroyal, US 3 920 438, 1973 (A. D. Brewer, R. W. Neidermyer, W. S. McIntire).
81. Schering, DE 2 214 632, 1972 (H. Schulz, F. Arndt).
82. M. C. Mok et al., *Phytochemistry* **21** (1982) 1509.

ULLMANN'S

Agrochemicals

Related Titles

Krämer, W., Schirmer, U. (eds.)
Modern Crop Protection Compounds
approx. 750 pages
in 2 volumes approx.
2007
Hardcover
ISBN-13: 978-3-527-31496-6
ISBN-10: 3-527-31496-2

Wiley-VCH (ed.)
Ullmann's Biotechnology and Biochemical Engineering
approx. 900 pages in 2 volumes
with approx. 500 figures
2007
Hardcover
ISBN-13: 978-3-527-31603-8
ISBN-10: 3-527-31603-5

Wiley-VCH (ed.)
Ullmann's Industrial Toxicology
1191 pages in 2 volumes
with 264 figures and 156 tables
2005
Hardcover
ISBN-13: 978-3-527-31247-4
ISBN-10: 3-527-31247-1

Plimmer, J. R. (ed.)
Encyclopedia of Agrochemicals
3 Volume Set
approx. 1648 pages
Hardcover
ISBN-13: 978-0-471-19363-0
ISBN-10: 0-471-19363-1

Wiley-VCH (ed.)
Ullmann's Encyclopedia of Industrial Chemistry
6th Edition in Print
30080 pages in 40 volumes with
22100 figures and 8500 tables
2003
Cloth
ISBN-13: 978-3-527-30385-4
ISBN-10: 3-527-30385-5

Voss, G., Ramos, G. (eds.)
**Chemistry of Crop Protection
Progress and Prospects in
Science and Regulation**
406 pages with 81 figures
and 27 tables
2003
Hardcover
ISBN-13: 978-3-527-30540-7
ISBN-10: 3-527-30540-8